How Does Earth Work?

How Does Earth Work?

Physical Geology and the Process of Science

Second Edition

GARY A. SMITH
UNIVERSITY OF NEW MEXICO

AURORA PUN
UNIVERSITY OF NEW MEXICO

Prentice Hall

New York   Boston   San Francisco

London   Toronto   Sydney   Tokyo   Singapore   Madrid

Mexico City   Munich   Paris   Cape Town   Hong Kong   Montreal

Library of Congress Cataloging-in-Publication Data

Smith, Gary A. (Gary Allen), 1958–
  How does earth work? : physical geology and the process of science / Gary A. Smith,
Aurora Pun.—2nd ed.
    p. cm.
  Includes bibliographical references and index.
  ISBN 978-0-13-600368-7 (alk. paper)
  1. Physical geology—Textbooks. I. Pun, Aurora. II. Title.
  QE28.2.S59 2010
  551—dc22

                                                                    2008048082

Editor in Chief, Science: *Nicole Folchetti*

Acquisitions Editor: *Drusilla Peters*
Project Manager, Editorial: *Crissy Dudonis*
Assistant Editor: *Sean Hale*
Editorial Assistant: *Kristen Sanchez*
Marketing Manager: *Amy Porubsky*
Marketing Assistant: *Keri Parcells*
Managing Editor, Chemistry and Geosciences:
  Gina M. Cheselka
Project Manger, Production: *Edward Thomas*
Full Service/Composition: *Macmillan Publishing
  Solutions*
Production Editor, Full Service: *Laura Lawrie*

Art Director: *Maureen Eide*
Interior and Cover Design: *Tamara Newnam*
Art Editor: *Connie Long*
Art Studio: *Precision Graphics*
Photo Research Manager: *Elaine Soares*
Senior Operations Supervisor: *Alan Fischer*
Senior Media Producer: *Angela Bernhardt*
Media Project Manager: *Natasha Wolfe*
Cover and Title Page Photo: *Marble Caverns of
  Lago Carrera, in Chilean Patagonia.
  (tbkmedia.de/Alamy)*

Credits and acknowledgments borrowed from other sources and reproduced, with permission, in this textbook
appear on page C-1.

Pearson Education Ltd., *London*
Pearson Education Australia Pty., Limited, *Sydney*
Pearson Education *Singapore,* Pte. Ltd
Pearson Education North Asia, Ltd., *Hong Kong*
Pearson Educacíon de Mexico, S.A. de C.V.
Pearson Education–Japan, *Tokyo*
Pearson Education Canada, Ltd., *Toronto*
Pearson Education Malaysia, Pte. Ltd.

ISBN-10 0-13-600368-0
ISBN-13 978-0-13-600368-7

Prentice Hall
is an imprint of

ABOUT THE AUTHORS

Gary A. Smith

Gary A. Smith is a Professor of Earth and Planetary Sciences at the University of New Mexico and Fellow of the Geological Society of America. He has an undergraduate geology degree with a specialty in geophysics from Bowling Green State University and a PhD in geology from Oregon State University. He has taught first-year geology courses, primarily to nonmajors, for 14 of the last 22 years. His research, reported in more than 100 publications, and his upper-division-teaching experience range widely across the geologic sciences, including sedimentology, volcanology, geomorphology, and hydrology. Gary's research has a strong field emphasis, and he has contributed to 17 published geologic maps in Oregon and New Mexico. Nonacademic employment experiences in the oil industry and at Department of Energy facilities engaged in environmental remediation further broaden his background. Gary has strong interests in science education through his membership in the National Association of Geoscience Teachers and the National Science Teachers Association and has published in the *Journal of College Science Teaching* and the *National Teaching and Learning Forum*. Gary is also Director of the Office of Support for Effective Teaching at the University of New Mexico. In this role he engages and empowers university faculty to employ the best practices in learner-centered course design and pedagogy that have been identified through research in education, cognitive psychology, and neuroscience.

Aurora Pun

Aurora Pun is a Lecturer in the Department of Earth and Planetary Sciences at the University of New Mexico. She holds an undergraduate degree in paleontology from the University of California, Berkeley and MS and PhD degrees in Geology from the University of New Mexico, Institute of Meteoritics. Aurora has taught physical geology for the last 14 years. Also a member of the National Association of Geoscience Teachers, Aurora taught, with Gary and an education instructor, a course for teachers on developing inquiry-based K–12 curricula in the earth and space sciences. Her research interests include meteorite geochemistry with an emphasis on planetary igneous systems, and the petrology, mineralogy, and chemistry of ultrafine-grained sub-millimeter particles, from both terrestrial and extraterrestrial environments. Aurora's research utilizes a variety of laboratory instruments (electron microprobe, ion-microprobe, and transmission electron microscope) that provide in-place chemical analyses of microscopic minerals.

Gary and Aurora have a variety of nonacademic interests. They enjoy hiking, traveling, and xeric gardening around their home in Albuquerque.

BRIEF CONTENTS

CONTENTS

PART III Earth Deformation 228

PART IV Surface and Near Surface Processes 324

CHAPTER 19 Shorelines: Changing Landscapes Where
Land Meets Sea **474**

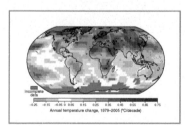

PREFACE

How Does This Textbook Relate to Your Teaching Objectives?

Our interest to write a physical geology textbook began with this question: What should students learn in a physical geology course? In most institutions this course serves primarily as a general-education requirement in the natural sciences, and secondarily as an introduction to a geoscience major. To us, the research in effective science education along with the commonly stated goals of science course requirements at colleges and universities are best served by courses that instill in students an understanding of how scientists come to know what they know about the natural world. Content information is available everywhere these days, and our students are more likely to seek information online than in our classes or their textbook. What the college experience should provide is the ability to do something with information and to understand the different discipline-based approaches to acquiring knowledge. When a student finishes a physical geology course they should certainly know something about the planet that they did not know beforehand. More importantly, they should have a better understanding of science and the attitudes and ways of knowing among scientists that they cultivated while using geology content as the vehicle to accomplish these goals.

The expectations that we place on our teaching and for those instructors this book is meant to support are best captured in the words of F. James Rutherford and Andrew Ahlgren in *Science for All Americans*, which was published 20 years ago by the American Association for the Advancement of Science. A partial excerpt reads as follows:

> Education has no higher purpose than preparing people to lead personally fulfilling and responsible lives. For its part, science education . . . should help students to develop the understanding and habits of mind they need to become compassionate human beings able to think for themselves and to face life head on. . . . Science, mathematics, and technology do not create curiosity. They accept it, foster it, incorporate it, reward it, and discipline it—and so does good science teaching. . . . Science prosper[s] because of the institutionalized scepticism of [its] practitioners. [The] central tenet is that one's evidence, logic, and claims will be questioned, and one's experiments will be subjected to replication. In science classrooms, it should be the normal practice for teachers to raise such questions as: *How do we know?* What is the evidence? What is the argument that interprets the evidence? Are there alternative explanations or other ways of solving the problem that could be better? The aim should be to get students into the habit of posing such questions and framing answers. (emphasis added)

4.5 How Do We Know . . . How Magma Is Made?

Picture the Problem

What Are the Necessary Conditions for Melting Rock? The lines in Figure 4.16 indicate whether peridotite is solid, partially melted, or completely melted at various combinations of temperature, pressure, and water content. These graphs illustrate the essential conditions necessary for rock to melt into magma. Geologists cannot go deep inside Earth to watch magma form naturally, so to understand the conditions required to melt rock, they must rely instead on experiments that reproduce in a laboratory the conditions of temperature, pressure, and water content inside Earth. These experiments are crucial to understanding how magma, and hence igneous rocks, form, so we should pause to understand how they are done.

For example, key laboratory investigations completed in the 1950s by O. F. Tuttle of Pennsylvania State University and N. L. Bowen of the Geophysical Laboratory at the Carnegie Institution of Washington considerably clarified geologists' understanding of the origins of felsic magma. Although later research significantly expanded on Tuttle and Bowen's findings, their study stands as a classic illustration of how geologists use experiments to learn about igneous processes within Earth. You might think of experiments as something to do simply to see what happens. Certainly, scientific experiments are conceived out of curiosity, but they are more carefully planned than a simple "What if?" activity. Bowen and Tuttle intended to provide data to support conclusions about the conditions necessary to make magma inside Earth.

Understand the Method

How Is Rock Melting Studied in the Lab? Tuttle and Bowen used the apparatus sketched in **Figure 4.17** to conduct their experiments. A rock sample is placed in a furnace and heated to a desired temperature while rods are squeezed together with a weight and lever to reproduce desired pressures. Water pumped into the chamber at very high pressure simulates the water present in felsic continental crust. In essence, Tuttle and Bowen created a very elaborate pressure cooker, but one capable of producing much higher temperatures and pressure than the familiar stovetop variety.

The "rock" materials for Tuttle and Bowen's experiments were synthetic mixtures of the elements known from laboratory chemical analyses to be present in granite. By combining these elements, the researchers created artificial granite. Dozens of melting experiments were undertaken. Here, we follow the results of only one series of experiments performed with elements found in quartz, potassium feldspar, and sodium-rich plagioclase feldspar. These elements—silicon, aluminum, potassium, sodium, and oxygen—typically comprise more than 95 percent of granite, so these synthetic samples are a reasonable substitute for natural granite.

Evaluate the Results

Under What Conditions Will Granite Melt? **Figure 4.18** graphs the results from the experiments. Each circle on the graph represents measured data from a single experiment. For each experiment, the sample was maintained at a selected temperature and pressure for several hours, and then the temperature was quickly dropped, so that any of the artificial granite that had melted into magma instantly solidified into obsidian-like glass.

The quickly cooled samples were then examined with a microscope. Three results are possible:

- *No melting:* If there was no glass in the sample and it was entirely crystalline, then it was clear that the selected temperature was not high enough at the experimental pressure to permit melting to begin.

- *Partial melting:* If the sample was a mixture of crystals and glass, then the experimental temperature was hot enough for partial melting to occur at the experimental pressure.

ACTIVE ART

Understanding Tuttle and Bowen's Data: See how animated and annotated graphs explain Tuttle and Bowen's data.

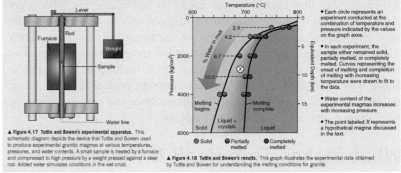

- Each circle represents an experiment conducted at the combination of temperature and pressure indicated by the values on the graph axes.
- In each experiment, the sample either remained solid, partially melted, or completely melted. Curves representing the onset of melting and completion of melting with increasing temperature were drawn to fit to the data.
- Water content of the experimental magmas increases with increasing pressure.
- The point labeled X represents a hypothetical magma discussed in the text.

▲ **Figure 4.17 Tuttle and Bowen's experimental apparatus.** This schematic diagram depicts the device that Tuttle and Bowen used to produce experimental granitic magmas at various temperatures, pressures, and water contents. A small sample is heated by a furnace and compressed to high pressure by a weight pressed against a steel rod. Added water simulates conditions in the wet crust.

▲ **Figure 4.18 Tuttle and Bowen's results.** This graph illustrates the experimental data obtained by Tuttle and Bowen for understanding the melting conditions for granite.

"How Do We Know?" sections in each chapter engage students in an appreciation for scientific methodology.

Our next question was what is the role of the textbook to achieve these results? To accomplish these objectives, which are aligned with learning outcomes for general-education science at most institutions of higher learning, requires a progressive pedagogy of which the students' textbook should be a part. The text is not the only thing that composes an effective science course but it should be supportive of the instructor's efforts to actively engage students in learning about Earth with a scientific perspective. In keeping with *Science for All Americans* we feel that it is critical for the text to focus on conceptual understanding of *how we know* rather than encyclopedic coverage of what we know. The text must provide models of how scientists do their work and further our knowledge if we are to expect students to understand scientific ways of thinking, and to appreciate that science is an ongoing endeavor that embraces curiosity and societal needs rather than existing as a revealed set of facts. The text must use examples of the questions that we expect students to ask, and use evidence to support the answers. And, consistent with the cognition research on how people learn, the text (as well as the course) should treat some subject material in depth in order to provide examples of concepts at work and to provide a foundation for factual knowledge and deeper, critical understanding that we want students to take away from the course. These expectations are the same for the student who never takes another science course and for the beginning geoscience student.

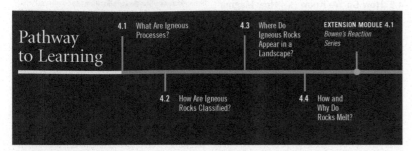

New "Pathway to Learning" features open each chapter with a visual guide to the questions students will seek to answer in order to achieve the learning outcomes of the chapter.

Equally important, the text must be readable, visually engaging, and set the stage for learning. *How Does Earth Work?* is designed to be read by your students rather than only summarized by your lectures. In our own teaching we expect, and receive, student engagement with the text before class so that class time can focus on active learning with peers along with instructor clarification of misconceptions and content that most challenges students. The instructor resources for *How Does Earth Work?* provide guides for how you can do the same. An essential goal of college education is to equip graduates to be lifelong learners. In the future they will advance in their careers through critical reading of text and interpretation of complex graphic information, and our teaching must build those expectations within our students.

What Makes This Book Different?

How Does Earth Work? emphasizes conceptual learning of processes, rather than immersion in facts, and adopts the credo "how we know" rather than only "what we know." The organization focuses on asking and then seeking answers to questions so as to:

1. Further engage the student to learn from reading, and
2. Encourage instructors to use a classroom pedagogy that facilitates interaction with students, regardless of class size.

The pedagogical spirit is:

1. Consistent with recently emphasized learner-centered paradigms, rather than simply as a companion to traditional teacher-centered lecturing, and
2. Intended to enhance science literacy through better appreciation of how scientific knowledge is obtained, and how it is relevant to societal concerns, needs, and decision-making.

These are the principal features of the text, some of which differ significantly from most comparable texts:

• **The presentation leans away from encyclopedic presentation of facts toward emphasizing active-process concepts**, and understanding how earth scientists acquire fundamental knowledge about our planet.

• **Each chapter opens with an essay that places a curious observer in realistic field or laboratory circumstances to make observations and ask questions about geological phenomena.** To the extent possible, these hypothetical experiences are of the type that students can readily relate to or may have experienced themselves. This opening vignette sets the stage with relevant questions that are addressed in the remaining pages of the chapter.

Each chapter begins with a story, from the field or lab, that gives students the opportunity to engage in the curiosity of scientific investigation and the questions that result.

• **Each chapter opening includes a list of measurable student learning outcomes about the topic content but that also ask the learner to develop abilities to apply, analyze, and synthesize knowledge rather than simply repeat information.** The companion instructor resources link the outcomes to specific parts of the chapter and show how to match end-of-chapter questions to these outcomes in order to assess student learning.

After Completing This Chapter, You Will Be Able to

- Apply established principles for placing geologic events, such as processes recorded by different rocks in an outcrop, in order from oldest to most recent.

- Compare and evaluate different approaches to estimating Earth's age.

- Describe how geologists measure ages of rocks and of Earth itself.

Learning outcomes are listed for each chapter and set goals for students.

• **The text uses questions as devices for learning.** Section headings are in the form of questions that intend to maintain reader interest, and to persistently illustrate relevancy. Students also encounter many questions in the text that are designed to cause them to stop, contemplate, and integrate what has been learned.

• **Observations are presented as a basis for developing underlying principles rather than as examples of already stated concepts.** "To state a theorem and then to show examples of it is literally to teach backwards" is an oft-cited quotation by mathematician E. Kim Nebeuts. Expert instructors organize knowledge by principles but most novice learners only understand the principles through an inductive process.

• **Key points are summarized at the end of each major section ("Putting it Together") rather than at the end of the chapter.** These bullets (1) relate directly to answering the question stated in the section heading, and (2) provide a learning check for students before moving on in the text.

Putting It Together— Why Are There Different Types of Volcanoes and Volcanic Eruptions?

• Volcanic phenomena and volcanic landforms closely depend on two properties of magma: gas content and viscosity.

• Gas content determines how explosive eruptions will be and the relative proportion of lava flows and pyroclastic deposits that eruptions will produce. More gas-rich magmas are more explosive.

• Viscosity determines whether lava flows are thin and widespread, or thick and short. Lower viscosity produces more fluid flows.

• Viscosity and gas content are in large part determined by magma composition, and both properties increase with increasing silica content.

"Putting It Together" summaries let students check their understanding of small sections before continuing their investigation.

• **Wherever appropriate, topics are linked to issues of societal concern, real-world experience, or both**, so as to increase relevancy and appreciation of the value of science learning. Topics of societal relevancy are not relegated to sidebars, boxes or separate chapters, but are integrated throughout the text.

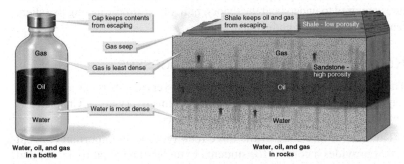

Real-world topics are integrated throughout the text, as are illustrations that teach and engage.

• **Illustrations are constructed to enhance learning.** Illustrations are as simple as possible, spatially associated with applicable text, and highly annotated in accordance with research demonstrating distinctions between effective versus ineffective use of textbook illustrations.

• **Animation is used to illustrate active processes.** The majority of the 63 "Active Art" animations use the same base illustration as the static art in the text, providing a clear link between text illustration and animation.

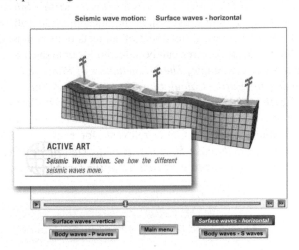

Text icons alert students to animation tutorials on the companion website that promote understanding of active processes.

• General concepts about Earth processes are presented with introduction of the absolute, bare minimum number of unfamiliar terms. **Classifications, where necessary, are inductively "derived" from "observation" rather than simply produced on the page as facts to be memorized.**

• **There is always consideration of "how it is that we know what we know," and a context for why it is important to understand these processes.** This focus ties to societal relevance in many cases, but also includes satisfaction of innate curiosity (i.e., the distinction, but equal importance, of strategic- and curiosity-based science).

• **Each chapter has a** *"How Do We Know?"* **section that provides more detail on the scientific accomplishments that provided a piece of key information in the topic for the chapter.** The *"How Do We Know?"* sections explicitly show how concepts about geological processes are developed, and implicitly serve (1) as a basis for appreciating how all concepts in the book have become scientific knowledge, and (2) to illustrate how geoscientists apply scientific methodology.

• **To facilitate text brevity, some background material (including that which we consider to be remedial), detailed examples, and supplemental information (which an instructor may or may not require students to know) are provided on the companion website as Extension Modules.** Unlike resources that are commonly treated as supplements or optional elements, and ignored by many instructors and students, we designed the extension modules as elements to be fully integrated with the text at the discretion of instructor assignments or student curiosity. Instructors may assign selected modules and hold students responsible for the material just as if it were a text section. The text itself reads continuously without consulting the modules. The modules are of two types:

1. Coverage of information that we assume students to have previously learned in high school chemistry or physical science. These modules (for example, *Basics of an Atom, Radioactivity and Radioactive Decay*)

serve as a refresher for students who previously learned this material but are also written as confidence-building first-time treatments for the less thoroughly prepared student.

2. Coverage of information that extends beyond the text. These modules include topics found in some physical geology texts, but which we feel are optional, in addition to topics that some books treat in "boxes" (for example, *Silicate Mineral Structures, Bowen's Reaction Series, When Will We Run Out of Oil?, Meteorites as Guides to Earth's Interior, The Geology of Caves*). Some extension modules provide substantially more depth or mathematical treatment of subjects than in most comparable textbooks; these features can be selected for courses designed for science majors (for example, *The Mathematics of Radioactive-Isotope Decay, Sizing up Earth, Calculating Magnitude and Energy Release from an Earthquake, Darcy's Law, Measuring Uplift Rates*). Other modules will be deemed essential for instructors teaching in regions where the subject matter is highly relevant (for example, *Mitigating and Forecasting Earthquake Hazards, How to Determine Recurrence Intervals of Floods, Changing Shorelines in the Great Lakes*).

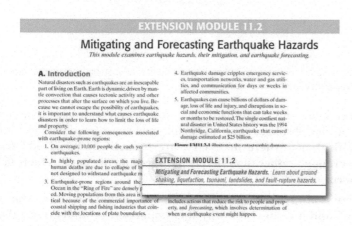

Additional content on the website extends topical coverage beyond the text to satisfy student curiosity and instructor flexibility.

What's New in the Second Edition?

The changes in the second edition resulted not only from helpful reviews by faculty adopters but also the specific feedback from more than 100 student readers. Teachers adopt texts but effective textbooks are written for learners. Student response to the first edition was overwhelmingly positive with a majority of those surveyed saying that *How Does Earth Work?* was superior to their other science texts because of its more engaging reading style, questions that piqued their interest, and explanatory graphics and animations that enhanced their learning. At the same time, the student feedback identified ways to improve understanding of conceptual arguments and where fewer details would enhance learning. As a result, the entire text has been either substantially edited or rewritten with a focus on readability and clarity. We have conscientiously worked to provide text that is assigned readability indices at the 10th to 12th grade level so that a moderately prepared college student is able to confidently read and understand the presented concepts.

A more obvious change is the addition of Chapter 21 on global warming. Working with our editors, we surveyed current and potential adopters of *How Does Earth Work?* and found that many of them desire to bring the global climate-change discussion into their geology classroom. Therefore, we provide in this edition an evidence-based approach that permits the learner to evaluate the most visible component of the climate-change debate—global warming. As the culmination of the book, Chapter 21 integrates knowledge from other chapters and, more importantly, integrates conceptual understanding of scientific arguments, hypothesis testing, and the roles of uncertainty and skepticism to further the science enterprise that are presented throughout the book. We also incorporate the important geological perspective of climate change over deep time.

How Does the Book Work?

Following, we detail our approach to the core topics of the course and the ways, if any, we depart from the standard treatment in most texts.

Overall Organization

The parts of the text follow a logical, progressive development of geological knowledge. Part I focuses on Earth materials seen at the surface: what they are and how they are used and studied. Part II moves on to what is known or inferred about the materials composing Earth's interior, how this knowledge is acquired, and the origin of Earth's internal structure. Part III takes what was learned about Earth's internal processes to understand descriptions of rock deformation, including plate tectonics. Part IV integrates knowledge of Earth materials and deformation with hydrosphere and atmosphere processes to determine the origin of diverse landscape features. The instructor's manual provides guidance on how to maintain coherency if the instructor desires to pursue only some topics or to present them in a different order.

Plate Tectonics

Plate tectonics is an underlying theme that can be threaded throughout the course. In writing a concept-oriented text, which encourages inquiry and application of scientific reasoning, we must first establish an understanding of the nature of geologic materials and phenomena before fully developing the theory by which they are explained; observations are first in scientific methodology. When introducing plate tectonics in Chapter 1, the emphasis is placed on the general phenomena at plate boundaries, and this background permits integration of plate tectonics into the chapters following. A thorough development of the theory is placed in Chapter 12, after students have a firm grasp of the geological materials and structural features required to understand the theory, rather than simply having its salient features presented as undisputed fact.

Rocks and Rock-Forming Processes

The dedication of an entire chapter to an overview of rocks is atypical for physical geology texts. We believe that this overview in Chapter 3 is important, however, for the following reasons:

- The inter-relationship of the three rock types is an important concept that provides a context for students to understand what rocks are and how they form.

- The classification of rocks, although typically encountered in grades 5–8 (and in some states in high school), is a nebulous concept to many students, many of whom are unclear on what classification schemes are or why they are developed.
- An overview of the three rock classes within a conceptual framework of genetic classifications, help students put rocks in perspective before considering each rock group in detail. Otherwise, students learn about each of the three classes separately, and rarely know where to begin identifying a rock unless they already know that it is igneous, sedimentary, or metamorphic.

We appreciate that many instructors may prefer to skip this chapter and sail onward into igneous rocks. For those inclined to do so, Chapters 4–6 are written independently of Chapter 3 so as to not impede its omission.

Weathering and Soil

Many physical geology textbooks insert a chapter on weathering and soil formation ahead of the chapter on sedimentary rocks. *How Does Earth Work?* takes a different approach. Weathering processes are included in the sedimentary rock chapter to emphasize how the components of sedimentary rock form. We find the consideration of soil to be a digression in the understanding of rocks, and provide a separate short chapter on soil in the context of landscape evolution in Part IV of the text.

Earth's Interior

Earth's interior is dealt with in a single short chapter in many texts. We believe that this brief treatment does an injustice to the current frontier science focusing on the interior of the planet. We dedicate Chapters 8–10 to the interior, one chapter to each of these concepts: what it is made of, how it got to be that way, and why it is in motion. Convection within Earth is worthy of more description and explanation than is found in most physical geology texts. Recent advances in geodynamics support the idea that plate tectonics can mostly be described as being mantle convection, rather than seeing convection as some nebulous, and unpopular, driving force. Convection is, therefore, given a prominent place in this text, but done at a level appropriate for the introductory student.

Earthquakes

The subject of earthquakes is divided into parts of two chapters rather than lumped into one. This differs from most physical geology texts where there is a single earthquake chapter simultaneously covering two completely different subjects: (a) The inelastic deformation mechanism of earthquakes and resulting effects and hazards; and (b) the use of elastic waves to deduce the internal structure of the planet. We integrate the surface effects and deformation caused by earthquakes with the discussion of faults and faulting in Chapter 11. Chapter 8 deals only with the properties of seismic waves that are fundamental to investigating the planetary interior.

Global Tectonics

Global tectonics (plate motions and plumes) is treated in a single chapter. In contrast to most text treatments of the subject, Chapter 12 does not provide significant historical development of the plate tectonics theory and plume hypothesis. Instead, our emphasis is on "showing the evidence." We think that it is more important for students to be able to explain why plate tectonics works, than to recite the history of critical developments. This approach is most consistent with the "how do we know" theme of the text, and keeps students focused on how scientific data successfully test hypotheses and theories. An underlying, subdued theme in this chapter is the distinction between plate tectonics as a theory and the mantle plume hypothesis; these two concepts do not enjoy the same level of supporting data and confirmation. We hope that students will come away with confidence in understanding why plate tectonics is entrenched as a fundamental concept in geoscience, while at the same time seeing the mantle-plume hypothesis as an area of continued fundamental testing, a continued frontier of exciting, ongoing science.

Surface Relief

We replace the typical chapter on mountain building with Chapter 13 on the origin of surface relief. While giving considerable attention to the formation of mountains, this chapter deals more generally with the tectonic foundation for the irregular surface of the planet as a fundamental feature that any introductory geology student should be able to explain. This chapter fully develops concepts of isostasy by making extensive use of simple analogy for complex Earth features, and the scientific modeling approach of starting simple and adding more complexity only as necessary to explain observed features.

Surficial Processes

Part IV, which emphasizes surficial processes, is written with an emphasis on understanding these processes and their visible results, rather than on developing an encyclopedic glossary of landform names. Our approach is somewhat more in depth than in most texts, because we think that students relate to studying and understanding landscapes that they see and have experienced more than any other topic in physical geology. We strive to develop understanding of how observed processes create observed landforms, resources, and hazards. A significant component of this understanding comes from understanding how surface processes impact humans, and how human modifications of the environment impact the surface processes in predictable ways. Chapters in this part of the text are largely written independently of one another, allowing the instructor to focus on topics that seem most relevant to the geologic setting of their institution, topics in the news, or both.

Modern Soil Treatment

The soil chapter (Chapter 14) uses modern soil descriptions and genetic interpretations. This chapter should be particularly engaging for students studying geology in those large regions of the country where agriculture plays an important role in the economy. This chapter also introduces the role of soils in evaluating landscape stability.

Mass Movements

Our discussion of mass movements in Chapter 15 emphasizes the characteristics of these phenomena that lead to a natural classification with genetic significance. As with most texts, we focus on the hazards of mass movements, but also provide insights into how these processes contribute to overall landscape development.

Rivers, Streams, and Lakes

Chapter 16 provides many of the familiar elements of learning about streams, but within the context of understanding how streams do work. With an understanding of how streams pick up, transport, and deposit sediment, it is possible to deduce the origins of a wide variety of landforms, and to understand why streams change along their courses and through time, in response to climate change, tectonic activity, sea-level fluctuation, and human activities. This chapter also includes a treatment of the origin and landscape-significance of lakes, which are rarely mentioned in physical geology texts despite being easily linked to the experiences of most students.

Ground Water

Chapter 17 discusses ground-water flow and its effects. Emphasis is placed on the simple mechanics of ground-water movement and particularly those aspects that influence the exploitation of this water resource. Ground water also plays a role in shaping landscapes. The formation of karst topography is discussed without undue emphasis on cave features, which are elaborated in an extension module. The role of ground-water sapping, rarely mentioned in other texts, to form canyon landscapes is also explained.

Glaciers

Chapter 18 considers the domain of frozen water—glaciers. Introducing a minimum of unfamiliar-sounding landform terms, the chapter first explains the nature of glacier flow and then relates glacier dynamics to erosional and depositional landforms. Glaciers are responsible for a large variety of landscapes and resources in North America, even though they are unseen by most students. Consideration of the late Pleistocene ice ages and the climatic driving forces for their occurrence is key to understanding these landscapes and resources. This chapter provides critical examples of how geologists interpret landscape origins by reference to vanished agents of landscape modification.

Shoreline Processes

Chapter 19 explains the role of waves, tides, and sediment supply to shape ocean shorelines. As with other chapters in this section, the shoreline chapter begins with grounding in the operative processes and then shows how these processes account for familiar shoreline features. Rapid, historical change of shorelines, by both natural and anthropogenic forces, is also a key part of the chapter and links to how scientists document and explain historical sea-level rise.

Wind

Chapter 20 evaluates wind as a geological process. Most introductory geological texts discuss wind within a chapter on deserts. *How Does Earth Work?* does not have a deserts chapter. The river chapter develops concepts of stream dynamics in both humid and arid landscapes and includes much of the information traditionally found in desert chapters. Wind is a critical geological agent everywhere on Earth, and not just in deserts—for this reason the text treats wind in a separate chapter. Although including significant information on desert dunes, wind is also featured as an effective landscape-forming agent along coastlines and as a globally important sediment-transport process as revealed by global dust movement.

Global Change

An instructor desiring to emphasize a global-change theme is provided with an opportunity to integrate content with several chapters. The glacial-processes chapter develops the science behind documenting and explaining the cyclic alternation of glacial and interglacial climates. The shoreline chapter examines the evidence of, causes for, and implications of sea-level change. The global-warming chapter builds on these concepts and allows the student to evaluate evidence for global change.

The Teaching and Learning Package

Instructors and students using *How Does Earth Work? 2e* have at their disposal a tremendous range of resources.

For the Instructor

The centerpiece of the instructor resources is the Instructor Resource Center (IRC), which aggregates online an **Instructor Manual with Test Item File**, lecture outlines, and suggestions for the classroom based on the latest pedagogical research; a **library of PowerPoint™ presentations**, including "art only," "classroom response system," and "lecture outline" versions; a library of Active Art animations set up for classroom projection, preloaded into PowerPoint slides; and an additional set of photos. To access these resources, visit www.mygeoscienceplace.com.

   The Instructor Resource Center online at www.mygeoscienceplace.com puts all of your lecture resources in one easily accessible site:

- Line art, tables, and photos from the text in .jpg format
- Art animations from textbook references for classroom projection
- *Images of Earth* photo gallery
- *Instructor Manual* in Microsoft Word
- *Test Item File* in Microsoft Word

The Prentice Hall Geoscience Animation Library (0-13-600377-X) includes over 100 animations illuminating the most difficult-to-visualize topics of physical geology. Created through a unique collaboration among five of Prentice Hall's leading geoscience authors, these animation represent a most significant leap forward in lecture presentation aids. They are provided both as Flash files and, for your convenience, pre-loaded into PowerPoint™ slides. At www.mygeoscienceplace.com, you may contact your local Pearson representative for the DVD containing the Prentice Hall Geoscience Animation Library.

PowerPoint™ Presentations found at www.mygeoscienceplace.com cut down on your preparation time no matter with four varieties to choose from for each chapter of the textbook:

- Exclusively art—all of the photos, art, and tables from the text, in order, loaded into PowerPoint slides.

- Lecture outline—this set of slides succinctly reviews the key content of each chapter.

- Animations—Active Art animations are loaded into PowerPoint to help students visualize Earth processes as you lecture.

- Classroom response—Questions for each chapter are included for Personal Response System (PRS) or H-ITT systems.

Environmental Geology Videos DVD (0-13-145666-0) contains more than 35 video segments featuring the results of a dynamic and changing Earth environment. These focused segments average less than a minute each to help you launch lectures and classroom discussion. Showing processes such as volcanic destruction, landslide devastation, and heartland tornadoes, this DVD is available to adopters through their local Pearson representative. Contact information is available at www.mygeoscienceplace.com.

For the Student

Designed to help focus students' efforts and maximize their results, the **Companion Website** (www.mygeoscienceplace.com) assesses student learning through multiple quizzes for each chapter of the text. It also delivers Active Art tutorials and Extension Modules for further study.

myeBook

*The integration of **myeBook** within mygeoscienceplace gives students easy access to the text when and wherever they have access to the Internet. **myeBook** pages look exactly like the printed text, offering powerful new functionality for students and instructors. Users can create notes, highlight text in different colors, create book marks, zoom, view in single-page or two-page view, etc. **myeBook** also links students to associated media files, enabling them to view an animation as they read the text.*

Encounter Earth

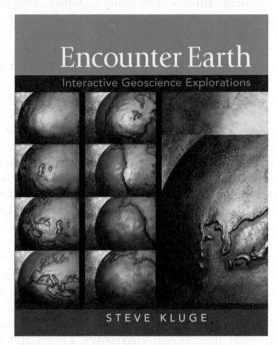

Encounter Earth: Interactive Geoscience Explorations offers 20 exercises using Google Earth™ mapping service .KMZ files to guide students through visual activities and homework assignments.

For a complete description of these resources or to request a desk copy, please contact your local Prentice Hall representative or visit www.pearsonhighered.com/.

Acknowledgments

We express our deep gratitude to more than 100 colleagues from schools across the country who provided us with hundreds of hours of their time during development of both editions of *How Does Earth Work?* They reviewed manuscript chapters, participated in focus groups, reviewed animations, and patiently answered our questions. Many of these reviewers shared their insights from using the first edition in their classes. *How Does Earth Work?* has benefited immeasurably from your effort, insights, objections, encouragements, and your selfless willingness to share your expertise as scientists and teachers. We also acknowledge the roughly 150 students at the University of New Mexico who provided anonymous critiques of those aspects of the text that promoted curiosity or challenge. Some of these students provided weekly reports of what they viewed as the positive and negative experiences with each chapter. Teachers select textbooks but they are written for learners. Student reviews have played the largest role in the changes reflected in the second edition.

Patrick Abbott (San Diego State University), Gerry Adams (Columbia College), John Anderson (Georgia Perimeter College), Jake Armour (University of North Carolina at Charlotte), Abdolali Babaei (Cleveland State University), Kathy Baldwin (Washington State University), Lisa Barlow (University of Colorado—Boulder), Jerry Bartholomew (University of Memphis), Raymond Beiersdorfer (Youngstown State University), Glen Blaylock (Utah Valley State

College), Robert Blodgett (Austin Community College), Kathleen Bower (Eastern Illinois University), James Brophy (Indiana University), Lloyd Burckle (Columbia University), Patrick Burkhart (Slippery Rock University), Patricia A. Campbell (Slippery Rock University), Susan Cashman (Humboldt State University), Beth Christensen, (Georgia State University), Jane Dawson (Iowa State University), John Dembosky (University of Akron), Halilang Dong (Miami University), William Dupre (University of Houston), Mark Everett (Texas A&M University), Roy Ewing (Aims Community College—Greeley), Tobias Fischer (University of New Mexico), Frank Florence (Jefferson Community College), Andrew Frank (Chemeketa Community College), Daniel Frederick (Austin Peay State University), Sharon Gabel (State University of New York—Oswego), Jacqueline Gallagher (Florida Atlantic University), Richard Gibson (Texas A&M University), Francisco Gomez (University of Missouri—Columbia), Yuri Gorokhovich (State University of New York—Purchase), Joseph Gould (University of South Florida), Nathan Green (University of Alabama), Lisa Greer (Pennsylvania State University), Paul Grogger (University of Colorado—Colorado Springs), Roy Hagerty (Oregon State University), Anne Larson Hall (Emory University), John Hanchar (George Washington University), William Heins (Lewis and Clark State College), Rosemary Hickey-Vargas (Florida International University), Christi Hill (Grossmont College), Scott Hippensteel (University of North Carolina—Charlotte), Lincoln Hollister (Princeton University), Daniel Holm (Kent State University), Mary Hubbard (Kansas State University), Scott Hughes (Idaho State University), Jamey Jones (University of Minnesota—Morris), Robert Jorstad (Eastern Illinois University), Richard Josephs (University of North Dakota), Amanda Julson (Blinn College), Haraldur R. Karlsson (Texas Tech University), Simon Kattenhorn (University of Idaho), David King (Auburn University), Katrien J. Kraft (Mesa Community College), Benjamin Laabs (Gustavus Adolphus College), Nicholas Lancaster (Desert Research Institute), Danita Maynard LaSage (Eastern Kentucky Unviersity), Patricia Lee (University of Hawaii), Robert Leighty (Mesa Community College), Alan Lester (University of Colorado—Boulder), Steven Lev (Towson University), Mian Liu (University of Missouri—Columbia), Harmon Maher (University of Nebraska, Omaha), I. Peter Martini (University of Guelph), Maureen McCurdy (Louisiana Tech University), Cheryl L. Metz (Blinn College), Dan McNally (Bryant College), Robert Meeks (St. Philip's College), Stanley Mertzman (Franklin & Marshall College), Barry Metz (Delaware County Community College), William Minarik (University of Maryland), William Mode (University of Wisconsin—Oshkosh), LeeAnn Munk (University of Alaska—Anchorage), Kathryn Nagy (University of Illinois—Chicago), Donald Neal (East Carolina University), Clive Neal (University of Notre Dame), Yuet-Ling O'Connor (Pasadena City College), Darryll T. Pederson (University of Nebraska, Lincoln), Raymond Pheifer (Eastern Illinois University), Nicholas Pinter (Southern Illinois University), Roy Plotnick (University of Illinois—Chicago), Jay Quade (University of Arizona), Monica Ramirez (Aims Community College—Greeley), John Renton (West Virginia University), Carl Richter (University of Louisiana—Lafayette), Jennifer Rivers (Northeastern University), Michael Roden (University of Georgia), Sarah Rogers (Arapahoe Community College), Laura Sanders (Northeastern Illinois University), William Sanford (Colorado State University), William Cullen Sherwood (James Madison University), Eric Small (University of Colorado at Boulder), Jay Sims (University of Arkansas (Little Rock), Jeffrey Snyder (Bowling Green University), John Paul Stimac (Eastern Illinois University), Karen Swanson (William Paterson University), Martha Sykes (Hawaii Pacific University), Charles Trupe (Georgia Southern University), Robert Kelly Vance (Georgia Southern University), Dorothy Vesper (West Virginia University), Joshua Villalobos (El Paso Community College), Chunzeng Wang (Hunter College of the City University of New York), David Warburton (Florida Atlantic University), Andrew Webber (Miami University—Hamilton), John Charles White (Eastern Kentucky University), James Wittke (Northern Arizona University), Ellen Wohl (Colorado State University), Wan Yang (Wichita State University), Aaron Yoshinobu (Texas Tech University), Brent Zaprowski (Salisbury University), Sally Zellers (Mississippi State University), Nick Zentner (Central Washington University)

Gary Smith and Aurora Pun
Albuquerque, New Mexico
January 2009

How Does Earth Work?

CHAPTER 1

Why Study Earth?

Why Study Geology?

You are beginning a study of **geology**—the science of Earth. Your goal is to learn how Earth works. Along the way, you will learn about the importance of geology to humans. You will consider the origin of essential resources and the forces that drive geologic processes that are hazardous to our lives and livelihoods. This text is not simply a collection of facts to memorize. It is an invitation to think about how scientists use creativity, diligence, and technology to arrive at the current body of knowledge about Earth.

The chapter title asks, why study Earth? Some reasons to consider are:

- Earth provides awesome, beautiful landscapes, such as the mountain view on the facing page, which inspire the mind to wonder how the planet came to be, what it is made up of, and how it works.

- Earth is an active planet that experiences cataclysmic earthquakes, volcanic eruptions, floods, landslides, and other processes that continually reshape the surface and sometimes threaten human lives and property. A geologist taking in the view in the photo to the right wonders, for example, if these rugged mountains are actively rising, driven by geologic forces that could cause earthquakes and landslides.

- Earth is a source of natural resources that provide for our quality of life: water, minerals, raw materials for manufacturing and building, precious metals, oil, natural gas, coal, and other materials used for generating energy. A geologist understands the origins of the construction materials for the buildings in this photograph and the sources of the energy consumed to produce electricity for the visible lights in the photo.

After Completing This Chapter, You Will Be Able to

- Define "geology" and describe the topics of study included within this scientific discipline.

- Describe examples of how geology is relevant to our society.

- Explain how geologists undertake scientific studies.

- Describe the basic elements of a key principle called uniformitarianism and of plate tectonics, an integrating theory that explains how Earth works.

- Explain how the transfer of energy drives work on and within Earth.

An evening view of Salt Lake City, Utah, and the Oquirrh Mountains

1.5 What Is the Theory of Plate Tectonics?

1.4 What Is the Principle of Uniformitarianism?

1.6 How Does the Concept of Work Apply to Earth?

IN THE FIELD

Your geologic inquiry begins with three imaginary field outings—outdoor excursions that focus on observing, describing, and asking questions about Earth processes, landscapes, and materials. Each chapter opens with such an experience, either in the field or in a laboratory.

In this chapter, your first stop takes place on a summer morning along the Snake River in Grand Teton National Park, Wyoming, shown in **Figure 1.1**a. An outdoors enthusiast taking in this view can envision countless opportunities in this landscape—hiking, rock climbing, canoeing, birding, and fishing. The diverse landscapes and seascapes of Earth provide widely different challenges, uses, and inspirational views.

As a resident of our planet, you may have wondered why there is so much variety in the appearance of Earth's surface. Why are mountains, such as the Teton Range, not found everywhere? How long have the mountains existed, and how did they form? Why do rivers, such as the Snake River, flow where they do? What determines how much water the river carries and how often the river floods? Geologists strive to answer such questions.

Your second stop is the island of Sicily, the "football" located off the "toe of the boot" of Italy. In Figure 1.1b, fountains of lava light up the night sky above the Mount Etna volcano. Molten-rock cinders fall back onto the slopes of a growing conical hill. Why is this eruption happening? Are the communities near the erupting volcano at risk of destruction?

The active processes on and within Earth—erupting volcanoes, shattering earthquakes, inundating floodwaters and tsunami, or devastating landslides—always make the news. These events clearly illustrate that the planet is dynamic, not stationary and static. As a student of geology, you probably are curious to know what processes lead to these hazardous events and whether geologists can forecast when and where they will happen.

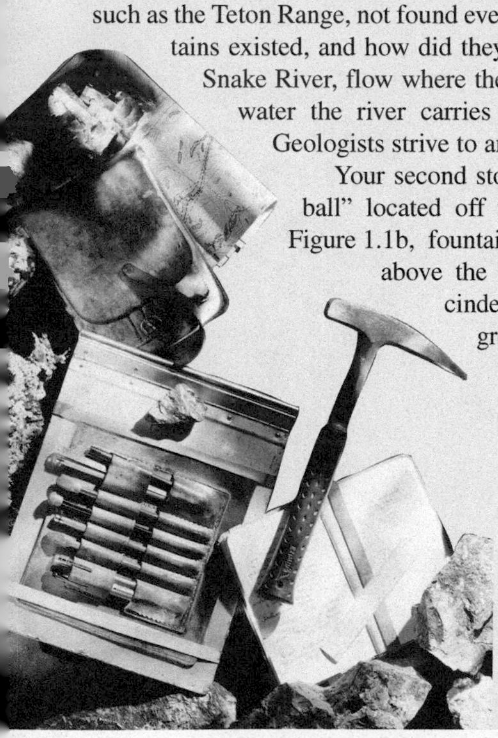

Your third stop is an overlook above a vast, open-pit coal mine in West Virginia. Giant shovels scoop out huge bites of the black rock shown in Figure 1.1c. The coal, carried by oversized trucks to conveyors, moves to equipment that crushes it, sorts the fragments by size, and then delivers them to waiting train cars. After being hidden underground for hundreds of millions of years, the coal will be burned in a power plant to generate electricity.

About half of the electricity generated in the United States comes from coal. What is coal? How does it form and where is it found? The same questions can be asked about a wide variety of Earth resources, from the rocks that are processed for iron, gold, or gemstones to the rocks that host oil, natural gas, and even water.

▶ **Figure 1.1 Why geologists study Earth.** (a) The Teton Range of Wyoming and the valley of the Snake River form one of countless scenic landscapes resulting from geologic processes that have long fascinated humans and inspired our curiosity to know more about how Earth works. (b) Volcanic eruptions, seen here at Mount Etna in Italy, are examples of dynamic Earth processes that are visually stunning but extremely hazardous. (c) The black rock in the bottom of this West Virginia mine is coal. The white rock visible in the walls of the mine is stripped away in order to reach and extract the coal.

1.1 What Is Geology?

Geology is the scientific study of Earth. Geology is commonly introduced in college classrooms as two broad fields—physical geology and historical geology.

Physical geology, the subject of this text, originated as the study of the appearance of Earth's surface and the processes that form surface features such as those seen in **Figure 1.2**. Geologists later concluded that heat energy within Earth drives many of these processes, so physical geology expanded to include the study of the physical and chemical interior workings of the planet.

Historical geology integrates understandings of physical, chemical, and biologic processes to interpret the history of Earth. Historical geology includes the ordering and timing of events in Earth history such as major mountain-building periods, changes in sea level and climate through time, and the evolution of organisms preserved in rocks as fossils.

This text emphasizes the physical geology of Earth. The objective is not only to describe how Earth looks, but also why it looks the way it does and how geologists acquire knowledge about its features and processes.

(a)

(b)

(c)

Geologists Work in Many Fields of Study

Geology integrates aspects of other sciences, such as chemistry, physics, and biology, and applies these diverse concepts to the study of Earth. Each geologic discipline has its own name, but all these studies fall broadly within the science of geology. Some examples of geologic disciplines are listed below.

- Geochemists bring chemical methods and theory to the study of rocks and water and also determine the ages of rocks and of Earth itself.

- Geophysicists apply the concepts of physics and the measurements of physical properties of rock to understand how processes within Earth shape its surface features.

- Paleontologists study fossilized organisms and the biology of past life.

- Planetary geologists work with astronomers, applying geologic knowledge of Earth to explain materials and surface features on other planets and moons.

- Geomorphologists study the origin of Earth's landscapes.

- Mineralogists study minerals, which are the materials that make up rocks.

(a)

(b)

(c)

▲ **Figure 1.2 Earth's surface is varied.** Dynamic geologic processes deform the surface and sculpt it into varied landscapes. (a) Mount Everest, in the Himalayas of southern Asia, is the highest point on Earth. (b) The Grand Canyon of the Colorado River, in Arizona, is but one of many examples of deep gorges eroded by rivers. The canyon is more than 1675 meters deep and carved through rock that formed as much as 1.7 billion years ago. (c) This Illinois landscape results from a combination of geologic processes acting over half a billion years of Earth history. These processes account not only for the flatness of the landscape, but also for the fertile soils that support economically important agriculture.

- Structural geologists study how rocks deform and are uplifted into mountains or depressed into lowlands.
- Resource geologists prospect for energy resources (e.g., coal, oil, gas, uranium) and other important resources, ranging from water to the constituents of concrete to metal ores and gemstones.
- Environmental geologists monitor and protect the environment by determining how human-caused contaminants move through soil and rock, and how to clean up contaminated areas.

During your course of study with this text, you will learn a bit about each of these disciplines within geology and some related scientific disciplines.

The Importance of Time in Geology

Geochemical studies of rocks show that Earth is approximately 4.5 billion years old. You will build your understanding of how we know that Earth is so old in Chapter 7. For now, simply consider how difficult it is to imagine time intervals measured in billions of years! Geologists strive to understand the variety of processes active on the planet at present and how these processes have varied over the enormous expanse of Earth's history.

The great antiquity of the planet requires an appreciation of the slow rates of many geologic processes, as illustrated in **Figure 1.3**. The rates of mountain building and processes that wear down mountainous landscapes are so slow that most aspects of these processes cannot be observed during a human lifetime. The slow rates of geologic processes invite us to contemplate the vast length of time necessary to produce geologic features such as the Himalayas, the Grand Canyon, the Teton Range, and other monumental landmarks on our planet.

Not all geologic processes are so imperceptibly slow. The volcanic eruption in Sicily (Figure 1.1b) altered the landscape as people watched, and large earthquakes can heave the ground surface many meters in an instant. This means that geologists need to understand the dynamics of both fast and slow Earth processes.

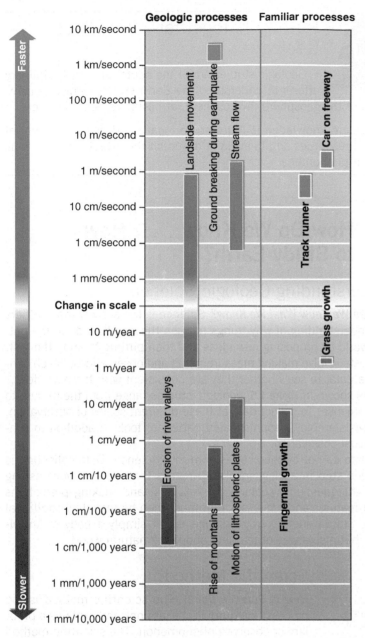

Geologic processes   **Familiar processes**

- Faster / Slower
- 10 km/second
- 1 km/second
- 100 m/second
- 10 m/second
- 1 m/second
- 10 cm/second
- 1 cm/second
- 1 mm/second
- **Change in scale**
- 10 m/year
- 1 m/year
- 10 cm/year
- 1 cm/year
- 1 cm/10 years
- 1 cm/100 years
- 1 cm/1,000 years
- 1 mm/1,000 years
- 1 mm/10,000 years

Labels: Landslide movement; Ground breaking during earthquake; Stream flow; Track runner; Car on freeway; Grass growth; Erosion of river valleys; Rise of mountains; Motion of lithospheric plates; Fingernail growth

◄ **Figure 1.3 Rates of geologic processes.** Many processes affecting Earth's surface take place rapidly, such as the flow of water in a stream or ground breaking during an earthquake (which occurs faster than the speed of sound). Most of the processes that have an enduring effect, however, take place very slowly. Each increment along the scale represents a tenfold increase in the rate of a process from top to bottom. Rate equals distance of movement per specified interval of time. In the metric measurement system, 10 millimeters equal 1 centimeter, 100 centimeters equal 1 meter, and 1000 meters equal a kilometer. Many of these processes occur at rates much slower than the growth of human fingernails. To account for the geologic features that result from such slow but persistent processes, such as mountain building, it is necessary to remember that Earth is 4.5 billion years old.

1.2 Why Study Geology?

We study Earth partly out of curiosity about our planet and our relationship with the Earth system. The Earth system encompasses the interactions among the geosphere, hydrosphere, biosphere, and atmosphere. The geosphere comprises the solid Earth, the sturdy, though ever-changing, foundation of the system. The other components of the Earth system include water that forms the hydrosphere; the gases surrounding Earth to form the atmosphere; and all living organisms, which form the biosphere. On a more practical level, geologic topics affect virtually every aspect of our daily lives, including such diverse issues as the economy, environmental health, and climate. The following discussion touches on just a few of the many reasons to study geology.

Geologists Study Earth Out of Curiosity

People commonly ask questions about their natural surroundings, and because the geosphere forms the foundation of our environment, even basic questions lead to geologic processes. Consider the question "Why does a particular type of tree grow only in certain places?" Answering this question requires more than biologic knowledge about plants, because the growth of specific trees partly depends on climate (e.g., temperature and precipitation) and the available nutrients in the soil. The climate, in turn, partly depends on how moisture from the ocean moves landward in the atmosphere and is released as rain and snow based on locations of the mountains and lowlands of continents molded by geologic processes. Soil forms by the breakdown of rocks resulting from biologic processes, the action of water, and chemical interactions of rocks with atmospheric gases. The disintegration of different rock types leads to the formation of different soils, each with unique nutrient properties. The slope of the land influences the thickness of the soil and its suitability as a viable foundation for trees, because water runoff from rainstorms erodes soil in steep areas. The slope of the land, in turn, results from geologic processes that form and modify the shape of the landscape. Answering the question "Why do these trees grow here?" requires an understanding of multiple aspects of the Earth system. The geosphere is central to the Earth system, so geologic studies are central to human curiosity about our surroundings.

Geologists Study Earth to Find Essential Resources

In addition to curiosity, geologists are motivated to understand the workings of the planet in order to locate and develop essential resources.

Putting It Together—What Is Geology?

- Geology is the scientific study of Earth. Geology includes the study of Earth materials (rocks, soil), the inner workings of the planet, and the origin and changes of surface features.

- Geology includes narrowly focused disciplines, many of which integrate knowledge from other sciences. Each discipline addresses specific aspects of the planet.

- Geologists not only strive to understand the cause and distribution of geologic processes on and within Earth today, but also to describe the 4.5-billion-year history of these processes on the planet.

Everything comes from planet Earth, from the food we grow to eat to the metal we use to build cars and the fuel that a car consumes. Geologists apply their knowledge to find economically viable deposits of natural resources, such as iron ore for making steel. The resources bound up in rocks and the geologic processes that concentrate them in particular places are illustrated in Chapters 2 through 6.

Water for drinking and irrigating crops is the most fundamental resource derived from the planet. We cannot survive without clean drinking water. Geologists use their knowledge of how water is stored and moves on the surface and in the shallow subsurface to locate usable supplies and to protect them from contamination. You will learn more about the geology of water in Chapters 16 and 17.

Geologists Study Earth to Reduce Hazards

Geologists work to diminish the detrimental impact of hazardous geologic processes. For example, an understanding of how different types of volcanoes erupt allows geologists to interpret potential hazards and respond to them. By examining the chemical composition of the fuming gases and the location and abundance of earthquakes below a volcano, it is sometimes possible to forecast the onset of an eruption and to determine the areas at risk in order to plan evacuations.

In addition, geologic information is essential to planning land use and responses to hazardous circumstances. Geologists routinely map the distribution of different materials at Earth's surface. These maps are useful to determine the suitability of a site for building or highway foundations or the instability of hillsides that could fail in disastrous landslides. **Figure 1.4** shows how geologists use their knowledge to produce maps that indicate the risk of damage from earthquakes and other hazardous phenomena.

All natural geologic disasters are expressions of the dynamic nature of the planet. Understanding the geologic processes that cause these events can reduce negative outcomes. This is especially important as current population growth makes it more and more difficult to keep humans out of harm's way.

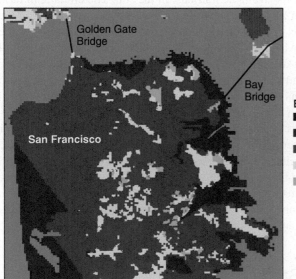

Geologists assign different shaking hazards to different areas based on knowledge of how different types of rock and soil respond to earthquakes.

Expected shaking severity
- ■ Very violent
- ■ Violent
- ■ Very strong
- ■ Strong
- ■ Moderate

▲ **Figure 1.4 Geologists map the risk of geologic hazards.** Many geologic processes are hazardous to humans, and as cities grow, it becomes ever more difficult to ensure people's safety. This map outlines the predicted severity of ground shaking during a possible future earthquake in San Francisco, California. Government agencies use hazard maps to determine what uses are best suited for different areas of land.

Putting It Together—Why Study Geology?

- Geologic studies are at the center of interdisciplinary efforts to understand the Earth system, which is composed of the geosphere, hydrosphere, atmosphere, and biosphere.

- Geologic knowledge is required to locate and develop essential natural resources and to avoid or diminish the effects of hazardous natural phenomena.

1.3 How Do We Know . . . How to Study Earth?

Understanding Geologic Science

How Do We Know What We Know? Science is the activity that furthers our understanding of why things happen the way they do in the natural world by proposing new ideas and then putting them to the test. The tests involve making measurements and observations, which constitute data, to see whether they are consistent with the new ideas.

As shown in **Figure 1.5**, geologic studies range from the minute to the immense. To collect data at these different scales of observation, geologists employ a sophisticated battery of tools in addition to simple visual observation.

Data do not by themselves comprise science. Data collection is just one step in a long, involved process that also includes asking the right questions, seeking the answers, and making predictions about features and processes that must then be tested by additional study. Stated another way, science is not simply a body of knowledge, but also a way of learning about the natural world.

Scientific Method

How Is Science Done? The scientific method is the process of inquiry that examines and explains a problem or observed phenomenon. The scientific method is not a simple, step-by-step procedure for doing science. Instead, the method describes how scientists measure natural features and processes and rigorously test new ideas about how the natural world works.

There is no single "recipe" for good science. Instead, as shown in **Figure 1.6**, scientists undertake many activities that collectively further our knowledge. If Figure 1.6 looks confusing, do not despair. This diagram is not supposed to be memorized; its function is to show the complex interaction of the activities that make up the process of science. A key factor to notice in this diagram is that scientific studies involve many activities and include repeating steps during the course of a research project. Another key item to note is that various paths can be followed to obtain valid conclusions. Unlike the steps of a recipe, the steps in the scientific method do not need to be taken in any particular order; scientific studies commonly take unexpected turns and can be triggered

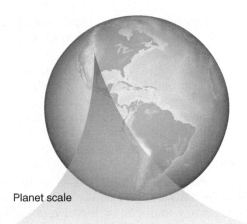

Planet scale

Mt. Rushmore, South Dakota

Landscape scale

Rock sample scale

Microscopic scale

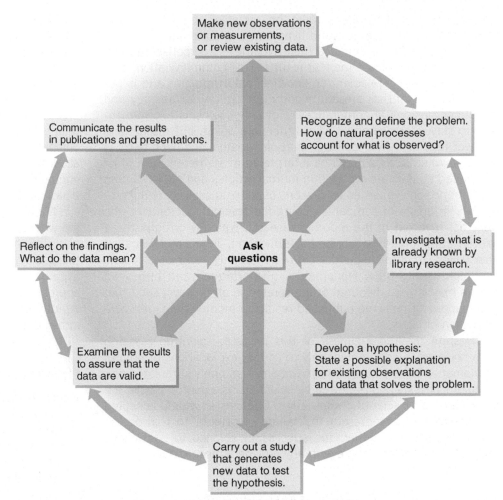

▲ **Figure 1.6 Activities reflect the scientific method.** This chart lists the activities that scientists undertake in their research. The scientific method emphasizes the development and testing of hypotheses. The arrows illustrate the many possible routes that a scientist might follow among these activities. Asking questions is central to every step of a scientific investigation, and questions always initiate an investigation.

◄ **Figure 1.5 Geology at different scales.** Geologic observations range from visualizing the whole planet with satellite images to observing the minute, such as the microscopic components of rocks.

by any number of initial observations or problems. To acquire knowledge we must ask many questions; the process of asking questions is central to the scientific activities depicted in Figure 1.6 and is reflected in the questions that serve as section titles throughout this text.

Scientists start their work by asking a question about something they do not understand. This question results from, or leads to, data collection. The data may include number of occurrences, locations of occurrences, appearances of features or processes, or changes of the features over time. A scientist needs to be imaginative and creative

in order to pose significant questions and to design the study that collects the data to answer the questions.

Once they have posed a question, scientists do library research to review previous work and the collected data on the topic. This research is essential during the early stages of a project, because it enables scientists to find out what the scientific community already knows about the problem and thus helps them to define more precisely the characteristics of the problem they are trying to solve. Data analysis also helps scientists to figure out what they still need to know to answer their original question. Last, the review of prior knowledge prevents the duplication of already completed efforts and commonly leads to a modification or refinement of the problem.

After, or sometimes while, data are collected, scientists propose a **hypothesis,** which is an explanation of the problem that accounts for existing data and predicts additional phenomena that should exist if the hypothesis is correct. A hypothesis is always testable, because it makes predictions about a natural process that can be checked by collecting more data. If these later tests refute the hypothesis, then scientists modify or abandon the hypothesis in favor of a new one that also can explain the new results. If tests undertaken by a number of scientists repeatedly support the hypothesis, then it gains recognition as the logical tested explanation for the studied phenomena. Hypotheses are critical to science, because all scientific ideas must be testable; if ideas are not testable, then they are not valid hypotheses and are not scientific.

After scientists have tested their hypothesis and collected data, they must communicate the test results. This communication is an important part of how scientific knowledge accumulates and is put to use. Most scientists present hypotheses, preliminary results, and initial conclusions at conferences. Not only does this allow scientists to stay informed about what others are doing, but researchers also get feedback from colleagues about whether the hypothesis is sound or whether data could be interpreted in a different way. Most completed studies are reported in journal articles and books, unless the information is guarded for reasons of national security or the proprietary interests of companies doing the research. Scientists who were not involved in the reported research typically critique these written communications prior to publication, in a process called peer review, to help ensure dissemination of only well-tested results and high-quality data. Nonetheless, the critiqued and published results may still be disputed or interpreted in other ways, such that later papers refute some published results or modify them into more robust hypotheses.

Despite the thoroughness of this process, scientific research does not always lead immediately to correct explanations of natural phenomena. Debate about interpretations commonly stimulates new questions or tests and is viewed by scientists as a central part of the advancement of scientific knowledge. Changes in widely accepted hypotheses do not represent the failure of science to provide the "right" result, but, rather, the success of science to continually improve the approach and eventually provide a meaningful solution.

Applying the Scientific Method in Geology

How Do Geologists Do Science? Geologists do not apply the scientific method the same way to every geologic problem. Many traditional scientific hypotheses are tested by laboratory experiments conducted under easily controlled conditions over periods ranging from minutes to several years. For many geologic questions, however, the number of variables is typically too large and the rates of processes much too slow (Figure 1.3) for direct laboratory experimentation. Earth is roughly 4.5 billion years old, so time plays a critical role in geologic processes. Clearly, the investigator must outlive the experiment, and geologists cannot reasonably run experiments that precisely duplicate processes that require centuries or millions of years in nature.

As a result of this problem, Earth is the laboratory for many geological research objectives. Rather than testing hypotheses solely in laboratories, geologists commonly address critical questions through careful observation of what happened on or within the planet. Hypothesis testing may include, for example, observing or measuring features in rocks or landscapes. If the features predicted by a particular hypothesis are present, then the hypothesis may be correct. If the predicted features do not exist, then the hypothesis must be modified or a new hypothesis must be stated and tested.

Geologists use many tools and methods to gather and analyze information about Earth. The simplest data are visual observations made in the field—similar to the observations you made "in the field" at the beginning of this chapter. Geologists also apply sophisticated instruments in laboratories to measure amounts of chemical elements in rocks. Field and laboratory measurements quantify physical properties of rocks, such as density. Other experimental labs reproduce the high temperatures and pressures within Earth's interior.

In addition to in-field observations and laboratory measurements, computer simulations are important tools in geologic studies. With computers, time can be "sped up" to reproduce Earth processes and variables can be changed one at a time to examine their effects on the outcome. Despite the tremendous usefulness of computers, necessary assumptions or simplifications can limit the accuracy of simulations. In addition, many simulations are constrained by computer-memory capacity or the uncertainty of mathematical formulas used to describe natural phenomena.

Many geologic investigations apply multiple working hypotheses whereby, instead of formulating a single hypothesis, the geologist simultaneously considers all conceivable alternatives. Then, the geologist identifies the data required to support a particular hypothesis while simultaneously refuting all other options. This approach reduces bias and enables hypothesis testing to progress rapidly, because many possibilities are evaluated simultaneously rather than one at a time.

Understanding the Big Picture

What Are Theories and the Laws of Nature? Missing so far from this discussion of scientific method are some terms that you have likely encountered in your past science classes, so we address them now. It is important to understand the roles of theories, laws, and principles to advance scientific knowledge.

Principles and **laws** of nature are synonymous terms referring to statements or mathematical formulas that always succeed in describing what is observed to happen. Principles and laws do not explain why things work the way they do, rather, they describe how humans have consistently observed that nature behaves. For example, seventeenth-century astronomer Johannes Kepler formulated the first law of planetary motion, which states that planets orbit the Sun along an elliptical path. The law derives from many undisputed observations but does not explain *why* planets follow elliptical paths around the Sun. This book describes many principles and laws applied by geologists, including the principle of uniformitarianism, which is explained in Section 1.4, and guides researchers to use observed processes to explain geologic features.

A **theory** is a widely applicable and generally accepted explanation for natural phenomena that explains all of the existing data. A scientific theory is a rigorously scrutinized and tested concept and not a hypothesis, a tentative explanation, or an opinion. In common usage, it is often misused in dismissive statements such as "It's only a theory." Laws and principles generalize *how* nature is observed to work; theories explain *why* nature works this way. For example, the theory describing gravitation forces between objects offers an explanation for Kepler's law of planetary motion. All theories undergo further testing by application in new studies and may eventually, although rarely, be found unsatisfactory. Unless data disprove or lead to modification of a theory, that theory serves as the valid accepted explanation of the observed phenomenon in the scientific community. Section 1.5 introduces the theory of plate tectonics, which is the most important unifying idea in geology and is, therefore, threaded as a common theme throughout this book.

Actually, even though the word "theory" does not appear often in this book, almost everything that you will read here is a theory, an observation that led to developing theories, or a prediction that is made by a theory. Theories, simply stated, are the best current explanations of natural phenomena that science has to offer. It is quite possible that some current theories will be discarded in the future if new evidence contradicts the current theory in favor of another. Likewise, existing theories do not explain everything, which is why scientific research continually progresses with a sense of excitement.

In general, principles and theories call for seeking the simplest explanation for natural phenomena. Unnecessarily complicated or overly elaborate explanations are not sought if the tested ideas at hand are adequate. Scientific theories also exclude supernatural explanations—explanations that cannot be observed or measured by scientific procedures. Science is concerned with seeking natural explanations for how the natural world works by posing hypotheses and advancing theories that are always testable through the observation and measurement of natural phenomena. Statements that are not testable in nature can be philosophically valuable, but they are not scientific.

Insights

How Do We Know? The knowledge of Earth delivered in the following chapters results from applying the scientific method. This body of knowledge results from two centuries of observation, experimentation, critique, challenge, reformulation, and testing of hypotheses documented in countless pages of reviewed scientific journals and books. What each generation of scientists takes for granted as established "fact" is actually the collective contribution of previous generations of hardworking researchers who proposed, tested, and established these theories as accepted knowledge. Throughout this book, we endeavor to convey an appreciation for how geologists have come to know what they know. It is impractical to assess every concept, but each chapter includes the scientific background for at least one fundamental geologic idea.

Putting It Together—How Do We Know . . . How to Study Earth?

• The scientific method seeks to understand natural phenomena by integrating the actions of asking questions, proposing explanatory hypotheses, and testing hypotheses.

• Geologists use laboratory experiments less often to test hypotheses than do other scientists, because the variables governing Earth processes usually are too many and the rates of the processes typically are too slow for direct experimental analysis.

• Principles (or laws) are generalizations about how nature is observed to work, whereas theories offer well-tested and accepted explanations of why natural systems work this way.

• To be scientific, hypotheses and theories must be testable by observing and measuring natural phenomena. Science remains an exciting and active human pursuit because all of nature is not explained by existing theory.

1.4 What Is the Principle of Uniformitarianism?

Geologists study observable processes happening on Earth today, and they interpret processes that occurred in the past based on features found in rocks and landscapes. Observing active processes is critical to a geologist's ability to interpret the ancient origins of rocks and landscapes. To understand this link between processes seen and unseen, let's consider two examples.

Compare the rippled sand shown in **Figure 1.7**, formed by the to-and-fro swash of water on a sandy beach, with the identical feature seen in a consolidated rock. It is possible to guess that the origin of the ripple features in the ancient rock is the same process observed today to form ripples on the beach. In reaching this conclusion, an observer infers the origin of the ripple marks in the ancient rock without having observed the process that formed them.

Likewise, although no one has witnessed the formation of a mountain range thousands of meters high, numerous documented instances exist of ground surfaces heaved upward many meters during single earthquakes.

▲ **Figure 1.7 An application of uniformitarianism.** Ripple marks on an ocean beach (top photo) are produced by the back-and-forth movement of waves. The surface of a rock layer hundreds of millions of years old shows similar ripple marks (bottom photo), and the rock contains sand grains much like those found on a modern beach. The similarities of the modern and ancient features indicate that the ancient rock formed in a beach environment. These photographs illustrate an application of the principle of uniformitarianism, whereby geologists interpret features preserved in ancient rocks in terms of observed processes.

Mountains uplifted during many earthquakes over a long time interval

Land surface broken and raised up 2 meters during earthquake

▲ **Figure 1.8 Uniformitarianism explains mountains created by uplift during repeated earthquakes.** Uplift during an earthquake in Taiwan in 1999 raised part of an athletic field and running track by two meters. Adding up the uplift caused by many earthquakes over long intervals of geologic time can explain the elevations of mountains.

Figure 1.8 illustrates an example of such instantaneous uplift during a recent earthquake. These observations lead us to readily accept the idea that mountains of broken rocks result from the combined effects of rock movement during innumerable earthquakes over long intervals of time.

These geologic interpretations are examples of the application of the **principle of uniformitarianism**. The principle states that observations of both a process *and* its result can apply to the interpretation of other, similar results where the process was not observed. In other words, if you can understand the geologic processes that are responsible for materials and features you presently see in nature or in the laboratory, then you can infer that similar materials and features found in ancient rocks and landscapes are the result of these same processes.

Two questions commonly arise when applying modern observations of geologic processes to interpreting events in the past: (1) Does uniformity of process also require that the rates of processes must be the same through time and limited to the values measured during the short human history of geologic study? (2) Is it possible that some processes acted in the past that humans have not witnessed during recorded history?

The immensity of geologic time plays an important role in answering questions about the application of uniformitarianism, because geologic studies of active processes have been ongoing for only a minuscule fraction of the 4.5 billion years of Earth history. Certainly, many geologic processes occur very infrequently (e.g., collisions of meteors and comets with Earth) or over a wide range of magnitudes from small to large (e.g., volcanic eruptions, floods). Two centuries of geologic investigation and a few thousand years of recorded history are insufficient for observing very rare or extremely large events. If the young Earth was hotter, for example, processes driven by thermal energy would have taken place at faster rates. Rates of erosion by wind and streams would have been higher before the appearance of rooted plants on land, which occurred 400 million years ago. With these examples in mind, and to address the questions posed above, geologists assert that we should not place too many restrictions on applying uniformitarianism, especially with regard to the rates and conditions of processes that might change over time.

The principle of uniformitarianism states the most fundamental concept in all sciences: Scientists use examined cases, where they observe both process and result, as a guide to explain features for which only the result, and not the active process of formation, is seen. Ancient rippled sandstone is only one of many examples where geologists see the result of a process without actually seeing the process at work. The application of uniformitarianism does not limit these examined causes to events that have actually been witnessed by humans. The results of laboratory experiments and mathematical calculations based on known natural phenomena also guide our interpretations of unseen processes concealed from view beneath the surface of Earth or that occurred in the ancient past. These experiments and calculations permit scientific simulation of processes that cannot be directly observed or that operate too slowly to study in nature.

If uniformitarianism guides interpretation of geologic history, then can it also be used to forecast future events? In a general way, yes, because scientists infer that the same natural processes operating today operated in the past and will operate in the future. This does not mean, however, that scientists can predict exactly when or where a geologic event will occur. For example, geologists can study the behavior of past eruptions of a volcano by observing the resulting lava flows and layers of volcanic ash, and then hypothesize that future eruptions will feature similar behaviors. This application of uniformitarianism does not, however, mean that the time and severity of the next eruption can be forecast far in advance.

Putting It Together—What Is the Principle of Uniformitarianism?

• The principle of uniformitarianism states that examined cases of both a process and its result can guide interpretation of visible results where the process was not witnessed. We can interpret ancient geologic features by understanding active processes observable today in nature or in the laboratory.

1.5 What Is the Theory of Plate Tectonics?

The theory of **plate tectonics** explains a vast array of geologic processes and features as being the result of the motion of the separate plates that make up the outermost 100 kilometers of Earth. The theory, developed during the 1960s, has had an immense impact on geology in recent decades, because it sparked revolutionary advances that simultaneously explained and integrated observed phenomena, while redirecting the focus of new research.

To appreciate the evidence for plate tectonics, you first must develop sufficient geologic knowledge to understand the strength of the supporting arguments. Here, we introduce some basic facts about the interior of Earth and the broad features of the theory; Chapter 12 fully develops the theory using knowledge gained from the intervening chapters.

The Interior of Earth

The interior of Earth, depicted in **Figure 1.9**, consists of three concentric layers—the **crust**, **mantle**, and **core**—that differ from one another in

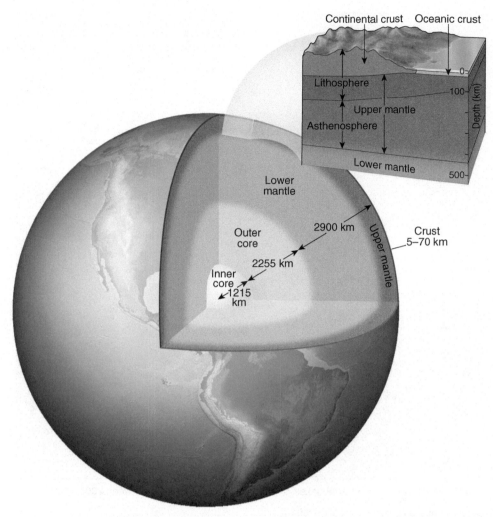

▲ **Figure 1.9 What Earth's interior looks like.** Earth consists of three major concentric layers, beginning at the center with a solid inner core and followed by a liquid outer core and a solid mantle and crust. The composition and physical properties of these layers, and the thinner sublayers within them, vary considerably. You regularly see only the upper part of the relatively thick continental crust, which differs from the thin crust under the oceans. The lithosphere is the firm, rigid outer skin of Earth consisting of the crust and uppermost mantle, which overlies a weaker layer of upper mantle called the asthenosphere.

composition and physical properties. Also, rock samples show that the crust on the continents differs from the crust beneath oceans. Knowledge of Earth's deeper layers comes primarily from measurements of how the planet shakes during earthquakes, which you will learn more about in Chapter 8. The mantle is Earth's thickest layer and consists of rocks that are distinct from the rocks that make up the crust. The deepest part of the planet, the core, consists of metal and is divided into the inner, solid core and the outer, liquid core. Geologists further distinguish between two zones whose differences in strength help to explain many aspects of how Earth works. The crust and uppermost mantle, down to about 100 kilometers, compose a strong outer lid on Earth's surface called the **lithosphere** (rocky sphere), which is labeled in Figure 1.9. The remaining upper mantle makes up the less-rigid, squishier, but mostly solid layer called the **asthenosphere** (weak layer).

Tectonic Plates

Plate tectonics theory explains that the outer shell of Earth, the lithosphere, is broken into many slabs called **plates**, which are roughly 100 kilometers thick. These lithospheric plates contain both continental and oceanic crust and the upper part of the immediately underlying mantle. **Figure 1.10**

depicts the outlines of plates on Earth today. The plates are rigid and strong, in contrast to the underlying, weaker asthenosphere.

Types of Plate Boundaries

Plate tectonics theory also describes the motion of the plates, which travel at speeds of a few centimeters per year, which is comparable to the rate of fingernail growth (Figure 1.3). The moving plates collide, move apart, or slide past one another at plate boundaries. **Figure 1.11** shows these three types of possible motions between two plates at their mutual boundary. The interactions between plates at these boundaries cause deformation of the lithosphere that is described by the term **tectonics**.

Plate edges move away from one another along **divergent plate boundaries** (Figure 1.11a). The Mid-Atlantic Ridge, for example, marks the divergence of the North American and Eurasian plates in the center of the Atlantic Ocean (Figure 1.10). As the plates separate, the resulting open gash in the lithosphere fills with molten material that rises from the underlying asthenosphere. The molten material solidifies to form new seafloor along a line of submarine volcanoes. This process causes the Atlantic Ocean to widen by about 5 centimeters each year as new lithosphere forms along the edges of the two plates at the mid-ocean ridge.

Plates collide at **convergent plate boundaries** (Figure 1.11b). Where the plates converge, one plunges into the deeper mantle in a process called **subduction**, while the overriding plate experiences volcanic activity and buckling that uplifts tall mountain ranges. Earthquakes, mountains, and active volcanoes in the Pacific Northwest result from a convergent boundary just offshore of Oregon and Washington (Figure 1.10).

Plates slide past one another along **transform plate boundaries**, without creating or destroying lithosphere (Figure 1.11c). An example of a transform boundary is the San Andreas Fault, which runs along nearly the entire length of California and marks where the Pacific plate slides past the North American plate (Figure 1.10).

Plate tectonics explains the tendency for earthquakes and active volcanoes to concentrate in long, narrow belts (Figure 1.10). The breaking and folding of rocks that accompany earthquakes are focused at plate boundaries, while there is much less deformation within plates. The largest and most damaging earthquakes happen at or near convergent and transform boundaries, whereas divergent-boundary earthquakes are small and occur mostly in the center of oceans away from cities and, therefore, rarely cause damage. Convergent and divergent plate boundaries are associated with processes that promote melting in the mantle, thus focusing volcanoes in narrow belts. We will examine the melting processes along plate boundaries in Chapter 4.

Hot Spots

Looking at the map in Figure 1.10, you may wonder, if volcanoes usually relate to plate-boundary processes, then how do geologists explain the volcanic Hawaiian Islands found near the center of the Pacific plate? Volcanoes such as those in Hawaii mark the locations of **hot spots**, where molten

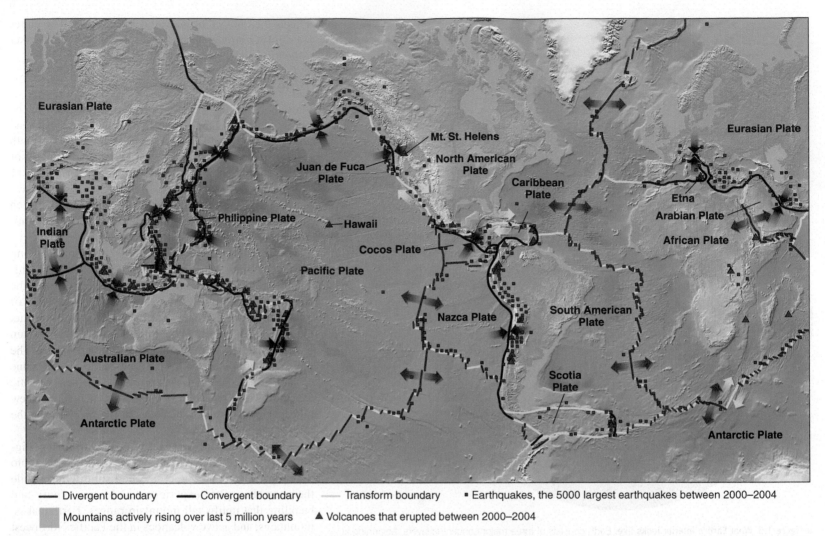

— Divergent boundary   — Convergent boundary   — Transform boundary   ▪ Earthquakes, the 5000 largest earthquakes between 2000–2004

▨ Mountains actively rising over last 5 million years   ▲ Volcanoes that erupted between 2000–2004

▲ **Figure 1.10 The world composed of plates.** This world map shows the outlines of the lithospheric plates whose motion is explained by the plate tectonics theory. Figure 1.11 explains the three types of boundaries between plates, which depend on how neighboring plates move relative to one another. Notice that nearly all historically recent active volcanoes, recent earthquakes, and geologically recent mountain building occur near plate boundaries.

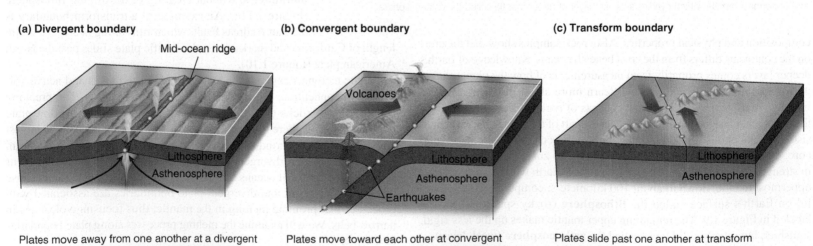

(a) Divergent boundary

Plates move away from one another at a divergent boundary, producing shallow earthquakes and volcanic activity where partly melted asthenosphere rises and solidifies to form new lithosphere in the gap. Most divergent plate boundaries coincide with mid-ocean ridges, which snake through the oceans.

(b) Convergent boundary

Plates move toward each other at convergent boundaries and one plate subducts beneath the other. Earthquakes occur over a wide range of depths where the rigid lithosphere thrusts downward into the weaker asthenosphere. Volcanic activity is triggered above the downward-moving slab to produce lines of volcanoes.

(c) Transform boundary

Plates slide past one another at transform boundaries. The sliding of one plate past its neighbor generates abundant earthquakes but is not typically associated with volcanic activity.

ACTIVE ART

Motion At Plate Boundaries: See how the plates move along their boundaries.

▲ **Figure 1.11 What happens at plate boundaries.**

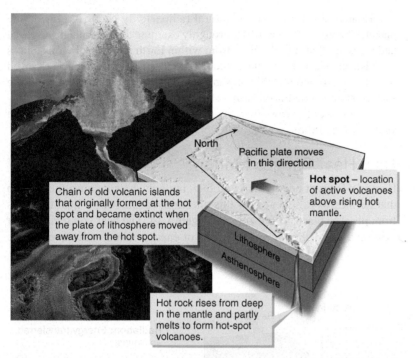

▲ **Figure 1.12 Hot spots are volcanic areas within plates.** The Hawaiian Islands are an example of a hot spot volcanic chain formed where rising deep-mantle material penetrates the interior of a lithospheric plate. The hot spot probably moves slightly through time but not as rapidly as does the plate. As a result, volcanoes are active only directly above the hot spot and become inactive as plate movement carries them away from the feeding conduit of hot mantle.

ACTIVE ART

Hot Spot Volcano Tracks: See how an island chain forms by plate motion across a hot spot.

material rises from deep in the mantle below the moving lithosphere as a result of processes not explained by plate tectonics. **Figure 1.12** shows how large volcanic islands build up as the Pacific plate moves over the Hawaiian hot spot. A volcano becomes extinct as plate motion carries it away from the hot spot, the area of rising hot mantle, and a new volcano appears in its wake. In Chapter 12, you will explore a hypothesis for the origin of hot spots and the likely relationship between hot spots and plate tectonics.

Anticipating the Evidence for Plate Tectonics

At this point in the text, perhaps the most compelling evidence to support plate tectonics is the ability of the theory to explain the restricted distribution of most earthquakes, active volcanoes, and young mountain belts. As we progress through the study of topics that include rocks, the internal workings of the planet, rock deformation, and the evolution of landscapes, you will see how plate tectonics provides the basis for understanding far more than earthquakes and volcanoes. It is the ability of plate tectonics to explain all of the topics listed above that makes it the centerpiece of geologic studies and earns it the title of "theory."

Putting It Together—What Is the Theory of Plate Tectonics?

• The plate tectonics theory states that the outer shell of Earth, the lithosphere, is divided into plates that move toward, away from, or past one another.

• The interaction of plates at their boundaries accounts for the distribution of earthquakes, volcanoes, and actively growing mountain belts and provides a basis for interpreting most geologic processes and products.

1.6 How Does the Concept of Work Apply to Earth?

Where does the energy come from to drive plate tectonics? Earth is an active place. The slow shifting of the lithospheric plates across the entire surface of Earth leads to ground-shaking earthquakes, violent volcanic eruptions, and gradually rising mountains. Water flows in streams and erodes rock particles that are carried to the ocean, where crashing waves move the particles onto sandy beaches. These are but a few of the observed Earth processes that move material mass from one place to another. These motions are evidence of work, just like the work you do to move an object from one place to another. **Work**, then, can be calculated by multiplying the distance something moves times the force required to move it. This text, as its title suggests, explains the many types of work that occur within Earth and on its surface.

Work Requires Energy

Energy is the measure of the ability to do work. This idea may be apparent from everyday experiences. A moving automobile is evidence of work. Burning gasoline provides the energy required for that work. If the automobile runs out of gas, no more work can be done, and the vehicle stops moving. You obtain energy from the food that you eat, which processes within your body convert to energy. Your body requires food as "fuel," and the accumulated body energy "burns" during your daily activities, which are a form of work.

Energy is most obvious when motion takes place, but energy also is stored. A battery is a familiar example of stored energy. Chemical reactions inside the battery are capable of producing electricity. If wires are connected to the battery, then the electricity lights a bulb, causes a toy to move, or produces other kinds of work. The work consumes the energy in the battery until it is drained.

Energy exists in a number of forms. In addition to stored energy and motion energy (also known as kinetic energy), there is radiant energy, such as the visible light or detectable heat emitted by a lightbulb. Energy transfers from one place to another and from one form to another. The stored chemical energy in a battery transfers through wires as electrical current, which may transform to motion energy in a toy or be converted to radiant energy in a bulb.

Work on and within Earth requires energy. This energy is stored, it transfers to cause motion and do work, and it moves as heat. While examining and understanding the active work represented by geologic processes described in this text, we need to also consider the energy required to do that work. **Figure 1.13** compares natural energy sources and energy expenditures with human energy consumption. It is challenging to think about all the energy consumed by humans across the entire planet, especially for transportation, generating heat and electricity, and manufacturing, yet natural energy producing and consuming processes are of comparable magnitude.

Heat Drives Geologic Processes

Heat represents the most important form of energy transfer for geologic processes. Heat from the Sun fuels Earth-surface processes, such as wind, ocean waves, and the evaporation and precipitation of water. Solar heat results from chemical reactions within the Sun, so in a sense, the Sun is like a huge battery of stored energy that constantly radiates heat and light energy through space.

The total energy output of the Sun each second is equivalent to the energy in about 4×10^{24} (4 followed by 24 zeros) 100-watt lightbulbs. Not all that energy reaches Earth, because the heat moves outward in all directions from the Sun and Earth is just a small speck in space about 150 million kilometers away from the Sun. So, Earth is in the path of only a tiny part of the heat from the Sun. In addition, not all the energy radiating from the Sun is heat. Some of the energy is visible light, and some of it is in the form of ultraviolet rays, which cause sunburn. Still, the warmth you feel on a sunny day clearly indicates the arrival of heat energy from the Sun. The amount of solar energy received by each square meter of Earth's surface is about one-tenth the amount of energy that emerges as light and heat from a 100-watt bulb. Some of the energy is consumed by work on Earth's surface and in the atmosphere, but most of it radiates back into space.

Heat is also produced by natural radioactivity occurring inside the planet. Chapters 7 through 10 thoroughly examine this internal heat source and the work that it fuels. Heat from within Earth is rarely obvious unless you visit an unusually hot place, such as a volcano, because the heat received from the Sun is about 4000 times greater than the heat reaching most of the surface from inside Earth. Nonetheless, this internal energy source is sufficient to power the work done by plate tectonics for more than 20 billion years into the future.

How Heat Causes Motion

Heat is a form of energy that you feel, but how does it cause the motion that defines work? To answer this question, consider how heat transfers from place to place, as illustrated in **Figure 1.14**. The heat you feel on your skin on a sunny day moves as waves of energy called **radiation**. As we

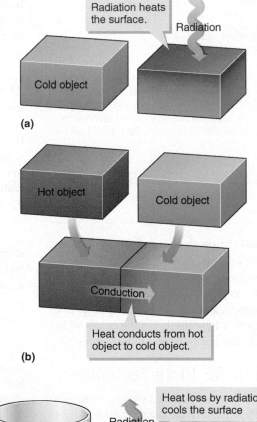

Radiation: Energy transferred as waves.

Example: Heat radiates through space from the Sun to Earth.

Conduction: Heat transferred by rapidly vibrating molecules when a hot object is placed against a cold object.

Example: Heat conducts from a stovetop burner to the bottom of a pan.

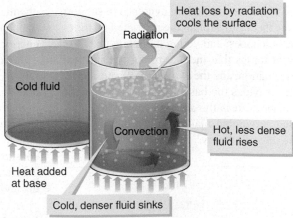

Convection: Simultaneous transport of material and heat caused by sinking of cold, dense material and rise of hot, less dense material. Radiation and conduction typically convey heat in or out of the convecting material.

Example: Heat convects when fluid turns over in a liquid heated on a stove.

▲ **Figure 1.14 How heat transfers from place to place.**

Annual world energy consumption multiplied by...

World energy consumption

Solar energy reaching Earth's surface

Internal heat energy reaching Earth's surface

Energy required to drive plate tectonics

Energy released by earthquakes

Energy consumed by river erosion

▲ **Figure 1.13 Visualizing energy numbers.** This bar graph illustrates the magnitude of geologically important energy sources and uses during one year as compared to humans' annual energy consumption. Human energy consumption includes the use of electricity and the burning of oil, coal, and natural gas. For example, annual solar energy reaching Earth's surface is almost 10,000 times greater than world energy consumption, whereas internal heat energy reaching Earth's surface each year is only three times greater.

have noted, light is another example of radiation, as are microwaves that heat food. The painful burn you feel when you touch a hot stove results from another type of energy transfer called conduction. **Conduction** is the transfer of heat from a hot object to a cooler object. Measurable heat in the hot object is caused when atoms that make up the object vibrate rapidly and collide. The rapidly vibrating atoms in the hot object collide with the less-active atoms in the colder substance, setting them in rapid motion, too.

Figure 1.15a illustrates how conduction of heat from a stove burner into a pot of liquid causes the liquid to move. The liquid heated by conduction at the bottom of the pot expands as the atoms of the liquid vibrate more energetically. These vibrations cause the hot liquid to expand and fill a larger volume. Expansion causes the hot liquid to become less dense than overlying, cooler, and unexpanded liquid. The denser, heavier liquid, therefore, sinks toward the bottom of the pot, which displaces the less dense, warmer liquid to the top of the pot. Transfer of heat energy into the liquid caused the liquid to move, and the moving mass of liquid carries heat with it. The motion that transports heat and material is called **convection**. When the warm liquid reaches the top of the pot, some of the heat radiates into the colder air above the pot. As this surface layer of liquid cools, it also becomes denser and sinks, while hotter liquid in the bottom of the pot rises to replace it—and the convection cycle begins again.

Convection in nature resembles convection on the stovetop. Where solar heating warms the atmosphere, the air expands and rises, while cooler air sweeps in to replace it, as shown in Figure 1.15b. We feel this motion in the atmosphere as wind. Liquid water evaporates where the Sun warms air, ground, and water surfaces. The warm air then moves upward by convection, where the air cools and the water in it condenses back into liquid. The liquid water falls as rain or snow that flows on the planet surface as streams, which do work eroding surface rock and soil. Therefore, atmosphere convection is an important observed process for shaping Earth's surface.

Convection also causes the motion of plates that drives plate tectonics, as shown in Figure 1.15c. Cold, dense rock near Earth's surface sinks, and less dense rock expanded by radioactive heating in the interior rises. Convection motion of the mostly solid mantle is extremely slow compared to the convection revealed by a bubbling pot on a stove or a windy day. The motion is sufficient, however, to move the rigid lithosphere in much the same way that patches of oil move on the surface of hot soup on a stove. Convection of rock may seem implausible, but Chapter 10 examines the scientific rationale for the phenomenon and explains convection in the molten outer core, which generates Earth's magnetic field.

Landscapes Have Potential Energy

Convection within Earth drives plate tectonics, and plate tectonics, in turn, causes uplift of mountains. Considerable

Heat conducts from a stovetop burner to a pot of soup. The heated liquid is less dense and rises upward as colder, denser liquid sinks to replace it. This circulation is convection and not only moves the soup up and down in the pot, but also transports heat upward with the warmer liquid.

Where solar heating warms the atmosphere, the air expands and rises while cooler air sweeps in to replace it. The air movement near the surface is wind.

Cold, dense lithosphere sinks at convergent plate boundaries. Sinking lithosphere pulls the plates apart at divergent plate boundaries. Hot spots are places where hot, low-density mantle convectively rises to the surface. Plate tectonics and hot spots, therefore are convection.

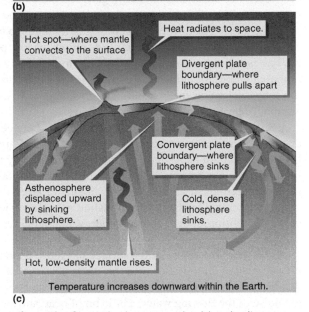

▲ **Figure 1.15 Visualizing how convection works.** Convection is a process involving simultaneous transport of heat and matter.

The higher the shelf you choose to lift the box to, the more work you do.

The potential energy of this box is less than the box on the higher shelf.

Potential energy is converted to motion energy, sound energy, and heat when the box falls to the floor.

▲ Figure 1.16 **Visualizing potential energy.**

work must be done to raise mountains, because the force of gravity pulls all materials toward Earth's interior. The higher that rocks are uplifted, the greater the work required to overcome the downward gravitational pull.

The effect is similar to lifting a heavy box from the floor to a shelf, as illustrated in **Figure 1.16**. The higher the shelf you choose to lift the box to, the more work you must do. Once placed on the shelf, the box may appear to just sit there, but it actually contains a form of stored energy called **potential energy**. Potential energy is the energy an object possesses because of its elevation and weight. The potential energy of the box on a high shelf is greater than the energy of the same box on a lower shelf or on the floor. The difference in energy between the box on the floor and the box on a high shelf is equal to the work done to move the box from the floor up to the shelf. The downward pull of gravity tends to move objects from positions of high potential energy to positions of low potential energy. If the box is not secured on the shelf, then it may fall down to the floor. The potential energy stored when the box was stationary on the shelf converts to motion energy as the box falls, and to sound-wave energy and a minor amount of heat when the box hits the floor.

Similarly, differences in potential energy between high elevations and low elevations in landscapes drive processes on Earth's surface. Rocks high on mountain peaks have higher potential energy than those on valley bottoms. The energy difference causes rocks to fall or slide down a slope, sometimes in impressive movements of large masses of material, commonly called landslides. A landslide is a conversion of the potential energy to motion, sound, and heat energy analogous to a box falling from a shelf.

Likewise, water on Earth's surface has more potential energy at high elevations than in lowlands, and this energy decreases as the water flows downhill in stream channels to the ocean. The potential energy converts to the noise of the flowing water, a little bit of heat, and mostly to motion energy that allows the stream to do work. Erosion and transport of rock material are the evidence of the work done by the stream. Each year, streams

erode about 14 cubic kilometers of sediment (gravel, sand, and mud) from the land surface, an amount that could fill enough railroad cars to form a train that encircles the equator 34 times. Clearly, considerable work is done to move this much mass for distances as great as several thousand kilometers. The annual energy consumed by stream work to erode Earth's surface is, however, less than one-half of 1 percent of the energy released each year by earthquakes (Figure 1.13), which continually deform Earth's surface, creating greater potential energy differences and ultimately causing even more stream erosion.

Putting It Together—How Does the Concept of Work Apply to Earth?

• Movement of mass within Earth, on Earth's surface, and in the atmosphere is evidence of work.

• Work requires energy.

• Geologically important energy sources are heat from the Sun and the internal heat of Earth, which results from natural radioactivity.

• Heat energy is transferred from place to place by radiation, conduction, and convection. Convection involves simultaneous movement of mass and heat, and powers geologically important motion in the gaseous atmosphere and in the solid and liquid parts of Earth's interior.

• Potential energy is an important form of stored energy that objects possess because of their elevation. Conversion of potential energy to motion energy occurs when materials move from high elevation to low elevation.

Where Are You and Where Are You Going?

This chapter has given you a basis for studying geology. You know the fields of study encompassed within geology and how these studies help us understand the Earth system, as well as some of its beneficial resources and perilous hazards. You also have learned that scientists design their work to test specific hypotheses and develop guiding theories and principles. You have been introduced to the principle of uniformitarianism and the plate tectonics theory. Uniformitarianism states that the observed link between processes and results in a few cases can be used to explain results in cases where the processes were not observed. Plate tectonics describes the slow motion of Earth's surface slabs, called plates, to produce earthquakes, volcanoes, mountain ranges, and other geologic features. In the chapters ahead, you will develop an even stronger appreciation of how these concepts explain geologic phenomena.

Earth is a dynamic, active planet. Active motion requires energy. Geologic processes are powered by heat energy from the Sun and from within the planet. These energy-fueled processes reveal how Earth works.

The chapters that follow are arranged into four parts. Before delving deeply into understanding routinely observed geologic features, you must first understand what the planet is made of. The first part of the book (Chapters 2–7) examines the solid Earth materials that are seen at the surface. The second part (Chapters 8–10) turns your attention to discovering the materials and processes hidden from view within the planet. This knowledge will help you appreciate internal processes that drive external deformation through plate tectonics. Deformation at and near the surface, and formulation of the plate tectonics theory compose the third part of the book (Chapters 11–13). The final part of the text (Chapters 14–21) explains the origin of landscapes by combining knowledge of Earth materials and rock deformation with observations of processes acting on the surface.

In essence, as its title states, this book aims to answer the question "How Does Earth Work?"

Active Art

Motion at Plate Boundaries: See how the plates move along their boundaries.

Hot Spot Volcano Tracks: See how an island chain forms by plate motion across a hot spot.

Confirm Your Knowledge

1. What is geology?
2. List two examples of fast Earth processes and two examples of slow Earth processes.
3. How old is Earth?
4. Geologists study Earth to find essential resources. What are some of these resources?
5. In many scientific disciplines, hypotheses are tested by direct laboratory experiments. Explain two reasons why direct laboratory experimentation is not always possible in geology.
6. What is the difference between a law and a theory? Give an example of each.
7. Why are supernatural explanations not included in scientific theories?
8. What is the principle of uniformitarianism, and can it be used to predict the future?
9. Earth is composed of three concentric layers. What are they, and how do they differ?
10. List and describe the three types of plate boundaries. Give one geographic example of each.
11. Energy drives Earth processes and when utilized, gives off heat. What are the two major sources for energy in and on Earth? How do they differ from potential energy?
12. List and describe the three types of energy. Give an example of each.
13. List the three types of heat transfer. Give an example of each.
14. Each year streams erode about 14 cubic kilometers of sediment. This amount could fill a train of railroad cars that would encircle Earth's equator 34 times. If this amount of sediment was spread over your home state, how thick would it be?

Confirm Your Understanding

1. Write an answer for the question in each section heading.
2. The text illustrates how answering a question such as "Why do these trees grow here?" requires an understanding of climate, soil formation, and slope development. What geologic information would you need to understand in order to answer the question "Where should we build a dam to provide water for our town?" List and explain at least three types of geologic information you will need.
3. Students in introductory science courses commonly are taught that the scientific method consists of formulating a hypothesis, testing that hypothesis, and developing a theory based on the results of the test(s). In practice, the scientific method is much more complex. Explain how.
4. If you were going to explore a planet orbiting around a star other than our Sun and you wanted to determine whether plate tectonics was occurring there, what would you look for? What instruments would you bring to test for the operation of plate tectonics?
5. You find fossil horseshoe crabs in a 100-million-year-old rock. You go to the coast and see horseshoe crabs living close to the shoreline. What can the principle of uniformitarianism tell you about the environment of deposition of the ancient rock?
6. How do hot spots, which are not explained by plate tectonics, help to confirm that lithospheric plates are moving?
7. Given what you know about convection, consider how global ocean circulation must work if the deep waters in the equatorial regions are known to be cold and surface waters are warm, but the latter cool when transported to cold polar regions.
8. Of the various reasons to study Earth, which is the most important to you? Why?

CHAPTER 2

Minerals: The Building Blocks of the Planet

Why Study Minerals?

Perhaps you have already wondered about the objects that geologists call minerals. Minerals include the beautiful natural gemstones that fill displays at jewelry stores, the astonishing specimens you see exhibited in museums, and the incredibly rare field examples such as those seen in the photo on the right. Like you, geologists admire the beauty of these mineral specimens, but geologists also study less spectacular examples, the ordinary and usually small minerals that appear in every rock on Earth.

Minerals are the building blocks of our rocky planet, because rocks consist of mixtures of minerals. Minerals form under a wide range of physical and chemical conditions that provide clues to the processes that generate Earth's lithosphere and deeper layers. Minerals containing iron, aluminum, lead, zinc, copper, silver, gold, and platinum are valuable economic resources. Other minerals are utilized in manufacturing useful commodities, from talcum powder for cosmetics to zeolite filtering agents for water purification and graphite for pencils. Biological processes also make minerals, ranging from seashells to pearls, and from bones to teeth.

After Completing This Chapter, You Will Be Able to

- Explain that each mineral has a definitive chemical composition and internal atomic structure.

- Relate physical properties of minerals to their chemical composition and structure.

- Describe examples of minerals that are important to your life and the products you use.

- Refer to the names, compositions, and properties of an important handful of the more than 4000 known minerals.

The largest natural crystals on Earth—gypsum crystals in the Cave of Swords, Mexico.

IN THE FIELD

To better appreciate a geologist's view of minerals, imagine hiking in unfamiliar mountains close to the home of friends. The view from afar, seen in the photo in **Figure 2.1**, reveals bold, rocky cliffs. "What kind of rocks are these?" you ask. "Granite," a friend answers. "You know, the stuff they use to make countertops, gravestones, and fancy walls inside office buildings and hotels." Prior to this, you may not have noticed these Earth materials, which are commonly used by humans, in their natural settings. At the trailhead, you take a closer look and notice that each solid outcrop and rock fragment lying on the ground is a mixture of many differently colored materials, visible in Figure 2.1.

The many-colored components of the granite interlock like the parts of a fine tile mosaic or inlaid jewelry. Each component varies in size, shape, and color, the latter ranging from black, pink, or white to translucent gray.

Hiking on the trail, you occasionally notice flashes of light from tiny particles on the ground; the flashes are caused by sunlight reflecting off the surfaces of small pink fragments. These fragments are the same color and shape as some of the components in the piece of granite you dropped into your backpack as a memento. The pink fragments apparently have crumbled away from the rest of the rock. Looking more carefully, you also find loose pieces of the white, black, and translucent components of the granite on the ground. A few of these grains have reflective surfaces, while others do not.

What are these differently colored, solid materials that somehow joined together to form a rock? These are minerals. In the simplest terms, minerals are the building blocks of rocks, because every rock is an aggregate of one or many minerals.

These minerals, however, are not the same "minerals" referred to in the nutrition labels of food, vitamin, and bottled water products. These labels list chemical elements, such as iron and calcium, whereas most geological minerals are compounds composed of two or more elements. However, not all compounds of elements are minerals, of course. To develop and understand the geologic definition of "mineral," which goes beyond simply saying that minerals are the materials that make up rocks, you need to learn more about their characteristics.

▶ Figure 2.1 **What you see in a rock.** Many rocks look uniform from a distance, but on closer examination consist of smaller components, called minerals.

2.1 What Are the Properties of Minerals?

A good way to understand what minerals are is to examine them closely and describe their properties. Consider the two minerals shown in **Figure 2.2**, which are named calcite and quartz. These are important minerals: calcite is the principal raw material for making cement (used to make concrete); quartz is used to make glass, as an abrasive on sandpaper, and as a gemstone (amethyst, carnelian, onyx), and it even has electrical properties that make it useful in the manufacture of radios, digital clocks, and watches. Calcite also forms some seashells and coral. At first glance, calcite and quartz look very similar. Both are transparent, six-sided, and have similarly shiny, reflective surfaces. **Luster** is the term used to describe how mineral surfaces reflect light.

Notable differences exist between the two minerals, however. They have distinct **crystal faces**, which are smooth, flat surfaces with regular geometric outlines. Although both minerals are six-sided, the shapes of their crystal faces differ. The calcite crystal resembles two six-sided pyramids placed together along jagged bases. Quartz, on the other hand, is a six-sided prism capped by a six-sided pyramid of unequally sized faces (Figure 2.2).

Minerals Have Characteristic Density

When similar-size specimens of calcite and quartz are held in each hand, they seem to weigh about the same, which suggests that they have similar density. **Density** is a measure of the mass of a material (the amount of matter an object contains) divided by its volume (how much space an object occupies). Density is typically measured in grams per cubic centimeter (g/cm^3), kilograms per cubic meter (kg/m^3) or pounds per cubic feet (lb/ft^3).

Minerals Have Characteristic Hardness

Hardness is a measure of the resistance of a mineral surface to scratching. Calcite and quartz exhibit different hardness. Forcefully rubbing quartz and calcite specimens against one another reveals that calcite is softer than quartz, as shown in **Figure 2.3**, because the calcite is scratched, while the quartz is unblemished.

Diamond, the hardest known mineral, is a cherished jewel, but its extraordinary hardness also makes it useful for a variety of industrial cutting and grinding purposes. How do geologists know that diamond is the hardest mineral? If you forcefully rub a diamond against any other

Distant view of outcrop of granite rock

Close view of granite reveals the minerals that compose the rock

Pink, box-shaped mineral grains

Black mineral in thin sheets

Gray, translucent, glassy looking mineral

White mineral

Calcite

Crystal form of calcite

Quartz

Crystal form of quartz

◄ **Figure 2.2 What do minerals look like?** Although many mineral samples appear shapeless, all minerals have distinctive crystal shapes. Calcite and quartz are two common minerals. Exceptional crystals of both minerals are transparent and six-sided, but as these drawings emphasize, they have different forms.

▲ **Figure 2.3 Comparing the hardness of minerals.** Scraping two minerals against one another tests their relative hardness. The quartz crystal remains unblemished, while the calcite is scratched and powdered, indicating that calcite is softer than quartz.

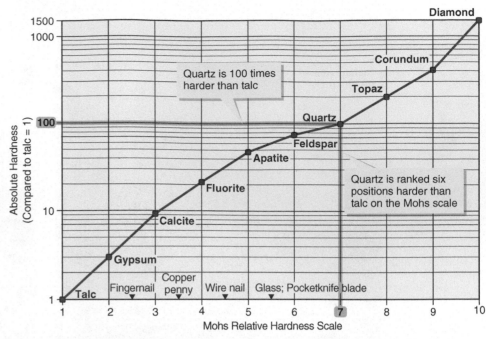

▲ **Figure 2.4 How mineral hardness is defined.** The Mohs hardness scale is a *relative* scale with values between 1 and 10 that simply rank minerals from softest to hardest. More quantitative measurement methods determine the *absolute* hardness of minerals. Calcite and quartz are four steps apart on Mohs scale, but quartz is more than 10 times harder than calcite. The black triangles show the Mohs hardness of common objects for comparison.

substance, the other material will be scratched, while the diamond is left unblemished, which reveals that diamond is harder. Rub two diamond crystals together and scratches appear on both, which indicates their equal hardness.

Geologists use **Mohs hardness scale** to describe mineral hardness. This scale, illustrated in **Figure 2.4**, has values from 1 to 10, with the hardest mineral, diamond, assigned a hardness value of 10. The scale defines the hardness of quartz as 7 and calcite with a hardness of 3. These values are consistent with the scratch test. The Mohs scale is relative and allows geologists to rank minerals according to their hardness; it is not an absolute scale, such as that for temperature, which is measured directly with a thermometer. Methods of measuring mineral hardness that are more sophisticated than the Mohs scale are more accurate and reveal that quartz is about 10 times harder than calcite.

Do geologists carry around pieces of the ten minerals on the Mohs scale in order to measure relative hardness of minerals? They could, but instead, field geologists use a few common objects, such as a fingernail, a penny, and the blade of a pocketknife to conduct

scratch tests. Comparing mineral hardness to that of these common objects permits a quick estimate of the Mohs-scale value for a mineral specimen.

Some Minerals Have Characteristic Cleavage

Figure 2.5 illustrates, for example, the different ways that calcite and quartz break when struck by a hammer. Quartz breaks into very irregularly shaped pieces that resemble broken glass. The broken surfaces display rounded indentations, and sharp edges, and each broken fragment has a different shape (Figure 2.5a). In contrast, the calcite breaks into six-sided pieces,

▲ **Figure 2.5 How minerals break.** Broken mineral fragments are differently shaped than mineral crystals. Broken quartz fractures like glass into irregular, sharp fragments. Calcite breaks into regularly shaped pieces that have six planar sides with rhomb shapes. The planes defining the calcite pieces are cleavage planes, while the pieces are called cleavage fragments.

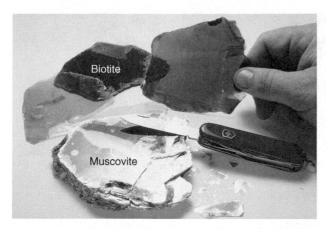

▲ **Figure 2.6 Micas cleave into thin sheets.** The common mica minerals, silvery muscovite and nearly black biotite, have one repeating cleavage plane. This means that a knife blade or fingernail easily separates samples into thin, transparent sheets.

each of which is shaped like a rhomb—sort of like a cardboard box that is partly squashed to one side (Figure 2.5b). Each surface of the calcite rhomb is a flat, smooth plane that brightly reflects light. Each rhomb-shaped piece can be broken into still-smaller rhombs. The term **cleavage** refers to the flat, smooth planes along which some minerals break and also to the shape of the resulting fragments. Calcite has three repeating cleavage planes that meet each other at 60-degree angles to form the rhomb-shaped fragments. As another example, **Figure 2.6** illustrates mica minerals that split into very thin transparent sheets parallel to a single cleavage. In contrast, quartz breaks along unpredictable, irregular surfaces so it does not have cleavage. The surface resulting from breakage that does not form smooth planes is called **fracture.**

▲ **Figure 2.7 Variations in mineral color.** Samples of quartz and calcite come in a variety of external forms and colors. One type of mineral can occur in a variety of colors, while different minerals can have the same color.

Mineral Color and Shape Can Vary

Centuries of measurements and observations reveal that density, luster, hardness, and cleavage are consistent characteristics of some minerals, such as quartz and calcite. But, for example, do *all* specimens of calcite and *all* specimens of quartz have the same properties? No, because other properties, such as color, are variable, as shown in **Figure 2.7**. Quartz (Figure 2.7a) is found in pink hues (called rose quartz), cloudy white (milky quartz), black (smoky quartz), purple (amethyst), and yellow (citrine). Calcite (Figure 2.7b) also comes in a variety of shades; cloudy white, gray, orange, red, and pink are the most common. You can conclude, therefore, that these variations mean that color is not always a reliable property for identifying a mineral.

A property related to color is **streak**, the color of the residue produced by scratching a mineral on a non-glazed porcelain plate, as illustrated in **Figure 2.8**. Different specimens of the same mineral may vary in color, but the streak color is always the same.

Mineral specimens also vary in shape. All calcite crystals have six sides, but the shapes and arrangement of crystal faces differ. Furthermore, some specimens of both calcite and quartz lack obvious crystal faces altogether (Figure 2.7). External crystal form, therefore, also is not always helpful when identifying minerals.

Red hematite streak

White calcite streak

▲ **Figure 2.8 Mineral streak colors.** Minerals softer than porcelain (about 6.5 on Mohs scale) leave a powdery residue, called streak, when scraped across a porcelain plate. Calcite streak is white, which is similar to the color of the mineral specimen. The iron mineral hematite always leaves a red-brown streak on a porcelain plate, even when the mineral specimen itself is not red-brown.

Putting It Together—What Are the Properties of Minerals?

• Minerals have observable or easily measured physical properties, some of which are more diagnostic than others for identifying a specific mineral.

• The principal physical properties used to describe minerals are color, luster, streak, hardness, cleavage (or fracture), density, and external crystal form.

2.2 What Makes Up Minerals?

Minerals, like all materials, consist of **elements**, which are substances that cannot be broken down chemically into simpler substances. **Appendix A** provides the periodic table of the elements for reference. Each element has properties that are determined by its component **atoms**, which are the smallest units of matter that can take part in chemical reactions. **Figure 2.9** shows that atoms consist of even smaller particles—positively charged protons, negatively charged electrons, and uncharged (or neutral) neutrons. The center, or nucleus, of an atom contains protons and neutrons, and electrons orbit its nucleus. The number of protons in an atom defines a particular element and is referred to as the element's atomic number. For example, the element oxygen has the atomic number 8, because its nucleus contains 8 protons and iron has the atomic number 23, because its nucleus contains 23 protons. Most minerals (including quartz and calcite) are compounds consisting of two or more elements, although a few minerals contain atoms of only one element (diamond, for example, contains only the single element carbon). Mineralogists perform laboratory chemical analyses to determine mineral compositions.

EXTENSION MODULE 2.1

Basics of an Atom. *Learn about the basic components of an atom.*

Our examples of calcite and quartz have distinctive chemical compositions. Calcite contains the elements calcium (represented by the symbol Ca), carbon (C), and oxygen (O), whereas quartz consists of silicon (Si) and oxygen (O). The chemical formula for calcite is $CaCO_3$, and quartz is SiO_2. The subscript numbers in these formulas tell us that three oxygen atoms exist for each calcium and carbon atom present in calcite, and that two oxygen atoms are present for each silicon atom found in quartz.

Figure 2.10 lists the results of laboratory determined chemical compositions of three quartz varieties identified by different colors. The analyses of the colored samples include tiny amounts of elements other than silicon

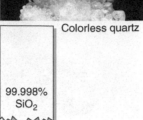
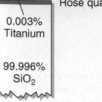

Colorless quartz

Rose quartz

Amethyst

99.998% SiO_2

0.003% Titanium
99.996% SiO_2

0.020% Iron
99.978% SiO_2

▲ **Figure 2.10 Why quartz has many colors.** All quartz specimens are almost entirely composed of silica (SiO_2), but the presence of even minute amounts of other elements can change its color. The purest quartz is colorless, but traces of titanium account for the pink color of rose quartz, whereas iron make amethyst purple.

and oxygen. The extremely small abundances of these extra elements are insufficient to affect the properties shared by all specimens of quartz, such as density, luster, hardness, and cleavage. The presence of trace constituents, however, may affect color. For example, rose quartz contains a tiny amount of titanium, and amethyst contains a tiny amount of iron.

Therefore, chemical composition, within narrow ranges of variation, is another defining characteristic of individual minerals. It turns out that composition determines the physical properties of minerals. These properties are determined not only by which elements are present in each mineral, but also by how these elements combine with one another. These points are the subjects of the next two sections of this chapter.

Putting It Together—What Makes Up Minerals?

• Minerals are chemical compounds consisting of combinations of atoms of one or more elements.

• Each mineral has a definitive, but possibly slightly varying, chemical composition.

2.3 How Do Elements Combine to Make Minerals?

In this section, we will explain how elements join to make minerals, what element combinations are possible in minerals, and how the element combinations produce an orderly arrangement of the atoms within minerals. If you can understand these almost invisible characteristics of minerals, then you should be able to understand the physical properties and uses of minerals.

Chemical Bonds Form Mineral Compounds

The arrangement of electrons around the nuclei of an atom determines how atoms **bond**, or combine to form molecules. Three bond types, and a fourth, weak force that attracts atoms to one another, are most common in minerals.

An atom of oxygen: A nucleus containing 8 protons and 8 neutrons surrounded by 8 orbiting electrons

Proton — Electron
Neutron — Nucleus

▲ **Figure 2.9 Parts of an atom.** An atom consists of a nucleus surrounded by orbiting electrons. Electrons are negatively charged. Within the nucleus are protons, which have positive charge, and neutrons, which have no charge.

The key to understanding bonding is to visualize the electrons orbiting the nucleus of an atom, sort of like planets moving around a star, but with several planets following the same orbit (Figure 2.9). In all atoms, only two electrons can occupy the innermost orbit, but most of the other orbits can accommodate up to eight electrons. In fact, atoms "prefer" to have eight electrons in their outer orbits and shed or gain electrons either by transferring or sharing electrons with other atoms to maintain eight electrons in that orbit. Transfer and share processes are responsible for the most common bonds.

Changing the number of electrons in the outer orbit of an atom produces an excess or deficiency of electrons in the atom compared to protons. When this happens atoms become positively or negatively charged particles called **ions**. Negative ions have more electrons than protons, whereas positive ions have fewer electrons than protons. **Appendix B** lists the ions for the elements that are most abundant in minerals.

The first type of chemical bond, an **ionic bond**, forms as a result of the attraction between negative and positive ions and occurs when electrons are transferred from one atom to another. **Figure 2.11** illustrates ionic bonding in table salt, which is also known as the mineral halite.

In other instances, two or more atoms share electrons to simultaneously fill the outer electron shell of the atoms. Sharing creates the second type of chemical bond, a **covalent bond**, as shown by the bonding of carbon atoms in diamond in **Figure 2.12**. Ionic bonds, covalent bonds, or a combination of the two are most common in minerals. Ionic bonds are usually weaker than covalent bonds.

Less commonly and typically of the same element, electrons freely roam among several atoms to form the third type of chemical bond, **metallic bonds**,

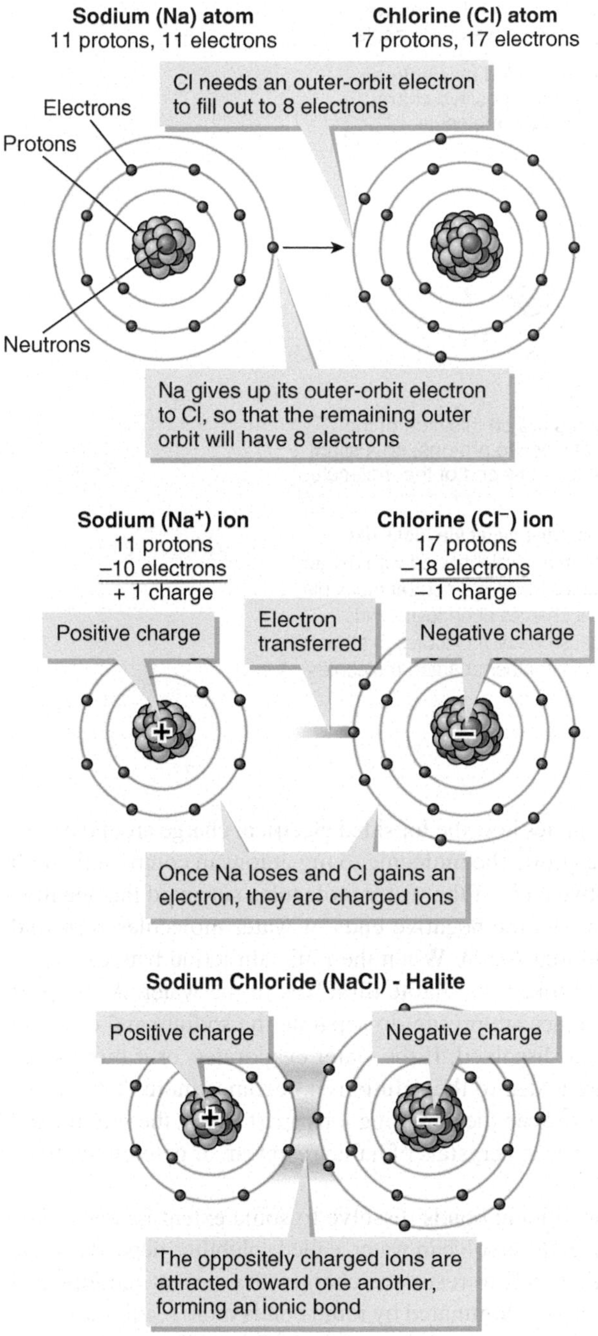

Sodium (Na) atom
11 protons, 11 electrons

Chlorine (Cl) atom
17 protons, 17 electrons

Cl needs an outer-orbit electron to fill out to 8 electrons

Electrons

Protons

Neutrons

Na gives up its outer-orbit electron to Cl, so that the remaining outer orbit will have 8 electrons

Sodium (Na⁺) ion
11 protons
−10 electrons
+ 1 charge

Chlorine (Cl⁻) ion
17 protons
−18 electrons
− 1 charge

Positive charge

Electron transferred

Negative charge

Once Na loses and Cl gains an electron, they are charged ions

Sodium Chloride (NaCl) - Halite

Positive charge

Negative charge

The oppositely charged ions are attracted toward one another, forming an ionic bond

▲ **Figure 2.11 How ionic bonding works.** A sodium atom has 11 electrons, with 1 electron in its outermost electron orbit. Chlorine has 17 electrons, with 7 electrons in its outermost orbit. The transfer of an electron from the sodium atom to the chlorine atom leaves each atom with 8 electrons in its outer orbit. The transfer of an electron creates a positive Na⁺ ion and a negative Cl⁻ ion. The two oppositely charged ions attract to each other to form an ionic bond. The resulting compound, NaCl (sodium chloride), is table salt; this same mineral compound is also known as halite.

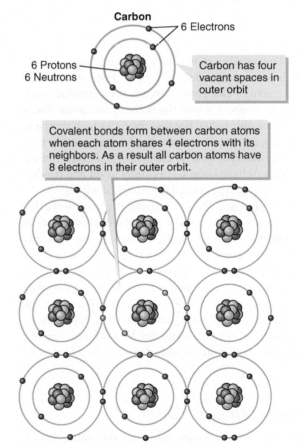

Carbon

6 Electrons

6 Protons
6 Neutrons

Carbon has four vacant spaces in outer orbit

Covalent bonds form between carbon atoms when each atom shares 4 electrons with its neighbors. As a result all carbon atoms have 8 electrons in their outer orbit.

▲ **Figure 2.12 How covalent bonding works.** Two or more uncharged atoms may bond by sharing electrons in their outermost orbits. The outer electron orbit surrounding a carbon nucleus has 4 electrons and needs 8 to be "full." A carbon atom, therefore, can share electrons with four other surrounding carbon atoms so that all of them have full outer orbits. Sharing electrons defines covalent bonds. Diamonds consist of covalently bonded carbon atoms.

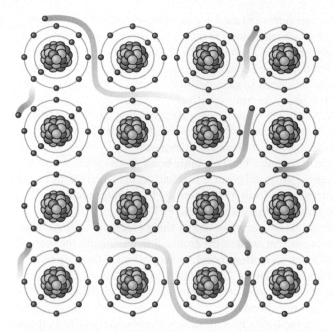

▲ **Figure 2.13 How metallic bonding works.** This diagram illustrates a cross section of the atomic structure of a metal. Metallic bonds form where electrons freely roam among atomic nuclei.

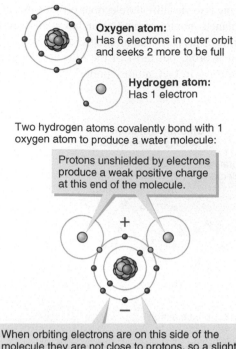

Oxygen atom: Has 6 electrons in outer orbit and seeks 2 more to be full

Hydrogen atom: Has 1 electron

Two hydrogen atoms covalently bond with 1 oxygen atom to produce a water molecule:

Protons unshielded by electrons produce a weak positive charge at this end of the molecule.

When orbiting electrons are on this side of the molecule they are not close to protons, so a slight negative charge exists at this end of the molecule.

▲ **Figure 2.14 What the water molecule looks like.** Two hydrogen atoms form a covalent bond with oxygen to make a water molecule. Water is a neutral molecule, but it has mild electrical charges at opposite ends. These weak charges cause water molecules to attract one another and also attract other charged molecules and ions.

as shown in **Figure 2.13**. The mobility of the electrons within metallic substances accounts for their ability to conduct electricity, because electrical currents involve the movement of electrons. Minerals with metallic bonds reflect light when rays of light interact with the roaming electrons. These minerals are said to have a metallic luster, because the rocks look similar to bright, shiny metal. Copper, gold, and silver are examples of minerals—each consisting exclusively of the atoms of a single element—that exhibit metallic bonds. All of these minerals have sought-after shiny luster and conduct electricity, which explains why copper is used for electrical wiring.

Last, the fourth type of chemical "bond" actually is a weak force that occurs because electrons are not always equally distributed around all sides of a molecule. This means that even a neutrally charged group of atoms, containing an equal number of protons and electrons, can behave like a weak miniature magnet, because there can be a weak negative charge on the side with more electrons and a weak positive charge on the side with fewer electrons. The slightly positive side of one molecule draws it close to the slightly negative side of a neighboring molecule. This weak attraction of neutrally charged particles is called a **van der Waals force**. You will soon see how this force is important to understanding some mineral properties.

Bonds Break when Minerals Dissolve

Water (H_2O) is a covalently bonded compound of considerable geologic importance. **Figure 2.14** shows that each of the single electrons encircling the two hydrogen atoms in a water molecule is shared with a single oxygen atom. The water molecule is electrically neutral (10 electrons and 10 protons), but it is lopsided. The molecule has a slight positive tendency on the side with the hydrogen atoms, while electrons spend more time on the oxygen side of the molecule, which gives this end a slight negative tendency.

Figure 2.15 illustrates how the lopsided electrical charge around a neutral water molecule allows the molecule to tug at ions in compounds such as halite. The positive ends of the water molecules surround the negative chloride ions (Cl^-), and the negative ends of water molecules surround the positive sodium ions (Na^+). When the ionic attraction between sodium and chloride is broken, the halite dissolves in the water. As long as enough water molecules are present to separate the sodium and chloride ions, the salt remains dissolved. If the water evaporates, or if more sodium and chloride are added to the solution, it becomes more difficult for the water molecules to keep the attracting ions apart. Then, the sodium and chloride ions combine, and crystals of halite reappear, or precipitate, from the solution.

All minerals with ionic bonds dissolve to some extent in water. The ease with which minerals dissolve in water, called solubility, depends on the strength of the ionic bonds to resist the charge attraction of the lopsided water molecule. Minerals dominated by ionic bonds more readily dissolve in water than those held together by covalent bonds. Adding acid to water further enhances solubility, because acids contain ions that draw atoms in minerals apart from one another.

The Internal Structure of Minerals

Figure 2.16 shows the presence of both ionic and covalent bonds in $CaCO_3$ molecules of calcite. Three oxygen atoms surround and share electrons

Halite molecular structure

Halite (rock salt) consists of ionic bonds between sodium (Na) and chloride (Cl) ions.

Water molecule

One end of the molecule has a weak positive charge while the other has a weak negative charge.

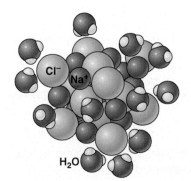

When halite is placed in water, the charged water molecules align with negative ends close to the sodium ions and positive ends next to the chloride ions. The weak electrical forces in the water molecules tug at the relatively weak ionic bonds and pull the mineral molecule apart.

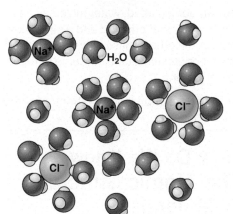

The halite completely dissolves in water when there are sufficient water molecules surrounding each ion to keep ionic bonds from forming between sodium and chloride.

▲ Figure 2.15 How bonds break when halite dissolves in water.

with each carbon atom to form covalent bonds. This group of carbon and oxygen atoms does, however, have two more electrons than protons, so the entire group of atoms behaves like a negative ion and is referred to as carbonate, CO_3^{2-}. The negatively charged carbonate and the positively charged calcium ions, Ca^{2+}, form ionic bonds. Like other ionic compounds, calcite dissolves in water, although much less readily so than halite, because the attraction between +2 and −2 ions of calcite is stronger than between the +1 and −1 ions of halite.

When calcite dissolves in water, the calcium ions separate from the carbonate group, but acid is required to break the strong covalent bond between carbon and oxygen within the

carbonate group. A common test for calcite and other carbonate minerals is to apply dilute hydrochloric acid (sold as muriatic acid in hardware and pool-supply stores) to a sample. The mineral dissolves and, as shown in **Figure 2.17**, violent bubbling of carbon dioxide (CO_2) gas reveals the breakdown of the carbonate.

Figure 2.18 illustrates the arrangement of calcium and carbonate ions within calcite. The diagram shows that each carbonate group, composed of covalently bonded carbon and oxygen atoms, is then ionically bonded to calcium ions. The structure is very orderly, with calcium and carbonate ions occupying alternating rows.

Bonding in quartz, depicted in **Figure 2.19**, is more complex than in calcite. The silicon and oxygen atoms share electrons, but the geometry of the covalent bonds are such that each silicon atom shares electrons with four adjacent oxygen atoms. The relative sizes of the atoms determine the four-to-one arrangement of oxygen atoms around each silicon atom. The much smaller silicon atom fits neatly in the small space between four oxygen atoms. The resulting negatively charged SiO_4^{4-} groups make ionic bonds with other nearby silica groups to form a linked, scaffold-like framework.

The diagrams of the atomic structure of calcite (Figure 2.18) and quartz (Figure 2.19) illustrate two important characteristics of all minerals. First, the size of atoms determines how they can be arranged, and the atomic structure determines the bonds between the atoms that compose the mineral. Second, orderly, repetitive patterns exist in the locations of atoms within a mineral.

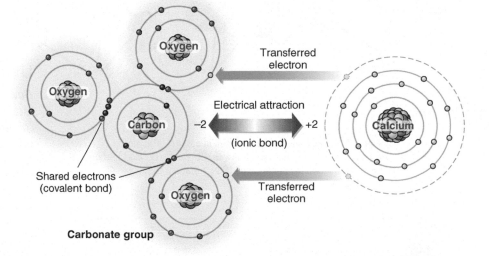

▲ Figure 2.16 How atoms bond in calcite. Calcite has both ionic and covalent bonds. Covalent bonding of carbon and oxygen atoms forms carbonate groups. Calcium ions (Ca^{2+}) form when electrons transfer to the carbonate groups (CO_3^{2-}). The oppositely charged calcium and carbonate attract to form ionic bonds.

▲ **Figure 2.17 Calcite reacts with hydrochloric acid.** A violent bubbling reaction occurs when hydrochloric acid drops contact this white calcite specimen. The bubbles are carbon dioxide gas released when the CO_3^{2-} groups in the carbonate mineral break down. This chemical reaction test distinguishes carbonate minerals from other, similar-appearing minerals.

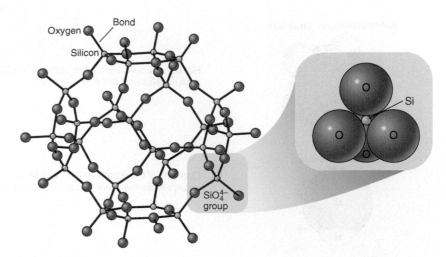

▲ **Figure 2.19 How atoms are arranged in quartz.** Quartz has both ionic and covalent bonds. The "balls" in this diagram represent the atoms and the black lines between the atoms represent the bonds. Four large oxygen atoms surround and covalently bond to each smaller silicon atom to make the basic SiO_4^{4-} group. This negatively charged group easily attracts positively charged Si ions. In quartz, each oxygen atom bonds to two different silicon atoms in a combination of covalent and ionic bonds. Oxygen atoms form the corners of a pyramid-like structure with a silicon atom in the center, as shown in the enlarged view.

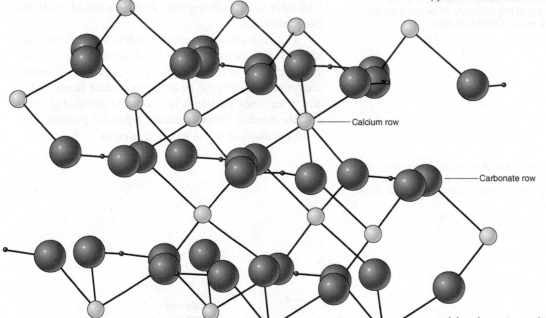

Oxygen

Calcium

Carbon

▲ **Figure 2.18 How atoms are arranged in calcite.** The atoms composing calcite are arranged in rows. Covalently bonded carbon and oxygen atoms form carbonate ions (CO_3^{2-}) that are then ionically bonded to calcium ions (Ca^{2+}).

• Ionic bonds are weaker than covalent bonds, in which atoms share electrons.

• The neutral, but lopsided, water molecule has a weak positive charge at one end and a slight negative charge at the other. These charges pull apart some weakly bonded ions, causing some minerals to dissolve in water.

• Many minerals exhibit combinations of ionic and covalent bonds.

2.4 How Do We Know . . . The Atomic Structure of Minerals?

Understand the Tool

How Do Geologists Visualize the Atoms Inside Minerals? Chemical analyses reveal that atoms of particular elements are present in minerals, but the analyses do not show where the atoms are located within a mineral. You may even wonder about the realism of diagrams such as those shown in Figures 2.18 and 2.19; after all, no one can actually see atoms when you hold a mineral in your hand. Comprehending the arrangements of atoms within minerals requires tools that allow geologists to "see" atoms.

One way to see small objects is to magnify them with a microscope. A common optical microscope uses glass lenses to focus light that reflects off of or passes through a mineral specimen. Under the best circumstances, an optical microscope magnifies a mineral nearly 1000 times and permits recognition of features approximately one-thousandth of a millimeter across. While this dimension seems very tiny, this level of magnification falls far short of seeing atoms.

Putting It Together—How Do Elements Combine to Make Minerals?

• Minerals consist of combinations of atoms held together by ionic, covalent, and less common metallic bonds, as well as the weak van der Waals forces.

► **Figure 2.20 How a TEM makes images of atoms inside minerals.** Electrons are fired at thin slices of minerals inside the electron microscope. Those electrons that transmit through the mineral sample, without encountering obstacles, reach and brightly illuminate a detector on the other side. Fired electrons that encounter dense electron clouds surrounding mineral atoms do not reach the detector, thus leaving a shadow on the image. Not all atoms cast a shadow, however, because some contain too few orbiting electrons that are too widely spaced to intercept all the electrons fired inside the microscope.

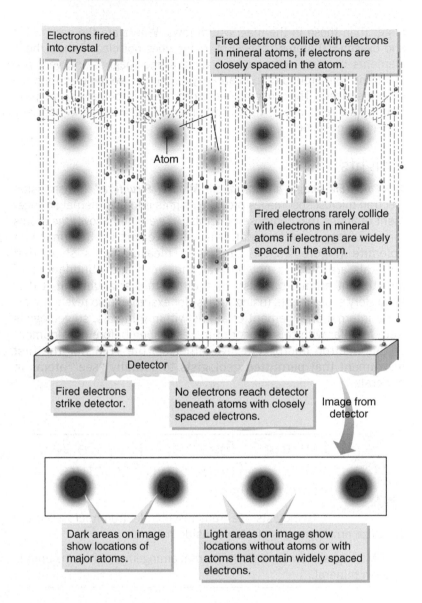

Electrons fired into crystal

Fired electrons collide with electrons in mineral atoms, if electrons are closely spaced in the atom.

Atom

Fired electrons rarely collide with electrons in mineral atoms if electrons are widely spaced in the atom.

Detector

Fired electrons strike detector.

No electrons reach detector beneath atoms with closely spaced electrons.

Image from detector

Dark areas on image show locations of major atoms.

Light areas on image show locations without atoms or with atoms that contain widely spaced electrons.

The much more sophisticated transmission electron microscope (TEM) uses magnets (rather than lenses) to focus streams of tiny electrons (rather than light) through a small part of a mineral sample. Some electrons pass through the sample without encountering any obstacles. Other electrons bounce off atoms within the mineral. A detector beneath the sample records the arrival of electrons that passed through unhindered. Electrons that encountered atomic obstacles do not make it to the detector, so they are not recorded. Mineralogists use the data from the detector to produce images of the interior of a mineral that are magnified up to 10 million times and that resolve features less than one-millionth of a millimeter across. This very high level of magnification is sufficient to see the outlines of most atoms.

Visualize the Result

What Does the Atomic Structure Look Like? **Figure 2.20** shows how a TEM produces an image of atoms in a mineral. The interaction of the streams of electrons fired inside the microscope with the atoms in the mineral produces an image of black dots and white dots. White areas indicate where electrons struck the detector beneath the mineral sample, and black areas occur where no electrons reached the detector. The black areas are like shadows formed where an object blocks light, or like a dry circle of pavement beneath an umbrella in a rainstorm. In the case of a TEM image, as noted above, the black areas form because electrons encountered atomic obstacles when passing through the mineral and never reached the detector.

Most of the size of an atom consists of the cloud of electrons that orbit the nucleus. If many electrons exist in a small area around the nucleus, then there is a good chance that electrons fired in the TEM will collide with an orbiting electron and bounce off it, rather than continuing through to the detector. If, however, the electrons are widely spaced around a nucleus, then the fired electrons may pass through the atom without hitting any obstacles. This means that the shadows in the TEM image may not represent all of the types of atoms in a mineral, but rather only the atoms with closely clustered electrons.

Figure 2.21 shows a TEM image of dolomite (CaMg[CO$_3$]$_2$), a mineral similar to calcite (CaCO$_3$) but in which magnesium (Mg) atoms replace half of the calcium (Ca) atoms. This magnification is high enough to see the shadows caused by the calcium and magnesium atoms in repeating rows of black dots. Calcium atoms are much bigger than magnesium atoms, so the dots in the calcium rows are

▼ **Figure 2.21 Visualizing atoms inside a mineral.** This TEM image of the mineral dolomite resembles a wallpaper of regularly spaced black dots and white dots. The black dots reveal locations of calcium and magnesium atoms that interfered with electrons fired inside the TEM. Dolomite also contains oxygen and carbon atoms that the TEM does not detect, but which occupy spaces between the calcium and magnesium atoms.

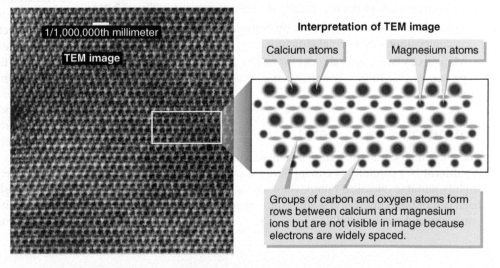

1/1,000,000th millimeter

TEM image

Interpretation of TEM image

Calcium atoms

Magnesium atoms

Groups of carbon and oxygen atoms form rows between calcium and magnesium ions but are not visible in image because electrons are widely spaced.

larger than those in the magnesium rows. Where are the carbon (C) and oxygen (O) atoms? The TEM image does not clearly show these elements because compared to calcium and magnesium, very few electrons spin around carbon and oxygen nuclei, so the electrons are widely spaced. The imaging electrons fired into the crystal pass through the relatively "empty" carbon and oxygen atoms without any deflecting collisions.

Insights

What Do TEM Images Reveal about Mineral Structure? Transmission electron microscope images provide insights into the internal arrangement of atoms that compose minerals. As we noted above, the atoms form orderly, repetitive patterns, as revealed by the image of dolomite (Figure 2.21). Laboratory measurements show that minerals are defined not only by the types and abundances of constituent atoms, but also by their atomic arrangement. Other laboratory experiments (using X-rays, for example) also reveal the orderly internal arrangement of atoms within minerals, so geologists were familiar with this aspect of minerals long before development of the electron microscope. Nonetheless, the TEM was the first instrument that permitted geologists to actually "see" atoms in minerals.

Putting It Together—How Do We Know … The Atomic Structure of Minerals?

• Geologists use transmission electron microscopes to visualize arrangements of atoms inside minerals.

• TEM images show the orderly internal arrangement of atoms unique to each mineral.

2.5 What Is a Mineral?

Pause to reflect on what you know about minerals so far:

• Each mineral has a chemical composition, which varies only slightly (e.g., Figure 2.10).
• Each mineral has a repetitive, orderly structure of atoms (e.g., Figure 2.21).
• Minerals have different physical properties, and these properties relate to composition and internal structure.
• Minerals compose natural rocks formed by geologic processes; biologic processes also form minerals, such as seashells and coral (also teeth and bone).

Definition of a Mineral

These listed observations lead to a comprehensive definition of a mineral. A **mineral** is a naturally occurring inorganic solid, with a definite, only slightly variable chemical composition and an ordered atomic structure. Each part of this definition is worth exploring more carefully to ensure that you completely understand it.

Only naturally formed substances are minerals; manufactured materials, such as synthetic gemstones (e.g., cubic zirconia), are not.

A mineral must be a solid with an orderly arrangement of atoms; it cannot be a gas or liquid. Opal, a popular gemstone, consists of silicon and oxygen, the same elements that compose quartz. Opal, however, lacks a highly ordered atomic structure and is not considered a mineral.

Organic compounds, those defined as containing mostly carbon and hydrogen atoms, are not minerals. Table sugar ($C_{12}H_{22}O_{11}$), therefore, is not a mineral, even though it is a naturally occurring solid with an orderly atomic structure.

The chemical composition of a particular type of mineral can vary slightly (as in the colored varieties of quartz illustrated in Figure 2.10), but the principal constituents are common to all specimens.

Putting It Together—What Is a Mineral?

• A mineral is a naturally occurring inorganic solid with a definite, only slightly variable chemical composition and an ordered atomic structure.

2.6 What Determines the Physical Properties of Minerals?

When comparing quartz and calcite (Section 2.1), for example, we examined their physical properties of color, external crystal form, cleavage, hardness, density, and luster (ability to reflect light). These properties are, to varying degrees, diagnostic of each mineral, in other words, they identify each mineral. These physical properties also determine whether a mineral has practical uses. These physical properties of minerals result from the chemical compositions and atomic structures of the minerals. The causes of these properties, at the atomic level, are summarized in **Table 2.1** and explained in the following paragraphs.

Composition Commonly Determines Color

When atoms of one (or even several) trace elements replace a few atoms of more abundant elements in a mineral crystal, this replacement has little effect on its composition and on most of its physical properties. It can have considerable effect on its color, however. Data depicted in Figure 2.10 show that trace amounts of some elements can affect color; for example, as we noted in Section 2.2, purple quartz (amethyst) contains minute amounts of iron, and the pink variety (rose quartz) contains a bit of titanium. **Figure 2.22** shows that the iron (Fe^{3+}) and titanium (Ti^{4+}) ions are larger than the silicon Si^{4+} ion, but Fe^{3+} and Ti^{4+} can still nestle between the four oxygen atoms in the quartz structure in place of Si^{4+}. The Fe^{3+}, Ti^{4+}, and Si^{4+} ions also have the same or nearly the same number of electrons to share.

Similarly, calcite is clear or white when pure, but its colors also vary because of elemental substitutions (Figure 2.7b). Iron (Fe^{2+}) and manganese (Mn^{2+}) ions have the same charge and are similarly sized as the calcium (Ca^{2+}) ion (Figure 2.22), permitting them to substitute for Ca^{2+} in the calcite atomic structure. Varying amounts of iron generate dark blue, green, and brown calcites, while manganese in calcite produces colors ranging from pale purple to deep red.

TABLE 2.1 Physical Properties of Minerals

Description of Property	Factors That Determine the Property
Luster describes how mineral surfaces reflect light.	Luster depends on the smoothness of the mineral surface at the atomic scale, which, in turn, depends on how mobile electrons are within the crystal.
Density is the measure of the mass of a substance contained within a particular volume of the substance.	Density depends on the types of atoms and how the atoms are arranged in the crystal structure. The heavier the atoms are and the more tightly packed they are, the higher the density of the mineral.
Crystal faces are flat, smooth surfaces on mineral exteriors with regular geometric forms.	Crystal faces reflect the atomic arrangement of atoms within the crystal structure and produce exterior geometric shapes during growth unless the crystal grows against another crystal.
Hardness is the resistance to scratching on a smooth surface.	Hardness reflects the atomic bond strength of a mineral, which depends on bond type and the spacing of atoms within the crystal.
Cleavage describes planes along which a mineral breaks and the shape of the resulting fragments.	Cleavage forms along regularly spaced internal planes where bonds are weakest in minerals.
Color results from the interaction of light with the mineral.	Atomic arrangement and composition determine how light passes through, or interacts with the atoms in, the crystal, determining the color.
Streak is the color of the residue remaining from scratching a mineral on a non-glazed porcelain plate.	The fine-grained nature of the powdered residue results in a more reliable observed color than the whole crystal.

Atomic Structure Determines Most Physical Properties

Diagnostic physical properties, especially hardness and cleavage, are more closely related to the arrangement of atoms within the crystal structure than to the types of atoms present (i.e., composition). A comparison of diamond and graphite, illustrated in **Figure 2.23**, provides an example of the importance of crystal structure in determining physical properties. Both minerals contain only carbon, but while diamond is the hardest mineral on Earth, graphite, with a Mohs hardness of 2, is soft enough to mark paper. Graphite readily cleaves into thin, scaly sheets, similar to mica (Figure 2.6), whereas diamond cleaves only with great difficulty into eight-sided fragments.

As we said at the beginning of the chapter, cleavage planes form where the weakest bonds align in the atomic structure of a mineral. Diamond consists of a tightly packed three-dimensional network of covalently bonded carbon atoms (Figure 2.23). Graphite consists of covalently bonded carbon atoms arranged in sheets, but only the weak van der Waals forces hold the sheets of carbon atoms together. So, whereas diamond is composed of carbon atoms bonded strongly in all directions, making that mineral difficult to break, graphite has weakly linked layers that make it easy to break.

You break bonds when scratching a mineral during a hardness test. Again, soft graphite and even softer talc (Mohs hardness of 1, Figure 2.4) are partly held together by van der Waals forces. Harder halite (Mohs hardness of 2.5) and calcite (Mohs hardness of 3) feature ionic bonds. Quartz and diamond contain mostly or entirely covalent bonds and are even harder. These examples show that bond type determines bond strength and, therefore, determines hardness. Covalent bonds are stronger than ionic bonds, and both are stronger than van der Waals forces.

The hardness of ionically bonded minerals also increases with increasing electrical charge on the ions. This means that an ionically bonded mineral with a +2 ion is harder than a mineral with a +1 ion. For example, calcite ($CaCO_3$) is harder than halite (NaCl), because the calcium ion in calcite has a +2 charge, whereas the sodium ion in halite has a +1 charge.

The density of minerals relates to the mass of the atoms that make up the mineral and also to how closely packed the atoms are within the crystal structure. For example, you can use a density measurement to distinguish gold from "fool's gold," the look-alike mineral pyrite, because the mass of gold atoms is many times greater than

▼ **Figure 2.22 Comparing the sizes of ions.** This chart schematically shows the electrical charges and diameters of several ions that are common in rock-forming minerals. Ions can substitute for one another in a mineral atomic structure when their charges and sizes are similar. Notice that iron (Fe) and manganese (Mn) each have two common ions.

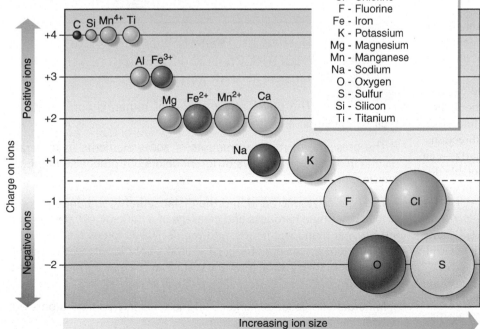

Key to chemical symbols:
Al - Aluminum
C - Carbon
Ca - Calcium
Cl - Chlorine
F - Fluorine
Fe - Iron
K - Potassium
Mg - Magnesium
Mn - Manganese
Na - Sodium
O - Oxygen
S - Sulfur
Si - Silicon
Ti - Titanium

▲ **Figure 2.23 Why diamond and graphite have different physical properties.** Diamond and graphite consist only of carbon atoms, but their physical properties differ. Carbon atoms in diamond are closely spaced and share strong covalent bonds in all directions; this configuration produces the hardest mineral on Earth. The smooth crystal faces defining the external form of diamond coincide with planes of carbon atoms in the crystal structure. In graphite, however, carbon atoms are more widely spaced than in diamond and are strongly bonded only in two dimensions. The covalently bonded carbon sheets are weakly held together by van der Waals forces. The weakly linked sheets readily separate in graphite, accounting for its softness and its tendency to cleave readily into thin, scaly plates.

than the masses of iron and sulfur atoms in pyrite. The contrasting density of the two carbon minerals in Figure 2.23, 3.51 g/cm³ for diamond, compared to 2.23 g/cm³ for graphite, is consistent with the more compact arrangement of the carbon atoms in diamond.

The arrangement of the carbon atoms also determines the external crystal forms. The covalent bonding of carbon in diamond commonly produces octahedron (eight-sided) crystals, which resemble two four-sided pyramids stuck together at their bases and pointing in opposite directions.

Think back to your initial experience with calcite and quartz—can you now explain their different physical properties? The most notable differences are hardness and cleavage. Covalent bonding dominates in quartz, whereas the ionic bond between calcium (Ca^{2+}) and carbonate (CO_3^{2-}) is the "weak link" in the calcite structure. The ionically bonded calcite (Mohs hardness of 3), therefore, is softer than the more covalently bonded quartz (Mohs hardness of 7) and is also much more soluble in water or dilute acid. The interlocking three-dimensional framework of equally spaced SiO_4^{4-} groups in quartz features no planes of preferential weakness; therefore, specimens fracture irregularly rather than cleave when broken. Calcite, however, breaks across ionic bonds, whose arrangement in the crystal structure produces rhomb-shaped cleavage fragments of various sizes.

Putting It Together—What Determines the Physical Properties of Minerals?

• The physical properties of minerals correspond to their composition and structure.

• The presence of minute amounts of some elements in minerals results in minerals that have different colors but otherwise identical physical properties.

• The type of atomic bond and the distance between bonded atoms or ions determine hardness and cleavage of a mineral. Covalent bonds are stronger than ionic bonds, and van der Waals forces are the weakest. Covalent bonds are strongest if atoms are closer together; ionic bonds are relatively stronger if the electrical charges on their ions are increased.

• Minerals composed of larger and more closely packed atoms are denser.

2.7 Which Minerals Are Most Important?

More than 4000 minerals are known, and more are discovered and named every year. So, how essential is it to know the details about all of them? What makes a mineral important to humans is based on one of two factors. First, the vast majority of rocks exposed at Earth's surface are composed of only a few dozen minerals; therefore, these substances have important geologic status as the "rock-forming minerals." Second, many other minerals, although not abundant, are essential economic resources. Some of these resource minerals occur in naturally pure forms and are used almost as they are when removed from the ground, for example, talc, sulfur, gold, and graphite. **Ore minerals**, on the other hand, are not pure, but contain economically important metallic elements that must be extracted from the minerals by metallurgical processes that break mineral bonds. Examples of elements obtained from ore minerals are iron (Fe), lead (Pb), zinc (Zn), copper (Cu), titanium (Ti), nickel (Ni), chromium (Cr), and uranium (U). Except for iron, all of these essential elements form far less than 1 percent of Earth's crust. This means that although these useful elements may be found within rock-forming minerals, they are typically present in such minute concentration that they cannot be extracted at acceptable cost. Fortunately, there are geologic processes (discussed in Chapters 4 through 6) that concentrate the economically valuable elements into mineral deposits that can be more easily mined.

Just 12 out of 89 naturally occurring elements account for 99.7 percent of the mass of the crust (see Appendix A for details), which explains why the list of common rock-forming minerals is quite short. **Figure 2.24** depicts geologists' estimates of the average elemental composition of the whole planet and of the crust alone. The composition of the crust is known from analyses of actual rocks and estimates of their various proportions based on maps of rock occurrences. The whole-Earth composition is clearly more difficult to determine, but can be accomplished using methods explained in Chapter 8. The two most common elements *in the whole Earth* are iron and oxygen. Although iron composes approximately 35 percent of the total Earth, it represents only 6 percent of the crust. The two most common elements *within the crust* are oxygen (46 percent by weight) and silicon (28 weight percent). This means that minerals containing silicon and oxygen, such as quartz, form the most common mineral group, known as silicates.

The remainder of this section and **Table 2.2** are designed as a reference for your continued study of minerals and rocks, not only in this chapter, but also in Chapters 3 through 7. For now, use these descriptions and explanations of important minerals to increase your familiarity with minerals names, properties, and uses. Then, refer back to these pages while pursuing later chapters in order to refresh and amplify your comprehension of why these minerals are important.

Silicate Minerals Are the Primary Rock-Forming Minerals in the Crust

Silicate minerals are those containing silicon and oxygen bonded in SiO_4^{4-} groups (Figure 2.19). Quartz is the most chemically simple silicate mineral, and it was the gray, glassy mineral in the granite outcrops you visited at the beginning of the chapter (Figure 2.1). In addition to silicon and oxygen, most silicates contain one to four elements in sufficient abundances to produce crystal structures that are quite different from quartz. Several of the dominant silicate minerals found in Earth's crust are listed and described in Table 2.2.

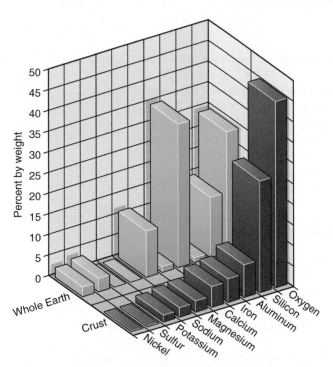

▲ **Figure 2.24 Visualize Earth's composition.** This graph depicts the abundances of the 10 most common elements within the entire Earth and within just the crust. Iron is the most abundant element in the whole planet, whereas oxygen, silicon, and aluminum are the most abundant elements in the crust.

EXTENSION MODULE 2.2

Silicate Mineral Structures. *Learn how the SiO_4^{4-} groups, called silica tetrahedra, bond together to form a variety of silicate-mineral structures.*

Many rock-forming minerals, including silicates, differ from one another because of element substitutions within their crystal structures. You have seen how the substitution of one element by even tiny amounts of another can cause variations in mineral color without substantially affecting composition or crystal structure (Figure 2.10). In many other cases, however, wholesale substitution of elements results in the formation of a series of minerals that do differ from one another in composition and structure. These gradual compositional variations correspond to equally gradual variations in physical properties, such as color and density.

Element substitutions can occur only under certain conditions. First, ions should be of similar size, so that a substituting ion fits into the same site in the crystal structure as the ion it replaces (Figure 2.22). Substitution of iron (Fe^{3+}) and titanium (Ti^{4+}) for silicon (Si^{4+}) in quartz does not happen readily, for example, because the substituting ions are nearly 50 percent larger than Si^{4+}, which creates a bad "fit." Consider aluminum, however, the third most abundant crystal element. The aluminum ion (Al^{3+}) is only approximately 22 percent larger than silicon and thus easily substitutes for Si^{4+} between four oxygen atoms. The aluminum substitution, however, is an unequal charge substitution, Al^{3+} for Si^{4+}. To

TABLE 2.2 Descriptions, Compositions, and Significance of Important Minerals

Mineral	Chemical Formula	Brief Description	Significance	Chapter(s) in Which You Can Learn More
Silicates				
Olivine Group	$(Mg, Fe)_2SiO_4$	Green to brown; equidimensional crystals	Important rock-forming mineral in the crust; most abundant mineral in the mantle. Gem form is peridot	3, 4, 5, 6, 8, 9, 11
Pyroxene Group	(Mg, Fe, Ca, Na) $(Mg, Fe, Al)Si_2O_6$	Blocky, black to green crystals; cleavage forms two planes that intersect at right angles	Important rock-forming mineral in the crust and mantle	3, 4, 5, 6, 8, 9
Amphibole Group	$(Na, Ca)_2 (Mg, Al, Fe)_5$ $(Si, Al)_8O_{22}(OH)_2$	Elongate to needle-shaped, black, green, and brown crystals; cleavage forms two planes that do not intersect at right angles	Important rock-forming mineral in the crust. Includes some forms of asbestos	4, 5, 6, 8
Mica Group	$KAl_2(AlSi_3O_{10})(OH)_2$ (Muscovite) $K(Mg, Fe)_3 (AlSi_3O_{10})(OH)_2$ (Biotite) $(Mg, Fe)_3 (Si, Al)_4O_{10}(OH)_2$ $(Mg, Fe)_3(OH)_6$ (Chlorite)	Silvery muscovite and dark-brown to black biotite; both have one plane of cleavage, so specimens split into thin sheets. Green chlorite usually forms very fine-grained crystal aggregates that form the green coloration in many rocks	Important rock-forming mineral in continental crust	4, 5, 6, 7, 8
Serpentine	$Mg_3Si_2O_5(OH)_4$	Fibrous, green crystals	Includes the most common form of asbestos, called chrysotile	6
Feldspar Group	$KAlSi_3O_8$ (Potassium feldspar) $(Ca, Na)Al_2Si_2O_8$ (Plagioclase feldspar)	Rectangular, blocky crystals; various colors from colorless to black; has two cleavage planes that intersect at right angles	The most common mineral group in rocks that form the crust	3, 4, 5, 6, 7, 8, 9, 14, 19
Quartz	SiO_2	Has six-sided crystal faces; various colors; fractures, but does not cleave	A very common mineral in rocks that form continental crust	3, 4, 5, 6, 8, 9, 11, 14, 19
Garnet	$A_3B_2(SiO_4)_3$ where "A" can be Ca, Mg, Fe^{2+} or Mn^{2+}, and "B" can be Al^{3+}, Fe^{3+} or Cr^{3+}	Varies in color from green to yellow to red; fractures rather than cleaves	Commonly used as a gemstone and as a commercial abrasive	6, 8, 19
Nonsilicates				
Calcite (Carbonate)	$CaCO_3$	Varies in color; has cleavage planes in three directions that break the mineral into rhombs; reacts with dilute hydrochloric acid	Used to make cement; composes marble, which is commonly used as an ornamental construction stone	3, 5, 6, 7, 13, 14, 17, 19, 20
Dolomite (Carbonate)	$CaMg(CO_3)_2$	Colorless to white and sometimes pink; has cleavage planes in three directions that breaks the mineral into rhombs; powdered dolomite reacts with dilute hydrochloric acid	Uses are similar to calcite	3, 5, 6, 17
Magnetite (Oxide)	Fe_3O_4	Black with metallic luster; magnetic	Important rock-forming mineral in the crust. An ore mineral for iron	3, 4, 10, 19
Hematite (Oxide)	Fe_2O_3	Black, red, or silvery gray with a red streak	Important rock-forming mineral in the crust. The most abundant ore mineral for iron	3, 5
Corundum (Oxide)	Al_2O_3	Variable in color; fractures rather than cleaves; insoluble in all acids	Commonly used in abrasives because of its hardness (9 on Mohs hardness scale). Includes the gemstones ruby and sapphire	6
Halite (Halide)	$NaCl$	Salty tasting; colorless to white cubic crystals; has three cleavage planes that intersect at right angles to form cubes	The most common mineral formed by evaporation of sea water; table salt	3, 5, 7, 14, 16, 17, 19, 20
Gypsum (Sulfate)	$CaSO_4 \cdot 2H_2O$	Colorless to white crystals; cleavage forms three planes that do not intersect at right angles	Commonly used in the manufacture of plaster and wallboard	3, 5, 14, 16, 17, 19, 20
Gold (Native element)	Au	Various shades of yellow; very malleable with a high density	Precious metal used in jewelry and many industrial applications because it does not tarnish	4, 11
Copper (Native element)	Cu	Copper-red to dark brown-green from tarnish; very malleable with a high specific density	Used for electrical wiring because of its high electrical conductivity	4, 6, 11, 17
Chalcopyrite (Sulfide)	$CuFeS_2$	Brass-yellow commonly tarnished with iridescent colors	Most common copper-ore mineral	4
Pyrite (Sulfide)	FeS_2	Pale brass-yellow cubic crystals. Also known as "fool's gold"	Most common sulfide mineral	5, 6, 17
Sphalerite (Sulfide)	ZnS	Various colors of yellow, brown	Most common zinc-ore mineral	17
Galena (Sulfide)	PbS	Lead-gray, commonly cubic crystals; has three cleavage planes that intersect at right angles to form cubes	Most common lead-ore mineral	17

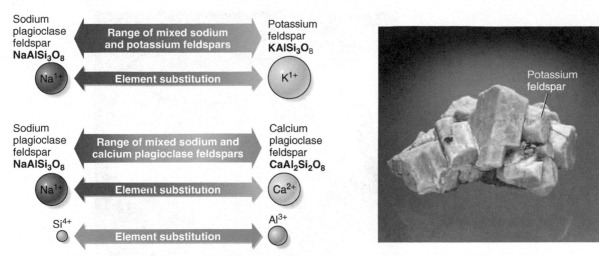

▲ **Figure 2.25 How composition varies among feldspars.** Feldspar minerals are aluminum-silicate minerals containing various amounts of sodium, potassium, and calcium. In some feldspars, sodium and potassium substitute for one another, because they have the same ion charge (+1), even though they are different sizes. A complete gradation from sodium-rich to calcium-rich compositions defines the plagioclase feldspars. Sodium and calcium ions are similar in size but have different charges. To compensate for the charge imbalance, aluminum substitutes for silicon at the same time that calcium substitutes for sodium.

(Ca^{2+}) ion is close in size to the sodium (Na^+) ion, permitting both ions to occupy the same sites in the feldspar structure. The charges of Ca^{2+} and Na^+ are not the same, however. This obstacle is overcome by the simultaneous substitution of Al^{3+} for Si^{4+} and Ca^{2+} for Na^+ in the atomic structure, which balances the charges. The gradual variation between relatively calcium- and aluminum-rich feldspars and sodium- and silicon-rich feldspars is characteristic of a mineral group called the plagioclase feldspars. Two feldspars appear in the granite illustrated in Figure 2.1: Potassium feldspar is the pink mineral, and sodium-rich plagioclase feldspar is the white mineral.

Magnesium and iron are two other significant components of the crust (Figure 2.24) that substitute for one another in the silicate minerals

maintain a neutral charge, other positive ions are added simultaneously, producing crystal structures (and, therefore, minerals) that differ from quartz.

Feldspars, illustrated in **Figure 2.25**, are the most common group of minerals in the crust; they result from the aluminum-for-silicon substitution described above. Potassium (K^+), sodium (Na^+), and calcium (Ca^{2+}) also are significant constituents of the crust (Figure 2.24) and are the most common positively charged ions added to offset the charge imbalance of the Al^{3+}-for-Si^{4+} substitution. Whereas sodium and potassium ions have the same positive charge, the potassium ion is much larger than the sodium ion. Therefore, substitution of K^+ for Na^+ does not easily happen, so potassium feldspars typically contain very little sodium, and sodium feldspar contains very little potassium. On the other hand, the calcium

pictured in **Figure 2.26**. This substitution happens because Mg^{2+} and Fe^{2+} ions have the same charge and are very similar in size (Figure 2.22). Four groups of mostly green and black silicate minerals—olivine, pyroxene, amphibole, and biotite—illustrate magnesium and iron substitutions. Although geologists casually refer to each of these four as single minerals, each name really refers to a closely related group of minerals whose crystal structures and physical properties vary depending on the amounts of magnesium and iron each contains. Biotite is a mica mineral (see Figure 2.6); it is present as the black mineral in the granite pictured in Figure 2.1.

Element substitutions also produce silicate mineral groups with widely varying composition but only slight differences in crystal structure. The term "garnet," for example, refers to members of six groups of common

▲ **Figure 2.26 Examples of iron- and magnesium-rich silicate minerals.** Iron and magnesium substitute for one another, and in some cases also with calcium, within the crystal structures of the dark silicate minerals olivine and amphibole.

▲ **Figure 2.27 Colorful garnets.** Garnets vary in color from green to red depending on the elements in their crystal structures. Color variations account for a wide variety of popular gemstones in jewelry.

▲ **Figure 2.28 Salt is a mineral.** Common table salt is the mineral halite, also called rock salt. Halite crystallizes as cubes and commonly forms from the evaporation of water, such as seawater, with abundant dissolved sodium and chloride ions.

silicate minerals with nearly identical atomic structure and indistinguishable crystal form. However, these minerals differ considerably in composition because of complex substitutions among calcium, iron, magnesium, manganese, chromium, and aluminum. Although the crystal structures of the various garnets differ only slightly, wide color variations occur, from wine-red to green. These variations not only aid in the identification of garnets, but also provide a variety of colored gemstones, as illustrated in **Figure 2.27.**

Nonsilicate Minerals Also Are Important

Not all major rock-forming minerals are silicates, as shown in Table 2.2. Carbonate minerals (so-called because they contain the carbonate ion complex ($[CO_3^{2-}]$), for instance, have a number of industrial uses. For example, they are important in the production of concrete and are present as the gentle abrasives in toothpaste. Carbonate minerals typically precipitate from water when the component ions dissolved in the water (e.g., Mg^{2+}, Ca^{2+}, and CO_3^{2-}) bond to make mineral solids. The carbonate group includes calcite and dolomite, which you have already learned about. Calcite is the white coating that you might find lining the inside of a pot in which you boiled water; the mineral precipitates from the water as the water boils off as steam.

Many other nonsilicate minerals precipitate from water besides carbonates. Halite, or rock salt, is a halide mineral shown in **Figure 2.28** that forms from evaporation of water. Sulfate minerals all contain the sulfate (SO_4^{2-}) group; the most common example is a calcium-sulfate mineral called gypsum. Gypsum is mined for the production of plaster used in manufacturing wallboard. Outrageously gigantic crystals of gypsum are illustrated on the opening pages of this chapter.

Oxide minerals contain elements other than silicon that bond to oxygen (Table 2.2). The related hydroxide minerals also contain the hydroxyl ion (OH^-). The most common oxide minerals are iron oxides, such as hematite (Fe_2O_3) and magnetite (Fe_3O_4), and are the primary

iron ores extracted for making steel. Hematite also is a source of red pigment. Magnetite, as the name suggests, is strongly magnetic, as shown in **Figure 2.29**.

Other oxide and hydroxide minerals are common ores of aluminum, chromium, uranium, tin, and titanium. The principal source of aluminum is bauxite, which is a mixture of very soft aluminum hydroxides. Aluminum is used to manufacture lightweight items, such as car parts, packaging and containers, and building products. Corundum, an aluminum oxide, is used as a polishing abrasive because of its high hardness (9 on the Mohs scale), and it is the source of the colorful, hard gemstones sapphire and ruby.

Atoms of many elements used to make important metal products do not fit within silicate crystal structures but readily bond to sulfur (S) to

▲ **Figure 2.29 Magnetite is magnetic.** Magnetite is a strongly magnetic oxide mineral. This particular property distinguishes magnetite from other, similar-looking minerals.

▲ **Figure 2.30 A collection of metallic minerals.** Galena and pyrite are examples of sulfide minerals in which metallic minerals bond with sulfur. Galena, a lead sulfide, is the most abundant source for industrial lead. Pyrite is iron sulfide and sometimes is called "fool's gold." Gold is an example of a mineral consisting of only a single element, similar to graphite and diamond.

form sulfide minerals, illustrated in **Figure 2.30**. Examples of these include galena (PbS), chalcopyrite ($CuFeS_2$), pyrite (FeS_2), and sphalerite (ZnS). Lead, copper, iron, and zinc, all are valuable natural resources that humans can easily separate from sulfur and use. There is a drawback to processing these minerals to obtain the metals, however. Separating lead, copper, zinc, and iron from the sulfur atoms produces sulfuric acid as a by-product. When released into the atmosphere from processing plants, sulfuric acid droplets combine with water vapor to form environmentally damaging acid rain. To limit these harmful by-products, pyrite, one of the most abundant iron-containing minerals, is not commonly used as an ore for iron. Instead, most iron is extracted from oxide minerals, such as hematite, which can be processed with less environmental impact.

Lead (Pb), obtained from galena, has many industrial uses because it can be easily cast, molded, and shaped into pipes and storage vessels, as well as be used in ceramic glazes, glassware, car batteries, ammunition, and even gasoline, to improve engine performance. Due to mounting evidence that toxic levels of lead have accumulated in soil, plants, and water, the use of lead has decreased. Lead is now less frequently added to gasoline, paint, and glassware.

Zinc (Zn) is obtained from sphalerite and is used primarily for galvanizing, a process that attaches a protective coating to steel and iron to prevent corrosion. Zinc also is commonly used to make white pigment and in ointments and lotions to prevent sunburn and infections. U.S. pennies are now made mostly of zinc and have only a thin copper coat, so as to conserve copper resources for other uses.

Copper (Cu) was one of the first metals used by humans, because it is a native metal in some places and, like gold (Figure 2.30), does not require chemical processing to separate the metal from other elements. Most copper production during the last century has, however, relied primarily on the mining of copper sulfide minerals, such as chalcopyrite. Copper is also mixed with tin to make bronze and with zinc to make brass. Copper is attractive, easily formed into many shapes, and corrosion resistant. Consequently, copper and its related bronze and brass have a long history of use

in weapons, tools, jewelry, pipes, and utensils. With the great expansion in the use of electricity beginning in the late nineteenth century, the abundance and highly conductive nature of copper made it ideally suited for the manufacture of electrical wiring.

Putting It Together—Which Minerals Are Most Important?

• Minerals are most important if they are common in rocks, provide essential resources, or both.

• More than 4000 minerals have been identified, but only a few dozen are important as rock-forming minerals, because only 12 of the 89 naturally occurring elements compose 99.7 percent of Earth's crust.

• Silicon and oxygen are the most abundant elements in the crust and, bonded with other elements, form silicates, the principal rock-forming minerals.

• Elements with similar ionic charge, size, or both substitute for each other within mineral structures. These substitutions cause minor changes in crystal structure that define groups of related minerals. Among the silicate minerals, the most important substitutions are Al^{3+} for Si^{4+} and the related interchange of K^+, Na^+, and Ca^{2+} in the feldspar group, and the exchange of Mg^{2+} and Fe^{2+} for each other in the olivine, pyroxene, amphibole, and biotite groups.

• Most economically valuable metals are processed from nonsilicate ore minerals, especially oxide and sulfide minerals.

EXTENSION MODULE 2.3

Gemstones. *Learn about the features valued in gems and the types of minerals and rocks that form gemstones.*

Where Are You and Where Are You Going?

Minerals are the building blocks of solid Earth. A mineral is a naturally occurring inorganic solid with a definite, only slightly variable chemical composition and an ordered atomic structure.

Each mineral possesses a unique set of observed or easily measured physical properties that directly relate to the elemental composition and the arrangement of the atoms of those elements within the mineral. Mineral color usually relates to composition and can vary significantly as a result of very small changes in composition. Mineral density is determined by the mass of the atoms and their spacing within the crystal structure. Scratching a mineral or breaking it into pieces involves breaking chemical bonds. If a mineral breaks along predetermined planes to form flat, shiny surfaces, it has cleavage; otherwise, it exhibits fracture. Both hardness and cleavage, therefore, vary depending on the type of bonds and the locations of repeated relatively weak bonds within the atomic structure of the mineral crystal.

Some groups of minerals differ only slightly in composition and structure because they are related to one another by the substitution of elements within the atomic structure. Elements of similar charge, size, or both can replace each other to cause minor changes in the structure, and in related physical properties, so as to produce distinctly different minerals.

Although there are more than 4000 known minerals on Earth, only a handful form most rocks, because just 12 elements account for 99.7 percent of the mass of the crust. The principal rock-forming-mineral class is the silicates. Quartz and feldspar are the most common silicate minerals in the crust, although others also are found in most rocks. A number of minerals precipitate from water, including most carbonates, sulfates, and some halides, such as halite (rock salt). Oxide and sulfide minerals are the primary ore sources of essential metals like iron, copper, lead, and zinc.

Now, you are ready to embark on the study of rocks—what they are, how they form, and why they have so many appearances. Rocks consist of one or more than one, and usually several, minerals. Clearly, part of your understanding of rocks must include the processes that determine which minerals form together, why and how they combine, and how the many different mineral grains become consolidated into a coherent rock. Geologists pursue answers to these questions by examining rocks in their natural settings. In Chapter 3, you will be introduced to the full array of rocks and the general relationships that exist among the three rock groups. We will consider the rock-forming processes for each group in greater detail in Chapters 4 through 6.

Extension Modules

Extension Module 2.1: Basics of an Atom. Learn about the basic components of an atom.

Extension Module 2.2: Silicate Mineral Structures. Learn how the SiO_4^{4-} groups, called silica tetrahedra, bond together to form a variety of silicate-mineral structures.

Extension Module 2.3: Gemstones. Learn about the features valued in gems and the types of minerals and rocks that form gemstones.

Confirm Your Knowledge

1. "Element," "mineral," and "rock" are important terms used in this chapter. Define each term using your own words, and then explain how the terms relate to one another.
2. Calcite and quartz are minerals. What properties do they have in common? How do they differ?
3. List and define the physical properties used to identify minerals.
4. Why is color not always a useful property in identifying a mineral?
5. Use the absolute hardness scale (Figure 2.4) to determine how many times harder fluorite is than gypsum.
6. List and describe the four types of bonds exhibited in minerals. Give an example of a mineral that exhibits each type of bond.
7. How does a geologist "see" atoms?
8. Does a TEM image detect all the elements in a mineral sample? Why or why not?
9. Why do some minerals exist in more than one color?
10. What factors determine the density of a mineral?
11. How does cleavage differ from fracture? How does cleavage relate to mineral structure?
12. What is the implication for the composition of Earth's interior if iron represents almost 35 percent of the entire Earth's composition, but less than 6 percent of Earth's crust?
13. Why are silicates the dominant rock-forming minerals?
14. Silicates are the most common mineral group and differences between some silicates are due to only slight variations in their compositions. Explain the conditions necessary for element substitution to occur for Si and Mg.
15. Give two uses for each of the following metals. Also list the minerals that are common ores for each metal.

    Iron   Lead   Zinc   Aluminum   Copper

Confirm Your Understanding

1. Write an answer for the question in each section heading.
2. For a substance to be a mineral, it must be a naturally occurring inorganic solid with a definite chemical composition and an ordered atomic arrangement. For each of the following substances, determine whether or not it is a mineral. If it is not a mineral, list all of the mineral properties that do not apply to it.
   a. amber
   b. beer
   c. calcite
   d. diamond
   e. emerald
   f. plagioclase feldspar
   g. glacial ice
   h. halite
   i. ice cubes from an ice machine
   j. rock candy
   k. kryptonite
   l. cubic zirconia
   m. mineral oil
   n. muscovite
   o. obsidian

3. Minerals have a wide variety of applications. Select five of the following substances and identify what minerals they contain.
   Baby powder
   Antacid
   Toothpaste
   Toothpaste with sparkles
   Epsom salt
   Red lipstick
   Red blush
   Table salt
   Kitty litter
4. Diamond and graphite have the same chemical composition, but are different minerals, because their atoms are arranged differently. Research and write a paragraph comparing another pair of minerals that share the same chemical composition.
5. Determine whether your state has a state gemstone or mineral. If it does, research that gemstone or state mineral and list its chemical formula, name two of its physical properties, and explain why it was chosen as the state gemstone or mineral.
6. Consider how easily halite (NaCl) dissolves in water. Why does quartz not dissolve readily in water?

CHAPTER 3

Rocks and Rock-Forming Processes

Why Study Rocks?

Ask a nongeologist what geology is all about, and the answer probably will be something like, "Geology is the study of rocks." Certainly, rock formation is a key topic in geology, because Earth is mostly solid rock, and rocks are the source of economically essential mineral ores, energy resources, and important building materials. Resources such as drinking water, oil and gas, fertile soil for agriculture, and building materials such as sand and gravel have provided geologists an incentive to study how rocks form in order to locate these resources. So, we study rocks to understand what they are and how they form.

Rocks form from minerals. How do minerals combine to make rocks and what do rocks reveal about how Earth works? You learned in the previous chapter that some minerals form by precipitation from solution (e.g., calcite) or are left when water completely evaporates (e.g., halite). Minerals also solidify from cooling molten lava and transform from one to another when temperature and pressure conditions change.

It also is important to understand rock-forming processes because they are the basis for how geologists classify rocks. The classification system for rocks introduced in this chapter is based on observations that reveal rock origins and fundamental scientific classification principles. This classification system provides a framework for further exploring the rock-forming processes in the chapters to come.

After Completing This Chapter, You Will Be Able to

- Describe the basic processes of rock formation and the relationships between rock types.
- Identify the factors that contribute to geologic classification systems and explain how to classify rocks.

Pathway to Learning

3.1 How and Where Do Rocks Form?

3.2 How Do We Classify Rocks?

Rock climber reaches for a hold on a New York cliff.

IN THE FIELD

To begin your study of rocks, we take a virtual rafting field trip in the Grand Canyon, as seen in **Figure 3.1**. The exposed bare rock stretches continuously for more than 200 kilometers. No wonder geologists flock to this place to study Earth processes and history.

Your eye quickly notices differences among the rocks. First, they vary considerably in color: Many are red, some outcrops are black, other rocks are off-white to gray, and a few have a greenish tint. You assume that these differences somehow relate to the combinations of different-colored minerals within the rocks. Also, you note that vertical cliffs alternate with more gradual slopes of loose rubble. These differences in slope steepness must have something to do with the durability of the different rocks when they are exposed to the weather—rain, snow, ice, and wind—over long periods.

The rocks also form striking patterns across the landscape. Most of the rocks form very even, nearly horizontal layers that continue as far as your eye can see (Figure 3.1a). In the dark depths at the bottom of the canyon, however, are rocks that lack layers and are streaked, here and there, with colorful bands (Figure 3.1b).

Up close, some rocks contain large mineral crystals (Figure 3.1c). Others contain no clearly visible minerals; you would need a microscope to determine their composition. Some contain remarkable fossils of ancient animals (Figure 3.1d).

A raft trip such as this is memorable for the spectacular scenery, but it also can whet your appetite for geology and make you want to learn more about rocks. You may be curious about why rocks exhibit such a huge variety of colors, patterns, degrees of durability, mineral compositions, and mineral textures, and wonder how these variations relate to the processes that form rocks. To find out more about rocks and their wonderful diversity, you will need to learn more about how rocks form.

▶ Figure 3.1 **Rocks in the Grand Canyon.**

3.1 How and Where Do Rocks Form?

Scientific explanations of natural processes depend on close observation of the natural world. For us to understand how and where rocks form, let's take three additional imaginary, but realistic, field trips, which can provide you with the observational opportunities you will need to begin to understand rock formation.

Observations on a Beach

Our first virtual environment involves walking along a beach, depicted in **Figure 3.2**a, where you feel gritty sand under your bare feet, punctuated by sharp-edged shell fragments. Waves crash against the shore. A small stream crosses the beach and enters the sea. A low cliff of layered rock rises above the landward side of the beach.

You may wonder what composes the sand and where it comes from. Figure 3.2b shows a handful of sand grains, seen with the aid of a magnifying glass. Some grains are colorless, glassy-looking quartz. The sand also contains many broken shell fragments; recalling what you learned about shell secretion in Chapter 2, you guess that the shell fragments are composed of calcite. If it were available, you could add a drop of dilute acid to test this idea. Identifying all the tiny grains is challenging, but the sand is clearly composed of small mineral grains.

The mineral grains and most of the shell fragments are smooth and round, with polished edges (Figure 3.2b). Breaking a shell, on the other hand, produces a sharp, jagged edge. These observations mean that the edges on the sand grains were somehow modified to form the smooth margins that you see. A possible solution comes to mind while watching sand and shell fragments move when waves surge across the beach—you wonder whether the motion of the water grinds the grains against one another, abrading away sharp edges and smoothing their outlines. You hypothesize that the tiny mineral grains that make up the sand began as larger, angular fragments from a rock outcrop that were gradually abraded into the small, smoothly rounded forms you now see.

The stream crossing the beach clearly delivers water to the sea, but a closer look (Figure 3.2c) indicates that the stream also transports sand,

(c) Some of the dark rocks in the bottom of the canyon have visible crystals, like this rock that has little dark-red garnets surrounded by shiny, silvery muscovite mica.

(a) These rocks form persistent layers. Some rock layers stand in vertical cliffs, whereas others form rubbly slopes.

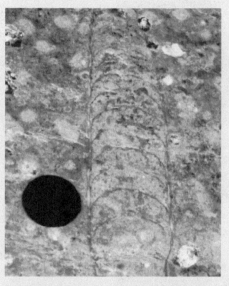

(b) These rocks in the bottom of the canyon do not exhibit layering like the rocks higher in the canyon and illustrated in (a).

(d) This rock has fossils in it. The camera lens cap rests next to a fossilized chambered nautilus.

small pebbles, and suspended silt. Some of these moving particles appear smoothly rounded, like the sand grains on the beach. Movement along the streambed may also reshape originally angular grains before they reach the ocean. Gusts of wind sting your face with blowing sand and remind you that wind also moves sediment particles and likely contributes to the abrasion process. Ocean waves crash into the mouth of the stream and spread the recently delivered sediment along the beach. At least some of the beach sand does not originate in the ocean, but is fed into it by streams and then redistributed by the ocean waves and currents, which add local shell fragments to the mix. The quartz grains in the sand came from inland areas through which the stream flows. In the face of all this observational evidence, you conclude that the mineral grains composing most of the sand must break off from rocks, during processes that are not obvious from this vantage point, and are then shaped and moved by wind and water.

Next, your attention turns to the low cliff of layered rock at the landward edge of the beach. Examining a piece of rock dislodged from the cliff, you notice that it shares some similarities with the beach sand. As seen in Figure 3.2d, the rock consists of rounded sand grains of quartz and fragments of broken shells. Applying the principle of uniformitarianism from Chapter 1, you infer that the grains in this rock were originally deposited in a setting like the modern beach. The most obvious difference is that the sand grains in the rock are not loose, as they are on the beach, but have somehow been stuck together (compare Figures 3.2b and 3.2d). Using a magnifying glass, you see a pale film along the margins of the sand grains that seems to act like glue to hold them together. Applying dilute acid causes this film to fizz and dissolve, suggesting that it is calcite. It seems reasonable to apply the name **sandstone** to this rock, because it is composed of sand grains.

You leave the beach with three important ideas about of rock-forming processes:

1. The mineral components of some rocks are pieces of other rocks and biologically produced minerals that form shells.

2. Abrasion—the grinding of objects together during transport by moving water or wind—can reshape rock and mineral fragments.

3. Fragments of older rock consolidate into new rock through the addition of mineral cement.

Waves move sand to and fro on a beach. The nearby sea cliff consists of sand grains consolidated into rock.

The beach sand consists of rounded mineral grains, and colorful, polished pieces of broken shells, all less than 2 millimeters across.

A stream crosses the beach and flows into the ocean. The current slowly rolls pebbles along the streambed. The water is cloudy with suspended grains of sand and silt.

The rock comprising the nearby sea cliff is a consolidated mixture of sand grains and fossil shell fragments. The components of the rock resemble the materials that make up the beach.

◄ **Figure 3.2 Geologic processes at a beach.**

Observations at a Spring

Our next stop is at a spring, illustrated in **Figure 3.3**a. Water issues from an opening atop a mound of light-colored rock. You examine this rock more closely and see thin layers of very small, clear and white crystals, as shown in Figure 3.3b. The rock reacts with mild acid, which suggests the presence of calcite. The calcite-rich rock is found only immediately surrounding the spring, suggesting a relationship between the rock and the spring water that exits from the top of the rock mound. Another striking feature visible in Figure 3.3a is that the rock encloses dead trees. This rock seems to form from the spring water by mineral growth at the surface. As the rock builds outward from the spring, it incorporates plants. You would be correct to hypothesize that the rock forms by a chemical reaction that precipitates calcite from the spring water. This process is similar to halite (table salt) crystals forming along the edge of a glass as salty water evaporates, and to mineral crusts forming in and clogging water pipes.

Remembering that the sandstone at the beach was composed of mineral grains cemented by calcite, you hypothesize that the calcite cement precipitated from water percolating through what was originally loose sediment. The rock formed at the spring resulted almost entirely from such precipitation, and the precipitated minerals are so closely intergrown that they form a coherent rock.

The key conclusion from visiting the spring is that another rock formation process is precipitation of minerals from water, including mineral precipitation that cements together loose sediment grains.

Observations at a Volcano

Perhaps no geologic phenomenon simultaneously causes so much awe and fear as an erupting volcano, such as the scene illustrated in **Figure 3.4a**. Fiery jets of molten material shoot into the sky and flow down the slopes of the volcano in rivers of lava. Unlike rivers of water, these molten streams solidify—they literally turn to stone, as seen in Figure 3.4b.

It is difficult to see what composes cooled volcanic rock, even if you look at it very closely (Figure 3.4c). A few blocky, green olivine crystals and rectangular, white feldspar are visible and are surrounded by a nondescript black background. Most of a volcanic rock consists of minerals too small to see with a magnifying glass; to observe these minerals, you must specially prepare the rock and view it under a microscope, where the minute crystals of olivine, plagioclase feldspar, and pyroxene finally become visible (Figure 3.4c). The crystals are intergrown with one another to form a coherent rock, much as interlocking ice crystals produce ice cubes.

Two features distinguish volcanic rock from the rocks observed at the beach and at the spring. First, the mineral grains in the volcanic rock have sharp edges and flat surfaces representing the crystal faces, whereas the grains in the sandstone are rounded and do not preserve crystal outlines. The reason for this difference is that

Sparkling spring water flows down over the surface of a mound of white rock. Notice the dead trees entombed in the rock.

The rock around the spring opening consists of thin layers of tiny crystals and sponge-like holes.

◀ **Figure 3.3 Rock-forming processes at a spring.**

An erupting volcano puts on a dazzling display as lava fountains into the air and then falls back to the ground to flow away in molten rivers.

Red-hot, fluid lava solidifies into hard black rock around a geologist's rock hammer.

▶ **Figure 3.4 Rock-forming processes at a volcano.**

minerals in the volcanic rock crystallized in place from the original molten material and were not tumbled and abraded during transport by moving water and wind. Second, although intergrowth of newly formed minerals explains the solid nature of the rock formed at both the spring and the volcano, the rock at the spring formed by precipitation of minerals out of the water, which remained liquid. In contrast, the volcanic rock formed by the complete crystallization of a molten liquid.

Volcano craters are surrounded not only by solidified lava, but also by rock fragments ripped from the walls of the volcanic conduit and blown out to the surface. These fragments reveal the kinds of rocks hidden below the volcano. One group of blown-out rock fragments, shown in Figure 3.4d, contains the same minerals seen in the solidified lava flows, but in this case, the minerals are readily visible to the naked eye. The intergrown mineral texture in the fragments, like in the lava rock, suggests solidification by freezing of a liquid. The larger crystal sizes in the fragments, however, suggest some different conditions during rock formation. A second group of blown-out rocks, seen in Figure 3.4d, contains angular silicate minerals in distinct layers or bands. The minerals in these fragments are intergrown in a fashion vaguely similar to the calcite-rich spring rock and the lava rock, but the minerals are curiously arranged in alternating bands of dark and light crystals, unlike anything you have seen forming at the surface.

Another important conclusion arises from these observations: Apparently, these two groups of rocks ejected from the volcano came from some depth below Earth's surface, because they are quite different from rocks observed forming at the surface. This implies that rock-forming processes occur within, as well as on, Earth.

Putting It Together—How and Where Do Rocks Form?

- Most natural rocks are aggregates of mineral grains.

- Many rocks originate from observable processes that take place at Earth's surface.

- Rocks that are not related to observable surface processes arise from processes active within Earth.

The volcanic rock contains a few visible crystals of green olivine and white plagioclase feldspar set against a dull, gray background.
A microscopic view shows that the gray background consists of interlocking crystals.

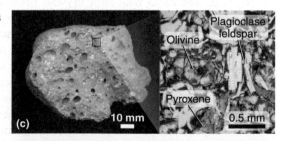

Olivine — Plagioclase feldspar
Pyroxene
(c) 10 mm 0.5 mm

The volcano ejected these blocks, which differ from the solidified lava. The rock on the left is a coarser-grained version of the lava rock, consisting entirely of visible crystals. The rock on the right contains different minerals in distinct bands. These rock types underlie the volcano and the pieces were blown out during the eruption.

Mosaic of feldspar, pyroxene, & olivine   Bands of biotite, quartz, & feldspar
(d) CENTIMETER

3.2 How Do We Classify Rocks?

How do the field observations and initial interpretations of rock origins permit you to classify and name rocks? Classification is an important part of natural science. Scientists commonly group similar objects, features, or phenomena in order to seek explanations for their origins. *Descriptive classifications* group items of similar appearance. *Genetic classifications* group items or phenomena by noting similarities in the processes that cause or create them. Chapter 2 utilized a descriptive classification to group minerals based on chemical composition and, in some cases, crystal structure. You may be familiar with other descriptive classifications of natural objects, such as animals and plants. Historically, the groupings of rocks have emphasized rock origin to form a genetic classification.

Which Is More Important: Where Rocks Form or How They Form?

How do geologists know how to group rocks? Let's try to classify the rocks we studied in the previous section. Our virtual field observations suggest that some rocks form by processes that leave visible evidence at Earth's surface. Other rocks, such as those found in loose pieces around the mouth of the volcano, seem to originate below the surface. Might the distinction between internal and external processes serve as a means to classify rocks? The sandstone from the beach cliff, the rock formed by precipitation of calcite from the spring water, and the rock formed by lava crystallization of can be viewed as rocks forming by external processes. The loose rocks around the volcano, which contain minerals similar to those in the volcanic rocks, but with much larger crystals, or contain bands of different minerals, can be categorized as those formed by interior processes.

Although this genetic division seems workable, it is not wholly satisfying. For example, consider the rocks composed of similar minerals intergrown in a fashion suggesting crystallization of a molten liquid (Figures 3.4c and 3.4d). Descriptively, these rocks differ only in the size of the mineral grains composing them. This fact suggests that they may belong in the same classification group. If we follow this line of reasoning, then all rocks formed from solidification of molten material can be grouped together. The rocks with large crystals result from crystallization below ground, whereas the volcanic rock with smaller crystals forms after the melt erupted at the surface. The distinction of internal and external processes is further blurred because although the volcanic rock formed on the surface, the molten liquid clearly originated below the surface. Volcanic activity is merely one of many links to be explored between the Earth's interior workings and its surface features. While some processes are wholly interior and others are wholly exterior to the planet, usually, the formation of rocks exposed at the surface is the result of complex relationships among

different processes in both of these two locations. It seems unwise, therefore, to use the site of formation as the primary criterion for genetic classification.

The Three Rock Classes

Rather than classify rocks according to *where* they form, geologists use a three-category rock genetic classification scheme, based on *how* rocks form. The three rock types are **sedimentary**, **igneous**, and **metamorphic**; they are briefly described in **Table 3.1**. This chapter provides a brief overview of these three rock classes, their origins, and their relationships to one another. Chapters 4, 5, and 6 are devoted, respectively, to one of these rock types. Genetic classification emphasizes the processes responsible for forming rocks in each group. Nonetheless, you will see in later chapters that further divisions within each rock class are descriptive and emphasize visible features rather than the origins of those features.

Sedimentary Rocks

Sedimentary rocks form by deposition and precipitation of mineral grains that originated from the breakdown of older rocks under the conditions found in Earth's surface environment. Rocks of any kind, when exposed at Earth's surface, slowly deteriorate due to the processes of **weathering**, schematically illustrated in **Figure 3.5**. Physical weathering causes rocks to disintegrate into rock fragments or mineral grains. Chemical weathering involves reactions among minerals, the atmosphere, and water, which produce dissolved ions and new minerals. Mineral or rock fragments produced by chemical and physical weathering, such as the sand you observed on the beach, are called **clastic sediment** (from the Greek *clastos:* "broken").

Ions dissolved in water by chemical weathering may later precipitate as new mineral crystals if conditions of water temperature, pressure, or chemical composition change. These crystals form **chemical sedimentary rocks** (Table 3.1), such as the rock you observed at the spring in Section 3.1. The minerals that precipitate from water are those whose chemical constituents most readily dissolve in water. Chemical sedimentary rocks are, therefore, most commonly composed of minerals such as calcite, dolomite, gypsum, and halite, which are strongly ionic and easily dissolve in water (Section 2.4). Although silicate minerals are the most abundant constituents of Earth's crust, they do not dissolve as easily as these other more strongly ionic compounds. Quartz is the only silicate mineral to form a common mineral precipitate from water.

Precipitation of minerals also plays an important role in making clastic sedimentary rocks, as you inferred from observation of the beachside sandstone in Section 3.1. Clastic sediment consists of loose particles generated by weathering and then transported to a final site of deposition by flowing water, blowing wind, or sliding glaciers. This loose sediment does

Table 3.1 The Genetic Classification of Rocks		
Origin	Rocks formed from the products of the breakdown of preexisting rock	Rocks formed by the solidification of molten rock material (magma)	Rocks formed by the transformation of minerals in a preexisting rock, without melting, because of the effects of elevated temperature, pressure, hot fluids, or all three variables
Rock Type	Sedimentary rocks	Igneous rocks	Metamorphic rocks

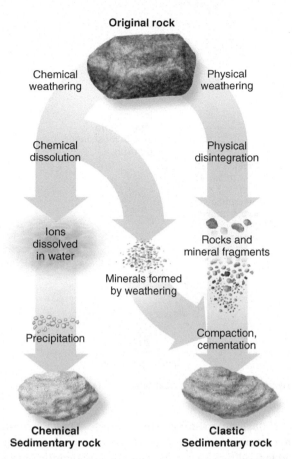

Original rock

Chemical weathering   Physical weathering

Chemical dissolution   Physical disintegration

Ions dissolved in water   Rocks and mineral fragments

Minerals formed by weathering

Precipitation   Compaction, cementation

Chemical Sedimentary rock   **Clastic Sedimentary rock**

▲ Figure 3.5 **Sedimentary rocks contain the weathering products of other rocks.** Rocks break down at the surface through the weathering processes of physical disintegration and chemical reactions. Clastic sedimentary rocks contain fragments set loose through physical weathering or newly formed by chemical weathering. Changing chemical conditions can cause ions dissolved in water to precipitate as interlocking mineral grains to form chemical sedimentary rocks.

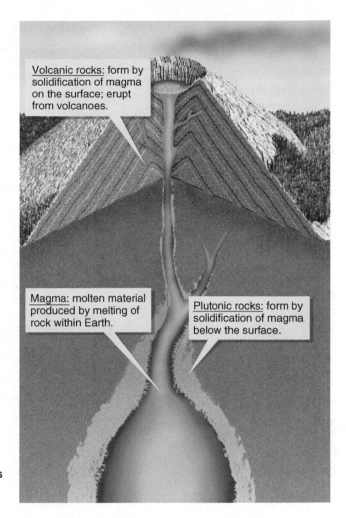

Volcanic rocks: form by solidification of magma on the surface; erupt from volcanoes.

Magma: molten material produced by melting of rock within Earth.

Plutonic rocks: form by solidification of magma below the surface.

▶ Figure 3.6 **Igneous rocks originate from magma.**

not become rock until it undergoes compaction under the weight of additional accumulating sediment and, especially, until the precipitation of cementing minerals between grains consolidates it into sedimentary rock. This transformation is **lithification**, which loosely translates as "making of rock" (Greek *lithos:* "rock").

Sedimentary rocks are most readily recognized from a distance by their distinctive layering, called **bedding** (Figure 3.1a). Bedding is caused by the deposition of different sedimentary materials, one layer above the other.

Igneous Rocks

Igneous rocks crystallize from molten material called **magma**, which originates from melting rock within Earth. The root for "igneous" is the Latin word *igneus,* meaning "fiery," an appropriate label for these rocks. **Figure 3.6** shows that magma may erupt at volcanoes, as lava and volcanic ash, or remain underground to solidify. Molten material is called "magma" below the surface and "lava" after it reaches the surface.

The two general kinds of igneous rocks are **volcanic rocks**, which solidify at the surface, and **plutonic rocks** (named for Pluto, the Greek

god of the underworld), which solidify beneath the surface. These rocks sometimes are referred to by their synonyms, **extrusive** (volcanic) and **intrusive** (plutonic) rock. Extrusive refers to lava, ash, and pumice extruded onto the surface at volcanoes. Intrusive describes magma intruded into pre-existing rocks, below the surface.

Experiments show that fast magma solidification usually produces small crystals, while slower crystallization forms larger crystals. This relationship between cooling rate and crystal size helps to account for the different appearances of volcanic and plutonic rock (Table 3.1). The generally fine-grained rocks result from rapid cooling following eruptions from volcanoes (see Figures 3.4b and c). The coarser rocks form from the slow solidification of magma beneath Earth's surface (an example is the rock on the left in Figure 3.4d). Magma is overwhelmingly dominated by the elements that make up silicate minerals. Igneous rocks, therefore, consist mostly of silicate minerals.

The accumulation of lava flows and ash deposits produces layers that superficially resemble sedimentary-rock bedding, as shown in **Figure 3.7**a. Plutonic rocks, however, rarely accumulate as successive layers (see Figure 3.7b) and as a result look very different from sedimentary rocks. Geologists use the term **massive** to describe rocks that lack layering.

Metamorphic Rocks

Metamorphic rocks result from changes to preexisting rocks (Greek *meta:* "change"; *morphe:* "form"); these changes occur when the minerals that

(a)

Successive lava flows stack up in layers, as in this view along the Columbia River in Washington. Notice how thick the lava layers are compared to the train.

Plutonic igneous rocks are usually massive without layers. Plutonic igneous rocks comprise many bold mountain landscapes, like the Wind River Mountains in Wyoming.

(b)

◄ Figure 3.7 What igneous rocks look like in the field.

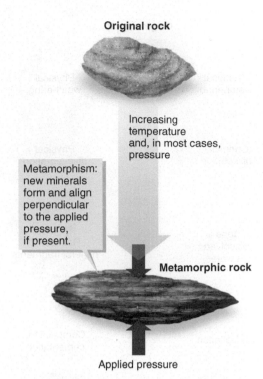

Original rock

Increasing temperature and, in most cases, pressure

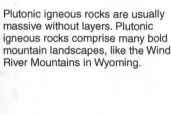

Metamorphism: new minerals form and align perpendicular to the applied pressure, if present.

Metamorphic rock

Applied pressure

▲ Figure 3.8 Metamorphic rocks form from preexisting rocks.

compose the rock are changed or rearranged or both. The preexisting rocks may be igneous, sedimentary, or older metamorphic rocks. The changes typically occur at temperatures in excess of 200°C, which distinguishes metamorphic rocks from the lower-temperature weathering and lithification processes that form sedimentary rocks. Metamorphic rocks are distinguished from igneous rocks because the temperatures that change the former are not high enough to cause the rock to melt, which defines the latter.

In nearly all cases, metamorphic processes, referred to as **metamorphism**, take place at temperatures and pressures found only within Earth and not at the surface. Field observations and laboratory experiments summarized in **Figure 3.8** indicate that metamorphism almost always occurs at elevated temperatures. Temperatures measured in deep wells and mines show that Earth's interior is hotter than the surface and that temperature increases progressively with depth. (Another sign of the fiery heat within Earth is magma that emerges from the depths at temperatures exceeding 1000°C.) Another primary metamorphic agent is high pressure. Pressure, like temperature, increases as we move deeper into Earth, because the weight of overlying rock increases with depth.

Chemical reactions between minerals and fluids at high temperatures are a third agent of metamorphic processes. Laboratory experiments simulating metamorphic temperatures and pressures reveal that minerals within rocks of any origin change under the conditions present at great depths within the crust or mantle. These changes are especially notable in the presence of watery fluids, which can remain liquid at high pressure

even if the temperature greatly exceeds 100°C, the boiling temperature of water at Earth's surface. The changes include modifications in the shape and orientation of existing crystals, and the production of new minerals. The new minerals result from chemical reactions among the original minerals and between the minerals and the hot watery liquid that passes through the rock.

Taken together, these three primary conditions found inside Earth—high temperate, high pressure, and chemical reactions that occur as a result of high temperate and pressure—are ideal for metamorphic transformation.

Metamorphic changes in rock can be local or affect vast regions. When local, the phenomenon usually is a straightforward case of the intrusion or extrusion of magma "cooking" the adjacent rocks. Widespread changes are particularly likely at convergent plate boundaries (Figure 1.12b). At these boundaries, rocks originating at the surface are forced down to great depth and, thus, experience very high pressures and temperatures. These metamorphic rocks include those with minerals oriented in layers or bands (right side of Figure 3.4d and Figure 3.8). The rocks seen deep in the Grand Canyon in Figure 3.1b, for example, are dark metamorphic rocks, cut through by pink, intrusive igneous rocks.

The minerals found in metamorphic rocks depend partly on the minerals in the original rock and partly on the nature of the metamorphic processes. Metamorphic rocks commonly contain abundant silicate minerals, but many of these silicate minerals are minor or not present in igneous rocks that crystallize from magma. This fact helps to distinguish metamorphic and igneous rocks.

The growth of metamorphic minerals in bands, illustrated in **Figure 3.9**a, only superficially resembles sedimentary bedding, as these bands are neither continuous nor uniform in thickness over large distances. In contrast, sedimentary beds are almost always readily traced across whole outcrop faces hundreds of meters long and typically can be traced for

Bands of different minerals are common in metamorphic rocks, such as the swirling dark and light-colored bands in this rock outcrop in Scotland.

The sheer cliffs of the Black Canyon of the Gunnison River, in Colorado, consist mostly of metamorphic rocks. The rocks appear massive from a distance but contain fine-scale mineral banding like that seen in photo (a).

◄ Figure 3.9 What metamorphic rocks look like in the field.

many kilometers. The banding of many metamorphic rocks may be apparent from great distances, but rarely is it as distinct and continuous as that of sedimentary bedding (compare Figure 3.9b to the photo at the beginning of the chapter).

Putting It Together—How Do We Classify Rocks?

• Rocks can be classified descriptively or genetically. The three principal rock types—igneous, sedimentary, or metamorphic—are defined genetically, according to the processes that form them.

• Sedimentary rocks are made up of minerals derived from the physical disintegration or chemical weathering of preexisting rocks. Clastic sedimentary rocks contain sedimentary particles of weathered rock cemented by minerals that precipitate from water. Chemical sedimentary rocks are intergrown mineral aggregates. Biological components may be significant constituents of some sedimentary rocks.

• Igneous rocks form from molten lava on the surface (volcanic, or extrusive, rocks) or from magma that remains and solidifies below ground (plutonic, or intrusive, rocks).

• Metamorphic rocks form when minerals in preexisting rocks change in response to temperature, pressure, fluid composition, or all three.

3.3 How Do We Know . . . How to Determine Rock Origins?

Picture the Problem

How Do Geologists Infer Process From Observations of Rocks? Do you feel confident that you could now identify a rock as belonging to one of the three principal genetic classes? Do not worry if you are not sure yet. It can be difficult to learn to use a genetic classification scheme, because descriptive classifications require only your powers of observation, whereas genetic divisions require the ability to interpret observations. Your imaginary experiences at the beach, spring, and volcano are modeled on the experiences of early geologists. In their quest to understand the world around them, they formulated questions and made inferences from their observations of rock-forming processes. The current threefold genetic classification scheme is less than 200 years old and represents the efforts of geologists of the late eighteenth and early nineteenth centuries.

In the late 1700s, geologists developed two competing views of rock-forming processes. Scientists in both groups appreciated that the wearing down of rocky landscapes results in loose sand, gravel, and mud. However, when it came to explaining how consolidated, coherent rocks form, the two groups reached very different conclusions. The supporters of each view argued at the scientific gatherings and in the written communications of the time. Neither group turned out to be entirely correct. Today, looking back at their debate and how their views arose, we can see how observations, inference, and critical review come together in evolving scientific knowledge.

The Neptunist View

Do Most Rocks Precipitate from Water? A group of geologists later known as the Neptunists (*Neptunist* is derived from Neptune, the Roman god of the sea) emphasized the existence of a cold early Earth where nearly all rocks formed by chemical precipitation from water. The Neptunist view was developed by Abraham Werner (1749–1817), a professor at a German mining academy whose stimulating instructional style attracted students from across Europe. At the academy, students learned about the formation and location of the economically important minerals used for manufacturing during the industrial revolution. Werner went beyond having students examine rock collections in laboratories and giving lectures that conveyed secondhand information: He also took his students to study rocks in the mines near the academy. This hands-on approach may have been the origin of the field instruction still required today in most university geoscience curricula.

Werner's concepts built upon prevailing views of eighteenth-century natural philosophers (the forerunners of modern scientists) that Earth originated as a vast blob of material uniformly dispersed in water. He envisioned this primordial ocean as a thick, dense mixture of solids and water rather than as a thin, dilute fluid containing dissolved salts, like the modern seas. From this mucky beginning, solids precipitated, and the liquid somehow evaporated into space.

Werner concluded that the sequence of precipitation of different minerals was readily determined by looking at the rocks the minerals form. He also concluded that successive layers of rock form in consecutive order, much like a pile of newspapers accumulates in your home—the oldest papers lie on the bottom, and the most recent edition is on top.

Werner noted that layered rocks usually rest above the more massive rocks that were exposed in deep river canyons or heaved upward in mountains. This observation led him to hypothesize that because the layered rocks are on top, the more massive rocks must have formed first. Accordingly, he named these thick, massive, mostly unlayered rocks, whose bottoms were never exposed to view, "Primitive Rocks." These rocks formed, according to Werner, by precipitation and settling of the least-soluble silicate minerals before dry land appeared. The Primitive Rocks are overlain by the so-called "Stratified Formations," distinctively layered rocks composed of precipitates of more-soluble minerals mixed in places with fragments worn from highlands of Primitive Rocks, which were gradually exposed as the enclosing ocean diminished in depth.

Werner also believed that rocks originating from volcanoes and unconsolidated gravel, sand, clay, and soil are of lesser significance and are found only at the surface because they were the most recently formed. The Neptunists referred to the unconsolidated sediment as "Washed Deposits." Werner viewed volcanoes as minor aberrations formed only very recently, and only at places where underground coal beds had somehow burst into flames and melted overlying rock, which then extruded at volcanoes. **Figure 3.10** summarizes the Neptunist origin of rocks.

To better grasp Werner's classification, you can relate his rock types to the classification presented in the previous section. Werner's Primitive Rocks include plutonic-igneous and most metamorphic rocks. His Stratified Formations include sedimentary rocks, as well as many volcanic-igneous rocks (e.g., layered lava flows such as those illustrated in Figure 3.7a) and some conspicuously banded metamorphic rocks (e.g., Figure 3.9a).

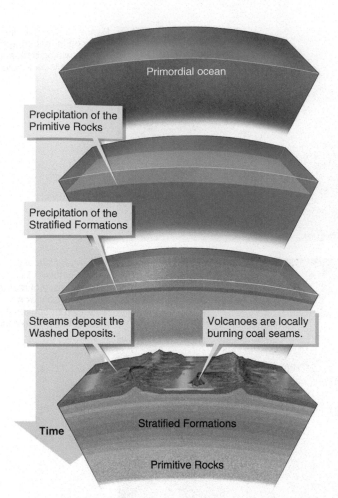

▲ **Figure 3.10 The Neptunist view of Earth.** Werner envisioned concentric shells of different types of rocks and minerals that precipitated in sequence from a "primordial ocean" solution that was extremely rich in dissolved materials. In his vision, the interior of the planet was cold, and volcanoes were rare, relatively recent phenomena related to the natural burning of subterranean coal beds.

The Vulcanist View

Do Most Rocks Result from Earth's Internal Heat? The Vulcanists (Vulcanist relates to Vulcan, the Roman god of fire) opposed the Neptunists and described an inherently hot planet. On this hot Earth, rocks formed from the fusing of particles by heat, or from solidification of magma rising from a molten interior. The man eventually labeled as the champion of the Vulcanists was James Hutton (1726–1797), a wealthy Scottish farmer who also helped to develop the principle of uniformitarianism (Section 1.4). Experienced and interested in scientific endeavors, Hutton sought to make his farm as productive and profitable as possible by studying farming methods across Britain and the European continent. It is easy to understand how his agricultural passion led to a curiosity about the characteristics and origin of soil and, eventually, to the rocks found beneath the soil.

In his travels, Hutton made observations, such as those documented in his illustration in **Figure 3.11**, that were vital to understanding the origin and classification of rocks. Hutton observed Stratified Formations above Primitive Rock, as had Werner. It was

▲ **Figure 3.11 An example of Hutton's field observations.** Hutton made this sketch of geologic features exposed alongside a road in Edinburgh, Scotland. Thinly layered rocks (Stratified Formations, now called sedimentary rocks) are cut across by a massive rock type (Primitive Rocks, now called igneous rocks). The massive rock appeared to Hutton to form by solidification of a molten liquid injected through the layered rocks.

quite clear to Hutton, however, that narrow veins of the Primitive Rock continued upward into the overlying Stratified Formations. The finger-like veins of the Primitive Rock suggested to Hutton the injection of a liquid along cracks in the Stratified Formations. This implied that the Primitive Rock could not always be the older of the two, because the Stratified Formations had to be present before the Primitive Rock could be injected into them. Furthermore, at the boundary of the two rock types, the layered formations were different in hardness and color, as if baked at high temperature.

Hutton demonstrated circumstances in which the Primitive Rocks could not have come before the Stratified Formations, even though they were found below them. In addition, he saw that the action of heat (what we now call metamorphism) was involved in forming the "fingers" of Primitive Rock (igneous rock solidified from magma) that interrupted the layers in the Stratified Formations (sedimentary rocks). This was contrary to the central Neptunian assumption that Earth was cold and rocks formed exclusively from the precipitation of minerals from water. Hutton cast further doubt on the notion that the silicate-rich Primitive Rocks could have precipitated from water by noting in his experiments that most silicate minerals are practically insoluble in water.

Hutton also noted that the Stratified Formations contained some rocks so closely resembling lava flows described at erupting volcanoes that they also must surely have resulted by crystallization of molten material (not precipitation from water). The presence of these ancient lava flows among rocks of various ages demonstrated that volcanic activity was not a minor aberration of modern time, but was an ancient and ongoing geologic process. So impressed was Hutton with the abundant evidence for subterranean heat that he went even further and suggested that the lithification of loose sediment was a consequence of heat causing sediment grains to fuse into rocks.

Analyze the Problem

How Did Geologists Resolve the Dispute? The Vulcanist view clearly clashed with the Neptunist doctrine not only by inferring a different sequence of events in Earth's history, but also by presenting contrasting explanations for the origins of rocks. Hutton's approach essentially followed the modern scientific method: His ideas were based on and derived from observations. In contrast, Werner explained observations in terms of a concept that was merely presumed to be factual and that lacked supporting scientific evidence. Hutton's conclusions were not all correct, however, because although he championed an internally hot Earth, as we do today, he took his Vulcanist views too far by also attributing a major role for heat in sedimentary rock formation. Today, we know that chemical sedimentary rocks and the cementing minerals in clastic sedimentary rocks owe their origin to precipitation from water and do not require heat. This is one of the few components of the Neptunian doctrine that has held up.

Scientists eventually accepted the Vulcanist explanation for the formation of what you now know as igneous rocks, when the Neptunist geologists changed their conclusions after being confronted by overwhelming evidence. French geologists had long suspected that volcanoes played a greater role in rock formation than the Neptunists allowed. So, they invited some of Werner's students to an area of recently extinct volcanoes in southern France. There, the Neptunists saw lava flows that clearly had issued from the volcanoes and were composed of a rock called **basalt**. In Werner's view, basalt was assigned to the stratified formations and was not volcanic, but that interpretation was inconsistent with the basalt lava flows seen at the volcanoes. In light of their field observations in France, Werner's students acknowledged that volcanic activity was more prevalent than the Neptunist view permitted. They agreed that it appeared that volcanic activity was responsible for some rocks that Werner had mistakenly attributed to chemical precipitation from water. One Neptunist undertook an experiment to further test the igneous origin of basalt by melting a basalt sample and allowing it to cool and solidify. The mineral content and crystal textures of the igneous rock formed in his laboratory so closely matched that of natural basalt that he shifted to the Vulcanist point of view.

This shift in thinking illustrates a critical part of being a scientist—the willingness to modify or abandon existing explanations when new observations or data are inconsistent with previous interpretations.

Insights

How Do We Apply Knowledge Gained from This Historical Debate? This recounting of the early debate about the origin of rocks underlines the complexities inherent in understanding the rock formation and establishing a genetic classification. However, the significance of the debate goes beyond its implications for classifying rocks. The dispute between those allied with Werner and those convinced otherwise by Hutton's arguments is an important milestone in the history of geology as a scientific discipline. Hutton and his peers demonstrated the utility of the scientific method using observations of Earth materials and features. The real-world observations made by these early geologists ultimately tested and confirmed hypotheses about the very origin of the planet.

Putting It Together—How Do We Know . . . How to Determine Rock Origins?

• The threefold genetic classification of rocks arose from debates between early geologists about the origin of rocks.

• Neptunists, led by Abraham Werner, asserted that nearly all rocks formed as chemical precipitates from water. This assertion was consistent with the prevailing, but not rigorously substantiated, view that Earth had always been internally cold and originated as a predominantly watery orb.

• Vulcanists, supporting the work of James Hutton, used careful field observations to refute the chemical-precipitate theory of the origin of many rocks. By documenting the prevalence of igneous rocks, Vulcanists demonstrated that the interior of Earth was, and probably always has been, very hot.

3.4 How Are the Rock Classes Related to One Another?

The three rock types—igneous, sedimentary, and metamorphic—are not isolated genetic groups; they are related to one another by processes that are active on and within the dynamic Earth. Sedimentary rocks result from the breakdown of *preexisting* igneous, metamorphic, or other sedimentary rocks. Metamorphic rocks owe their origin to changes in the form and nature of *preexisting* igneous, sedimentary, or other metamorphic rocks. Igneous rocks crystallize from magmas that form when *preexisting* rocks, typically of metamorphic or igneous origin, melt. Each rock type originates from preexisting rock, which demonstrates the relationships among the groups.

The interrelationship of the three rock types is illustrated as the **rock cycle** in **Figure 3.12**. The rock cycle is a useful depiction of the processes that form rocks and relate rock types to each other. Every process has a product, and some of those products are consolidated aggregates of minerals that fit the definition of a rock. Take a moment and carefully trace the routes of rock origins by following the different arrowed paths.

The rock cycle is a reminder of the dynamic nature of Earth and its constituent rocks. Individual rocks on Earth are continually changing, but the overall number and type of components do not change in any significant way. The minerals or constituent elements in an igneous rock may ultimately be recycled as components of a sedimentary or metamorphic rock, but they remain on the planet as part of the rock cycle.

The process links in the rock cycle also relate to plate tectonics. For example, the conditions required to produce magma correspond to processes nearly unique to divergent and convergent plate boundaries and hot spots. The limited distribution of active volcanoes demonstrates this point (look back at Figure 1.10). Sedimentary rocks also are not found everywhere across the surface of the planet. Clastic sediment erodes from high-standing areas and is transported primarily by water to be deposited in low areas, called sedimentary basins, where accumulations of rock several kilometers thick are found. Plate motions drive the uplifting of mountainous sediment sources and the downward sagging of basins where sediment is deposited. You learned in Section 3.2 that plate-boundary processes are also key to the formation of metamorphic rocks.

Another important point to reflect on is that plutonic-igneous and metamorphic rocks form below the surface and yet are commonly exposed at the surface. In order for this to happen, these rocks must rise toward the surface, and the material that originally resided above them must erode away. Tectonic forces drive this process of uplift and form steep mountain slopes prone to erosion. Plutonic-igneous and metamorphic rocks, therefore, not only form in tectonically active regions, but also become exposed at the surface as a result of tectonic activity.

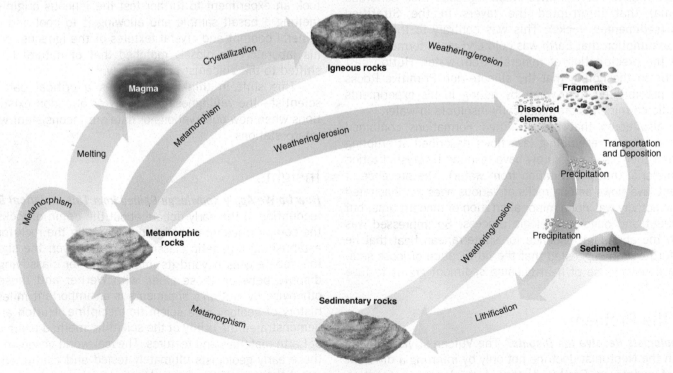

▲ **Figure 3.12 The rock cycle.** The rock cycle is a conceptual illustration of the relationships between the three rock types. A variety of process paths convert the components of one class of rocks into another rock class, with intermediate products forming along the way.

Putting It Together—How Are the Rock Classes Related to One Another?

• The three rock classes are linked through processes that convert the materials in one class of rock into new rocks of the same or a different class. This series of processes is referred to as the rock cycle. Despite the changes individual rocks undergo, the chemical composition of Earth remains essentially constant.

• Rock-forming processes and the relationships among rock classes are closely related to processes at plate boundaries. Plate tectonics is, therefore, a large-scale driving force in the rock cycle.

Where Are You and Where Are You Going?

You now have a framework for understanding the origins, properties, and uses of rocks. Geologists have used the threefold genetic classification to understand the origins, properties, and uses of rocks for nearly two

centuries. Sedimentary rocks form from the deposition and precipitation of products formed by the breakdown of older rocks. Minerals and rock fragments produced from weathering produce clastic sediment, which when lithified, forms clastic sedimentary rocks. Chemical sedimentary rocks are aggregates of minerals that precipitate from ion-rich water previously dissolved from other rocks by weathering. Biologic materials, such as shell fragments, also are components of sedimentary rocks. Igneous rocks form from the crystallization of molten magma. Volcanic, or extrusive, igneous rocks crystallize on Earth's surface, whereas plutonic, or intrusive, igneous rocks crystallize below the surface. Metamorphic rocks result from changes in preexisting rocks resulting from elevated temperatures and pressures and fluid-rock interactions. The rock cycle relates the rock types to one another. Plate tectonics provides most of the driving force for the transformations in the rock cycle.

This chapter offers only a glimpse into the understanding of the three rock groups. Many important questions remain. In the next three chapters, we will discuss how magma is created and how it reaches the surface at volcanoes. We also will look at how weathering breaks down rocks and what constituents of weathering end up in sedimentary rocks. We will explore what temperatures and pressures are required for metamorphism and how economically important resources, such as oil, coal, metal ores, and gems, form in rocks.

Confirm Your Knowledge

1. Define the three types of rocks.
2. Where does the sand on a beach come from?
3. Why do some rocks consist of mineral grains with sharp edges and flat surfaces, while others consist of mineral grains with rounded edges and surfaces?
4. A rock exposure consists of layers of fine-grained, rounded particles that are cemented together. Which type of rock is it: igneous, metamorphic, or sedimentary?
5. A rock consists of small but well-formed, randomly oriented, intergrown silicate-mineral crystals. Which type of rock is it: igneous, metamorphic, or sedimentary?
6. A shiny rock contains minerals oriented preferentially in discontinuous layers. Which type of rock is it: igneous, metamorphic, or sedimentary?

7. A rock contains fragments of shell and coral. Which type of rock is it: igneous, metamorphic, or sedimentary?
8. Contrast the Neptunist and Vulcanist explanations for the origins of rocks.
9. Which part of the Neptunist doctrine on the formation of rocks was correct?
10. Which part of the Vulcanist doctrine on the formation of rocks was incorrect?
11. Which conditions are required for metamorphism?
12. Draw you own diagram of the rock cycle and label the processes that change one rock type to another rock type.

Confirm Your Understanding

1. Write an answer for each question in the section headings.
2. Look ahead to the full-page photos at the beginnings of Chapters 5 and 6. Which features of these rock outcrops are similar? Which features are different? If you were to take samples from these two outcrops and examine them closely, what would you look for to determine that one rock is sedimentary, whereas the other is metamorphic?
3. Distinguish the differences you would see between intrusive and extrusive igneous rocks formed from the same magma.
4. How could you test the rate of weathering in your area by observing local gravestones?
5. The classification of rocks into igneous, metamorphic, and sedimentary rocks is a genetic classification. Create a descriptive classification

of rocks. Compare that classification system to the genetic one. Assess the strengths and weaknesses of each system.
6. Develop genetic and descriptive classifications for the objects on your desk at home.
7. If you were visiting Germany in 1800, what would you do to convince a geology student that not all rocks formed by precipitating from water?
8. The rock cycle describes recycling on a planetary scale. How might our 4.5-billion-year-old Earth differ if there was no rock cycle? Here are some ideas to consider in your answer: land elevation, depth of the oceans, and variability of rocks.

CHAPTER 4

The Formation of Magma and Igneous Rocks

Why Study Igneous Rocks?

Why is it important for us to study igneous rocks and processes? Self-preservation is one reason. Igneous processes—the partial melting of rock into very hot magma, which is a mixture of molten silicate minerals, or liquid rock, and the solidification of that liquid into new igneous rock—occur during volcanic eruptions. If we understand the workings of volcanoes, then we can perhaps reduce the incidences of volcanic hazards that devastate people and property. When the magma does not reach Earth's surface, solidification of igneous rock also occurs at depth, forming Earth's crust.

Also, understanding the formation of igneous rocks provides clues to finding and recovering some of Earth's economic resources, including valuable metal ores and building materials. Many metals are concentrated by igneous processes and occur within igneous rocks that solidify below the surface. Geologists use their understanding of igneous processes to locate these resources. Magma formation provides insights into processes inside Earth and teaches us more about the processes related to plate tectonics.

After Completing This Chapter, You Will Be Able to

- Explain the conditions that cause rocks to partly melt into magma and how these conditions relate to plate tectonics.

- Describe the processes that produce magmas of different compositions and show how differences in magma composition relate to the many types of volcanic eruptions and to the shapes of volcanoes.

- Apply your comprehension of igneous processes and products to understand how some important economic resources form and where they are found.

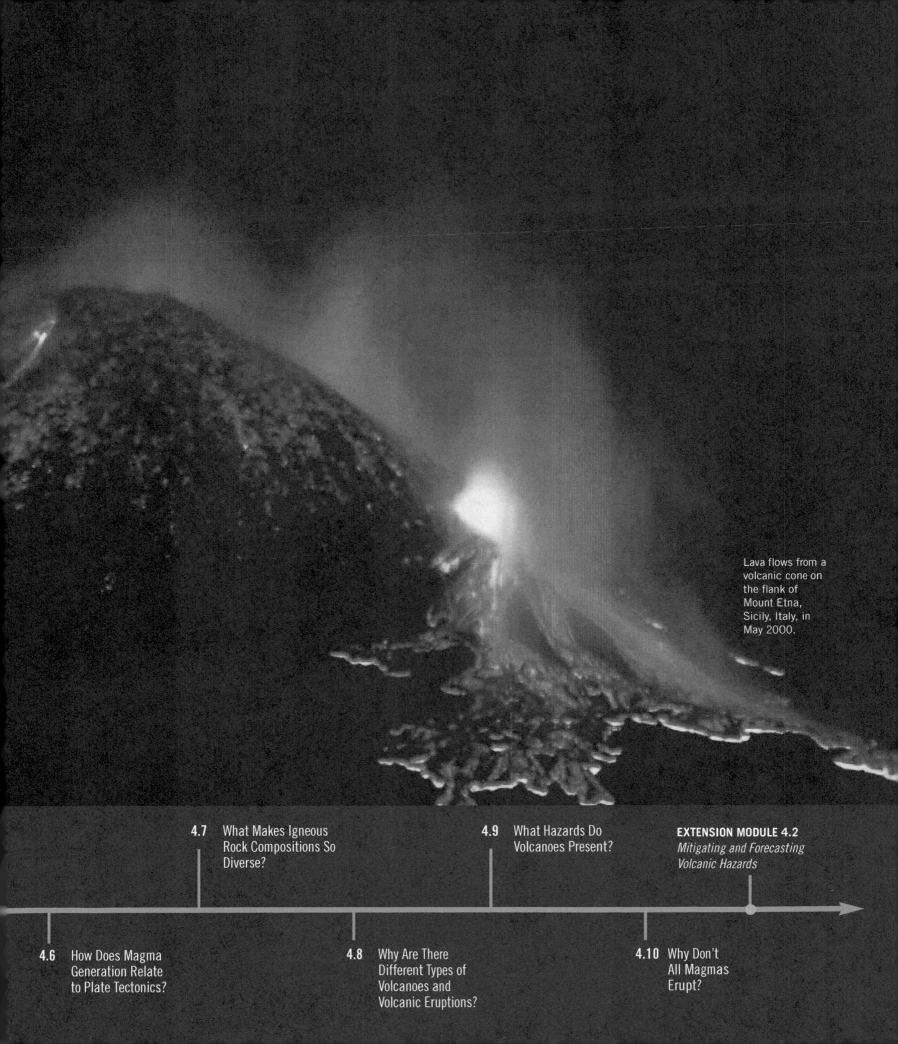

Lava flows from a volcanic cone on the flank of Mount Etna, Sicily, Italy, in May 2000.

4.7 What Makes Igneous Rock Compositions So Diverse?

4.9 What Hazards Do Volcanoes Present?

EXTENSION MODULE 4.2
Mitigating and Forecasting Volcanic Hazards

4.6 How Does Magma Generation Relate to Plate Tectonics?

4.8 Why Are There Different Types of Volcanoes and Volcanic Eruptions?

4.10 Why Don't All Magmas Erupt?

IN THE FIELD

Let's set out on three virtual field trips during which we will explore the nature of igneous rocks and the processes that form them. First, we visit the slope of the Kilauea volcano in Hawaii, illustrated in **Figure 4.1a**. We see a fountain of fiery, red-hot liquid shooting straight up from the crater within a black cone of rock. The liquid flows down the sides of the cone almost as quickly as water from a fountain—but this liquid is not water, it is molten **lava**, molten material ejected from a volcano at Earth's surface. Some of the lava, whose temperature is 1000°C, falls around the cone, congealing and building the black volcano even higher. But most of the liquid streams away as lava flows, which move downslope at speeds of 1 to 10 meters per second. Volcanologists dressed in heat-resistant clothing carefully approach the flow. Stopping within a few meters of it, they insert tools into the lava to sample it for analysis back at the lab.

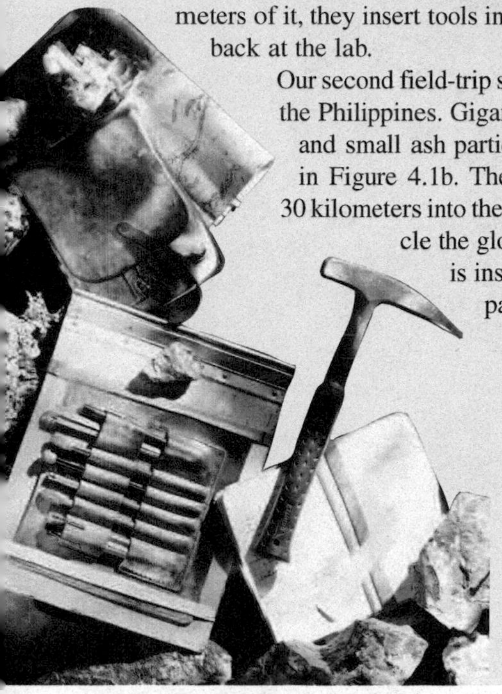

Our second field-trip stop is Mount Pinatubo, a volcano in the Philippines. Gigantic explosions hurl porous pumice and small ash particles high into the air, as illustrated in Figure 4.1b. The cloud of ash will rise more than 30 kilometers into the sky and drift with the wind to encircle the globe. But the force of the explosions is insufficient to carry all of the erupting particles into the atmosphere, which results in avalanches of hot pumice and ash rushing down the slopes of the volcano, obliterating everything in their paths. Unlike in Hawaii, we do not witness spectacular lava fountains and mesmerizing streams of flowing lava. No volcanologist will dare venture near this dangerous volcano until days after the explosions cease.

Figure 4.1c shows the spectacular landscape of Yosemite National Park in California, our final stop. The park contains deep valleys, high waterfalls, and vast expanses of rock laid bare by erosion. From a distance, the rock appears uniformly gray, without contrasting colors or layering. Up close, however, the rocks are salt-and-pepper mosaics of mineral grains colored white, black, and light pink (Figure 4.1d). A geologist recognizes these rocks as products cooled from molten magma. Whereas lava is ejected from a volcano at Earth's surface, magma is molten material below the surface. At Yosemite, the magma solidified into rock deep below ground. Uplift and erosion raised and wore away the rocks above the solidified magma, and more than 100 million years later, the igneous rocks are exposed for you to study.

Why are lava fountains and lava flows the most clearly visible products of the eruptions at Kilauea, while pumice and ash, flung high into the sky or avalanching rapidly down slopes, are the main products of the Mount Pinatubo eruption? Why did the once-molten magma that solidified into the rocks at Yosemite never reach the surface at a volcano? How did the molten material form at these three locations to begin with? As you read through this chapter, the answers to these and other questions will reveal a cohesive picture of igneous processes and their powerful results.

▶ **Figure 4.1 Contrasting views of igneous processes and rocks.** (a) The slope of the Kilauea volcano. Red-hot lava fountains high above the cone and feeds lava flows with a dark, solidified crust. (b) This is a distant view of a 1991 eruption of Mount Pinatubo, in the Philippines. The volcano is hidden behind ash clouds rising from incandescent avalanches of pumice and ash that rush at hurricane speeds down all sides of the volcano. (c) The rocks in this scenic landscape at Yosemite National Park, California, formed when molten magma solidified deep underground. Erosion exposed these rocks to view at Earth's surface. (d) A closer view of this igneous rock from Yosemite reveals the intergrown mosaic of minerals that crystallized from the molten magma as it cooled. The camera lens cap provides a sense of scale.

4.1 What Are Igneous Processes?

The three field trips described above illustrate different phenomena, but each offers insights into the processes that form igneous rocks. In order to form magma, rock first must melt at high temperature inside Earth. The liquid magma then moves upward, in some cases reaching the surface to erupt as volcanoes (such as Kilauea and Pinatubo), and in other instances remaining trapped underground (as at Yosemite). The properties of magma determine how volcanoes erupt, either with fiery fountains and fluid lava, like at Kilauea, or with mind-boggling explosive force, as at Mount Pinatubo.

Geologists learn about igneous rocks and the processes that form them by combining knowledge gained from field observations, geochemical analyses, and laboratory experiments. Field studies of ancient volcanoes provide clues to volcanic processes that cannot safely be studied at active volcanoes. Studies of locations such as Yosemite Valley allow geologists to infer how magma moves and solidifies below Earth's surface. Geologists also conduct microscopic studies of rocks to identify the minerals found in rocks and to determine how magmas crystallize to form rocks. Geochemists identify and measure the abundance of different elements in rock samples. Volcanologists remotely or even directly analyze gases from volcanoes. Scientists combine the results of all these analyses to infer the processes involved in the evolution of igneous rocks.

In this chapter, you will first study igneous rocks more closely and learn how they are encountered in the field. Then, you will work to understand the processes that create the features of igneous rocks.

(a) Kilauea, Hawaii – Red-hot lava fountaining above a growing volcanic cone.

(b) Pinatubo, Philippines – Explosions hurl pumice and ash more than 24 km above the volcano, which is hidden by clouds rising from incandescent avalanches of debris rushing at hurricane speed down all sides of the mountain.

(c) Yosemite National Park, California – Erosion has exposed igneous rock that solidified from magma deep below Earth's surface.

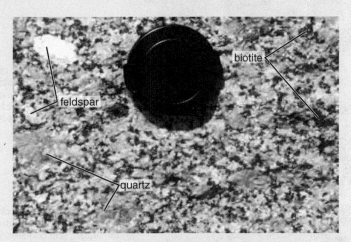

biotite

feldspar

quartz

(d) A closer look at the Yosemite rock reveals the intergrown mosaic of minerals that crystallized from the magma when it cooled; camera lens cap for scale.

Putting It Together—What Are Igneous Processes?

• Igneous processes involve the melting of rock to form magma and the solidification of magma into new rock.

• Igneous products form where magma crystallizes below Earth's surface or erupts as lava and other materials, like volcanic ash and pumice, onto Earth's surface.

• Geologists study processes forming igneous rocks through field observations and laboratory studies that include geochemical analyses and experiments.

4.2 How Are Igneous Rocks Classified?

The features of igneous products observed in Hawaii, the Philippines, and at Yosemite National Park are the result of the processes that form them. Thus, careful description and classification of the rocks are essential to understanding igneous processes.

Field observations support the logic of dividing igneous rocks into two general categories based on where they formed. Rocks originating from the eruption of molten material at the surface are volcanic (also called extrusive) rocks. Volcanic rocks form from flowing lava, as in Hawaii, or when explosions break apart the magma and eject sticky blobs that quickly solidify into particles of natural glass and fall to the ground. These fragmented **pyroclastic** materials (from the Greek *pyros:* fire, and *clastos:* broken)

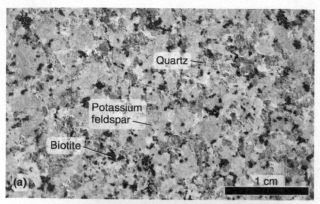

This light-colored rock contains large, readily visible crystals of quartz, feldspar, and biotite. These minerals and coarse crystal sizes characterize the felsic, coarse-grained (phaneritic) rock called granite.

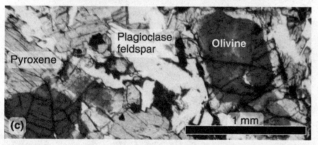

This rock from a lava flow is dark and contains no visible crystals. These are features of the mafic, fine-grained (aphanitic) rock called basalt. The vesicle holes represent former gas bubbles in the magma.

A microscopic view of basalt reveals the tiny crystals of olivine, pyroxene, and plagioclase feldspar formed by crystallization of the lava flow.

This rock is intermediate in color between granite and basalt, and contains visible plagioclase feldspar and hornblende in a gray background of microscopic crystals. This rock is porphyritic andesite.

Obsidian is volcanic glass that lacks crystals. Notice the curving fractures that are typical of broken glass.

◀ **Figure 4.2 Igneous rocks have different mineral contents and textures.**

include the lava-fountain blobs building the black cone at Kilauea and the pumice and ash that formed the high eruption cloud and hot avalanches at Mount Pinatubo. Igneous rocks that solidify below the surface are plutonic (also called intrusive) rocks. The word "plutonic" derives from Pluto, the Roman name for the god of the underworld. The rocks exposed at Yosemite are plutonic rocks.

Figure 4.2 shows a collection of igneous rocks that illustrate the differences in appearance that form a basis for descriptive classification of these rocks. The most common igneous rocks vary in color from black to very pale gray or nearly white with tinges of pink or red. These color variations are explained by the many kinds of minerals that make up the different rocks. Darker rocks have high abundances of the dark, iron- and magnesium-rich minerals, such as olivine and pyroxene, whereas the light-colored rocks consist almost entirely of feldspar and quartz. All the mineral grains are visible to the unaided eye in some rocks (Figure 4.2a), but a microscope is needed to see the crystals in other samples (Figures 4.2b and c). In still other cases, the rock sample consists of both large, visible mineral grains and microscopic ones (Figure 4.2d). Some igneous rocks lack mineral crystals of any size and instead consist of natural glass (Figure 4.2e).

Composition (minerals present) and texture (size of mineral grains) provide a framework for naming igneous rocks. Composition and texture in igneous rocks directly result from the processes that formed them.

The First Component in Classification: Composition

The composition of magma determines which minerals form in an igneous rock and is the first component of classification. Magma cools when it loses heat to its lower-temperature surroundings. Each mineral in magma crystallizes, or forms solid mineral crystals, at a particular temperature, which is the crystallization temperature of that mineral. As the magma cools, each mineral appears when its crystallization temperature is reached, a process similar to ice forming from liquid water. Ice is a single mineral, however, whereas many minerals crystallize from any particular magma.

With only rare exceptions, magma is rich in silica, the compound of the two most common elements in Earth's crust, silicon and oxygen (expressed as SiO_2). By weight, typical magma ranges from 45 to almost 80 percent silica. The remainder of magma consists of aluminum, magnesium, iron, calcium, sodium, potassium, and other elements in very minor concentrations. Given the dominance of silica, it is not surprising that igneous rocks are overwhelmingly composed of silicate minerals—feldspars, pyroxene, olivine, and quartz are the most common.

Gases also are present in all magmas; water vapor, carbon dioxide, and odorous sulfur compounds are the most abundant of these. When magma nears or reaches the surface, the gases escape from the liquid to form bubbles, called **vesicles**, which are preserved in some volcanic rocks when the liquid solidifies around the bubble and then the gas leaks out (Figure 4.2b). A critical difference between magma and lava is the presence or absence of gas. Magma is a mixture of liquid melt, crystals, and dissolved gases below the surface, whereas lava is the same liquid melt and solid crystal mixture at the surface from which the gases have mostly escaped.

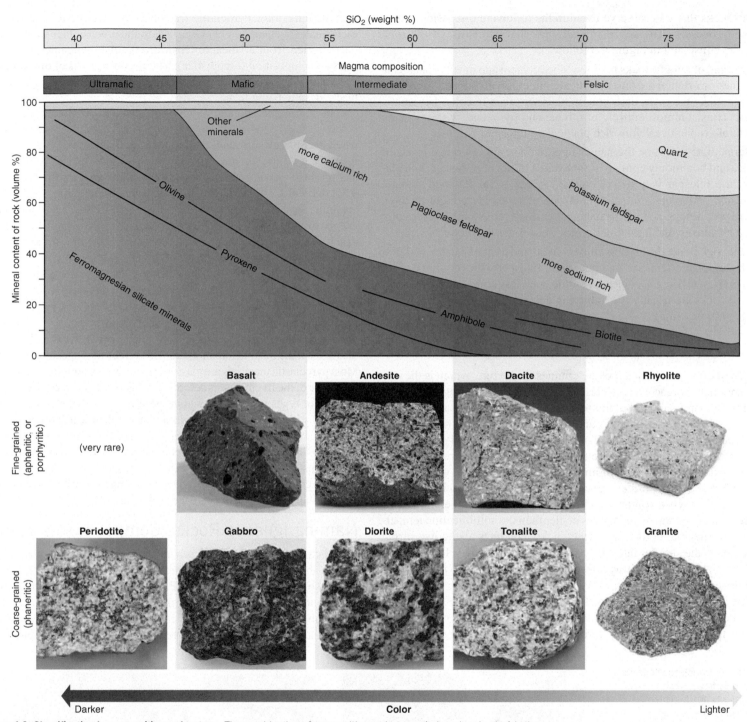

▲ **Figure 4.3 Classification by composition and texture.** The combination of composition and texture (mineral grain size) is the basis for classification and naming of igneous rocks. Each rock in the figure has a predominantly aphanitic or phaneritic texture. They are further distinguished by the relative abundance of their mineral constituents. Light-colored felsic rocks are rich in silica (SiO_2) and contain mostly quartz and feldspar. Dark-colored mafic and ultramafic rocks have a lower silica content and are richer in iron and magnesium, permitting the formation of dark silicate minerals, such as olivine and pyroxene.

Figure 4.3 shows how geologists divide magma and igneous-rock compositions into **felsic**, **intermediate**, **mafic**, or **ultramafic** categories according to the silica abundance of the magma or rock. The term "felsic" describes rock and magma that contain mostly *fel*dspar and *sili*ca (in the form of quartz). The term "mafic" refers to the elements *ma*gnesium and iron (*ferric*), which are more abundant in magma and rocks with lower silica content. Igneous rocks and magmas gradually vary from one compositional group into the next. Magma composition determines which minerals crystallize to form rock. So, different abundances and combinations of minerals describe igneous rocks.

The terms "felsic," "intermediate," "mafic," or "ultramafic" occur throughout this book, so please take a moment to familiarize yourself with their meanings and use. In fact, learning all the terms and rocks shown in Figure 4.3 will help you to better navigate and understand the contents of this chapter. As you study this figure, you will find that you can make a number of observations and generalizations about the classifications of igneous rocks that will serve you well in the following discussions about igneous processes. The key compositional characteristics of the rock categories illustrated in Figure 4.3 follow:

- Mafic and ultramafic rocks have low silica content and mostly contain the magnesium-rich and iron-rich silicate minerals olivine and pyroxene, which make the rocks dark gray to black or green in color. Ultramafic rocks consist almost entirely of olivine and pyroxene, whereas mafic rocks also contain calcium-rich plagioclase feldspar.
- Intermediate rocks, as the name suggests, contain midrange levels of silica and commonly contain hornblende, biotite, or pyroxene, and plagioclase feldspar with nearly equal amounts of calcium and sodium. Feldspar and magnesium- and iron-rich minerals are present in about equal proportions, so intermediate-composition rocks are lighter in color than ultramafic and mafic rocks; gray shades are most typical.
- Felsic rocks have the highest silica content and contain quartz, sodium-rich plagioclase feldspar, and potassium feldspar with very minor dark minerals, including hornblende, biotite, and, less commonly, pyroxene. These rocks are usually the lightest in color of the igneous rocks (e.g., light gray or white).

The Second Component in Classification: Texture

The size of crystals within a rock determines the texture, which is the second component of igneous rock classification. Crystal size in igneous rocks depends on many factors, but the magma cooling rate is the most important one. (We will consider yet another factor in Section 4.10.) When magma cools down quickly, the resulting rock consists of very small crystals, which are generally invisible to the unaided eye (Figure 4.2b). This fine-grained texture is called **aphanitic** (Figures 4.2b, 4.3), a word derived from the Greek word *aphanes*, meaning "invisible." Very rapid crystallization can happen when volcanic rocks form because the temperature at the surface is many hundreds of degrees cooler than the solidification temperature of the magma. In this case, heat quickly conducts and radiates from the magma to the surroundings. Intrusive rocks may also develop aphanitic texture in cases where small volumes of magma intrude into much colder rocks close to Earth's surface.

The liquid melt solidifies into glass if the cooling rate is so fast that few, if any, crystals have time to form. Glassy texture is common in rocks formed at the surfaces of lava flows (Figure 4.2e) and along the margins of some igneous intrusions. Pyroclastic fragments are commonly glassy because each small fragment quickly loses heat to the much colder surrounding air once ejected from a volcano. **Obsidian** is felsic volcanic glass with an unusual composition (Figure 4.2e): Microscopic grains of black magnetite are widely dispersed through obsidian, making it opaque and dark gray to black, despite being felsic.

In contrast to volcanic (above-surface) and shallow intrusive (just-below-surface) conditions, magma cools slowly deep beneath the surface, where warm rock surrounds and insulates it. Because magma cools slowly in this deep intrusive environment, mineral crystals also grow slowly and produce a coarse-grained texture known as **phaneritic** (Figures 4.2a, 4.3), derived from the Greek word *phaneros*, meaning "visible." Thin, aphanitic lava flows may cool within hours on Earth's surface, whereas deeply formed phaneritic intrusive rocks may require tens of thousands of years, or longer, to solidify completely deep below the surface.

In some instances, magma starts to crystallize slowly at depth, producing some large crystals, before moving into colder environments near or at the surface, where the remaining melt crystallizes quickly. The resulting **porphyritic** igneous rock consists of some large crystals surrounded by smaller crystals (Figure 4.2d).

Pyroclastic materials erupted from volcanoes when expanding gas bubbles disrupt the magma into blobs and drops require a special classification. Most pyroclastic materials are glassy, so in order to distinguish them we look at the size of the fragments, as described in **Figure 4.4** and **Table 4.1**, rather than the size or types of minerals within the fragments. **Bombs** are the largest fragments, **lapilli** are intermediate in size, and **volcanic ash** fragments are the smallest. In contrast to the more common meanings of the words "bombs" and "ash," volcanic bombs do not explode, and volcanic ash is not a product of combustion. When the pyroclastic fragments are consolidated into hard rock, the comparable grain-size related terms, agglomerate, lapillistone, tuff, and lapilli tuff, are used to name the rocks (Table 4.1).

Naming Igneous Rocks: Composition + Texture

The classification and naming scheme for igneous rocks combines information about the composition and texture of the magma or rock. Let's look at some of the rocks introduced in Figure 4.3. Ultramafic, phaneritic **peridotite** composes Earth's mantle, and pieces of this rock are sometimes carried to the surface at erupting volcanoes. **Basalt** and **gabbro** form from mafic magma and contain the same minerals but have different textures;

▶ **Figure 4.4 Pyroclastic deposits classified by fragment size.** All of the fragments pictured here fell to the ground after powerful volcanic explosions ejected them into the sky. The largest fragments accumulate on the slope of the volcano, whereas wind carries smaller particles progressively greater distances. Table 4.1 shows how size determines the name applied to the individual fragments and the resulting deposits.

(a) Basaltic bombs have aerodynamic ribbon and spindle shapes formed as the solidifying blobs of lava traveled through the air.

(b) Dark basaltic and andesitic lapilli are called cinder or scoria.

(c) Light dacitic and rhyolitic lapilli are called pumice.

1 cm

(d) Fine, light-colored volcanic ash collected 140 kilometers east of Mt. St. Helens, Washington during the May 1980 eruption.

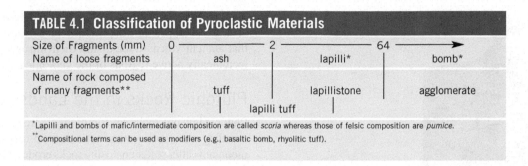

TABLE 4.1 Classification of Pyroclastic Materials

Size of Fragments (mm)	0 —	2 —	64 ——>
Name of loose fragments	ash	lapilli*	bomb*
Name of rock composed of many fragments**	tuff	lapillistone	agglomerate
		lapilli tuff	

*Lapilli and bombs of mafic/intermediate composition are called *scoria* whereas those of felsic composition are *pumice*.

**Compositional terms can be used as modifiers (e.g., basaltic bomb, rhyolitic tuff).

gabbro is the coarse-grained equivalent of basalt. The oceanic crust is composed mostly of these two mafic rocks. **Andesite** and **diorite** solidify from intermediate-composition magmas. **Rhyolite** and **granite** form from the most felsic magmas. **Dacite** and **tonalite** have compositions that fall between andesite and rhyolite. Many recently active volcanoes, including Mount St. Helens (Washington, active 1980–present day), Mount Pinatubo (Philippines, 1991), and Mount Unzen (Japan, 1991), erupted dacitic lava flows and pyroclastic fragments. A wide range of intermediate to felsic igneous rocks composes the continental crust. A combination of the various rocks that have been sampled from the upper crust results in an average crust of tonalite composition.

Pyroclastic deposits are named on the basis of fragment size and of whether the fragments are loose or consolidated into rock (Table 4.1). Compositional labels complete a descriptive naming of pyroclastic deposits (Table 4.1). Lightweight, highly vesicular lapilli is called **pumice**, if dacitic or rhyolitic in composition, and **cinder** (or **scoria**) if basaltic or andesitic.

Igneous Rocks Provide Essential Economic Resources

Igneous rocks are widely used in construction and industry. Many volcanic rocks, for example, are strong because their finely intergrown mineral crystals are difficult to fracture along grain boundaries. These rocks make good building stone and, when crushed, form excellent aggregate for highway and railroad-bed construction. Tuff is a strong, yet lightweight, rock used as building stone. Volcanic ash and pumice are mixed into lightweight concrete. Pumice also is mined for use as an abrasive in toothpaste and soap. Obsidian is easily chipped and flaked into incredibly sharp tools and weapons. Although mostly replaced by later manufacture of metal implements, the use of obsidian for weaponry and tools made it an essential resource in many ancient cultures, and its availability determined migratory and trading routes. Obsidian knives are still preferred by many plastic surgeons. Phaneritic igneous rocks are commonly quarried for decorative purposes, including for such uses as building facades, monuments, and gravestones.

Putting It Together—How Are Igneous Rocks Classified?

- Composition (minerals present) and texture (crystal size) are used to classify and name igneous rocks.

- Ultramafic, mafic, intermediate, and felsic are compositional categories for magma and igneous rocks. Light-colored silicate minerals dominate in silica-rich felsic rocks, whereas darker silicate minerals form mafic rocks.

- Rapidly cooled volcanic and shallow plutonic rocks are fine-grained (aphanitic) whereas slowly cooling deep magma intrusions produces coarse-grained (phaneritic) rocks. Porphyritic texture describes rocks containing at least two distinct crystal sizes.

- Pyroclastic deposits, igneous products formed by volcanic explosions into the atmosphere, are classified on the basis of fragment size.

- Igneous rocks lend themselves to a variety of uses in construction and industrial processes.

4.3 Where Do Igneous Rocks Appear in a Landscape?

Understanding the origin of igneous rocks requires knowing where they are found and how they appear in natural landscapes. Volcanoes are built of lava flows and pyroclastic deposits and form where magma reaches the surface. When magma solidifies before reaching the surface, it is called plutonic rock. Plutonic rock bodies, sometimes called **plutons**, exist in varying shapes and sizes. These rocks are revealed at the surface only after the rock that originally covered them has been eroded (at Yosemite, for example; Figure 4.1c).

Modern volcanoes show where magmas form and rise to the surface today. Geologists see only the extruded volcanic materials, but they infer simultaneous magma intrusion in the plumbing beneath the volcanoes. Geographic coincidences of extrusive and intrusive processes are revealed by:

- Earthquakes caused by magma movement during intrusion beneath volcanoes.

- Pieces of plutonic rocks ejected from volcanoes (Figure 3.4d, for example).

- The presence of plutonic rocks exposed by erosion of the overlying volcanic rocks.

Igneous processes clearly relate to plate tectonics, because most active volcanoes exist at or near divergent and convergent plate boundaries, and only a few active volcanoes are found within plates (confirm by looking again at Figure 1.10). Volcanoes are especially numerous along continental margins and island chains above subduction zones in the Pacific Ocean, forming the "Ring of Fire." Within the United States, subduction-related volcanoes form the Cascade Range, (which stretches from northern California to northern Washington) and the peaks and islands of southern Alaska and the Aleutian Islands. Seafloor volcanoes compose the mid-ocean ridges (Figure 1.11), and some rise to form islands, such as Hawaii, where magma rises below plate interiors at hot spots (Figure 1.12). Volcanoes appear within continents at locations where plate divergence is beginning (e.g., East Africa) or at hot spot localities, such as Yellowstone National

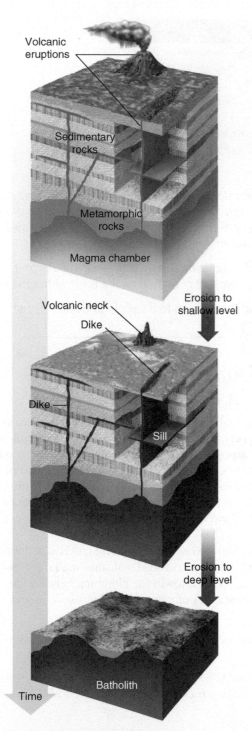

▲ **Figure 4.5 Visualize intrusive features.** Names assigned to igneous intrusions relate to shape and size of each intrusion. Large magma chambers have very irregular shapes because they form over long periods from multiple injections of magma deep in the crust or mantle. Smaller bodies of magma rise through vertical or near-vertical fractures to form dikes, which cut across sedimentary rock layers. Magma injected between layers forms sills. A volcanic neck is rock that solidifies in the volcano throat. Dikes, sills, and volcanic necks form close to the surface, so they are the first features revealed at the surface by later erosion. Erosion to deeper levels reveals the larger, deeply rooted solidified magma chambers, called batholiths.

Park, in Wyoming. Observed volcanic activity occurs in active tectonic settings, so applying the principle of uniformitarianism, geologists infer that ancient volcanic rocks and their intrusive equivalents formed in the tectonically active regions of the geologic past.

Plutonic Rocks in the Landscape

Plutonic igneous rocks are commonly more resistant to erosion than the surrounding rocks, so they stand out in the landscape as hills, ridges, and even large mountains after softer enclosing rocks erode away. **Figure 4.5** shows the landforms that result when igneous intrusions of various sizes and shapes, which relate to the volume of magma intruded and the mechanism of intrusion.

Figure 4.6 shows tabular intrusions called **dikes** and **sills**, which, like a tabletop, are extensive in two dimensions but relatively thin in the third dimension. In some cases, magma rises to form long, steep (sometimes vertical), tabular dikes along fractures in the surrounding rocks (Figure 4.6a). Dikes cut across layers if the magma intrudes sedimentary rocks. In other cases, magma moves horizontally and forms sills, which may be injected between sedimentary layers (Figure 4.6b).

Large magma chambers form deep within Earth's crust where the enclosing metamorphic rock is very hot and deforms in a plastic, squishy fashion rather than by being brittle and breaking. In these places, the rising magma squeezes aside the surrounding rocks. Adjacent rocks melt into the magma, and large blocks of rock fall from the roof of the magma

ACTIVE ART

Forming Igneous Features and Landforms: See how intrusions form and how they produce unique landforms when exposed by erosion.

Dikes are steeply inclined intrusions that cut across sedimentary layers. This basalt dike in the Grand Canyon, Arizona, cuts across red sedimentary layers. Notice the geologists (circled) standing on the dike for scale.

Sills are gently inclined or horizontal intrusions that are parallel to sedimentary layers. This gabbro sill in Glacier National Park, Montana, is between horizontal sedimentary layers. The sill is more than 50 m thick.

◄ **Figure 4.6 What dikes and sills look like.**

◀ **Figure 4.7 What a volcanic neck looks like.** Shiprock, New Mexico, is a volcanic neck, composed of tuff and intrusive rock that filled the conduit of an old volcano. Erosion of the surrounding soft sedimentary rock exposed the neck.

volcanic eruptions along long fissures. In other cases, cylindrical pipes of magma feed volcanoes; when erosion later exposes the rocks left in the pipes, **volcanic necks** are formed, as shown in **Figure 4.7**.

Volcanic Rocks in the Landscape

Volcanic rocks appear in the landscape on a small scale as the products of single eruptions and on a larger scale as volcanoes form over hundreds of thousands of years by many eruptions. Lava flows are the most common volcanic landscape features. **Figure 4.8** shows basaltic lava flows in Hawaii that are only meters thick. **Figure 4.9** illustrates basaltic lava flows that cover very large areas. In contrast, intermediate and felsic lava flows, shown in **Figure 4.10**, are many tens to hundreds of meters thick but extend only a few kilometers from the vent where they erupted. **Lava domes** are an extreme feature created when lava seemingly does not flow at all but merely squeezes out to form large mounds directly over the volcanic vent

chamber, which allows the magma to move farther upward. These large plutons, called **batholiths**, solidify at depths from a few kilometers to more than 10 kilometers below Earth's surface, and are exposed only where substantial uplift and erosion of surrounding rocks has occurred (Figure 4.5). The intrusive rocks exposed at Yosemite National Park (Figure 4.1c) are merely a small part of the vast Sierra Nevada batholith. This batholith is exposed over nearly 100,000 square kilometers and was formed by countless intrusions during a period of more than 50 million years.

Erosion in volcanic regions reveals that volcanoes form above shallowly intruded dikes, sills, and small upward protrusions of large batholiths (Figure 4.5). In some cases, dikes of magma extend to the surface to cause

Basaltic lava flows poured across the Pacific Northwest about 15 million years ago. The Grande Ronde River canyon in Washington exposes hundreds of meters of basalt lava. Each layer in the photo is a separate lava flow. Some flows cover more than 100,000 square kilometers.

Lava flows

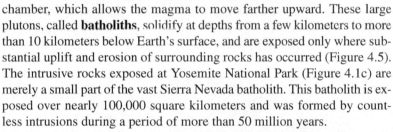

◀ **Figure 4.8 Flowing lava in Hawaii.**

Some lava flows are very thin, attesting to the very fluid character of basaltic lava.

A'a

Pahoehoe

Smooth and wrinkled lava flow surfaces, called pahoehoe in Hawaii, form where the lava stops flowing while it is still very hot. If the flow continues to move after the crust hardens, then the rock crust breaks into pieces and forms rubbly lava flows, called a'a.

Some basalt flows fracture into columns when the rock contracts while cooling. This thick lava flow is the highlight for visitors to Devil's Postpile, California.

◀ **Figure 4.9 What mafic lava flows look like.**

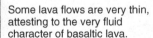

A thick rhyolite-obsidian lava flow erupted about 1300 years ago to form Paulina Peak in Oregon.

A large dacite lava dome formed in the crater of Mt. St. Helens, Washington, between 2004 and 2008. The dome is about 1 km across and the volume of erupted lava would fill about 200 large sport stadiums.

◄ **Figure 4.10 What felsic lava flows look like.**

the slope of a volcano. Moving pyroclastic flows at Mount Pinatubo are visible in Figure 4.1b. The deposits contain a wide range of ash- and lapilli-sized particles that form lapilli tuffs (Figure 4.11a, Table 4.1). Ash and lapilli fragments may be sticky and plastic if pyroclastic-flow deposits remain very hot after coming to rest. In these cases, particles within the hot interior of the deposit are squashed and stuck together by the weight of the overlying material, producing **welded tuff** (Figure 4.11b).

Types of Volcanoes

Volcanoes are hills, ridges, or high mountains formed by the accumulation of lava flows and pyroclastic deposits around the conduit, or crater depression, from which they erupted. The size of a volcano relates to the volume of extruded volcanic materials. The shape of the conduit and the types of eruptions determine the shape of a volcano. Volcanoes with a classic cone shape form from eruptions through a single, centrally located crater. More irregular or elliptical shapes result from eruptions through many craters or along fissures where dikes break through to the surface. Deeply eroded river canyons reveal the internal structure of volcanoes, enabling geologists to determine how volcanoes grow over time. **Figure 4.12** presents a classification of volcanoes according to size, shape, and erupted volcanic materials.

(Figure 4.10b). Some lava flows move like cooling maple syrup and have smooth to slightly wrinkled surfaces. Other flows are encased in jumbled, broken blocks of rock that break away from the cooler solidified outer margin of the lava as the still-fluid interior continues to flow. Geologists apply the Hawaiian term *pahoehoe* to describe the smooth or wrinkled flows, and they call the rubbly lava *a'a* (Figure 4.8).

Pyroclastic deposits are another type of volcanic landscape feature. The characteristics of these deposits depend on whether the exploded fragments fall from the sky or flow down the slopes of volcanoes. When particles fall from the sky, they are referred to as **pyroclastic-fall deposits**; when they flow across the land surface, they are called **pyroclastic-flow deposits**.

Pyroclastic-fall deposits may extend more than 1000 kilometers from a volcano if the explosions carry the particles high into the atmosphere. Very violent explosions, such as the one that occurred at Mount Pinatubo in 1991 (Figure 4.1b), eject fine-ash particles so high into the atmosphere that they form a dusty ash veil that travels on prevailing winds and encircles the planet. Wind carries pyroclastic fragments away from the volcano, so the ash deposits are thickest in the downwind direction from the volcano. Larger fragments settle out of the air first (Figure 4.4), so pyroclastic-fall deposits exhibit a uniform decrease in particle size with increasing distance from the volcano.

Pyroclastic-flow deposits, illustrated in **Figure 4.11**, form when avalanches of incandescent pumice and ash flow down

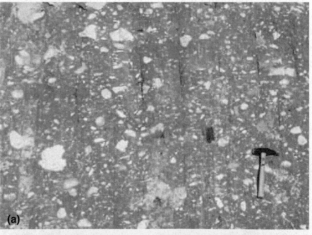

A nonwelded lapilli tuff, in New Zealand, consists of white pumice lapilli and bombs enclosed in a matrix of ash.

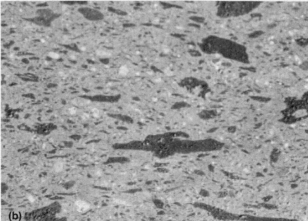

A welded tuff from Mexico formed when the pyroclastic-flow deposit remained sufficiently hot for the glassy particles to flatten under the weight of overlying pyroclastic debris. The black fragments, as much as 6 centimeters long, are pumice lapilli that compacted into black obsidian during welding.

◄ **Figure 4.11 What pyroclastic-flow deposits look like.** Pyroclastic-flow deposits are composed of an unsorted mixture of pumice and ash.

Cinder cone. Small volcanoes, usually less than 600 m high, composed of basaltic to andesitic cinder (scoria). Slopes are near 30–35 degrees, the angle defined by the loose bombs and lapilli when they come to rest. Lava flows may issue from the base of the cone. Typically produced by single, prolonged eruptions lasting 1–20 years.

Example: SP Crater, near Flagstaff, Arizona.

Shield volcano. Small (300 m high) to giant (10,000 m high) volcanoes composed of many thin and widespread basaltic lava flows. Definitive shield shape is characterized by gentle slopes typically less than 15 degrees. Active for centuries to a few million years.

Example: Mauna Loa, Hawaii. The summit of Mauna Loa is nearly 4300 m above sea level; this volcano rises nearly 10,000 m above the sea floor. The summit of the Kilauea shield volcano is in the foreground.

Composite volcano. Modest (100 m high) to large (3000 m high) volcanoes composed of interlayered lava flows, lava-flow rubble, and pyroclastic deposits. Lava flows are typically thicker and shorter than those on shield volcanoes. Volcanic rocks may range in composition from basalt to rhyolite at single volcanoes although most are dominated by basalt, andesite, and/or dacite. Slopes are generally greater than 25 degrees. Formed by rare to frequent eruptions over several hundred thousand years.

Example: Mount Fuji, Japan, rises to 3776 m above sea level and last erupted in 1707.

Dome complexes. Modest (500–2000 m high) volcanoes composed of multiple, overlapping volcanic domes. Usually of dacitic and rhyolitic composition. Formed by several eruptions over thousands to hundreds of thousands of years.

Example: Chaos Crags, northern California, is a series of dacitic domes that formed during eruptions 1100 years ago.

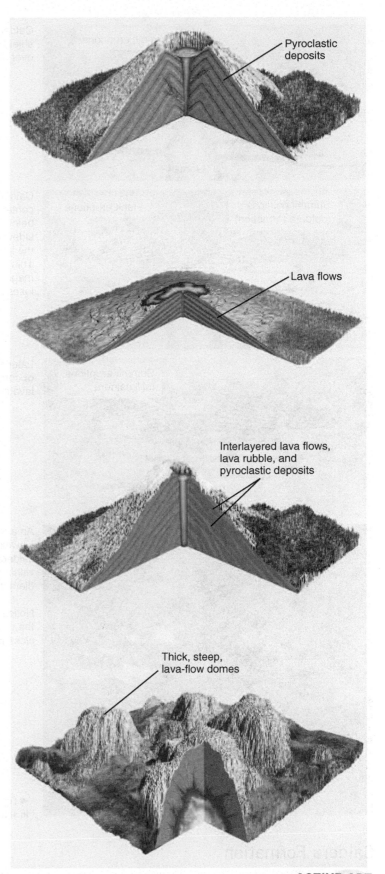

Pyroclastic deposits

Lava flows

Interlayered lava flows, lava rubble, and pyroclastic deposits

Thick, steep, lava-flow domes

▲ **Figure 4.12 How volcanoes are formed and classified.** Volcano size and shape relates to the types of eruptions that produced them and the types of volcanic materials that compose them.

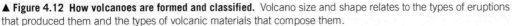

ACTIVE ART

Forming Volcanoes: See how different types of volcanoes form.

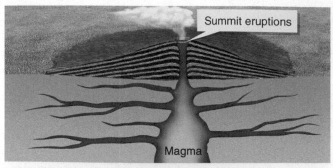

Caldera formation at Hawaiian shield volcanoes occurs gradually over thousands of years.

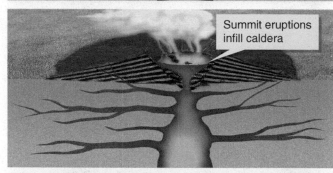

Caldera collapse occurs when the contents of the magma chamber below the volcano summit intrude sideways to supply eruptions on the lower flank of the caldera. This causes the roof of the magma chamber to subside as it loses support.

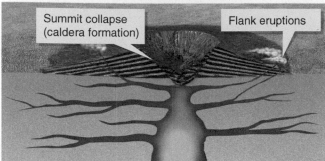

Later summit eruptions may partly or completely fill the caldera with lava flows.

An aerial view of Kilauea volcano, Hawaii, shows the summit caldera. Halemaumau Crater within the caldera is about 1 km in diameter.

Notice the eruption on the flank of the volcano, which is seen up close in Figure 4.1a.

◄ **Figure 4.13 How calderas form in shield volcanoes.**

magma chambers form within a few kilometers of the surface and the magma is supporting the weight of the overlying rocks. If large volumes of magma erupt from the chamber, then the unsupported roof of the magma chamber collapses. This collapse produces a large circular or elliptical pit at the surface, called a **caldera**. Calderas are distinguished from smaller craters by being more than one kilometer in diameter.

Both of the volcanoes that you virtually visited at the beginning of this chapter (Kilauea and Mount Pinatubo) have calderas. **Figure 4.13** shows how calderas form in shield volcanoes when magma moves away from beneath the volcano summit to feed flank eruptions on the slopes. Since 1790, several periods of subsidence by this process at Kilauea have formed a caldera 3.2 kilometers long, 2.6 kilometers wide, and hundreds of meters deep, but this depression was partially filled with lava flows during later eruptions. In contrast, **Figure 4.14** illustrates how other calderas form dramatically within hours or days during explosive eruptions of pumice and ash that evacuate many cubic kilometers of magma from shallow magma chambers. The Mount Pinatubo caldera formed in 1991 during the eruption of about 10 cubic kilometers of magma during less than 24 hours of eruptions. The caldera is approximately 2.5 kilometers in diameter and lowered the elevation of the mountain by 250 meters. Crater Lake National Park (Figure 4.14) more accurately encloses a caldera lake, rather than a crater lake, produced by the collapse of Mount Mazama about 7600 years ago during eruption of 50 cubic kilometers of magma as pumice and ash.

Monstrous calderas, now partly filled in and surrounded by extraordinarily thick pyroclastic-flow deposits, are found across the western United States and central and northern Mexico. Dozens of caldera-forming eruptions occurred in this region, beginning approximately 35 million years ago. The most recent such eruption extruded 1000 cubic kilometers of pumice and ash at Yellowstone, Wyoming, about 640,000 years ago. The largest caldera in the United States is the 28-million-year-old La Garita caldera in the San Juan Mountains of Colorado. It is approximately 60 by 30 kilometers and formed during eruption of more than 4000 cubic

Caldera Formation

Volcanic eruptions are usually thought of as processes that increase ground elevation by adding layers of lava and pyroclastic material, but some eruptions actually decrease surface elevation. This can occur when large kilometers of magma as pyroclastic flows. Very large calderas such as those at Yellowstone and La Garita encompass areas much larger than single volcanoes and represent the collapsing roofs of unusually shallow batholiths.

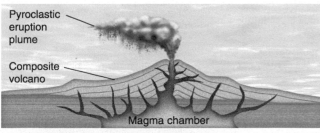

Calderas form in composite volcanoes when large volumes of magma explosively erupt in just hours to a few days to form thick, widespread pyroclastic deposits.

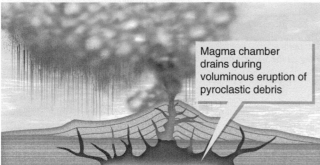

The eruption of magma causes the roof of the magma chamber to subside as it loses support.

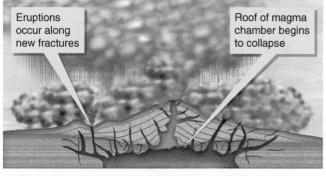

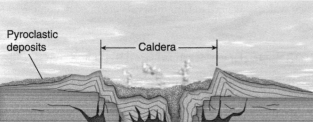

Crater Lake, Oregon, is an 8-km diameter caldera formed at the top of ancient Mount Mazama volcano. Later eruptions created Wizard Island cinder cone, seen in the center. The caldera partly filled with water to form the deepest lake in North America.

▲ Figure 4.14 How calderas form in composite volcanoes.

ACTIVE ART

How Calderas Form: See how explosive eruptions form calderas in composite volcanoes.

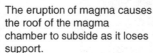

Putting It Together— Where Do Igneous Rocks Appear in a Landscape?

• Plutonic rocks at Earth's surface are exposed when erosion removes the rocks that originally covered the solidifying magma.

• Once plutonic rocks are exposed to view, the resulting landforms display a variety of sizes and shapes corresponding to the processes of intrusion and how erosion exposed the rocks.

• Volcanic rocks include lava flows, pyroclastic-fall deposits, and pyroclastic-flow deposits.

• Different magma compositions and types of eruptions, and varying proportions of lava flows to pyroclastic deposits determine the sizes and shapes of volcanoes.

• Calderas form when volcanoes collapse into shallow magma chambers that partially drain during eruptions.

4.4 How and Why Do Rocks Melt?

At this point, you are familiar with the descriptions and classification of igneous rocks and with the landscapes that igneous processes create. Now, we discuss how magma forms in the first place. In other words, under what conditions does something as hard as rock melt?

Keep in mind that most of the interior of Earth is solid (Figure 4.10), as illustrated by the way Earth shakes during earthquakes (a subject explored in detail in Chapter 8). Volcanoes, therefore, are not fed from vast underground oceans of magma. Therefore, the most important point to understand about igneous processes is how the largely solid interior of the planet becomes locally molten. This is intriguing because all of the oceanic crust and most of the continental crust consists of igneous rocks. In other words, Earth would not have a crust if magma did not somehow form out of the solid rock below the planet's surface.

Now that you know there is no ocean of magma simmering under the Earth's surface that feeds through cracks to make volcanoes, you are likely wondering how magma forms.

How Rocks Melt—Melting Temperature

Rock begins to melt if its temperature is raised high enough to cause any of its constituent minerals to melt. Laboratory studies indicate that melting temperatures of rock depend on the combination of minerals in the rock. Minerals undergo chemical reactions with each other at high temperatures, and these chemical reactions affect melting temperatures. Specifically, mixtures of minerals melt at lower temperatures than the constituent minerals melt individually. For example, olivine present in basalt that also

Stages in rock melting

1. Solid basalt at room temperature.

2. Rock remains solid until heated to 1100°C, at which point all three minerals begin to melt.

3. Pyroxene completely melts first and plagioclase and olivine continue to melt as temperature is increased.

4. At 1200°C the last of the plagioclase crystals melts, leaving only olivine crystals within the liquid.

5. Olivine continues to melt until about 1225°C, when the last traces of crystals disappear into the liquid.

6. Above 1225°C the sample is completely molten.

Lower temperature

Olivine
Pyroxene
Plagioclase

1100°C

1150°C

1200°C

1225°C

1230°C

Increasing temperature during melting

Decreasing temperature during crystallization

Higher temperature

Stages in rock crystallization

6. Below 1100°C the sample is completely solid basalt.

5. At 1100°C pyroxene crystallizes along with remaining olivine and plagioclase.

4. Olivine and plagioclase crystallize simultaneously as the temperature drops lower.

3. Plagioclase begins to crystallize (in addition to olivine) as the temperature reaches 1200°C.

2. Olivine begins to crystallize in the liquid at about 1225°C.

1. At a temperature above 1225°C the sample is completely molten.

◄ **Figure 4.15 How melting and crystallization work.** These diagrams summarize results of the experimental melting and crystallization of a basalt sample. The rock does not completely melt at a single temperature, but gradually transforms from solid rock to complete liquid over a range of more than 100°C. Similarly, the molten liquid gradually solidifies through the sequential appearance of different minerals over the same temperature range during cooling experiments.

contains pyroxene and plagioclase feldspar begins to melt at about 1200°C, a melting temperature almost 700°C lower than a sample composed only of olivine.

Figure 4.15 illustrates the results of laboratory rock-melting and magma-crystallization experiments. These experiments showed that nearly all the minerals *begin* to melt at a single temperature, but that once one of the minerals completely melts, the temperature must increase to cause further melting of the remaining minerals. Also, it was determined that each mineral completely transforms to liquid at a different temperature. The process is somewhat analogous to melting a bowl of chocolate-chip ice cream. At room temperature, the ice cream completely melts to a white liquid. For the solid chocolate chips to melt, however, the temperature must be increased, possibly by heating the bowl on a stove.

Every mineral has a melting temperature that also depends on pressure, and pressure equates to the depth below the surface and the weight of overlying rock. The effect of pressure on rock melting is analogous to the effect of pressure on the boiling temperature of water. Water boils at 100°C at sea level, but it boils at a lower temperature on top of a high mountain, where air pressure is lower, and it boils at a higher temperature in a pressure cooker, where pressure is higher. Similarly, rocks require a higher temperature to melt at high pressure than at low pressure. For example, laboratory experiments show that olivine, which is common in basalt, melts at 1890°C at Earth's surface. At a pressure equivalent to 100 kilometers below the surface, however, its melting temperature increases to 2050°C.

Water also influences melting temperature. At the high pressure deep in Earth where rocks melt, water does not boil off as steam. Instead, water remains in liquid form or exists as water molecules bound within minerals such as amphibole and mica. Experiments show that rocks melt at lower temperature when water or water-bearing minerals are present compared to when rock is dry.

How Magma Composition Affects Rock Melting

Only when rock melts completely does the resulting magma have the same composition as the original rock. If the temperature of the rock is not higher than the final melting temperature for all of the constituent minerals, the rock has only partially melted. Going back to the ice-cream analogy, partial melting produces a vanilla-flavored milky liquid that lacks the chocolate, which remains in the unmelted chocolate chips.

The composition of partial melts depends on the proportions of minerals that melt and the chemical reactions that occur between the melted minerals and the remaining crystals. In Figure 4.15, for example, the partial melt formed by melting basalt at 1150°C contains all of the chemical components of the melted pyroxene, but the olivine and plagioclase are mostly still solid. In most cases, partial melts contain a greater abundance of silica than the original rock. Experiments demonstrate, therefore, that partial melting of ultramafic peridotite produces more silica-rich mafic magma, that partial melts of mafic rocks become even more silica-rich intermediate magma, and so forth (Figure 4.3). Therefore, partial melting of rocks creates many magma compositions from similar starting materials.

Why Rocks Melt

As we know, rocks melt when temperatures below the surface rise to certain levels, causing the minerals in rocks to liquefy. Temperatures increase, in part, with increasing depth; this relationship of increasing temperature with increasing depth is called the **geothermal gradient**. Measurements obtained from drilled wells and mine shafts show temperature increasing at greater depth below Earth's surface. Below continents, the temperature increases about 25° to 30°C with each kilometer below the surface, while beneath oceans, the geothermal gradient averages 60°C per kilometer. Generally, however, these high gradients are the case only for the first few kilometers beneath the surface. Below 10 to 50 kilometers, the temperature does not increase so dramatically, for reasons we will explore in Chapter 8.

Figure 4.16 graphs the geothermal gradients along with the laboratory measurements of the melting temperature of the peridotite that makes up Earth's mantle. Remember, although temperature increases at depth, melting temperatures of silicate minerals in peridotite also rise because of increasing pressure. Figure 4.16a shows that average continental geothermal gradients are not hot enough to cause melting in the mantle, and only small amounts of partial melting are likely at depths of about 100 kilometers beneath the oceans.

Now, we understand why an ocean of magma does not simmer under the Earth's surface and feeds through cracks to make volcanoes: We can conclude that magma does not readily form under the conditions graphed in Figure 4.16a and, therefore, volcanoes are not abundant on Earth's surface.

Despite this, the relationships we earlier explored among melting temperature, pressure, and water content of rock reveal two processes that can cause partial melting of mantle peridotite. The first process is to decrease pressure; the second is to add water.

- *Decrease the pressure:* Figure 4.16b shows that rock does not cool off quickly if it rises toward the surface. Very hot, but solid, peridotite deep in the mantle melts if it rises to a shallower depth, and therefore lower pressure, where the melting temperature is less than the rock temperature.

- *Add water:* Figure 4.16c shows that wet peridotite melts at a lower temperature than dry peridotite. Hot, solid peridotite may melt if water is present in the mantle.

Keep these two points in mind because in Section 4.6 you will explore plate-tectonic processes that follow the scenarios illustrated in Figure 4.16 to generate magma.

EXTENSION MODULE 4.1

Bowen's Reaction Series. Learn how early formed crystals react with a cooling, crystallizing magma.

Putting It Together—How and Why Do Rocks Melt?

- Rocks melt gradually as the melting temperatures of their constituent minerals are exceeded.

- At higher pressures, rocks melt at higher temperatures. This means that very hot rocks rising from great depth will melt as pressure decreases provided that temperature remains high.

- Increasing water content in rock decreases the melting temperature. Melting can be induced, therefore, by adding water to hot rock at high pressure.

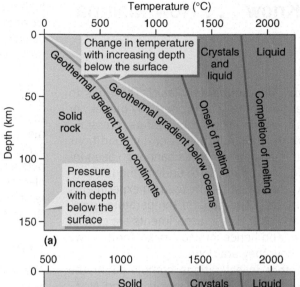

(a)

Melting along geothermal gradient: The average geothermal gradient beneath continents does not intersect the onset-of-melting curve for peridotite. The gradient below oceans intersects the region of partial melting at about 100 km which accounts for the partly molten asthenosphere at that level in the mantle.

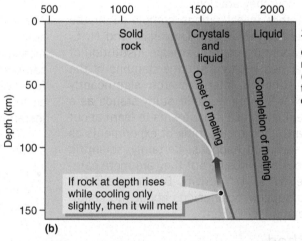

(b)

Melting by decreasing pressure: The black circle denotes a location deep in the mantle. If that mantle material moves toward the surface (experiences a pressure decrease), then it crosses the onset-of-melting curve and partially melts.

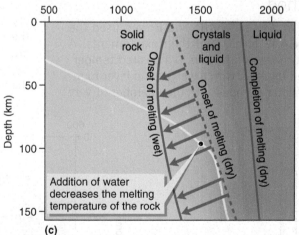

(c)

Melting by adding water: Rocks with water-bearing minerals melt at lower temperature than dry rocks. Water lowers the melting temperature only at elevated pressure, so the onset-of-melting curves under wet and dry conditions are the same at the surface. The black circle denotes a location deep in the mantle where dry peridotite is solid. If water is introduced, then this location will be hotter than the melting temperature.

▲ **Figure 4.16 How mantle rocks melt.** These graphs use results from experiments where peridotite rock was melted at different temperatures and pressures to illustrate conditions within Earth where magma forms. Look at the graph axes carefully—temperature increases from left to right and pressure *increases downward* on the vertical axis. Geologists make the graphs this way because pressure increases the farther below Earth's surface one goes.

ACTIVE ART

Using Graphs to Understand Mantle Melting: Animated and annotated graphs explain how magma forms in the mantle.

4.5 How Do We Know . . . How Magma Is Made?

Picture the Problem

What Are the Necessary Conditions for Melting Rock? The lines in Figure 4.16 indicate whether peridotite is solid, partially melted, or completely melted at various combinations of temperature, pressure, and water content. These graphs illustrate the essential conditions necessary for rock to melt into magma. Geologists cannot go deep inside Earth to watch magma form naturally, so to understand the conditions required to melt rock, they must rely instead on experiments that reproduce in a laboratory the conditions of temperature, pressure, and water content inside Earth. These experiments are crucial to understanding how magma, and hence igneous rocks, form, so we should pause to understand how they are done.

For example, key laboratory investigations completed in the 1950s by O. F. Tuttle of Pennsylvania State University and N. L. Bowen of the Geophysical Laboratory at the Carnegie Institution of Washington considerably clarified geologists' understanding of the origins of felsic magma. Although later research significantly expanded on Tuttle and Bowen's findings, their study stands as a classic illustration of how geologists use experiments to learn about igneous processes within Earth. You might think of experiments as something to do simply to see what happens. Certainly, scientific experiments are conceived out of curiosity, but they are more carefully planned than a simple "What if?" activity. Bowen and Tuttle intended to provide data to support conclusions about the conditions necessary to make magma inside Earth.

Understand the Method

How Is Rock Melting Studied in the Lab? Tuttle and Bowen used the apparatus sketched in **Figure 4.17** to conduct their experiments. A rock sample is placed in a furnace and heated to a desired temperature while rods are squeezed together with a weight and lever to reproduce desired pressures. Water pumped into the chamber at very high pressure simulates the water present in felsic continental crust. In essence, Tuttle and Bowen created a very elaborate pressure cooker, but one capable of producing much higher temperatures and pressure than the familiar stovetop variety.

The "rock" materials for Tuttle and Bowen's experiments were synthetic mixtures of the elements known from laboratory chemical analyses to be present in granite. By combining these elements, the researchers created artificial granite. Dozens of melting experiments were undertaken. Here, we follow the results of only one series of experiments performed with elements found in quartz, potassium feldspar, and sodium-rich plagioclase feldspar. These elements—silicon, aluminum, potassium, sodium, and oxygen—typically comprise more than 95 percent of granite, so these synthetic samples are a reasonable substitute for natural granite.

Evaluate the Results

Under What Conditions Will Granite Melt? **Figure 4.18** graphs the results from the experiments. Each circle on the graph represents measured data from a single experiment. For each experiment, the sample was maintained at a selected temperature and pressure for several hours, and then the temperature was quickly dropped, so that any of the artificial granite that had melted into magma instantly solidified into obsidian-like glass.

The quickly cooled samples were then examined with a microscope. Three results are possible:

- *No melting:* If there was no glass in the sample and it was entirely crystalline, then it was clear that the selected temperature was not high enough at the experimental pressure to permit melting to begin.

- *Partial melting:* If the sample was a mixture of crystals and glass, then the experimental temperature was hot enough for partial melting to occur at the experimental pressure.

ACTIVE ART

Understanding Tuttle and Bowen's Data: *See how animated and annotated graphs explain Tuttle and Bowen's data.*

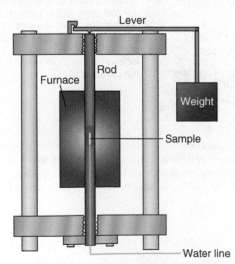

▲ **Figure 4.17 Tuttle and Bowen's experimental apparatus.** This schematic diagram depicts the device that Tuttle and Bowen used to produce experimental granitic magmas at various temperatures, pressures, and water contents. A small sample is heated by a furnace and compressed to high pressure by a weight pressed against a steel rod. Added water simulates conditions in the wet crust.

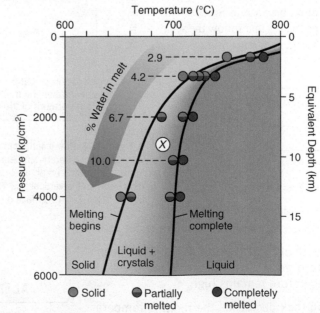

- Each circle represents an experiment conducted at the combination of temperature and pressure indicated by the values on the graph axes.

- In each experiment, the sample either remained solid, partially melted, or completely melted. Curves representing the onset of melting and completion of melting with increasing temperature were drawn to fit the data.

- Water content of the experimental magmas increases with increasing pressure.

- The point labeled *X* represents a hypothetical magma discussed in the text.

▲ **Figure 4.18 Tuttle and Bowen's results.** This graph illustrates the experimental data obtained by Tuttle and Bowen for understanding the melting conditions for granite.

- *Complete melting:* If the sample was entirely glass, then the temperature was sufficiently high to cause complete melting.

Tuttle and Bowen then drew curves between, *not through*, their graphed data points to visualize the approximate conditions of temperature and pressure for the onset and completion of melting (Figure 4.18). They also analyzed the experimental glasses for water content to determine how much water was incorporated into the artificial magma at different pressures and temperatures.

Examination of the results (Figure 4.18) reveals several critical implications.

- The melting temperature of granite is much lower than that of peridotite (compare Figures 4.16 and 4.18b).
- Partial melting in the presence of water begins at lower temperature as the pressure increases. This result also is seen for peridotite (Figure 4.16c) and is, indeed, a characteristic shared by all magmas.
- As pressure increases, more water dissolves into the magma, which means that magma can hold more water at high pressure than at low pressure.

Insights

How Do Laboratory Experiments Explain Field Geology? Laboratory experiments enable geologists to determine the combinations of temperature, pressure, and water content that permit granite to melt (Figure 4.18), which, in turn, allows scientists to understand this process in the field. Notice that the melting temperature of *wet* granite decreases when pressure increases. *Dry* rock behaves in an opposite fashion in experiments, as illustrated for the peridotite results in Figure 4.16. Because water-bearing minerals such as amphibole and biotite are common in granite, these experiments suggest that granite in the continental crust might melt at pressures equivalent to depths of 25 to 40 kilometers below the surface.

The experimental results also allow geologists to predict what happens when felsic magma moves toward the surface. Consider a felsic magma at a temperature and pressure equivalent to point "*x*" in Figure 4.18. If this magma rises toward the surface, then the point moves up on the graph. The magma crosses the "melting begins" curve and becomes solid granite at a depth of about 5 kilometers, even if it does not cool to a lower temperature. In this way, wet magma solidifies into rock without losing heat. Also, when at the conditions of point "*x*," the magma contains a little less than 10 percent water, but just before it crystallizes at 5 kilometers, it contains only 5 percent water. Somehow, the magma loses water as it rises. Where does the water go? Sections 4.8 and 4.9 consider igneous phenomena that are explained by this behavior of water in magma formation and crystallization that was originally revealed by Tuttle and Bowen's experiments.

Putting It Together—How Do We Know . . . How Magma Is Made?

- The ability to simulate the temperature, pressure, and compositional characteristics of the deep crust and mantle in the laboratory allows geologists to conduct experiments that teach us more about the melting of rock and crystallization of magma.

- Tuttle and Bowen's experiments showed that magma under higher pressure can contain more water, and that the presence of water lowers the melting temperature of rock.

4.6 How Does Magma Generation Relate to Plate Tectonics?

Sections 4.4 and 4.5 explained that variations in temperature, pressure, and water content determine melting conditions for different rock compositions. We also know that these conditions are not met in many places under the surface because nearly all of the silicate Earth (crust and mantle) from which magmas must come is solid, not molten. Thus, we must ask, where does rock melting consistent with the graphs in Figures 4.16 and 4.18 occur within Earth? The association of igneous rocks with tectonically active areas, especially plate boundaries (Figure 4.10), indicates a link between plate tectonics and magma generation, which we explore in this section.

Decompression: Melting at Divergent Plate Boundaries and Hot Spots

Figure 4.16b describes how mantle peridotite can partially melt if it rises toward Earth's surface without losing much heat. **Figure 4.19a** shows that this process happens at mid-ocean ridges (divergent plate boundaries) where the hot asthenosphere rises and fills the spaces where the oceanic lithosphere belonging to different plates separates. When this occurs, the mantle peridotite is exposed to lower pressure, or decompression, as it rises. Decompression results in a combination of temperature and pressure that allows partial melting (Figure 4.16b). Experiments show that partial melting of the peridotite produces mafic magma. The mafic magma is less dense than the surrounding mantle, and so it rises to erupt onto the seafloor as basalt.

Decompression melting also happens where unusually hot mantle rock rises at hot spots (Figure 4.19a). This process builds many volcanic islands such as Hawaii and Iceland (Figure 4.12), and explains the eruption of lava at Kilauea, which we visited at the beginning of the chapter.

Adding Water: Melting at Convergent Plate Boundaries

The close relationship between the locations of volcanoes and convergent plate boundaries shown in Figure 1.10 indicates that melting takes place in the vicinity of subducting plates. However, subduction causes the insertion of cold lithosphere from the surface into the deeper mantle, which has a chilling effect. So, if subduction zones are relatively cool, how does the mantle melt to make the magma that feeds the volcanoes?

Figure 4.19b shows how geologists explain magma formation near convergent plate boundaries. The upper part of the subducted plate is seafloor basalt that has resided in seawater for tens to hundreds of millions of years. By the time the seafloor basalt reaches the subduction zone, it contains many water-bearing minerals because of chemical reactions between the basalt and seawater. Water also is present in the sediment that accumulates on top of the basalt during its slow journey to the subduction zone. In this way, subduction carries water deep into the mantle within minerals present in basalt and sediment.

At a depth of approximately 125 kilometers, the temperature and pressure causes metamorphic reactions in which water-bearing minerals (amphibole, for example) in the subducting seafloor turn into minerals lacking water (such as pyroxene). The water is released into the overlying

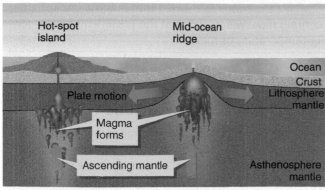

(a)

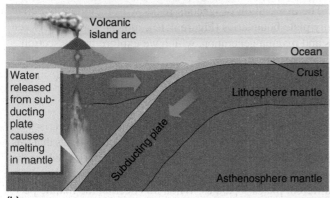

(b)

Mantle rises toward the surface at mid-ocean ridges and at mid-plate hot spots. If the mantle ascent occurs with minimal heat loss, then the mantle peridotite begins to melt as it rises and the pressure decreases (this is the process illustrated in Figure 4.16b).

Subduction carries water-bearing minerals into the asthenosphere. There, metamorphic reactions release the water. The presence of water causes melting of the mantle peridotite, as illustrated in Figure 4.16c. This process explains why volcanoes are present at convergent plate boundaries.

▲ **Figure 4.19 How plate tectonics cause mantle melting.** The experimental insights illustrated by graphs in Figure 4.16 explain the relationships between plate tectonic processes and magma generation at divergent plate boundaries (mid-ocean ridges), hot spots (such as Hawaii), and convergent plate boundaries (subduction zones).

ACTIVE ART

Plate Tectonics and Magma Generation: See how magmas form at plate boundaries and hot spots.

asthenosphere. Recall that laboratory experiments show that wet rocks melt at lower temperatures than dry rocks (Figure 4.16c). The expulsion of water from the subducting plate adds water to the asthenosphere and causes partial melting of the peridotite, as illustrated in Figure 4.19b.

Decompression partial melting of peridotite forms mafic magma, so you may hypothesize that mafic magma forms when peridotite partially melts near subduction zones. Indeed, basalt is found on volcanoes near subduction zones, but andesite and dacite also are common. The greater diversity of magma composition at subduction zones relates to processes described in Section 4.7.

Putting It Together—How Does Magma Generation Relate to Plate Tectonics?

• Magma forms only where unique circumstances exist that permit melting to occur in the mantle or lower crust.

• Partial melting of the mantle happens where rising mantle remains hot but is exposed to decreasing pressure at divergent plate boundaries and hot spots.

• Partial melting of the mantle also occurs due to the addition of water from metamorphic reactions during plate subduction at convergent plate boundaries.

4.7 What Makes Igneous Rock Compositions So Diverse?

The igneous rock classification system illustrated in Figure 4.3 demonstrates the remarkable diversity of igneous rock compositions. How can such a diverse range of magmas form in the first place? Geologic studies show that the diversity of magmas results from four processes:

1. Different types of rock can melt or partially melt by different amounts.

2. Magmas of different composition can mix to produce new, in-between compositions.

3. Rock surrounding magma may partially melt and mix with, or be assimilated by, the original magma.

4. One magma composition can be derived from another during crystallization.

The Melting of Different Rocks

Recall the observation from Section 4.4 that partial melts are more silica rich than the original rock. When examining the plate-tectonic environments of magma formation (Section 4.6), you learned that melting mantle peridotite generates mafic magma, which is consistent with mafic magma being more silica rich than ultramafic peridotite. Carrying this logic a bit further, it becomes apparent that if you start with different rock types and melt them to different amounts, you will end up with different magma compositions. For example, geologists have determined that the subducted mafic crust at convergent plate boundaries does, in a few places, partially melt to produce magma that solidifies as dacite or tonalite. Also, partial melting of intermediate-composition continental crust rocks results in even more felsic magma. This type of melting commonly happens where mafic magma produced by partial melting in the mantle rises into continental crust.

Magma Mixing

Another way to form a compositionally distinct magma is to mix two or more magmas of different composition. Many large batholiths form from numerous injections of magma into a single magma chamber. As a new batch of magma arrives in the chamber, it encounters the resident magma, which may be of a different composition. In some cases, geologists have found convincing evidence of magmas mixing to form a melt with a new geochemical composition.

Magma Assimilation

When magma rises toward the surface, it comes in contact with, and may incorporate pieces of, the surrounding rock. These surrounding rocks may partially or completely melt and mix into the magma, which changes the magma composition. This process, called **assimilation**, is suggested by

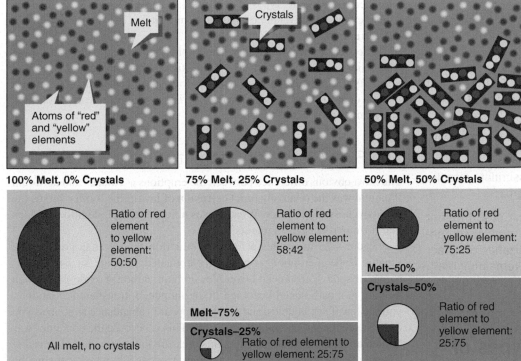

▲ **Figure 4.20 How crystallization changes melt composition.** The sample melt in this diagram includes equal amounts of two elements represented by red and yellow dots. The first minerals to crystallize when the melt cools contain three times more of the "yellow" element than the "red" element. Crystallization preferentially removes the "yellow" element from the melt so that the remaining liquid becomes progressively richer in the "red" element. Melt composition changes, therefore, as crystallization takes place.

ACTIVE ART

Fractional Crystallization: Animation illustrates how magma composition changes during crystallization.

Deriving One Magma Composition from Another

A less common, but very important, process begins with one magma composition and produces another composition during crystallization. If, for example, you start with a mafic melt in the laboratory, then a basalt rock will form when all of the liquid crystallizes into minerals (see Figure 4.15). What would happen, however, if each mineral grain is taken out of the magma as soon as it formed? Referring back to our partially melted chocolate-chip ice cream, this would be analogous to removing the solid chocolate chips and refreezing the remaining liquid into vanilla ice cream that lacks the original chocolate.

Figure 4.20 illustrates how melt composition changes during magma crystallization. No single mineral has the exact same composition as the liquid melt. The first mineral crystals that form when magma begins to cool will contain more of some elements and less of other elements

outcrops of some pluton margins, where adjacent rocks are seen to have melted, and by certain chemical peculiarities in many igneous rocks.

compared to the original magma. In a case such as the one illustrated in Figure 4.20, a particular element (represented by the yellow dots) is preferentially incorporated within the mineral crystals. As a result, that element is less abundant in the remaining molten liquid after crystallization begins. Likewise, other elements (such as the element represented by the red dots in Figure 4.20) now compose a greater proportion of the remaining liquid than was the case initially. In natural silicate magmas, the melt remaining after a fraction of magma crystallizes is almost always more felsic (more silica rich) than the original, completely molten liquid because minerals with relatively low silica contents tend to crystallize first. If the first-formed mineral crystals separate from the melt, then the remaining liquid magma has a different composition from the original magma. This process is called **fractional crystallization**, because each fraction of the magma that crystallizes leaves behind a melt of new composition.

Figure 4.21 shows how fractional crystallization can occur by mineral separation. In some cases, the minerals are denser than the liquid and simply sink to the bottom of the magma chamber. Mineral accumulations such as these can form significant natural economic resources, including the ore minerals for chromium and platinum.

Some plutons (magma solidified into rock bodies before reaching the surface) consist of progressively more felsic rocks the nearer you get to their center. This suggests the process for separating crystals and melt illustrated on the right side of Figure 4.23. As intruded magma solidifies inward from colder surrounding rocks, the remaining magma is progressively enriched in silica and other elements that are less abundant in the early formed minerals near the edge of the pluton.

Fractional crystallization can result in intermediate and felsic magmas that are derived from an original mafic magma by early crystallization of minerals that are rich in iron and magnesium but contain less silica than the original magma. The compositional diversity of rocks at some volcanoes is evidence that this process occurs. Fractional crystallization of mafic magma, for instance, accounts for many of the abundant andesitic and dacitic volcanic rocks found near subduction zones.

▶ **Figure 4.21 Two ways crystals separate from magma.** Crystals that form early during crystallization of the magma can settle to the bottom of the chamber if they are denser than the melt (left). In some cases, crystallization takes place first on the wall of the magma chamber, where the temperature is coolest, and progresses inward (right). In both cases, the crystals separate from the remaining molten liquid so that magma composition changes.

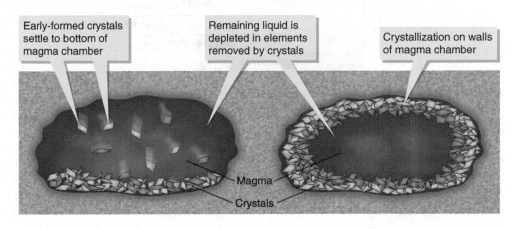

Putting It Together—What Makes Igneous Rock Compositions So Diverse?

- Melting different types of rock produces magmas of different composition. The extent to which the rocks melt also determines the composition of the magma.

- Two or more magmas may mingle to produce a hybrid magma with a new composition.

- Rocks surrounding magma may partly melt and assimilate into the liquid to change the overall composition of the magma.

- Magma of one composition can form from magma of a different composition by fractional crystallization. No single mineral composition is exactly that of the whole magma, so crystallization of a fraction of the magma changes the composition of the remaining melt. If the liquid melt separates from the crystals, then a new magma composition results.

4.8 Why Are There Different Types of Volcanoes and Volcanic Eruptions?

You now have answers to many questions about magma and igneous rock compositions and the relationship between igneous rocks and plate tectonics. But we still do not know why such different eruption styles occur, and

Composition	Rhyolitic	Dacitic	Andesitic	Basaltic
Silica content	70%		60%	50%
Eruption temperature	750°–900°C		900°–1000°C	1100°–1200°C
Viscosity (water = 0.01 poise)	1×10^7 poise		3×10^4 poise	500 poise
Gas content	5.0%		2.0%	0.5%
"Explosiveness"	More explosive			Less explosive
Volcanic products	Lava domes / Pyroclastic deposits			Lava flows
Volcano types			Shield volcanoes / Cinder cones / Composite volcanoes / Dome complexes	

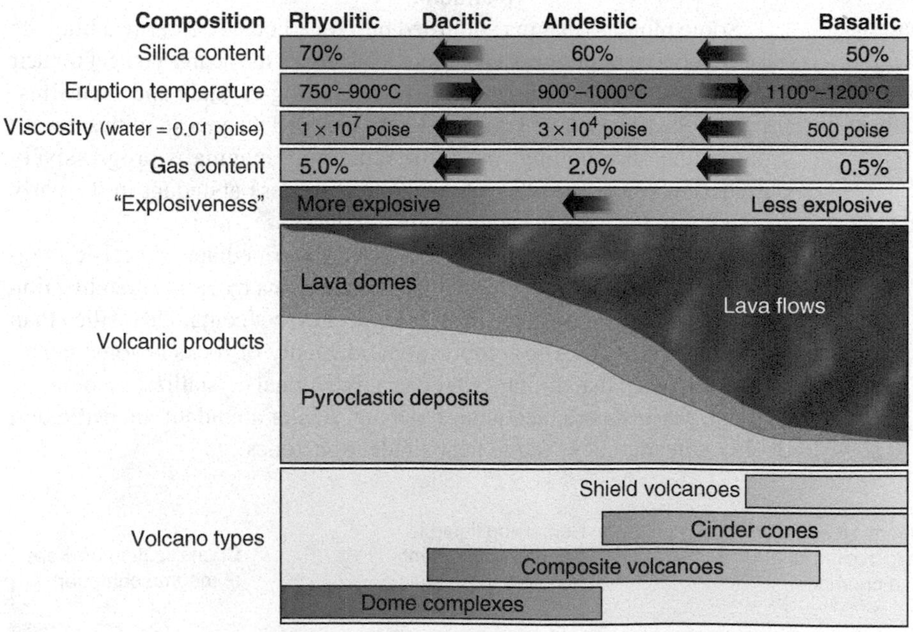

▲ **Figure 4.22 How magma composition determines volcano types and eruption style.** Volcano types relate to the proportions of lava and pyroclastic materials composing them and the fluidity of lava flows, if present. Laboratory experiments demonstrate that gas content and viscosity vary according to the silica content of magma. The more silica content in a magma, the higher its viscosity and the more gas it can hold. Black arrows show how a property increases in value depending on magma composition. Higher gas content favors more explosive eruptions, which generate more pyroclastic deposits. Viscosity determines whether lava flows readily or accumulates in steep-sided domes. These measurable magma characteristics provide explanations for the commonly observed relationships among magma composition, explosiveness of eruptions, types of volcanic products, and types of volcanoes.

why there are so many types of volcanic landforms (Figure 4.12). Is there a link between the diversity of magma compositions and the variety of volcanic eruption styles and landforms?

The key to understanding volcanic phenomena and addressing these issues lies in understanding two properties of magma: gas content and viscosity. **Figure 4.22** shows how gas content and viscosity depend on magma composition and also relate to variations in volcanic eruptions and landforms.

The First Clue: Gas Content

The most obvious difference between the eruptions at Kilauea and at Mount Pinatubo was the relatively quiet extrusion of lava at the former versus the explosiveness of the latter. What causes a liquid to explode into the blobs that form pyroclastic fragments? Gas.

Figure 4.23 illustrates a familiar situation that is analogous to the Pinatubo type of eruption—opening a carbonated beverage, such as Champagne or soda. Carbonated beverages contain carbon dioxide gas. When looking at a carbonated beverage in an unopened, transparent container, you see few if any bubbles, and yet they appear in abundance as soon as you open the container and pour its contents. This is because the container was bottled under high pressure, and at this pressure the carbon dioxide gas dissolves into the liquid. When the container is opened, however, causing the pressure to release and drop, the gas comes out of the solution in bubbles.

Like a carbonated beverage, magmas also contain dissolved gases, as mentioned in Section 4.2. Water vapor is the most abundant gas found in magma. Like in a carbonated beverage, gas remains dissolved at the elevated pressure deep below ground. When magma moves toward the surface, however, that pressure decreases and the gases cannot remain dissolved in the magma. Recall from Tuttle and Bowen's experimental results (Figure 4.18) that magmas at lower pressure cannot hold as much water as magma at high pressure. In nature, bubbles form in the magma and stream toward the surface when water vapor and other gases come out of the magma. The bubbles and surrounding liquid rapidly rise and explode out of the volcano just like foam spraying from a suddenly opened Champagne bottle.

There is, however, a major difference between magma and Champagne: Champagne spray remains liquid, but magma solidifies when it explodes out of the volcano because the temperature in the atmosphere is very cold, well below the solidification temperature of the melt. Glassy pyroclastic particles are frozen fragments of bubbly magma, and highly vesicular pumice and cinder are chunks of the frozen foam that preserve the outlines of the bubbles (Figure 4.4).

The amount of gas dissolved in magma strongly depends on the composition of the magma. Water vapor is the most abundant magmatic gas, as we have noted, and experiments show that water vapor is four to five times more soluble in felsic magma than in mafic magma at the same temperature and pressure (Figure 4.22). Basaltic volcanoes, therefore, rarely erupt violently, but dacitic and rhyolitic volcanoes almost always do so. This also explains why basaltic volcanoes have a very high ratio of lava flows to pyroclastic deposits; fewer explosions result in fewer pyroclastic deposits. On the other hand, pyroclastic materials are more abundant among andesitic, dacitic, and rhyolitic volcanoes (Figure 4.22).

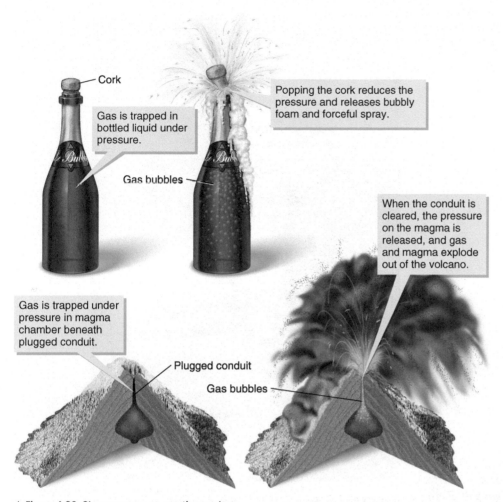

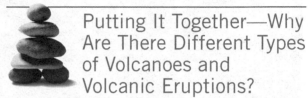

stubby and build high, steep-sided volcanic domes (Figure 4.12). Gas bubbles also separate more easily from low-viscosity mafic magma than from high-viscosity felsic magma. Not only is the gas content high in a felsic magma, but the gas pressure rises dramatically when the bubbles cannot rise through the viscous melt to the surface.

Putting It Together—Why Are There Different Types of Volcanoes and Volcanic Eruptions?

• Volcanic phenomena and volcanic landforms closely depend on two properties of magma: gas content and viscosity.

• Gas content determines how explosive eruptions will be and the relative proportion of lava flows and pyroclastic deposits that eruptions will produce. More gas-rich magmas are more explosive.

• Viscosity determines whether lava flows are thin and widespread, or thick and short. Lower viscosity produces more fluid flows.

• Viscosity and gas content are in large part determined by magma composition, and both properties increase with increasing silica content.

▲ Figure 4.23 Champagne as an eruption analogy.

Of course, you may wonder how water vapor gets into magma in the first place. Minerals containing water molecules exist in small quantities throughout the mantle, so when the minerals melt, they release the water into magma. Also, recall from Section 4.6 that subduction carries water into the mantle and triggers melting at convergent plate boundaries (Figure 4.19b). It is not surprising, therefore, that subduction-zone magma has a high content of dissolved water and that volcanoes above subduction zones typically produce the most violently explosive eruptions.

The Second Clue: Viscosity

Another critical property of magma is its **viscosity**, or resistance to flow. Water, for example, flows readily on an inclined surface and thus has a very low viscosity. Other fluids, such as cake batter or tar, flow very slowly as thick, pasty masses, which indicate greater resistance to flow. Thus, slow-moving, thick fluids are highly viscous.

Viscosity partly relates to the molecular bonds within a liquid, so viscosity varies with magma composition. Magmas with higher silica content tend to be more viscous. For comparison, at a typical eruption temperature, basaltic lava has a viscosity comparable to ketchup, whereas rhyolitic lava is as viscous as creamy peanut butter. Shield volcanoes, therefore, consist of thin, laterally extensive lava flows of mafic (low-silica) composition and low viscosity (Figure 4.12). On the other hand, silica-rich dacitic and rhyolitic lava flows have a high viscosity, so they are thick and

4.9 What Hazards Do Volcanoes Present?

The serious hazards of volcanic eruptions are compelling reasons for geologists—and you—to understand igneous processes. More than 85,000 people died worldwide in the twentieth century as a result of volcanic activity. In the United States, volcanic hazards are particularly notable along the convergent plate boundaries of the Pacific Northwest and Alaska, and on the hot-spot island of Hawaii.

Volcanoes have caused some of the most dramatic and destructive natural disasters in history. From the destruction of the Italian cities of Pompeii and Herculaneum by the eruption of Vesuvius in 79 CE to the obliteration of Armero, Colombia, by erupting Nevada del Ruiz in 1985, history is full of accounts of destructive eruptions. Lava flows cover large areas of arable land and destroy buildings and crops, but fortunately, they usually advance slowly enough for people to evacuate from harm's way. The lava flow depicted in **Figure 4.24**, for example, traveled so slowly that people could simply avoid it. Volcanic phenomena other than lava flows, such as pyroclastic flows and lahars, account for most eruption casualties.

The Hazards of Pyroclastic Flows

Fast-moving pyroclastic flows commonly cause horrific casualty tolls. **Figure 4.25** shows billowing ash from a pyroclastic flow at Unzen volcano, Japan, in 1991. Moving at velocities in excess of 100 kilometers per hour,

▲ **Figure 4.24 A river of lava.** Basaltic lava streams from a crater on the flank of Mauna Loa volcano in Hawaii. Although lava flows destroy buildings and bury agricultural land, they rarely flow fast enough to prevent people from escaping their path.

these blasts offer the people in their path little chance of survival. This fact is driven home by the story of Mont Pelée, which is located near St. Pierre, Martinique, French West Indies. Pelée erupted in 1902, destroying St. Pierre with a pyroclastic flow. The city was flattened, and all but two of eighteen boats anchored in the harbor sank. The photograph in **Figure 4.26** was taken after that disaster and still remains a striking record of this event, which killed nearly all of the 28,000 residents in the city. Pyroclastic flows are so dangerous because anything that is not knocked over by the fast-moving avalanche of pumice and ash usually burns after contacting pyroclastic fragments and gases at temperatures greater than 300°C. Most human deaths, however, result from the rapid asphyxiation that occurs in the choking cloud of ash and gas generated by these flows.

The Hazards of Lahars

The loss of nearly 25,000 people at Armero, illustrated in **Figure 4.27**, is the most tragic of many examples of destruction caused by the rapid flow of water and loose debris down steep volcanic slopes. The Indonesian term **lahar** describes this phenomenon. The Armero catastrophe was caused by the melting of snow and glacier ice by hot pyroclastic fragments. Lahars also can be triggered by the rapid erosion of loose, pyroclastic deposits from steep hillsides during heavy rainstorms. During the annual rainy seasons in the Philippines, lahars led to the burial of entire villages in the years following the 1991 eruption of Mount Pinatubo.

The Hazards of Far-Flung Pumice and Ash

Ash and lapilli ejected high above volcanoes would, at first glance, seem more likely to be a nuisance than a hazard, especially given that volcanic

▲ **Figure 4.25 A deadly pyroclastic flow.** Ash billows above a fast-moving pyroclastic flow unleashed from the obscured summit of Mount Unzen, Japan, in 1991. The combination of rapidly moving fragments, high temperature, and suffocating dust and gases makes pyroclastic flows potentially the most destructive and deadly hazards of volcanic eruptions. This pyroclastic flow destroyed 50 homes and killed 28 people, including three volcanologists.

▲ **Figure 4.26 Pyroclastic-flow devastation.** In May 1902 a pyroclastic flow from Mont Pelée on the West Indies island of Martinique, visible in the background, thoroughly devastated the city of St. Pierre, only 10 km away. Approximately 28,000 people died.

▲ **Figure 4.27 Lahar devastation.** A rapidly flowing mixture of melted snow, volcanic ash, and other debris from the Nevada del Ruiz volcano buried the city of Armero, Colombia, in 1985 and killed at least 23,000 people. The buildings visible in the photo are all that remains of what was once a hilly part of the city, most of which is completely buried beneath gray volcanic mud.

ash produces highly fertile soils in tropical regions. This beneficial aspect of volcanic eruptions is sometimes offset, however, by the encroachment of agricultural villages too close to active volcanoes. Not only are these settlements at increased risk of destruction by lava flows, pyroclastic flows, and lahars, but also by pyroclastic-fall deposits that at such close proximity can be heavy enough to cause buildings to collapse. Nearly all of the 350 people who lost their lives during the 1991 eruption of Mount Pinatubo were victims of falling roofs weighed down by the ash of pyroclastic-fall deposits.

Volcanic ash is strongly abrasive and can cause machinery failures hundreds of kilometers downwind of a volcano. Ash clouds also present risks to jet aircraft, because ash ingested by unfiltered jet engines is subjected to temperatures above the ash melting point. When this happens, the ash converts to lava, which coats the turbine blades. Eventually, the engines fail because of the stress of the added weight. Following several near catastrophes with jumbo aircraft in the 1980s and 1990s, new precautions were enacted to prevent such air disasters.

Indirect Volcanic Hazards

In some cases, property destruction and loss of life are less directly related to an eruption itself. One example of this is the death of more than 36,000 Indonesian people as a result of the 1883 eruption of Krakatau. The volcano was on a remote, uninhabited island where impact on humans was predicted to be negligible. The eruption culminated, however, in caldera collapse, creating a deep depression on the ocean floor. The collapse disturbed the water surface and formed a huge wave more than 35 meters high that swept outward from the ruins of the volcanic island and devastated coastal villages on nearby Java and Sumatra. Another eruption in Indonesia, at Tambora in 1815, initially led to few casualties, but the devastation of farmland by pyroclastic flows ultimately led to famine and the loss of more

than 117,000 lives in an era before international relief efforts could be rapidly deployed.

Reducing Volcanic Hazards

Destruction by volcanic eruptions is substantially diminished when people practically apply knowledge about eruption types. As we have learned, the composition of erupted magma is the primary factor determining eruption style, as illustrated in Figure 4.22. It is possible, therefore, to loosely forecast the hazards of an eruption at its outset by analyzing the first magma that is erupted. It is also possible to ascertain what sort of eruption will occur by studying the lava flows and pyroclastic deposits from previous eruptions. If geologists can determine the type of eruption, then they can identify the geographic areas most susceptible and evacuate the people most likely to be at risk. Increasing sophistication of volcanological studies in the latter part of the twentieth century has substantially reduced the casualties from volcanic activity.

EXTENSION MODULE 4.2

Mitigating and Forecasting Volcanic Hazards: Learn about the types of volcanic hazards.

Putting It Together—What Hazards Do Volcanoes Present?

• The nature of eruption hazards (lava flows, pyroclastic flows, distant fallout of volcanic ash) corresponds to the composition of the erupted material.

• The most deadly and destructive volcanic phenomena are fast-moving, far-traveling pyroclastic flows and lahars.

4.10 Why Don't All Magmas Erupt?

A particularly nagging question about volcanoes remains: Why does some magma erupt at volcanoes, while other magma solidifies below ground? Earth's many large outcrops of plutonic rocks, such as those exposed in Yosemite National Park (Figure 4.1b), indicate that a lot of magma never finds a way to the surface to erupt at a volcano.

How to Make a Pluton

As we know, magma forms at high temperatures within Earth and rises into progressively shallower, cooler environments closer to the surface. As it rises, the hot magma loses heat by conduction into the surrounding, cooler rocks. This means that the magma temperature decreases as it rises and the magma solidifies if it is not ascending rapidly enough to counteract this cooling effect. This is why no ultramafic volcanic rocks form on Earth today. On modern Earth, the high temperatures (about 1600°C) required to keep ultramafic magmas in a largely molten state cannot be sustained at shallow depths, and the magma solidifies before it reaches the surface (notice the geothermal gradient in Figure 4.16). The presence of ultramafic volcanic rocks older than 2.5 billion years old, however, is powerful evidence that our planet was formerly much hotter than it is today.

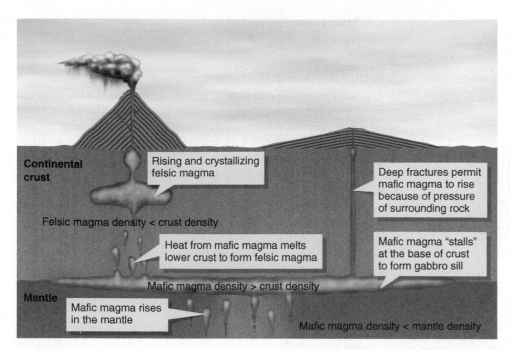

Continental crust

Rising and crystallizing felsic magma

Deep fractures permit mafic magma to rise because of pressure of surrounding rock

Felsic magma density < crust density

Heat from mafic magma melts lower crust to form felsic magma

Mafic magma "stalls" at the base of crust to form gabbro sill

Mafic magma density > crust density

Mantle

Mafic magma rises in the mantle

Mafic magma density < mantle density

◀ **Figure 4.28 How magma rises.** Magma rises only if it is less dense than surrounding rock. Comparing magma and rock densities determines whether magma reaches the surface to erupt at a volcano or remains underground to form a pluton.

ACTIVE ART

Density and Magma Movement: See how magma rises or stalls because of density contrast between magma and rock.

Why does magma rise to begin with? For most compounds, including silicates, the liquid phase is less dense than the solid form of the same composition. Once magma forms, therefore, it is less dense than the surrounding rock and rises, as shown in **Figure 4.28**.

Mafic magma generated by partial melting in the mantle rises through, and shoulders aside, the surrounding denser peridotite and solid basalt in the oceanic crust, but experiences greater difficulty rising in continental regions. This is because the intermediate and felsic rocks of the continental crust, by virtue of their different composition, are actually less dense than the mafic magma. The mafic magma, therefore, stalls at the base of the continental crust (Figure 4.28). If tectonic processes deeply fracture the crust, the melt may make it through to the surface. Otherwise, the magma is stuck near the base of the crust and solidifies to form a gabbro sill. The mafic intrusion may, however, conduct enough heat into the crust to partially melt the crust and form felsic magma. This new-formed felsic magma can rise farther because it is less dense than the surrounding crustal rocks (Figure 4.28).

However, even most felsic magmas do not make it all the way to the surface because of the behavior of water in the melt. Intermediate and felsic magmas tend to be water rich. The implications of water content for magma ascent were discussed in Section 4.5, and it is worthwhile to revisit Figure 4.18.

The hypothetical water-rich magma at point "x" in Figure 4.18 is partially melted granite. Tracing a vertical line upward from "x" helps us to visualize what happens when the felsic magma moves toward the surface. As the pressure decreases, the water content of the magma also decreases, and the temperature for the onset of melting, which is the same as the temperature for the completion of crystallization of the rising magma,

increases. In the real world, the temperature of magma cannot increase as it rises through progressively cooler rocks toward the surface, so the path of the magma must, at best, remain a vertical line on the graph. Now, notice the key point—this magma crosses the completion-of-crystallization (also the onset of melting) curve before reaching the surface. Therefore, the hypothetical felsic magma is destined to solidify below ground and become a granitic intrusion, not a rhyolitic volcanic deposit.

Indeed, it is a wonder that any magma reaches the surface to be erupted at all. The hurdles presented by lower temperatures near the surface, density barriers (such as that presented by the crust), and the release of gases during ascent ensure that all magma crystallizes to at least some extent before reaching the surface at a volcano. The largest proportion of Earth's magma volume ultimately solidifies as plutonic rocks. For every cubic kilometer of lava and pyroclastic debris that erupts at the surface, probably at least 50 cubic kilometers of igneous rock solidify below the surface. These facts explain why Earth's crust overwhelmingly consists of plutonic igneous rocks.

Pluton Crystallization Produces Economic Resources

The overwhelming majority of gold, silver, copper, and other metal resources are exploited from rocks within or near igneous intrusions. A wide variety of useful ore deposits form as a result of fractional crystallization or of gases (water and other vapors) being released from crystallizing magma.

Many economically important metallic elements (e.g., gold, silver, copper, lead, zinc, molybdenum, and nickel) exist in tiny amounts in magma. Nonetheless, these elements are naturally concentrated in sufficient abundance for profitable mining because they do not fit within the crystal structure of the dominant silicate minerals that form when magma solidifies. This occurs because, although these elements make up a small percentage of the melt and are disseminated throughout the original magma, they become concentrated by fractional crystallization into the last bit of the magma that remains after most of the silicate minerals have formed. It is also at this stage that gas-vapor pressure is highest (Section 4.8) in the melt, so that fractures form within the solidified parts of the intrusion and in the surrounding rock. The remaining metal- and vapor-rich magma intrudes into these fractures, producing ore veins.

Many metallic elements readily bond with sulfur, which also is commonly present at this gas-rich magma phase. When the sulfur-rich fluids separate from the magma during crystallization of the silicate minerals, the metals go with the sulfur. Then, the sulfur and metal dissolve in hot ground water that encircles the intrusion. Subsequent changes in temperature, pressure, or water composition cause the metals to precipitate as sulfide minerals.

▲ **Figure 4.29 The world's largest open-pit mine.** The Bingham Canyon Mine, Utah, is roughly 3.5 km by 2.5 km across and more than 1 km deep. Sulfide ore minerals of copper and other metals are extracted from a felsic pluton and surrounding rocks that were mineralized by fluids released from the magma as it crystallized. Note the circled buildings for scale.

Especially notable in this regard are the huge copper mines in southern Arizona, Bingham, Utah, and Butte, Montana, which produced much of the world's copper from copper sulfides such as chalcopyrite. These mines are located in plutons that range in composition from diorite to granite. The Bingham mine, shown in **Figure 4.29**, is the world's largest open pit, nearly 3.5 kilometers across and 1 kilometer deep. At peak productivity in the mid-twentieth century, more than 400,000 metric tons of rock was excavated *each day* to recover 1200 tons of copper.

Heated ground water near shallow intrusions produces hot springs and geysers at the surface. Although most of the hot water and steam are simply heated local ground water, some of the water vapor and other gases originate from magma, accounting for the sulfurous odors around many thermal springs. These are the same fluids that produce the voluminous metal ores at deeper levels below the ground. Where these hot fluids are located close to the surface, they can be used directly for heating purposes or to generate electricity.

Pegmatite, shown in **Figure 4.30**, is another economically important igneous rock that forms very late in the magma crystallization process. Very large crystals, approaching several meters in size, define these extraordinary coarse-grained rocks. The large crystals form not because of slow cooling but because the high fluid content of the magma inhibits the beginning of crystal formation but then favors rapid crystal growth once crystals start to form. The size of the crystals, while spectacular in itself, is not the principal economic value of pegmatites. In vapor-rich magma, pegmatite is the last material to crystallize. Thus, as pegmatite crystals form, they are enriched in elements that were excluded from typical rock-forming silicates. In certain pegmatites, these uncommon elements form silicate and oxide minerals when concentrations are hugely enriched in the melt due to fractional crystallization. These pegmatites include ore minerals of rather exotic elements such as beryllium, lithium, and tantalum, which have important uses in metallurgy and the production of ceramics. Some pegmatites also are the source of spectacular gemstones such as tourmaline, kunzite, and aquamarine.

Spodumene crystals

▲ **Figure 4.30 Finding valuable minerals in the field.** Huge crystals of spodumene, a silicate mineral containing the element lithium, form most of this pegmatite at the Etta Mine in the Black Hills of South Dakota. Compare the crystal size to the pine trees, which are approximately 2 m high. The bottom photos show a beautiful raw tourmaline crystal and several cut tourmaline gemstones from pegmatite mines.

Putting It Together—Why Don't All Magmas Erupt?

• When rocks partially melt, the resulting liquid magma is less dense than the remaining solids and rises. If, however, the magma encounters less dense rock as it ascends toward the surface, then the magma stalls and crystallizes.

• Rock temperature decreases as magma moves near the surface. If magma does not rise quickly, then its heat is conducted to surrounding rocks and the melt cools and crystallizes before it can erupt on the surface.

• Most water-rich magma crystallizes below ground because the magma solidifies as it releases dissolved water at shallow depths.

• Economically important elements present in minor amounts in magma melts are concentrated into pegmatites and sulfide-mineral deposits by fractional crystallization.

Where Are You and Where Are You Going?

The beginning of this chapter included descriptions of the igneous rocks and landscapes formed at three very different geographic localities: Hawaii, Mount Pinatubo, and Yosemite. Each landscape relates to the melting of rock to produce magma and its subsequent solidification into igneous rock. Igneous rock types are classified by mineral composition and texture. Pyroclastic deposits, such as those observed at Mount Pinatubo, are classified by fragment size and degree of consolidation.

Magma crystallizes above the surface to produce volcanic rocks, as observed in Hawaii and at Mount Pinatubo, or it solidifies below the surface to form plutonic rocks that may be exposed by later erosion, as at Yosemite. Plutonic igneous rocks form plutons with various shapes and sizes depending on how much magma intrudes into surrounding rocks and the manner in which it intrudes. Volcanoes vary in shape and size depending on the amount of material they erupt and the style of the eruption, which is the result of different compositions of magma. Gas content and viscosity of magma play a major role in eruption style and the abundance of lava flows and pyroclastic deposits. These variations in eruption style, in turn, determine the nature and extent of volcanic hazards.

The different compositions of magma relate to how rocks melt and how the magma crystallizes. Temperature, pressure, and water content are the key variables in these processes. Experiments show that rocks melt in areas where pressure decreases while temperatures remain high, or where water is added to hot rock at high pressure. Partial melting of mantle peridotite to produce mafic magma at divergent plate boundaries and hot spots results from decompression of the mantle as it rises. Magma forms at convergent boundaries where subduction carries water into the mantle and lowers the melting temperature of peridotite. The different varieties of igneous rocks form as a result of four processes. Different source rocks melt to form different types of magma. Magmas of different composition may mix to produce another composition of magma. Magma can change its composition by assimilating nearby rocks. Fractional crystallization changes the composition of the remaining melt as the magma gradually crystallizes.

Magma forms deep in the crust and most commonly in the mantle. Only a small fraction of magma erupts from the surface through volcanoes—most magma solidifies within Earth's crust. This process occurs where magma encounters less dense rock on its way to the surface and is blocked, causing it to solidify in place, or when the heat from the magma is conducted to its cooler surroundings and crystallizes before reaching the surface. Dissolved water lowers the melting temperature of magma, and most water-rich magma crystallizes below ground because the water releases from the magma at a shallow depth where the temperature is too low to maintain a dry melt. Some unusual elements are concentrated into mineral deposits during the crystallization of intrusions, creating valuable economic resources, including pegmatites (a source for gemstones and rare elements) and metal ores.

Now, you should understand the formation of magma and igneous rocks. In the following chapters, you will learn how other rock types form and how sedimentary and metamorphic rocks are related to igneous rocks. In Chapter 5, you will learn that most minerals composing igneous rocks disintegrate or dissolve at Earth's surface. Igneous and other rocks are subjected to a series of surface and near-surface processes that recast their constituents as sediment and sedimentary rocks.

Active Art

Forming Igneous Features and Landforms. See how intrusions form and how they produce unique landforms when exposed by erosion.

Forming Volcanoes. See how different types of volcanoes form.

How Calderas Form. See how explosive eruptions form calderas.

Using Graphs to Understand Mantle Melting. See how animated and annotated graphs explain how magma forms in the mantle.

Understanding Tuttle and Bowen's Data. See how animated and annotated graphs explain Tuttle and Bowen's data.

Plate Tectonics and Magma Generation. See how magmas form at plate boundaries and hot spots.

Fractional Crystallization. See how magma composition changes during crystallization.

Density and Magma Movement. See how magma rises or stalls because of density contrast between magma and rock.

Extension Modules

Extension Module 4.1: Bowen's Reaction Series. Learn how early formed crystals react with a cooling, crystallizing magma.

Extension Module 4.2: Mitigating and Forecasting Volcanic Hazards. Learn more about the types of volcanic hazards.

Confirm Your Knowledge

1. How does magma differ from lava?
2. Geologists gain understanding of the formation of magma and igneous rocks by making field observations and performing laboratory experiments. Give an example of each.
3. What are the four basic types of magma and rock compositions, and how do they differ in their respective abundance of silica, iron, and magnesium?
4. What aspect of igneous rock formation is best illustrated by crystal grain size? Explain the relationship between rock formation and grain size.
5. How are pyroclastic deposits classified?
6. Which of the nine igneous rock types shown in Figure 4.3 were formed during last century's eruptions of Mount St. Helens and Mount Pinatubo?
7. List the three pieces of evidence for how we know that there is a magma chamber and the formation of plutonic rocks below an active volcano?
8. What factors determine the size and shape of a volcano?
9. Explain how peridotite can melt at a divergent plate boundary without an increase in temperature.
10. How do magmas form at convergent plate boundaries?
11. Explain how calderas form.
12. Define "partial melting" and explain how it affects magma composition.
13. What processes can form intermediate and felsic magmas from a melt that is originally mafic?
14. How does the type of volcanic eruption relate to gas content and magma viscosity?
15. Obsidian is a felsic rock, yet it has a black color. Explain.
16. How does fractional crystallization contribute to the formation of economically important metal ores?
17. Explain why some lava flows cover large areas and others do not.
18. Referring to Figure 4.3, identify the following igneous rocks based on their texture and mineral content:
    a. Fine-grained rock with 0% quartz, 0% potassium feldspar, 61% plagioclase feldspar, 13% biotite, 17% amphibole, 9% pyroxene, 0% olivine
    b. Fine-grained rock with 0% quartz, 0% potassium feldspar, 67% calcium-rich plagioclase feldspar, 0% biotite, 0% amphibole, 25% pyroxene, 8% olivine
    c. Coarse-grained rock with 0% quartz, 0% potassium feldspar, 52% calcium-rich plagioclase feldspar, 0% biotite, 0% amphibole, 32% pyroxene, 16% olivine
    d. Coarse-grained rock with 0% quartz, 0% potassium feldspar, 0% plagioclase feldspar, 0% biotite, 0% muscovite, 0% amphibole, 56% pyroxene, 44% olivine
    e. Fine-grained rock with 32% quartz, 26% potassium feldspar, 26% sodium-rich plagioclase feldspar, 5% biotite, 11% amphibole, 0% pyroxene, 0% olivine

Confirm Your Understanding

1. Write an answer for each question in the section headings.
2. What is a geothermal gradient? Given an average geothermal gradient for continents and a surface temperature of 15°C, what is the temperature at a depth of 10 kilometers?
3. What evidence is necessary to prove that batholiths form over a long period of time by multiple intrusions of differing composition?
4. Mafic magma at a temperature of 1200°C intrudes into felsic continental-crust rocks 15 km below the surface. Will the granite melt? If so, what will happen to the resulting magma? Use information presented in the chapter to support your answer.
5. What type of volcanic products (lava flows, lava domes, or pyroclastic material) and types of volcanoes would you expect from an eruption of basalt? of andesite? of dacite? of rhyolite? If there is more than one type of volcanic product, rank them from most abundant to least abundant.
6. Write a paragraph explaining the processes that create the various igneous rock compositions on Earth.
7. You are planning to move to a volcanic island and you want to choose the safest location. You have to decide between an island with a composite volcano consisting of dacite and andesite or an island with a shield volcano consisting of basaltic lava flows. Which island do you choose? Why?

The Formation of Sediment and Sedimentary Rocks

Why Study Sedimentary Rocks?

Why study sedimentary rocks and processes? These rocks provide information about Earth's history. Sediment results from the disintegration of rocks exposed to weather and is then redistributed on Earth's surface by water, wind, and glaciers. As sediment accumulates, it gets buried and transforms to hard rock. The photograph on the facing page shows sedimentary rocks that began more than 100 million years ago as desert sand dunes. How is it that geologists know that this area was once a desert and that these rock layers started out as sand dunes? Sedimentary rocks reveal a record of shifting seas, of changing climate, of the raising and eroding of mountains, and of life. These rocks preserve the remains of once-living organisms and record the emergence and evolution of life on Earth.

Sedimentary rocks also hold oil, natural gas, and coal, which are essential to society. Over time, buried biologic matter is deposited with sediment, and eventually, this matter chemically transforms into energy resources.

After Completing This Chapter, You Will Be Able to

- Describe how rocks break down by physical and chemical weathering to produce sediment, the raw materials of sedimentary rocks.

- Explain how loose sediment converts into hard sedimentary rock.

- Examine sedimentary rocks and interpret the materials from which the sediment was made and the environment in which the sediment was deposited.

- Explain the economic importance of sedimentary rocks, particularly as energy and industrial resources.

Pathway to Learning

5.1 How and Why Do Rocks Disintegrate to Form Sediment?

EXTENSION MODULE 5.2 *Why Is Seawater Salty?*

5.2 What Is the Link Between Weathering and Sediment?

EXTENSION MODULE 5.1 *Chemical Reactions and Chemical Equations*

EXTENSION MODULE 5.3 *Geochemistry of Calcite*

5.3 How Does Loose Sediment Become Sedimentary Rock?

Scenic sandstone in Paria Canyon, Arizona, formed when desert sand dunes blew across the area more than 150 million years ago.

IN THE FIELD

Our study of sedimentary rocks begins in the field. Imagine that you are hiking along the dry streambed shown in **Figure 5.1**a. With each step, your feet sink slightly into the pea-size gravel and dry sand, leaving slight impressions; you occasionally step over larger cobbles. Grass and shrubs grow sparsely along the banks, rooted in layers of sand and mud. Pausing to rest and examining a fist-sized stone from the streambed, you quickly recognize the coarse-grained mosaic of quartz, pink and white feldspars, and less abundant flakes of black biotite—minerals that together form the igneous rock granite, which you learned about in Chapter 4. You guess that flowing water must have transported the cobble from another location. You leave the stream channel and head up the nearby slope, looking for granite outcrops similar in appearance to the cobble.

A short distance above the stream, you discover a large granite outcrop (Figure 5.1b), with characteristic large crystals of quartz, feldspar, and biotite. This granite is crumbly and brown, however, unlike the solid, pinkish-gray granite samples you have seen elsewhere in the field and lab. Chipping with your rock hammer, you notice that mineral fragments readily fall off to join the similar loose fragments surrounding the base of the outcrop. A more forceful swing knocks loose a large fragment from the stronger rock that resides below the crumbly surface. You look carefully at this larger chunk (Figure 5.1c) and, indeed, you see that the hard rock has a mosaic of quartz, feldspar, and biotite crystals that resembles the granite specimens you are more accustomed to seeing.

So, what is the story with the brownish outer part of the rocky outcrop and the pile of rock crumbs at your feet? The fragile brown rock also is granite; an unsurprising observation, as there is no sharp boundary between the crumbly and hard materials to suggest that two different rock types compose the outcrop. The brown rock fragments (Figure 5.1d) clearly contain quartz and feldspar. Biotite also is present, but not as abundantly as in the hard granite. Yellow-brown stains surround the biotite crystals, and the cleavage surfaces, instead of being shiny black, flash brassy yellow reflections in the sunlight.

Grains of all three minerals crumble and fall to the ground as you handle the fragile brown rock. Stooping, you pick up a large handful of the loose mineral and rock particles littering the slope below the outcrop. These smaller particles, visible in Figure 5.1e, range in size from brown granite pebbles, a few centimeters across, to dust. Most particles are single crystals of quartz and feldspar, a millimeter or two across. Recognizing quartz and feldspar as the dominant components of the granite, you look through several handfuls in search of biotite. You pick out a few biotite flakes, but they are not as easy to find in the loose particles as they are in the granite. Your hands become increasingly stained with a dusty brown residue, which turns to sticky mud when you spare some drinking water to wash it off.

Returning to the streambed, you see the sediment deposited there during past floods in a new light. Now, you know that the sediment—not only the cobbles of recognizable granite, but also the sand grains composed mostly of quartz and feldspar—originated from the nearby outcrops of granite. The sediment on the stream bank includes the same brownish, muddy dirt that is mixed with the deteriorated granite on the hillside above, along with scattered flakes of brassy biotite.

This experience may only scratch the surface of your curiosity about sediment and sedimentary rocks. All sorts of other questions arise: How does the granite disintegrate? Why do the quartz and feldspar seemingly survive this process more readily than the biotite? What is the brown dust that turns to mud when moistened? How does loose sediment like that seen in the dry wash today become hard sedimentary rock in the future? Will some of the loose sediment eventually wash into larger rivers and ultimately travel to the sea? Will the sediment carried by larger rivers, found on a beach, or lying on the ocean floor differ in any significant way from the deposits in the dry streambed? Will leaves falling onto the sediment from the streamside plants also become part of the sedimentary rocks?

In this chapter, you will learn how geologists seek answers to these and other questions. The methods you will study in this chapter are broadly similar to those used to study igneous rocks in Chapter 4. These include field and microscopic examination of sediment and sedimentary rocks, chemical analyses of the minerals composing the rocks, and experiments designed to duplicate sedimentary processes.

▶ **Figure 5.1 Stages in the transformation of granite to sediment.** These field photos illustrate the transformation of hard granite to crumbly granite and then to loose mineral fragments. The sand and gravel along the stream consist of loose mineral fragments similar to those found at the base of the granite outcrop.

stream bank

dry stream bed

(a)

Sand and gravel are deposited on this dry stream bed during rare floods. Similar sediment, along with some mud, forms the eroded stream bank.

This crumbly granite forms outcrops uphill from the stream.

(b)

(c)

The interior of the granite outcrop consists of hard igneous rock, with light-colored feldspar and quartz, and dark biotite.

(d)

The outer part of the outcrop is crumbly granite, notably browner than the interior sample and with less biotite.

(e)

Fragments that accumulated at the base of the outcrop are almost entirely quartz and feldspar, coated in fine brown dust.

5.1 How and Why Do Rocks Disintegrate to Form Sediment?

By studying the disintegrating granite outcrop and the stream sediment, you observed the products of weathering. Weathering is the processes that break down preexisting rocks at Earth's surface, reducing them to loose particles, dissolving some minerals, and producing new minerals. Keep in mind that weathering is different from erosion. Erosion refers to the processes that pick up sediment particles, which is the first step in transporting the particles to locations where they ultimately come to rest to form a sedimentary deposit. Sedimentary rocks seen in the field today record the processes of weathering to produce the sediment; the processes of sediment erosion, transport, and deposition; and the processes that convert deposited sediment into hard rock. With so many processes to keep in mind, it is important from the outset to keep them separated. The best place to start is with the formation of sediment by weathering processes.

Weathering occurs when the geosphere interacts with the atmosphere, hydrosphere, and biosphere. Rocks weather through both physical and chemical means. The mechanical processes, called **physical weathering**, break large rocks into smaller fragments. The chemical processes, called **chemical weathering**, involve reactions among minerals and water and atmospheric gases that dissolve some minerals and produce new ones.

Physical Weathering

Natural rock outcrops do not have smooth, continuous faces. Cracks, such as those seen in **Figure 5.2**, always break the rock surface. These fractures form in many ways:

- In some igneous rocks, fractures and the separation of rock into blocks relate to the processes that formed the rock. Think of the cracks formed by cooling and breakage associated with flowing lava (look back to Figures 4.8 and 4.9).
- Plate tectonics stresses the crust and causes many fractures. You will learn more about these features in Chapter 11.
- Cracks open up parallel to the ground surface when rock expands slightly outward after overlying materials erode away (Figure 5.2b). This process of forming rock sheets parallel to the ground surface is called **exfoliation** (from the Latin *exfoliatus*, which describes stripping leaves or bark from trees).
- Bedding (layering) in sedimentary rocks causes planar breaks in rock (see the chapter-opening photo in Chapter 3).

No rock, therefore, is a continuous, unbroken mass of intergrown mineral crystals. Weathering processes at or near Earth's surface exploit the original breaks in rock to create more fractures while dislodging smaller and smaller fragments.

Figure 5.3a shows how water, when frozen in the cracks of rocks, becomes an effective agent of physical weathering. Water enters rock along existing fractures and then moves along boundaries between mineral grains or into open spaces, called **pores**, between minerals. When the liquid water freezes, it expands and acts like a wedge, exerting sufficient force to break rock and opening new cracks that, in turn, provide new openings for more water to enter after the ice thaws. In regions where the climate causes many freeze-thaw cycles during a single year, repeated freezing and thawing effectively break rocks apart.

Other weathering processes also take advantage of existing fractures and further break down rock. One example is the common nuisance of crumbling pavement that results when communities apply salt to melt ice during winter months. Freezing and thawing of water alone can contribute to this problem, but salting greatly speeds up the process. Salt dissolves in water, and the salty water percolates into cracks. When the water evaporates, salt crystals form in the cracks and break the rock apart as they grow. This same salt weathering process occurs along shorelines of oceans and salty lakes and when rainwater evaporates in deserts (Figure 5.3b).

A closer look at Figure 5.2a reveals trees growing in the rock; this is another process that exploits cracks. As roots grow to a larger diameter within an existing crack, they wedge the rock apart, which forms new cracks for additional root growth.

Expansion and contraction of minerals in rocks also cause rock disintegration. Some rocks contain clay, a mineral group with crystal structure similar to mica (see Chapter 2). These minerals generally form at or near Earth's surface, mostly by chemical weathering processes

Cracks are pathways for water to enter into rocks. Water causes physical weathering–when it repeatedly expands when freezing and contracts when thawing–and chemical weathering–when reactions occur between water and minerals. Physical weathering can also result from the growing roots of plants within fractures. Fractured granite dominates this view of the Sandia Mountains in New Mexico.

Enlarged view

(a)

Fractures can open parallel to the ground surface, as seen here in granite at Yosemite National Park, California. The fractures form when erosion removes the weight of overlying rock and permits the once buried rock to expand upward and outward.

◀ Figure 5.2 Natural exposures of rocks are always fractured.

(b)

Water enters rock along fractures and mineral-grain boundaries.

Water freezes forming ice, which forces the rock apart.

Repeated freezing of water to ice and thawing pries the rock apart.

(a) **Time**

Salty water enters rock along fractures and mineral-grain boundaries.

Water evaporates causing crystallization of soluble minerals, which forces the rock apart.

Repeated wetting and drying loosens mineral grains that separate from the original rock.

(b) **Time**

▲ **Figure 5.3 How physical weathering works.** (a) The most important physical weathering process is the freezing and thawing of water and ice that pries rocks apart along existing fractures and mineral-grain boundaries. Repeated freeze-and-thaw cycles break rock outcrops into angular fragments of various sizes, as seen here at Beartooth Pass, Montana. (b) Salt crystals that grow as water dries in cracks make rock disintegrate along the shorelines of the ocean and salty lakes and in arid deserts. The pitted sandstone surface along the California coast results from salt weathering.

ACTIVE ART

Physical Weathering. *See how the freezing and thawing of water and the evaporation of salty water pry rocks apart.*

that you will explore in upcoming paragraphs. Clay minerals incorporate water within their crystal structures. During cycles of wetting and drying, water molecules are added to and then drawn away from the clay grains, causing the mineral crystals to expand and contract. The repeated wetting and drying causes grains to separate from one another and substantially decreases the strength of the rock to the point where it eventually falls apart.

Chemical Weathering

The fact that rocks, such as the streamside granite in Figure 5.1, formed in the past under one set of conditions and deteriorate later under another set of conditions may make you wonder: What unique conditions at Earth's surface cause rocks to weather? Most igneous and metamorphic rocks (Chapter 3) form within Earth where oxygen gas (O_2) is nonexistent and water (H_2O) is a very minor constituent. Even lava flows that reach the surface solidify without including oxygen and water, because atmospheric gases and water do not incorporate into lava at low surface pressure.

Thus, the key to understanding chemical weathering lies in figuring out how minerals react with oxygen and water. The reactions that take place with oxygen and water are important not only in rock weathering at the surface, but also in the shallow subsurface where ground water interacts with minerals.

Dissolution reactions break apart mineral molecules, and these molecules then disperse in water. Many minerals readily dissolve in water, especially those with ionic bonds (Section 2.3).

EXTENSION MODULE 5.1

Chemical Reactions and Chemical Equations. Learn about chemical reactions and how to write chemical equations.

An example of dissolution is halite crystals (NaCl) dissolving in water. The solid crystals completely disappear and are replaced by Na^+ and Cl^- ions, which exist invisibly between the H_2O molecules (Figure 2.15). It is easy to determine this process has occurred if you add a tablespoon of salt to a glass of tap water and note the salty taste of the originally flavorless water. A simple chemical equation to represent this reaction is:

$$NaCl(s) + H_2O \rightleftharpoons Na^+(aq) + Cl^-(aq) + H_2O \qquad (5.1)$$
halite　　water　　　sodium ion　chloride ion　water

In a chemical equation, "(s)" indicates compounds that are solid (in this case, the mineral halite), while "(aq)" indicates which components are present in watery (aqueous) solution. Equation 5.1 states that solid halite reacts with water to produce sodium and chloride ions that are dissolved in water.

For a mineral such as halite, dissolution happens as soon as the mineral becomes wet, as seen in **Figure 5.4**a. For other minerals, such as calcite, dissolution is not instantly noticeable, but is detectable over a period of time, as illustrated in Figure 5.4b. Dissolution of still other minerals is even slower. Experiments and field observations show, however, that given enough time and the proper conditions of temperature and water chemistry, *all* the principal rock-forming minerals will dissolve in water.

More complicated **hydrolysis** reactions consume both the mineral and some of the water molecules, and reorganize the elements into new solid minerals as well as dissolved ions. An example of hydrolysis is the weathering of potassium feldspar to make kaolinite, a clay mineral (Figure 5.4c). The following chemical equation represents this reaction:

$$2\ KAlSi_3O_8(s)\ +\ 11\ H_2O\ \longleftrightarrow\ 2\ K^+(aq)\ +\ 2\ OH^-(aq)\ +\ 4\ H_4SiO_4(aq)\ +\ Al_2Si_2O_5(OH)_4(s) \qquad (5.2)$$

potassium feldspar   water   potassium ion   hydroxyl ion   silicic acid   kaolinite

Dissolution of halite (rock salt) happens quickly enough to witness in the laboratory or kitchen.

(a)

Calcite dissolves very slowly in mildly acidic water, but the dissolution is noticeable over long time periods. The letters on this gravestone were clearly defined when first engraved, but weathering dissolved the outer surface of the calcite-rich rock (limestone) so that no letters remain legible.

(b)

0.02 mm

Kaolinite

Potassium feldspar

(c)

Microscopic examination shows the results of hydrolysis reactions between feldspar and water. New crystals of the clay mineral kaolinite formed where the corroded potassium feldspar reacted with water.

◀ **Figure 5.4 Minerals undergo chemical reactions with water.**

Equation 5.2 states that potassium feldspar reacts with water to produce kaolinite, plus a solution of ions and compounds that include potassium and some of the silica originally present in the feldspar. The hydrogen and oxygen atoms in the original water molecules are redistributed among the solid kaolinite and the dissolved compounds.

The hydrolysis weathering reaction represented by Equation 5.2 has two important implications:

1. Chemical weathering does not just destroy minerals (e.g., potassium feldspar). It also forms new minerals (e.g., kaolinite). **Table 5.1** lists the products of the weathering of some common minerals. The dust particles you saw among the fragments of weathered granite (Figure 5.1e) above the streambed are made up mostly of clay minerals. The clay formed not only from weathering of feldspar but also from the weathering of the biotite (Table 5.1). This fact is supported by your observation that the amount of biotite in the weathered crumbly granite was diminished compared to the amount in the unweathered, solid granite (Figure 5.1d).

2. Although the reaction began with pure water, the result is an aqueous solution containing other dissolved components (e.g., K^+, OH^-, H_4SiO_4 in Equation 5.2).

No natural water is pure H_2O. Water always reacts with minerals in the natural world and, as a result, contains dissolved ions. Salty seawater contains the dissolved products of rock-weathering on land combined with ions generated by reactions of water with seafloor rocks and the gases rising from submerged volcanoes. Dissolved ions in water eventually bond to form new minerals that precipitate from the water. The presence of dissolved ions also can enhance the ability of the water to react with other minerals. This is especially true if reactions cause the water to become acidic, because acids enhance the weathering of most minerals.

EXTENSION MODULE 5.2

Why Is Seawater Salty? *Learn why some ions become concentrated in seawater and give it its salty taste.*

TABLE 5.1 Weathering Products of Some Common Rock-Forming Minerals

Original Mineral	Mineral Weathering Products	Ions in Solution
Halite (NaCl)		Na^+, Cl^-
Gypsum ($CaSO_4$ plus water)		Ca^{2+}, SO_4^{2-}
Calcite ($CaCO_3$)		Ca^{2+}, HCO_3^-
Quartz (SiO_2)		SiO_4^{4-}
Plagioclase feldspar (Ca, Na, Al silicate)	Clay	Ca^{2+}, Na^+, SiO_4^{4-}
Potassium feldspar (K, Al silicate)	Clay	K^+, SiO_4^{4-}
Olivine (Mg, Fe silicate)	Limonite, clay	Mg^{2+}, SiO_4^{4-}
Pyroxene (Ca, Mg, Fe silicate)	Limonite, clay	Ca^{2+}, Mg^{2+}, SiO_4^{4-}
Amphibole (Ca, Mg, Fe silicate)	Limonite, clay	K^+, Mg^{2+}, SiO_4^{4-}
Biotite (Fe, Mg, K, Al silicate)	Limonite, clay	K^+, Mg^{2+}, SiO_4^{4-}
Muscovite (K, Al silicate)	Clay	K^+, SiO_4^{4-}

The most abundant acid active in chemical weathering forms from the natural mixing of carbon dioxide (CO_2) and water. Carbon dioxide is present in the atmosphere and in soil. Water vapor in the atmosphere and water percolating through soil reacts with CO_2 to produce weak carbonic acid (H_2CO_3). The mixing reaction is written as:

$$\underset{\text{water}}{H_2O} \; + \; \underset{\text{carbon dioxide}}{CO_2\,(g)} \; \longleftrightarrow \; \underset{\text{carbonic acid}}{H_2CO_3\,(aq)} \qquad (5.3)$$

The "(g)" in the equation indicates that the carbon dioxide is a gas. The resulting aqueous carbonic acid solution is especially effective in dissolving carbonate minerals, such as calcite (Figure 5.4b). Other acids released from plant tissues (e.g., citric and ascorbic acids in fruit), and a variety of acids produced by bacteria (e.g., acetic acid, commonly known as vinegar) also contribute to the chemical weathering of minerals.

EXTENSION MODULE 5.3

Geochemistry of Calcite. Learn the factors that determine whether calcite dissolves or precipitates in water, which explains many features of rocks and landscapes.

Chemical reactions between minerals and atmospheric oxygen (O_2) also weather rocks. **Oxidation** is the process in which substances react with oxygen to form new substances by exchanging electrons. You observe the results of oxidation whenever you see rusted metal. Iron, for example, reacts with oxygen and oxidizes to form a new, less reactive, but much weaker compound, commonly known as rust. Recall that iron is one of the most abundant elements in Earth's crust (Figure 2.24) and is well represented in rock-forming minerals, such as biotite in granite. Plutonic and metamorphic rocks form below Earth's surface in the absence of oxygen gas, so they are especially prone to oxidation reactions on the surface.

During oxidation reactions, electrons transfer from a substance to the O_2 molecule, which causes transformation of O_2 into O^{2-} ions. The most common iron ion present in minerals that make up igneous and metamorphic rocks is Fe^{2+}. In the presence of O_2, the iron ion donates an electron to oxygen and becomes the Fe^{3+} ion. Fe^{3+} is smaller than Fe^{2+} and, clearly, has a different charge. These size and charge changes destroy mineral structures.

Minerals containing abundant Fe^{2+} (such as olivine, pyroxene, amphibole, biotite, and pyrite) readily weather, because the oxidation of Fe^{2+} breaks down the original crystal structure. Some ions released from the disintegrating minerals, including Fe^{3+}, combine with O^{2-} to form new minerals.

Oxidation weathering produces oxide and hydroxide minerals. Of these minerals, those containing iron are the most common. The iron hydroxides, loosely referred to as "limonite" (Table 5.1), form yellowish to brownish grains or stains, as seen in **Figure 5.5**. The brown hue of the weathered granite (Figure 5.1d, e) and soil reveals the presence of limonite produced when the iron within the biotite and hornblende oxidized. Another telltale indication that oxidation has occurred is the brassy color of the biotite flakes in the weathered granite. This coloration reveals that some of the biotite has been converted to the mineral vermiculite (also used in potting soil), which contains oxidized iron.

Hematite is the most common iron oxide mineral. Hematite forms during rock weathering and when oxygen-rich ground water reacts with sediment containing iron-bearing minerals. Hematite colors rocks red, even when present in very small amounts, as shown in Figure 5.5b and c.

What Minerals Survive Weathering?

Chemical weathering processes do not equally affect minerals. However, it is challenging to rank common minerals according to resistance to weathering, because the abundance of moisture and water chemistry, the factors that drive weathering reactions, varies widely from place to place. Nonetheless, some generalizations can be made. Minerals dominated by ionic bonds (e.g., halite, gypsum, calcite) most readily dissolve in water, especially if the water is acidic. Silicate minerals are generally more resistant to chemical weathering; those with a greater abundance of strong Si–O bonds (e.g., quartz and feldspar) weather more slowly than minerals with a greater abundance of ionic bonds, especially when those ions include easily oxidized Fe^{2+} (e.g., olivine, pyroxene, biotite, and amphibole). This explains why the quartz and feldspar particles remained and the biotite grains were sparse in the sediment that resulted from weathering of the granite at the field site. You might take a moment and consider whether an igneous rock with a composition other than granite—gabbro, for example—would weather more or less readily than granite. To complete this application of what you have learned about weathering, you may want to refresh your memory of the mineralogical differences between granite and gabbro shown in Figure 4.3.

Pyrite oxidizes to form iron hydroxide minerals called limonite. Notice that the typical cube shape of pyrite crystals is preserved in the limonite.

Many sedimentary rocks, such as these sandstones at Arches National Park, Utah, have red coloration from small quantities of hematite (iron oxide), and limonite (iron hydroxide).

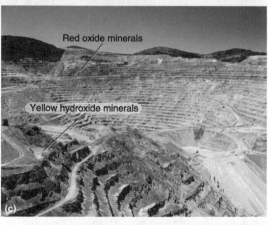

Iron-bearing sulfide minerals present in rocks near the top of this copper mine in New Mexico are weathered to oxide and hydroxide minerals by oxidizing ground water. The pale igneous rocks exposed deeper in the open pit are not oxidized. Geologists use the presence of oxidation minerals to locate metal ores for mining.

▲ **Figure 5.5 What the results of oxidation reactions look like.**

Putting It Together—How and Why Do Rocks Disintegrate to Form Sediment?

• Weathering is the interaction of the geosphere with the atmosphere, hydrosphere, and biosphere.

• Physical weathering disaggregates rocks by mechanical means. The most effective agent of physical weathering is the freezing and thawing of water. Other processes include salt weathering, cracks formed by temperature-induced expansion and contraction, and tree roots growing in rocks.

• Chemical weathering occurs when minerals react with water and oxygen. The two products of chemical weathering are ions dissolved in aqueous solutions and new minerals formed by hydrolysis and oxidation.

5.2 What Is the Link Between Weathering and Sediment?

In the rock cycle (Figure 3.12), weathering is the process that forms sediment, the raw material for sedimentary rocks. Weathering produces two products—solid minerals and dissolved ions. Each of these products relates to a different type of sediment.

Making Clastic Sediment

Clastic sediment is solid mineral particles that are left behind as rocks weather. These residual particles are the physically weathered parts of the original rock combined with minerals, such as clays, that form by chemical weathering. Weathered-rock residue that remains more or less where it forms is a principal component of soil, along with decaying organic matter and windblown dust. You will learn more about soil and the weathering processes that form it in the context of landscape evolution in Chapter 14.

Among common minerals, quartz is the most resistant to weathering because of its strong, mostly covalent bonds and lack of iron. Recall that the iron-rich biotite in your granite sample (Section 5.1) was the most readily weathered. If your sample is exposed to more time and water, all of the biotite will be destroyed, along with most or all of the feldspar. Only quartz, clay, and limonite will remain, and a large mass of ions will have been carried away in solution (Table 5.1). Quartz and clay are, in fact, the most abundant constituents of clastic sediment and sedimentary rock.

Currents of wind or water, and in some places the slow movement of glaciers, erode (pick up) clastic particles and transport them from where they formed. Water and wind currents exert sufficient force on the particles to roll or even pick up and suspend them. Thus begins an odyssey whereby sediment grains may ultimately be transported thousands of kilometers, mixed with sediment from countless other weathered rock outcrops, and ultimately deposited to form thick layers of sedimentary rock. The quartz, feldspar, and clay grains and larger fragments of granite you observed in your virtual trip to the streambed below the granite outcrop represent the beginning of that journey (Figure 5.1).

Making Chemical Sediment

Not all of a weathered rock is represented by clastic particles, however. Remember that chemical weathering also forms dissolved ions that move away in solution. Most dissolved ions eventually precipitate as solid ionic compounds called **chemical sediment**. Precipitation is the opposite of dissolution: Ions in solution bond to form mineral grains, such as halite and calcite, when water chemistry or temperature change. (You will learn more about the process of precipitation later in this chapter.) Most precipitation of chemical sediment takes place in lakes and oceans, which are the ultimate destinations for the dissolved weathering products of continents. In many cases, biochemical processes also cause mineral precipitation, such as the formation of shells and bones. Minerals also precipitate from the water in open pores between sediment grains; this process ultimately cements the grains together into coherent sedimentary rock (you will read more on this in Section 5.3).

Putting It Together—What Is the Link Between Weathering and Sediment?

• Weathering breaks down rock into the components that form sediment.

• Clastic sediment consists of the minerals and rock fragments remaining from physical weathering and newly formed mineral grains produced by chemical weathering.

• Chemical sediment consists of minerals precipitated from water. The ions composing these minerals are generated mostly by chemical weathering reactions.

5.3 How Does Loose Sediment Become Sedimentary Rock?

Sediment mostly is a mixture of residues from weathering and precipitated ionic compounds, which have been transported from the weathering site and deposited, usually by flowing water or blowing wind. For most igneous rocks, the original intergrowth of crystals forms a consolidated rock immediately. In a similar fashion, some chemical sediment forms as rock by the intergrowth of mineral crystals that precipitate from water. In contrast, clastic sediment consists of loose grains, and some chemical sediment particles also form independently of neighboring grains and are not intergrown with them. Let's examine how loose clastic or chemical sediment transform into rock.

The First Step in Forming Sedimentary Rock: Compaction

Lithification is the process that converts sediment into rock. The most common first step in lithification is **compaction**, illustrated in **Figure 5.6**. At the beginning of the chapter, when you walked along the dry streambed full of weathered granitic sediment, your feet sank into the sand, leaving footprints. This happened because the sand grains were inefficiently packed and had large open spaces between them. As you walked, your weight caused the grains to slide past one another and repack into a smaller space (Figure 5.6). This caused some fragile mineral grains (thin flakes of biotite or weathered feldspar, for example) to break. Natural sediment compaction takes place when additional layers of sediment accumulate and press down on the layers underneath them. The total volume of solid material remains the same during compaction, but the air- or water-filled pore space between grains decreases.

Does compaction of sediment grains cause them to stick together enough to form rocks? Usually, compaction simply decreases the volume of pore space and packs the grains closer together. Compacted sediment may not form a hard rock, but it can still become somewhat consolidated. Have you squeezed slightly moist sand in your hand and noticed the grains clumping into delicate clods? Add more water (or finer clay particles), and you can make more coherent clods, such as the ones you possibly threw at siblings or friends when you were younger. This consolidation results from weak electrical attractions both among the mineral grains themselves and between the mineral grains and pore water.

The stickiness of compacted sediment is explained by the observation that natural broken surfaces on minerals expose atoms with unbalanced charges that tend to attract oppositely charged atoms in adjacent grains. Clay minerals contain many stray charges on their outer surfaces, and the attraction between particles enhances the clumping of clay-rich sediment. The lopsidedness of water molecules (Figure 2.15) also creates weak electrical charges that help to hold grains together. The greater the compaction, the greater the mineral-surface contacts within the sediment and the greater the amount of mild electrical attraction between atoms in adjacent particles. Still, although the compacted sediment may not strictly be loose, it is a long way from being a hammer-ringing-hard rock.

The Second Step in Forming Sedimentary Rock: Cementation

To make hard rock from compacted sediment, the pore spaces need to fill, partly or completely, with the precipitated minerals. This is the **cementation** process illustrated in **Figure 5.7**. Mineral grains not only precipitate from water and accumulate as chemical sediment, but minerals also precipitate from water in the pore spaces between sediment particles and cement them together into rock.

Loosely packed sediment

Compacted sediment

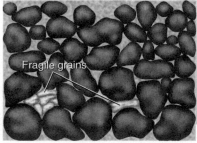

Fragile grains

Time

The weight of accumulating sediment provides the pressure to compact sedimentary layers.

During compaction, sediment grains rotate and repack closely together, which reduces the pore space between the grains.

Some fragile grains may break or squash between stronger grains.

Weak electrical forces at grain boundaries are more effective at holding compacted sediment together because the grains are in close contact and touch along larger surfaces in contrast to uncompacted sediment.

◀ Figure 5.6 How sediment compacts.

Time

▲ **Figure 5.7 How cementation works.** Loose sediment forms sedimentary rock when cementing minerals precipitate in the pore spaces between the sediment grains.

Mineral cements are not like glue. Cementing minerals are not sticky, and in only a few cases do they form chemical bonds with the sediment particles. Instead, cement minerals grow to fill pore spaces and surround the sediment particles. Even though the sediment grains and cement minerals are not chemically bonded, they are so intimately interlocked—like pieces in a complex three-dimensional puzzle—that neither the sediment grains nor the cement minerals can move.

The most common cementing agents are calcite, quartz, clay minerals, and hematite. The elements composing most of these minerals are available in solution as a result of chemical weathering (Table 5.1). Additional dissolved ions come from continued reactions between the pore water

and mineral grains in the sediment. For example, pore water may react with the feldspar grains in sand to form clay cement or with iron-silicate grains to form hematite cement.

Precipitation of mineral cements is caused partly by an increasing ion concentration in the water, but it also is strongly affected by temperature. Calcite, for example, dissolves more readily in cold water and precipitates in warmer water. So, pore water containing abundant calcium and carbonate ions could precipitate calcite as binding cement if it were to experience higher temperature. Recall from Chapter 4 that inside Earth, temperatures increase at depth (Section 4.4). This means that sediment is exposed to higher temperatures as it is buried deeper and deeper. As a result, calcite precipitates and forms cement readily at depth. Quartz, on the other hand, dissolves slightly in warmer water, but precipitates in cooler water. Consider warm pore water at depth that becomes increasingly enriched with dissolved silicic acid ions (SiO_4^{4-}; see equation 5.2) as it reacts with feldspar and other silicate grains in the sediment. If the silicon-rich water is squeezed upward as the sediment compacts, then it may cool and precipitate quartz as cement.

Compaction and cementation lithify unconsolidated sediment into sedimentary rock as pore spaces squeeze shut or fill with cement. It follows that the ability of fluid to move through the rock diminishes as pores get smaller and disappear. Ground water, oil, and natural gas move easily through loose sediment and slightly consolidated sedimentary rocks, but do not readily move through rocks, where most or all of the pore spaces are either closed off or filled with cement. Geologists determining the extent of fluid resources must, therefore, have knowledge of how lithification varies from place to place and at different depths within a sedimentary deposit.

Putting It Together—How Does Loose Sediment Become Sedimentary Rock?

• As sediment accumulates, older sediment compacts under the weight of overlying younger sediment. Compaction packs sediment grains closer together, facilitating particle clumping because of weak electrical attractions between mineral-grain surfaces.

• Strong, hard sedimentary rocks form when cementing minerals precipitate in the spaces between sediment grains. Calcite, quartz, clay minerals, and hematite are the most important mineral cements.

5.4 How Are Sedimentary Rocks Classified?

The classification scheme for sedimentary rocks incorporates the concepts we have just discussed: how sediment forms and how it lithifies into rock. It starts with the two most common types of sediment grains—clastic (weathered rock residue) and chemical (dissolved ions that have precipitated as solid ionic compounds).

Sediment accumulations at Earth's surface also incorporate biogenic (sometimes called organic) particles. For example, in some circumstances, animal shells and plant leaves form most of the sedimentary deposits, so these biogenic particles also are part of the classification scheme. Most biogenic sediment particles are chemically precipitated inorganic compounds (such as calcite seashells), making for an often fuzzy distinction between chemical and biogenic sediment. Lumping chemical and biogenic sediment into a single class allows us to avoid this problem. So, this book uses a two-part classification system whose primary categories are:

1. *Clastic sedimentary rocks* composed primarily of mineral grains remaining from or produced by the weathering of preexisting rocks and cemented by minerals that precipitated from pore water.
2. *Chemical and biogenic sedimentary rocks* composed of minerals that were precipitated from water or that are simply the remains of organisms.

As we have discussed previously, texture (grain size) and composition (minerals present) are easily observed features of rocks and thus used for classification and naming. These two characteristics are just as suitable for classifying sedimentary rocks as they are for classifying igneous rocks. Geologists emphasize texture when describing clastic sedimentary rocks and composition when describing chemical and biogenic sedimentary rocks.

Clastic Sedimentary Texture

The most important aspect of clastic-sediment texture, explained in **Figure 5.8**, is grain size. Weathering produces a wide range of sediment grain sizes, from tiny sand particles to large boulders more than 1 meter across. However, the vast majority of sedimentary rocks consist of particles that are small enough to be eroded and then transported by water or wind to a site of deposition. These fluid currents exert forces on the sediment grains to transport them. Thus, the forces required to move particles are correspondingly weaker or stronger: Stronger forces, such as fast-flowing water or strongly blowing wind, are necessary to move large grains, while only the slightest current or gentle breeze may move tiny dust particles. A fast-moving flood in a stream may, for example, move grains ranging in size from dust to large boulders. As a flood current slows, the large boulders come to rest first, while sufficient force remains to move the smaller pebbles, sand, and mud farther downstream. As the current power further decreases, successively smaller particles come to rest at greater and greater distances downstream. The result is that any given sedimentary deposit usually consists of a relatively narrow range of grain sizes.

Sorting describes this process that arranges sediment according to grain size, with well-sorted specimens containing mostly one grain size and poorly sorted sediment containing a wide range of grain sizes (Figure 5.8). The more steady the transporting current and the more frequent the events that pick up and move the sediment grains, the more sorted the resulting deposit.

Particle **rounding** also plays a significant role in describing clastic rock texture. Most clastic grains have sharp edges and corners when initially produced by weathering and are described as angular. However, these edges and corners abrade, or are smoothed away, when the particles

collide with each other during transport by wind and water. As a result, sediment grains become progressively more rounded with increasing distance and frequency of transport, as depicted in Figure 5.8.

The classification of clastic sediment and sedimentary rocks, illustrated in **Figure 5.9**, is based on three grain-size categories—gravel, sand, and mud. The mud-sized category is further broken down into silt and clay subcategories. You probably already use the terms in Figure 5.9 to refer to the relative sizes of sedimentary particles, with gravel being largest, or coarsest, and mud being smallest, or finest. Notice in Figure 5.9, however, that geologists attribute specific particle sizes to each of the categories and to subcategories within them. Figure 5.9 shows the different types of

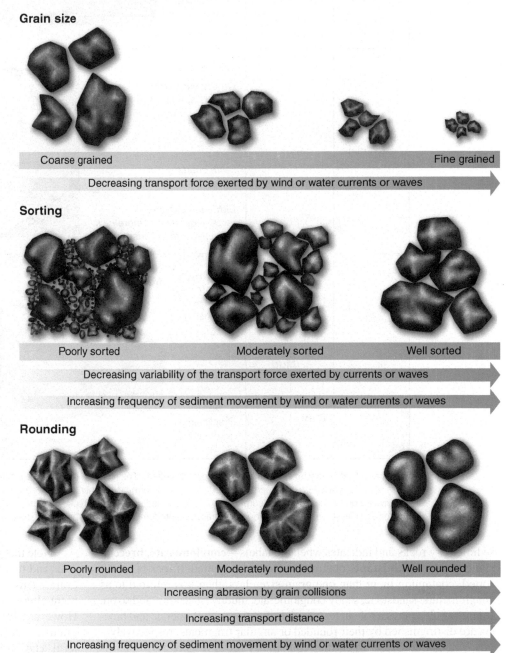

Grain size

Coarse grained   Fine grained

Decreasing transport force exerted by wind or water currents or waves

Sorting

Poorly sorted   Moderately sorted   Well sorted

Decreasing variability of the transport force exerted by currents or waves

Increasing frequency of sediment movement by wind or water currents or waves

Rounding

Poorly rounded   Moderately rounded   Well rounded

Increasing abrasion by grain collisions

Increasing transport distance

Increasing frequency of sediment movement by wind or water currents or waves

▲ **Figure 5.8 The texture of clastic sediment.** Grain size, sorting (the range of grain sizes present), and the degree to which grains are rounded describe the texture of clastic sediment. Sediment deposited by flash floods in the vicinity of a steep mountain stream likely will be coarse grained, poorly sorted, and poorly rounded. In contrast, sediment deposited on a beach will be finer grained, better sorted, and better rounded.

Grain size	Sediment name	Rock name
Larger than 2 mm	**Gravel** (pebble, cobble, boulder)	**Breccia** (if fragments are angular) **Conglomerate** (if fragments are rounded)
$\frac{1}{16}$ to 2mm	**Sand**	**Sandstone** *Quartz sandstone:* > 95% quartz grains *Arkose:* > 25% feldspar grains with quartz *Lithic sandstone:* < 90% quartz and more rock fragments than feldspar
$\frac{1}{16}$ to $\frac{1}{256}$ mm — **Silt** **Mud** Smaller than $\frac{1}{256}$ mm — **Clay**		**Mudstone, Claystone, Siltstone** (if blocky) **Shale** (if splits into sheets)

▲ **Figure 5.9 How clastic sedimentary rocks are classified.** The first step in naming a clastic sedimentary rock is to determine the dominant grain size. For gravel-size sediment, the next step is to determine whether the grains are rounded or angular. Sandstones are further subdivided on the basis of grain composition. Fine-grained sedimentary rocks can simply be called mudstone. (They also can be called siltstone or claystone, depending on what grain size predominates, or can be called shale if the rock shows a tendency to split into thin sheets.)

sedimentary rocks and indicates when the labels—**conglomerate**, **breccia**, **sandstone**, and **mudstone** (or **shale**)—should be applied. If rock is poorly sorted, containing more than one grain-size class, then it is referred to as conglomeratic sandstone, sandy conglomerate, muddy sandstone, and so on. Rounding also plays a role in texture classification; conglomerate and breccia are distinguished by their rounded or angular fragments, respectively.

You can now refer to clastic sedimentary rocks (Figure 5.9) in terms of the three ingredients of sedimentary-rock texture:

1. The dominant grain size
2. The range of grain size (degree of sorting)
3. The extent to which the grains are rounded

Note that geologists use the term "clay" to refer both to a group of minerals and to a grain size (Figure 5.9), regardless of the composition of the grains. This sounds confusing, but fortunately, clay-mineral grains in sediment almost always are also clay size, which diminishes some confusion. However, although most clay mineral grains are clay size, not all clay-size grains are clay minerals. Therefore, it is important to understand whether the term "clay" is being used to describe the composition or texture of sediment.

Clastic Sedimentary Rock Composition

Composition plays a secondary role to texture in the classification of clastic sedimentary rocks, and it is most commonly applied to sandstones

(Figure 5.9). **Quartz sandstone** consists almost entirely of quartz. Quartz is abundant in sedimentary rocks because it is the common rock-forming mineral most resistant to chemical weathering. Quartz also is hard and difficult to break, because it lacks cleavage, which makes it well suited to surviving long-distance transport by flowing water or blowing wind. **Arkose** is sandstone containing at least 25 percent feldspar. The remaining 75 percent of arkose is mostly quartz, usually with minor amounts of mica and other silicate minerals. **Lithic sandstone** consists of sand-size rock fragments. Lithic sands form by weathering of very fine-grained rocks, such as basalt. When these fine-grained rocks weather, the resulting individual sand grains are considered to be miniature rocks because they contain many mineral crystals.

The composition of clastic sedimentary rocks reveals the types of rocks that weathered to produce the sediment and the amount of weathering that occurred. For example, both quartz sandstone and arkose can result from weathering and erosion of granite, which also contains quartz and feldspar. A greater amount of weathering or breakage of cleavable feldspar during transport, might account for the greater quartz content of quartz sandstone than of arkose.

Chemical and Biogenic Sedimentary Rock Composition

While composition plays a secondary role in classifying clastic sedimentary rocks, it plays a primary role in classifying and naming the chemical and biogenic sedimentary rocks. It is very unusual for the conditions to be just right for two or more minerals to precipitate from water simultaneously, so most chemical sedimentary rocks consist overwhelmingly of only one mineral, as depicted in **Figure 5.10.**

Limestone, composed mainly of calcite, and **chert**, composed mostly of quartz, are the most abundant rocks formed by chemical precipitation from water. These common rocks form by both inorganic and biologic processes. The chemical components of calcite and quartz precipitate from water as a result of inorganic chemical reactions that occur when water chemistry or temperature change slightly. Many invertebrate animals, protozoa, and algae secrete shells and other hard tissues composed of calcite. Other protozoa and algae, such as diatoms, and some invertebrates, including sponges, secrete hard tissues composed of silica. Most limestone and chert form primarily by these biologic precipitation processes. Additional inorganic precipitation of calcite or quartz as cement occurs after the biogenic sediment is buried, ultimately forming limestone or chert, depending on composition.

Dolostone is a chemical sedimentary rock composed of the calcium-magnesium carbonate mineral dolomite. No organisms secrete shells of dolomite, and dolomite precipitates from seawater only under conditions of unusual water chemistry in some nearshore lagoons. Nonetheless,

dolostone is abundant among ancient sedimentary rocks although notably sparse in more modern ones. How does it form? Most dolostone forms long after the original sediment has been deposited, when ground water chemically reacts with calcite in limestone to produce dolomite. The Mg^{2+} ion in the dolomite is smaller than the Ca^{2+} ion in the calcite, resulting in

Composition	Rock name
Calcite [CaCO$_3$]	Limestone
Dolomite [CaMg(CO$_3$)$_2$]	Dolostone
Quartz (microscopic) [SiO$_2$]	Chert
Halite [NaCl]	Rock Salt — Evaporite
Gypsum [CaSO$_4$•H$_2$O]	Rock Gypsum
Plant organic matter	Coal

▲ **Figure 5.10 How chemical and biogenic sedimentary rocks are classified.** To name a chemical or biogenic sedimentary rock, it is necessary to know the composition of the sediment, because a single, dominant component forms each rock.

tighter packing of atoms in the dolomite crystal structure compared to calcite. This means that when calcite reacts with water to produce dolomite, the resulting mass of dolostone occupies a smaller volume than its limestone predecessor, leaving behind open pores that may fill with oil or natural gas.

Other minerals, such as halite and gypsum, are very soluble in water. This means that the water must undergo considerable evaporation before the appropriate ions can bond to make these minerals. Rock salt and rock gypsum are, therefore, commonly referred to as **evaporites** to emphasize the fact that they form in arid lagoons or drying desert lakes, where extensive evaporation allows precipitation of halite and gypsum.

The remaining rock in this category is coal, which consists entirely of plant organic matter compressed into rock under elevated heat and pressure during burial. Coal geologists apply sophisticated classifications of coal based on the degree of transformation of the original plant tissue into new compounds. The greater the extent of transformation, the greater the combustibility or heat production from the coal when it burns. The term **peat** refers to plant material before it transforms into the lightweight, black rock that you normally think of as **coal**. Most coal is relatively soft and is specifically called **bituminous coal**. If metamorphosed, coal transforms to **anthracite**. Anthracite is more accurately classified as a metamorphic rock than as a sedimentary rock, although it is still commonly called coal. Purity also determines the quality of coal as a fuel. If the plant material accumulated in the nearly complete absence of other sediment, then a coal sample will burn almost completely and leave little waste material (called fly ash). If substantial clay and silt washed into the coal swamp, then the quality of the resulting coal is compromised.

The rock names you encounter in Figures 5.9 and 5.10 are referred to throughout the rest of Chapter 5, so please take a moment to familiarize yourself with their meanings and usage. Doing so will help you to better navigate and understand the content of this chapter. As you spend time with these figures, you will find that you can make a number of observations and generalizations about the classifications of sedimentary rocks that will serve you well in the following discussions about sedimentary process.

Putting It Together—How Are Sedimentary Rocks Classified?

- Origin, texture, and composition of sediment grains form the basis for sedimentary rock classification.

- Clastic rocks are classified primarily according to sediment texture, especially grain size, and secondarily according to composition.

- Clastic sediment texture relates to the strength and steadiness of currents transporting the sediment and the distance of transport.

- Clastic sediment composition reveals the source rock that weathered to create the sediment and the extent of weathering.

- Chemical and biogenic sedimentary rocks are classified primarily according to dominant composition and consist of sediment produced by inorganic chemical precipitation, biologic processes, or both. Some rocks, such as limestone, form by both inorganic and biologic chemical processes.

5.5 Why Are Fossils and Fossil Fuels Found in Sedimentary Rocks?

Sediment forms and accumulates at the surface of the planet. Organisms live and die in the same environments where sediment forms and accumulates. As a result, an archive of the biosphere is sometimes preserved within sedimentary rocks as they form.

Fossilized Organisms

No doubt you have long been aware of fossils; dinosaurs and their remains may have even been one of your childhood obsessions. **Fossils** are the remains of organisms from the prehistoric geologic past. There are three basic ways for fossils to form:

1. As the fossilized remains of shelled organisms and of vertebrate teeth and bone, shown in **Figure 5.11**a and b, which consist of the original mineral matter secreted by the organisms.

2. From minerals that replace the original carbon-rich organic compounds buried in sediment. These minerals precipitate from ions dissolved in ground water. As they precipitate, the new-formed minerals conform to and preserve the original form of the fossilized organism. Calcite and quartz are the most common minerals that "petrify" organic remains (Figure 5.11c).

3. As an impression in the sediment that remains where some or all of the original organic material was completely destroyed as a result of organic decay or mineral dissolution, as illustrated in Figure 5.11d.

Very few organisms become fossils, because weathering destroys soft tissue and even the hard minerals that form shell, bone, and teeth. Nonetheless, sufficient numbers of organisms become fossils to provide insights into the history of life on Earth. Fossils also provide clues for geologists about the environment that existed when the surrounding sediment was deposited. Some volcanic rocks, especially those formed from rapidly accumulating ash layers, also contain fossils, but sedimentary rocks are the primary hosts for fossils because they are the most abundant rocks formed at Earth's surface.

Fossil Fuels

Coal, oil, and natural gas, the combustible energy sources called **fossil fuels**, originate from the organic matter deposited with clastic, chemical, and biogenic sediment millions to hundreds of millions of years ago. The energy in fossil fuels is a type of stored energy. This energy came from the Sun and is utilized by plants during photosynthesis, a chemical reaction whereby plants (and a few microscopic organisms) convert the Sun's energy into carbon-rich organic molecules and oxygen. Photosynthesis allows plants and some microscopic organisms to produce organic molecules full of energy. Typically, non-photosynthetic organisms (like us) consume photosynthesizers and utilize this converted solar energy to fuel our daily activities.

When organic molecules are buried in sedimentary rock, the stored energy is still in them. That energy is liberated as heat when the organic molecules—in the form of fossil fuels—burn. The liberated heat can, in turn, be converted to motion energy that rotates turbines to generate electricity or drives pistons inside vehicle engines. Every aspect of modern living involves the use of fossil fuels, and humans use an immense amount of energy, but this consumption totals a surprisingly small amount of Earth's

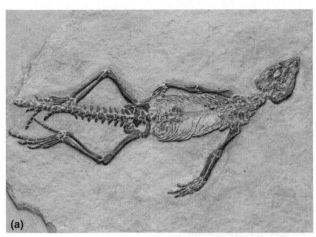

Bones are hard mineral matter commonly preserved as fossils, as we see in this fossil reptile skeleton. In other cases, minerals precipitate from ground water and replace the bone minerals to form fossils.

(a)

Shells, such as these fossil clams from California, consist of hard minerals secreted by invertebrate animals.

(b)

This petrified wood in eastern Arizona formed when quartz and hematite replaced the original woody tissue.

(c)

Soft-bodied animals and plants such as this fossil fern, are preserved as coaly organic films and impressions between layers of sedimentary rock.

◄ Figure 5.11 What fossils look like.

(d)

energy budget. All the known fossil-fuel reserves on Earth add up to an amount of energy equivalent to the energy delivered to the planet from the Sun in just 10 days.

Organic matter becomes buried in sediment under only certain circumstances. Usually, when an organism dies, its carbon-rich organic tissue is consumed by decay or reacts with oxygen to form carbon dioxide and other gases before the organism can be buried. However, in some deep lakes and ocean basins, swamps, and nearshore lagoons, the oxygen content of the water is very low, because it was previously used up in the chemical reactions that oxidize organic matter and Fe^{2+}-bearing minerals. In these situations, organic matter cannot be oxidized and, instead, is buried along with inorganic sediment. So, sediment accumulates on top of the organism, increasing the pressure and temperature in the organic-rich sediment, which promotes chemical reactions that change the organic molecules into new carbon- and hydrogen-rich compounds—these compounds are fossil fuels.

Coal, introduced in Figure 5.10, is a sedimentary rock composed almost entirely of the compacted remains of fossil plants, such as that shown in **Figure 5.12**a. Oil and natural gas are organic compounds produced when molecules that are mostly the remains of plants and organic aquatic microorganisms are altered by exposure to high temperatures in deeply buried sedimentary layers. These important fluid energy resources accumulate in the pore spaces within sediment and sedimentary rock (Figure 5.12b).

To locate precious fossil fuels, geologists must apply their knowledge of sedimentary rocks. The scientists need to know where, in the geologic past, organic matter probably was not destroyed by oxidation and is likely to be buried with sediment. They also need to know where to find the rocks with abundant, interconnected pore spaces through which oil and gas move and accumulate in sufficient volume to make extraction economically viable. This combination of appropriate depositional conditions and rock types to host the fossil fuels is rare. As a result, vast fossil-fuel resources exist in some regions and countries, such as oil in the Middle East and coal in the United States and China, whereas these resources are virtually nonexistent in other regions. **Figure 5.13** indicates the amount of fossil-fuel energy that is produced by and consumed in various countries. The need of high-consuming countries to import fossil fuels from high-producing countries is a driving force of the global economy.

We treat fossil fuels as nonrenewable resources because it takes an extremely long time for buried organic matter to convert into coal, oil, and gas. Logged forests regrow in decades, but fossil fuels take millions of years to form. For example, coal mined in the Appalachian region of the eastern United States originated as deposits of coastal swamps about 300 million years ago.

EXTENSION MODULE 5.4

When Will We Run Out of Oil? Learn the facts about oil reserves and consumption that determine the uncertain future of this important energy resource.

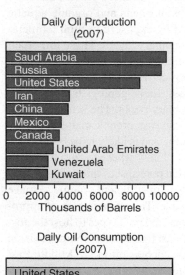

◀ **Figure 5.12 Fossil fuels come from sedimentary rocks.**

(a) Coal consists of compressed plant remains. Microscopic examination (inset) reveals the original cell structure of the plant matter.

(b) Oil and natural gas are organic compounds formed from the remains of microscopic aquatic organisms. The gray color of this sandstone in northern Alaska is caused by crude oil within the pore spaces between the sediment grains.

Daily Oil Production
(2007)

Saudi Arabia	
Russia	
United States	
Iran	
China	
Mexico	
Canada	
United Arab Emirates	
Venezuela	
Kuwait	

0 2000 4000 6000 8000 10000
Thousands of Barrels

Annual Natural Gas Production
(2007)

| Russia |
| United States |
| Canada |
| Iran |
| Algeria |
| Norway |
| United Kingdom |
| Netherlands |
| Indonesia |
| Saudi Arabia |

0   5   10   15   20   25
Trillion Cubic Feet

Annual Coal Production
(2006)

| China |
| United States |
| India |
| Australia |
| Russia |
| South Africa |
| Germany |
| Indonesia |
| Poland |
| Kazakhstan |

0 500 1000 1500 2000 2500
Million Tons

◀ **Figure 5.13 Who produces and consumes the most fossil fuels?** The bar graphs illustrate the top 10 producers and consumers of fossil fuels. The United States is a global leader in oil, natural gas, and coal production, but the country consumes more oil and natural gas than it produces.

Daily Oil Consumption
(2007)

| United States |
| China |
| Japan |
| Russia |
| India |
| Germany |
| Canada |
| Brazil |
| Korea, South |
| Mexico |

0   5000 10000 15000 20000
Thousands of Barrels

Annual Natural Gas Consumption
(2006)

| United States |
| Russia |
| Iran |
| Germany |
| Canada |
| United Kingdom |
| Japan |
| Ukraine |
| Italy |
| Saudi Arabia |

0   5   10   15   20   25
Trillion Cubic Feet

Annual Coal Consumption
(2006)

| China |
| United States |
| India |
| Germany |
| Russia |
| Japan |
| South Africa |
| Poland |
| Australia |
| Korea, South |

0 500 1000 1500 2000 2500
Million Tons

Putting It Together—Why Are Fossils Found in Sedimentary Rocks?

• Many sedimentary rocks also contain biologic remains called fossils.

• Living organisms abound in the same environments where sediment accumulates, and their remains are buried with the sediment.

Geologists interpret ancient environments with the help of biologic remains preserved as fossils.

• Organic matter buried with sediment and sedimentary rock changes at elevated temperature and pressure to form combustible organic compounds that yield large amounts of energy when burned. These organic materials are the fossil fuels—coal, oil, and natural gas.

5.6 How Do Sedimentary Rocks Reveal Ancient Environments?

Sediment can end up at a streambed, a beach, a desert sand dune, the seafloor, or any number of Earth-surface environments where sediment comes to rest or precipitates from water. Geologists study the features of rocks formed from this sediment to determine the environmental conditions on Earth's surface where and when the sediment accumulated. In effect, sedimentary rocks contain a "memory" of where deposition occurred, so it is possible for geologists to reconstruct ancient geography by mapping the extent of different recorded environments. Ancient geography is fascinating, indicating the past presence of mountains, shorelines, coral reefs, and other surface features, commonly in locations that feature dramatically different geography today.

This environmental and geographic information can help geologists do more than just imagine how places looked in the distant geologic past, however. Ancient geography has economic relevance when it allows geologists to locate environments that were conducive to fossil-fuel production and storage. For example, some environments favored the accumulation of organic material that transformed into coal. Other environments contained the type of sediment that eventually formed pore spaces, which may have filled with ground water, oil, or natural gas resources.

Interpreting ancient depositional environments is an example of the principle of uniformitarianism, introduced in Chapter 1. The principle states if you can understand the geologic processes responsible for materials and features you see in nature or in the laboratory today, then you can infer that similar materials and features found in ancient rocks are the result of these same processes.

What Fossils Reveal

Fossils (Figure 5.11) provide the most easily interpreted record of past environments. Organisms adapt to living in particular conditions: land versus water, shallow lagoons versus deep ocean, dry deserts versus tropical rainforests, and so forth. By identifying the environmental requirements for the organisms preserved as fossils, paleontologists—scientists who study fossils—can infer key information about the ancient depositional environment.

What Rock Types Reveal

To decipher details about past depositional environments from sedimentary rocks, geologists begin by figuring out where certain types of sediment accumulate today. **Figure 5.14** illustrates examples of how modern

◄ **Figure 5.14 Using modern environments to interpret ancient environments.** These illustration pairs are examples of how geologists interpret the ancient depositional environments recorded in sedimentary rocks by comparing them to where sediments of similar texture and composition are observed to accumulate today.

The photo on the left shows very coarse, angular, poorly sorted gravel that accumulates during flash floods at the base of steep present-day mountains in Death Valley, California. The photo on the right shows a coarse, angular, poorly sorted breccia and conglomerate that geologists interpret to have formed by flash floods near steep mountains in Death Valley, millions of years ago.

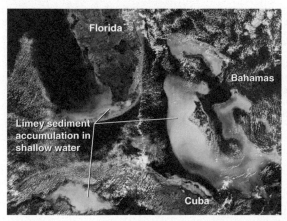

The satellite image on the left shows light-colored areas where limey sediment accumulates in shallow, tropically warm water near Florida, the Bahamas, and Cuba. Geologists interpret the limestone layers on the New Mexico mountainside in the right photo to have formed in a similar warm, shallow sea hundreds of millions of years ago.

The horizontal rock layers in this view of the Grand Canyon, Arizona, are large-scale sedimentary beds. Changes in depositional environment through time caused deposition of different sediment types in successive beds, one above the other. Hard sandstone and limestone form cliffs because these rocks are more resistant to weathering than the slope-forming mudstone and shale.

Small-scale bedding, as seen here in old lake beds in Washington (with a pocket knife for scale), commonly forms by changes in depositional processes or the type of sediment deposited. Each pair of these dark and light beds represents a single year of deposits in a glacier-fed lake. Light-colored silt settled during spring and summer snow-melt runoff. Dark organic matter settled to the bottom during the winter when the lake surface froze. Can you determine how many years were required to deposit the sediment visible in the photo?

▲ Figure 5.15 **What bedding looks like.**

the sediment becomes sedimentary rock. Sedimentologists can observe sedimentary structures forming in modern environments or in simulated laboratory environments and apply the knowledge they gain to reconstruct ancient environments.

Figure 5.15 illustrates a fundamental sedimentary structure, called **bedding** or **stratification**, which simply refers to the layering that always exists in sedimentary rocks. Bedding reflects changes in depositional processes. Some bedding planes represent pauses in deposition, or even erosion, whereas others represent changes in the sediment-depositing processes. Bedding shows that accumulation of sediment in any environment is not a monotonously continuous process. The energy of moving currents and waves varies with time, sometimes eroding sediment and sometimes depositing it. Recall from Section 5.4 that fluctuation in the strength of the currents that deposit sediment also causes variations in the grain size and sorting of deposited sediment (Figure 5.8). Some depositional environments may receive clastic sediment only during powerful river floods or strong oceanic storms, or the volume and composition of sediment may vary with the seasons. All these variations produce separate beds in the resulting rock.

Figure 5.16 illustrates the sedimentary structures referred to as **mud cracks**. You probably have seen mud cracks; they form wherever wet mud dries up. Muddy sediment usually contains abundant clay minerals that absorb water and swell when wet, but then lose water, shrink, and cause the sediment to crack while drying. If sediment washes or blows into open cracks, then the polygon shapes of the cracked ground are preserved. The presence of mud cracks in sedimentary rocks

sedimentary rock composition and texture allow us to figure out where ancient sediments formed. Coarse-grained clastic rocks, such as conglomerate, form in environments with strong transporting currents, whereas fine-grained mudstone requires still water in which fine silt and clay settle slowly from suspension. Limestone reveals the past presence of submerged lake bottoms or seafloor where limey (calcite-rich) sediment accumulated. Evaporites indicate arid lakes and shoreline lagoons where water evaporated and soluble ions precipitated as minerals.

What Sedimentary Structures Reveal

The physical features of modern sediment relate to the processes that deposit it. By extension, if we see similar features in ancient rocks, we can assume that the same processes occurred in the past to form the rock. These physical features are called **sedimentary structures**. Sedimentary structures form during sediment deposition or shortly after deposition before

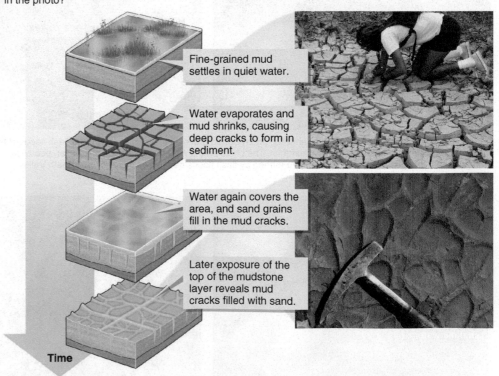

Fine-grained mud settles in quiet water.

Water evaporates and mud shrinks, causing deep cracks to form in sediment.

Water again covers the area, and sand grains fill in the mud cracks.

Later exposure of the top of the mudstone layer reveals mud cracks filled with sand.

Time

▲ Figure 5.16 **What mud cracks reveal about past environments.** This illustration shows how mud shrinks and cracks when it dries. Mud cracks, commonly filled in with sand, are preserved in some sedimentary rocks. The appearance of mud cracks indicates an environment that was alternately wet and dry.

implies a depositional environment exposed to the air where sediment occasionally dries, like a river floodplain, rather than an environment always submerged in water, like the seafloor.

Figure 5.17 illustrates **cross-beds**, which are inclined layers in sediment or sedimentary rock that reveal current or wave transport of sediment. Cross-beds form by the movement of sediment **dunes** and **ripples**. Dunes and ripples are curving ridges of loose sediment that move along with water or wind currents, or move back and forth beneath oscillating water waves. The detailed physics of dune and ripple formation differ, but generally, you can think of ripples as small features, less than

ACTIVE ART

Forming Cross-beds. See how cross-beds form and how sand dunes migrate through time.

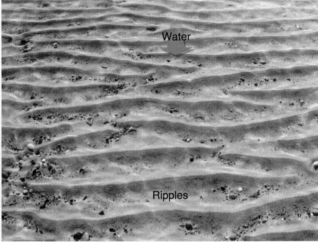

Cross-bedding forms by movement of dunes and ripples, which are ridges of loose sediment shaped by wind or water currents.

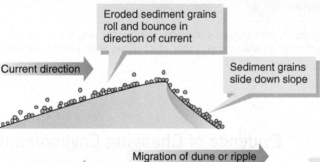

Eroded sediment grains roll and bounce in direction of current

Current direction

Sediment grains slide down slope

Sediment grains move with the current to accumulate on the down-current side of the dune or ripple. Successive positions of this down-current inclined slope form cross-beds as the dune or ripple migrates in the direction of the current.

Migration of dune or ripple

Current direction

Cross-beds mark shifting position of down-current side of dune

Current

Cross-bedding

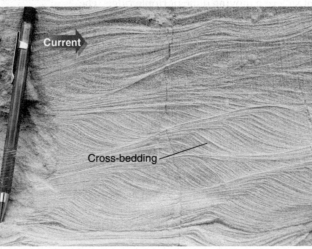

Current

Cross-bedding

Large-scale dune (left) and small-scale ripple (right) cross-bedding appear in ancient sedimentary deposits. The cross-beds incline downward in the direction of the transporting current.

◄ Figure 5.17 How cross-bedding forms.

Sediment grains of many sizes suspend in a rapidly moving current.

Grains settle as current slows down; larger grains settle faster than small grains.

A graded bed has larger grains at the bottom and smaller grains at the top.

Graded bed of gravel and sand deposited by a flood.

It is easy to make graded beds in the laboratory by stirring a sand and mud mixture in water and letting the sediment grains settle.

▲ **Figure 5.18 How graded bedding forms.** Graded beds are recognized by concentrations of coarser grains at the base of the bed and finer grains at the top (for the most part). This sorting of grains by size occurs when currents slow, which allows larger grains to settle to the resulting deposit before smaller grains accumulate.

3 centimeters high, and dunes as similar in shape but larger, ranging in height from several centimeters to many meters. Although the term "dune" usually conjures up images of windblown sand dunes, flowing water also forms submerged dunes, and both wind and water generate ripples (Figure 5.17).

Figure 5.16 illustrates how cross-beds form when sediment accumulates on the steep, down-current side of a moving dune or ripple ridge. The recognition of cross-beds in sedimentary rock not only indicates an environment where water or wind move across the surface, but also reveals the direction the currents were traveling.

Figure 5.18 illustrates the formation of **graded beds**, in which sediment grain size uniformly changes from coarser at the base of a bed to finer at the top. Graded beds reveal strong currents that transport a wide mixture of sediment grain sizes and then slow, allowing large particles to settle first and smaller particles to accumulate progressively on top of the larger ones. Graded beds commonly result from rapid sediment deposition during river floods, the settling of sediment stirred up by big storm waves on the seafloor, or in the wake of episodic currents of sediment-laden water, called **turbidity currents**, that sweep across lake bottoms and the seafloor. (You will learn more about turbidity currents in Section 5.7.)

Evidence of Changing Environments

Vertical transitions from one sedimentary rock type to another represent a change in the depositional environment. **Figure 5.19**, for example, illustrates an upward progression from sandstone, to shale, to limestone. All these layers contain fossils of marine organisms. How did the marine environment change through time in order to cause deposition of sandstone, followed by deposition of shale, and then precipitation of limestone?

Figure 5.20 shows how geologists interpret this progression of rocks by thinking about where these rock types typically form in relationship to a shoreline. Clastic sediment is produced by weathering on continents and delivered to the coastline by rivers. Coarser-grained sediment, such as sand, is deposited closest to shore as the river currents slow when entering the ocean. Wave agitation suspends the silt and clay, and these fine grains settle to the seafloor farther offshore. Calcite-secreting organisms live throughout the area of sand and mud deposition, where their hard mineral parts are buried in the sediment that is washed from the continent. Far from shore, however, there is less continent-derived clastic debris settling to the seafloor, so abundant calcite fossil remains and precipitated calcite crystals accumulate on the seafloor in the near absence of clastic sediment. This scenario illustrates how sediments that will someday become sandstone, shale, and

▲ **Figure 5.19 Rocks record changing environment.** This upward change in sedimentary rock types in the Grand Canyon records changing environments in a shallow sea 500 million years ago. The transitions between rock types are not sharp. The lower part of the shale is slightly sandy, and there are some thin limestone layers in the upper part of the shale before it merges upward into layers of mostly limestone interspersed with a few shale layers. These observations suggest that the changes in the depositional environment were gradual rather than abrupt.

limestone simultaneously accumulate in different environments defined by proximity to a shoreline (Figure 5.20).

This information allows geologists to explain the vertical arrangement of sandstone, shale, and limestone at a single location. Figure 5.20 shows that if sea level rises and the shoreline position changes, then the area where sand was once deposited is buried in mud, and the previous area of mud deposition becomes covered with an accumulation of limey sediment. The vertical change in rock types, therefore, records changing environmental conditions—in this case, a rise in sea level.

Putting It Together—How Do Sedimentary Rocks Reveal Ancient Environments?

• Different sedimentary rock types indicate the physical and chemical processes active in ancient depositional environments.

• Sedimentary structures reveal details about physical processes in the depositional environment. Cross-beds also reveal the direction of moving water and air currents that transport sediment.

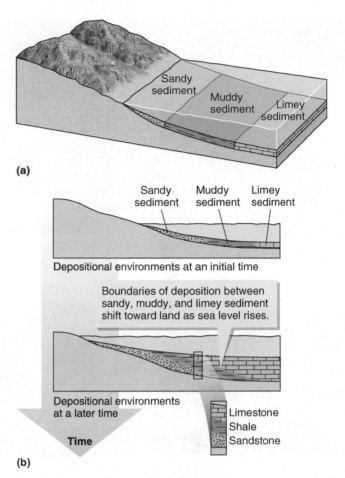

(a)

(b)

▲ **Figure 5.20 How shifting environments form different rocks.** Each sedimentary layer accumulates on the layer below, so vertical changes in rock type represent environmental changes during the time of sediment deposition. (a) Different sediment types accumulate offshore, with coarser clastic sediment deposited closest to shore and limey (calcite-rich) sediment deposited farther from shore. (b) If sea level rises, then the environments represented by each rock type shift toward land. Compare this diagrammatic sequence of sedimentary rocks with real rocks illustrated in Figure 5.19.

ACTIVE ART

Changing Shorelines and Sedimentation. *See how sedimentation near shorelines differs with changing sea levels.*

• Changes in environments or processes produce boundaries between beds of sedimentary rock. Observations of vertical successions of rocks with different features allow geologists to interpret environmental change over time.

5.7 How Do We Know . . . How to Interpret Unseen Deep-Ocean Currents?

Picture the Problem

How Is Coarse Sediment Deposited in the Deep Sea? **Figure 5.21** illustrates a common sedimentary sequence of alternating sandstone and shale beds. On the basis of their fossil content, geologists interpret these beds to have been deposited on the seafloor. Even before 1900,

◀ **Figure 5.21 What deep-sea sandstone and shale look like.** Tilted layers of light-colored sandstone and dark shale crop out along the coast near San Francisco, California. The field photo shows a close view of a sandstone layer sandwiched in shale. The enhanced view of the sandstone bed labels the features that a geologist sees:

- The layer is a graded bed, with the sizes of the grains decreasing upward until the distinction between sandstone and shale is unclear.
- The upper part of the bed contains a thin interval of small-scale cross-bedding produced by ripples moving with a current from right to left.

sedimentologists were able to look at sequences such as this and make a number of observations and interpretations, highlighted in Figure 5.21.

- The most conspicuous sedimentary structure is the presence of graded beds. The graded beds imply sediment settling out of suspension in the water, with larger particles reaching the seafloor first, followed by successively smaller grains. The base of each graded bed is sharp and typically not flat, indicating that the underlying mud was scoured by erosion before sand was deposited on it.

- The next most notable structures within the graded bed are ripple cross-beds. The creation of cross-bedding required moving currents to deposit the sediment.

- The scoured base of the graded bed and the current-formed cross-beds indicate that the sediment did not simply settle quietly to the seafloor, but also was swept along by a current. The current eroded the muddy seafloor before the sand began to accumulate. The intervening shale layers reveal that the depositional environment was characterized by slow accumulation of fine clay particles between the times when the sandy graded beds were deposited.

Paleontologists contributed further observations and interpretations during the early twentieth century:

- The shale layers contain fossil microorganisms that secreted microscopic shells of calcite and silica. These fossils accumulate only in sediment deposited in deep water far from shore. There are no fossils in the shale layers, which prefer shallow, nearshore conditions.

- The graded sandstone beds, on the other hand, contain fossil remains of nearshore animals and shreds of land-plant debris. These

fossils reveal that the sediment composing the graded sandstone beds was swept away from the shoreline into deep water, where it was deposited with fine mud and settling plankton carcasses on the deep ocean floor.

Many geologists treated the paleontologists' interpretation with disbelief—how could sandy sediment that typically accumulates close to shorelines be transported into deep water? These relationships are inconsistent with the pattern illustrated in Figure 5.20.

Clues from the Field

How Does Sediment Travel on Lake Bottoms? As it turned out, geologists working on problems far removed from the origin of deep-sea graded beds provided observations that supported the existence of deep-water currents. The geologists noted that muddy river water disappears from the surface of otherwise clear lakes very close to river mouths. Furthermore, they observed that muddy river water seemingly disappeared at the upstream end of a clear-water reservoir, only to reappear at the base of a dam 20 kilometers farther downstream. Geologists explained this phenomenon by suggesting that sediment-laden river water is denser than the clear lake water and sinks downward to flow along the lake bottom. This turbid water, thick with suspended sediment, flows in a current along the sloping bottom of the lake, below the clear, still water, for the entire length of the reservoir. The name "turbidity current" was applied to this unseen, hypothesized process.

Could turbidity currents, caused by the higher density of sediment-laden water moving below less dense clear water, also form in the ocean? If they existed, what would be the erosional and

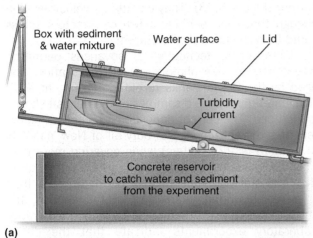

Box with sediment & water mixture   Water surface   Lid

Turbidity current

Concrete reservoir to catch water and sediment from the experiment

(a)

This sketch shows Kuenen's experimental apparatus for producing small turbidity currents. A tilted glass-walled tank simulates the sloping seafloor. Sediment and water are mixed in a box suspended above the upslope end of the tank. A trap door opens in the bottom of the box and releases the sediment and water mixture into the tank, where it flows along the bottom as a turbidity current.

(b)

This photograph was taken through the glass sidewall of the 2-m-long tank during one of Kuenen's 1950 experiments. The milky white, sediment-laden turbidity current moves along the bottom of the tank below less dense, clear water.

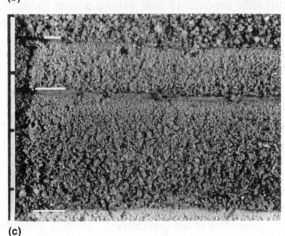

(c)

Each experimental current produced a graded bed. The white bars mark the base of three graded beds. The black marks on the left are 1 cm apart.

◀ **Figure 5.22 How to make graded beds in the laboratory.**

depositional effects that might be recognized in sedimentary rocks? Could these currents produce graded beds such as those documented in Figure 5.21?

Clues from Experiments

How Does Sediment Move on Submerged Slopes? Oceanographer Philip Kuenen conducted laboratory experiments in the late 1940s to understand the erosional properties of turbidity currents. He constructed a large, aquarium-like tank with a gently sloping bottom, shown in **Figure 5.22**. He filled the tank with seawater and mixed fine sand, mud, and seawater in a separate container. The sediment-water mixture was vigorously stirred to suspend the sediment and then quickly dumped through a trap door into the upslope end of the large tank. Figure 5.22 shows the result of such an experiment. The turbid sediment-water mixture does not mix with the clear seawater, but instead produces a turbulent, swirling current that sweeps rapidly down the sloping bottom of the tank to the opposite end. In this experiment, Kuenen had created laboratory-scale turbidity currents.

Integrating Experiments and Field Descriptions into a Hypothesis

Do Turbidity Currents Deposit Deep-Sea Graded Beds? Many field geologists read the results of Kuenen's experiments with great interest. Until this point, there had been no clear link between the hypothesized turbidity currents in lakes and the deep-marine graded beds of sandstone and conglomerate. The results of the laboratory turbidity currents suggested a connection. The hypothesis goes like this:

- River floods, storm waves, or submarine landslides triggered by earthquakes stir up nearshore sediment.
- The sediment and water mixture sweeps downslope along the ocean bottom into deep water as a turbidity current, carrying along its cargo of nearshore organic remains.
- The energetic current erodes some of the seabed clay, and then, as the current energy wanes, the sediment grains settle to form the graded beds.
- Continued motion of the slowing current moves the settled sediment to produce ripple cross-beds.
- Peaceful, slow accumulation of clay and microorganism carcasses resumes between these episodic events.

Geologist C. I. Migliorini seized the opportunity to link field observation and experiment, and invited Kuenen to visit his Italian field-study locations with graded beds. Kuenen saw in this invitation the potential to further expand scientific understanding of how sedimentary deposits form. He returned to his laboratory tanks and adjusted his experiments and measurements to focus on the deposits left by artificial turbidity currents. Indeed, these deposits, shown in Figure 5.22c, closely resembled Migliorini's graded beds, and the two scientists teamed up to publish their results. The deposits from turbidity currents became known as **turbidites.**

Completing a Uniformitarian Analysis

Do Ocean Turbidity Currents Happen in Nature? You learned in Section 5.6 (Figures 5.14–5.18) that geologists explain the formation of ancient sedimentary rocks by watching sediment accumulate in modern environments and by experiments with sediment in the lab. Clearly, it is a tall order to watch a real, natural, rare, submarine turbidity current and then compare the resulting deposit to ancient rocks with graded beds. Watching waves on a beach to interpret beach sandstone or diving off a modern coral reef to interpret an ancient limestone is not the same as trying to be in the right place at the right time to witness (and survive!) a catastrophic turbidity current on the deep-sea floor. Nonetheless, a rare opportunity presented itself in the last century for sedimentologists and oceanographers to link turbidite beds with historical turbidity current events.

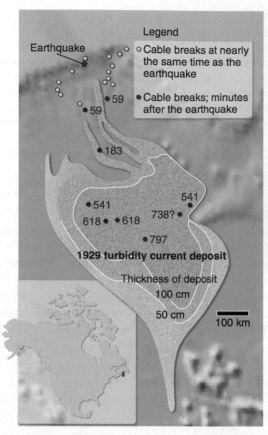

Geologists and oceanographers used remotely operated cameras and sophisticated sonar to map the outline of the 1929 Grand Banks, Canada, turbidity current deposit.

The positions of telegraph cables broken following the earthquake, and the time elapsed between the earthquake and the break, are also depicted on the map.

When the earthquake occurred along the steep edge of the continental shelf, it triggered a landslide that broke some cables instantaneously. Subsequently, the landslide sediment mixed into seawater and formed a turbidity current, which broke more cables as it swept across the seafloor.

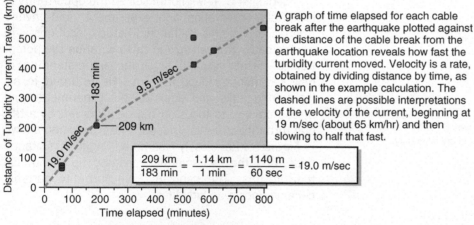

A graph of time elapsed for each cable break after the earthquake plotted against the distance of the cable break from the earthquake location reveals how fast the turbidity current moved. Velocity is a rate, obtained by dividing distance by time, as shown in the example calculation. The dashed lines are possible interpretations of the velocity of the current, beginning at 19 m/sec (about 65 km/hr) and then slowing to half that fast.

$$\frac{209 \text{ km}}{183 \text{ min}} = \frac{1.14 \text{ km}}{1 \text{ min}} = \frac{1140 \text{ m}}{60 \text{ sec}} = 19.0 \text{ m/sec}$$

▲ Figure 5.23 **Documenting a real turbidity current.**

A 1929 earthquake off the Grand Banks of southeastern Canada generated the best-documented historic turbidity current, as illustrated in **Figure 5.23**. This part of the North Atlantic Ocean was crisscrossed with telegraph cables, several of which broke almost instantaneously when the earthquake occurred. Another cable broke about an hour later at some distance from Grand Banks, and then cables at progressively greater distances broke in succession over the next 17 hours. Oceanographers reexamined the cable-break times in the 1950s when Kuenen and others developed the turbidity-current concept. They hypothesized that the successive cable breaks were caused by a southerly moving turbidity current generated by the earthquake. They used the known times of the earthquake and the cable breaks to estimate the velocity of the

current (Figure 5.23). They calculated velocities high enough that even pebbles a few centimeters across could be suspended in the current.

More recent technological advances permitted researchers to take photographs of the seafloor ravaged by the Grand Banks turbidity current. In addition, carefully retrieved samples of the deposit show a graded bed. The current deposited 175 cubic kilometers of sediment (enough to bury all of New York City to a depth of 225 meters) while covering more than 150,000 square kilometers (an area a little larger than the state of Indiana) to a depth of more than 1 meter. In the northern part of the deposit, the surface undulates with dunes of cross-bedded gravel. Laboratory experiments indicate that the dunes formed at a velocity of about 15 meters per second. This is consistent with the velocity data extrapolated from the cable breaks (Figure 5.23) if the dunes formed as the current slowed from its peak velocity of 19 meters per second.

No person has observed an actively moving underwater turbidity current of the size required to produce turbidite-sandstones that extend across thousands of square kilometers. However, it is apparent from the story of the Grand Banks turbidity current that modern observation plays an important role in the uniformitarian interpretation of ancient sedimentary rocks.

Insights

How Do Geologists Use Interdisciplinary Studies? There are many examples similar to the interpretation of sedimentary structures and depositional environments described above. All these interpretations build from observations of ancient rocks, modern environmental processes, and laboratory experimentation.

This story of the discovery of turbidity currents and turbidites also reveals that the scientific process of resolving a question is not always systematic. Although field sedimentologists and paleontologists hypothesized turbidity currents, Kuenen's initial experiments were not designed to test that hypothesis. Like many geologic problems, this one was solved by successfully integrating various observations by scientists in different disciplines. Scientists with broad perspectives who appreciate the interrelationships of different research results and are willing to enter into collaborative projects usually make the greatest contributions to scientific knowledge.

Putting It Together—How Do We Know . . . How to Interpret Unseen Deep Ocean Currents?

• Combining field observation of rocks and modern processes with laboratory experiments revealed that certain graded sandstone beds interlayered with deep-water marine shale are the result of turbidity currents.

• Turbidity currents are mixtures of sediment and water that are denser than clear water and rush along the seafloor or lake bottoms.

• Solving some geologic problems requires integrating modern field observation and laboratory experimentation with study of ancient rocks and modern processes.

5.8 How Are Plate Tectonics and Sedimentary Rocks Connected?

Modern global geography corresponds closely to plate tectonic processes (see Section 1.5): High mountain ranges form where plates converge, while wide, deep oceans flank divergent boundaries along mid-ocean ridges. The patterns of ancient environments preserved in sedimentary rocks permit interpretation of ancient geography, which in turn provides a history of plate tectonics.

Sediment Accumulates in Tectonic Basins

Sedimentary rocks on continents are thickest where depressions, called **basins**, form on Earth's surface. Tectonic processes raise mountains and cause basins to subside. The presence of thick sedimentary-rock accumulations, therefore, tells geologists where tectonic processes actively formed basins at various times during the course of geologic history. Depositional environments described by the sedimentary rocks provide insights into the size of the basin. The direction toward, and proximity to, mountainous highlands is revealed by the locations of the coarsest clastic sediment. The composition of clastic-sediment grains indicates the types of rocks in those ancient mountains. These observations integrate to reveal which tectonic processes affected a region in the distant geologic past.

Example of Sedimentary Rocks and Plate Tectonics

Figure 5.24 illustrates an ancient geography interpreted from sedimentary rocks. The map on the left shows distributions of sedimentary rocks that accumulated about 425 million years ago in what is now the eastern United States. Notice the changes in sediment type from place to place across the map. These rocks outline a sedimentary basin that subsided deep enough for seawater to submerge the middle of the continent. The distribution of rock types shows the location of a shoreline along the eastern edge of the basin. The absence of sedimentary rocks of this age farther east implies the presence of a sediment source area—an upland area where rocks weathered and eroded to provide the clastic sedimentary materials deposited to the west. The presence of conglomerate in the northeast suggests steep, mountainous slopes on which fast-moving streams transported coarse gravel. Some conglomerate fragments include volcanic and plutonic rocks, which suggest erosion of igneous rocks near a subduction zone.

The ancient geographic map in Figure 5.24 depicts an ancient landscape as interpreted by geologists from the sedimentary-rock evidence. In this map, you can see a mountainous, volcanically active landmass along a convergent plate boundary (near where the Atlantic Ocean is today). Farther to the west, you can see the submerged sedimentary basin forming a shallow sea in the mid-continent, which is presently high and dry. The submerged basin is the result of tectonic subsistence.

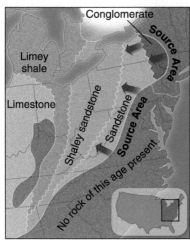
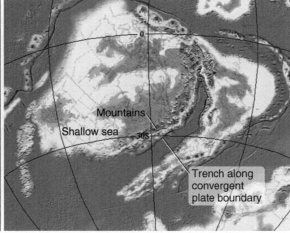

▲ Figure 5.24 **Making a paleogeographic reconstruction.** The map on the left depicts the presence of different sedimentary rock types deposited in the eastern United States about 425 million years ago. The sandstone and conglomerate contain fossils mostly of land organisms, whereas some of the sandstone and all of the other rocks contain marine fossils. These sedimentary rocks permits the following interpretation of the ancient geography:

• The westward change from sandstone through shale to limestone is the pattern expected along a westward-deepening, shallow sea (compare to Figure 5.20). This suggests the presence of dry land to the east and rivers flowing westward to a warm, shallow sea, which is consistent with current directions indicated by cross-bedding (shown by the arrows). The area of missing rock near the present Atlantic coast likely represents a relatively high area, which was undergoing erosion to supply the sediment found farther west.

• The presence of conglomerate in the northeast suggests that streams flowed on steeper slopes to transport the coarser fragments, implying more mountainous topography in that area.

The geographic picture of this region reconstructed from the sedimentary rocks on the right depicts a mountainous landmass adjacent to a convergent plate boundary that existed near where the Atlantic Ocean is today and a shallow sea in the mid-continent, which in modern times is high and dry.

Putting It Together—How Are Plate Tectonics and Sedimentary Rocks Connected?

• Plate tectonic forces form basins that accumulate the sediment eroded from mountains.

• The composition of the sediment reveals the nature of the rocks in the original sediment source region, possibly uplifted by tectonic activity.

• The depositional environments of the sedimentary rock reveal ancient landscapes and seascapes that resulted from tectonic processes.

Where Are You and Where Are You Going?

With the ending of this chapter, you have completed two-thirds of your circuit around the rock cycle.

The processes that form sediment and sedimentary rocks act as geologic recyclers. Older rocks, physically and chemically broken down into solid and dissolved components, are shuffled, redistributed, and formed into new rocks. Clastic sedimentary rocks contain the fragments created from rocks during physical weathering, as well as new minerals produced by chemical-weathering reactions. Chemical sedimentary rocks form when minerals precipitate from ionic solutions generated when other rocks dissolved via chemical weathering. Chemical precipitation in the pore spaces between sediment grains cements both clastic and chemical sediment into rock.

Observing modern processes provides a useful basis for understanding the origin of sedimentary rocks. You can witness the weathering processes that break down rocks and analyze water to track the ions carried away in solution. You can observe sediment transport and deposition by a variety of processes and in countless environments, and analyze key features in the modern sediment that can then be used to interpret ancient depositional environments. These features include sedimentary structures such as cross-beds, which also reveal the direction that currents moved while transporting sediment.

Sedimentary rocks are the historical archive of past processes at Earth's surface and of the organisms that live there. Vertical changes in rock types record changing processes of sediment deposition and shifting environments, such as those caused by fluctuating sea levels. Horizontal changes across a map in rocks of the same age reveal clues about ancient landscapes and seascapes from which geologists interpret tectonic processes. Reconstructing past views of our planet does not require a fanciful imagination once you understand how to translate the rich archives of sedimentary rocks.

Now, you are ready to complete your trip around the rock cycle. You appreciate that rocks formed well below the surface of the Earth weather at the surface because they are not stable there. Likewise, rocks formed at or near the surface, at relatively low temperature and pressure, are unstable at the higher temperature and pressure found at greater depth. This is especially true where tectonic forces also exert strong horizontal pressure and where water is available to assist chemical reactions that transform one group of minerals to another. You are now ready to investigate metamorphic rocks.

Active Art

Physical Weathering. See how the freezing and thawing of water and the evaporation of salty water pry rocks apart.

Forming Cross-beds. See how cross-beds form and how sand dunes migrate through time.

Changing Shorelines and Sedimentation. See how sedimentation near shorelines differs with changing sea levels.

Extension Modules

Chemical Reactions and Chemical Equations. Learn about chemical reactions and how to write chemical equations.

Why Is Seawater Salty? Learn why some ions become concentrated in seawater and give it its salty taste.

Geochemistry of Calcite. Learn the factors that determine whether calcite dissolves or precipitates in water, which explains many features of rocks and landscapes.

When Will We Run Out of Oil? Learn the facts about oil reserves and consumption that determine the uncertain future of this important energy resource.

Confirm Your Knowledge

1. Define "weathering" and provide examples of the two types of weathering processes.
2. Why do rocks weather?
3. In learning about chemical reactions involving geologic materials, keep in mind the differences among elements, minerals, and rocks. For each word listed below, indicate whether it refers to an element, a mineral, or a rock. If it refers to a mineral, list the elements that compose the mineral. If it refers to a rock, list the common minerals that compose the rock, and the elements that compose each mineral.

silicon	**quartz**	**granite**	**basalt**
iron	**hematite**	**calcium**	**calcite**
limestone	**halite**	**rock salt**	**biotite**
aluminum	**kaolinite**	**feldspar**	**sandstone**

4. What is the role of carbon dioxide (CO_2) in weathering? Explain Equation (5.3).
5. Oxidation is a common weathering process. Which group of minerals does it affect most? What minerals do oxidation reactions commonly form? How can you quickly recognize in the field the most common minerals formed by oxidation?
6. The processes of weathering, erosion, transport, deposition, and lithification are important in the formation of sedimentary rocks. Briefly explain each of these processes in your own words, being sure to distinguish each from the others.
7. Explain the processes involved in lithification of sediment into sedimentary rock.

8. What three characteristics do we use to describe the textures of clastic sedimentary rocks? Describe how one of the characteristics can be used to provide information about the environment of deposition.

9. What features are used to name chemical and biogenic sedimentary rocks?

10. Define a fossil and list the ways that fossils form. Explain why fossils are usually found in sedimentary rocks rather than in igneous or metamorphic rocks.

11. What are fossil fuels, and why are they given this name?

12. List the most common biogenic and chemical sedimentary rock types, and describe the depositional environments that favor deposition of each sediment type.

13. How do geologists reconstruct an ancient environment from study of sedimentary rocks?

14. List three sedimentary structures and explain what each one indicates about the depositional environment.

15. How are plate tectonics and sediment deposition linked?

Confirm Your Understanding

1. Write an answer for each question in the section headings.

2. Predict whether limestone exposed in a dry desert with sparse vegetation would chemically weather faster or slower than a limestone exposed in a very wet, tropical forest. Explain how you arrived at your prediction.

3. While hiking with two of your classmates, you observe an outcrop of sandstone rich in quartz and feldspar. Friend A says that the sand grains in the rock were eroded from a nearby basaltic volcano. Friend B says that the sandstone was derived from a more distant granite outcrop. Which one of your friends is right and why?

4. Imagine a river that flows through steep mountain valleys and then across a gently sloping coastal plain to the ocean. Write a prediction of how grain size, rounding, and sorting of the sediment deposited by the river will change from the mountains to the coastline. Explain why these changes will take place.

5. Explain why the sediment in windblown sand dunes is better sorted than the sediment deposited by a flooding river.

6. Look ahead to Figure 19.24, which shows three beach deposits. If each deposit became a rock, what would be the name for each rock formed? Be as specific as possible in selecting a name.

7. What sedimentary structure is illustrated in this photograph? What direction (toward the left or the right) were sediment-transporting currents flowing when the sediment was deposited?

8. This photo shows an outcrop on Mars; the image was transmitted to Earth by a robotic rover traveling over the Martian surface. When geologists saw this image, some of them suggested that the rocks could be evidence of flowing water at some point in Martian history. What do you see to support that hypothesis? What process other than flowing water might produce the features visible in the photo? What additional information would you use to distinguish between these two hypotheses?

9. Think about a modern coastal lagoon with mudflats that are alternately submerged and exposed each day by rising and falling tides. With this environment and tidal processes in mind, what clues would you look for to identify the deposits of an ancient lagoon in the geologic record? How would you distinguish an ancient coastal lagoon mudstone from a floodplain deposit, where mud collects during river floods?

10. At a field outcrop, you notice a sequence of sedimentary layers. All the rock layers contain fossils of marine organisms. From bottom to top, the layers consist of sandstone, mudstone, limestone, mudstone, and sandstone. What can you interpret about the environment during the deposition of these layers of rock?

11. You are a geologist assigned to identify possible coal resources in an area of central Asia where little is known about the geology. Preliminary work indicates that area A consists primarily of outcrops of volcanic rocks. Area B contains sedimentary rocks that have been interpreted as forming in a desert. Area C contains sedimentary rocks that probably were deposited along a river delta near a coastline. You have time and money to explore for coal in only one of these regions. Which one do you choose? Explain why you chose the area you did, as well as why you did not choose the other two.

12. Reexamine the map on the left side of Figure 5.24. Rocks of this age in southern Michigan and northwestern Ohio provided large quantities of oil and natural gas in the early 1900s. However, no oil is found in rocks of this age in eastern Pennsylvania. Use the information in the map to propose a hypothesis for why oil and natural gas are found in rocks of this age in one location but not the other.

CHAPTER 6

The Formation of Metamorphic Rocks

Why Study Metamorphic Rocks?

Some rocks do not form by igneous or sedimentary processes, but are transformed by undergoing changes in mineral content, texture, or both—these are metamorphic rocks. In fact, these rocks have been so changed that their original minerals, and the story of how they first formed, can be hard for geologists to determine.

Metamorphic processes cannot be seen in nature because almost no metamorphic rocks form on Earth's surface. Metamorphic processes require higher temperatures, and usually higher pressures, than exist at the surface. Therefore, geologists use experimental and theoretical approaches to recognize and explain these processes that form metamorphic rocks deep within Earth.

Metamorphic rocks are exposed in actively forming mountains and in eroded ancient mountain belts in the interior of continents. Uplift and erosion bring these rocks, originally formed deep below ground, to the surface and are key to reconstructing the history of processes inside Earth and the making of mountains.

Metamorphic rocks and minerals also are used as economic resources, such as talc for talcum powder, graphite for pencils, marble for building material, garnet and corundum for industrial abrasives, and metamorphosed coal for fuel. Metamorphic processes also form some metal ores. Understanding the metamorphic history of a region helps geologists discover and extract economic resources.

After Completing This Chapter, You Will Be Able to

- Explain the distinctive characteristics of metamorphic rocks.
- Explain how metamorphic rocks form.
- Explain where metamorphic rocks form.
- Use metamorphic rocks to interpret the geologic history of a region.

Pathway to Learning

6.1 What Is Metamorphism?

6.2 What Role Does Temperature Play in Metamorphism?

6.3 What Role Does Pressure Play in Metamorphism?

6.4 What Role Does Fluid Play in Metamorphism?

6.5 Why Do Metamorphic Rocks Exist at the Surface?

Swirling mineral bands characterize this metamorphic-rock outcrop in the Shining Rock Wilderness Area of the Appalachian Mountains of western North Carolina.

IN THE FIELD

You are on a virtual road trip when you notice an interesting road cut, seen in **Figure 6.1**a. A shiny outcrop catches your eye from the car, so you pull over to take a look, thinking this may be a chance to identify some of the rocks you are learning about in your geology class. When you explore the outcrop, you notice a very subtle layering in it that runs parallel to the hillside but does not possess obvious signs of the bedding common in sedimentary rocks.

A closer examination reveals that the rock consists mostly of muscovite mica, along with scattered crystals of two darker minerals. The thin, flaky mica crystals are several millimeters across and arranged like sheets of paper strewn across the floor. The cleavage surfaces of the mica crystals are parallel, so light reflects off the mineral surfaces to produce the shine you saw from the car (Figure 6.1b). The darker minerals do not resemble the common ingredients you recall from your study of igneous or sedimentary rocks. One mineral type, forming equidimensional crystals, with a deep, red-wine color and well-formed crystal faces (Figure 6.1c), does remind you of a mineral illustrated in Chapter 2; this is garnet (see Figure 2.27). The brown crystals stump you, so you return to your car and retrieve your dog-eared field-guide-to-minerals book. The brown crystals match the description for a silicate mineral called staurolite. A key identifying feature is the tendency for staurolite crystals to grow through one another to form crosses (the mineral name derives from the Greek word *stauros:* "cross").

What kind of rock is this? Muscovite can be a minor component in igneous rocks, playing a supporting role for much more abundant quartz and feldspar, but in this case, muscovite forms most of the rock in the outcrop. You also recall that garnet and staurolite are not typical igneous-rock minerals. Athough the outcrop has a subtle layering formed by the parallel orientation of the thin, flat muscovite crystals, it bears no other resemblance to a sedimentary rock. The garnet and staurolite crystals have well-defined crystal faces with sharp edges and corners, so they cannot be clastic sediment grains, which would be rounded by transport in flowing water. Also, a rock composed of silicate minerals such as these is not likely to be a chemical sedimentary rock, because silicate minerals are barely soluble in water.

The texture and mineral content of this rock are very different from what you have come to expect of sedimentary and igneous rocks. If this seems like new territory to you, do not worry; it should. You have encountered an example of a third group of rocks—metamorphic rocks. Metamorphic rocks result from changes in preexisting rocks. To change the appearance of a rock, the minerals that compose it must rearrange in some fashion, must change to new ones, or both. How do metamorphic rocks form? What was the rock before it was metamorphosed? Where do metamorphic rocks form? How do you name this rock you have picked up?

▶ Figure 6.1 Field observations of metamorphic rock.

6.1 What Is Metamorphism?

As a youngster, you probably learned about and possibly witnessed the metamorphosis of caterpillars into butterflies. The word "metamorphosis" derives from the Greek terms *meta* for "change" and *morphe* for "form," which appropriately describe the change in the form of the same insect from an immature caterpillar to a mature butterfly. Similarly, geologists see evidence for changes in rocks that do not relate to the igneous and sedimentary processes of melting, dissolving, or physically disintegrating. Geologists, therefore, define **metamorphism** as the process by which pre-existing rocks undergo changes in chemical composition, mineral content, or physical texture while remaining in a solid state. Metamorphic changes do not involve melting, as required for forming igneous rocks, or the break-down and recombination of minerals and rocks that relate to the formation of sedimentary rocks. This means that metamorphic rocks are intuitively more difficult to understand. We gain important insights into the formation of igneous and sedimentary rocks when we watch lava cooling on the slope of a volcano or watch weathered rock residue wash down a stream, but we cannot see rock metamorphism taking place.

The compositional and mineralogical changes during metamorphism are chemical reactions that transform one association of minerals in the original rock into new minerals that then compose the metamorphic rock. Many of the minerals found in metamorphic rocks, such as the garnet and staurolite in the roadside outcrop, can be formed out of other mineral mixtures in the laboratory through reactions that require high temperatures and pressures. From this important observation, we can infer that metamorphism occurs in Earth's crust and mantle under conditions that differ from the conditions in which the original rock formed. Chemically reactive fluids and rock deformation processes within our dynamic planet also contribute to metamorphic transformations.

Under what conditions of temperature and pressure does metamorphism happen? If temperatures are high enough to cause substantial melting, igneous rocks will form by crystallization of the resulting melt. Relatively low temperatures and pressures, on the other hand, can favor chemical reactions that cement sedimentary rocks, but not enough to change the minerals and textures in the sediment. These observations reveal that the conditions for metamorphism fill a range of temperatures and pressures between those necessary to lithify sediment and those necessary

(a) Subtle layering parallel to hillside is caused by parallel orientation of flaky mica crystals

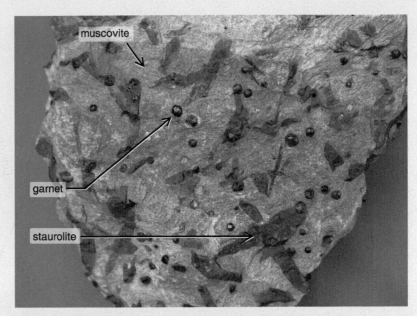

(c) Close examination shows the presence of three minerals: shiny flakes of muscovite, equidimensional red crystals of garnet, and elongate, sometimes crossing brown crystals of staurolite.

(b) A closer view of the rock, with car keys to indicate scale. The rock is shiny because light reflects off of the flat cleavage surfaces of large muscovite mica crystals.

to melt rock into magma, as illustrated in **Figure 6.2**. Metamorphism occurs to varying degrees that you might think of as ranging from slight to extreme. Geologists describe the intensity of temperature and pressure during metamorphism as ranging from **low grade** to **medium grade** to **high grade**. Figure 6.2 illustrates the approximate temperature and pressure conditions for the different metamorphic grades.

What Happens During Metamorphism

We know that during metamorphism, high temperature and pressure deep below the surface transform preexisting rocks into metamorphic rocks. So, how do geologists know what happens during metamorphic processes if these processes cannot be observed?

Geologists gain insights into metamorphic processes through laboratory experiments. Recall from Chapter 4 that geologists can use laboratory equipment to recreate the conditions deep inside Earth and to control magma composition, temperature, pressure, and abundance of water in experiments (see Section 4.5) in order to simulate igneous-rock forming processes. Similarly, to reproduce the minerals found in metamorphic rocks,

geologists conduct laboratory experiments with these four variables: (1) a starting rock composition, (2) temperature, (3) pressure, and (4) abundance and composition of fluid. Experiments indicate that changes in any of these four parameters affect the mineral composition and texture of the resulting metamorphic rocks. Furthermore, experiments and field observations indicate that rocks exhibit two types of metamorphic changes: (1) New minerals can form at the expense of the original ones as a result of chemical reactions, or (2) the rock texture can be altered by changes in the size, shape, and orientation of the constituent minerals.

The Significance of the Original Rock

Although changes in temperature, pressure, and fluid composition cause metamorphism, the original composition of the rock—commonly called the parent rock—is key to determining which type of metamorphic rock forms. For some metamorphic rocks, the overall chemical composition of the metamorphosed rock can be quite similar to the starting material, with only minor modification. In these cases, the composition you begin with is very similar to the composition you end up with, except that the

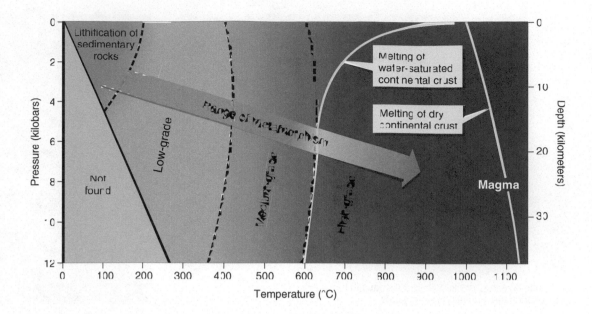

► **Figure 6.2 Defining the conditions for metamorphism.** The range of temperatures and pressures where metamorphism occurs exceeds the values for lithification of sedimentary rocks but falls short of the values at which rock melts to form magma. Grades of metamorphism approximately correspond to the labeled areas of temperature and pressure on the graph. The temperatures and pressures where melting begins depend partly on the water content of the rock, so potential melting curves differ for wet and dry, granite-rich crustal rocks. Note that the conditions for high-grade metamorphism of dry crust overlap with the melting conditions for water-rich rocks.

minerals present, the rock texture, or both, will be different. In some cases, features in the original rock, such as igneous crystals and sedimentary bedding, may still be recognizable.

Think about making a cake. After mixing the ingredients and baking the mixture, the cake has a very different appearance and texture than the original bowl of ingredients. Yet, the bulk chemical composition of the cake is nearly the same as the mixture of ingredients in the batter. Chemical analysis informs us that the metamorphic rock you picked up along the road in your virtual field trip at the beginning of the chapter (Figure 6.1c) has the same chemical composition as shale. So, geologists infer that shale was the parent rock for your field sample.

In other cases, however, the composition of the rock undergoes significant changes during metamorphism, because a large amount of chemically active water that is rich in dissolved ions participates in the metamorphic reactions. This happens, for example, when hot, water-rich solutions from magma move into adjacent rocks during metamorphism. Metamorphic minerals incorporate elements from the hot fluid and produce a rock with different composition from the parent rock.

The conditions of metamorphism determine the minerals formed in the rock (composition), or the shapes and arrangement of minerals in the rock (texture), or both. Metamorphic changes in composition and texture are the responses to changes in temperature, pressure, and fluid availability and composition. What role does each factor play in forming the metamorphic rock? Addressing each factor independently of the others is tricky because it is very rare that only one factor explains a metamorphic reaction. Still, it is useful to examine the role of each variable separately to the extent that is possible, in order to learn more about how these factors affect metamorphic processes.

Putting It Together—What Is Metamorphism?

• Metamorphism is the process that results in mineral, chemical, and texture changes to preexisting rocks in a solid state. The variables that affect these changes include temperature, pressure, fluid availability, and the original rock composition.

• Metamorphic rock may retain the overall chemical composition of its original (parent) rock, or metamorphism may significantly change the composition of the rock. Whether the composition of the metamorphic rock changes from that of the parent rock depends on the availability of chemically active fluids.

6.2 What Role Does Temperature Play in Metamorphism?

The cake-baking analogy in the last section offers insights into the role of temperature in metamorphism. No matter how long batter sits in the bowl, it will never become a cake unless it is put into the oven and baked, that is, exposed to high temperature. Similarly, most metamorphic reactions require heat. Thus, how does temperature increase, allowing metamorphism to take place?

How Rock Temperature Increases

We measure the intensity of heat with temperature. Temperature increases with depth below Earth's surface along the geothermal gradient (Section 4.4). Some metamorphic reactions occur when minerals that formed at low temperature transform into different minerals at higher temperature. The rates of metamorphic reactions are also faster at higher temperature than at lower temperature.

Three processes most commonly cause heat transfer to rocks and result in increasing temperature.

• Sediment burial, depicted in **Figure 6.3**. Temperature increases with depth along the geothermal gradient. When sedimentary rocks accumulate in sinking basins, the early-deposited sediment experiences progressively higher temperature as more sediment accumulates above it. Eventually, the sedimentary rocks are buried deep enough to experience the heat required for metamorphism.

• Tectonic burial, depicted in **Figure 6.4**. Where one block of crust is forced over another block, the lower block experiences the higher temperatures associated with greater depths below the surface.

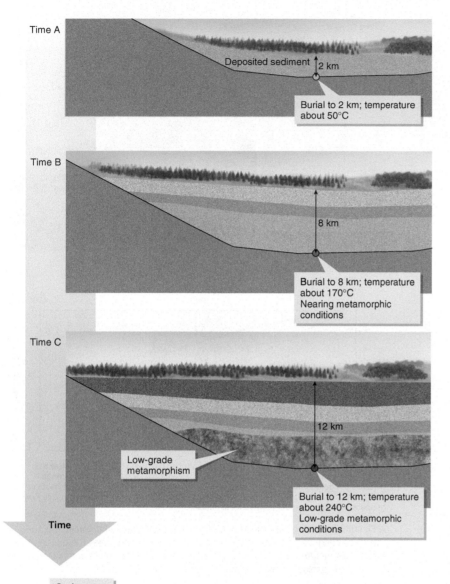

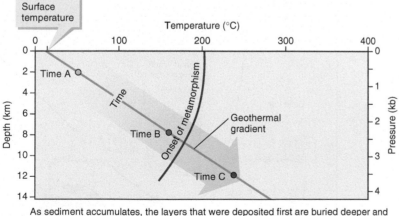

As sediment accumulates, the layers that were deposited first are buried deeper and experience higher temperature.

▲ Figure 6.3 Raising the temperature of rocks through sediment burial. Sedimentary rocks experience increasing temperature and pressure during sediment accumulation. If the rock burial is sufficiently deep, then low-grade metamorphism occurs. The graph traces the changing conditions at the point shown in the geologic cross section.

- Magma intrusion, depicted in **Figure 6.5**. The injection of hot magma into rocks changes the local geothermal gradient by increasing the temperature. The transfer of heat from magma metamorphoses the surrounding rocks.

Notice that increasing temperature by burial (Figures 6.3 and 6.4) happens simultaneously with increasing pressure exerted by the weight of overlying rock. In contrast, increasing temperature near an intrusion (Figure 6.5) need not be accompanied by a change in pressure.

Heat Drives Away Water and Gases

Observation from laboratory experiments show that minerals containing water or gas molecules at low temperature release those molecules at high temperature, which causes transformations to different minerals. These reactions are dehydration (loss of water) and degassing (loss of gas). An example of a high-temperature dehydration chemical reaction is the transformation of muscovite and quartz into sillimanite and potassium feldspar. During this chemical reaction water molecules escape the crystal structure of muscovite at high temperature, forming two new minerals that lack water. The reaction is written like this:

$$\text{Add Heat}$$
$$\downarrow$$
$$\underset{\textit{muscovite}}{KAl_3Si_3O_{10}(OH)_2} + \underset{\textit{quartz}}{SiO_2} \rightarrow \underset{\textit{sillimanite}}{Al_2SiO_5} + \underset{\substack{\textit{potassium-} \\ \textit{feldspar}}}{KAlSi_3O_8} + \underset{\textit{water}}{H_2O}$$

You can read this reaction like this: "When a rock containing muscovite and quartz is heated, the minerals react to form sillimanite and potassium feldspar, and liquid water is released." (You will learn more about sillimanite, a common mineral in high-temperature metamorphic rock, in Section 6.6.)

Dehydration reactions release liquid water or water vapor into the spaces between the mineral grains and may speed up further metamorphic reactions that depend on the presence of water. Recall from Chapter 4 that the presence of water also can decrease the melting temperature of silicate minerals and thus enhance the formation of magma (see Figures 4.18c and 4.21b). Dehydration reactions during metamorphism of subducted oceanic crust cause melting of mantle peridotite to form magma near convergent plate boundaries (see Section 4.6).

The most common degassing reactions release carbon dioxide from carbonate minerals. Cement producers make cement from limestone, which consists of calcite ($CaCO_3$), by employing one such reaction:

$$\text{Add Heat}$$
$$\downarrow$$
$$\underset{\textit{calcite}}{CaCO_3} \rightarrow \underset{\textit{lime}}{CaO} + \underset{\textit{carbon dioxide}}{CO_{2(gas)}}$$

Roasting the limestone in a kiln drives carbon dioxide gas out of the calcite and produces lime (CaO), which is the primary ingredient in cement. A natural example of a degassing reaction is the "baking" of dolostone near a hot magma intrusion. The dolomite releases carbon dioxide gas at high temperature, and the mineral

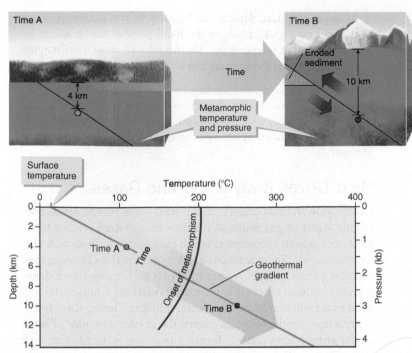

At time A, a rock (represented by the green dot) is 4 km below the surface at a temperature of 100°C. After tectonic processes displace the blocks of crust, the same rock (now shown by a red dot) is at 10 km depth and experiences a temperature of 260°C, resulting in low-grade metamorphism.

▲ **Figure 6.4 Raising the temperature of rocks through tectonic burial.** Tectonic processes can move rocks deeper into Earth, where they experience higher temperatures and pressures and metamorphism occurs. The graph traces the changing conditions of the point shown in the geologic cross section.

transforms into a combination of calcite and periclase (a magnesium-oxide mineral):

$$\text{Add Heat}$$
$$\downarrow$$
$$\underset{\textit{dolomite}}{CaMg(CO_3)_2} \rightarrow \underset{\textit{calcite}}{CaCO_3} + \underset{\textit{periclase}}{MgO} + \underset{\textit{carbon dioxide}}{CO_2 \, (gas)}$$

Mineral Stability

These example dehydration and degassing reactions describe the concept of mineral stability, which is the tendency of an object to remain the same rather than to change. Most minerals are stable only in particular ranges of temperature and pressure. If these conditions change, then the minerals become unstable and react to form different minerals that are stable at the new temperature and pressure conditions. The production of a new mineral requires that the chemical bonds in the original mineral break and that the newly free atoms rearrange into a new mineral structure. Dolomite, for example, is stable at low temperature, but is unstable and degasses, breaking down to other minerals and gas, at higher temperature.

In summary, rocks are stable under certain conditions, but once those conditions change, they may become unstable. It may help you to understand this concept if you think of water and its various states—liquid, solid, or steam. **Figure 6.6** shows the temperature and pressure conditions at which liquid water, water vapor (steam), and solid ice are each stable. At sea level, water boils at 100°C to become steam and

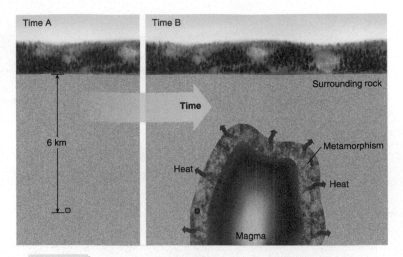

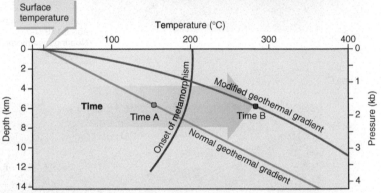

Prior to the magma intrusion, the rock (represented by the green dot) is 6 km below the surface where the normal temperature is 160°C. Heat conducted from intruding magma modifies the geothermal gradient, and the rock (now represented by the red dot) heats to 285°C and metamorphoses to low grade. Rocks closer to the magma experience even higher temperature, higher-grade, metamorphic effects.

▲ **Figure 6.5 Raising the temperature of rocks through magma intrusion.** Magma intrusions conduct heat into cooler surrounding rock, which raises the temperature and causes metamorphism. The graph traces the changing conditions of the point shown in the geologic cross section.

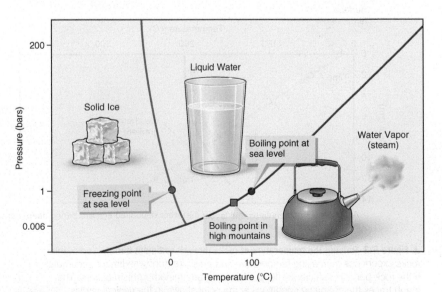

▲ **Figure 6.6 Visualizing the pressure and temperature stability conditions of water.** This diagram illustrates the temperatures and pressures at which solid, liquid, and vapor water are stable. Air pressure is measured in bars; one bar is atmospheric pressure at sea level. When pressure drops below atmospheric pressure, such as in mountains high above sea level, the temperature at which water converts to steam also drops.

freezes at 0°C to become ice. If pressure decreases, the boiling temperature of water drops. This is why water heated on a stove in the mountains boils sooner than at lower elevations. Similarly, when you put an ice-cube tray of water into the freezer, the temperature drops significantly, and the liquid water converts to ice. Once the temperature drops to the freezing point of water, the atoms reorganize into the more orderly arrangement of atoms found in ice, because this configuration is more stable than liquid at the lower temperature in your freezer. Because the conditions are different inside the freezer than they are outside of it, the water changes, by freezing, to achieve stability in its new environment. Similarly, rocks under different environmental conditions change to achieve the most stable conditions. The difference is that metamorphism involves different stable forms of solid minerals rather than the stability conditions of solids, liquids, and gases.

Note that at certain combinations of temperature and pressure, indicated by the blue and red lines on the graph in Figure 6.6, more than one phase can coexist: liquid water and steam, or ice and liquid water. At the point where the blue and red lines converge, liquid, ice, *and* steam can exist at the same time.

Putting It Together—What Is the Role of Temperature in Metamorphism?

• Temperature increases along the geothermal gradient as rocks are buried deeper in Earth's crust by tectonic processes or by the weight of the rocks above them. Rocks also experience higher temperatures near igneous intrusions.

• A temperature increase can drive the chemical reactions of metamorphism. These reactions include dehydration and degassing of minerals that contain water or gas molecules in their structure.

• Minerals are stable within a defined range of temperature and pressure. Outside of this stability range, original minerals break down and form new minerals.

6.3 What Role Does Pressure Play in Metamorphism?

A striking feature of many outcrops of metamorphic rocks, such as the photo at the beginning of the chapter and the road-cut exposure in Figure 6.1, is the subtle layering caused by parallel arrangement of minerals or alternating bands of different minerals. This layering results from physical and chemical changes that relate to high pressure.

Pressure affects not only mineral stability, but also crystal size and orientation, which, as we know, determine rock texture. Pressure relates to the concept of **stress**. Both stress and pressure are defined as the magnitude of a force divided by the area of the surface on which the force is applied. When you are standing you exert a pressure, or stress, that is equal to your weight, which is a force, divided by the area of your feet. **Figure 6.7** illustrates two types of stress. **Normal stress** is force applied perpendicular to a surface (Figure 6.7a), and **shear stress** is force applied parallel to a surface (Figure 6.7b). When you stand still, you exert a normal stress on the floor. If you slide your feet across the floor, then you are exerting a shear stress. In Earth, the increasing force exerted by the weight of overlying rock causes pressure to increase at greater depth. **Pressure** is a special type of normal stress where the forces are equal in all three dimensions (Figure 6.7c).

Stress can change the shape of the stressed material. Geologists use the term **strain** to describe the deformation of rock that occurs as a result of an applied stress. The changes in shape and size of the stressed cubes depicted in Figure 6.7 are examples of strain.

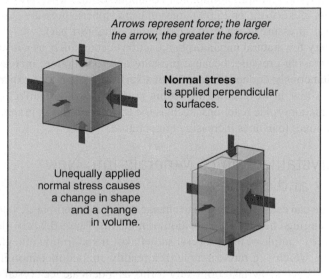

(a)

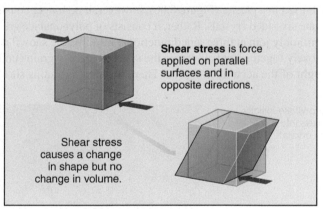

(b)

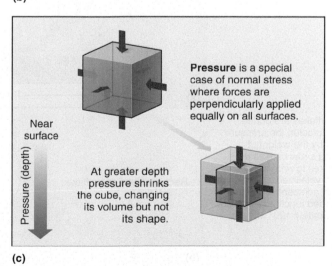

(c)

▲ Figure 6.7 How normal and shear stresses deform rock.

Changes in Mineral Stability

Mineral stability relates not only to temperature, but also to applied stress, especially equal-dimensional normal stress, or pressure (Figure 6.7c). Atoms rearrange into denser, more closely packed crystal structures at high pressure, as illustrated by the conversion of graphite to diamond (refer to Figure 2.23). Laboratory experiments show that the metamorphic reaction of graphite to diamond occurs when pressure exceeds 25 kilobars, or 25,000 times the atmospheric pressure at sea level (air pressure at sea level is defined as equal to 1 bar).

Very few natural metamorphic reactions occur solely as a result of increasing pressure, because pressure and temperature increase simultaneously at depth below Earth's surface. Laboratory experiments show, however, that increasing pressure has more effect on some metamorphic reactions, such as the transformation of graphite to diamond, than does increasing temperature.

Recrystallization of Minerals into New Sizes and Shapes

Pressure can cause rock texture to change during metamorphism without changing chemical composition or minerals. **Figure 6.8** provides a simple example of this process: snowflakes transforming into glacial ice, which is a rare example of a readily studied metamorphic change. Each snowflake originally forms as a delicate ice crystal at low pressure in the atmosphere (Figure 6.8a). When snow becomes deeply buried in a glacier (Figure 6.8b), however, it no longer consists of delicate six-sided crystals. Rather, it consists of polygonal ice grains that commonly join at three-sided junctions. Initially, the snowflakes pack closely together and eliminate the spaces between grains under the weight of the accumulating ice. Then, the smaller grains shrink,

Notice the rounded quartz grains surrounded by deep-red hematite cement in this microscope photo of sandstone.

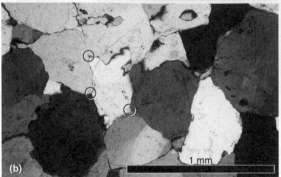

When metamorphosed, the quartz grains recrystallize into grains with straighter-sides that seem to penetrate into one another, and locally intersect at three-sided junctions. (Example junctions are circled).

▲ **Figure 6.9 The transformation from sandstone to metamorphic rock.**

and the larger grains grow as a result of **recrystallization**, which involves the transfer of atoms from one part of a crystal to another part of the crystal or to an adjacent crystal. The atoms within the crystals rearrange as a result of the increasing pressure from the accumulating weight of the glacier; they fit tighter together, making the minerals more compact and denser. Recrystallization is a metamorphic process that changes the size and shape of existing minerals rather than making new minerals.

Another example of recrystallization occurs when quartz sandstone metamorphoses as shown in **Figure 6.9**. You can see the individual, round quartz grains within the sedimentary rock (Figure 6.9a). During metamorphism, however, the rounded quartz grains recrystallize to form a texture of straight-sided crystals that form three-sided junctions (Figure 6.9b).

Foliation: How Rock Textures Record Strain and Recrystallization

Strain and recrystallization change rock texture. **Figures 6.10**, **6.11**, and **6.12** illustrate how crystals may rearrange into layered planes, a process called **foliation**, depending on the type and orientation of the stress during metamorphism. The term "foliation" comes from the Latin word *folium*, meaning "*leaf*." Foliation planes are recognized by:

- the preferred orientation of minerals (Figure 6.10).
- alternating bands of different minerals (Figure 6.11).
- the flattening and stretching of minerals (Figure 6.12).

Foliation forms by physical rotation of preexisting minerals, recrystallization, or dissolution and new mineral growth along a preferred orientation rather than a random orientation. Rotation or growth of crystals parallel to the

Each snowflake initially crystallizes into a six-pronged crystal.

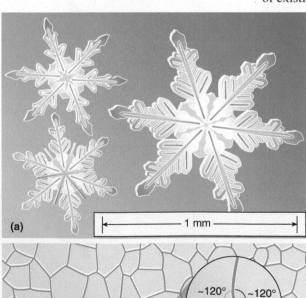

As snowflakes are buried within a glacier, the pressure exerted by the weight of overlying snow causes the snowflakes to recrystallize. The recrystallized ice crystals join with adjacent crystals at three-sided junctions, forming approximately 120° angles.

Recrystallized snow flakes

▲ **Figure 6.8 The metamorphic change from snowflake to ice.**

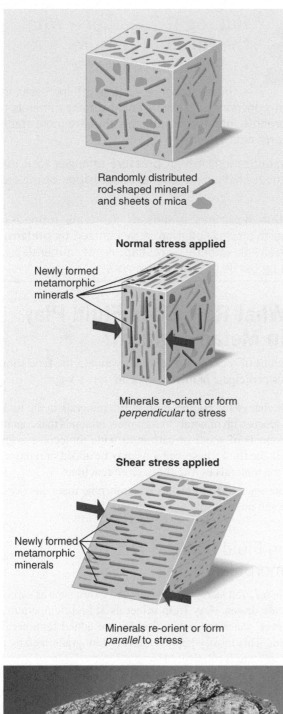

Randomly distributed rod-shaped mineral and sheets of mica

Normal stress applied

Newly formed metamorphic minerals

Minerals re-orient or form *perpendicular* to stress

Shear stress applied

Newly formed metamorphic minerals

Minerals re-orient or form *parallel* to stress

Metamorphic rock with foliation defined by oriented mica crystals

▲ Figure 6.10 How minerals reorient and crystallize to form foliation.

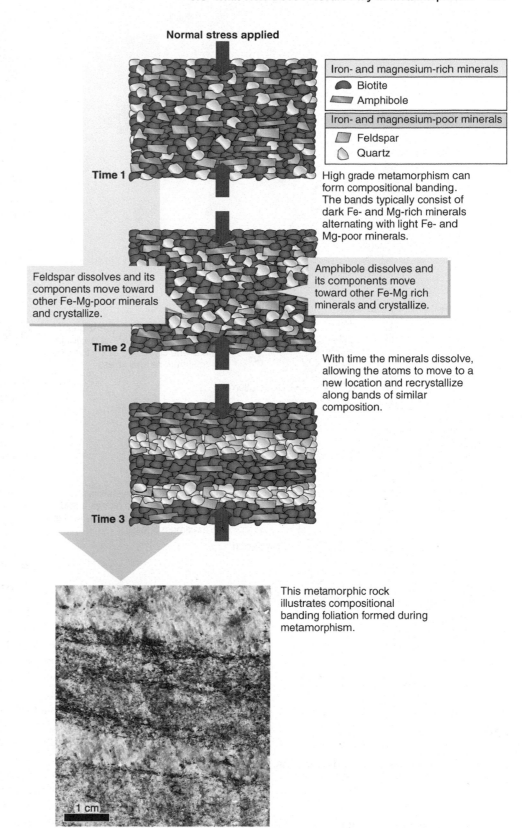

Normal stress applied

Iron- and magnesium-rich minerals	
⬭	Biotite
▱	Amphibole

Iron- and magnesium-poor minerals	
▱	Feldspar
⬠	Quartz

Time 1

High grade metamorphism can form compositional banding. The bands typically consist of dark Fe- and Mg-rich minerals alternating with light Fe- and Mg-poor minerals.

Feldspar dissolves and its components move toward other Fe-Mg-poor minerals and crystallize.

Amphibole dissolves and its components move toward other Fe-Mg rich minerals and crystallize.

Time 2

With time the minerals dissolve, allowing the atoms to move to a new location and recrystallize along bands of similar composition.

Time 3

This metamorphic rock illustrates compositional banding foliation formed during metamorphism.

1 cm

▲ Figure 6.11 How compositional banding forms foliation.

ACTIVE ART

Forming Foliation. See how the three types of foliation form.

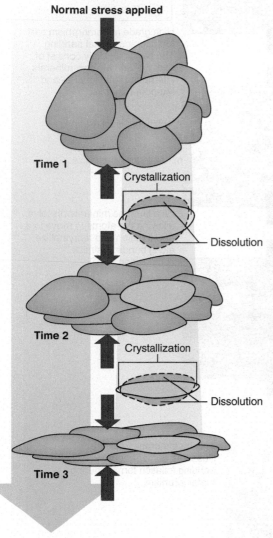

Normal stress applied

Time 1

Crystallization

Under normal stress, quartz grains dissolve along the axis of maximum stress and crystallize along the axis of minimum normal stress.

Dissolution

Time 2

Crystallization

This process of recrystallization causes the grains to elongate and flatten along the axis perpendicular to maximum stress.

Dissolution

Time 3

▲ **Figure 6.12 How minerals flatten to form foliation.**

direction of least stress typically forms foliation in rocks with platy minerals, such as mica (Figure 6.10). Minerals also can dissolve and recrystallize, or their atoms can rearrange into new minerals that form parallel to the direction of the smallest normal stress (Figure 6.12). In other cases, the minerals segregate into compositional layers that are oriented according to the stress orientation (Figure 6.11). All types of foliation are planes in the rock that are perpendicular to the greatest normal stress or parallel to shear stress.

Putting It Together—What Is the Role of Pressure in Metamorphism?

• Pressure increases as depth increases inside Earth. Increasing pressure causes many minerals to metamorphose into denser minerals that are more stable at high pressure.

• Recrystallization at high pressure changes rock texture without changing chemical composition or mineral content.

• Foliation describes planes of minerals formed in response to stress. Foliation is recognized by preferred orientation of minerals, reshaping of minerals, or alternating bands of different minerals.

6.4 What Role Does Fluid Play in Metamorphism?

Observations of rocks in the laboratory and in the field show that fluids participate in metamorphism in two ways:

1. Fluids change the composition of the minerals in the rock. Fluid reacts with minerals to form new minerals that contain components of water or carbon dioxide molecules originally in the fluid. Dissolved ions may be added or removed from the minerals during these same reactions.

2. The presence of fluid makes metamorphic reactions occur faster and more easily.

Adding Fluid and Gas During Metamorphism

In Section 6.2, you learned that water and gases, such as carbon dioxide, are driven away from minerals at high temperature. In some cases, however, water and gas molecules are added to minerals during metamorphism. This mostly happens during low-grade metamorphism when the temperature is relatively low.

You already know about processes of chemical weathering, in which water is added to minerals (Section 5.1). Recall the example of feldspar, which lacks water. When feldspar weathers, it turns into clay, which contains water molecules in its crystal structure. In contrast, metamorphic reactions that produce different minerals by adding water or carbon dioxide occur at much higher temperature and pressure than chemical weathering.

The metamorphism of olivine in mafic and ultramafic igneous rocks is a common metamorphic reaction involving water and carbon dioxide. Under low-grade metamorphic conditions, water reacts readily with olivine, as does the small amount of carbon dioxide that is almost always dissolved in the water. This reaction is written as:

$$2Mg_2SiO_4 + 2H_2O + CO_{2(gas)} \rightarrow Mg_3Si_2O_5(OH)_4 + MgCO_3$$
Mg-olivine   water   carbon dioxide   serpentine   magnesite

Magnesite is a carbonate mineral with a crystal structure similar to calcite. Serpentine is a complex silicate mineral (see Table 2.2) that includes the most abundant form of asbestos, which was once widely used to make insulation and fireproof fabric.

Fluid Enhances Metamorphism

Without water, chemical reactions between solids take place extremely slowly or never even begin. **Figure 6.13** illustrates the role water plays in making reactions occur faster and more easily. If you mix powdered aspirin and baking soda in a shallow dish, nothing happens. If you add water, however, the aspirin dissolves to form an acid that reacts with the baking soda and releases bubbles of carbon dioxide gas.

Laboratory experiments repeatedly show that metamorphic reactions take place much more quickly in the presence of water, and commonly at lower temperature, than they do in dry environments. Experimental mineral reactions that require many days at temperatures around 1000°C may take place in minutes at half the temperature in the presence of water.

Clearly, water plays an impressive role in enhancing metamorphic reactions, but why? The most important factor is the role of water as a solvent (also see Section 5.1). Water dissolves minerals, and as a result, the ions in the minerals are able to move through the water to react with other ions. Also, experiments show that while crystals readily precipitate out of aqueous solutions, crystals are much slower to form from a combination of solids.

The Origin of Fluids

Where do the fluids that drive metamorphism come from? Fluid may either be part of the original (parent) rock or be introduced into the

▲ **Figure 6.13 Adding water to trigger a reaction.** A dry mixture of crushed aspirin and baking soda does not react. Adding water to the mixture, however, causes a visible, bubbling reaction between the two compounds.

metamorphic environment. All rocks that form at or near Earth's surface contain water in open pore spaces or fractures, or along boundaries between mineral grains. When these rocks are buried to low-grade metamorphic conditions, the fluids are already there to participate in the metamorphic reactions. The high-grade dehydration and degassing reactions described in Section 6.2 also generate fluids that rise toward regions of lower temperature and pressure. These fluids participate in and help drive metamorphic reactions at less extreme temperatures and pressures.

Igneous intrusions are another important source of fluids for metamorphism. Hot, watery solutions commonly form during the late stages of magma crystallization (see Section 4.10). These fluids are rich in reactive ions and move into the surrounding rock as it metamorphoses because of heating from the intrusion. Fluid from the magma not only delivers reaction-enhancing water, but also heat (which speeds up reactions) and ions from the magma (which participate in reactions). These solutions may introduce economically valuable metal ions that form ores of metal sulfide and oxide minerals in metamorphic rocks. Even if the magma lacks abundant water to be forced into the surrounding rocks, its high temperature causes water in the surrounding rock to heat up. This process moves warm, reactive water into rocks that already contain the ingredients for metamorphism. At that point, all that is needed is a bit more heat and water from the migrating fluid to initiate the reactions.

Fluid Changes the Composition of the Rocks

The ability of water to transport chemical components means that metamorphism in the presence of abundant fluid produces a metamorphic rock with a bulk composition very different from that of the parent rock. Ions in the water deliver new components, and elements liberated by mineral breakdown are transported away in solution.

Putting It Together—What Is the Role of Fluid in Metamorphism?

• Fluids, present in the original rock or introduced from magma or high-temperature dehydration and degassing reactions, create metamorphic minerals that incorporate components from the fluid molecules.

• Metamorphic reactions take place faster and at lower temperature in the presence of water than they do under dry conditions.

• The movement of fluid during metamorphism delivers some ions and removes others, so that the metamorphic rock has a different bulk composition than that of the parent rock.

6.5 Why Do Metamorphic Rocks Exist at the Surface?

If temperature and pressure are critical factors of metamorphism and can convert graphite to diamond, then why is diamond not converted back (reverted) to graphite at Earth's surface (i.e., much lower) temperature and pressure? In other words, do you need to worry that the diamond ring you inherited will eventually become a piece of graphite? How do metamorphic rocks, composed of minerals that are stable at high temperature and

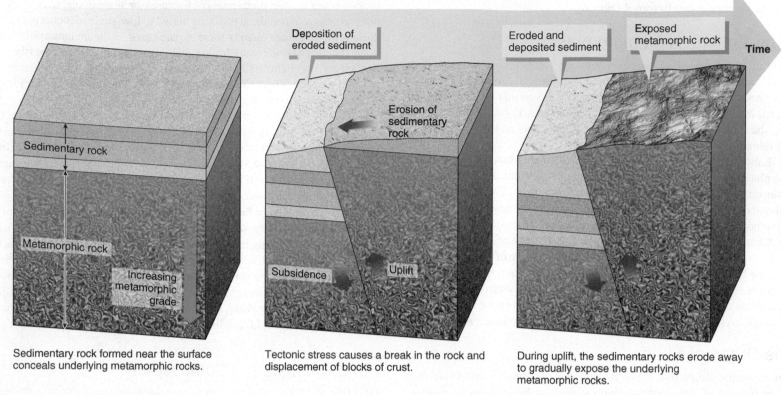

▲ **Figure 6.14 How metamorphic rocks are exposed at Earth's surface.**

Deposition of eroded sediment

Erosion of sedimentary rock

Sedimentary rock

Metamorphic rock

Increasing metamorphic grade

Sedimentary rock formed near the surface conceals underlying metamorphic rocks.

Subsidence   **Uplift**

Tectonic stress causes a break in the rock and displacement of blocks of crust.

Eroded and deposited sediment

Exposed metamorphic rock

Time

During uplift, the sedimentary rocks erode away to gradually expose the underlying metamorphic rocks.

ACTIVE ART

Exposing Metamorphic Rocks. *See how metamorphic rocks end up at the surface after forming at great depth.*

pressure, even end up being exposed at the surface where temperature and pressure are low? Also, once exposed, how can they retain their metamorphosed form?

Most metamorphic rocks form kilometers underground (see Figures 6.3, 6.4, and 6.5) and are exposed to view for study only when overlying rock has been removed. This usually occurs during mountain building and erosion, which uplifts rock and reveals what was once below the surface. That is the primary reason that outcrops of metamorphic rocks are most commonly found in active or ancient mountain belts. We will explore the process of mountain building more thoroughly in Chapters 11–13. For now, we focus on the primary processes that expose metamorphic rocks. **Figure 6.14** illustrates that as mountains rise, erosion removes the overlying uplifted rock, eventually exposing metamorphic rocks that formed at great depth.

The reason metamorphic minerals are not transformed back into the original pre-metamorphic minerals at surface temperatures and pressure is explained by the same factors that account for the original metamorphism. Heat is commonly needed for chemical reactions to take place, and because the temperature decreases as the rocks are uplifted toward the surface, there is not enough heat to drive the reverse reactions. The loss of fluid during metamorphic dehydration also prevents metamorphic minerals from reverting to their original minerals. The reactions cannot reverse if fluids moved away from the rock and are no longer available to participate in the reactions. For these reasons, high-grade metamorphic rocks retain their form at the surface and thus, can be studied.

However, it is important to remember that metamorphic rocks are not only transformed sedimentary and igneous rocks. Metamorphic rocks themselves can be metamorphosed. In certain cases, high-grade metamorphic rocks may rise to depths associated with low-grade metamorphism and reside there for a long time. Before they are further uplifted and exposed, they may be re-metamorphosed, especially if fluids are abundant.

Therefore, your diamond will not revert to graphite on Earth's surface because the transformation of diamond to graphite requires high temperature to initiate the reaction. The temperature at Earth's surface is not sufficiently high for the reaction to take place. As a result, not only are most metamorphic rocks not converted back to their original form at Earth's surface, but your family heirloom will remain a diamond for many, many, many generations to come.

Putting It Together—Why Do Metamorphic Rocks Exist at the Surface?

• Mountain-building processes uplift metamorphic rocks, which are later exposed at the surface when erosion removes overlying rocks.

• Once formed, metamorphic rocks do not revert to their original minerals at Earth's surface, because sufficient temperature, pressure, fluid, or time to promote the necessary reverse reactions do not exist at the surface.

6.6 How Do We Know . . . How to Determine the Stability of Minerals?

Picture the Problem

Under What Conditions Are Metamorphic Minerals Stable? How do geologists know what minerals are stable under specific metamorphic conditions? How high does the temperature have to be for muscovite mica to dehydrate? How much water and carbon dioxide needs to be present before olivine metamorphoses to hydrous serpentine and the carbonate mineral magnesite? These mineral reactions served as examples of metamorphic changes in the previous three sections, but how are these reactions known to occur?

Geologists infer the reactions from observations. Look closely at the microscopic view of a metamorphic rock shown in **Figure 6.15**. The mineral sillimanite has formed within the mineral andalusite. This observation suggests that the andalusite grew first in the metamorphic rock. Then, the andalusite stopped growing, and sillimanite formed by metamorphic reactions that consumed andalusite. What caused andalusite to become unstable, while sillimanite became stable? If the necessary variations in temperature, pressure, or fluid composition for different mineral stabilities can be determined, then it is possible to reconstruct the history of changing metamorphic conditions that produced this rock.

Experimental Setup

How Are Metamorphic Conditions Reproduced in the Lab? Geologists explore metamorphic mineral stability by conducting laboratory experiments. As an example, you can easily determine the stability conditions of liquid water at Earth's surface depicted in Figure 6.6. You do not need to enter a lab to find out that water freezes to ice on winter days when temperatures descend below 0°C, and that water turns to steam when heated to temperatures above 100°C. But to fully reproduce the stability relationships represented in Figure 6.6, you would need laboratory facilities that produce pressures higher or lower than those observed on Earth's surface. In a lab setting, the temperature could be varied for each chosen pressure, and you could observe exactly when the liquid water would freeze or boil. In a similar way, geologists investigate mineral stability in laboratory experiments by progressively changing temperature, pressure, and fluid content to simulate metamorphic conditions within Earth. This approach resembles Tuttle and Bowen's experiments to determine the conditions required to melt granite (Section 4.5).

The changing mineral stability illustrated by the rock shown in Figure 6.15 was explained by laboratory experiments conducted in the 1970s by Michael Holdaway at Southern Methodist University, in Dallas, Texas. Holdaway worked with andalusite, sillimanite, and a third mineral, kyanite, all of which are polymorph minerals. The term "polymorph" indicates that all three minerals share the same chemical composition, Al_2SiO_5, but have different atomic structures. One or more of these aluminum-silicate polymorphs commonly appear in aluminum-rich metamorphic rocks, so Holdaway hoped to determine their stability conditions, which would in turn provide tremendous insight into the temperature and pressure conditions that rocks experience during metamorphism.

It was important for the experiments that these minerals have the identical chemical composition and do not contain water in their crystal structures, because these features make the reaction relationships simple compared to chemical reactions involving multiple minerals of different composition or minerals that contain water. In other words, only temperature and pressure needed to be measured in the laboratory to determine the stability of these three minerals.

Figure 6.16 illustrates an experimental setup used by Holdaway for determining the stability of minerals at different pressures and

▲ **Figure 6.15 Microscopic evidence of changing metamorphic conditions.** In this metamorphic rock, the mineral sillimanite crystallized, while the mineral andalusite dissolved, so that the sillimanite replaced the andalusite. This observation reveals that the metamorphic conditions changed from circumstances in which andalusite is stable to conditions in which sillimanite is stable. (Polarized lighting produces unnatural colors in this microscope photo. The color of both minerals actually is tan to brown.)

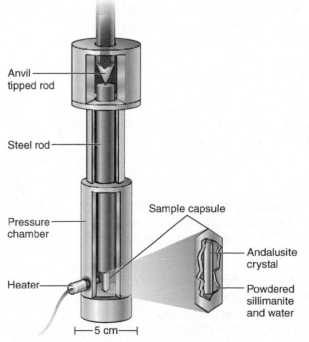

Mineral-stability studies use a cylindrical pressure chamber.

A capsule containing the mineral samples is placed at the bottom of the chamber. A steel rod fills the remainder of the chamber above the sample capsule.

The top of the chamber closes by turning a nut that lowers an anvil tip onto the steel rod to apply a measured pressure against the sample capsule.

A heater heats the chamber to the temperature required for the experiment.

Anvil tipped rod

Steel rod

Pressure chamber

Heater

Sample capsule

Andalusite crystal

Powdered sillimanite and water

⊢ 5 cm ⊣

▲ **Figure 6.16 An experimental apparatus for determining mineral stability.**

temperatures. A heated pressure chamber encloses a sample capsule containing the reaction components plus water to speed up the reaction. Temperature and pressure are changed in a controlled manner and measured.

To explore the stability of andalusite compared to sillimanite, Holdaway placed a single andalusite crystal of known mass into the sample capsule and surrounded it with powdered sillimanite and water. After heating the capsule at a measured temperature and pressure for several weeks, he turned off the heater. The removal of heat energy stopped whatever reactions may have been taking place in the capsule. Then Holdaway removed and weighed the andalusite crystal.

The mass of the andalusite reveals which mineral is more stable. If andalusite mass increases during the reaction, then andalusite is stable and sillimanite is not. The instability of sillimanite is inferred in this case because the breakdown of unstable sillimanite is the only source of the aluminum and silica that are needed for andalusite growth. On the other hand, if the mass of the andalusite crystal decreases, then andalusite is not stable, and sillimanite is the stable polymorph of Al_2SiO_5.

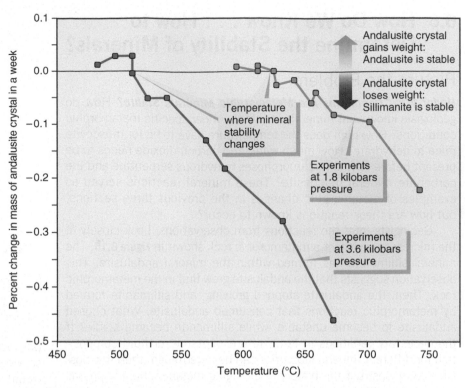

▲ **Figure 6.17 Visualize the experimental results.** Andalusite is stable at temperatures and pressures in which the andalusite crystal grew, and therefore gained mass during the experiment. Loss of mass in the andalusite reveals conditions in which andalusite is unstable and sillimanite is stable. The temperatures at which mineral stability changes differ according to pressure.

Visualize the Results

At What Temperatures and Pressures Are Aluminum Silicate Minerals Stable? **Figure 6.17** graphically illustrates the results of several of Holdaway's experiments. The changing mass of the starting andalusite crystal reveals whether andalusite or sillimanite is the more stable mineral at the recorded temperature and pressure. At the relatively low pressure of 1.8 kilobars (equivalent to the pressure that occurs at a depth of approximately 6 kilometers below Earth's surface), andalusite is stable to a temperature of approximately 625°C. Double the pressure to 3.6 kilobars, however, and andalusite is stable only to approximately 520°C.

Figure 6.18 summarizes the results of many similar experiments, including experiments with kyanite. Holdaway's results, along with those of other geologists, outline the stability conditions of the three aluminum-silicate minerals in terms of the experimentally varied temperature and pressure. The experiments reveal that kyanite is more stable at higher pressure and lower temperature than andalusite or sillimanite; that sillimanite is more stable at higher temperature than either kyanite or andalusite; and that andalusite is more stable at lower pressure over a wide range in temperature.

Insights

How Do Lab Results Increase Understanding of Metamorphism? What do the laboratory results mean for interpreting real metamorphic rocks? Take another look at Figure 6.15 and examine the graph in Figure 6.18. Two conclusions emerge:

1. The metamorphic reaction that formed the rock in Figure 6.15 requires an increase in temperature, or an increase in pressure, or both, in order for andalusite to become unstable and sillimanite to be stable. The rock, therefore, experienced an increase in metamorphic grade.

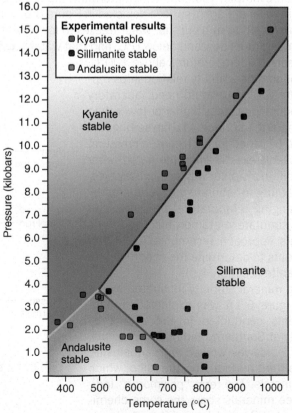

Each dot on this graph represents an experimental result that indicates which of the three aluminum-silicate minerals is stable at the indicated temperature and pressure.

Lines drawn between the data points separate the graph into three areas representing the ranges of temperatures and pressures where kyanite, sillimanite, and andalusite are each stable.

▲ **Figure 6.18 Aluminum-silicate mineral stability.**

2. The metamorphic temperature at which this rock formed had to exceed 500°C, which is the lowest experimental temperature where sillimanite is stable.

Nearly all metamorphic rocks contain more than one mineral. A variety of experimental results have allowed geologists to define the stability conditions for most of the minerals that are commonly seen. By applying this information, it is possible to ascertain the temperature of metamorphism to within 50°C and the pressure to within 0.5 kilobar. Knowledge of mineral stability, therefore, allows scientists to reconstruct metamorphic conditions and determine under what conditions most rock samples were formed.

Putting It Together—How Do We Know How to Determine the Stability of Minerals?

• Many minerals are stable over limited ranges in temperature and pressure. Changes in these two variables cause reactions that result in the formation of new mineral or minerals.

• Laboratory experiments conducted with different minerals at measured temperatures and pressures reveal the specific conditions under which common metamorphic minerals are stable.

6.7 How Are the Conditions of Metamorphism Determined?

Once geologists recognize that a rock is metamorphic, whether from recrystallization, foliation, or the presence of unique metamorphic minerals, they set out to find out how strongly it has been metamorphosed. Why do geologists want to know what temperature and pressure metamorphic rock experienced? So that they can reconstruct how deep it once was inside Earth. If the rock contains andalusite, kyanite, or sillimanite, then the data graphed in Figure 6.18 can answer that question. Similar stability-field graphs documenting metamorphic reactions have been made for most of the commonly observed metamorphic minerals.

Using Index Minerals

Minerals used to estimate the pressure and temperature conditions where metamorphic rocks form are known as **index minerals**, which reveal metamorphic grade. Geologists have combined field observations with knowledge gained from experiments on mineral stability to develop a generalized guide to the minerals whose presence indicates specific grades of metamorphism.

Figure 6.19 summarizes the relationship between the minerals present in a rock and metamorphic grade. Only minerals with limited ranges of stability serve as useful index minerals. Other minerals, such as quartz and feldspar, do not indicate metamorphic grade because they are stable over large temperature and pressure ranges (Figure 6.19).

Consider again the rock you picked up along the road (shown in Figure 6.1c) in your virtual field trip, and refer to Figure 6.19. The presence of muscovite is not very diagnostic of metamorphic conditions. Garnet reveals that the rock could have experienced medium- or high-grade metamorphism. Staurolite is the most useful index mineral in your rock, because its presence indicates metamorphism toward the upper limits of medium-grade conditions. The stability conditions of staurolite determined in experiments suggest that metamorphism for your rock took place at a minimum temperature of approximately 550°C and at pressure of 2 to 7 kilobars. Remember that pressure depends on depth, so if the pressure range is known, you also know the depth below Earth's surface where metamorphism took place. Because your rock metamorphosed at a pressure of 2 to 7 kilobars, the equivalent depth is approximately 6.5 to 23 kilometers beneath Earth's surface.

Putting It Together—How Are the Conditions of Metamorphism Determined?

• Index minerals reveal metamorphic temperature and pressure because they are stable over limited ranges of temperature, pressure, or both

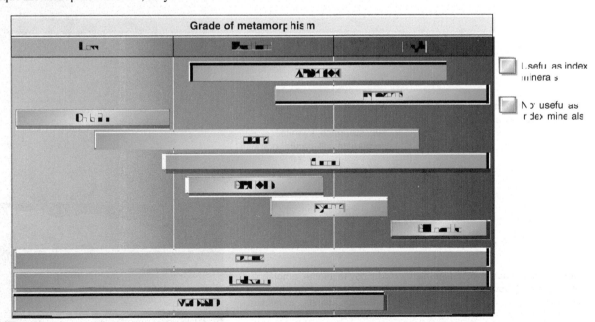

▲ **Figure 6.19 Using index minerals.** Stability of some metamorphic minerals relates to relatively narrow ranges of temperature and pressure that correspond to different metamorphic grades. For example, chlorite, a green, iron- and magnesium-rich mica, indicates low-grade metamorphism whereas sillimanite indicates high-grade conditions. Some minerals, such as quartz, feldspar, and muscovite are not good indicators of grade, because they are stable over a wide range of metamorphic grades. This diagram emphasizes minerals whose stability is affected by pressure and temperature. Other stability relationships are determined for minerals whose stability is mostly a function of temperature or pressure alone.

6.8 How Are Metamorphic Rocks Classified?

We now know that observed variations in the mineral content and texture of metamorphic rocks reveal the temperature and pressure of metamorphism, as well as the composition of parent rocks and reactive fluids. Composition and texture are, therefore, useful criteria for classifying metamorphic rocks. In metamorphic rocks, the primary textural attributes are the presence or absence of foliation and mineral grain size.

Using Composition and Texture to Classify Metamorphic Rocks

Figure 6.20 illustrates how to use composition and texture to classify and name metamorphic rocks. The first distinction is whether the rock contains a well-developed foliation. If it does, then the rock is named primarily by the type of foliation and whether the minerals are too small (fine grained) to distinguish with the naked eye or are large (coarse grained) and easily visible. Notice that each foliated rock type potentially contains many different minerals, so mineral content is not distinctive for naming the rock.

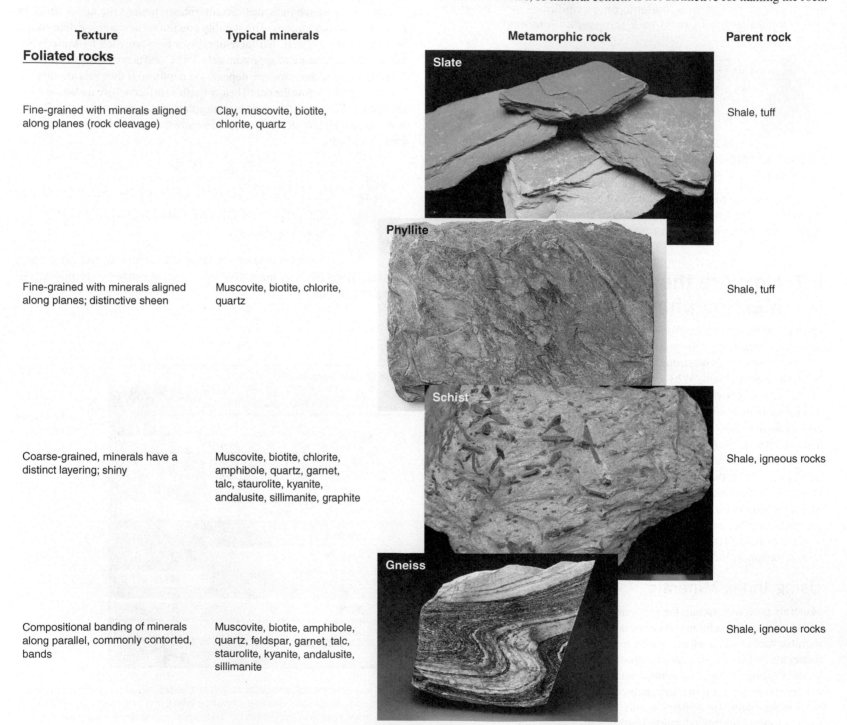

Texture	Typical minerals	Metamorphic rock	Parent rock
Foliated rocks			
Fine-grained with minerals aligned along planes (rock cleavage)	Clay, muscovite, biotite, chlorite, quartz	Slate	Shale, tuff
Fine-grained with minerals aligned along planes; distinctive sheen	Muscovite, biotite, chlorite, quartz	Phyllite	Shale, tuff
Coarse-grained, minerals have a distinct layering; shiny	Muscovite, biotite, chlorite, amphibole, quartz, garnet, talc, staurolite, kyanite, andalusite, sillimanite, graphite	Schist	Shale, igneous rocks
Compositional banding of minerals along parallel, commonly contorted, bands	Muscovite, biotite, amphibole, quartz, feldspar, garnet, talc, staurolite, kyanite, andalusite, sillimanite	Gneiss	Shale, igneous rocks

▲ **Figure 6.20 The classification of metamorphic rocks.** Texture and mineral content form the basis of this classification scheme of metamorphic rocks. Foliated rocks are divided primarily according to texture (grain size), whereas non-foliated rocks are named primarily according to composition (minerals present).

Texture	Typical minerals	Metamorphic rock	Parent rock
Non-foliated rocks or weakly foliated rocks			
Mosaic of coarse grains	Calcite, dolomite		Limestone, dolostone
Mosaic of coarse grains	Quartz		Quartz, sandstone
Mosaic of fine, microscopic grains	Quartz, feldspar, muscovite, biotite, garnet, andalusite		Any fine-grained rock
Fine-grained, commonly fibrous or greasy	Serpentine		Peridotite
Shiny, curving broken surfaces	Carbon (not a mineral)		Coal
Fine-grained	Chlorite, amphibole, feldspar, quartz		Mafic igneous rocks
Coarse-grained; minerals commonly aligned	Amphibole, feldspar, garnet, quartz		Mafic and intermediate igneous rocks
Mosaic of coarse grains	Pyroxene, garnet		Mafic igneous rocks

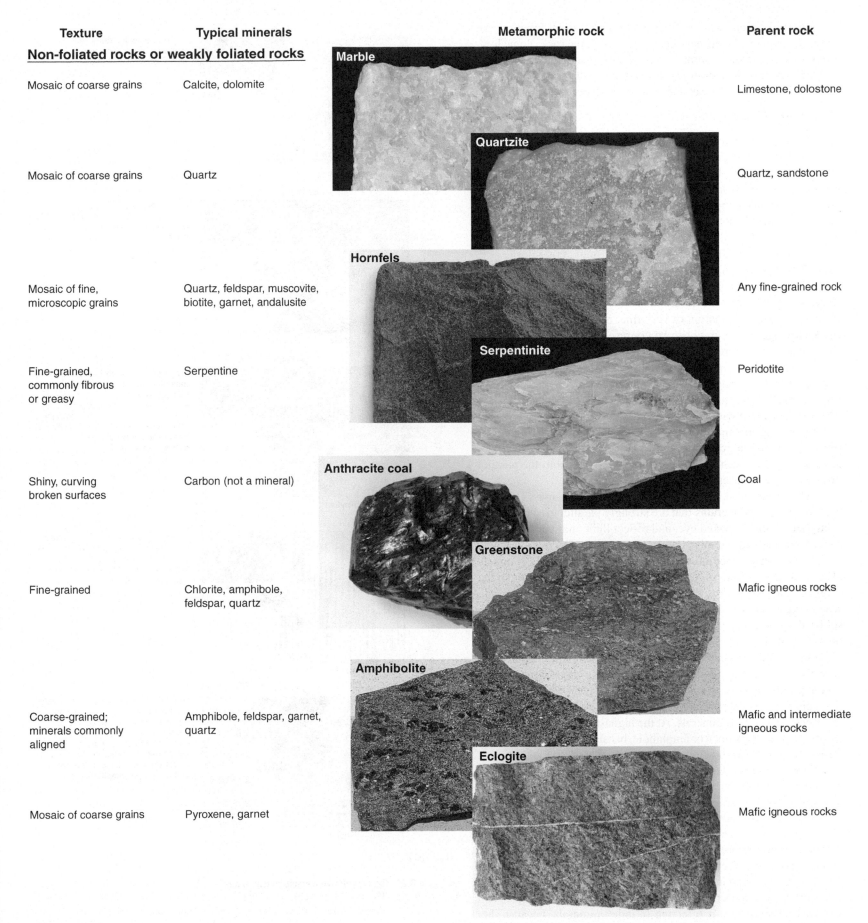

Metamorphic rock images: Marble, Quartzite, Hornfels, Serpentinite, Anthracite coal, Greenstone, Amphibolite, Eclogite

▲ Figure 6.20 (Continued)

In contrast, for rocks with no foliation or an indistinct foliation, the mineral content is the primary basis for naming the rock. Some low- to medium-grade metamorphic rocks retain many of their original features. These rocks are simply described by putting the prefix *meta* in front of the parent rock name, i.e., metabasalt or metaconglomerate.

Foliated Rocks

Recall from Section 6.3 that many rocks that undergo metamorphic strain become foliated. **Figure 6.21** illustrates the rock names associated with changing foliation and grain size produced by increasing metamorphism of shale, the most common sedimentary rock. At progressively higher temperature and pressure, the predominant clay minerals present in shale at relatively low temperature and pressure become unstable. Initially, the clay minerals convert to micas. Muscovite is the most abundant low-grade mica, but chlorite (see Table 2.2) may form if the original clays contain iron or magnesium.

At low-grade metamorphic conditions, shale first transforms into **slate**, which exhibits well-developed rock cleavage planes produced from the parallel orientation of very fine, microscopic mica grains. Rock-cleavage foliation, not to be confused with mineral cleavage (Section 2.1), is the preferential splitting of rock along planes of parallel microscopic layers of mica. The property of slate that causes it to break into thin sheets of hard rock makes it ideally suited as a roofing material, flooring, and even for making pool tables (the slate tabletop is covered with felt).

Increasing metamorphic grade leads next to the formation of **phyllite**. The mica grains in phyllite are coarser than in slate and generate a silky sheen caused when light reflects from the parallel mica cleavage surfaces that define its foliation.

Schist forms at still higher temperatures and pressures. This rock has even larger mica grains that are strongly parallel to one another, easily seen with the naked eye, and reflect light, making for a shiny rock. Your road-trip sample is schist (Figure 6.1). The mica grains in schist commonly surround scattered, well-formed crystals of minerals such as garnet, staurolite, and kyanite. If chlorite is present in the original slate or phyllite, then it reacts to form biotite at the grade where schist forms. If the parent rock was organic-rich shale, then the schist may contain graphite produced by the metamorphism of organic matter.

Gneiss (pronounced "nice") forms when temperature and pressure rise to high grade. Gneiss is defined by its characteristic foliation of parallel compositional layers of readily visible, light-colored (e.g., quartz and feldspar) and dark-colored (e.g., biotite, amphibole, pyroxene, and garnet) minerals. At the highest metamorphic temperatures, gneisses lack mica or amphibole, because these water-bearing minerals break down by dehydration reactions.

If the high-grade temperature, pressure, and fluid conditions are appropriate for melting to begin, then the resulting rock contains the textures of both metamorphic and igneous rock. An example of such a rock, called **migmatite**, is shown in **Figure 6.22**. The migmatite resembles gneiss except that the light-colored bands have the igneous-crystallization texture of granite, whereas the dark layers feature metamorphic crystal growth and recrystallization.

The most abundant coarse-grained minerals in foliated rocks are commonly used as modifiers in the rock name. These adjectives are particularly useful if referring to index minerals, because the name of the rock then conveys information about grade independent

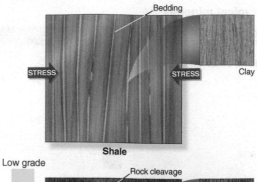

Bedding

Shale

Shale is a sedimentary rock composed of compacted clay-size, clay minerals.

Clay

Low grade

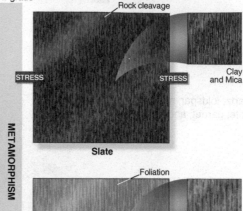

Rock cleavage

Slate

Clay and Mica

New minerals such as muscovite and chlorite micas start to form. This produces fine banding or cleavage in the rock that is perpendicular to the maximum direction of normal stress.

METAMORPHISM

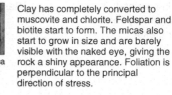

Foliation

Phyllite

Mica

Clay has completely converted to muscovite and chlorite. Feldspar and biotite start to form. The micas also start to grow in size and are barely visible with the naked eye, giving the rock a shiny appearance. Foliation is perpendicular to the principal direction of stress.

METAMORPHISM

Foliation

Schist

Mica and Garnet

Micas recrystallize into large, easily seen crystals. New minerals such as garnet and staurolite form. Foliation is perpendicular to the principal direction of stress.

METAMORPHISM

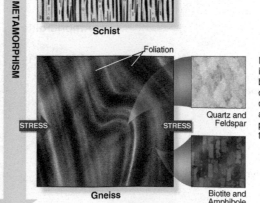

Foliation

Gneiss

Quartz and Feldspar

Biotite and Amphibole

Minerals recrystallize and segregate into bands or layers. The foliation bands typically are feldspar and quartz, alternating with bands of darker biotite or amphibole. Minerals are coarse-grained and show preferred orientations perpendicular to the principal direction of stress.

High grade

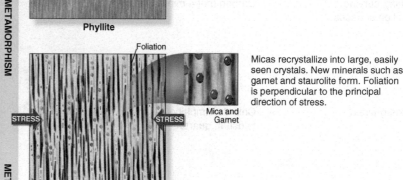

▲ **Figure 6.21 The progressive metamorphism of shale.** Visible changes in rock texture and mineral content take place as the grade of metamorphism increases. These are some of the possible changes observed in the progressive metamorphism of shale to gneiss.

▲ **Figure 6.22 A rock at the transition from metamorphism to melting.** Migmatites possess the textures of both metamorphic and igneous rocks, as seen in this outcrop near Tucson, Arizona. When this gneiss began to melt and form granitic magma, the melt accumulated in bands parallel to the foliation.

Hornfels is a particular type of non-foliated rock—any very hard, non-foliated, metamorphic rock composed mostly or entirely of microscopically small crystals, regardless of composition. Typically, the newly formed metamorphic minerals are those whose origin more closely relates to changes in temperature rather than changes in pressure.

Other non-foliated or weakly foliated metamorphic rocks containing many minerals include **greenstone** and **eclogite**, which are formed by metamorphism of mafic igneous rocks. Greenstone is the descriptive name of the metamorphic rock that contains abundant green minerals. Usually, its primary constituent is chlorite, an iron-magnesium mica, although green amphibole, feldspar, and quartz usually are present in it, too. Eclogite is a very high-grade metamorphic rock that lacks water-bearing minerals. Its composition is dominated by garnet and pyroxene, sometimes with a minor quartz. Eclogite is denser than peridotite. This means that when basaltic oceanic crust subducts at convergent plate boundaries, it metamorphoses to dense eclogite, which drags the subducting plate deeper into the mantle.

Please take the time to familiarize yourself with the terms and rocks shown in Figures 6.20 and 6.21. Doing so will help you to better navigate and understand the contents of this chapter as well as future chapters.

of the type of foliation. For example, your schist sample has staurolite and garnet surrounded by muscovite. You name this rock staurolite-garnet-muscovite schist, indicating the dominant component is muscovite, followed in abundance by garnet and staurolite.

Non-Foliated or Weakly Foliated Rocks

If rock is non-foliated or only weakly foliated, then its mineral content is more crucial to naming it. For instance, rock composed mostly of metamorphically recrystallized calcite is **marble**, whereas **quartzite** consists of metamorphically recrystallized quartz. These non-foliated rocks are distinct from limestone and sandstone because they have metamorphic recrystallization textures, such as large grain size, crystals meeting along straight edges (Figure 6.9), or both. You likely have seen marble, because it is a commonly used decorative building stone. Calcite is a fairly soft mineral, so marble also commonly is used for carving statues. Other metamorphic rocks also are composed primarily of a single mineral. These include **amphibolite** (amphibole minerals with plagioclase feldspar) and **serpentinite** (typically composed almost entirely of serpentine).

Organic compounds also can experience metamorphic reactions. Coal contains few minerals, but as temperature increases, hydrogen- and nitrogen-rich organic compounds in it break down and escape in the form of gas, converting the organic matter to 90 percent or more pure carbon. Pyrite, commonly found in coal (Section 5.6), also breaks down, and the resulting sulfur is likewise released as gaseous compounds. Metamorphosed coal, called anthracite, is the highest-quality coal because it burns with the most heat and contains almost none of the sulfur compounds that pollute the atmosphere when sedimentary coal burns. At high pressure and high temperature, the organic carbon converts entirely to graphite, which, being a platy mineral, such as mica, forms a dark, greasy schist.

Putting It Together—How Are Metamorphic Rocks Classified?

• The classification of metamorphic rocks is based on texture (foliation and grain size) and composition (mineral content). The classification of foliated rocks is based on texture. Modifying adjectives indicate which index minerals are present. Most non-foliated rocks are named on the basis of their mineral content.

6.9 What Was the Rock Before It Was Metamorphosed?

You know that all metamorphic rocks start out as something else. But, how do you determine the identity of the parent rock? This can be very challenging, despite knowing how to name a metamorphic rock and how to estimate the grade of metamorphism that occurred. Recall from Section 6.4 that if a lot of fluid moves through the rock during metamorphism and transports ions to and away from the reaction sites, then the end result is a metamorphic rock containing minerals that are inconsistent with the starting bulk composition of the parent rock.

Metamorphosed Sedimentary Rocks

Figure 6.23 illustrates some possible metamorphic pathways for different parent sedimentary and igneous rocks. The rock transformations depicted in

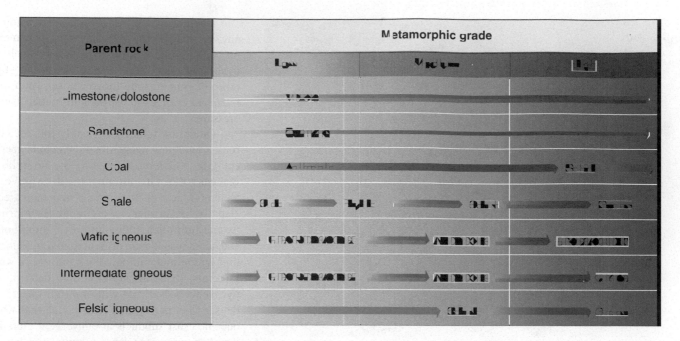

▲ **Figure 6.23 Possible metamorphic pathways for parent rocks.** This diagram outlines some of the metamorphic rocks that may form when various parent rocks metamorphose to different grades. Some parent rocks, such as limestone, produce the same metamorphic product, such as marble, regardless of the grade of metamorphism. Other parent rocks, such as the mafic igneous rock basalt, produce a variety of metamorphic rocks depending on the minerals that are stable at different metamorphic grades.

Figure 6.23 assume that water participates primarily to make water-bearing minerals and does not substantially change the bulk composition of the rock. The metamorphic changes for shale are illustrated in Figure 6.21.

Quartz sandstone and limestone generally metamorphose to non-foliated rocks composed of a single mineral because a single, non-platy mineral dominates the original rock. Arkose or lithic sandstone containing significant non-quartz grains may form micas at the expense of feldspar and other reactive minerals during low-grade metamorphism in the presence of water. The resulting rocks may be weakly foliated mica-rich quartzite or quartz-rich schist.

Metamorphosed Igneous Rocks

Low- to medium-grade metamorphism of mafic and intermediate igneous rocks in the presence of water generates dark-colored metamorphic rocks with abundant iron- and magnesium-bearing mica, such as chlorite or biotite. If foliated, these rocks are chlorite or biotite schist; if not foliated, they are greenstones. At higher metamorphic grade, the mineral hornblende becomes increasingly stable and forms amphibolites. At highest grade, the metamorphic product is gneiss or eclogite (Figure 6.23).

Felsic igneous rocks metamorphose to schist if water is available to convert feldspar to muscovite. Otherwise, there may not be substantial textural or mineralogical changes, especially for coarse-grained granite, until high-grade conditions cause segregation of minerals into bands to form gneiss.

Gneiss formed by metamorphism of igneous rock may be difficult to distinguish from highly metamorphosed shale. Igneous gneisses are less likely, however, to contain the very aluminum-rich minerals, such as muscovite, staurolite, corundum, kyanite, and sillimanite, that form by metamorphism of clay.

Putting It Together—What Was the Rock Before It Was Metamorphosed?

• The composition of the parent rock is interpreted from the mineral composition of the metamorphic rock. This can be done as long as fluids have not extensively removed or delivered chemical components during metamorphism.

• The major elements found in the constituent metamorphic minerals, such as Fe and Mg, are similar to those found in the minerals from the parent rock. Felsic igneous rocks, therefore, metamorphose to rocks containing abundant quartz, feldspar, and muscovite, whereas mafic igneous rocks contain abundant amphibole, biotite, and pyroxene.

• If the parent rock contains mostly one mineral, as is the case with limestone and quartz sandstone, then the metamorphic equivalent may also consist solely of one mineral. We see this in marble and quartzite.

6.10 Where Does Metamorphism Take Place?

Variations in temperature, pressure, and fluid availability determine metamorphic reactions, and the increased heat and pressure inside Earth provide suitable conditions for these reactions. Is it possible, however, to interpret more precisely where metamorphism might occur? In other word, can we determine where the processes that account for the specific changes in

Table 6.1 The General Types and Settings of Metamorphism

Type of Metamorphism	How Metamorphism Occurs	Where Metamorphism Occurs	Characteristics of Metamorphic Rocks
Contact	Increase in temperature and, in some cases, migration of hot fluids	Contact metamorphism occurs in tectonically active areas adjacent to igneous intrusions and, to a lesser extent, below lava flows. Metamorphism extends over distances of less than a meter to as much as 20 kilometers from the igneous intrusion, depending on size, temperature, the composition of the intrusion, and the composition of the surrounding rocks.	Non-foliated rocks, of which hornfels is typical, form around near-surface intrusions. Contact metamorphic zones around deeper intrusions are identified by unusually high temperature minerals within regional-metamorphic rocks.
Hydrothermal	Large-volume interaction of hot fluid with rocks	Hydrothermal metamorphism occurs where water is abundant and temperature is high, especially below mid-ocean ridges and near some igneous intrusions in both continental and oceanic crust. Metamorphism may affect areas from as small as hundreds of square meters to larger than hundreds of square kilometers.	Non-foliated rocks containing water-rich minerals, including micas and amphiboles, form this way. Hydrothermal metamorphic rocks commonly occur along fractures that cut across other rocks. These rocks commonly contain economically valuable sulfide minerals.
Regional	Rock transformation over large regions affected by high tectonic stress and geothermal gradients typically higher or lower than average	Regional metamorphism occurs in rocks near convergent plate boundaries where magma forms and mountain building takes place. Regional metamorphic rocks typically cover tens of thousands of square kilometers, or more.	Rocks are foliated except in cases where parent rock types do not metamorphose to platy minerals.

temperature, pressure, and fluid availability that are crucial to metamorphism occur?

A variety of observations help us determine the location of metamorphic environments. Geologists start by observing the relationship between different metamorphic rock types in the field. If you observe transitions over some distance between rocks representing different metamorphic grades, then you know the location where temperature and pressure were elevated to metamorphic conditions or the direction from where fluid entered the rock. In many cases, it is even possible to follow the rock through diminished grades of metamorphism to the unmodified parent rock. This observation not only permits a straightforward way of determining the parent rock of a given metamorphic rock, but it also helps to determine how metamorphism occurred.

Field studies of the distribution of metamorphic rocks and their relationship with igneous and sedimentary rocks indicate three general settings for metamorphism, as shown in **Table 6.1**—contact, hydrothermal, and regional metamorphism.

Contact Metamorphism

The geologic map in **Figure 6.24**a shows igneous intrusions in Nevada surrounded by concentric zones of metamorphic rocks. The close association of these metamorphic zones with the outlines of the igneous rocks indicates a relationship between metamorphism and the intrusion of magma into rock. The field relationships suggest that the intruded magma cooled by conveying heat to the surrounding rock. The heat from the magma caused the temperature in the surrounding rock to rise considerably. These rocks display a progressive series of changes beginning from the igneous contact and moving outward. The mineral constituents closest to the igneous contact are the highest-temperature (higher-grade) metamorphic minerals, while the outer zone contains low-temperature (lower-grade) minerals. The

metamorphic rocks closest to the intrusions contain andalusite and sillimanite, whereas the rocks farther away have andalusite but no sillimanite. This indicates that the temperature and pressure conditions in the inner zone were stable for both andalusite and sillimanite or that andalusite was converting to sillimanite (Figure 6.18). At even greater distance from the stocks, the surrounding rocks do not appear to be modified because as the heat dissipated, the grade of metamorphism dropped.

The metamorphic rocks in Nevada are an example of **contact metamorphism**, which occurs near igneous intrusions (Figure 6.24b), or less significantly beneath lava flows. Heat causes the metamorphism, which is restricted to the region adjacent to the magma or lava. The amount of heat introduced by the magma or lava and the amount of fluid movement determines the volume of rock that is metamorphosed. The heat associated with lava flows is trivial compared to that associated with large subsurface intrusions, so here we focus on metamorphism adjacent to intrusions.

Because heat (not pressure) is the dominant factor in contact metamorphism, the resulting rocks are generally non-foliated. Hornfels is the most common contact-metamorphic rock and forms adjacent to magma bodies that intruded into sedimentary and volcanic rocks within a few kilometers of the surface. Where limestone and dolostone are interbedded with chert or shale, the heat from the intrusion causes reactions between the carbonate and silicate minerals and forms coarse-grained rocks composed of calcium and magnesium silicates, such as garnet, talc, and pyroxene. At greater depths, the surrounding rocks already are metamorphosed. In this case, contact metamorphism simply causes higher-temperature metamorphic minerals to form near the intrusion than are found in the metamorphic rocks that are farther from where the magma solidified.

Metamorphic grade is highest close to the contact with the igneous rock and grade decreases away from the intrusion, as seen in the example from Nevada. If geologists observe a decreasing grade of metamorphism as they move away from an intrusion, then that indicates that heat from the

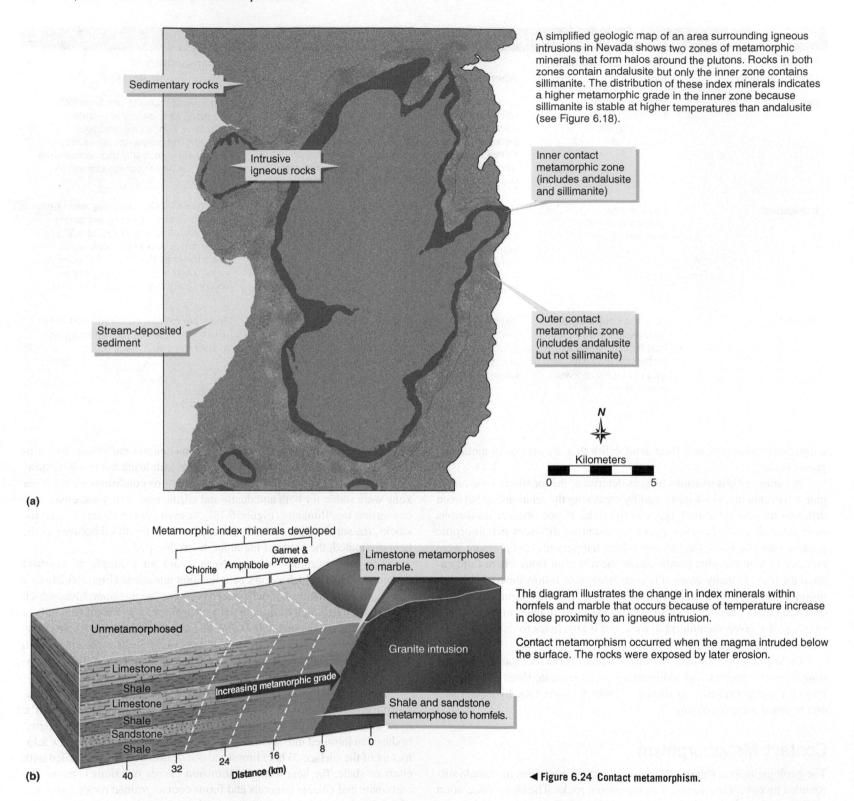

(a)

A simplified geologic map of an area surrounding igneous intrusions in Nevada shows two zones of metamorphic minerals that form halos around the plutons. Rocks in both zones contain andalusite but only the inner zone contains sillimanite. The distribution of these index minerals indicates a higher metamorphic grade in the inner zone because sillimanite is stable at higher temperatures than andalusite (see Figure 6.18).

Sedimentary rocks

Intrusive igneous rocks

Stream-deposited sediment

Inner contact metamorphic zone (includes andalusite and sillimanite)

Outer contact metamorphic zone (includes andalusite but not sillimanite)

N

Kilometers
0   5

(b)

Metamorphic index minerals developed

Chlorite   Amphibole   Garnet & pyroxene

Unmetamorphosed

Limestone metamorphoses to marble.

Limestone
Shale
Limestone
Shale
Sandstone
Shale

Increasing metamorphic grade

Granite intrusion

Shale and sandstone metamorphose to hornfels.

Distance (km)
40   32   24   16   8   0

This diagram illustrates the change in index minerals within hornfels and marble that occurs because of temperature increase in close proximity to an igneous intrusion.

Contact metamorphism occurred when the magma intruded below the surface. The rocks were exposed by later erosion.

◀ Figure 6.24 Contact metamorphism.

intruding magma body caused the metamorphism. This is typically the way contact metamorphism is recognized in the field.

Hydrothermal Metamorphism

An ongoing example of hydrothermal metamorphism can be studied below the Mid-Atlantic Ridge. Here, geologists retrieved rocks from holes drilled into the seafloor near a submarine hot spring. These samples provide insights into the anatomy of an actively forming seafloor mineral deposit that includes sulfide minerals that are metal ores. **Figure 6.25** shows a cross-sectional view of the ore deposit. The bulk of the ore deposit consists of pyrite and quartz along with sulfide minerals containing valuable copper, nickel, and zinc. Similar metal-ore deposits are mined where ancient seafloor is uplifted onto the margins of continents.

Direct observation of a submarine hot spring, shown in **Figure 6.26**, shows how the sulfide minerals form. Where the hot fluids, nearing 360°C,

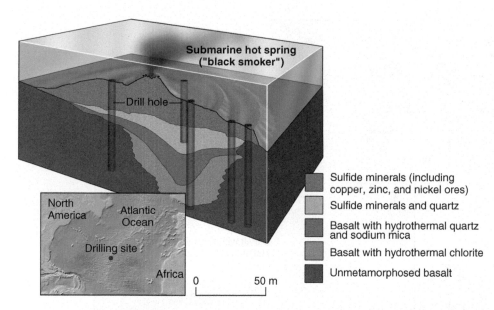

▲ **Figure 6.25 Metamorphosed seafloor rocks with an ore deposit.** A cross section of rocks encountered by drilling into rocks near a submarine hot spring along the Mid-Atlantic Ridge. Deposits of sulfide ore minerals and quartz form near the surface. Below these deposits are metamorphosed basalts containing sodium-rich mica and chlorite.

▲ **Figure 6.26 The creation of ore deposits.** This photograph, taken from a research submarine, shows hot, mineral-rich fluid emerging from the seafloor. Geologists call this feature a "black smoker," but the black material is not smoke. The "cloud" is fine particles of pyrite, galena, and sphalerite (iron, lead, and zinc sulfides) carried along in a rapidly rising column of hot water. The mineral components dissolve in the high-temperature (>300°C) fluid below the seafloor, but then precipitate when they contact cold seawater. Metal sulfides fill fractures and pore spaces in the seafloor basalt and form the chimney through which the hot fluid erupts.

emerge onto the seafloor and mix with the cold seawater, metal-sulfide minerals precipitate to form a "cloud" in the water. The dark, mineral-rich fluid earns these hot springs the name, "black smoker."

The observations shown in Figures 6.25 and 6.26 are evidence for **hydrothermal metamorphism**, which involves the migration and reaction of hot, ion-rich fluids with rock. This metamorphism results in chemical change to the rock because of the substantial role hot water plays as it circulate through pore spaces and cracks. The term "hydrothermal" conveys the equal importance of water and elevated temperature to cause the metamorphic reactions ("hydro" refers to water; "thermal" refers to heat). Hydrothermal metamorphism is commonly associated with contact metamorphism on a local scale in addition to taking place at volcanically active mid-ocean ridges.

Figure 6.27 shows how hydrothermal metamorphism takes place at a divergent plate boundary. Hot magma rising to form new seafloor also heats up seawater that circulates through cracks in the rock. The resulting hot fluid reacts with the basaltic crust and peridotitic mantle and metamorphoses them. The hot fluid carries away dissolved metal and sulfur ions that originated in the rock, the seawater, or both. Because the fluid cools as it rises toward the seafloor and mixes with cold seawater, the metal and sulfur ions combine to form the sulfide-mineral ore deposits.

The rocks encountered below the ore deposit shown in Figure 6.25 are chlorite-rich basalts metamorphosed by hot fluids flowing through the cracks and fissures of the seafloor. Further metamorphism converted some of the chlorite-rich basalt into sodium-rich mica. The sodium came from reactions of minerals and salty seawater. Below the metamorphosed basalt is original, unmodified basalt.

Regional Metamorphism

Metamorphic rocks on continents in the vicinity of modern or ancient convergent plate boundaries commonly cover tens of thousands of square kilometers. **Figure 6.28** shows an example of such widespread metamorphism in the modern subduction-zone setting in Japan. Two features stand out on this map. First, the metamorphic rocks form a long but relatively narrow region that is parallel to the deep-sea trenches where subduction is taking place. Second, there are two different belts of metamorphic rocks. The belt closest to the trench contains minerals that record metamorphism at high pressure but relatively low temperature. In contrast, the belt farther inland indicates lower pressures but higher temperatures. The presence of both belts with an orientation that is parallel to the subduction zone implies that the paired metamorphic belts form because of processes that are unique to convergent plate boundaries.

Figure 6.29 illustrates how paired metamorphic belts can be explained by plate tectonics. The diagram includes a schematic view of the temperature variations near a subduction zone between two lithospheric plates, and it highlights regions of various temperature and pressure conditions of metamorphism. The plate that subducts into the mantle stays relatively cool, even at great depth. This produces a region of relatively low-temperature and high-pressure metamorphism. Magma forms deeper in the subduction zone where the mantle melts above the subducting plate. The water produced by metamorphic dehydration reactions in the subducted plate induces this melting (see Section 4.6). Close to the surface, the rising magma produces high-temperature, low-pressure metamorphic conditions in the crust above the subducted plate. At greater depth below the volcanic arc, the pressure and temperature are high enough for high-temperature and high-pressure metamorphism, but these rocks remain out of view in Japan.

The paired metamorphic belts of Japan are an example of **regional metamorphism**, which refers to metamorphism over large areas not

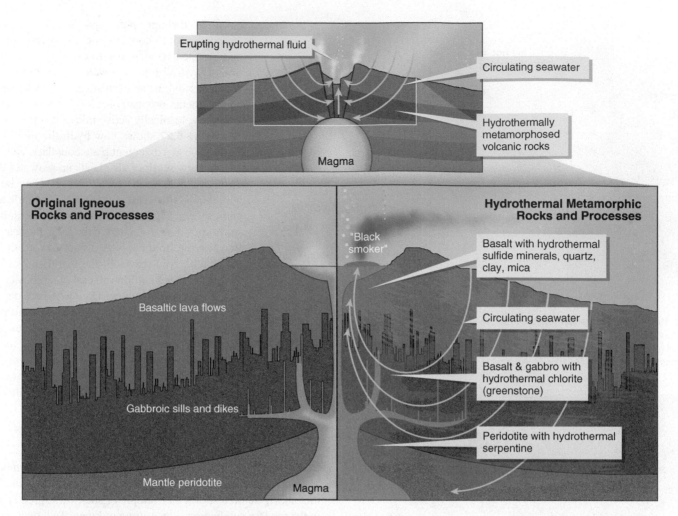

▲ Figure 6.27 Hydrothermal metamorphism at divergent plate boundaries. Hydrothermal metamorphism occurs when magma heats circulating seawater. Gases from the magma also dissolve in the hot fluid. The fluid provides sources of ions and pathways for ions to travel, and the magma provides heat to drive the metamorphic reactions. The enlarged view shows the original igneous rocks (on the left) and how they are modified by metamorphism (on the right).

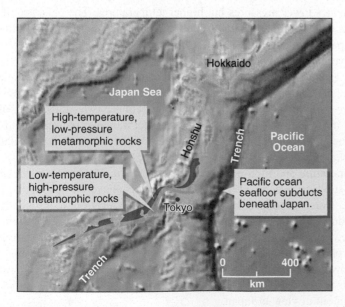

▲ Figure 6.28 Paired metamorphic belts in Japan. This map shows the location of paired metamorphic belts near the subduction zone beneath Japan.

related to specific igneous intrusions or sources of hydrothermal fluid. This metamorphism typically is associated with the formation of mountain belts along subduction zones. Regional metamorphism involves progressively increasing temperature- and pressure-driven mineralogical and textural changes to rock. It encompasses large volumes of continental crust, and in some cases oceanic crust and mantle. This kind of metamorphism occurs when the influence of tectonic stresses is combined with the pressure exerted by overlying rock. These stresses include those associated with the horizontal convergence of plates at subduction zones (Figure 6.29). Regional-metamorphic rocks are almost always foliated, but the foliation may be vertical or at some intermediate angle between horizontal and vertical depending on the orientation of stresses. The heat driving the regional metamorphic reactions is enhanced in many cases by rising magma.

Figure 6.30 summarizes the different metamorphic-environment conditions in terms of the pressure-temperature graph that you first examined in Figure 6.2. Take a moment to compare these two illustrations. In terms of pressure and temperature, different types of metamorphism (contact, hydrothermal, regional) relate to the pathways of changing temperature and pressure that rocks experience.

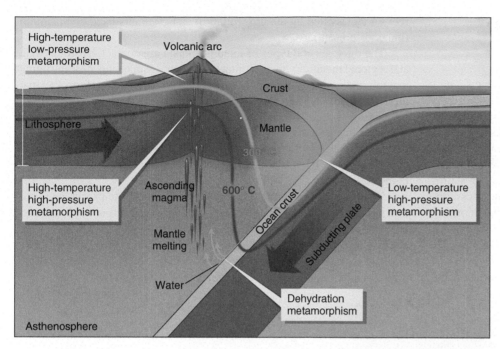

◀ **Figure 6.29 The tectonic setting for regional metamorphism at a convergent plate boundary.** This cross section illustrates the different temperature and pressure conditions that exist near a convergent plate boundary. The 300°C and 600°C temperature curves indicate that conditions are cooler near the cold subducting slab and warmer under the volcanic arc, where hot magma rises toward the surface. Relatively low-temperature but high-pressure metamorphism occurs near the subduction zone in both the subducted and overriding plate. As the subducted plate releases water by metamorphic dehydration reactions, magma forms and rises into the overriding plate. The heat from the rising magma causes high-temperature, and near the surface, also low-pressure, metamorphism in the crust. Below the volcanic arc, at greater depths and higher pressures, high-temperature and high-pressure metamorphism takes place. Erosion of uplifted rocks near convergent plate boundaries exposes these different types of metamorphic rocks.

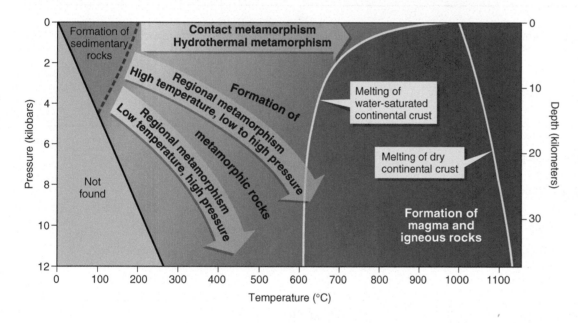

◀ **Figure 6.30 Summarizing metamorphic environments on a graph.** The temperature and pressure conditions under which metamorphism will occur lie between the conditions in which sedimentary rocks and magma form. The paths of changing temperature and pressure that rocks experience during progressive metamorphism further distinguish the metamorphic environments. Most observed examples of contact and hydrothermal metamorphism occur near the surface under conditions of low pressure and increasing temperature. Regional metamorphism follows different paths of increasing temperature and pressure.

Putting It Together—Where Does Metamorphism Occur?

• The three primary types of metamorphism are contact, hydrothermal, and regional.

• Contact metamorphism occurs along igneous intrusions. It is local metamorphism with reactions driven primarily by heat. Contact metamorphism is most intense closest to the magma and decreases away from the intrusion.

• Hydrothermal metamorphism on a local scale involves hot fluids, derived from intruding igneous bodies or infiltrating ground water, which circulate through the rock. This type of metamorphism occurs on a large scale at divergent plate boundaries, where large volumes of circulating hot seawater promote metamorphism and ore production.

• Regional metamorphism depends on temperature and pressure increases that occur over very large volumes of crust and produces extensive tracts of foliated rocks. It typically occurs near convergent plate boundaries.

EXTENSION MODULE 6.1

Metamorphic Isograds, Zones, and Facies. *Learn how geologists use metamorphic minerals and the chemical reactions that form them to determine the metamorphic history of a region.*

Where Are You and Where Are You Going?

This chapter introduced you to the last of the three major rock groups found on Earth: metamorphic rocks. These rocks form from any of the major rock groups, including previously metamorphosed rocks, when the conditions under which the rocks originally formed change. Increases in temperature or pressure, or both, and the introduction of chemically reactive fluids modify the mineral contents and textures of rock. Temperature is a measure of the heat energy required for metamorphic reactions. Pressure, resulting from the weight of overlying rocks or directed tectonic stress, determines the orientation and atomic structure of minerals in metamorphic rocks. Fluids enhance metamorphic reactions by transporting ions in solution. Experiments define the temperature and pressure conditions at which metamorphic minerals are stable. Key metamorphic index minerals reveal the pressure and temperature conditions of metamorphism. Metamorphic rocks are classified by texture (foliation and grain size) and composition (minerals present).

The three general types of metamorphism are contact, hydrothermal, and regional metamorphism. Contact metamorphism occurs near intruding magma and is driven primarily by elevated temperature. Hydrothermal metamorphism involves hot fluids derived from intruding magma or from infiltrating ground water and seawater heated by the magma. Regional metamorphism occurs where variations in temperature and pressure affect very large volumes of crust to produce vast regions of foliated rocks near convergent plate boundaries. Uplift and erosion bring metamorphic rocks to the surface.

You have examined the three major rock types. The processes forming these rocks provide the *what*, *why*, and *how* for interpreting Earth history. The remaining question we must answer is *when* these processes were active. The next chapter wraps up your study of Earth materials by revealing how rocks and minerals help to reconstruct the sequence of geologic events and determine when these events took place.

Active Art

Forming Foliation. See how the three types of foliation form.

Exposing Metamorphic Rocks. See how metamorphic rocks arrive at the surface.

Extension Module

Metamorphic Isograds, Zones, and Facies. Learn how geologists use metamorphic minerals and the chemical reactions that form them to determine the metamorphic history of a region.

Confirm Your Knowledge

1. What is metamorphism?
2. Why is it impossible to observe the processes that produce metamorphic rocks when they happen?
3. What features can you observe in a rock sample or outcrop that would permit you to recognize a rock as being metamorphic, rather than igneous or sedimentary?
4. Which four factors determine the mineral content and texture of a metamorphic rock?
5. List and describe the three common processes that cause rocks to experience increasing temperature after they form.
6. What are the three ways foliation can form in a metamorphic rock?
7. Describe why fluids are important in the formation of a metamorphic rock.
8. Determine whether the change caused by the mineral reactions listed below results in hydration or dehydration.
   • Muscovite and quartz react to form sillimanite and potassium feldspar
   • Olivine reacts with water and carbon dioxide to form serpentine and magnesite

9. Given that diamonds form from graphite at high pressure, why don't they revert to graphite at Earth's surface where pressure is no longer high?
10. Why are some minerals useful as metamorphic index minerals, while other minerals are not?
11. What is mineral stability? How does mineral stability change with temperature and pressure?
12. Define "pressure," "stress," and "strain." How do they relate to one another?
13. Explain why metamorphic rocks, which require high temperature, pressure, or both to form, are commonly exposed at Earth's surface.
14. Identify a parent rock for each of these metamorphic rocks: slate, marble, and gneiss.
15. Explain how you could use field observations to distinguish between rocks produced by contact metamorphism and regional metamorphism.
16. Turn to Figure 3.9a in your textbook. What is the name for the pictured metamorphic rock?

Confirm Your Understanding

1. Write an answer for each question in the section headings.
2. Calculate the geothermal gradient depicted by the graph in Figure 6.3.
3. Determine whether a rock formed at the following pressures and temperatures typically would be sedimentary, metamorphic, or igneous. If metamorphic, determine whether the rock would be considered low-, medium-, or high-grade.
   - Pressure = 1 kb, Temperature = 700 °C
   - Pressure = 2 kb, Temperature = 100 °C
   - Pressure = 3 kb, Temperature = 300 °C
   - Pressure = 4 kb, Temperature = 500 °C
   - Pressure = 6 kb, Temperature = 1100 °C
   - Pressure = 7 kb, Temperature = 1000 °C
4. Stress is defined as a force exerted on an area. Weight is an example of a force. Describe how the stress of a person's weight on the ground surface is different if that person is wearing flat-soled shoes, high heels, or snow shoes.
5. Metamorphic changes include a change in mineral content, or a change in texture, or both. Determine the type of metamorphic change for each of the metamorphic changes listed below.
   - Shale metamorphoses into a schist
   - Dolostone metamorphoses near an intrusion to a rock containing calcite and periclase
   - Graphite metamorphoses into diamond
   - Quartz sandstone metamorphoses into quartzite
   - Limestone metamorphoses into marble
6. Use Figure 6.18 to determine whether andalusite, kyanite, or sillimanite is stable under the following pressure (P) and temperature (T) conditions:
   - P = 1 kb, T = 500 °C
   - P = 2 kb, T = 650 °C
   - P = 4 kb, T = 450 °C
   - P = 5 kb, T = 650 °C
7. Classify the following metamorphic rocks into low-, medium-, or high-grade based on the minerals present in the rock.
   - Quartz, biotite, chlorite
   - Quartz, feldspar, biotite, garnet, sillimanite
   - Quartz, feldspar, biotite, garnet, staurolite
   - Quartz, feldspar, muscovite, biotite, chlorite
   - Quartz, feldspar, muscovite, biotite, garnet
   - Quartz, feldspar, pyroxene
8. What would you name the following metamorphic rocks?
   - Foliated rock with alternating compositional bands rich in quartz and feldspar or biotite, garnet, and sillimanite
   - Shiny, foliated rock rich in muscovite and biotite, with numerous crystals of coarse-grained garnet
   - Weakly foliated rock consisting entirely of coarse-grained recrystallized calcite
   - Weakly foliated rock consisting entirely of fine-grained recrystallized quartz
   - Non-foliated green rock consisting of quartz, feldspar, chlorite, and amphibole
9. Define "recrystallization." Describe the changes a crystal undergoes during recrystallization. How does recrystallization differ from cementation of sedimentary particles? How might you distinguish between cementation and recrystallization?
10. Mountain building and erosion near convergent plate boundaries commonly expose rocks that formed in the middle crust at depths of 15 to 20 km below the surface. These rocks include plutonic-igneous rocks and metamorphic rocks. The metamorphic rocks formed under both regional and contact metamorphic conditions. What combination of field and laboratory observations could you make to distinguish the effects of the two sets of conditions? In developing your answer, assume that the rocks contain aluminum silicate polymorphs, and use the graphs in Figures 6.5 and 6.18 to guide your thinking.

CHAPTER 7

Earth Materials as Time Keepers

Why Study the Ages of Rocks?

How useful is a recipe that lists ingredients but neglects the preparation and cooking time, or product-assembly instructions that do not list each step in order? Like a successful recipe, rock-forming processes also take place in a certain order and last for established intervals. This chapter introduces you to the methods geologists use to unlock the history of Earth archived in rocks. Here we ask, how can you look at a complicated rock outcrop, such as the one featured in the photo at right, and determine the sequence of events that formed it and how long it took for those events to take place? However, if geologists want to know how old the rocks are or how long it took for the events to occur, they must employ complicated laboratory procedures.

Many humans before us had an avid curiosity about Earth's age. To figure out the age of the planet, you first need to understand how the age of a rock is measured and why there is a high degree of confidence in the results.

Earth formed about 4.5 billion years ago. Can you even begin to fathom how long 4.5 billion years is? Geologists must put their amazement aside, for they always work within the context of the vast length of Earth's history that makes the average human lifespan, the rise and fall of nations, even the evolution and extinction of individual species, seem insignificantly short.

After Completing This Chapter, You Will Be Able to

- Apply established principles for placing geologic events, such as processes recorded by different rocks in an outcrop, in order from oldest to most recent.

- Compare and evaluate different approaches to estimating Earth's age.

- Describe how geologists measure ages of rocks and of Earth itself.

Pathway to Learning

7.1 How Do We Describe Rock Age?

7.2 How Are Geologic Events Placed in Relative Order?

7.3 How Do Geologists Determine the Relative Ages of Rocks That Are Found Far Apart?

7.4 How Was the Geologic Time Scale Constructed?

7.5 How Do You Recognize Gaps in the Rock Record?

A geologist's rock hammer rests on steeply tilted sedimentary layers that are overlain by less steeply tilted layers. This outcrop, at Siccar Point on the east coast of Britain, was central to James Hutton's interpretation of an extraordinarily long Earth history.

IN THE FIELD

Enjoying a virtual stroll along the seashore, you pause and contemplate the rocks that form the steep cliff rising above a sandy beach. The rocks display a variety of colors and an eye-catching pattern of horizontal, vertical, and inclined transitions from one rock type to another. You examine these rocks a bit more closely and, as field geologists typically do, sketch the scene in your notebook.

Your sketch, reproduced in **Figure 7.1**, shows red sandstone and shale beds, inclined at an angle downward to the right, that are overlain by horizontal layers of sandstone, limestone, and shale. A band of basalt cuts across all of the sedimentary layers, forming a dike. Discolored and hardened sedimentary rock borders the basalt, implying that contact metamorphism occurred adjacent to the dike.

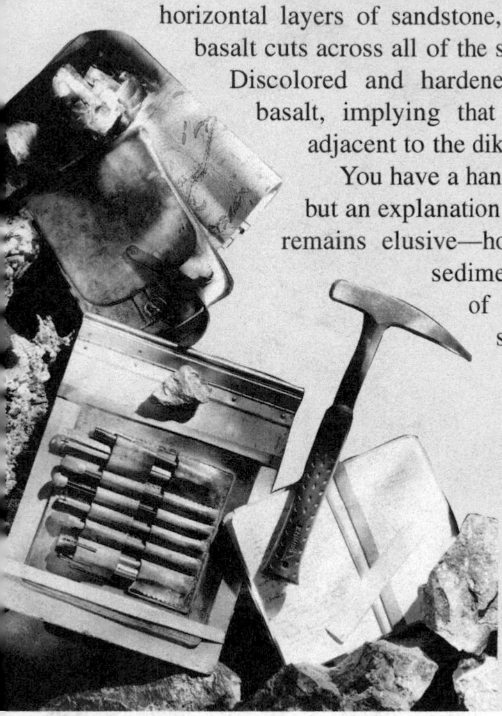

You have a handle on the origin of each rock type, but an explanation for the origin of the whole outcrop remains elusive—how was it all put together? The sedimentary rocks indicate deposition of clastic and chemical sediment in sedimentary basins. In some layers, cross-bedding reveals the direction of currents that deposited the sediment. Different types of fossils reveal that the reddish strata were deposited on land and that the overlying layers were deposited beneath the sea. The dike implies a tectonically active area where magma formed and moved upward, perhaps even reaching the surface at volcanoes that have eroded away. The heat of the rising magma metamorphosed the adjacent sedimentary rocks. You can identify and explain each rock type by itself because you are familiar with rock-forming processes. However, knowing what each rock is and what forces created it, does not mean you understand why the rocks are located how they are in relation to each other. Another dimension is at play—time.

The sea-cliff outcrop reveals a history. Understanding that history requires that you sort out the order of geologic events and seek an explanation for why the rocks are oriented at different angles. This chapter provides the tools for understanding the history of the seaside outcrop.

▶ Figure 7.1 **How do geologists interpret Earth's history?** A field sketch illustrates rocks exposed along a sea cliff. Each rock reveals its own record of the processes that formed it. With careful study you can understand the order in which these processes took place and even when they happened.

7.1 How Do We Describe Rock Age?

Ancient rocks reveal the deep time history of our planet much like the way documents reveal recorded human history. Historians track the sequence of human events by studying written papers, art, photographs, and oral recordings, which form the historical record. Geologists use the "rock record" to decipher Earth's long history. Sedimentary and volcanic rocks record processes that occur on Earth's surface, whereas plutonic and metamorphic rocks reveal processes that take place in the interior.

To establish the order of events within the historical record, historians use specific clues. For example, a historian might put a pile of undated photographs of a city skyline in sequence from oldest to most recent by documenting the appearance of new buildings and the disappearance of older buildings in the pictures. The ordering of objects or features from oldest to most recent establishes the **relative age** of each; this process determines whether one thing is older or younger than another.

Historians also work to establish exactly when an event took place; they may verify the date when a document was written, a photograph taken, or an art object completed. Establishing the date of an event provides its **absolute age**. In history, this means the calendar date and possibly even the hour of the event.

These steps are a good place to start in order to learn more about your seaside cliff. To describe the geologic history of the events preserved in this particular rock record, you need to establish two things: (1) the order in which the events occurred, and (2) when they occurred. In other words, you need to determine either the relative or absolute age, or both, of each rock type exposed in the cliff.

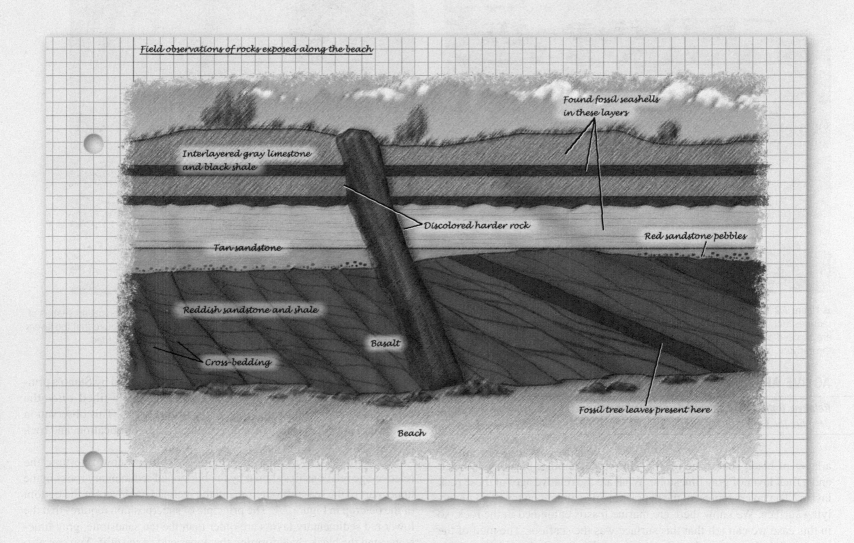

Field observations of rocks exposed along the beach

Found fossil seashells in these layers

Interlayered gray limestone and black shale

Discolored harder rock

Red sandstone pebbles

Tan sandstone

Reddish sandstone and shale

Basalt

Cross-bedding

Fossil tree leaves present here

Beach

Putting It Together—How Do We Describe Rock Age?

• Relative ages establish the sequence of a series of events without establishing exactly when each event occurred. To establish relative age, items or events are placed in order of what happened first, what happened next, and what happened last.

• Absolute ages indicate exactly when an event took place. An absolute age is a specific age in years, or a statement that an event happened a particular number of years in the past.

7.2 How Are Geologic Events Placed in Relative Order?

Geologists have established a set of rules to determine the relative ages of geologic events preserved in the rock record of Earth's history. These rules, also called principles, are mostly self-evident yet very powerful. Together, the four principles of superposition, original horizontality, cross-cutting relationships, and inclusions provide a system for ordering geologic events.

The Principle of Superposition

Let's start by looking at the horizontal sedimentary layers exposed high on the sea cliff in Figure 7.1. From bottom to top we see tan sandstone,

▲ **Figure 7.2 Applying the principle of superposition.** The principle of superposition requires that the oldest visible sedimentary rocks along the Colorado River, in Utah, are exposed at the bottom of the canyon while the youngest rocks form the top rim of the canyon.

ACTIVE ART

Relative Dating Principles. *See how the relative dating principles are used to decipher the sequence of geologic events.*

a layer of black shale, one of gray limestone, another black shale layer, and a final layer of gray limestone. What are the relative ages of these rock layers? The layers represent the deposition of sediment on top of an underlying surface. We know there are marine fossils entombed in the rock, so in this case we can tell that this surface was the seafloor. The mud of the lowest shale layer was deposited on top of the sandy layer below. This means that the sand making up the tan sandstone was deposited before the shale, which indicates that the sandstone is older than the shale.

With this simple exercise, we have introduced one principle for determining relative ages. Within a sequence of horizontal rock layers formed at Earth's surface, those lower in the sequence are older than those found above, as illustrated in **Figure 7.2**. This **principle of superposition** was formally written down in 1669 by Danish physician Niels Steensen, who is better known by his Latinized name, Nicolaus Steno.

The Principle of Original Horizontality

Steno also noticed that the surfaces where sediment usually accumulates— the seafloor, a riverbed, the bottom of a lake—are nearly flat. Steno's **principle of original horizontality** states that sediment tends to be deposited in horizontal layers. As a result, you expect the bedding planes in sedimentary rock also to be horizontal or nearly so, as seen in Figure 7.2. Sedimentary layers are usually not *precisely* horizontal because the seafloor, riverbeds, and other depositional surfaces are not perfectly flat. Nonetheless, these surfaces rarely slope at angles of more than 0.5 degree, which appears horizontal to the human eye.

Non-horizontal sedimentary layers, therefore, require an explanation. Consider, for example, the sedimentary rocks illustrated in **Figure 7.3**, where

▲ **Figure 7.3 Applying the principle of original horizontality.** The vertical orientation of the beds in these sedimentary layers found in Quebec, Canada, require that the layers were tilted from an initial horizontal orientation after they were deposited.

the bedding is vertical, rather than horizontal, as described by Steno's principle. These sedimentary rock layers were somehow tilted *after* the deposited sediment lithified to rock in horizontal layers. This observation establishes the relative ages of sediment deposition, lithification, and rock deformation.

Combining these first two principles, you can now fine-tune the relative-age relationships shown by the horizontal sedimentary rocks in the upper part of the sea-cliff outcrop and the inclined red layers at the bottom of the outcrop in Figure 7.1. The principle of superposition requires that the lower red sedimentary layers are older than the tan sandstone, gray limestone, and the black shale forming the upper part of the cliff. You can also safely assert that the red layers were originally horizontal and then tilted at a later time. The sequence of events is (1) deposition of the red sand and mud; (2) lithification of the red sediment; (3) tilting and erosion of the red sedimentary rock layers; and (4) deposition of horizontal sediment layers that subsequently lithified to tan sandstone, black shale, and gray limestone.

The Principle of Cross-Cutting Relationships

The basaltic feature in Figure 7.1 is a dike, a tabular mass of igneous rock that cuts across sedimentary layers (see Section 4.3). When the dike intruded into the surrounding rock, the magmatic heat caused contact metamorphism. This tells us that the sedimentary rocks existed before the dike-forming magma intruded; otherwise there would not have been surrounding rock for the magma to intrude.

The **principle of cross-cutting relationships**, applied by James Hutton in Scotland (look back to Figure 3.11), states that geologic features that cut across rocks must form after the rocks that they cut through. **Figure 7.4** illustrates applications of this principle. Besides igneous intrusions, another common cross-cutting feature is **faults**—fractures across which rocks are displaced as a result of tectonic forces. Faults clearly must develop after the formation of the rocks they displace.

► **Figure 7.4 Applying the principle of cross-cutting relationships.** In the photo on the left, the geologist examines where a relatively young basalt dike cuts across part of a relatively older granite batholith in northern Michigan. In the photo on the right, a fault offsets sedimentary layers. The dashed lines show how much one layer was displaced by fault movement. The sedimentary deposits formed prior to movement along the fault.

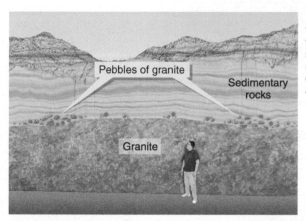

In this diagram, geologists can tell that the sediment was deposited on top of the granite because the lowest sedimentary bed contains pebbles eroded from the granite. The principle of superposition applies in this case and tells us that the sedimentary rocks are younger than the granite.

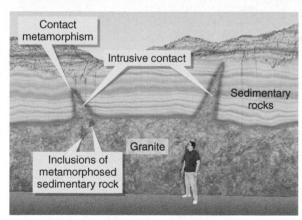

In this diagram, the granite cuts across the sedimentary strata along an intrusive contact and the sedimentary rock is metamorphosed near the contact with the granite. Therefore, the granite must be younger than the sedimentary rock.

▲ **Figure 7.5 Determining which rock is older.** Granite underlies sedimentary rocks in both of these outcrop sketches, but the age relationships of granite and sedimentary rocks are different in each example.

The Principle of Inclusions

As simple as they seem, care must be taken in applying the principles of superposition and cross-cutting relationships. Igneous intrusions form by magma rising from great depth; therefore, plutonic igneous rocks commonly appear underneath sedimentary rocks into which the magma was rising

when it crystallized. A closer look at the bottom image of **Figure 7.5** shows that if a geologist does not carefully examine the contact between the plutonic-igneous rock and the overlying rock, he or she can mistakenly apply the principle of superposition and conclude that the underlying plutonic igneous rock is older. However, in this case, the plutonic rock is actually younger, a relationship that is demonstrated by the fact that the plutonic rock cuts across layers in the sedimentary rock, which experience metamorphism along the contact. Keep in mind that the principle of superposition infers that rocks at the bottom of an outcrop are older than those above *only if all* the rocks formed at the surface. Because plutonic-igneous and metamorphic rocks form below the surface, a geologist needs to apply the principle of superposition cautiously when these rocks are present in an outcrop.

Another concept originating with Steno applies to determining the relative-age relationships illustrated in Figure 7.5. The **principle of inclusions** states that objects enclosed in rock must be older than the time of rock formation. Granite pebbles embedded in sedimentary rock reveal that the granite is older than deposition of the sediment that became sedimentary rock. Metamorphosed inclusions of sedimentary rock within granite indicate that sedimentary rock is older than the granite.

Putting It Together—How Are Geologic Events Placed in Relative Order?

• The principle of superposition states that when rocks form at Earth's surface in layers, the lowest layer formed first and each successively higher layer is younger than the one below.

• The principle of original horizontality states that sedimentary layers are horizontal, or nearly so, when they are deposited. Non-horizontal layering indicates disruption of the beds at some time after deposition.

• The principle of cross-cutting relationships states that geologic features, such as dikes and faults, that cut across otherwise continuous rocks formed after the rocks that they cut across.

• The principle of inclusions states that objects enclosed by rock formed prior to inclusion within the rock.

Individual rock layers in the Grand Canyon can be continuously traced for more than 200 km.

7.3 How Do Geologists Determine the Relative Ages of Rocks That Are Found Far Apart?

The principles of superposition, original horizontality, cross-cutting relationships, and inclusions allow geologists to order the events archived in the rock record at a single location, such as the sea cliff shown in Figure 7.1. These principles do not, however, reveal the relative ages of these rocks compared to rocks observed elsewhere. To construct the geologic history of a wide region, or even across the entire Earth, geologists must determine relative ages without relying on features seen only at one location. Two additional principles, lateral continuity and faunal succession, are applied to this problem.

The diagram below shows how geologists use the lateral continuity of sedimentary layers to interpret the extent of layers concealed beneath the surface, and the original continuity of layers interrupted by erosional irregularities of the surface.

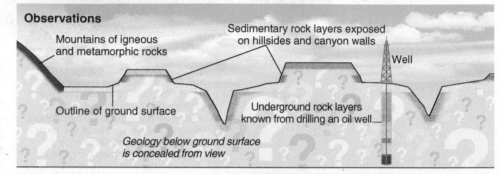

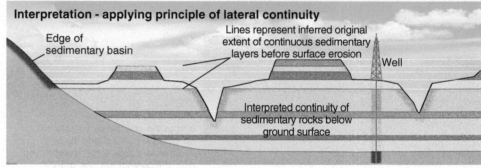

▲ Figure 7.6 Lateral continuity of sedimentary rocks.

The Principle of Lateral Continuity

In some regions, sedimentary rock layers continue in outcrops for long distances. This observation is especially common in regions with dry climates where dense vegetation does not obscure the rocks, and in places where deep canyons lay rocks bare for many kilometers. Perhaps the most dramatic example of such continuous exposures is seen in the Grand Canyon of the Colorado River in northern Arizona. **Figure 7.6** shows a part of the canyon where intervals of sandstone, shale, and limestone can be traced continuously for 200 kilometers, establishing (via the principles of superposition, cross-cutting relationships, and original horizontality) the chronological order of events for a large area.

Steno anticipated that sedimentary rock layers would be continuous for long distances because he assumed that sediment beds accumulate in a continuous pattern until encountering some obstruction. This concept is the **principle of lateral continuity** of beds. The bottom diagram in Figure 7.6 illustrates the application of this principle.

In some places sedimentary layers may erode, while in other localities they may be buried beneath younger rock. The fact that ancient sediment was continuously deposited across large areas does not necessarily mean that the resulting rocks are now exposed everywhere in that region. The principle of lateral continuity, however, encourages geologists to relate rocks in isolated outcrops to one another (Figure 7.6)—this procedure is called **correlation**. Combining principles of lateral continuity and superposition extends relative age relationships over larger areas.

The Principle of Faunal Succession

How can geologists determine the relative ages of rocks in widely separated regions where the rocks bear little resemblance to one another? **Figure 7.7** illustrates this dilemma. Englishman William Smith solved this puzzle with careful observations in the late 1700s and early 1800s.

Smith made key observations while surveying coal mines and canal excavations during the industrial revolution in Britain. He recognized that the coal seams in different underground mines were found in predictable positions between other sedimentary layers. He reached this conclusion not only by using superposition to put the different rock layers in relative order within each mine, but also by noting that some layers contained unique fossils. For example, if he had established that two different coal layers in one mine were each overlain by limestone layers that contained distinctly different fossils, then when he encountered coal and limestone layers in another mine, he was able to confidently correlate the layers in the two mines by comparing the fossil types in the corresponding limestone layers.

▶ **Figure 7.7 The problem of determining relative ages between distant outcrops.** These two outcrops, 1000 kilometers apart, are similar but are not the same. Which rocks are older and which are younger?

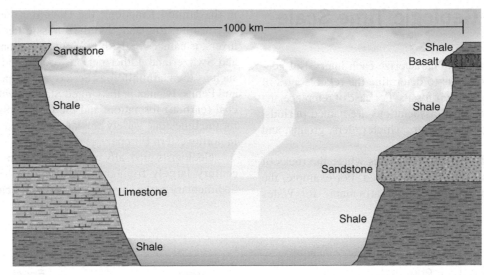

• Are the two rock sequences of different ages? If so, which set of rocks is older and which is younger?

• Are the rock sequences the same age but appear as different rock types because they formed in different depositional environments? If so, how can their similar age be proven?

Smith saw long, high exposures of rock while he was surveying canal excavations dug for transporting coal from the mines into cities. He carefully noted the vertical order of rock types and the fossils found within the layers at each artificial and natural exposure. **Figure 7.8** illustrates how Smith used fossils to correlate the rocks at various localities to establish a complete vertical sequence of the beds, ordered chronologically from oldest to youngest. Smith became so knowledgeable at placing fossil-bearing rocks in the appropriate relative order that amateur fossil collectors could show him specimens and he would tell them exactly which rock layers each had come from and where those rock layers were exposed.

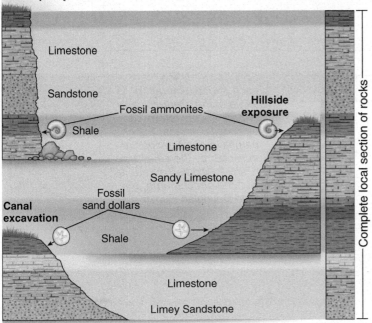

▲ **Figure 7.8 How William Smith used fossils to correlate rocks.** Fossils reveal how to relate the rocks in one place to those found at another location. In this illustration, one shale layer contains fossil ammonites (relatives of the modern chambered nautilus) and another contains a fossilized, now extinct, sand dollar. Combining information on rock types and fossils, it is possible to correlate the rock sections, as shown by the color shading, into a single complete section of rocks that is not actually wholly exposed at any one place.

Using this knowledge, Smith established the **principle of faunal succession**, which states that fossil plants and animals appear in the rock record according to definite chronological patterns. Over the course of Earth's history, each species of organism existed for a limited time interval. Thus, the presence of each type of fossilized organism represents a definite interval of geologic time, which means that the relative ages of fossil-bearing rocks are defined by the fossils they include. The principle of faunal succession asserts that:

• fossils of different organisms, such as clams and fishes, appear at distinct times.

• fossils of related organisms, such as different fishes, appear in the same order every place they occur.

• fossil species disappear from the rock record everywhere when they become extinct, as happened to the dinosaurs, and do not reappear in younger rocks.

Geologists have confirmed this principle by examining thick successions of fossil-bearing sedimentary rock in many different places. After using superposition to place the rocks in relative order, they have established that the different types of fossils *always* appear and disappear in the same order.

Putting It Together—How Do Geologists Determine the Relative Ages of Rocks That Are Found Far Apart?

• The principle of lateral continuity states that sedimentary beds are continuously deposited over large areas until some sort of barrier limits deposition.

• The principle of faunal succession states that fossils found in rocks change through time as some species become extinct and new ones appear. Each species of organism has a limited time interval of existence. Fossil-bearing rocks are placed in relative-age order by determining the interval of geologic time represented by the fossils that the rocks contain.

7.4 How Was the Geologic Time Scale Constructed?

Unique groupings of fossil organisms characterize each interval of geologic history, which enables geologists to determine the relative ages of rocks in distant places by comparing their fossils. For ease of reference, the intervals represented by different fossil associations are named **periods**. **Figure 7.9** illustrates the **geologic time scale**, which orders, groups, and subdivides the periods.

The names of many periods derive from places where the rocks of that age, based on their fossils, were first described. For example, the Cambrian Period is named for Cambria, the Latin name for Wales in the United Kingdom; the Devonian Period is named for rocks in Devonshire, England; and the Jurassic Period, well known to dinosaur lovers, is named for the Jura Mountains along the border between France and Switzerland. Other periods are named after dominant rock types, such as the Carboniferous Period in Europe (divided into Mississippian and Pennsylvanian in the United States), which was a prominent time for coal (carbon) formation; and the Cretaceous Period, which was named for the limestone variety chalk (*creta* in Latin) found in rocks of this age in northwestern Europe.

Geologists gradually constructed the time scale during the nineteenth century largely from the study of fossil-bearing sedimentary rocks. Sedimentary rocks anywhere in the world can be assigned to a particular

▶ **Figure 7.9 The geologic time scale.** The geologic time scale is a chronological listing of time intervals of varying duration. The intervals are arranged within a hierarchy. **Periods** are the fundamental time interval. Periods include shorter intervals called **epochs**. Periods are grouped into **eras**, which are grouped into **eons**. Fossils contained in sedimentary rocks define each Phanerozoic time interval. Age boundaries between Precambrian time intervals are adopted by international convention. Quaternary and Tertiary are traditionally defined periods of the Cenozoic Era, although Neogene and Paleogene are alternative, and currently preferred, names. The Pennsylvanian and Mississippian Periods are referred to as the Carboniferous Period outside of the United States.

Eon	Era	Period		Epoch	Age (millions of years)
Phanerozoic	Cenozoic	Quaternary	Neogene	Holocene (Recent)	
				Pleistocene	0.01
					1.8
				Pliocene	
					5
				Miocene	
		Tertiary			23
			Paleogene	Oligocene	
					34
				Eocene	
					56
				Paleocene	
					65
	Mesozoic	Cretaceous			145
		Jurassic			200
		Triassic			
					251
	Paleozoic	Permian		*Epochs are defined for each period although only those of the Cenozoic era are commonly referred to by specific names. Epoch names in other periods are indicated by the adjectives "Early," "Middle," and "Late" with the period name; e.g., Late Devonian Epoch.*	300
		Carboniferous	Pennsylvanian		318
			Mississippian		
		Devonian			359
					416
		Silurian			444
		Ordovician			488
		Cambrian			
					542
Precambrian	Proterozoic	Neoproterozoic			1000
		Mesoproterozoic			1600
		Paleoproterozoic			2500
	Archean	Neoarchean			2800
		Mesoarchean			3200
		Paleoarchean			3600
		Eoarchean			4500

interval on the time scale based on the fossils in the rock. Geologists solve the puzzle illustrated in Figure 7.7 by establishing the period of deposition for each layer of rocks through the study of the fossils that they contain. Fossils define the periods, so ages of igneous and metamorphic rocks can be inferred by relative-age relationships to datable sedimentary rocks. The time scale in Figure 7.9 also lists absolute ages, in millions of years, for the boundaries between periods. These numerical ages were added to the time scale starting about 1960, through the use of techniques that will be introduced in Section 7.6. This is an important fact to consider—the length of time in each geologic period and when that period actually took place were unknown when the time scale was first formulated; instead, the intervals were defined as the times of existence of particular organisms. This led to the use of labels such as "the age of dinosaurs" for the Mesozoic Era because the fossils of dinosaurs are known only in rocks assigned to the Triassic, Jurassic, and Cretaceous Periods.

With all this in mind, take another look at the sketch in Figure 7.1. The rocks in the seaside cliff contain fossils, so if you identify those fossilized organisms, you can match them up to the appropriate periods on the geologic time scale. Of course, the assistance of a professional paleontologist, a geologist who studies fossils and the history of life on Earth, will help!

Take a close look at **Figure 7.10**, which shows the result of the paleontological investigation at your seaside cliff. Only the sedimentary rocks are assigned to geologic periods at this stage because the basalt, an igneous rock, does not contain fossils. The periods of deposition of the sedimentary rocks are consistent with your earlier application of the principle of superposition. The red layers at the bottom of the cliff are oldest, the limestone and black shale are the youngest, and the tan sandstone is of intermediate age.

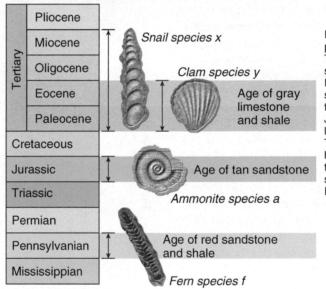

Fossils are collected from the sedimentary rock layers along the sea cliff.

Each fossil is diagnostic of a particular geologic time interval. The fossil fern from the tilted red strata lived during the Pennsylvanian Period. Ammonite species *a* indicates deposition of the tan sandstone during the Jurassic Period. Snail species *x* lived during many epochs of the Tertiary Period. Clam species *y*, however, restricts deposition of the gray limestone and black shale to one or both of the Eocene and Paleocene Epochs.

▲ Figure 7.10 Assigning depositional ages to rocks.

Putting It Together—How Was the Geologic Time Scale Constructed?

• The geologic time scale is a series of intervals with an established chronological order. The fossils that are found in rocks deposited during that time define each named interval.

7.5 How Do You Recognize Gaps in the Rock Record?

Take another look at the ages established for sedimentary rocks at the virtual sea cliff (Figure 7.10). Although these ages are consistent with the principle of superposition, the three intervals of deposition are not adjacent to each other on the time scale. For example, there are no rocks of Permian and Triassic age in between the Pennsylvanian and Jurassic rocks in this outcrop. This incomplete geologic record is like a biography constructed from a diary that is missing many pages.

Why is the rock record incomplete? Because the once-horizontal red sediment required some amount of time to lithify tilt, and erode to produce a nearly flat surface before deposition of the tan sand began. The time it took to complete this process is not represented by rocks at the sea cliff. Therefore, the discovery that the tan sandstone is substantially younger (Jurassic) than the red sandstone and shale (Pennsylvanian) is not surprising.

The time gap above the tan sandstone is more surprising. Referring to your sketch, however, leads you to realize that the contact between the tan sandstone and the overlying black shale is irregular rather than flat, which suggests that a period of erosion separated the two intervals of sediment deposition. Using fossils to determine the age of sedimentary rocks commonly reveals gaps in deposition; a single cross section of rock rarely reveals a complete geologic history.

Unconformities

Gaps in the rock record, when erosion rather than deposition happened, are called **unconformities**. There are two unconformities in the sea-cliff section, one above and one below the tan sandstone. Unconformities are significant for two reasons. First, to construct the geologic history of an area, it is important to know which part of the rock record does not exist, just like a biographer must know which years are missing from an incomplete diary. Second, geologists seek to identify any event or succession of events that caused the break in the rock record, because these events also are part of the geologic history. Geologists generally recognize three types of unconformities.

An **angular unconformity**, illustrated in **Figure 7.11**, is found between intervals of layered rocks (sedimentary beds or lava flows) that are inclined at different angles. There must be an interval of time when the lower layers of rock were tilted and eroded at Earth's surface prior to deposition of the overlying rocks. An angular unconformity separates the red sandstone and shale from the overlying tan sandstone at the seashore in Figure 7.1.

A **disconformity**, illustrated in **Figure 7.12**, also is found between intervals of layered rock, but in this instance all layers are either horizontal or inclined at the same angle. A surface of erosion, which might be a deeply eroded channel or a subtle and almost planar feature, marks a disconformity. The geologic record is incomplete across the disconformity because no rock exists to show the interval of erosion or non-deposition separating the deposited layers. A disconformity separates the tan sandstone and black shale on the sea cliff (Figure 7.1).

A **nonconformity**, illustrated in **Figure 7.13**, is found where sedimentary or volcanic rocks accumulate on top of eroded plutonic-igneous or metamorphic rocks. This contact must be an unconformity because plutonic-igneous and metamorphic rocks form *beneath* Earth's surface, whereas sedimentary and volcanic rocks accumulate *on* the surface. When metamorphic and plutonic rocks form, there must be other rock above them that extends to the surface. All that other rock, possibly many kilometers thick, must erode in order to expose the metamorphic and plutonic rocks at the surface. Only then can the metamorphic and plutonic rocks be reburied by sediment or volcanic deposits. The exposed rocks in the photograph in Figure 7.13, then, do not record the time required to erode the rock that was *originally* above the metamorphic and plutonic materials.

The rock record at a single locality is usually incomplete and may be riddled with unconformities, but no one unconformity extends completely around the planet. Erosion in one locality produces sediment that is deposited at another place during the same time.

The black line traces part of an angular unconformity in the Grand Canyon. Beds above the unconformity are horizontal, whereas beds below the unconformity incline down to the right. Approximately 500 million years of geologic history are missing along the unconformity.

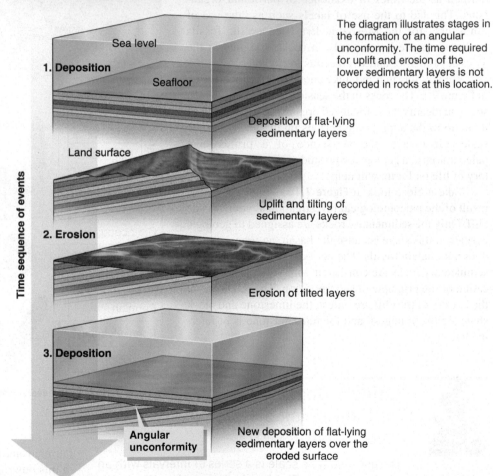

The diagram illustrates stages in the formation of an angular unconformity. The time required for uplift and erosion of the lower sedimentary layers is not recorded in rocks at this location.

Time sequence of events

1. **Deposition**

Sea level

Seafloor

Deposition of flat-lying sedimentary layers

Land surface

Uplift and tilting of sedimentary layers

2. **Erosion**

Erosion of tilted layers

3. **Deposition**

Angular unconformity

New deposition of flat-lying sedimentary layers over the eroded surface

▲ Figure 7.11 What an angular unconformity looks like.

ACTIVE ART

Unconformities. See how the three types of unconformities form.

The black line traces part of a disconformity in the Grand Canyon. Fossils indicate that several million years are missing from the rock record along this sharp, eroded boundary between sedimentary rock layers.

▶ Figure 7.12 What a disconformity looks like.

Time sequence of events

1. Deposition

Sea level

Seafloor

Deposition of flat-lying sedimentary layers

2. Erosion

Land surface

Erosion of sedimentary layers

3. Deposition

Disconformity

New deposition of flat-lying sedimentary layers

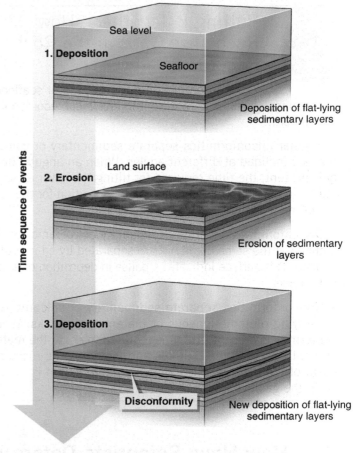

The diagram illustrates how a disconformity forms.

Time sequence of events

1. Deposition

Land surface

Deposition of flat-lying sedimentary layers

2. Plutonism / Metamorphism

Intrusion of magma; metamorphism

3. Erosion

Igneous and metamorphic rocks

Erosion of sedimentary rocks to expose plutonic-igneous and metamorphic rocks

4. Deposition

Sea level

Seafloor

Sedimentary rocks

Nonconformity

New deposition of flat-lying sedimentary layers

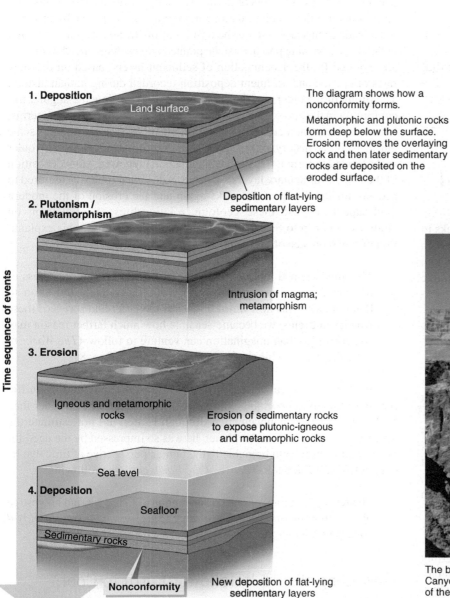

The diagram shows how a nonconformity forms.

Metamorphic and plutonic rocks form deep below the surface. Erosion removes the overlaying rock and then later sedimentary rocks are deposited on the eroded surface.

The black line traces part of a nonconformity in the Grand Canyon. Metamorphic and plutonic-igneous rocks in the bottom of the canyon are overlain by sedimentary rocks. Approximately one billion years of Earth history is missing along the unconformity.

◀ Figure 7.13 What a nonconformity looks like.

Putting It Together—How Do You Recognize Gaps in the Rock Record?

• Unconformities are former land or seafloor surfaces between rock layers that represent intervals *not* recorded in the local rock record.

• Angular unconformities separate sedimentary or volcanic rocks that are inclined at different angles. When an angular unconformity is present, the time required for tilting the lower rocks and eroding them to produce a new surface of deposition for the upper rocks is *not* represented in the geologic record.

• Disconformities separate sedimentary or volcanic rocks that are inclined at the same angle but are separated by an irregular erosion surface. This surface indicates a pause in deposition or erosion of the older rocks.

• Nonconformities separate sedimentary or volcanic rocks from underlying plutonic-igneous or metamorphic rocks. When a nonconformity is present, the time required to erode the materials that originally covered the plutonic and metamorphic rocks before the onset of the more recent sediment accumulation is *not* represented in the geologic record.

7.6   How Have Scientists Determined the Age of Earth?

How long has it taken for all the geologic events recorded in the rocks to take place—how old is Earth? Over hundreds of years, people have taken both scientific and nonscientific approaches to answer this question. These efforts show the varied ways scientists try to reach quantitative answers. Geologists now have strong evidence that Earth is about 4.5 billion years old.

Nonscientific Assessments of Earth's Age

Earth's age engaged people long before scientists developed methods for addressing the problem. These nonscientific approaches did not present testable hypotheses or rely on observations of rocks and geologic processes. Ancient Greek philosophers, for example, envisioned the universe as continually cycling in a circular fashion without a beginning or an end. Hindus believe in cycles of cosmic destruction and renewal that do not close in a circle. Based on Hindu traditional accounts of these cycles, Earth is about 2 billion years old. Numerous efforts were made between the second and seventeenth centuries to determine Earth's age from biblical chronologies. The best-known effort was by Irish Bishop James Ussher during the 1650s. Bishop Ussher adopted a completely literal interpretation of time on the basis of the Bible's first book, Genesis, asserting that God created the planet and all that is on it in six days. Ussher used biblical genealogies to conclude that creation occurred in 4004 B.C.E., which corresponds to a Universe and an Earth that would have been 6013 years old in 2009.

Evidence of an Older Earth

During the eighteenth century several people attempted to determine Earth's age by applying scientific reasoning to observations of natural processes. The French diplomat Benoit de Maillet, for example, found marine fossils in rocks high above sea level. He used historical records of measured small variations in sea level to conclude that a period of 2.4 million years would account for the locations of these fossils if they were left behind by the fall of a once all-encompassing deep sea. Geologists now know that sea level has risen and fallen countless times during Earth history, and that both falling sea level or uplift of land can explain the presence of marine fossils found above present sea level. Nonetheless, de Maillet's scientific estimate was based on realistic observation and stated assumptions that were widely accepted at the time.

James Hutton also challenged the notion that Earth was only a few thousand years old. He was particularly struck by the significance of unconformities as indicators of Earth's antiquity. Hutton supported his conclusion that Earth was older than contemporary assumptions by observations made at Siccar Point, on the east coast of Britain, which is pictured in the chapter opening photo. First, he pointed out that huge stretches of time are required for the accumulation of sediment layers, based on the slow rates of erosion and sediment deposition recorded during human history. The tilting of the older rock layers at Siccar Point, below the angular unconformity in the photo, also required an immense period, in human terms, to occur, given that there is no indication of single upheavals on this scale during recorded history. Next, a great deal of time passed while these older rocks eroded to form a flat surface and then the renewed, slow deposition of the upper sedimentary layers occurred. Additional time was required to account for the tilting of the rocks again and the erosion of the modern landscape to expose these geologic relationships. A participant on Hutton's field trip to Siccar Point later recalled listening to his explanation of this sequence of events:

> The mind seemed to grow giddy by looking so far into the abyss of time; and while we listened with earnestness and admiration to [Hutton] who was now unfolding to us the order and series of these wonderful events, we became sensible how much farther reason may sometimes go than imagination can venture to follow. (*The Works of John Playfair*, Archibald Constable and Co., 1822, p. 81)

Hutton's argument based on real-life observations and scientific reasoning clearly support the hypothesis that Earth must be far older than Bishop Ussher's pronouncement. Although Hutton did not estimate the absolute numerical age of the planet, he was so impressed by the need for an antiquity nearly unfathomable to the human perspective of time that he concluded his 1788 book, *Theory of the Earth*, with the following:

> The result, therefore, of our present enquiry is, that we find no vestige of a beginning—no prospect of an end. (*Transactions of the Royal Society of Edinburgh*, vol. 1, p. 304)

How Long Would It Take for Earth to Cool?

British physicist William Thomson, better known as Lord Kelvin, made a rigorous, quantitative attempt to calculate the duration of Earth's history in the late nineteenth century. Kelvin assumed that Earth started out as a molten sphere and then cooled over time by conduction. He then reasoned that if Earth was originally hotter than at present, then its age can be calculated by knowing (a) the original temperature, (b) the current distribution of temperatures within the planet, and (c) the rate at which heat conducts through rock. He developed what modern scientists call a "conceptual model." **Figure 7.14** illustrates Kelvin's concept.

How did Kelvin obtain the values that he used for his calculation? There was abundant evidence from temperature measurements in deep coal and metal-ore mines that temperature increases with depth into the planet. This temperature change is the geothermal gradient described in Section 4.4. The rate of heat conduction through rock is readily measured in the laboratory. An estimate of Earth's original temperature, however, is a matter of greater uncertainty. Kelvin assumed that Earth coalesced from the amalgamation of smaller objects in the early history of our solar system (this topic will be explored further in Chapter 9). When fast-moving objects collide, the energy of motion converts to heat when one or both objects instantaneously decelerate to zero velocity (slap your hands together to experience this effect on a small scale). The challenge is determining how hot early Earth was as a result of this heating from collisions. Kelvin and others assumed initial surface temperatures ranging from 1200°C to 3900°C.

Lord Kelvin admitted the uncertainty of the values used in the calculations and the nature of his assumptions. His biggest and most important assumption was that Earth cools only by conduction. Based on this assumption and his conceptual model, Kelvin argued that the most likely age of Earth was between 20 million and 40 million years. He concluded that if Earth was much older than that, conduction of heat to Earth's surface and then radiation into space would leave crustal rocks much colder than the warm temperatures measured in deep mines.

Most geologists at the time felt that, in light of the slowness of observed rates of Earth processes, the complex rock record required an even longer Earth history than Kelvin calculated. Of course, a final calculated answer is only as good as the certainty of the values used in the calculation and the assumptions used to construct the formula. This does not mean that mathematical solutions are not worth seeking, but that these solutions, like any scientific result, must be carefully scrutinized. In Kelvin's case there was no error in the math. Instead, it turned out that the most important of Kelvin's assumptions was suspect. If, as geologists now think, Earth's mantle convects (refer back to Figures 1.14 and 1.15 to refresh your understanding of the differences between conduction and convection), then heat moves upward in the convecting mantle to maintain warm geothermal gradients near the surface for billions of years. Therefore, the geothermal gradient required to maintain high temperatures in deep mines can be maintained much longer than the 40 million years calculated by Kelvin. Kelvin's conduction assumption, therefore, yields an imprecise minimum age for Earth.

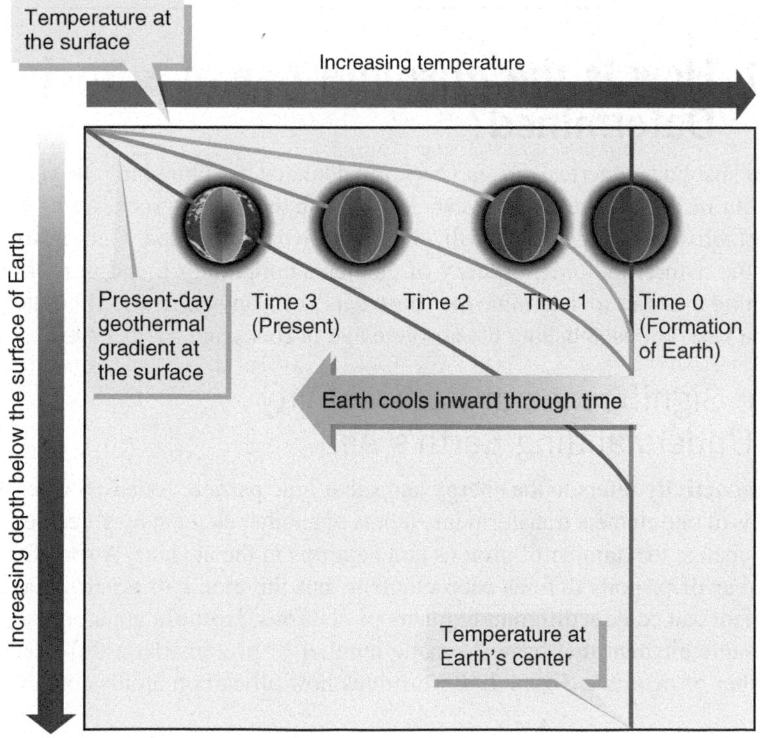

Using the Salty Ocean as a Clock

Around 1900, Irish geology professor John Joly developed a different conceptual model to determine Earth's age. His idea was that if the saltiness of the ocean results from dissolved salt carried to the ocean by rivers, then the age of the oceans, as an approximation of Earth's age, can be calculated if (a) the amount of salt in the ocean and (b) the quantity of salt carried to the sea by rivers are known. **Figure 7.15** illustrates Joly's logic.

What are the key assumptions and sources of uncertainty in Joly's calculation? In Joly's time there were many measurements of the amount of salt (NaCl) dissolved in seawater, and this value did not vary much from place to place. There also were measurements of the salt content of river water flowing to the ocean, but the values ranged considerably from river to river, so estimates of the total salt delivery to oceans from weathering on continents was highly uncertain. The *area* of Earth's oceans was well known at the time, but there were very few measurements of ocean depth, so any selection of a value for seawater *volume* was uncertain. The volume value is essential because the total amount of ocean salt is calculated by multiplying the average salt concentration in seawater by the total volume of seawater. Joly knew that some salt in seawater chemically precipitated during evaporation as rock salt (halite), so the volume of ancient rock salt also was estimated, but also with great uncertainty because the amount of rock salt buried from view in subsurface layers was unknown. Joly assumed that his value for salt input from rivers is constant through geologic time and estimated an initial saltiness for the primeval ocean instead of assuming that seawater started out completely fresh. Acknowledging his assumptions and the uncertainty of the values he used, Joly estimated the antiquity of the oceans to be on the order of 80–100 million years.

Temperature at the surface

Increasing temperature

Increasing depth below the surface of Earth

Present-day geothermal gradient at the surface

Time 3 (Present)   Time 2   Time 1   Time 0 (Formation of Earth)

Earth cools inward through time

Temperature at Earth's center

Kelvin assumed that Earth was equally hot from the center to the surface when it formed. Surface layers cooled down as heat radiated into space. Curves relating temperature to depth show cooling from formation of Earth to the present. Kelvin estimated the age of Earth by calculating the time required for the temperature curve to change shape from the vertical line, *Time 0*, to that reflected by the *Time 3* curve.

▲ Figure 7.14 How Lord Kelvin calculated Earth's age.

ACTIVE ART

Kelvin's Calculation of Earth's Age. See how Kelvin determined Earth's age.

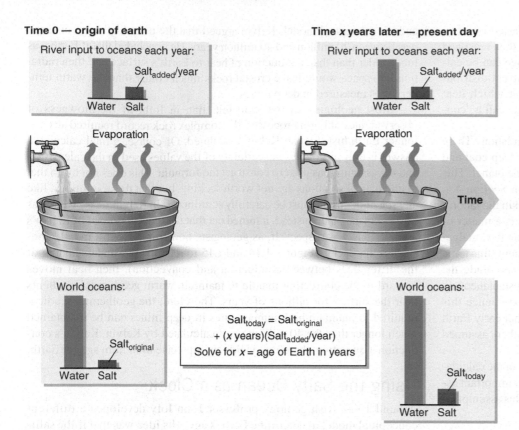

$$Salt_{today} = Salt_{original}$$
$$+ (x \text{ years})(Salt_{added}/\text{year})$$
Solve for x = age of Earth in years

▲ **Figure 7.15 How Joly calculated Earth's age.** Evaporation removes water from the ocean but the salt stays behind and accumulates over time. Joly assumed that the amount of salt in the ocean today, $salt_{today}$, was simply that amount that was originally present, $salt_{original}$, plus the annual input of salt by rivers, $salt_{added}$, multiplied by the number of years elapsed since Earth formed.

Putting It Together—How Have Scientists Determined the Age of Earth?

• Eighteenth-century geologists, including James Hutton, applied scientific observations and principles to conclude that the geologic rock record requires that Earth be extraordinarily ancient.

• Lord Kelvin calculated Earth's age based on a conceptual model of how much the planet has cooled since its origin. His calculations suggested an age in the range of 20–40 million years.

• John Joly calculated the age of the oceans based on a conceptual model of how long it would take Earth's rivers to transport enough salt to account for the current saltiness of the ocean. His calculations suggested an age of 80–100 million years.

• Kelvin's and Joly's calculations were correct but included assumptions that could not be verified at the time of their work. Reconsideration of their assumptions shows that the ages they calculated are minimum values and, therefore, Earth must be still older.

Although Joly calculated a greater age than that settled on by Lord Kelvin, many geologists still believed Earth to be older. Modern geologists also see many problems with Joly's calculation. Oceanographers later explored the ocean depths and now know that oceans hold a much larger volume of seawater than estimated by Joly. Subsurface explorations for oil and gas and field excursions in remote lands reveal that rock-salt deposits also are much larger than known by Joly. Applying Joly's logic to these newer data would result in a substantially older age for Earth.

Kelvin's and Joly's scientific reasoning led to the conclusion, by the early 1900s, that Earth was at least tens of millions of years old. As large as this number seems, it also was clear that Kelvin's and Joly's calculations must be minimum ages and that Earth *must be still older*. These early efforts at determining Earth's age set the stage, however, for twentieth-century innovations that led geologists to techniques for measuring the absolute ages of rocks.

7.7 How Is the Absolute Age of a Rock Determined?

Your absolute numerical age, in years, is calculated by subtracting the year of your birth from the current year. To establish the age of a rock, we need to establish the year of its "birth"—the date when it formed. Geologists use the natural radioactive decay of elements commonly found in rock-forming minerals to determine the "birth date" of many minerals. This date is the basis for establishing the absolute age of rocks and Earth's age.

The Significance of Radioactivity to Understanding Earth's Age

Radioactivity refers to the energy and subatomic particles released when atoms of one element transform into atoms of another element by processes that change the number of protons and neutrons in the nucleus. A specific number of protons defines each element, but the atoms of a particular element can contain different numbers of neutrons. **Isotopes** are atoms of the same element that have the same number of protons but a different number of neutrons. **Figure 7.16** illustrates how all carbon atoms contain

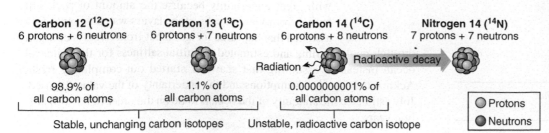

| **Carbon 12 (^{12}C)** | **Carbon 13 (^{13}C)** | **Carbon 14 (^{14}C)** | **Nitrogen 14 (^{14}N)** |
| 6 protons + 6 neutrons | 6 protons + 7 neutrons | 6 protons + 8 neutrons | 7 protons + 7 neutrons |

Radiation ← Radioactive decay →

98.9% of all carbon atoms | 1.1% of all carbon atoms | 0.0000000001% of all carbon atoms

○ Protons
● Neutrons

Stable, unchanging carbon isotopes — Unstable, radioactive carbon isotope

◀ **Figure 7.16 Visualizing isotope examples.** Carbon exists as three isotopes. All carbon atoms contain six protons, and most of these atoms also contain 6 neutrons so that the total atomic mass number is 12. One carbon isotope, ^{13}C, has 7 neutrons, for a total mass number of 13. Another very rare isotope, ^{14}C, has 8 neutrons. ^{12}C and ^{13}C are stable isotopes whose abundances do not change over time. ^{14}C, however, is an unstable isotope that undergoes radioactive decay and becomes the most common isotope of nitrogen. During this radioactive decay one carbon neutron converts to a proton.

six protons but may have six, seven, or eight neutrons; this means that there are three isotopes of carbon. All isotopes of the same element have the same atomic number (the number of protons) but different atomic mass numbers (the number of protons plus the number of neutrons). Scientists have identified 339 isotopes among the 84 naturally occurring elements. **Appendix B** lists isotopes that are important in geologic studies.

Only certain combinations of the number of protons and neutrons can exist in an atomic nucleus without causing instability between forces acting within the nucleus. This instability causes a transformation of an unstable isotope to a stable one by changing the number of protons, neutrons, or both. This transformation is called **radioactive decay**. For example, unstable carbon 14 (^{14}C) radioactively decays to become stable nitrogen. Only 70 of the 339 natural isotopes are unstable, or radioactive, isotopes.

Radioactive decay permits geologists to date the age of minerals because the abundances of isotopes within minerals change at known rates as time passes. The change occurs because the radioactive-isotope abundance decreases over time, whereas the abundance of the related stable isotope created by decay increases. If these isotope abundances are measured in a mineral, and the rate of decay is known, then the time that has elapsed since the mineral formed can be calculated.

EXTENSION MODULE 7.1

Radioactivity and Radioactive Decay. *Learn the different ways that radioactive isotopes decay and the resulting levels of natural radioactivity.*

Measure the Isotope Abundances

Figure 7.17 illustrates the logic used by scientists to determine the absolute age of a geologic sample by measuring the abundances of selected isotopes. The hypothetical mineral sample contains atoms of many elements, but it is simpler to focus on the abundance of only a particular radioactive **parent isotope** that decays through time to produce a stable **daughter isotope** of another element. (In the example of radioactive decay from Figure 7.16, ^{14}C is the parent isotope, and nitrogen is the daughter isotope.) In the hypothetical example in Figure 7.17, no daughter isotope is originally present, but over a period of time, radioactive decay causes the number of daughter isotope atoms to increase, while the abundance of the parent isotope decreases. Each daughter atom originates from the decay of a parent atom, so the total number of daughter plus parent isotopes is always the same. The ratio of daughter isotope to parent isotope relates directly to how much time has elapsed since radioactive decay began in the sample.

ACTIVE ART

Radioactive Decay. *See how the abundance of parent and daughter isotopes changes over time because of radioactive decay.*

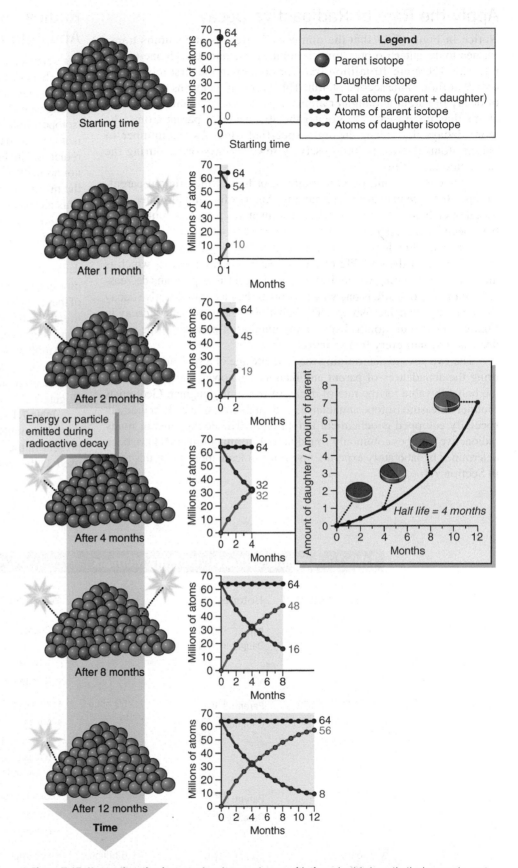

▲ **Figure 7.17 How radioactive-isotope abundances change with time.** In this hypothetical example, red atoms of a parent isotope decay to blue daughter-isotope atoms. The number of parent atoms decreases by half during every four-month interval, indicating a half-life of four months. The inset graph shows how the ratio of daughter to parent atoms changes during radioactive decay.

Apply the Rate of Radioactive Decay

Notice in Figure 7.17 that the number of parent-isotope atoms transformed to daughter-isotope atoms is not the same during each one-month interval. There are 10 million transformations in the first month, but only 9 million in the second month. This observation means that the rate of decay of parent to daughter atoms depends on the number of parent atoms that are present. The greater the abundance of parent atoms, the more decays that occur during a specified time. As the number of parent atoms decreases, progressively fewer decays occur during the same amount of time.

Notice that 32 million parent atoms, or half of the original parent-isotope atoms, remain after four months. Another four months later, after a total of eight months, the 32 million parent atoms present at four months have been halved again, to 16 million. After 12 months this number once again decreased by half, to 8 million atoms. This observation leads us to the definition of the **half-life** of the radioactive decay process, which is the time interval during which the number of parent-isotope atoms decreases by half. The half-life is one way to express the rate of radioactive decay from parent to daughter isotope. The half-life of the hypothetical decay in Figure 7.17 is four months because the number of parent-isotope atoms decreases by half every four months.

The two keys to determining the absolute age of a rock are (1) measuring the abundances of parent and daughter isotopes, and (2) knowing the half-life value for the rate of decay of parent to daughter. Geologists routinely measure isotope abundances with high precision and accuracy in specially equipped geochemical laboratories. The decay rates of many radioactive isotopes commonly found in rock-forming minerals have been determined by laboratory experiments (you will learn more about this topic in Section 7.8).

Radioactive-Isotope Decay Provides Absolute Ages

Let's consider an example of calculating the age of a mineral by application of the hypothetical radioactive-decay scheme shown in Figure 7.17. This requires some arithmetic but the math is not difficult; just take the time to work through it with confidence. Begin by analyzing the number of atoms of each isotope within a mineral sample: There are 10 million atoms of the parent isotope and 30 million atoms of the daughter isotope. An important thing to remember in this case is that the atoms of the daughter-isotope element do not normally bond with other elements found in the mineral, so that when the mineral first formed it did not contain atoms of the daughter isotope. This means that the 30 million atoms of the daughter isotope that we now measure in the mineral started out as atoms of the parent isotope. So, we now know that when the mineral first formed it contained 40 million atoms of the parent isotope (the 10 million atoms that are in it now plus the 30 million that decayed to the daughter-isotope atoms that are also in the mineral now). The time elapsed since the mineral crystallized is the time required for 30 million of these original 40 million radioactive parent atoms to decay. After one half-life, the number of parent atoms would decrease from 40 million to 20 million. After a second half-life, the number of parent atoms would further decline to 10 million, which is the number of parent atoms that have been measured in the rock. These simple calculations lead to the conclusion that two half-lives have elapsed since the mineral formed; each half-life is 4 months in duration (Figure 7.17), so the mineral is 8 months old.

Table 7.1 summarizes four radioactive-decay relationships that geologists commonly use to determine the age of Earth materials. The minerals present in the rock are a critical factor for determining which method is used. Only certain minerals contain the useful radioactive isotopes.

Table 7.1 Example Radioactive-Dating Methods

Name of Method	Isotopes	Half-life	What Can Be Dated?	Age Range of Application
Carbon 14	Parent: ^{14}C Daughter: ^{14}N	5730 years	• Charcoal • Organic matter (including wood and bone) • Calcite (including shells)	1 to 75,000 years
Potassium-argon	Parent: ^{40}K Daughters: ^{40}Ar ^{40}Ca	1.250 billion years	• Feldspars • Micas • K-bearing clay minerals • Whole rocks that contain K minerals	5,000 to 20 billion years[1]
Uranium-lead	Parent: ^{238}U Daughter: ^{206}Pb	4.468 billion years	• Uranium-ore minerals • Zircon • Calcite	10,000 to 40 billion years[1]
Rubidium-strontium	Parent: ^{87}Rb Daughter: ^{87}Sr	48.6 billion years	• Potassium feldspar • Micas	100,000 to 500 billion years[1]

[1] These methods could date materials that are older than the age of our solar system.

Table 7.1 also includes the practical age limits for using each method. Why do these practical limits exist? When the half-life is long and the decay has only recently begun, the abundance of parent atoms has barely changed. In this case the abundance of daughter isotopes is very small, and the precision of the geochemical instruments is insufficient to determine that any decay has actually occurred. Likewise, when nearly the entire parent isotope has decayed, there is too little remaining to be accurately detected by the instruments. For this reason, the carbon-14 method is ideal for dating items that are less than 75,000 years old because its decay to ^{14}N is fast, as indicated by a half-life of only 5730 years. To date very old rocks, however, requires application of the other radioactive-decay methods shown in the table, which have half-lives on the order of billions of years.

The validity of the radioactive-decay method is easily tested. **Figure 7.18**a shows an early test of radioactive-isotope dating that used the carbon-14 method to calculate ages for materials of known age. In some cases, the half-lives of two or more methods are appropriate for dating the same sample, and multiple age determinations provide a further verification of the approach (Figure 7.18b). Consistency of relative and absolute ages also validates the radioactive-isotope method.

Understanding the Geologic Conditions

Determining when a rock formed requires more than simply analyzing its constituent minerals for parent and daughter isotopes. Understanding the geologic origins of the rock and how to apply the isotope-dating methods is fundamental. To apply the method illustrated in Figure 7.17, it is essential that

a. no daughter-isotope atoms are present when the rock forms,

b. no parent-isotope atoms are gained or lost after the rock forms, and

c. no daughter-isotope atoms are lost or gained from the rock once it forms and radioactive decay begins.

▶ **Figure 7.18 How radioactive-isotope ages are confirmed.**

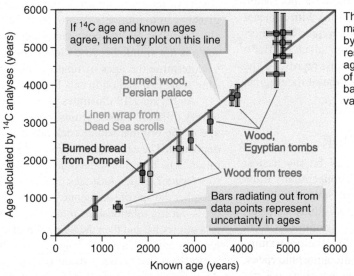

The ^{14}C dating method is tested by dating archaeological materials of known age, and tree wood whose age is known by counting the annual growth rings. The graph shows the results. For a perfect match between known and isotope ages, all data points should lie on the purple line, as nearly all of them do within the known uncertainties indicated by the bars on the data points. This consistency demonstrates the validity of the method.

(a)

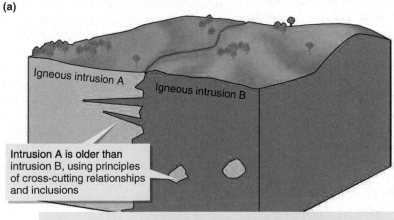

This diagram shows the relative-age relationships between two igneous intrusions exposed in Australia and isotope ages determined for rocks from each intrusion. The absolute ages are consistent with the relative ages. Ages obtained by three different methods for three different minerals in intrusion B are also identical within measurement uncertainty.

Absolute ages determined from laboratory measurements:

Intrusion A:
417 ± 1.0 million years

Intrusion B (ages obtained using different methods to date different minerals in the rock):
412.8 ± 1.9 million years
413.6 ± 3.7 million years
412.2 ± 2.3 million years

(b)

Condition (a) is not always met; and in that case additional measurements and calculations are needed to determine how much of the measured daughter isotope was present prior to the decay of the parent. Conditions (b) and (c) are met in most cases but can be affected by weathering or metamorphism, which can add or remove atoms of parent isotopes, daughter isotopes, or both. Before attempting to determine the age of a rock, a geologist must undertake other field, chemical, and microscopic studies of the rock to determine whether it is significantly weathered or metamorphosed.

It is also important to keep in mind that a radioactive-isotope age indicates when the *minerals* in the rock formed. When a mineral crystallizes from magma, precipitates from a watery fluid, or forms from metamorphic reactions, its radioactive "clock" starts to "run." At that point atoms of daughter isotopes accumulate in the crystal structure as parent atoms decay. For example, the age of a feldspar crystal collected from granite usually records when the granitic magma crystallized. If a feldspar crystal from that granite erodes as a sand grain that ends up in sandstone many millions of years later, the age of the feldspar crystal remains the age *when it formed as part of the granite*, and this is not the age of deposition of the sandstone. If the granite metamorphoses to gneiss, the potassium feldspar crystals in the granite heat up and exchange atoms with adjacent crystals and fluids, and recrystallize into new feldspar crystals. Metamorphism, therefore, usually resets the radioactive-isotope "clock." After a rock metamorphoses, the age determined by a geochemist represents the *age of metamorphism* rather than the age of crystallization of the original granite.

An Example—Potassium-Argon Dating

The potassium-argon (^{40}K-^{40}Ar) dating method may be the easiest radioactive-isotope dating method to understand; it is conceptually illustrated in relationship to the rock cycle in **Figure 7.19**. Several common minerals contain abundant potassium, including potassium feldspar, muscovite, and biotite. These potassium-bearing minerals commonly form when magma crystallizes or during metamorphism, so the potassium-argon method applies handily to dating igneous and metamorphic rocks. Unlike reactive potassium ions, argon is a nonreactive gas that does not readily bond to other elements to form compounds. For this reason, the crystal structures of growing minerals do not incorporate argon.

The potassium-argon method is, therefore, similar to the hypothetical example in Figure 7.17, which specifies that the starting material contains parent isotope, in this case the ^{40}K atoms, but no daughter isotope, such as ^{40}Ar (Figure 7.19). A difference from the hypothetical example in Figure 7.17 is that ^{40}K simultaneously decays to ^{40}Ar *and* ^{40}Ca. The geological dating method monitors only the decay of ^{40}K to ^{40}Ar because ^{40}Ca is the most common isotope of calcium and is initially present in almost all potassium-bearing minerals even before decay of ^{40}K begins.

Argon is a nonreactive element, and therefore it is not always retained inside a mineral once it forms by radioactive decay. If the mineral heats up during metamorphism, expansion of the crystal structure may permit the loosely caged argon atoms to escape. In this case, the isotope clock resets and the calculated age from measurements of ^{40}K and ^{40}Ar reflects the time of metamorphic reheating.

We can use the potassium-argon method to determine when the basaltic dike exposed at the beach crystallized (Figure 7.1). **Figure 7.20** graphs

the results of the laboratory analyses. Chemical analyses of a 100-gram sample of the basalt detect 7.54×10^{21} atoms of ^{40}K and 1.13×10^{19} atoms of ^{40}Ar produced by decay of ^{40}K. This means that the ^{40}Ar/^{40}K ratio is 0.0015. On the graph in Figure 7.20, based on knowledge of the decay rate of ^{40}K to ^{40}Ar, you can see that a ratio of 0.0015 corresponds to an age of 25.8 million years.

EXTENSION MODULE 7.2

The Mathematics of Radioactive-Isotope Decay. *Learn how to use mathematical equations, instead of graphs, to calculate mineral ages.*

Putting Absolute Ages on the Geologic Time Scale

The distribution of fossils within sedimentary rocks defines the geologic time scale (Figure 7.9). The scale was established decades before the discovery of radioactivity and about a century prior to the time when geologists began to routinely determine the absolute ages for rocks. Obtaining absolute ages for the boundaries between periods and eras on the time scale is not simple because radioactive-isotope methods are used mostly to determine the ages of igneous and metamorphic rocks, whereas the time scale is based on fossils found in sedimentary rocks.

As Figure 7.19 illustrates, direct isotope dating of sedimentary rocks is very difficult. Clastic grains eroded from older rocks yield the absolute age of the original source rock and not the time when the sediment was deposited. The ages of cementing minerals reflect only the time of cementation, which occurs some time after deposition.

Figure 7.21 demonstrates how to combine absolute and relative ages to decipher the age of sedimentary rocks. The principles of superposition and cross-cutting relationships relate the relative age of sedimentary rocks to igneous rocks, and then the igneous rocks provide samples for absolute-age measurements. Then, the range in possible absolute ages of sedimentary rocks deposited between dated volcanic ash or lava layers or cross-cut by dikes is estimated. **Figure 7.22** applies this approach to obtain ages for boundaries on the geologic time scale. It is extraordinarily rare to find dateable igneous rocks right at a boundary in the rocks between two periods defined by fossils, so the ages of boundaries are estimated. Geologists revise the ages on the time-scale boundaries to new values with less uncertainty as they obtain more absolute dates that are relevant to establishing boundary ages. For this reason you may find that the ages on the time scale in this book (Figure 7.9) differ from those in older texts, and the ages listed here doubtless will be revised in the future. The names of the intervals on the time scale do not change, and the fossils that define each time interval that have been agreed upon by international convention remain the same, but the boundary ages continually fluctuate by small amounts as new absolute-age data become available.

The Oldest Rocks and the Age of Earth

Finding the oldest rocks on Earth is a daunting task because our planet is a dynamic place. Metamorphism resets most isotope-dating clocks in rocks. Weathering and erosion recycle material from old rocks into younger ones.

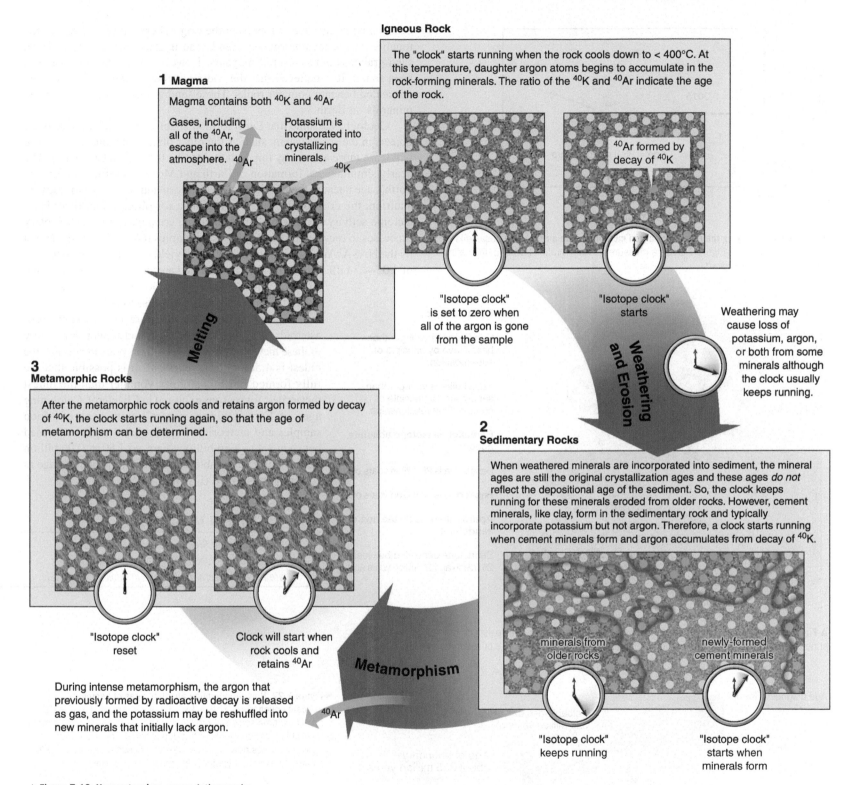

Igneous Rock

The "clock" starts running when the rock cools down to < 400°C. At this temperature, daughter argon atoms begins to accumulate in the rock-forming minerals. The ratio of the ^{40}K and ^{40}Ar indicate the age of the rock.

1 Magma

Magma contains both ^{40}K and ^{40}Ar

Gases, including all of the ^{40}Ar, escape into the atmosphere. ^{40}Ar

Potassium is incorporated into crystallizing minerals. ^{40}K

^{40}Ar formed by decay of ^{40}K

Melting

"Isotope clock" is set to zero when all of the argon is gone from the sample

"Isotope clock" starts

Weathering may cause loss of potassium, argon, or both from some minerals although the clock usually keeps running.

Weathering and Erosion

3
Metamorphic Rocks

After the metamorphic rock cools and retains argon formed by decay of ^{40}K, the clock starts running again, so that the age of metamorphism can be determined.

2
Sedimentary Rocks

When weathered minerals are incorporated into sediment, the mineral ages are still the original crystallization ages and these ages *do not* reflect the depositional age of the sediment. So, the clock keeps running for these minerals eroded from older rocks. However, cement minerals, like clay, form in the sedimentary rock and typically incorporate potassium but not argon. Therefore, a clock starts running when cement minerals form and argon accumulates from decay of ^{40}K.

"Isotope clock" reset

Clock will start when rock cools and retains ^{40}Ar

Metamorphism

minerals from older rocks

newly-formed cement minerals

During intense metamorphism, the argon that previously formed by radioactive decay is released as gas, and the potassium may be reshuffled into new minerals that initially lack argon. ^{40}Ar

"Isotope clock" keeps running

"Isotope clock" starts when minerals form

▲ Figure 7.19 How potassium-argon dating works.

The accumulation of sedimentary and volcanic rocks at the surface buries older rock, which is exposed again only when and where tectonic forces and erosion conspire to uplift and remove the covering strata. Earth's oldest rock may lay buried deep below the surface and out of view.

The oldest dated materials on Earth are more than 4 billion years old. The oldest rock found so far is gneiss that resulted from the metamorphism of a tonalite intrusion in northwestern Canada. The tonalitic gneiss contains the silicate mineral zircon, which incorporates radioactive uranium when it crystallizes. Furthermore, zircon is unusual for holding onto parent uranium isotopes and daughter lead isotopes during metamorphism. Uranium-lead isotope measurements reveal a 4.03-billion-year age for some of the zircon crystals in the gneiss. These

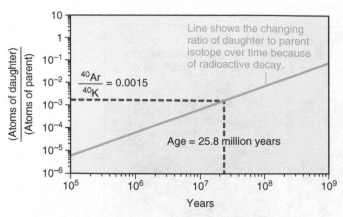

▲ **Figure 7.20 Applying the potassium-argon method to date a rock.** The basalt dike at the seaside outcrop in Figure 7.1 has a measured $^{40}Ar/^{40}K$ ratio of 0.0015, indicating an age of 25.8 million years.

data can be interpreted to represent the original crystallization age of the tonalite before metamorphism. Sandstone in Australia contains zircon sand grains as old as 4.4 billion years. However, as no rock this old has been found, it is believed that the source of these zircon grains may be concealed beneath younger rocks. These sand grains are the oldest dated minerals on Earth.

Geologists assume that Earth formed at about the same time as other objects in the solar system, such as the other planets and our Moon. (You will learn more about why this is thought to be the case in Chapter 9.) The nearly simultaneous formation of Earth and Moon is useful to determine Earth's age because the Moon is relatively close and accessible to study. In addition, the crust of the Moon is laid bare for observation; there is no tectonic activity to cause metamorphism, no erosion, and no sedimentary processes to recycle and bury old rocks. Samples returned from the Moon by the NASA Apollo missions in the 1970s yield radioactive-isotope ages of 3.05–4.3 billion years for the dark-colored areas visible from Earth, and 3.65–4.53 billion years for rocks collected in the light-colored areas. There are also radioactive-decay dates from meteorites found on Earth, which are mostly fragments of asteroids that orbit between Mars and Jupiter. A majority of these meteorites exceed 4.4 billion years in age, and the oldest is 4.568 billion years old. It is possible that the fully formed Earth is slightly younger than the oldest meteorites, for reasons explored in Chapter 9. Combining ages of Earth's oldest minerals with ages of Moon samples and meteorites implies that the planet formed between 4.4 to 4.56 billion years, so 4.50 ± 0.06 billion years old is a reasonable expression of the planet's age as currently interpreted (Figure 7.9).

EXTENSION MODULE 7.3

Using Geologic Clocks. *Learn the different radioactive-isotope dating methods and how they are used to determine the age of geologic materials.*

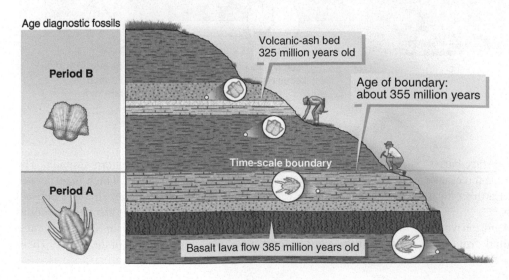

Relative ages:

Sandstone is younger than basalt lava by principle of superposition.

Basalt dike is younger than sandstone by principle of cross-cutting relationships.

Radioactive isotope absolute ages:

Basalt lava is 25 million years old.

Basalt dike is 20 million years old.

Approximate absolute age of sandstone:

Sandstone deposited between 25 million and 20 million years ago.

▲ **Figure 7.21 How to estimate absolute ages of sedimentary rocks.** Geologists combine relative-dating principles and absolute-dating methods to determine the depositional age of sedimentary rocks.

◄ **Figure 7.22 How to determine absolute ages of geologic-time-scale boundaries.** Fossils found in sedimentary deposits define the geologic time scale but radioactive-isotope dating methods rarely give reliable measurements of the depositional age of sediment. This diagram illustrates how absolute ages of igneous rocks are used to estimate the age of a hypothetical time-scale boundary.

Putting It Together—How Is the Absolute Age of a Rock Determined?

• Radioactive-dating methods provide absolute ages of minerals. Naturally occurring radioactive parent isotopes decay to stable daughter isotopes at known measured rates. Measurement of parent and daughter isotope abundances permits calculation of rock age.

• Absolute ages are determined for the geologic time scale by combining radioactive-dating and relative-dating methods. Fossils present in sedimentary rocks define the time-scale boundaries. Time-scale boundary ages are estimated from absolute ages of nearby igneous rocks whose relative age relationships to fossiliferous sedimentary layers are known.

• The oldest rock measured thus far on Earth is 4.03 billion years old, and the oldest mineral is 4.4 billion years old. Even older rocks are found on the Moon and among meteorites that have landed on Earth. The current estimated of Earth's age is 4.50 ± 0.06 billion years.

7.8 How Do We Know . . . How to Determine Half-Lives and Decay Rates?

Understand the Problem

Why Is It Important to Measure Radioactive-Decay Rates? Chemical analyses of the abundance of elements, and individual isotopes of elements, have been routine for several decades. The abundances of most isotopes are measured down to minute fractions of 1 percent.

Unless the rate is known for the decay of the parent to the daughter isotope, however, measuring the abundances of parent and daughter isotopes in a mineral is insufficient for determining the age of the mineral. Geochemists use carefully designed experiments to measure these rates.

Make the Measurements

What Data Are Required to Calculate Decay Rate? The potassium-argon dating method provides an example of how geologists design experiments to measure radioactive decay. The half-life for the decay of ^{40}K to ^{40}Ar plus ^{40}Ca is 1.25 billion years (Table 7.1). This extremely long half-life implies that the decay rate is very slow. Clearly, no one is able to conduct an experiment lasting more than 1 billion years to confirm that half of the original ^{40}K atoms converted to daughter isotopes. Instead, the rate of decay is determined by measuring the radiation emitted when each parent atom decays. The number of decays during a particular time interval depends on how many parent atoms exist in the sample, as noted from the example in Figure 7.17. The decay-rate calculation, then, requires two types of data:

1. Measurements of the amount of radioactive parent isotope in the sample, which is possible to far better than 1 percent accuracy.

2. Measurements of the amounts and types of radiation emitted during decay. This second step requires further consideration.

Two radiation-producing processes take place during decay of ^{40}K:

1. Conversion of ^{40}K to ^{40}Ar occurs when an electron orbiting near the nucleus of ^{40}K enters into the nucleus, combines with a proton, and produces a neutron. This causes rearrangement of electrons circling the nucleus, which releases measurable x-rays.

2. Conversion of ^{40}K to ^{40}Ca happens when a neutron in the nucleus of ^{40}K changes to a proton, which in order to maintain charge balance causes simultaneous production of a detectable negatively charged electron that exits the nucleus.

Geochemists use detectors to measure the release of x-rays and electrons from highly purified samples containing a measured amount of potassium. Each detection of an x-ray or an electron represents the decay of a ^{40}K atom. Detection of an x-ray records decay of a ^{40}K atom to an ^{40}Ar atom. Detection of an electron records decay of a ^{40}K atom to a ^{40}Ca atom.

Visualize the Results

How Fast Does ^{40}K Decay? **Figure 7.23** summarizes many laboratory measurements of the radiation released during decay of ^{40}K. Notice that different measurements made at different times produced different decay rates, and some experiments produced more precise values than others. These data show that as measurements continue to improve, more recent experiments provide more reliable values. Many different scientific research groups, in different laboratories using different starting samples and different types of radiation-counting instruments, obtained very similar values. Reproducibility is an essential part of establishing the merits of scientific results.

In 1976 an international commission of geochemists adopted values for the two decay paths of ^{40}K to its daughter isotopes. Figure 7.23 depicts these selected values. The geochemists did not simply average all of the available numbers, but gave greater importance to the more recently obtained values as these were found to reflect smaller uncertainty. Adopting specific values for the decay rates ensures that every scientist will calculate the same mineral age from a single set of measurements of ^{40}K, ^{40}Ar, and ^{40}Ca abundance.

The data confirm a very slow radioactive decay as already suspected from the long half-life reported in Table 7.1. Adding the two decay paths together reveals that between 31 and 32 ^{40}K atoms decay each second in every gram of potassium. This may seem like a lot of decays, but one gram of potassium contains 1.8×10^{18} atoms of ^{40}K, so the ^{40}K abundance decreases very, very slowly; 1.25 billion years must go by before half of the ^{40}K decays to its two daughter isotopes, ^{40}Ar and ^{40}Ca.

Is the half-life of 1.25 billion years *exactly* correct? Not in the strictest sense, because the experimental results yield small differences. The differences in decay rates, however, are so slight that you can have a high degree of confidence that the half-life is uncertain by no more than one-half of one percent (or about six million years).

Insights

How Reliable Are Decay Rates? The consistent reproduction of experimental determinations of decay rates leave geologists very

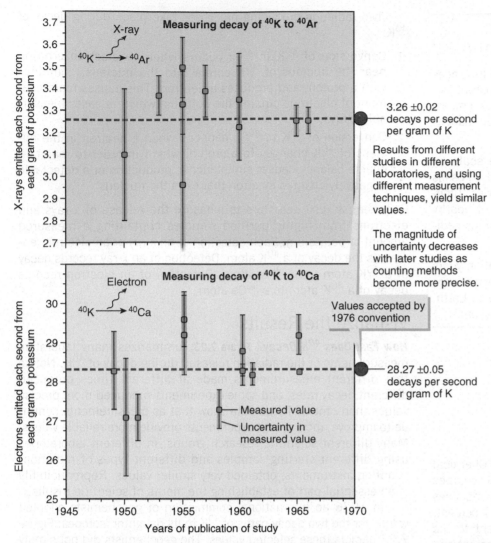

▲ **Figure 7.23 Measured decay rates.** Measurements of radioactivity emitted from potassium reveal the rate of decay of ⁴⁰K to its two daughter isotopes. These graphs show different measurement results of both decay rates over a 20-year period.

confident of the conclusion that Earth is close to 4.5 billion years old. Measurements of the abundances of the parent and daughter isotopes in minerals and the rate of decay of parent to daughter isotopes have been confirmed by repeated laboratory measurements. Another key question, however, is whether these decay rates are always the same or whether they might change under different conditions. If decay rates are not constant, then the laboratory measurements are not valid for calculating mineral ages.

Many experiments have been conducted to verify the constancy of decay rates, typically for radioactive isotopes that disintegrate faster than ⁴⁰K. These experiments tested for changes in decay rates by varying temperature from −250°C to 1550°C, by applying pressure as great as 2000 times that experienced at Earth's surface, by varying the force of gravity, and by employing magnetic fields more than 160,000 times as strong as the natural field at Earth's surface. No results showed variations in the rate of decay that would have a noticeable effect on the calculation of a mineral's age.

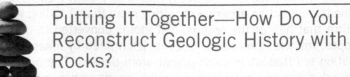

Putting It Together—How Do We Know . . . How to Determine Half-Lives and Decay Rates?

• Radioactive-decay rates are measured in the laboratory by detecting the energy released by each decay event. Decay rates are expressed as the number of decays per unit of time per gram of parent isotope.

• Decay rates measured in different laboratories are the same within the uncertainty of the measurement. International committees establish the universally used decay rates based on the most reliable and up-to-date laboratory values.

• Laboratory experiments show that decay rates are unaffected by changes in temperature, pressure, variations in gravitational field, or different magnetic fields.

7.9 How Do You Reconstruct Geologic History with Rocks?

Return to your field sketch of the rocks exposed in the sea cliff (Figure 7.1). The picture is now literally more complete. You are now able to do more than explain the origin of the rocks, one at a time.

• You can explain the *order* in which these rock-forming processes took place, using the principles of relative dating.

• You can, with the aid of fossils and radioactive-isotope dates, specify *when* these processes took place.

Stop here and write down the geologic history of the outcrop, and then compare your history to the version presented in **Figure 7.24**.

Putting It Together—How Do You Reconstruct Geologic History with Rocks?

• Both relative and absolute dating methods, which reveal the order and the age of events, allow us to determine the geologic history of an area.

• Combined with knowledge of rock-forming processes and the origins of other geologic features, relative and absolute dating methods permit narrative descriptions of geologic history.

Rivers deposit sediment during the Pennsylvanian Period

Application of the principle of superposition requires that the lowest exposed layers reveal the earliest known history. The red sandstone and shale layers contain fossils of nonmarine organisms characteristic of the Pennsylvanian Period (see Figure 7.10). The cross-bedded sandstone records deposition by rivers, with fine silt and clay accumulating on floodplains now represented by the shale. The geologic time scale (Figure 7.9) indicates that the Pennsylvanian sedimentary strata formed during some part of the interval between 300 and 318 million years ago.

Late Paleozoic or early Mesozoic Tilting

Tectonic forces tilted the rocks during the time following Pennsylvanian deposition but before accumulation of overlying Jurassic tan sandstone. You know this because the inclined red Pennsylvanian strata were originally nearly horizontal. Tilting of the Pennsylvanian rocks and erosion to produce a new surface for later deposition took place during some part of the late Paleozoic or early Mesozoic Eras, between about 300 and 200 million years ago (see Figures 7.9 and 7.10).

Sedimentation in the Jurassic Sea

Submergence of the tilted and eroded red sandstone and shale caused deposition of tan beach sandstone containing Jurassic fossils. Fragments of Pennsylvanian sedimentary rocks were included in the Jurassic beach sand. Based on the time scale (Figure 7.9) this sea existed during part or all of the Jurassic Period between 146 and 200 million years ago.

A pause in deposition

A disconformity forms the top of the Jurassic rocks (see Figure 7.10), so there was a pause in deposition and, quite likely, some erosion of rock. It is not evident how long this period of nondeposition and erosion persisted.

Deposition in the early Tertiary Sea

The area submerged again during the early Tertiary, which accounts for deposition of gray limestone and black shale. The fossils in these rocks suggest that deposition took place during some part of the time interval between 36 and 65 million years ago (see Figure 7.10).

Oligocene igneous activity

During the late part of the Oligocene Epoch, at about 25.8 million years ago based on a ^{40}K-^{40}Ar age (Figure 7.20), tectonic processes formed basaltic magma that intruded as a dike. We can speculate that the dike fed volcanoes at the surface, but if this is true, those volcanic rocks eroded away and there is no remaining evidence.

Erosion to form the present landscape

Erosion resumed sometime after 25.8 million years ago. We can tell this because the basaltic dike is exposed at the surface. Intrusive igneous rocks are only exposed at the surface if overlying rock has eroded.

Time

▲ **Figure 7.24 The geologic history recorded in the sea cliff.** Here is a reasonable historical narrative for the sea-cliff outcrop illustrated in Figure 7.1. This history is consistent with principles for establishing the relative order of events recorded in rocks, the knowledge of how to use fossils to establish the geologic-time-scale age of sedimentary strata, and the use of radioactive-isotopes to establish the absolute age of rock-forming minerals. Only events occurring during time represented by rocks preserved in the outcrop are described with great confidence. The processes occurring during the intervening times are inferred only by the presence of unconformities.

Where Are You and Where Are You Going?

Rocks record the antiquity of Earth's history and the multitude of events, slow and fast processes, and ever-changing environments that define this history. Not only do the features in a rock reveal the processes that formed that rock, but they also make it possible to determine the relative and absolute ages of the rock-forming processes.

Relative ages describe a sequence of events. The principles of superposition, original horizontality, cross-cutting relationships, and inclusions allow you to place the events recorded in rocks into a sequential order from oldest to most recent. The principle of lateral continuity of beds permits you to apply knowledge gained from one location to another, where more rocks may be exposed, in order to gain a more complete picture of geologic history. Fossils play an important role in determining the relative ages of rocks over large areas and even globally. Because most organisms existed for only short intervals of the vast history of the planet, the types of fossils in a rock reveal that the rock formed during a discrete interval of Earth's history. The geologic time scale formally defines and names these intervals.

Absolute ages of rocks refer to how much time has passed since the rocks formed. The natural radioactive decay of elements found in rock-forming minerals provides scientists with a way to record the absolute ages of rocks. Parent isotopes decay to daughter isotopes at rates that are measured and verified by laboratory experiments. Half-life refers to the elapsed time required for half of the parent to convert into an equal amount of daughter products. Half-life is a convenient way to express the rate of decay. The abundances of the parent-to-daughter isotopes can be measured with a high degree of precision. The measurements of isotope abundances are combined with knowledge of the decay rate, sometimes expressed as the half-life, in order to determine when component minerals formed in the rock.

Radioactive-isotope dating allows scientists to estimate boundary ages on the geologic time scale. This work is required because isotope-dating techniques are best suited for providing crystallization ages for igneous and metamorphic rocks, but fossils found in sedimentary rocks define the geologic time scale. Applying the principles of superposition and cross-cutting relationships, it is possible to estimate the age of the time-scale boundaries by obtaining isotope ages on interlayered and cross-cutting igneous rocks. Both relative-age relationships and absolute-age measurements are key to deciphering geologic history.

Radioactive-isotope ages allow scientists to estimate Earth's age. The oldest absolute-age determination for minerals formed on Earth is 4.4 billion years, but older rocks have been returned from the Moon and are found among meteorites that have fallen to Earth. These observations have been combined to suggest that Earth originated about 4.5 billion years ago.

The history of measurements pertinent to determining Earth's age illustrates the changing nature of scientific knowledge caused by challenging established interpretations and by incorporating new knowledge and technology when they become available. These measurements also indicate the importance of uncertainty, which is present in nearly all calculations and numerical models. Joly's and Kelvin's calculations of Earth's age were not specific single numbers, but instead were expressed as ranges in values because of uncertainties in the pair's calculations. Likewise, radioactive-isotope ages have uncertainties that result from variations in the measurements of isotope abundances and decay rates.

Your knowledge of Earth materials, extensive as it now is, still pertains only to what is known about the outermost skin of the planet. In the next three chapters, you will find out how geologists learn about the materials and processes within Earth's interior.

Active Art

Relative Dating Principles. See how the relative dating principles are used to decipher the sequence of geologic events.

Unconformities. See how the three types of unconformities form.

Kelvin's Calculation of Earth's Age. See how Kelvin determined Earth's age.

Radioactive Decay. See how the abundance of parent and daughter isotopes changes over time because of radioactive decay.

Extension Modules

Extension Module 7.1: Radioactivity and Radioactive Decay. Learn the different ways that radioactive isotopes decay and the resulting levels of natural radioactivity.

Extension Module 7.2: The Mathematics of Radioactive-Isotope Decay. Learn how to use mathematical equations, instead of graphs, to calculate mineral ages.

Extension Module 7.3: Using Geologic Clocks. Learn the different radioactive-isotope dating methods and how they are used to determine the age of geologic materials.

Confirm Your Knowledge

1. Distinguish between relative age and absolute age. Give an example of each type of age.
2. Which principle(s) would you use to determine the relative ages of beds in a sequence of flat-lying sedimentary rocks? Why?
3. Which principle(s) would you use to determine the relative sequence of events that accounts for an exposure of non-horizontal beds of sedimentary rocks? Why?
4. Which principle(s) would you use to determine the relative ages of an intrusive igneous rock, such as a dike, and the rock adjacent to the dike? Why?
5. What principle(s) would you use to determine the relative ages of a granitic intrusion if there are granite pebbles in the sedimentary rock overlying the granite? Why?

6. What is correlation? How can you be sure you are correlating the same layers?
7. What are the three types of unconformities? How does each form?
8. What is an isotope?
9. What percentage of the known naturally occurring isotopes is radioactive?
10. What is the meaning of the half-life of a radioactive-decay process?
11. In order to apply the isotope-dating method, what must be known or measured, and what conditions must be met?

Confirm Your Understanding

1. Write an answer for each question in the section headings.
2. Turn to Figure 4.6a. Pretend that you are leading a geology field trip in the Grand Canyon. Explain the sequence of events recorded by the rocks that are visible in this photograph.
3. Explain a hypothetical example in which the oldest rock exposed in a deep canyon is *not* at the bottom of the canyon.
4. Examine the diagram below, which is a diagrammatic sketch of geologic features exposed in cross section in a deep river canyon. Write a paragraph that summarizes the geologic history at this locality, using the letter labels to describe the different rock units. Your history should be chronological, starting with the first event and ending with the most recent. Identify unconformities where they exist.

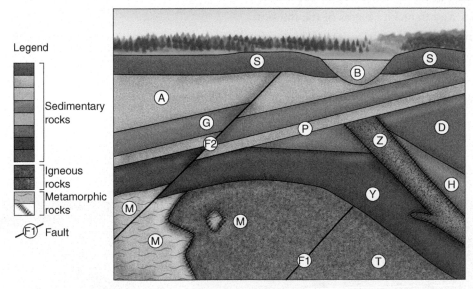

Legend

Sedimentary rocks

Igneous rocks

Metamorphic rocks

F1 Fault

5. Approximately what percent of geologic time is represented by the Precambrian? Which period within the Phanerozoic eon represents the longest interval of time? How long is it?

6. Early scientific estimates of Earth's age used observations of exposed marine fossils requiring a drop in sea level (Benoit de Maillet), cooling of a hotter Earth (Lord Kelvin), and the salinity of the ocean (John Joly). What age did each of these estimates come up with? What assumptions were made for each of these estimates that we now know to be incorrect?
7. How do we know that the ^{14}C radioactive dating method is valid for dating archaeological and geological materials?
8. You need to know the age of a rock collected during a field research project. How do you select which radioactive dating method to use?
9. Age determinations from radioactive-isotope decay methods are usually performed on a mineral crystal. Assume that a rounded grain of feldspar is plucked from a sandstone, analyzed, and determined to be 200 million year old. What does this tell you about the age of the sandstone? What other methods can be used to determine when the sand in the sandstone was deposited?
10. Why do geologists sometimes change the absolute ages of the boundaries in the geologic time scale?
11. The potassium-argon dating method is applied to dating potassium feldspar in rocks. But it is important to keep in mind just exactly what the resulting numbers mean. For each example, indicate which geologic process or event is being dated:
   • A potassium-argon date of 2.3 million years on potassium feldspar in a volcanic rock
   • A potassium-argon date of 535 million years on potassium feldspar in a metamorphosed volcanic rock
   • A potassium-argon date of 164 million years on potassium feldspar in a clastic sedimentary rock
12. Assume that a particular radioactive-isotope dating method has a half-life of 150 million years. Also assume that the parent isotope is commonly incorporated in igneous-rock minerals but that the daughter isotope is not. Measurements show that one crystal in an igneous rock contains 5 million atoms of the parent isotope and 75 million atoms of the daughter isotope. How old is this crystal?

CHAPTER 8

Journey to the Center of Earth

Why Study Earth's Interior?

Humans have always been curious about Earth's interior. Magma rises from underground and erupts at volcanoes. Earthquakes, apparently driven by unseen forces from below, heave and split the land surface. These events reveal an active interior where energy drives processes, some visible on the surface, that profoundly affect human lives. Understanding earthquakes, volcanoes, and the persistent motion of continents provides motivation for studying Earth's interior.

Even if you visited every country on every continent and sailed the seven seas, you would see only a very small percentage of Earth, because most of the planet is beneath you. Even trips into Earth access only a small fraction of the planet. The deepest diamond mine in South Africa, for example, is 3.6 kilometers deep. The deepest well, in northeastern Russia, penetrates to 12 kilometers, a scratch into Earth as compared to its radius of 6371 kilometers. How can geologists know anything about Earth's deep interior? Earthquakes are key to our knowledge of inner Earth because earthquake energy provides a way to make images of the interior. These images are analogous to doctors using X-rays to reveal an image of a person's interior. Countless images are combined to reveal the various layers inside Earth and provide clues to the nature and composition of the material that makes up Earth's interior. This ingenuity of geoscientists allows them to decipher Earth's interior without being able to physically visit Earth's interior.

After Completing This Chapter, You Will Be Able to

- Describe Earth's interior as layers arranged concentrically, in nearly spherical shells of different composition and physical properties.

- Explain how geoscientists use information from earthquakes to construct this view of the interior.

- Explain how geoscientists determine temperatures inside Earth and the origin of the planet's internal heat.

Pathway to Learning

8.1 How Do Geologists Learn about Earth's Interior?

8.2 How Do Earthquakes Make Images of Earth's Interior?

8.3 How Do We Know . . . How to Determine Velocities of Seismic Waves in Rocks?

EXTENSION MODULE 8.1
Sizing Up Earth

EXTENSION MODULE 8.2
How to Locate an Earthquake

The diamonds in these specimens from Africa and Asia formed 150 kilometers below Earth's surface.

IN THE LAB

Determining the internal structure and composition of Earth is similar to trying to figure out what is inside a baseball. Imagine picking up a baseball in your geology laboratory and being told to determine what it consists of—without cutting it in half. In fact, you are not allowed to cut deeper than through the leather covering. This exercise in "scratching the surface" is analogous to studying Earth because geologists' direct knowledge of the interior arises from studying little more than the outer skin of the planet.

It makes sense to start making observations as deeply as you can. When you take a knife to the outer layer of the baseball, you see that the stitched leather cover is only a millimeter thick, with tightly wound wool yarn beneath it, as seen in **Figure 8.1**a. You now know that the ball is inhomogeneous—that is, it consists of more than one type of material. Does the yarn continue to the center of the ball? The rules forbid you to dig deeper, so how can you tell?

One option is to compare some property of the entire ball to the properties of the observed leather and yarn. The easiest property to measure is density. Your instructor tells you the accepted densities of the tightly bound yarn and cowhide. If the density of the whole ball is similar to the density of the yarn, then perhaps the remainder of the ball is also made of yarn. It takes just a minute to weigh the ball and measure its outer dimensions. From these measurements, written in Figure 8.1, you determine that the density of the ball is more than the stated densities of yarn and leather. There must be something in the interior that is denser than the materials you found in your shallow excavation at the surface.

What next? Is there a way to look inside and learn about the invisible interior? You might think of x-rays. Doctors figure out much about the interior processes of a living person by using x-rays and other imaging technology. Security officers determine the contents of airline luggage with scanning devices. One way to interpret the deeper interior of the baseball simply would be to use an x-ray machine, but none is available in your geology lab.

Your instructor ends the suspense and allows you to survey more deeply into the ball. As shown in Figure 8.1b, you cut open the ball. Sure enough, the center is higher-density rubber, which fits your conclusion that something else composes the ball besides cowhide and yarn.

How does this exercise help you understand the nature of the inaccessible Earth interior? Extending the application of density, can you determine the density of the whole Earth and compare it to the densities of some rocks found near the surface? This would be similar to comparing the density of your baseball to the density of its near-surface components. You will soon discover that this process provides some clues but not a complete picture. Extending the idea of imaging to study Earth's interior, can you zap the planet with x-rays? No—aside from the fact that no x-ray machine is large enough, the x-rays are not likely to be helpful for imaging Earth because rock, like bone, is mineral matter and nearly opaque to x-rays. How then have geologists developed cross sectional views of Earth such as the one sketched in Figure 8.1c? The answer to that question is a fundamental learning objective for this chapter.

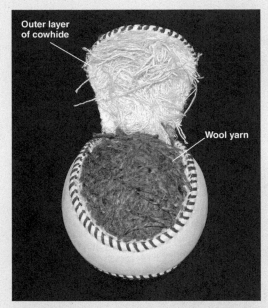

(a) Inside of baseball revealed with a shallow knife cut, and inititial calculations.

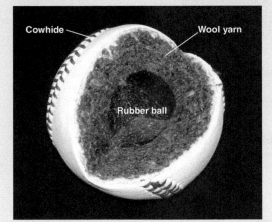

(b) Interior of baseball revealed by cutting the ball completely open, along with follow-up calculations.

▲ **Figure 8.1 The internal structure of a baseball and Earth.**

Information provided:

Overall density of cowhide leather and tightly wound wool yarn is:
   0.59 g/cm³

Is the rest of the ball composed of yarn?

Calculating the density of the baseball:

$$Density = \frac{mass}{volume}$$

Mass measurement = 146 g
Radius measurement = 3.81 cm

$$Sphere\ volume = \frac{4}{3}\pi\,(radius)^3$$

$$Volume\ of\ ball = \frac{4}{3}(3.14)(3.81\ cm)^3 = 231\ cm^3$$

$$Density = \frac{146\ g}{231\ cm^3} = 0.632\ g/cm^3$$

Baseball is denser than yarn, therefore there must be some denser material deeper inside the baseball.

Determing the density of the rubber-ball center of the baseball:

Radius of rubber ball = 1.61 cm

Calculated volume = 17 cm³

Mass of rubber ball = 22.4 g

$$Therefore\ density\ of\ rubber\ ball = \frac{22.4\ g}{17\ cm^3}$$
$$= 1.32\ g/cm^3$$

As predicted in (a), the center of the baseball consists of denser material than the outer layers of cowhide and yarn.

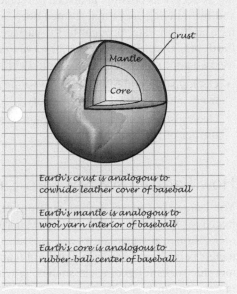

(c) Drawing a comparison between the concentric layers that compose Earth and the baseball.

Earth's crust is analogous to cowhide leather cover of baseball

Earth's mantle is analogous to wool yarn interior of baseball

Earth's core is analogous to rubber-ball center of baseball

8.1 How Do Geologists Learn About Earth's Interior?

Myths tell of gods, demons, or whole civilizations that reside in its inner world. R. D. Oldham, an early twentieth-century geologist, captured this fascination on the part of scientists and nonscientists, alike: "Of all the regions of the earth none invites speculation more than that which lies beneath our feet" (*Quarterly Journal of the Geological Society,* 1906, vol. 62, p. 456). Perhaps no one captured that speculative imagination better than nineteenth-century author Jules Verne in his epic *Journey to the Center of the Earth,* in which explorers descend through a volcano in Iceland and undertake a perilous journey through a hypothetical underworld before emerging, from another crater, in Italy. Because geologists cannot take journeys such as the one described in Verne's novel, how do we know so much about Earth's interior?

Geologists gain some knowledge about Earth's interior in a manner similar to your analysis of the baseball. They make real observations as far as they can—by looking as deeply as possible in drilled boreholes and mines, and by studying rocks that formed deep below the surface and were later brought to the surface by dynamic Earth processes. Beyond these observations, geologists use two approaches similar to the study of the baseball:

- Geologists infer the interior composition at deep levels from calculations of Earth's density.
- Geologists use energy from earthquakes to construct images of Earth's interior.

How Far Can We See into Earth?

You learned in Chapter 6 about outcrops of metamorphic rocks that originated far below the surface. The metamorphic minerals in some of these rocks formed at temperatures and pressures that exist as much as 50 kilometers below Earth's surface. Later uplift and erosion exposed these metamorphic rocks for geological studies, providing insights into the makeup of deep continental crust.

Surface outcrops and rocks from deep drill holes reveal that continental crust consists primarily of quartz-rich igneous and metamorphic rocks, depicted in **Figure 8.2**a. Sedimentary layers are also common at the surface, but these rocks are rarely more than 5 kilometers thick. Felsic igneous rocks, such as tonalite and granite, along with gneiss containing high abundances of quartz, feldspar, and mica, compose most of the upper crust. Metamorphic rocks formed deeper than about 15–25 kilometers are typically amphibolite or eclogite, which contain less quartz and more amphibole, pyroxene, and garnet. These deeper rocks probably are metamorphosed mafic igneous rocks.

Volcanoes provide additional clues about the interior because some eruptions eject pieces of rock torn loose by rising magma. **Figure 8.3** shows two examples of these rocks. Most common are pieces of granite, gneiss, and other materials similar to those found in outcrops of plutonic and metamorphic rocks that also are exposed at the surface by uplift and erosion. More exotic samples, such as those pictured at the beginning of this chapter, contain diamonds, which form at depths of about 150 kilometers.

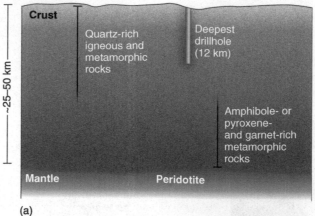

Crust

Quartz-rich igneous and metamorphic rocks

Deepest drillhole (12 km)

Amphibole- or pyroxene- and garnet-rich metamorphic rocks

Mantle   Peridotite

~25–50 km

(a)

Continental crust:
- Typically ranges from 25–50 km thick.
- Mostly a mix of plutonic and metamorphic rocks.
- Rock compositions are more felsic in the upper crust, and more mafic in the lower crust.

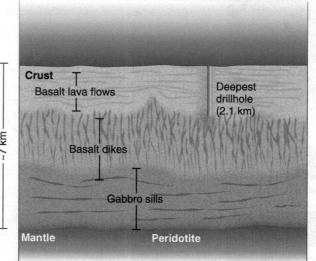

Crust

Basalt lava flows

Deepest drillhole (2.1 km)

Basalt dikes

Gabbro sills

Mantle   Peridotite

~7 km

(b)

Oceanic crust:
- Usually about 7 km thick.
- Composed of mafic igneous rocks.
- Gabbro sills pass upward into closely spaced basaltic dikes, which were the volcanic feeders to basalt lava flows found on the sea floor.

▲ **Figure 8.2 Comparing continental and oceanic crusts.** Continental and oceanic crusts, with different thickness and composition, overlie peridotite mantle. Notice that the vertical scale is different for the two illustrations—continental crust is thicker than oceanic crust.

▼ **Figure 8.3 What volcanoes bring up from inside Earth.** Volcanoes sometimes eject pieces of rock from deep inside Earth that were torn loose as magma moved toward the surface. The rock on the left is a fragment of green peridotite enclosed in basalt. The peridotite is interpreted to be a piece of the mantle because it contains an association of minerals not seen in rocks from Earth's crust. The fragment of metamorphic rock on the right contains minerals consistent with formation more than 15 km below the surface and, therefore, represents a piece of deep continental crust.

Other ejected rocks are chunks of peridotite (Figure 8.3) that contain garnet in addition to olivine and pyroxene. Laboratory experiments indicate that garnet peridotite forms at depths greater than 50 kilometers. Taken together, these pieces of evidence imply that below the typical quartz-bearing rocks of continental crust there must be rocks of strongly contrasting ultramafic composition. This is the boundary between crust and mantle pictured in Figure 8.2a.

Geologists also have substantial knowledge of rocks found beneath the oceans. In the nineteenth century, oceangoing scientists explored the composition of the seafloor by dragging buckets across the bottom of the ocean at the end of long chains and hauling them back to the deck of the ship. These buckets contained muddy oozes composed primarily of the remains of microscopic plankton and chunks of basalt. More sophisticated exploration beneath the seafloor began in the 1960s, when ship-based drill rigs began operation. An international scientific partnership has drilled holes in the sea bottom at more than 2000 locations worldwide. Many of these wells penetrate far into the seafloor basalt, as deep as 2.1 kilometers. Rocks recovered from the wells reveal crust that is nearly devoid of quartz and overwhelmingly dominated by basaltic lava flows and intrusions, as shown in Figure 8.2b. These mafic oceanic rocks contrast sharply with the felsic composition of most continental rocks at similar depths.

Oceanic crust also is present within mountain belts, in places where subduction-zone processes shoved slivers of seafloor onto the edge of continents. **Figure 8.4** illustrates a landscape made of this kind of uplifted seafloor. Sedimentary layers of chalk and chert embedded with deep-sea fossils rest on basaltic lava flows and reveal the oceanic origin of these rocks. Stacks of lava flows, such as those encountered in the deep-ocean drill holes, are commonly more than 2 kilometers thick, and they are increasingly interrupted at greater depths by dikes, which rise from a yet lower zone of gabbro sills. In the thickest uplifted seafloor slices, the mafic igneous rocks are underlain by peridotite or its metamorphosed equivalent, serpentinite. The total thickness of the uplifted mafic oceanic-crust rocks is about 7 kilometers, resting on peridotite mantle. These field observations of uplifted oceanic crust contribute substantially to the schematic illustration in Figure 8.2b.

To summarize these observations, the crusts of oceans and continents differ in both composition and thickness (Figure 8.2). Continental crust contains mostly quartz-bearing igneous and metamorphic rocks, with an average composition of diorite to tonalite, and it is commonly greater than 25 kilometers thick. Oceanic crust consists of mafic igneous rocks and is typically about 7 kilometers thick. Both types of crust rest on top of peridotite mantle.

Remarkable as it is that we have samples of rocks that formed tens of kilometers below the surface, geologists have a rather poor sampling of the Earth overall. Rocks available for study represent less than the outermost 1 percent of the radius of the planet. They tell us no more about the whole Earth than the leather covering of the baseball tells you about the entire ball. Density measurements show that the baseball is made up of different layers; similarly, Earth's visible rocks reveal inhomogeneity, with a variety of rocks in the crust that rest on very different mantle peridotite. Does this peridotite continue on to the center of Earth? If not, what else is there?

What Is the Density of Earth?

You determined that the baseball's center could not be composed entirely of yarn by comparing the density of the entire baseball to the materials you could see. Is it possible to conduct a similar test to help learn more about the composition of the interior of Earth? To calculate the density of Earth you need to know its shape, its size, and its mass.

In the fifth century B.C.E., Greek observers established that Earth was nearly spherical in shape. They reached this conclusion by observing how ships gradually disappear from view when sailing beyond the horizon, and by noting the curved outline of Earth's shadow on the Moon during lunar eclipses.

The astronomer Eratosthenes used simple surveying and measurements of shadows in sunlight to estimate Earth's size, placing the planet's radius at about 6350 kilometers. In the seventeenth century Sir Isaac Newton calculated that a rotating planet would not be perfectly spherical, but must bulge at the equator and be somewhat flatter at the poles. Refinement of surveying methods led to the current calculation that Earth has a radius of 6378 kilometers at the equator, 6357 kilometers through the poles, and an average radius of 6371 kilometers. Using these values, one can readily calculate the volume, or size, of Earth.

Determining the mass of Earth is much more challenging than placing a baseball on a scale, and this mass-measurement process has played out over centuries. Newton formulated the laws of gravity that explain the mutual

◄ **Figure 8.4 What uplifted mantle and oceanic crust looks like.** The dark rocks in the photo are part of a large sliver of mantle peridotite and basaltic oceanic crust that was shoved onto the Arabian Peninsula in the country of Oman by plate tectonic forces. A close-up of one outcrop of these rocks shows peridotite sliced through with closely spaced gabbro sills.

attraction of objects and how planets orbit around the Sun. Other scientists combined Newton's mathematical expression of gravitational force with in-genious experiments during the late eighteenth and early nineteenth centuries to make the first calculations of Earth's mass. More precise measurements during the twentieth century, which calculate the effects of Earth's mass on satellite orbits, provide the currently accepted value of 6×10^{24} kilograms.

Combining these measurements of shape, size, and mass leads to a calculation of Earth's overall density that is equal to about 5.5 g/cm^3. Typical mantle peridotite has a density of only about 3.2 g/cm^3. These data tell us that, like the baseball, Earth's near-surface materials have a different density from that of the whole sphere. The deeper interior must be much denser than the rocks that form the outermost layers.

EXTENSION MODULE 8.1

Sizing Up Earth. Learn how Eratosthenes determined the radius of Earth and how eighteenth-century scientists calculated the mass of Earth.

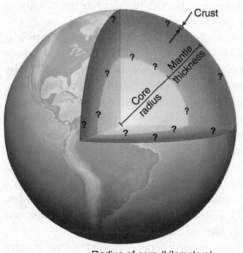

The average Earth density is 5.5 g/cm^3, and calculations based on the planet's wobbly rotation require that most of the mass is concentrated toward the center. If Earth is mostly composed of mantle and a denser core, how thick are each of these layers?

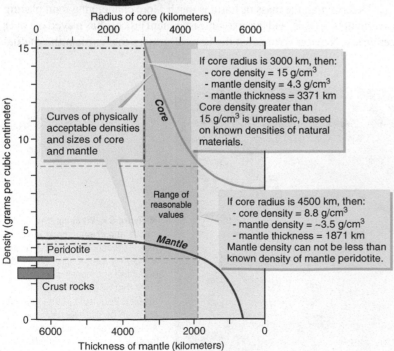

If core radius is 3000 km, then:
- core density = 15 g/cm^3
- mantle density = 4.3 g/cm^3
- mantle thickness = 3371 km
Core density greater than 15 g/cm^3 is unrealistic, based on known densities of natural materials.

Curves of physically acceptable densities and sizes of core and mantle

Range of reasonable values

If core radius is 4500 km, then:
- core density = 8.8 g/cm^3
- mantle density = ~3.5 g/cm^3
- mantle thickness = 1871 km
Mantle density can not be less than known density of mantle peridotite.

▲ Figure 8.5 **Acceptable dimensions of the mantle and core.**

Applying what we have learned in previous chapters, we can begin to formulate some interesting questions. Is it possible that peridotite is denser at greater depth due to the weight of overlying rock compressing mineral structures more tightly? On the other hand, would not minerals expand as temperature increases deeper and deeper into Earth? Further laboratory explorations show that the density increase caused by increasing pressure is greater than the expansion due to higher temperatures. These experiments reveal, however, that the slightly higher density of hypothetical peridotite at the center of Earth still would be insufficient to explain the density of the whole planet.

Another piece of the puzzle fell into place when measurements of Earth's wobbly rotation on its axis allowed astronomers and geophysicists to conclude that the planet has a distinct concentration of high-density material near the center rather than a gradual inward transition from low-density to high-density rock. The notion of a very dense core surrounded by a less dense mantle and a thin covering of crust has been a part of scientific knowledge since the late nineteenth century.

Combining knowledge of Earth's density and the mechanics of rotation identifies limits on the density of mantle and core, as shown in **Figure 8.5**. The exact dimensions and densities of these concentric layers remained speculative, however, until the early twentieth century. What was missing until then was a way to create an image of the interior in much the same way you thought of x-raying your baseball to determine its internal structure.

Putting It Together—How Do Geologists Learn About Earth's Interior?

• Some rocks seen at Earh's surface formed as much as 50 kilometers or (rarely) as far as 150 kilometers below the surface. These rocks reveal mostly felsic continental crust and mafic oceanic crust, both of which overlie peridotite mantle.

• The rocks found near the surface cannot account for the density of the whole Earth. Along with wobbles in Earth's rotation, the high overall density of the planet implies a central core of material that is much denser than mantle peridotite.

8.2 How Do Earthquakes Make Images of Earth's Interior?

Earthquakes happen around the globe and occurrences are particularly concentrated near plate boundaries (see Figure 1.10). News media headlines focus on infrequent large earthquakes that cause tremendous destruction and loss of life, but thousands of smaller earthquakes occur every day. Earthquakes are important geological events, and thus this book examines them in detail. Chapter 11 explores the causes and effects of earthquakes, while this chapter focuses on a scientifically beneficial aspect of earthquakes—the use of earthquake waves to make images of Earth's interior.

ACTIVE ART

Dimensions of the Mantle and Core. See how to determine physically possible thicknesses of Earth's mantle and core.

Visualizing Waves

Consider what happens when you throw a pebble into a pool of water. **Figure 8.6** shows ripples forming on the surface of the water and moving outward in ever-widening circles. These ripples are waves. **Waves** are disruptions that move through a medium, such as rock, air, or, in this case, water, without any overall transport of the medium in the direction that the wave moves. The height of the water waves relates to the amount of energy released at the water surface when the pebble lands. Movement of water by the waves is greatest at the surface and decreases downward in the pool (Figure 8.6 inset). If you are underwater when a pebble plunks into the pool, then you will not feel the surface waves pass over you, but you may hear a sound. In contrast to surface waves, sound waves move radially outward in three dimensions away from the splash point and not just along the surface (Figure 8.6).

Water molecules move when the waves pass by but then return to their original position. For this reason the waves are described as **elastic**, because they do not permanently deform the water. Elastic deformation also happens when you stretch a rubber band and then release it, because the rubber band returns to its original shape after being deformed. In contrast, if you stretch a piece of clay, it does not return to its original shape when you let go of it; this type of deformation is **plastic**, and it is permanent.

Earthquake Waves

You can relate the experience of the pebble landing in the pool to the waves produced by earthquakes. An **earthquake** is a nearly instantaneous release of stored energy resulting from the breaking and sudden movement of rock under stress. The release of energy, analogous to the pebble hitting the water, creates three types of elastic waves, illustrated in **Figure 8.7**: surface waves, primary (P) waves, and secondary (S) waves.

The **surface waves** are similar to the ripples on a pond surface, except that they not only disrupt water surfaces, but also cause motion in rock and soil when they pass. Some surface waves cause Earth's surface to roll up and down like ocean waves, whereas others cause sideways motion like a slithering snake. Similar to water waves, Earth surface-wave disturbance decreases downward into Earth (Figure 8.7a). Most destruction by earthquakes results from the passage of surface waves, which cause the ground to shake up, down, and sideways by as much as a meter or more.

The primary (P) and secondary (S) waves are **body waves** that move through Earth below the surface, somewhat like the sound waves produced by the pebble hitting the water. **P waves** elastically displace material in the same direction that the wave is moving, which causes alternating squeezing and stretching of the material as the wave passes, similar to squeezing and stretching a spring (Figure 8.7b). **S waves** twist material at right angles to the direction of wave motion (Figure 8.7c). Unlike surface waves, body waves are confined by the pressure of overlying rock, so they move the rock only a few millimeters or centimeters. The

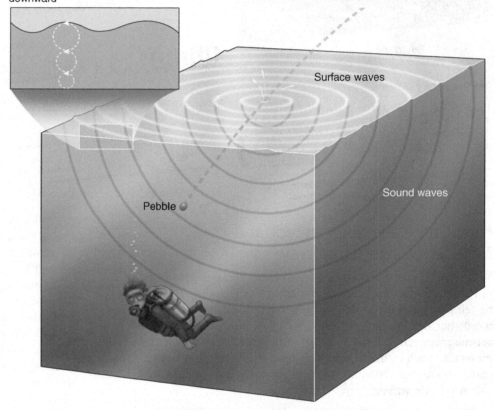

Circular particle motion caused by passing surface wave decreases downward

Surface waves

Sound waves

Pebble

▲ **Figure 8.6 Waves in water.** Circular ripples expand outward from where a pebble impacts the water surface. When the pebble hits the water, waves move on the water surface, and sound waves move outward and downward through the water. The inset diagram shows that the water motion caused by the passing surface waves diminishes downward and is not noticed by the diver. Sound waves, however, move as a pulse of changing water pressure through the water, so the diver hears the sound of the pebble hitting the surface.

magnitude of this motion abruptly decreases as the wave spreads out through Earth. This means that P and S waves are not usually as damaging as surface waves but, because they pass through Earth, they are key to understanding the interior of our planet.

The different motions of P and S waves determine what the body waves can pass through. Gases, liquids, and solids can all deform by the squeezing and stretching caused by P waves. Therefore, P waves can travel through all materials, although they move fastest through rocks. However, only solids can twist, so S waves do not travel through gases or liquids.

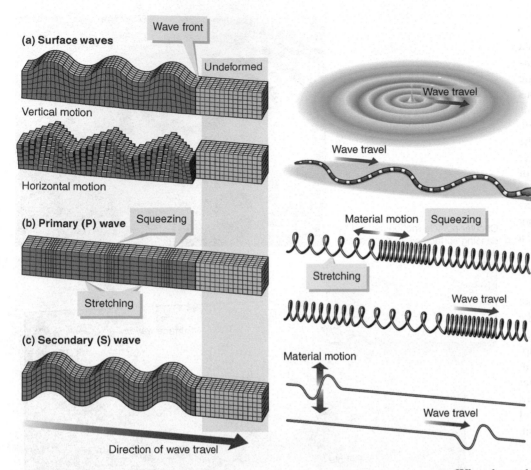

(a) Surface waves

Wave front

Undeformed

Vertical motion

Horizontal motion

(b) Primary (P) wave

Squeezing

Stretching

(c) Secondary (S) wave

Direction of wave travel

Wave travel

Wave travel

Material motion

Squeezing

Stretching

Wave travel

Material motion

Wave travel

◀ **Figure 8.7 How earthquake waves move through rock.** Each diagram portrays how one type of earthquake wave deforms originally cubic blocks of rock, along with an analogy. The grid lines help you to visualize how the rock deforms as the waves pass.

a. Surface waves distort the rock vertically, like ripples in water, or horizontally, like a slithering snake. The deformation decreases downward into Earth.

b. The P body wave moves through rock as a pressure pulse, alternately squeezing and stretching the blocks. Pushing on a spring produces a similar kind of wave.

c. The S body wave moves through rock by vertical twisting distortions of the blocks. Shaking a string produces a similar kind of wave.

ACTIVE ART

Seismic Wave Motion. *See how the different seismic waves move.*

How Can Geologists Detect and Measure Earthquake Waves?

Seismometers are extremely sensitive instruments designed to detect, amplify, and record surface and body wave motion, some of which may not be felt by humans. The instrument name derives from *seismos*, a Greek word that means "ground shaking." This word root also appears in **seismogram**, which is the record of earthquake waves detected by a "seismometer," such as that shown in **Figure 8.8**, and in "seismologist," a scientist who studies earthquake records. Similarly, earthquake waves also are called **seismic waves**.

▲ **Figure 8.8 Seismometers record earthquakes.** These seismometers consist of pens suspended above rotating drums that are covered with a sheet of paper. Earthquake waves move the drums, causing the pens to trace out the movement on the paper while the drum turns. This paper record of the earthquake is a seismogram.

What do earthquakes have to do with deciphering the internal structure of Earth? Records of earthquake body waves allow us to construct an image of Earth's interior, not unlike x-rays reveal an image of the interior of your body. The key to deciphering what the seismograms tell us is understanding the factors that determine how fast seismic waves travel, especially body waves that penetrate deeply into the planet. The velocity of seismic waves depends on the properties of the material through which the waves move. The velocity also determines how long it takes between when an earthquake occurs and when a distant seismometer detects the resulting seismic waves. Seismologists measure these elapsed times at different locations on Earth's surface in order to learn about the properties of materials within the planet through which the waves passed. This knowledge permits imaging of Earth's interior.

A Simple Experiment for Measuring Seismic Waves

Seismologists gain knowledge about seismic-wave properties not only from earthquakes, but also from energy released by artificial explosions. Worldwide networks of seismic stations were established during the 1950s and 1960s not to measure earthquakes, but instead to monitor underground nuclear-bomb tests during the Cold War. Explosions in rock quarries and mines, and even disasters such as the collapse of the World Trade Center towers in September 2001, generate seismic energy, which can be detected by seismometers.

Figure 8.9 illustrates how a seismometer records a dynamite explosion at a gold mine. The seismogram records the arrival of three waves. The primary (P) and secondary (S) body waves arrive first and second; now you can see that these names relate to the order of wave arrival at

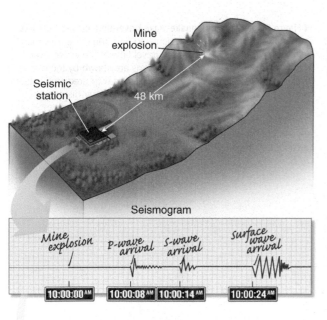

A seismic station records body and surface waves produced by a precisely timed mine explosion.

The seismogram shows the recorded arrival times of P, S, and surface waves from the mine explosion.

The velocities of the waves are calculated by dividing the travel distance by the travel time.

$$\text{P-wave velocity} = \frac{48 \text{ km}}{8 \text{ sec}} = 6.0 \text{ km/sec}$$

$$\text{S-wave velocity} = \frac{48 \text{ km}}{14 \text{ sec}} = 3.4 \text{ km/sec}$$

$$\text{Surface-wave velocity} = \frac{48 \text{ km}}{24 \text{ sec}} = 2 \text{ km/sec}$$

▲ Figure 8.9 Determining the velocity of earthquake waves.

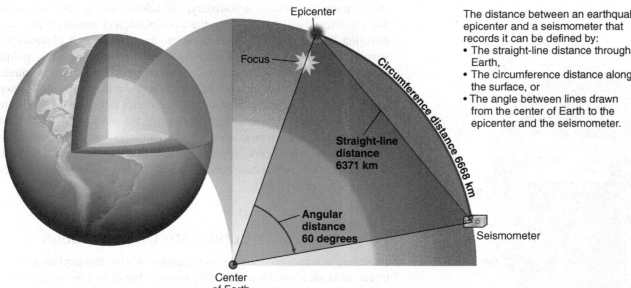

The distance between an earthquake epicenter and a seismometer that records it can be defined by:
• The straight-line distance through Earth,
• The circumference distance along the surface, or
• The angle between lines drawn from the center of Earth to the epicenter and the seismometer.

▲ Figure 8.10 Describing the distance from an earthquake.

earthquake within Earth, and the point on the surface directly above the focus is the **epicenter**. The circumference distance along the curving surface between the epicenter and a distant seismometer is much longer than the straight-line distance through Earth. Geologists usually describe the distance between the epicenter and a seismometer distance as an angle, called the angular distance, rather than as a length.

Assume for a moment that the P- and S-wave velocities measured during the mine-explosion experiment (Figure 8.9) also apply to P- and S-wave motion everywhere within Earth. This assumption may or may not be true, but it provides a starting point for investigating Earth's interior. *If* this assumption is true, then you can predict the time when the P and S waves from an earthquake will be recorded anywhere around the world. The travel-time curves graphed in **Figure 8.11** show this prediction of P- and S-wave travel times between the focus and seismometers. The graph shows that the predicted curves compare poorly to the real travel times determined from decades of earthquake records from all over the globe. The predicted travel times, based on the mine-blast experiment, match up with the actual data only where the travel distance is small. At distances greater than about 500 km the real waves arrive much earlier than predicted. The assumption that P- and S-wave velocities are the same throughout Earth does not hold true. What does this tell us about Earth's interior?

Why Seismic Waves Speed Up

Earthquake P and S waves that have traveled long distances arrive at earlier times than predicted from the mine-blast data (Figure 8.9), which means that seismic waves speed up as they travel to more distant locations. **Figure 8.12** illustrates the paths of P and S waves through Earth and shows that waves detected at progressively greater distances from the focus pass through progressively deeper parts of the planet. Because the average wave velocity increases with increasing distance between the focus and the seismic station, geologists conclude that seismic-waves travel faster at greater depths.

Careful measurements show that seismic waves move at different velocities through rocks with different properties. Therefore, faster wave velocities in Earth's interior, compared to near-surface rocks, imply that rocks in the interior differ from those seen at the surface.

The seismic-wave travel times also help us understand how waves move through rocks with different properties. **Reflection** and **refraction**, illustrated in **Figure 8.13**, describe important processes that occur where waves

a seismic station. Later-arriving surface waves cause the remaining disturbances on the seismogram. Figure 8.9 shows how to calculate the speeds of the seismic waves because the mine geologist recorded the precise time of the dynamite detonation. A key observation is that the three waves move at different velocities: P waves are fastest, surface waves are slowest, and velocity of S waves falls in between those of the other two.

Figure 8.10 illustrates different ways of describing the distance between a seismometer and an earthquake. The **focus** is the location of an

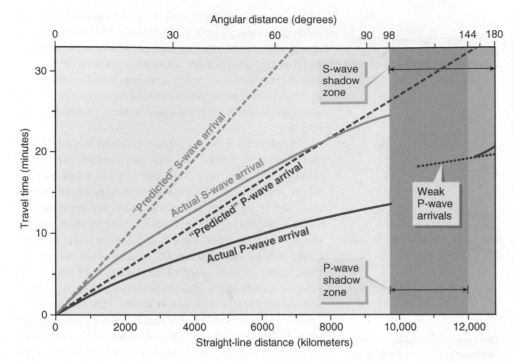

◄ **Figure 8.11 What earthquake-wave travel-time curves look like.** The P- and S-wave velocities calculated from the mine-explosion data (from Figure 8.9) *predict* the travel time of the waves over any distance on Earth; these predictions are shown by the dotted lines. *Actual* arrival times of the waves agree with predicted values only when the travel distances are short. At greater distances, the waves arrive at seismic stations increasingly earlier than predicted. There are also "shadow zones" where the waves are not detected at all, or where P waves are weak and arrive at unexpected times.

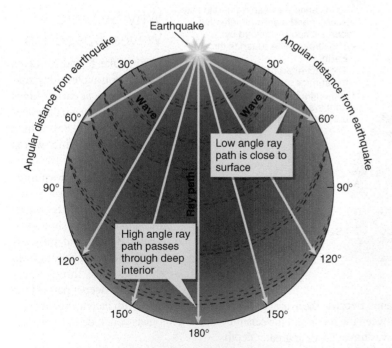

▲ **Figure 8.12 Waves that travel farthest also penetrate deepest into Earth.** This cutaway view of Earth illustrates how a seismic wave moves through the planet from a hypothetical earthquake focus at the North Pole. It is sometimes easier to visualize the ray path, which are arrows drawn perpendicular to the wave. Ray paths to seismometers at large angular distance from the epicenter travel through deeper parts of the planet than those that travel small angular distances.

encounter boundaries between materials with different properties. When a wave encounters a boundary, such as that between layers of different rock types, some of the energy reflects, which means that a new wave forms and bounces back from the boundary. That part of the original wave energy that crosses the boundary changes velocity, which is consistent with the change in rock properties across the boundary. The velocity change also causes the wave to refract, which means that the wave bends and moves off in a new direction. To convert these descriptions into images, Figure 8.13 shows the **wave front**, which is the continuous line or surface including all the points in space reached by a wave as it travels, and the **ray path**, which is direction of wave travel drawn perpendicular to the wave front (and also shown in Figure 8.12). The wave-front and ray-path lines in Figure 8.13 indicate the new directions that the reflected and refracted waves take. **Figure 8.14** demonstrates that as seismic-wave velocity increases with depth, refraction causes a wave path to curve upward and eventually emerge at the surface. So the shape of the travel-time curve (Figure 8.11) is consistent with changes in seismic velocity.

Why Travel-Time Curves Are Discontinuous

Increasing seismic-wave velocities within Earth's interior explain why the travel times are shorter than initially expected, but increasing velocities do not explain the breaks in the curves depicted in Figure 8.11. Particularly notable are the regions called **shadow zones** where seismometers do not record P or S (or both) waves. In other words, seismic waves from a particular earthquake are not recorded by seismometers everywhere in the world. A seismic station located in a shadow zone does not record the earthquake waves.

The shadow zones are very important to the interpretation of Earth's interior because they tell geologists that something inside Earth either stops the waves or deflects them away from the shadow zone. **Figure 8.15** shows that the S-wave shadow zone begins at locations that

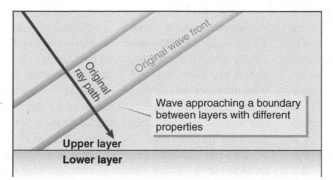

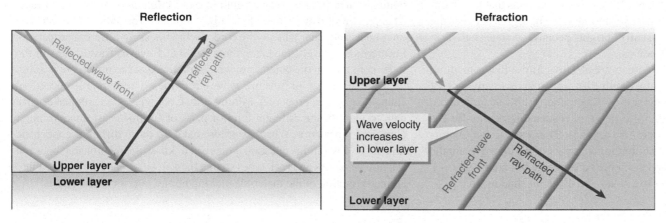

◀ **Figure 8.13 How waves reflect and refract.** Two things happen when waves encounter a boundary between layers with different physical properties. (1) Some of the wave energy reflects off of the boundary like a ball bouncing off of the floor, as shown in the left diagram. (2) Some of the wave energy continues into the lower layer but speeds up or slows down depending on the properties of the two layers. In the example shown in the right diagram, the wave speeds up in the lower layer causing a change in the direction of the ray path. This change in the direction of wave motion is called refraction.

Reflection

Refraction

ACTIVE ART

Wave Reflection and Refraction. *See how waves reflect and refract through rock.*

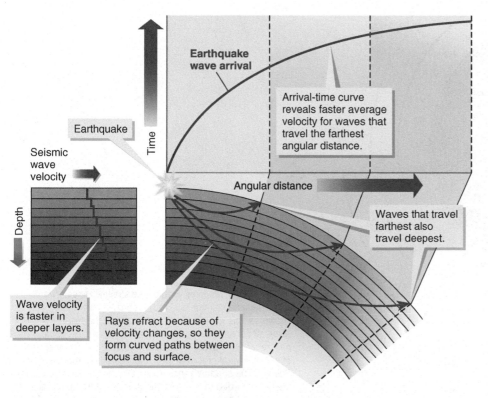

◀ **Figure 8.14 Earthquake-wave velocity increases at depth.** This diagram schematically shows how waves move through a planet where the wave velocity increases downward in each deeper layer. Refraction bends the ray paths back to the surface. The farthest traveled ray paths also travel through the deeper layers where velocity is fastest. This means that ray paths through deeper layers have a higher average velocity than waves traveling only through shallower and slower layers. As a result, earthquake-wave arrival times at greater distances are progressively earlier than predicted using constant wave velocities measured over short distances in shallow layers (compare this graph to Figure 8.11).

are 98° around the globe in all directions from an epicenter. The P-wave shadow zone is more complex. Similar to the S-wave case, there also is an abrupt break in the P-wave arrivals at an angular distance of 98° from the epicenter, but then the P waves reappear at 144° (Figure 8.15). The reappearing waves arrive much later than expected by simple extrapolation of the arrival times at lower angles (see Figure 8.11), which implies that something slowed the waves down. There also is a second set of P waves that show up as weak arrivals within parts of the shadow zone, and these waves overlap with the more typical P waves that reappear at 144°.

An analogy to light and shadows helps to illustrate these complexities. The S-wave shadow zone is completely "dark" and receives no S waves. The two P-wave shadow zones are "dimly lit" by P-wave arrivals that are inconspicuous except on the most sensitive seismometers. The region between the two P-wave shadow zones is unusually "bright" because seismic stations in this region receive two sets of P waves.

Looking at these results, geologists are able to surmise that something dramatic happens inside Earth at a depth that corresponds to an angle of 98° between epicenter and seismic station. Calculations show that this depth is about 2900 kilometers, within the range of reasonable depths for the boundary between mantle and core (Figure 8.5). This leads to two important conclusions: (1) something about the properties of the core explain the shadow zones, and (2) what happens at the core-mantle boundary affects P and S waves differently in order to explain the different shadow zones.

Figure 8.16 shows how wave refraction at boundaries within Earth explains the shadow zones for a hypothetical earthquake at the North Pole. The absence of S waves on the opposite side of Earth from the epicenter indicates that these waves do not pass through the core. Recall that S waves do not pass through liquid, so this observation leads us to conclude that at least the outer part of the core is liquid. Inward refraction of P-wave paths at the core-mantle boundary accounts for the P-wave shadow zones. The inward refraction indicates a large decrease in wave velocity where P waves pass into the outer core. This is consistent with a liquid outer core because P waves move more slowly through liquids than solids. The weak P waves recorded in the P-wave shadow zone are explained by refraction from a boundary *within* the core at a depth of 5155 kilometers below the surface. This pattern of wave refraction can be accounted for only if there are distinct inner and outer parts to the core, each possessing different properties. Furthermore, travel-time curves indicate that P waves travel faster in the inner core than in the outer core, which implies that the inner core is solid.

The behavior of seismic waves provides a remarkable record of changing physical properties within Earth. Geologists use earthquake-wave data to identify the different interior parts of Earth—crust, mantle, outer core, and inner core. What materials make up these different concentric layers? To answer this question, you need to know the relationship between seismic velocities and rock types so that the velocities can be translated into an interpretation of the materials making up the interior of the planet.

EXTENSION MODULE 8.2

How to Locate an Earthquake. *Learn how to use seismographs and travel-time curves to determine where an earthquake happened.*

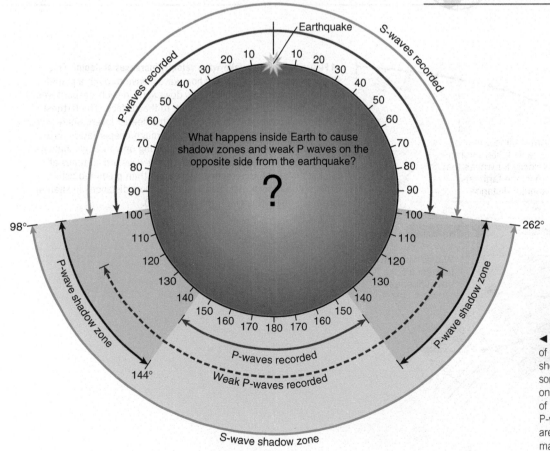

◀ **Figure 8.15 Mapping earthquake shadow zones.** Not all of the seismic stations on Earth record an earthquake. As shown for this hypothetical earthquake at the North Pole, some stations do not receive S waves or P waves, some only receive P waves, while others receive two P waves, one of which is weaker and arrives later than the stronger P-wave. The shadow zones where S waves, P waves, or both are not recorded indicate major changes in the properties of materials deep inside Earth.

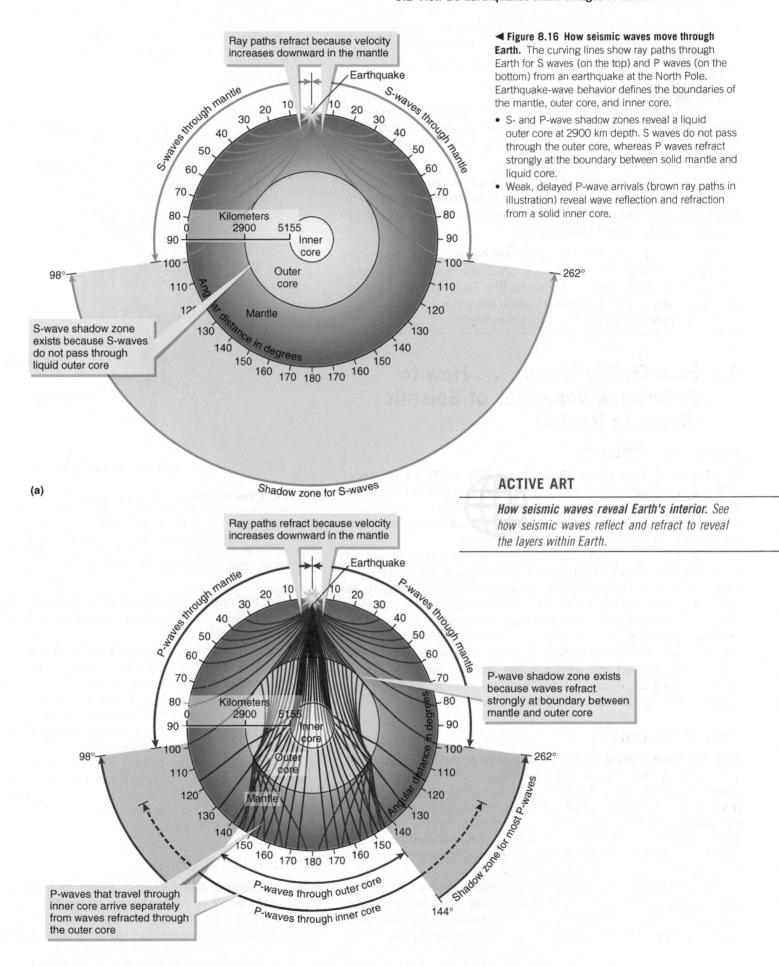

◄ Figure 8.16 How seismic waves move through Earth. The curving lines show ray paths through Earth for S waves (on the top) and P waves (on the bottom) from an earthquake at the North Pole. Earthquake-wave behavior defines the boundaries of the mantle, outer core, and inner core.

- S- and P-wave shadow zones reveal a liquid outer core at 2900 km depth. S waves do not pass through the outer core, whereas P waves refract strongly at the boundary between solid mantle and liquid core.
- Weak, delayed P-wave arrivals (brown ray paths in illustration) reveal wave reflection and refraction from a solid inner core.

ACTIVE ART

How seismic waves reveal Earth's interior. See how seismic waves reflect and refract to reveal the layers within Earth.

Putting It Together—How Do Earthquakes Make Images of Earth's Interior?

- Earthquakes produce surface waves and two distinct body waves, called primary (P) and secondary (S) waves, which travel through the planet. P and S waves reflect and refract when encountering boundaries between materials.

- Seismograms are records of earthquake waves detected by seismometers located around the globe. The time required for different earthquake waves to reach these instruments reveals the velocity at which the waves move through Earth.

- Travel-time curves reveal that, in general, seismic-wave velocity increases at deeper levels in the mantle. Shadow zones where S waves, P waves, or both are not recorded reveal that Earth has a liquid outer core, beginning at 2900 kilometers below the surface, surrounding a solid inner core at 5155–6371 kilometers depth.

8.3 How Do We Know . . . How to Determine Velocities of Seismic Waves in Rocks?

Define the Problem

How Fast Do Seismic Waves Move through Different Rocks and under Different Conditions? When geologists understood that seismic-wave arrivals required changing wave velocities within Earth, they understandably wanted to know what is different about the various internal layers to account for the changing velocities.

Most of the rocks that earthquake body waves travel through are concealed from view beneath Earth's surface. In addition, temperature and pressure conditions vary dramatically inside Earth. These changing conditions may cause variations in seismic-wave velocities compared to velocity measurements made at Earth's surface temperature and pressure. To relate rock types to different seismic velocities, and to know how velocities vary with increasing temperature and pressure inside Earth, geophysicists designed laboratory experiments to measure P- and S-wave velocities in different Earth materials and under different conditions.

Gather the Data

How Are Seismic-Wave Velocities Measured in the Lab? **Figure 8.17** schematically illustrates how a typical seismic-velocity experiment works. A mechanical signal enters at one end of a cut cylinder of mineral or rock. The time that passes by before the signal arrives at the opposite end is measured in order to calculate the velocity of wave travel. The setup shown in Figure 8.17 is modified for experiments at different temperatures and pressures to permit measurements at conditions such as those inside Earth.

Seismic waves travel at velocities of several kilometers per second, so they pass through a 5-centimeter-long rock sample in mere hundredths of thousandths of a second. Electrical devices are much more accurate than mechanical devices for measuring such short

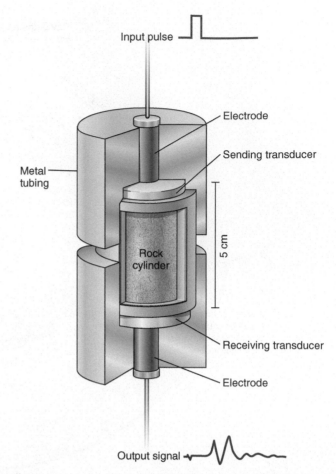

▲ **Figure 8.17 How seismic-wave velocities are measured.** This diagram illustrates an experimental setup used to measure seismic velocities in the laboratory. Metal tubing confines a cylinder of rock. A transducer converts an electrical impulse to a body wave. When the wave exits the lower edge of the rock, another transducer converts the body wave into another electrical signal. The time elapsed between the input and output signals reveals the body-wave velocity through the rock.

times. For this reason, the original input pulse of energy is electrical. A transducer converts the electrical pulse to a mechanical thump on the rock.

A transducer is simply a device that converts one form of energy to another. You use transducers every day when you make a telephone call. The vibrating sound waves of your voice enter the phone as mechanical energy. A transducer converts the sound waves to electrical energy that transmits by wire or microwave signal to a receiver. Another transducer in the receiver converts the transmitted energy back to mechanical sound waves.

The transducer at the bottom of the rock sample works like a receiving telephone and converts the received mechanical vibration into an electrical signal that is recorded by an instrument. The time between the input pulse and the recorded output pulse reveals the seismic velocity of the rock.

Evaluate the Results

How Does Seismic-Wave Velocity Relate to Rock Properties and Conditions? **Figure 8.18** depicts typical ranges of P-wave velocities in a variety of materials. Notice that unconsolidated sediment has lower seismic velocity than sedimentary rock. This is because, as the graph

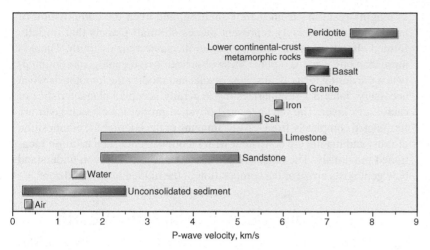

▲ **Figure 8.18 How fast P waves move in Earth materials.** This bar graph summarizes laboratory and field measurements of P-wave velocities for materials at surface temperature and pressure. Velocities will be different for the temperature and pressure conditions within Earth. In all rocks there are natural variations in mineral abundance and open pore space that affect wave velocity, so the lengths of the bars reveal typical velocity ranges rather than a single value.

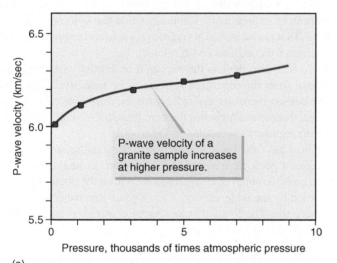

(a)

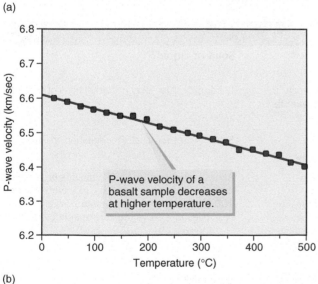

(b)

▲ **Figure 8.19 How pressure and temperature affect seismic velocity.** Experimental measurements illustrate how P-wave velocity changes at different pressures and temperatures.

shows, waves move very slowly through the air or water that fills the pore spaces between the solid sediment grains. Seismic waves in igneous rocks are generally fast, and they move faster in mafic gabbro than felsic granite and even faster in ultramafic rocks; these differences between rock types indicate the importance of mineral types on seismic velocity.

Seismic velocity also varies with changes in temperature and pressure, as the data graphed in **Figure 8.19** show. The experiments show that velocity increases with increasing pressure (Figure 8.19a) and decreases with increasing temperature (Figure 8.19b).

Insights

How Do Rock Properties Determine Seismic Velocities within Earth?

Velocity varies considerably for several of the materials illustrated in Figure 8.18. Some of these variations result from differences in the abundance of open fractures or other pore spaces where the seismic waves move very slowly through air or water. Variation in mineral abundances within different samples of the same rock also causes variations in seismic velocity. A related observation from Figure 8.18 is that different materials can have the same seismic velocity. A measured P-wave velocity of 4.75 kilometers per second, for example, is compatible with granite, salt, limestone, or sandstone at surface temperature and pressure (see Figure 8.18). Seismic velocity, alone, therefore does not allow geologists to identify rock type.

Figure 8.19 illustrates that increasing pressure and temperature have significant, but opposite, effects on seismic velocity. How temperature and pressure affect seismic velocity relates to how minerals deform when a body wave passes through. If the mineral twists or squeezes easily when body waves pass, then some of the wave energy is consumed by these elastic deformations, which slow the wave down. If, on the other hand, there is not much twisting or squeezing, then the wave travels fast.

As an analogy, consider what would happen if you were holding one end of a rubber pole in one hand and one end of a steel pole in the other hand while someone struck the opposite ends of the poles with a hammer. You would feel much stronger vibrations through the steel pole than through the rubber pole. This happens because the body waves initiated by the hammer blows cause more squeezing and twisting within the rubber than within the steel, so the waves travel more efficiently, and thus faster, through the steel. In rocks, high pressure keeps mineral structures from deforming very much when body waves pass through, so seismic velocity increases as pressure increases. As temperature increases, without changing pressure, the mineral structure expands a tiny bit so that more twisting and squeezing can take place than in colder rock. This explains why seismic velocity decreases when temperature increases. Both temperature and pressure increase in value at greater depth within Earth, so experiments at various combinations of pressure and temperature are critical for determining how seismic velocity changes for different materials at different depths below Earth's surface.

How does density affect velocity? You might be tempted to speculate that seismic velocity increases when density increases, given that high pressure causes minerals to compact more closely in rocks. Experiments show, however, that although density affects velocities, velocities are even more strongly influenced by the details of how the minerals in the rock squeeze and twist when the waves pass through. These squeezing and twisting properties, along with density, are more accurately measured in the lab than are the rapid seismic velocities, so seismic velocities are commonly calculated from these other results rather than measured directly.

EXTENSION MODULE 8.3

Velocity of Seismic Waves. *Learn about the physical properties of rocks that determine how fast P and S waves move.*

Putting It Together—How Do We Know . . . How to Determine Velocities of Seismic Waves in Rocks?

• Laboratory experiments relate seismic-wave velocities to different types of rocks, under temperature and pressure conditions comparable to those in the deep interior of Earth.

• Seismic waves move more slowly in unconsolidated sediment than through sedimentary rocks, and they move even faster in igneous and metamorphic rocks. Within igneous rocks, felsic rocks transmit waves more slowly than mafic rocks, and ultramafic rocks display the most rapid velocities.

8.4 What Composes the Interior of Earth?

A hypothesis about the composition of Earth's interior must take into account all available data. These include seismic data that image the interior, experimental data on seismic velocities of minerals and rocks, the geochemistry

of magma resulting from mantle melting, and even the composition of meteorites that arguably represent pieces of small planets that initially formed elsewhere in the solar system at the same time as Earth. There is more than one way to combine these observations, so geologists continually test prevailing ideas against new results and modify the hypothesis when necessary. **Table 8.1** summarizes the generally accepted characteristics of each Earth layer. The thickness and physical properties of each layer are formulated completely from seismic imaging of the interior; the composition of crust and mantle are consistent with laboratory data such as those mentioned previously. The objective of this section is to help you understand how geologists arrive at the composition of the hidden mantle and core.

The Velocity and Density Structure of Earth

Figure 8.20 depicts Earth's interior using seismic-wave velocity data. This mapping of the interior using the wave data generated by earthquakes and explosions is the starting point for identifying the rocks and other materials present in the interior.

It is not possible to get a clear picture of what rocks are present at what depth with just a quick comparison of Figures 8.18 and 8.20, because the seismic velocities within Earth are much faster than those predicted from laboratory experiments. Remember that the velocities summarized in Figure 8.18 were measured at surface pressure and temperature and need to be adjusted for conditions within Earth.

Figure 8.20 also depicts the variation in density with depth. Density is estimated from the measured seismic-wave velocity. These estimates show that density increases as depth increases. Furthermore, the density is, on average, three times higher in the core than in the mantle, which is consistent with arguments presented in Figure 8.5.

You now have the basic knowledge to translate seismic velocity into a prediction of geologic materials and complete an image of the interior. There are considerable velocity variations within the crust, which is readily explained by the wide variety of rock types that we actually see composing the crust. The P-wave velocities in continental and oceanic crust

TABLE 8.1 Summary of Characteristics of Earth's Inner Layers			
Layer	Thickness	Composition	Solid or Liquid?
Crust			
Continental	Range: 25–85 km; mostly 30–40 km	Mostly igneous and metamorphic rocks with average composition similar to diorite; upper crust is similar to tonalite, and lower crust is probably gabbroic	Solid, except for local accumulations of magma
Oceanic	Range: 5–25 km; mostly 5–10 km.	Mafic igneous rocks (basalt and gabbro)	Solid, except for local accumulations of magma
Mantle	~2900 km	Ultramafic igneous composition. Upper mantle is olivine-rich peridotite with pyroxene and garnet. Transition zone (410–660 km depth) and lower mantle (below 660 km) probably have peridotite chemical composition, but with silicate minerals that are stable at high pressure and not present in the upper mantle	Almost entirely solid. Small amounts of partially molten rock are likely present at base of the lithosphere (around 100 km depth, on average), and possibly in discontinuous zones just above the core-mantle boundary
Core			
Outer	2255 km	Iron and nickel with as much as 10% lighter elements	Liquid melt
Inner	1215 km	Very iron rich, iron-nickel alloy	Solid

▼ **Figure 8.20 How seismic waves reveal the interior structure of Earth.** Abrupt changes in seismic-wave velocity reveal boundaries between different materials within the planet. Densities of the different layers are estimated from the seismic-wave velocities. Besides the boundaries between crust, mantle, and inner and outer core, there also are abrupt discontinuities in velocity and density within the mantle at about 100 km, 410 km, and 660 km. The low-velocity zone below 100 km defines the boundary between the lithosphere and asthenosphere. The expanded part of the diagram shows the contrasting seismic properties and thickness of oceanic and continental crust.

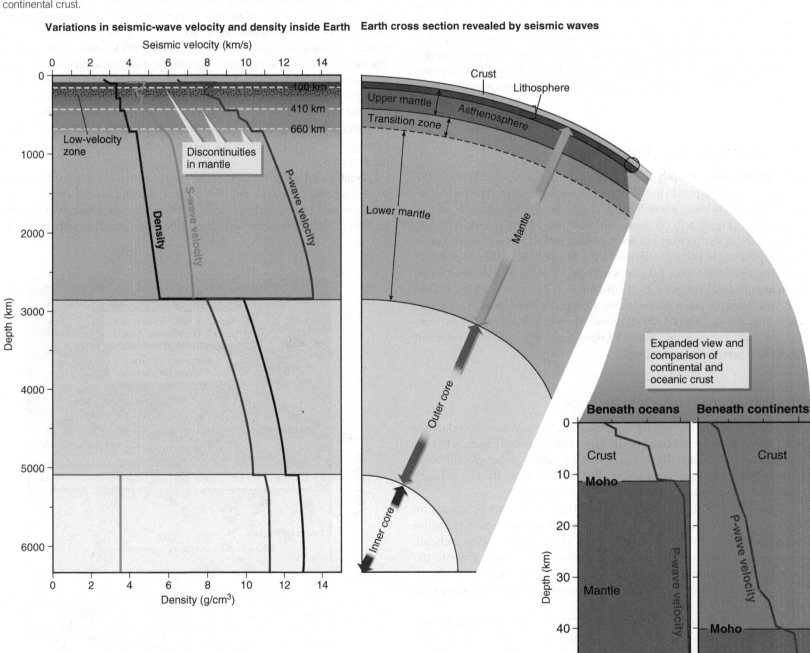

differ in an expected way because of the different rock types composing each type of crust (Figure 8.20; also see Figure 8.2). Velocity increases downward within continental crust, consistent with data from outcrops, deep drill holes, and fragments from volcanoes that suggest a downward transition from felsic to more mafic rocks (Section 8.1). Mafic rocks have higher seismic velocities than felsic rocks (Figure 8.18). Relatively high velocity within oceanic crust is consistent with the abundance of basalt and gabbro found in drill holes in the seafloor and in slivers of uplifted oceanic crust found on continents. The gradual increases in density and velocity within the mantle and core might simply relate to increasing pressure. The more critical features to explain are the abrupt step-like changes in velocity, density, or both that show up in Figure 8.20. These changes take place not only at the base and top of the mantle, but also within mantle and core.

The Crust-Mantle Boundary

Velocity increases abruptly at the crust-mantle boundary (Figure 8.20). This change corresponds to the abrupt downward change to peridotite, which geologists know about from geologic data collected at the surface (Section 8.1, Figure 8.2). Yugoslavian seismologist Andrija Mohorovičić discovered this sharp change in seismic velocity in 1909 and, in his honor, the crust-mantle boundary is known as the **Mohorovičić discontinuity**, or more simply as the Moho. Estimates of depth to the base of the crust based on field-collected geologic data are available only in very few places. Seismic data, however, allow geophysicists to map the Moho around the globe. The boundary is typically 5–20 kilometers deep beneath ocean basins. Continental crust varies from 25 to 85 kilometers thick, although values of 30–40 kilometers are most typical.

The Mantle Low-Velocity Zone

At a depth of approximately 100 km below the surface, seismic velocities slow; thus, this area is defined as the **low-velocity zone** (Figure 8.19). At depths between 220 and 400 kilometers the P- and S-wave velocities typically increase to values comparable to those just above the low-velocity zone, and then continue to increase at still greater depth.

Laboratory studies show that decreasing velocity can be caused by the presence of rocks that are less rigid, meaning that they yield more readily to stress and twist and squeeze more readily when seismic waves pass through. Rocks become less rigid when heated close to the melting point, and magma has no rigidity. The low-velocity zone, therefore, can be interpreted as a somewhat squishy, less rigid layer in the mantle.

Is this layer actually molten? You have the information needed to answer this question. S waves do not travel through liquids but they do pass through the low-velocity zone, so this zone *cannot* be entirely molten. At most, there is a small amount of melt in the low-velocity zone, which along with the presence of rocks close to their melting point accounts for the slowing of seismic waves.

The abrupt change at the top of the low-velocity zone marks the boundary within the mantle between the strong **lithosphere** (rocky sphere) consisting of crust and uppermost mantle and the underlying less rigid **asthenosphere** (weak sphere). These layers play an important role in plate tectonics, as mentioned in Section 1.6 and explored more thoroughly in Chapter 12.

The Mantle Transition Zone

P- and S-wave velocities abruptly increase globally at depths of about 410 and 660 kilometers (Figure 8.20). Given the similar changes in seismic velocities at the crust-mantle boundary, you may suspect that peridotite in the upper mantle is underlain by some other rock type at greater depth. This is not necessarily an accurate conclusion, however.

Minor changes in chemical composition within the mantle almost certainly exist, but an effect of increasing pressure can also account for these step-like discontinuities in seismic-wave velocity. Geologists explain the increases in seismic wave velocity at 410 and 660 kilometers by changes that occur to the crystal structure of peridotite minerals at pressures that are more than 100,000 times greater than at Earth's surface. At this extremely high pressure, atoms within minerals reconfigure with a tighter fit to produce new crystal structures. **Figure 8.21** illustrates that reconfiguration of atoms within peridotite minerals, rather than substantial changes in rock composition, readily explains the observed changes in seismic velocity at these depths.

Nearly all the upper-mantle peridotite recovered at Earth's surface consists of olivine, pyroxene, and a little bit of garnet. Lab experiments show that dominant olivine and pyroxene transform to new minerals at the progressively higher pressures and temperatures that coincide with the depths

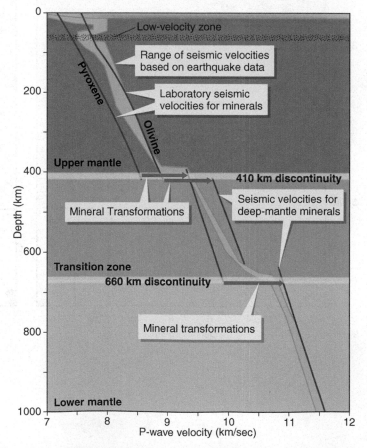

▲ **Figure 8.21 Why there are abrupt increases in seismic velocity within the mantle.** High-pressure repacking of the crystal structures of olivine and pyroxene explain sharp discontinuities in seismic-wave velocity at about 410 and 660 km below the surface. These metamorphic transformations change the mineral structure but not the mantle chemical composition. The observed seismic-wave velocities are consistent with laboratory measurements of velocity in olivine and pyroxene, and in the minerals that they transform to at greater pressure.

of 410 and 660 kilometers below Earth's surface. These deep-mantle minerals have the same composition as olivine and pyroxene but have different crystal structures. Experimental measurements of seismic velocity within these high-pressure olivine and pyroxene equivalents (Figure 8.21) indicate substantial velocity increases that account for the velocity discontinuities illustrated in Figure 8.20. This region of the mantle between 410 and 660 kilometers below Earth's surface is a **transition zone** where minerals undergo changes between peridotite upper mantle and a remarkably uniform lower mantle of high-pressure minerals.

EXTENSION MODULE 8.4

Mantle Minerals. Learn about the changes in mantle minerals that occur at high pressures and temperatures deep within Earth.

Changes in the Lowermost Mantle

In the lowest part of the mantle, just above the core-mantle boundary, there is a zone as much as 300 kilometers thick where body-wave velocities abruptly increase a few percent. Figure 8.20 does not include a label for this zone because it varies considerably in thickness from place to place around the core. The reason for the seismic velocity variations of this poorly defined zone is uncertain and the subject of considerable ongoing research. Recently acquired laboratory data support the possibility of another mineral transformation in the lowermost mantle that accounts for these seismic properties.

About one-tenth of the mantle that is in contact with the core exhibits a narrow zone, 5 to 50 km thick, where P- and S-wave velocities abruptly decrease by 10–15 percent. These spots in the lowermost mantle may be partly molten or a compositional mixture of mantle and core components. The presence of this unusual lowermost mantle plays a role in understanding dynamic motion within Earth, and you will revisit its significance in Chapters 10 and 12.

Core Composition

Recall that P- and S-wave behaviors imply a solid inner core surrounded by a liquid outer core, both of which are substantially more dense than the mantle (Figure 8.20). Three observations imply that iron composes both the solid and liquid parts of the core:

1. Iron is the only common chemical element that is dense enough to account for the high density of the core.
2. Liquid iron in the core can explain Earth's magnetic field (to be examined more thoroughly in Chapter 10).
3. Some meteorites consist mostly of iron and are interpreted to be the cores of broken-up planets.

The density of the inner core matches that of nearly pure iron, perhaps containing some nickel, at the high pressure near Earth's center. The outer-core density is a little lower than expected for liquid iron by itself. The density of the outer core calculated from the seismic-wave velocities can be explained by the presence of about 10 percent elements that are less dense than iron and nickel. Different geologists argue that this minor, lighter component could alternatively be oxygen, sulfur, silicon, potassium, or hydrogen. Without actual samples from Earth's core, it is difficult to know which of these elements is present.

EXTENSION MODULE 8.5

Meteorites as Guides to Earth's Interior. Learn about meteorites and why some meteorites likely represent analogs of Earth's interior.

Putting It Together—What Composes the Interior of Earth?

• The internal composition of Earth is inferred by relating calculated seismic-wave velocities to properties of rocks and minerals at temperatures and pressures equivalent to Earth's interior.

• The crust-mantle boundary, called the Moho, displays a sharp increase in seismic velocity where mafic to felsic igneous and metamorphosed igneous rocks of the crust are underlain by mantle peridotite.

• The low-velocity zone that exists about 100 kilometers below the surface is evidence that part of the upper mantle is less rigid and may be slightly molten. This less rigid zone defines the top of the asthenosphere and is below the more rigid lithosphere consisting of the crust and uppermost mantle.

• Crystal-structure transformations of peridotite minerals at high pressure and temperature explain the abrupt increases in seismic velocity at about 410 and 660 kilometers below the surface. The interval between 410 and 660 kilometers depth is the transition zone between the upper mantle (shallower than 410 kilometers) and the lower mantle (deeper than 660 kilometers).

• The core probably is composed of iron or of mostly iron with some nickel in it. The inner core is solid, while the outer core is molten. The outer core must contain about 10 percent lighter elements—likely oxygen, sulfur, silicon, potassium, or hydrogen—in addition to iron to account for the calculated seismic velocities.

8.5 How Hot Is the Interior of Earth?

Seismic-wave data also provide insights into internal temperatures within Earth. Seismic-wave properties distinguish solids and liquids, and therefore indicate whether material is above or below its melting temperature. You have frequently encountered statements about the temperature inside Earth in this text, beginning in Chapter 3, but you have not learned (a) how geologists determine the planet's internal temperature, or (b) why the temperature is so high inside the planet. You are now equipped with enough information to address both of these questions.

Determining the Geothermal Gradient

The geothermal gradient is simply the increase in temperature with increasing depth; at shallow depths it is calculated from temperature measurements in deep wells and mines (see Section 4.4). These measured geothermal gradients vary considerably from place to place, but temperatures typically increase between 25 and 30°C for each kilometer of increasing depth on continents and average about 60°C per kilometer beneath the oceans.

Is it reasonable to use geothermal gradients, which are based on near-surface measurements, to determine temperatures deep within Earth? Let's explore the possibilities. A geothermal gradient of 30°C per kilometer near the surface under continents would, if extended deeper, reach the melting temperature of mantle peridotite at 50 kilometers and an outrageous value approaching 200,000°C at Earth's center. The travel of S waves throughout the mantle indicates, however, that the peridotite is mostly solid. We must conclude, therefore, that the geothermal gradient measured near the surface must decrease at greater depth in the mantle.

Integration of different geologic data sets provides more realistic estimates of the geothermal gradient deep within the mantle.

- If, as seems likely, the low-velocity zone in the mantle relates to small degrees of partial melting at the top of the asthenosphere, then temperatures of about 1300°C (the melting temperature of peridotite in the upper mantle) must be reached at depths of 75–100 kilometers.

- Laboratory experiments with diamond-bearing mantle rocks brought to the surface through continental volcanoes indicate that these rocks formed at depths of 100–150 kilometers below the surface but only at temperatures of ~1000–1200°C. So the overall gradient must decrease to less than 10°C per kilometer deep below the continents.

- Laboratory experiments on the transformation of olivine to a denser mineral at 410 kilometers suggest a temperature of only 1450 ± 150°C. This result suggests a further decrease in the geothermal gradient to about 3.5°C per kilometer beneath both continents and oceans.

Figure 8.22a summarizes these data and provides a sketch of the geothermal gradient in the upper mantle based on them.

▶ **Figure 8.22 How temperature increases below the surface.**
This diagram illustrates how geologists interpret temperature variation within Earth. Here are the key points:

- Typical near-surface geothermal gradients of 25–60°C/km (dashed lines) cannot extend downward very far or else molten mantle would exist at shallow depths, where S waves instead reveal solid rock.
- The interpreted geothermal gradient below oceans intersects the dry-peridotite-melting curve to account for the low-velocity zone at 75–100 km.
- Laboratory experiments provide ranges of possible temperatures and pressures for the formation of diamond-bearing rocks found in continental volcanoes and for the mineral changes at the top of the mantle transition zone.
- The blue, speculative geothermal gradient for the deeper Earth is consistent with estimated melting conditions for lower-mantle and core materials. This profile explains small amounts of melting in the upper-mantle low-velocity zone, the possibility of some melting just above the core-mantle boundary, and complete melting in the outer core.

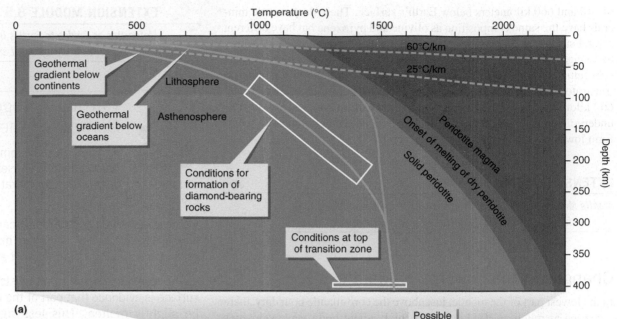

(a)

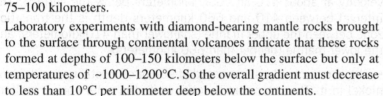

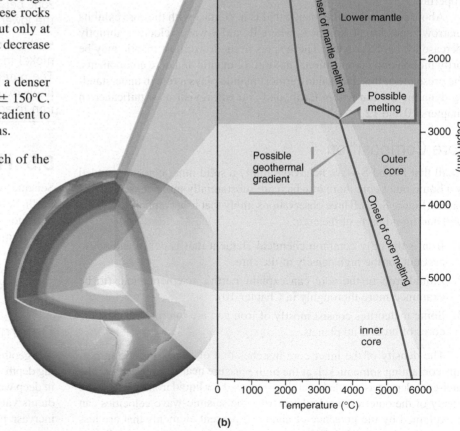

(b)

Continuing the geothermal gradient deeper into the mantle and core requires considerable speculation, because the precise melting temperatures are unknown for rocks in the lower mantle and core. Figure 8.22b portrays a geothermal gradient that is consistent with what geologists do know.

The geothermal gradient depicted in Figure 8.22 shows temperature rising very gradually through most of the lower mantle and then more abruptly to a value of 3500°C just above the core-mantle boundary. We know the temperature in the outer core exceeds the melting temperature of the iron-nickel mixture present there because S waves do not pass through it. In the uppermost outer core, current best estimates based on laboratory experiments suggest a melting temperature of 3500 ± 500°C. A temperature of approximately 3500°C is also close to the estimated melting temperature for high-pressure mantle rock. Remember that a small degree of melting in the lowermost mantle is one explanation for the curiously low seismic velocities observed in some spots just above the core-mantle boundary.

Extending the meager information on the temperature gradient into the deeper core requires further speculation. The inner core must be cooler than this melting temperature because seismic data are consistent with nearly pure and solid iron in the inner core. This does not mean that temperature decreases into the inner core, however—remember that melting temperature increases as pressure increases. The possible geothermal gradient illustrated in Figure 8.21 proposes a temperature of about 5000°C at the boundary between outer and inner core and a temperature at the center of Earth between 5500 and 6000°C.

Why the Interior Is So Hot

The geothermal gradient suggested by seismic-wave data and laboratory experiments requires heat generation inside Earth in order to account for the inferred high internal temperatures. Chapter 7 introduced the primary heat source—radioactive decay of unstable isotopes.

The most abundant radioactive elements within Earth are potassium, uranium, and thorium. All three elements have a strong affinity to accumulate in the crystal structures of silicate minerals, especially those found in felsic igneous rocks and in mineralogically similar sedimentary and metamorphic rocks. As shown in **Figure 8.23**, this means that a cubic meter of continental-crust rock generates far more heat from radioactive decay than does an equal volume of basaltic oceanic crust or peridotitic mantle. Although some geologists speculate on the presence of significant potassium in the core, most scientists think this is unlikely. If potassium is absent, then there is probably little or no radioactive-decay heat generated in the core.

How do geologists explain the high temperature of the mantle if radioactive elements concentrate into the continental crust? The total heat production from the mantle is huge because the mantle composes 83 percent of the volume of Earth. The pie charts at the bottom of Figure 8.23 show that even though a small sample of mantle rock contains miniscule concentrations of radioactive elements, 75 percent of Earth's radioactive

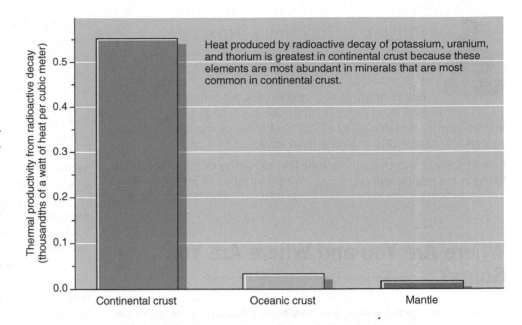

Heat produced by radioactive decay of potassium, uranium, and thorium is greatest in continental crust because these elements are most abundant in minerals that are most common in continental crust.

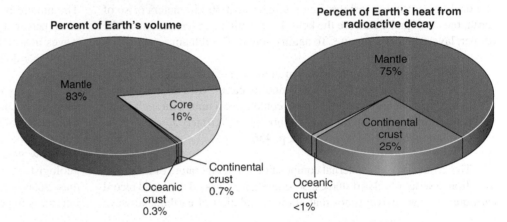

Percent of Earth's volume

The mantle overwhelmingly composes the majority of the volume of Earth, even though it contains low abundances of radioactive, heat-generating elements.

Percent of Earth's heat from radioactive decay

Multiplying heat production by the volume of rocks reveals that about 75% of all of Earth's radioactive heat originates in the mantle.

▲ **Figure 8.23 Radioactive decay heats the interior of Earth.** Radioactive elements are most abundant in continental crust. Nonetheless, most radioactive heat comes from the mantle simply because the mantle composes so much of Earth's volume.

heat originates in the mantle simply because the mantle occupies such a vast volume of the planet. Even this enormous source of radioactive heat, however, is insufficient to account for temperatures near 3500°C at the base of mantle, and it does not explain the higher temperature in the core, where radioactive isotopes are rare or lacking altogether.

Why, then, is the core so hot? The additional heat required to keep the core so hot probably originates from crystallization of iron from the molten outer core to form the solid inner core. Just as you must add heat to melt something, the opposite occurs whenever melt crystallizes—heat is released. This may seem strange at first, because you know that decreasing temperature causes crystallization. Yet crystallization releases heat that increases temperature. Therefore, heat released by crystallization actually slows the cooling of magma. Calculations based on laboratory measurements suggest that crystallization of iron might account for the high temperature of the core as well as the high temperatures in the lowermost mantle adjacent to the core.

Putting It Together—How Hot Is the Interior of Earth?

- Geothermal gradients measured near the surface are higher than in the interior. Earth's internal temperature is on the order of 1000–1200°C at 100–150 kilometers depth, perhaps 3000–4000°C at the core-mantle boundary, and greater than 5000°C in the inner core.

- Heat sources within Earth include (a) radioactive decay of elements found in the mantle and crust; and (b) crystallization of iron in the core.

Where Are You and Where Are You Going?

Thanks to earthquake waves, you now have a descriptive image of the interior of the planet. Rocks found at the surface or retrieved from deep mines and drill holes provide insights into the outermost 50 kilometers or so of Earth, but earthquake data are the keys that permit an understanding of the deeper, layered inner structure. To again quote R. D. Oldham,

> The seismogram, recording the unfelt motion of earthquakes, enables us to see into the earth and determine its nature with as great a certainty, up to a certain point, as if we could drive a tunnel through it and take samples of the matter passed through. (*Quarterly Journal of the Geological Society*, 1906, vol. 62, p. 456)

The deduction of the internal composition and temperature of Earth follows from a series of related studies. Seismic stations around the world record earthquake-wave arrival times that allow calculation of earthquake-wave velocities. These data demonstrate that wave velocity, for the most part, increases with depth. A shadow zone exists where S waves, which do not travel through liquid, do not reach Earth's surface because they encounter a liquid outer core. The otherwise persistent S waves indicate that the remainder of the inner planet is solid and contains only minor amounts of melting. Laboratory measurements reveal relationships between seismic velocity and rock type. These relationships allow geologists to use seismic-velocity data to predict what physical materials occur at specific depths within the planet.

Earth's crust usually is 25–50 kilometers thick beneath continents and averages 7 kilometers thick below the oceans. Continental crust is largely quartz-bearing igneous and metamorphosed igneous rocks that become progressively more mafic at greater depth. Oceanic crust is composed almost entirely of mafic lava flows and intrusive rocks. The base of the crust is a boundary called the Moho, where a sharp increase in seismic velocity indicates the existence of peridotite in the upper mantle.

The mantle continues downward to about 2900 kilometers below the surface. Discontinuities in seismic velocity reveal a number of internal boundaries within the mantle. Seismic velocity increases downward within the crust and uppermost mantle, which together compose the lithosphere, but then it decreases in the low-velocity zone at the top of the asthenosphere. The mantle is less rigid and the low-velocity zone probably contains a small percentage of magma. Sharp increases in the velocities of earthquake waves near 410 and 660 kilometers depth imply repacking of atoms in mantle-peridotite minerals into more compact and seismically faster crystal structures.

The very dense core consists of two parts, a liquid outer core and a solid inner core. The high density of the core is most consistent with a dominance of iron or a combination of iron and nickel. The outer core probably also contains small amounts of lighter elements.

Now you know about Earth's layered internal structure and how geologists discovered this structure. The next chapter investigates how these zones formed, and Chapter 10 shows why geologists infer that Earth's interior is in persistent motion.

Active Art

Dimensions of the Mantle and Core. See how to determine physically possible thicknesses of Earth's mantle and core.

Seismic Wave Motion. See how the different seismic waves move.

Wave Reflection and Refraction. See how waves reflect and refract through rock.

How seismic waves reveal Earth's interior. See how seismic waves reflect and refract to reveal the layers within Earth.

Extension Modules

Extension Module 8.1: Sizing Up Earth. Learn how Eratosthenes determined the radius of Earth and how eighteenth-century scientists calculated the mass of Earth.

Extension Module 8.2: How to Locate an Earthquake. Learn how to use seismographs and travel-time curves to determine where an earthquake happened.

Extension Module 8.3: Velocity of Seismic Waves. Learn about the physical properties of rocks that determine how fast P and S waves move.

Extension Module 8.4: Mantle Minerals. Learn about the changes in mantle minerals that occur at high pressures and temperatures deep within Earth.

Extension Module 8.5: Meteorites as Guides to Earth's Interior. Learn about meteorites and why some meteorites likely represent analogs of Earth's interior.

Confirm Your Knowledge

1. How was the radius of Earth determined? Is Earth a perfect sphere?
2. How do geologists determine the composition of Earth's interior? Why not examine it directly?
3. How do we get samples of rocks from the crust that formed deeper than 5 kilometers below Earth's surface?
4. How do oceanic and continental crusts differ? How do scientists know that they are different?
5. What evidence tells us that Earth has a distinct concentration of high-density material near the center and does not gradually transition from low-density to high-density rock?
6. What is a wave? How does a wave move through rock?
7. What instrument records seismic waves?
8. What is the difference between elastic and plastic deformation? What type of deformation does a rock experience when an earthquake wave passes through it?
9. What is the difference between an earthquake's focus and its epicenter?
10. Describe the three types of earthquake waves in your own words.
11. Which type of wave arrives first at an earthquake recording station? Why do the waves not arrive at the same time?

12. The illustration below shows a seismogram record of an earthquake. How far away did the earthquake occur from the seismic station?

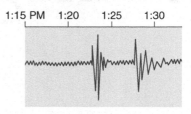

13. What causes the mantle low-velocity zone?
14. What is the most likely cause of the mantle seismic velocity transition zones at 410 and 660 kilometers?
15. What causes the S-wave shadow zone? What causes the P-wave shadow zone?
16. What is the evidence that the outer core consists mostly of liquid iron? What is the evidence that the outer core contains elements other than iron?
17. What is the source of heat responsible for the high temperatures in Earth's mantle? What is the source of heat responsible for the high temperatures in Earth's core?

Confirm Your Understanding

1. Write an answer for each question in the section headings.
2. How do we know that Earth is inhomogeneous?
3. How do we know that P and S waves do not travel through Earth at constant velocities?
4. How could you use seismic-wave travel velocities to determine if magma is present below a restless volcano?
5. You have collected data on seismic-wave velocities in rocks that exist a few hundred meters below the surface. Use information in Figure 8.18 to determine what kind of rock(s) is suggested by these P-wave velocities: 2.25 km/s; 4 km/s; 6.25 km/s; 6.75 km/s; 8 km/s.
6. Use Figure 8.20 to compare the P-wave velocities of oceanic and continental crust at a depth of 10 km. Why are these velocities different?
7. Use data graphed in Figures 8.18 and 8.20 to write a paragraph explaining why geologists interpret the lower continental crust to consist of igneous rocks that are more mafic than the rocks composing the upper continental crust.

CHAPTER 9
Making Earth

Why Study the Origin of Earth?

When astronauts view Earth from space, they see rocky continents and wide oceans blanketed by an atmosphere. Have you ever wondered how Earth's geosphere, hydrosphere, and atmosphere originated? We have ventured deep into Earth using seismic waves and geophysical principles to reveal concentric layers distinguished by different compositions and physical properties. Our curiosity about Earth now leads to the next question—*how* did Earth become layered?

How the layers formed, in turn, leads to other questions: How did Earth form? How did the atmosphere and hydrosphere form? How did Earth end up with distinctly different continental and oceanic crusts? A complete look at the complicated chemistry and physics behind the hypotheses for the formation of the planet, its layers, and the origin of its water are beyond our scope, but a look at the fundamental ideas for how Earth formed and evolved in its youth can be explored here. Observations from other worlds, such as the Moon and Mars, as seen on the adjacent page, will help us to answer questions about the origin of our own planet. In this chapter you will learn how geologists have drawn on the research achievements of astronomers studying other worlds in order to better understand our own world.

After Completing This Chapter, You Will Be Able to

- Explain the prevailing theory for the construction of our solar system.
- Explain the origin of Earth's layered internal structure.
- Contrast the processes required to form two distinct types of crust.
- Apply understanding of geologic processes to explain the origin of Earth's atmosphere and hydrosphere.

Pathway to Learning

9.1 How Did Earth Form?

EXTENSION MODULE 9.1
A Geologic Tour of the Solar System

EXTENSION MODULE 9.2
The Origin of the Moon

9.2 How Did the Core and Mantle Form?

Mars Exploration Rover Spirit captured this stunning view of a Martian sunset in 2005. Because Mars is farther from the Sun than Earth is, the Sun appears only about two-thirds the size that it appears in a sunset seen from Earth.

9.3 How Does the Crust Form?

9.5 How Do We Know . . . the Hydrosphere Came from the Geosphere?

9.4 How Did the Hydrosphere and Atmosphere Form?

IN THE FIELD

In this virtual field trip, you are outside on a clear night brightened by the light of a full Moon. A store-bought telescope enlarges your view of the Moon, and you take a picture for your notebook, as seen in **Figure 9.1**a. As you admire the Moon, you ask yourself how it is that this incredible planet formed.

Sometimes people call Earth the "Goldilocks planet" because it is "just right." As you probably remember, in the fairytale of Goldilocks and the three bears, Goldilocks ate the baby bear's porridge because it was neither too hot nor too cold—it was just right. Likewise, our planet is the perfect distance from the Sun—Earth is neither too hot nor too cold to sustain liquid water, an atmosphere, and life. None of the other planets currently have liquid water. Also unique to Earth is a crust containing quartz-rich rock, such as granite. In addition, no other planet is as geologically active as Earth, with its moving, deforming lithosphere, and its ever-changing landscape crafted by weathering and erosion.

As you peer at the Moon, you easily recognize the contrasting whitish-gray and black colors of the Moon's rocky surface. Imagine what it would be like to take a geology field trip to the Moon, sample the rocks and bring them back to Earth for careful study. One geologist has done field work on the Moon: Dr. Harrison Schmitt was part of the crew for Apollo 17 in 1972 (Figure 9.1b). Dr. Schmitt and astronauts on NASA's five earlier lunar landing missions brought back many rocks and made key field observations during their brief visits to the Moon. Among the rocks returned from these visits, the youngest are 2.9 billion years old and the oldest are about 4.56 billion years old. The lack of rocks younger than 2.9 billion years old implies that the Moon has not changed much since very early in its history. Unlike Earth, the Moon is geologically "dead"; it has no oceans, no atmosphere, no active volcanoes, and only rare moonquakes. Earth is dynamic; it has oceans, an atmosphere, volcanoes, and abundant earthquakes. These changing features and processes cause Earth to look very different today from the way it looked when it first formed. The Earth and the Moon are two planetary bodies in complete contrast, yet they are close to one another in the solar system and likely share some common history.

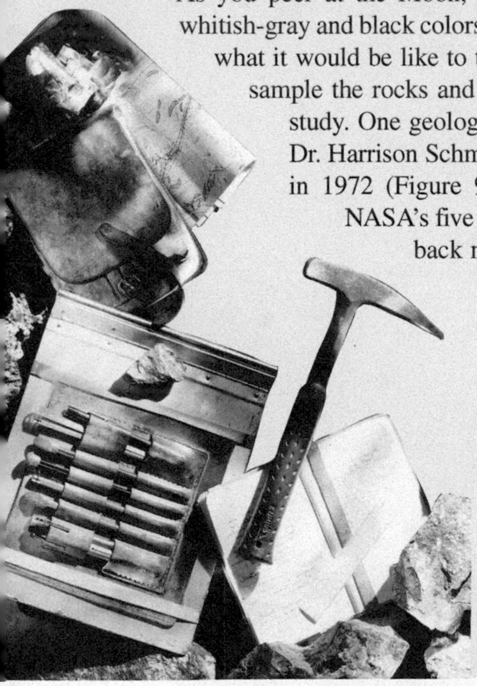

Did Earth and Moon form at the same time, from the same material? In Chapter 7 you learned how scientists infer that Earth and Moon probably formed at roughly the same time as the other planetary bodies of the solar system. Geologists also think that the Moon likely originated from the same material as Earth. To understand how Earth and Moon formed, it makes sense to broaden your view to contemplate the origin of the Sun and our planetary neighbors.

Hypotheses about the origin of Earth must be consistent with the seismic data that reveal the three distinct layers: core, mantle, and crust. These hypotheses should also address whether the layers formed during planet formation or at a later time. Additional puzzles include why the iron core separated from the silicate materials that form the mantle and crust, and why there are distinct continental and oceanic types of crust.

Any hypothesis about Earth's origins also must explain the fact that Earth is unique compared to other planets because it has the right temperature and pressure conditions to contain water as solid, liquid, and gas. You know that water is critical for geologic processes because it explains many metamorphic reactions, aids in forming magma near convergent plate boundaries, causes explosive volcanic eruptions, weathers rocks to make sediment, and erodes landscapes. It is important to understand, therefore, how the hydrosphere formed, as well as the atmosphere that we breathe.

On any clear night, as you gaze at the Moon and think about Earth and the solar system, you might imagine the variety of suggested hypotheses for the formation of Earth, its layers, and water. How Earth formed is one of the most fundamental questions in geology. Geologists combine their knowledge with that of astronomers and astrophysicists to investigate and explain how Earth formed.

▶ Figure 9.1 Examining Earth's Moon.

9.1 How Did Earth Form?

One of the greatest challenges in geology is figuring out the physical and chemical processes and events that formed Earth. Understanding the origin of planets involves consideration of a wide variety of physical and chemical processes. Various theories describe scientific understanding of these processes. Rather than examine each elaborate theory separately, this book considers the condensed and integrated highlights, which we simply call the planetary-origin theories. Many of the fundamental ideas about Earth's formation originated in the seventeenth and eighteenth centuries and remained hypothetical until space-age observations enabled scientists to test them.

Finding the Evidence

To understand the origin of Earth, geologists first examine its rocks. The oldest known rocks on Earth provide clues, but even these ancient rocks are not as old as the planet. Earth is about 4.5 billion years old, but the oldest dated minerals on Earth are slightly younger, approximately 4.4 billion years old (Section 7.6). This means that the rocks that originated on Earth during or immediately after planetary formation are not preserved for us to study. They were recycled by weathering, erosion, metamorphism, and melting in the rock cycle, including the dynamic resurfacing of the planet by plate tectonics. The geologic record is not a complete dead end,

(a) Telescope observations of Earth's Moon: The surface consists of areas of dark and light rocks. Most of the dark rock occupies more or less circular areas, which are large meteor-impact craters.

(b) In the field on the Moon: Geologist-Astronaut Harrison Schmitt sampling a large boulder of igneous rock at the Taurus-Littrow in 1972. Notice the "Rover" field vehicle in the left foreground.

however, because knowledge of Earth's composition and internal layered structure are factors that any theory of planet formation must explain.

In addition to the information gained from Earth rocks, data from elsewhere in the cosmos provide critical evidence of planet formation. Meteorite samples and rocks from the Moon that are older than Earth's oldest rocks help to fill in the gap in Earth's earliest history. More puzzle pieces are supplied by studies of the other planets in our solar system, along with their moons and other small objects such as rocky asteroids and icy comets. Information on our planetary neighbors partly comes from Earth-based telescope observations but mostly from data-collecting instruments on spacecraft that pass close to these other bodies. Also offering key pieces of evidence are telescopic observations of other solar systems, including those where planets are probably forming as we watch.

Whatever processes account for how Earth formed must explain the similarities and the differences among all of the objects that formed in our solar system. **Table 9.1** summarizes the most important statistics about the planets and emphasizes their differences. The inner planets (Mercury, Venus, Earth, and Mars) are relatively small and dense, which implies that they are mostly solid rock. The outer planets are huge but have low densities, which imply that they mostly consist of compressed gases with relatively small centers of rock or ice. **Figure 9.2** provides recent images of the surface of Mars, a rocky world that resembles an earthly desert, in contrast

Table 9.1 Basic Planet Properties

Planet	Distance from the Sun (millions of km)	Diameter (km)	Density (g/cm³)	Composition
Inner Planets				
Mercury	58	4,878	5.4	Silicate crusts and mantles
Venus	108	12,104	5.2	with iron-nickel cores
Earth	150	12,756	5.5	Atmospheres are thin or
Mars	228	6,794	3.9	lacking altogether
Outer Planets				
Jupiter	778	143,884	1.3	Mostly hydrogen and helium
Saturn	1,427	120,536	0.7	gases, liquids, and ices with
Uranus	2,870	51,118	1.2	small rock and iron cores
Neptune	4,497	50,530	1.7	

▶ **Figure 9.2 Views of other worlds.** The inner, rocky planets contrast with the gas giants in the outer solar system. NASA's Mars Exploration Rover *Spirit* took the image on the left in June 2004 while traveling the rocky desert plains of Mars. NASA spacecraft *Cassini* took the image of Saturn, on the right, in March 2004 as it approached the planet for closer study. Ice and rock fragments encircle the great planetary ball consisting mostly of gas and ice.

with Saturn, a gas giant. The distinction between the inner rocky planets and the gas giants must be explained by any theory of planet formation.

Pluto, recently demoted from true planet status, is a small ice-and-rock oddball in comparison to its large, gas-rich neighbors. To understand why astronomers now classify Pluto, along with less familiar Ceres and Eris, as dwarf planets and plutoids requires understanding how planets form.

EXTENSION MODULE 9.1

A Geologic Tour of the Solar System. Learn the geologic basics of the planets, moons, asteroids, and comets of the solar system.

Starting with a Cloud of Gas and Dust

The planetary-origin theories explain the formation of planets by the gradual accumulation of small dust particles and gas molecules to form increasingly larger objects that eventually achieve the status of planets. **Figure 9.3** illustrates the key stages of this process.

Data and images collected at astronomical observatories show that stars and planets originate from **nebulae** (singular: nebula), which are

clouds of gas and dust in space. In these clouds gas molecules, mostly hydrogen, strongly outnumber dust particles, which compose less than 1 percent of a nebula's mass. Uneven distribution of matter in a nebula causes uneven gravitational forces that attract particles and molecules to one another and to the center of the nebular cloud. The gas and dust rotate around the center point at increasing speed as mass concentrates more and more toward the center, a process called nebular contraction. This speed-up effect is similar to the accelerating spin of an ice skater who pulls his or her arms in close to the body. Forces of motion associated with the increasingly faster spin also flatten the cloud into a pancake-shaped disk (Figure 9.3), much the same way that a pizza chef makes a thinner crust by spinning the dough in the air.

Forming the Protosun

The disk-shaped cloud that formed our solar system is called the **solar nebula** because the concentrated mass near the center eventually formed into the Sun. **Figure 9.4** illustrates recent observations of nebular clouds and disks that are the forerunners of distant stars and planets. The density and temperature within a spinning nebular disk are highest at the center. The

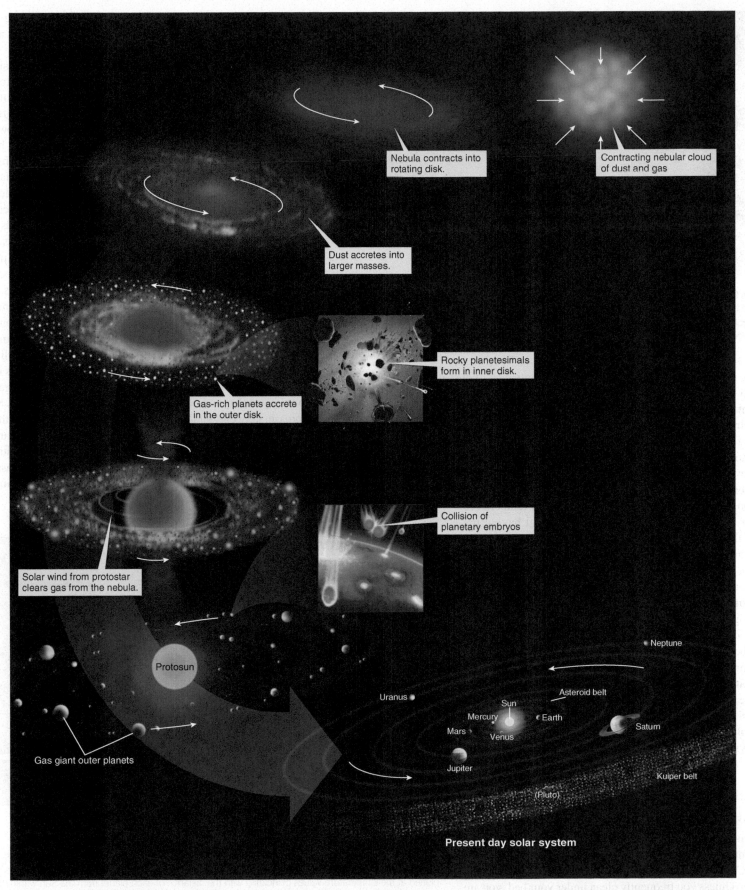

Nebula contracts into rotating disk.

Contracting nebular cloud of dust and gas

Dust accretes into larger masses.

Rocky planetesimals form in inner disk.

Gas-rich planets accrete in the outer disk.

Collision of planetary embryos

Solar wind from protostar clears gas from the nebula.

Protosun

Neptune

Uranus

Asteroid belt

Sun

Mercury   Earth

Mars   Venus   Saturn

Gas giant outer planets

Jupiter

(Pluto)

Kuiper belt

Present day solar system

▲ **Figure 9.3 How the solar system probably formed.** Planetary-origin theories explain the origin of the planets by accretion of progressively larger masses of dust and gas into planetesimals, planetary embryos, and finally into planets. The outer, gas-giant planets formed before the inner, rocky planets. Smaller objects left over from planet formation form the asteroid and Kuiper belts.

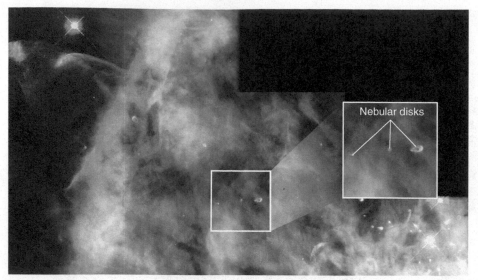

▲ **Figure 9.4 Witnessing star and solar system formation.** This Hubble Space Telescope image of the Orion Nebula shows a giant cloud of dust and gas illuminated by countless bright, young hot stars. The closer view of a small part of Orion shows faint stars surrounded by dust and gas. These solar nebulae are the beginnings of planetary systems.

density is high because gravity pulls an increasing mass of dust and gas into a small volume of space. Heat energy, related to the temperature, can be thought of as the motion energy of atoms and their internal particles, such as electrons. When a large number of moving atoms occupy a small volume, the heat energy is concentrated in that small volume and causes the temperature to increase. The accumulation of more mass in an increasingly smaller volume also increases the density.

About 100,000 years after the beginning of nebular contraction, the dense, hot center of a disk has a temperature greater than 1 million degrees. While not yet having the internally fueled furnace of a true star, this central mass is bright and constitutes a **protostar**, or in the case of our solar system, the protosun. The heat radiating from a protostar is sufficient to split the dust particles in the inner part of the spinning nebular disk into their component atoms. Where spinning forces carry these atoms farther from the protostar, they cool and condense into new dust particles.

The original nebula was peppered with dust grains of various compositions. However, when the dust reformed after being atomized near the protosun, the composition of each particle was determined by its location and temperature within the disk. **Figure 9.5** shows how ice particles of various frozen gas molecules formed in the relatively cool outer part of the solar system, whereas only silicate, oxide, and metal dust particles condensed in the hotter inner part of the solar system.

Making Planetesimals and Planetary Embryos

The crowding of matter along the Frisbee-shaped plane of the spinning nebular disk caused collisions between dust particles. Colliding particles stuck to one another to make larger clumps. Unless you frequently clean under your bed, you are well aware of how effectively tiny dust particles clump into larger aggregates. This process in space, called **planetary accretion**, gradually accumulates small particles into large

objects that we recognize as the planets. Judging from radioactive-isotope ages of minerals in meteorites, this accretion process began at 4.567 billion years ago.

The initial small objects, called **planetesimals** (Figure 9.3), ranged in size from only football stadiums to large cities. Planetesimal growth resembles the construction of a snowman. When you roll a small snowball along the ground, more small snowflakes stick to it, and the snowball grows larger and heavier. Planetesimals did not roll on a surface, however, so the better analogy might be throwing a snowball through a heavy blizzard of snowflakes, with the snowball enlarging with every accreted flake. The attraction of water molecules binds a snowball together, whereas during planetary accretion gravity is the force that pulls particles together and holds them together to form a growing spherical planetary ball. As planetesimals grew, their larger mass generated a stronger pull of gravity to shape the mass into a smoothly rounded object.

Once planetesimals were a few kilometers across, their gravity forces also perturbed the motion of their neighbors, pushing and pulling them into collision courses. Some of the resulting collisions undoubtedly pulverized planetesimals into smaller chunks. Computer modeling of the physics of collisions suggest, however, that most collisions produced overall accretion of the planetesimals into larger bodies. This is sort of like piling up large snowballs to construct a snowman. These enlarged objects are **planetary embryos** (see Figure 9.3), with sizes as large as Moon or Mars, approximately one-hundredth to one-tenth the mass of Earth.

Each embryo staked out a swath of space near its orbital path around the protosun, from which it swept up remaining dust while also gravitationally pulling in nearby planetesimals. These orbital zones that fed mass into the growing embryos contained materials of different compositions in

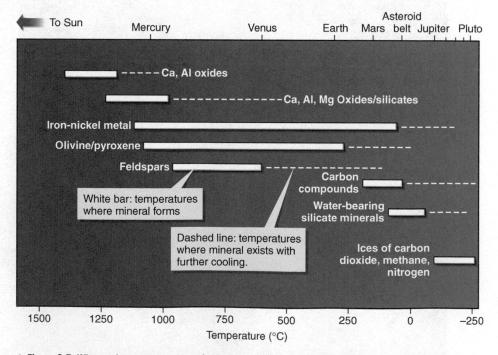

▲ **Figure 9.5 Where minerals crystallized in the solar nebula.** This chart illustrates where minerals crystallized in the solar nebula depending on temperature and distance from the Sun. The planets are plotted at the temperature values coinciding with the distance of each planet from the Sun within the early nebula. Condensed dust particles that accreted into planetesimals in the vicinity of Earth contained iron-nickel metal and silicate minerals such as olivine, pyroxene, and feldspar. Water-bearing silicate minerals, carbon compounds, and ices of gas-rich elements accumulated farther out in the solar system where nebular temperatures were much cooler.

different parts of the nebular disk, because of the different condensation temperatures of particles at various distances from the hot protosun.

The first planetesimals formed in the outer part of the disk where cooler temperatures permitted condensation of ice particles, while the inner nebula was still too hot for significant formation of any solids. Eventually, the icy planetesimals and embryos in the outer disk became sufficiently large for their gravitational forces to pull in surrounding envelopes of gas molecules from the nebula. In this fashion the gas giants were born first in the outer solar system (see Figure 9.3).

Planetesimals probably were just starting to combine into embryos in the hotter inner solar system when the protosun underwent a violent stage of its own history that altered the composition of the solar nebula. This is what is thought to have happened: A few million years after the nebular disk formed, the internal temperatures in the continually contracting protosun exceeded 5 million degrees. At this extraordinarily high temperature, the energetic motion of atoms tore the atoms apart into their component protons and electrons. These charged particles exploded out into space in spectacular displays. Today, astronomers sometimes witness such events in other nebulae. The streams of charged particles produce a **solar wind**, which is not moving air but is, instead, fast-moving protons and electrons blasting outward from the Sun. The solar wind remains a feature of the modern-day Sun but today the wind lacks the intensity of its early history.

The early solar wind had little effect on the condensed dust particles and planetesimals, but it effectively blew the lightweight gas molecules out of the nebula except where they were strongly bound by gravitational forces around the gas giants in the outer solar system. This clearing out of the nebula (see Figure 9.3)—the removal of all the gaseous molecules from this region of the solar system—affected the continued growth of the inner planets, where high temperatures had not permitted formation of ices or water-bearing minerals. This meant that the planets forming in the inner solar system lacked abundant gases. The distinction between the inner, rocky planets and the outer, icy and gas-rich planets was a result of this early event in the history of planet and Sun formation.

Planets and the Sun

The final stages in the formation of the inner planets were complicated by the increasingly strong gravitational interactions between the enlarging inner planetary embryos and the gravitational pull of the early formed gas giants in the outer part of emerging solar system. Gravitational push-and-pull directed embryos into different orbits, which had two effects.

1. The embryos began to gather particles from outside of their original accretion zones. This blurred the original compositional gradients between the inner to outer parts of the solar system. Wider accretion zones resulted in embryos that contained mixtures of materials that originally formed at varying condensation temperatures.
2. The embryo orbits crossed one another, setting up collisions that continued the planetary accretion process to the next stage—the making of the inner, rocky planets (Figure 9.3).

As Mercury, Venus, Earth, and Mars started to form perhaps a few dozen embryos and an uncountable number of planetesimals remained in the inner solar system. Over a period of perhaps as short as 5 million years the embryos collided to form the planets. Earth and Venus probably formed from a dozen or more embryos, while continuing to sweep up smaller planetesimals that were pulled in by the gravitational attraction of the growing planets. Smaller Mars may consist of only a few embryos, and Mercury may be a single embryo.

The final accretion of Earth is thought to have included giant embryo collisions. The scientifically popular giant-impactor hypothesis states that the Moon formed when a catastrophic collision occurred between a large embryo and Earth. Vaporized debris from the impacting embryo, along with chunks from Earth, spun into a ring around the planet that then accreted to form the Moon. This collision added the last 10 percent of Earth's current mass and took place about 30 million years after the accretionary process began.

Until recently, the term "planet" lacked a rigorous definition, causing controversy over which solar system objects qualified as planets. An understanding of how planets form provides a basis for clarifying a definition. In 2006, the International Astronomical Union adopted a definition that requires a planet to meet three criteria. A **planet** is a celestial body that (1) orbits the Sun, (2) has sufficient mass and therefore gravitational pull to form into a nearly spherical shape, and (3) has cleared all other objects from the neighborhood of its orbit, hence completing the accretion process.

We need to backtrack slightly to point out that the protosun matured into a star early during the formation of the inner planets. The key final step occurred when continued contraction of the dense, gas-rich ball drove the temperature at the interior of the protosun above 10 million degrees. The rapid motion of atoms, necessary to account for so much heat, is sufficient to fuse them together when they collide. **Nuclear fusion** describes the process whereby protons fuse together into larger atoms and release huge amounts of energy. When nuclear fusion began, the nuclear furnace ignited and our Sun became a true star.

EXTENSION MODULE 9.2

The Origin of the Moon. Learn about the different hypotheses for the origin of the Moon.

The Leftovers

Not all planetesimals and embryos accreted to form planets. Many of the small objects in the inner solar system fell into the Sun. Hundreds of other objects were flung to the outer edges of the solar system to form sort of a planetary junkyard called the Kuiper belt (Figure 9.3). Pluto is one such castoff embryo that has a spherical shape but has not cleared its orbit of other debris. Eris is another spherical Kuiper-belt object that, like Pluto, does not qualify as a planet. Pluto and Eris are examples of dwarf planets, and dwarf planets that exist beyond the orbit of Neptune are specifically called plutoids. Several dozen embryos escaped further collisions to become the moons trapped by the gravitational field of various planets. Some of the leftover debris still spins as rings around the outer planets, most notably Saturn (Figure 9.2).

In addition, a cluster of irregularly shaped, rocky planetesimals called **asteroids** form an orbiting belt between Mars and Jupiter (Figure 9.3). One object within this crowded belt, Ceres, qualifies as a dwarf planet because of its spherical shape. Some small asteroids and pieces of rock flung off by asteroid collisions are pulled into highly stretched orbits by the gravitational force of gigantic Jupiter. These oddly shaped orbits bring the asteroids and smaller fragments on potential collision courses with Earth. These objects are called **meteors** as they streak through the atmosphere and **meteorites** when they land on the surface.

Comets are mixtures of ice and rock that astronomers compare to dirty snowballs. Most comets originate beyond the Kuiper belt, but some have orbits that cross into the inner solar system where they sometimes collide with planets.

Meteorites and smaller dust particles provide critical evidence for the planetary-origin theories. Some meteorites have mineral textures, suggesting that they are pieces left over from very early stages of dust accumulation

▲ **Figure 9.6 What a meteorite-impact crater looks like.** Meteor Crater in northern Arizona formed about 50,000 years ago when a meteor crashed into Earth. The crater is 1.2 kilometers wide and 180 meters deep. The meteorite fragments are iron metal, which suggests that the meteor was a piece of the core from a shattered planetary embryo in the asteroid belt.

that never formed into larger embryos. These meteorites are artifacts that preserve the materials from which the planets formed but without the chemical and physical modifications that accompanied planetary accretion. Other meteorites seem to be chunks of embryos that were smashed to bits during collisions, and they provide pieces of planetary interiors for geologists to study.

The accretion process that built Earth is not entirely over. Small meteors and dust particles continue to add about 100,000 kilograms to Earth each day, which is negligible compared to the mass of Earth. Most of the incoming mass is fine space dust, and even the somewhat larger colliding objects are too small to notice, other than the occasional bright fireballs and meteor streaks (shooting stars) visible in the night sky.

Impact craters, such as the one illustrated in **Figure 9.6**, are slightly eroded records of large collisions in Earth's recent geologic past. Abundant evidence for one such collision points to an impact of an asteroid about 20 kilometers in diameter, which crashed into Earth 65 million years ago. This event, which happened in Mexico, caused catastrophic environmental change on Earth including extinction of the last of the dinosaurs and approximately 75 percent of other species then present on the planet. Such catastrophic collisions will happen in the future, as the orbits of some remaining asteroids cross Earth's orbit. Every few years, one of these objects passes Earth at a distance closer than the distance from Earth to the Moon. These occurrences show that planetary accretion is still occurring, although nearly all the mass in the original solar nebula now resides within the Sun and planets, and very little additional growth of these bodies is possible. Earth accreted to more or less its current volume and mass by about 4.5 billion years ago (Section 7.6 discusses of the age of Earth).

Planetary Differentiation

The collision of planetesimals and embryos generates a tremendous amount of heat. When fast-moving objects with large masses collide, their motion energy instantaneously converts to enormous heat energy. Clap your bare hands together several times with some force and then press them to your face to feel the warmth. Now, imagine the heat generated when objects the size of Moon collide with one another.

| Oil and Vinegar | Oil |
| Vinegar |

Oil and vinegar are well mixed in vigorously shaken salad dressing.

Over time, oil and vinegar separate from one another. Oil is less dense than the vinegar, so oil occupies the top of the bottle while vinegar segregates on the bottom.

▲ **Figure 9.7 Oil and vinegar are analogies for planetary differentiation.** The chemical characteristics of oil and vinegar cause them to remain separate as different fluids rather than mixing. Vinegar is denser than oil, so when a bottle of oil-and-vinegar dressing is shaken then set down, gravity pulls the vinegar to the bottom of the bottle, which forces the oil to occupy the top of the bottle. In a similar fashion, molten iron metal separated from molten silicates in planetary embryos: Gravity pulled the denser iron toward the embryo centers to form metallic cores surrounded by silicate minerals.

Collisional heating is sufficient to melt planetary embryos. When this occurs, molten metallic compounds do not mix with molten silicates. Instead, gravity pulls the denser molten metal toward the center of the embryo, leaving less dense molten silicates near the surface. **Figure 9.7** illustrates the somewhat similar process of unmixing and separation observed in a bottle of oil-and-vinegar salad dressing. The denser molten metallic compounds are analogous to the vinegar in the bottle and are drawn by gravity toward the center of Earth.

Planetary differentiation is the name applied to the planetary-scale process of separation of matter based on density. Differentiation within embryos could readily produce denser metallic cores and less-dense silicate mantles. Many meteorites collected on Earth consist only of silicate minerals or iron-nickel metal, which suggests that these meteorites are fragments of ancient differentiated planetesimals or planetary embryos.

Putting It Together—How Did Earth Form?

• Earth, as well as other solar system bodies, gradually accreted from collisions of progressively larger objects formed from dust and gases composing a nebular disk surrounding the central Sun.

• Earth formed in a part of the solar system where silicate minerals and iron-metal alloys crystallized. Gaseous compounds formed icy planets in the colder, outer solar system.

• Several moon-sized embryos collided to form Earth. The embryos may have been differentiated, with metal-rich cores and silicate mantles. One such collision probably spun off debris that formed the Moon.

9.2 How Did the Core and Mantle Form?

The major chemical and physical boundaries within Earth are between the mantle and the core and between the mantle and the crust. The contrast between the metallic core and the mostly silicate mantle is the more distinctive of these two boundaries, and the core and mantle compose almost all of Earth's mass. Therefore, it makes sense to consider first the origin of the distinct mantle and core layers.

Core Formation from a Magma Ocean

Figure 9.8 illustrates the current geologic thinking, called "the magma-ocean hypothesis," which asserts that Earth was once partly or largely molten. Within this magma ocean, dense, iron-rich molten metal separated from extensively melted silicate minerals to form Earth's core and mantle, respectively.

In a liquid state, materials easily differentiate by density as described in the previous section. High-density metallic elements settle inward to form a planetary core, while remaining low-density silicate compounds compose the enveloping mantle. This is a *physical* process and, when thought of in terms of the oil-and-vinegar analogy, is fairly intuitive. Scientists who study core formation and test the magma-ocean hypothesis, however, concentrate on the *chemical* process of differentiation, in particular the behavior of elements with chemical properties similar to iron. We can call these the iron-seeking elements, because they prefer to bond with metallic iron (Fe) rather than silicon. Do not confuse the metallic, uncharged version of iron with its common ions (Fe^{2+} and Fe^{3+}), which do readily fit into silicate mineral structures.

Analyses of a class of meteorites not modified by accretion into planetesimals provide estimates of the starting abundances of these iron-seeking elements in the solar system and are a window into the chemical composition of early Earth. In these meteorites, geochemists find an undifferentiated soup of elements with varying affinities to metallic iron or to silicate minerals. There are many iron-seeking elements and their affinity for bonding with metallic iron ranges from "strong" to "moderate" to "slight."

If the magma-ocean hypothesis is correct, then two things must have happened as Earth differentiated.

1. Under the physical influence of gravity, the heavy, dense iron sank to form the core, leaving the less dense silicates to form the mantle.

2. As the iron sank, one can reasonably expect that it took with it most of the "strongly" iron-seeking elements and some of the "moderately" iron-seeking elements and left behind some of the "moderate" and most or all of the "slightly" iron-seeking elements.

Unfortunately, we cannot test the core to see whether it is full of strongly iron-seeking elements. On the other hand, we can test the other side of the assertion—if the strongly iron-seeking elements went to the core with the iron, there should be only small amounts, if any, of them left behind in the mantle.

To test the magma-ocean hypothesis, geochemists compare Earth rocks with the unmodified meteorites. Differences between elemental abundances in mantle rocks compared to the abundances in meteorites, which are assumed to represent the original overall Earth composition, will then imply what elements moved into the core.

Figure 9.9 illustrates what the geochemists find when they compare the element abundances. The measured abundances of the iron-seeking elements in mantle peridotite form a remarkable "stair step" pattern: Slightly iron-seeking elements are most abundant in the mantle, moderately iron-seeking elements are less abundant, and strongly iron-seeking elements are present at less than 1 percent of the expected abundances found in the meteorites.

Many geochemists argue that the data plot in Figure 9.9 supports the magma-ocean hypothesis. Slightly and moderately

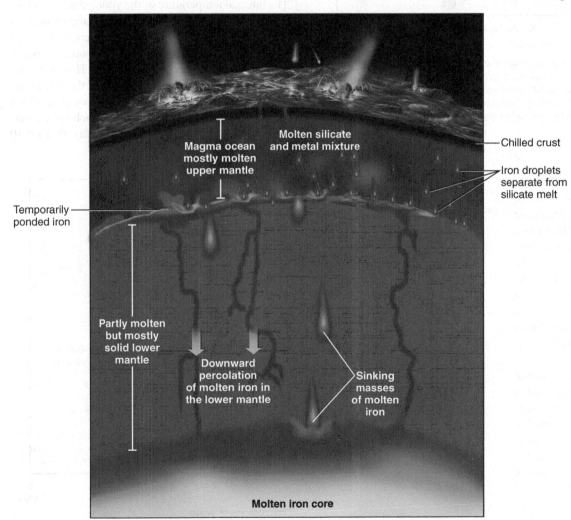

▲ **Figure 9.8 Visualizing core formation in a magma ocean.** This diagram summarizes how the magma-ocean hypothesis accounts for core formation on Earth. Dense, molten, metallic iron sinks through the silicate magma ocean in the upper mantle and temporarily accumulates at a depth of about 400 km, where the lower mantle is only partly melted. The molten-iron ponds enlarge until they are heavy enough to sink through the partially melted lower mantle to form the central core of Earth. Smaller volumes of molten iron and other metals also percolate downward through the mostly solid lower mantle.

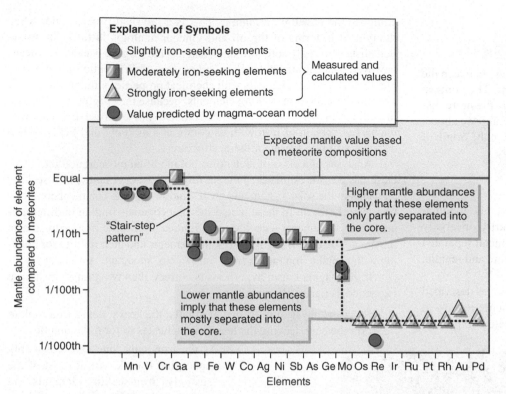

▲ **Figure 9.9 How element abundances support the magma-ocean hypothesis.** Graphed abundances of the iron-seeking elements in Earth's upper mantle exhibit a stair-step pattern. The pattern supports the hypothesis that strongly iron-seeking elements more completely separated from the mantle into the core compared to slightly iron-seeking elements. The actual mantle abundances for all these elements are close to what is predicted by the separation of molten iron and silicate compounds in a magma ocean.

important conclusion when he tried to calculate the age of Earth (Section 7.6). In addition to data supporting the magma-ocean hypothesis, geologists have other evidence to indicate that early Earth was much hotter than today.

1. The current heat flow from within Earth is approximately twice as high as can be accounted for by present-day radioactive decay in the crust and mantle. This suggests excess heat that remains from processes active in the past when Earth was hotter.

2. Ultramafic lava flows are found on Earth only among rocks more than 3 billion years old. Experiments reveal that these unique lava flows, the only known extrusive equivalents of peridotite, could only form at mantle temperatures several hundred degrees hotter than those estimated today.

So where did the heat come from to make the magma ocean that permitted separation of core and mantle? Radioactive decay and collisional heating are the likely heat sources.

The internal temperature of the young Earth was higher than today partly because of higher abundances of radioactive elements. Decay of radioactive potassium, uranium, and thorium isotopes provide the source for much of Earth's current heat energy (Section 8.5). However, over the last 4.5 billion years, large amounts of these radioactive isotopes have decayed to daughter isotopes, which means that the radioactive parents were much more abundant early in Earth history than they are today. **Figure 9.10**

iron-seeking elements only partly sank with metallic iron into the core, leaving behind residual abundances in the mantle. Elements with stronger affinity to iron metal have extremely low mantle abundances because they more efficiently separated from the mantle when the core formed. Experiments successfully reproduce the step-like pattern of elemental abundances in the mantle shown in Figure 9.9 and suggest that the metallic core components first separated from a silicate-rich magma ocean that was at least 400 kilometers deep. The dense core material sank to the center of Earth through the mushy, and partly melted, silicate lower mantle.

As Earth cooled, iron metal crystallized and settled under the force of gravity toward the initial central core, which was entirely molten. Crystallization of the core from the inside out gradually led to the distinction between the solid inner core and the liquid outer core. The inner core continues to enlarge very slowly at the expense of the outer core because of this crystallization. Cooling also led to almost complete solidification of the enveloping mantle.

Heat Sources for Magma Ocean and Core Formation

Geologists are confident that early Earth was much hotter than it is at present. Lord Kelvin also reached this

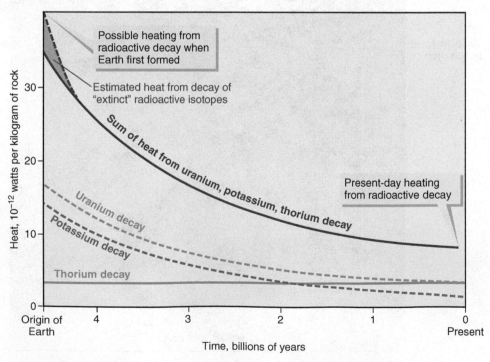

▲ **Figure 9.10 How hot was Earth in the past?** Geologists estimate the amount of heat energy released by radioactive decay of uranium, thorium, and potassium in Earth by combining measurements of the present-day abundances of the parent and daughter isotopes with calculations of their half-lives. When Earth formed, the total heat produced by decay should have been about four times greater than it is today. Many "extinct" isotopes with short half-lives have completely decayed to become their daughter isotopes but contributed heat very early in Earth history.

illustrates how geologists use the decay rates of these three isotopes to calculate that radioactive heat production in the first two billion years on Earth was four times higher than it is today.

The total heat production by radioactive decay must, however, have been even greater than that produced by decay of potassium, uranium, and thorium. Experiments produce other radioactive isotopes in the laboratory that are not found on Earth today. These isotopes are missing in nature because they have very short half-lives compared to the ancient age of Earth, so the isotopes long ago decayed completely to their daughter isotopes. The evidence of their earlier existence comes from detecting these daughter isotopes in Earth and Moon rocks and in meteorites. This means that the "extinct" radioactive isotopes were present during the early history of Earth and must have contributed additional heat as they decayed.

Heating by collision during accretion, added to the heat resulting from radioactive decay, probably was sufficient to melt the early Earth to produce a magma ocean. If we accept the giant-impactor hypothesis for the origin of the Moon, then we can assume that the impact of that collision would have provided heat energy to account for the simultaneous formation of magma oceans on both the Moon and Earth. Wholesale melting by the impact also was made possible by the already hot state of the planet resulting from radioactive decay.

Putting It Together—How Did the Core and Mantle Form?

• Earth's metallic core separated from the mantle by planetary differentiation in a global magma ocean. Very dense metallic compounds separated from lower-density silicate compounds to produce a metallic core and a silicate-mineral mantle.

• As Earth cooled, crystallization and inward settling of iron metal caused distinction of a solid inner core and a remaining molten outer core. The mantle solidified into silicate minerals with a chemical composition similar to peridotite.

• Very high temperatures within early Earth resulted from collisions with accreting embryos and higher heat production by radioactive decay.

9.3 How Does the Crust Form?

With a tested hypothesis at hand for explaining the separation of the core from the mantle, we can now move on to the question of how Earth's crust formed. There is no direct geologic information about the earliest crust, simply because none has survived the active tectonic processes that reshape the surface of the planet. Geochemical characteristics of silicate-mineral crust, which is composed mostly of igneous rock, suggest that it is formed from partial melting in the mantle. The resulting magmas are less dense than the remaining solid peridotite, so the magmas migrate upward and solidify to form the crust.

Oceanic Crust Compared to Continental Crust

Sampling in the field, experimental studies, and seismic data provide the basic information about Earth's crust (Table 8.1). Oceanic crust, of basaltic composition, covers roughly 60 percent of the Earth's surface, and the more

silica-rich continental crust forms the remainder. Oceanic crust reaches a maximum age of about 180 million years under the western Pacific Ocean, and the average age of oceanic crust is about 100 million years. Continental crust, on the other hand, has a maximum recorded age of approximately 4.4 billion years and an average age of about 2.2 billion years.

Plate tectonics explains the greater antiquity and permanence of continental crust compared to oceanic crust. Subducting basaltic oceanic crust metamorphoses in the mantle to very dense eclogite (Section 6.8). This allows the dense oceanic lithosphere to sink into the asthenosphere at subduction zones, whereas the less dense continents, consisting mostly of intermediate and felsic igneous rocks, do not subduct (you will learn more about this concept in Chapter 12). The density comparison makes continental crust more permanent and explains why continental crust is much older than oceanic crust and why the oldest history of Earth is found in continental crust.

The Origin of Oceanic Crust

It is perhaps a bit surprising that the formation of the basaltic oceanic crust, despite its relative inaccessibility beneath the oceans, is better understood than that of continental crust. Section 4.6 explained the recipe for making basalt—partial melting of mantle peridotite produces basaltic magma. The non-melted peridotite minerals remain behind to make up the mantle part of the lithosphere. Oceanic crust, therefore, simply is extracted from the mantle by partial melting and crystallization. This melting and extraction process is witnessed today at divergent plate boundaries and where mantle plumes generate thick accumulations of basalt at places such as Hawaii.

The Origin of Continental Crust

Deciphering the origin of continental crust is much more challenging. Experiments show that partial melting of the mantle produces basaltic magma such as that which makes up the oceanic crust. On the other hand, the average composition of the continental crust is more similar to andesite, and many continental igneous rocks are still more felsic. These observations imply different processes of magma generation (and hence crust formation) than are witnessed at mid-ocean ridges.

Geologists are still testing two hypotheses about how the earliest continental crust formed. One hypothesis turns to modern convergent plate boundaries for clues, because we see intermediate and felsic composition magmas erupt at volcanoes at these locations today. Mantle melting near convergent plate boundaries happens because fluids expelled by dehydration metamorphism of the subducted plate cause adjacent peridotite to melt (Section 4.6 describes this melting process). Fractional crystallization (explained in Section 4.7) of these fluid-rich partial melts causes early removal of low-silica, water-bearing minerals (such as amphibole) to produce the more silica-rich intermediate and felsic magmas that are commonly erupted near modern convergent boundaries and more closely resemble continental crust. However, the early Precambrian magma-forming processes that operated when much of the continental crust formed were probably different than those observed today. Recall from Section 4.6 that partial melting of a rock produces magma that is more felsic than the original rock. Experiments show that partial melting of the water-bearing minerals in subducting mafic crust yields dioritic and tonalitic magma rather than basalt. This process for making large volumes of felsic magma is not very important today because plate melting occurs at very few modern subduction zones. However, melting of subducted basaltic crust could have been more important in early Earth history when the mantle was much hotter. According to this hypothesis, the first melting of the mantle at divergent plate boundaries formed basaltic oceanic crust, which later melted at convergent plate boundaries to form felsic magma.

The alternative hypothesis suggests that while recently formed intermediate and felsic crust owes its origin to subduction-zone processes, the oldest continental crust possibly originated before the distribution of heat within Earth would even permit plate tectonics to work. In this view, the early formed basaltic crust resulted from partial melting of mantle peridotite at mantle plumes. Once this initial oceanic crust formed, it was subjected to further partial melting when later plume magma heated the crust from below. This melting of the basaltic crust, by rising plumes rather than near convergent plate boundaries, could then generate the more silica-rich igneous rocks that are most abundant in continental crust.

Regardless of which hypothesis, or perhaps another option, is eventually best supported by ongoing research, we can make some statements with certainty. First, the formation of continental crust requires a more complicated recipe than oceanic crust formation, because the oceanic crust is partially melted or metamorphosed in order to generate the intermediate and felsic igneous rocks that dominate continental crust. This means that oceanic crust forms first and then is used as starting material to create continental crust. Second, both oceanic and continental crust, whether formed from a one-step or multiple-step recipe, ultimately were derived from the mantle by igneous processes.

The third key statement about the formation process is that the generation of the commonly seen granite in continental crust requires an additional melting step and the presence of water. Both hypotheses described above form dioritic and tonalitic rocks but not much granite during the early steps in the crust-forming recipe. In the required next step, these early formed igneous rocks partially melt when heated by still more rising magma. Experimental data indicate that the partial melting of diorite or tonalite that contains water-bearing minerals produces most of the granitic magma in the continental crust.

A tiny, 4.4-billion-year-old sand grain composed mostly of zircon is the oldest known geologic record of crust (Section 7.6). The zircon crystal encloses tiny quartz and feldspar crystals. This mineral association suggests that the zircon crystallized from felsic magma. Other details of the mineral chemistry suggest, but do not prove, the presence of dissolved water in the magma. From this evidence, some geologists suspect that less than 150 million years after Earth formed, rock resembling the continental crust seen today was forming and water probably was present. The next problem to contemplate is how water originated on Earth.

Putting It Together—How Does the Crust Form?

• Basaltic oceanic crust is the direct result of the partial melting of peridotite in the mantle.

• Intermediate and felsic continental crust cannot form by simple mantle melting. Partial melting of basaltic oceanic crust in hot early Earth may have generated magmas with continental-crust composition.

• All processes for producing felsic magma require melting in the presence of water. Water was present in minerals melted to form granitic magma 4.4 billion years ago.

9.4 How Did the Hydrosphere and Atmosphere Form?

According to the planet-origin theories, water and gaseous compounds had not yet condensed in the hotter inner solar system when solar winds from the protosun cleared the light elements out of that region of the system. The cold temperatures necessary for gaseous compounds such as water vapor to condense and to form ices existed only in the vicinity of the outer planets, far distant from Earth, prior to this violent phase of solar evolution. Thus Earth (along with Mercury, Venus, and Mars) formed in a part of the solar system where water and gaseous compounds did not exist (Figure 9.5).

Two contrasting hypotheses attempt to explain the paradox of water existing on Earth even though the planet formed in a part of the solar system where water probably did not exist. Both hypotheses have merit and strong supporters, so it is appropriate to examine both.

Later Delivery of Water to an Initially Dry Earth

One hypothesis for the origin of Earth's water assumes that most of Earth formed without water and that the planet acquired water late in the planetary accretion process. Planetesimals growing in the orbits of the inner rocky planets accreted in a "dry" state because they were too close to the hot early Sun for water molecules to exist. This hypothesis requires later delivery of water to Earth from comets and meteors that originated farther out in the cooler, gaseous part of the solar system. Earth-based and spacecraft analyses indicate that comets, such as those seen in **Figure 9.11**, are composed mainly of water ice. Some varieties of meteorites also have water-bearing minerals. The first hypothesis states that these types of water-bearing objects accreted to Earth during the very latest stages of planet formation, mixed water into the upper mantle, and produced the water-bearing minerals found in Earth rocks today.

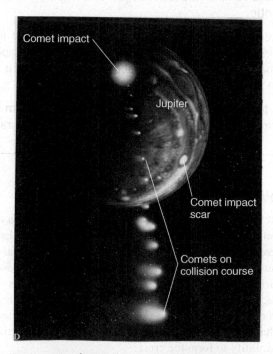

▲ **Figure 9.11 Watery comets impact Jupiter.** This painting, based on actual telescope images, shows fragments of Comet P/Shoemaker-Levy 9 slamming into Jupiter. This first-ever observation of two colliding solar system bodies occurred in July 1994 when 20 comet fragments, as large as 2 kilometers across, collided with Jupiter. The icy comets delivered at least 2 million metric tons of water to Jupiter. Similar impacts of icy comets with early Earth are one hypothesized source of water for our planet.

The heavily cratered surface of the Moon (Figure 9.1a) attests to a large number of encounters with stray comets and asteroids. Studies of Moon rocks indicate that most of these impacts occurred between 3.8 and 4.4 billion years ago. Given the close proximity of Earth and Moon, and given Earth's larger size, it seems likely that millions of planetesimals left over from planetary accretion likewise pummeled Earth at this time. These objects may have included icy comets and meteorites with water-bearing minerals, either of which could deliver water to Earth.

Planetary Embryos Form an Initially Wet Earth

In a more recently proposed and evolving hypothesis, researchers suggest that planetary embryos rich in water-ice or water-bearing minerals accreted to form the planet—that is, the planet initially accreted "wet," rather than having water added near the end of planet formation.

According to this second hypothesis, some planetesimals and embryos that contributed to Earth's growth did not originate in the nearby water-absent, orbital feeding zone for the growing planet. Instead, some of these accreting objects condensed farther from the Sun, near the present-day asteroid belt, where we know that water-bearing minerals formed (Figure 9.5) because we find them in meteorites that have reached Earth. According to this second hypothesis, erratic orbits influenced by the proximity of giant Jupiter sent one or more these water-bearing embryos on a collision course with Earth. Late-arriving comets are of minor significance in this hypothesis, perhaps providing only 10 percent of Earth's total water.

What is the evidence for accretion of a wet planet? For one thing, the predicted abundances of iron-seeking elements depicted in Figure 9.9 are based on the assumption that water existed in the mantle. The predicted abundances are so close to actual measured elemental abundances that they imply the presence of a hydrous magma ocean during core formation or, in other words, when Earth initially formed. Supporters of the early water hypothesis also point to the Moon as evidence that Earth's water did not arrive as late as 3.8–4.4 billion years ago. If Earth received most of its water late in accretion from watery bodies that impacted it, then the Moon should have also received a supply of water at or around that time. Moon rocks, however, do not have water-bearing minerals. In addition, the 4.4-billion-year-old Australian zircon eroded from granite indicates the presence of water before the late heavy bombardment had hardly gotten underway.

Insights on Water Delivery from Comets and Meteorites

One way to test whether comets and meteorites originating far from Earth could deliver water to Earth is to look more carefully at chemical data. To do this, geochemists commonly compare the abundances of two non-radioactive isotopes of hydrogen that are found in water molecules on Earth, as well as in other planetary objects. Most hydrogen atoms have no neutrons in the nucleus, and this common form of hydrogen is almost 10,000 times more abundant on Earth than deuterium, a hydrogen isotope that does contain one neutron. If the ratio of deuterium and normal hydrogen in water molecules on Earth matches that measured in comets, this would support the hypothesis that comets delivered Earth's water after planetary accretion was nearly completed.

Figure 9.12 compares the ratios of deuterium to common hydrogen from comets and meteorites to that of water measured on Earth. The existing data, which include measurements from only three comets, do not support the hypothesis of comets delivering much water to Earth because comets

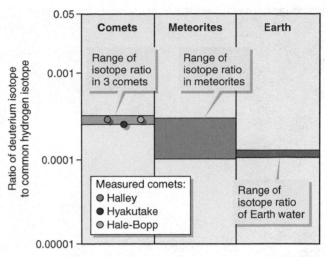

▲ **Figure 9.12 Where did Earth's water come from?** This chart compares the deuterium/hydrogen isotope ratios from three comets, various meteorites, and Earth. The deuterium/hydrogen ratio of Earth does not resemble that of measured comets so comets may not be the primary source of water on Earth. Some meteorites have a similar deuterium/hydrogen ratio to water on Earth, but meteorites contain less water than comets.

have a different ratio of the hydrogen isotopes from Earth water. Some meteorites, however, do have similar deuterium/hydrogen ratios as Earth water.

Taken together, the data graphed in Figure 9.12 suggest that Earth's water mostly arrived during the original accretion process within embryos that had compositions similar to those represented by the analyzed meteorites. In this case, the later heavy bombardment of objects whose scars remain visible on the Moon delivered very little, if any, water. However, supporters of the late-delivery hypothesis point out that data from only three comets may not be sufficient to represent most potential comet contributions of water to early Earth. As a result, both hypotheses remain possible. Intensive research continues to investigate when and how Earth obtained the water for its hydrosphere.

Making the Early Atmosphere

Regardless of when water arrived on Earth, it is thought that it initially mixed into the molten mantle and then incorporated as water molecules into crystallizing minerals, rather than accumulating as liquid on the surface. Experiments show that water immediately releases into magma at the onset of partial melting of rocks containing these water-bearing minerals. Therefore, both hypotheses for the origin of Earth's water propose that magma formation exported steam and other gases from the mantle through volcanoes to Earth's surface. **Figure 9.13** portrays the volcanically active, hot early Earth spewing out gases through innumerable volcanoes. These gases accumulated as the first atmosphere, and steam condensed to water and formed the hydrosphere.

Gases from currently active volcanoes are potential examples of the gases that formed the early atmosphere. **Figure 9.14** compares the composition of the modern atmosphere to the composition of volcanic gas. Water vapor is the most important constituent in volcanic gas. This observation is consistent with the hypothesis that water degassed from the mantle to form the atmosphere and hydrosphere.

The analyzed abundances compared in Figure 9.14 are very different. Perhaps the initial atmosphere composition more closely resembled the composition of the volcanic gases, but the abundances of compounds in the

▲ **Figure 9.13 Artistic rendition of early Earth.** Scientific illustrators use geological data and hypotheses to portray the early Earth surface. The initial atmosphere consisted of gases emitted from volcanoes that must have been much more numerous on the hot early Earth than at present. Meteors and comets plunged through the atmosphere to the surface.

atmosphere have changed through time. Weathering reactions between atmospheric gases and surface minerals, as well as biological processes after life first appeared on the planet, are called upon to explain these changes.

Most importantly, Figure 9.14 shows that oxygen gas (O_2), an obviously critical ingredient for most life on present-day Earth, does not erupt from volcanoes. Abundant geological evidence also exists for a lack of oxygen in the early atmosphere. This evidence supports the hypothesis that

volcano degassing produced the atmosphere. For example, the vast iron-ore resources of the Great Lakes region, South America, and Australia indicate the lack of atmospheric oxygen during early Earth history. These iron ores are peculiar chemical sedimentary rocks that cannot form from water that contains oxygen (as the modern oceans do). Almost all of these iron-rich sedimentary rocks formed prior to 2.3 billion years ago, indicating low oxygen levels on the early Earth. Oxygen most likely accumulated in the atmosphere very gradually after the appearance of photosynthetic organisms, which consume carbon dioxide CO_2 while manufacturing oxygen.

Putting It Together—How Did the Hydrosphere and Atmosphere Form?

• Two contrasting hypotheses explain the origin of Earth's atmosphere and hydrosphere. One hypothesis suggests delivery of water-rich materials from meteorites and comets that accreted to the initially dry Earth very late in the formation of the planet. The other hypothesis suggests that Earth accreted wet, from planetesimals or embryos already containing water-bearing minerals.

• Water vapor is the most abundant gas erupted from volcanoes. This observation is consistent with the hypothesis that the hydrosphere derived from melting water-bearing mantle minerals and erupting water vapor through volcanoes.

• Atmosphere composition changed through Earth history as a result of weathering reactions and the presence of photosynthetic organisms, which consume carbon dioxide and create oxygen.

▶ **Figure 9.14 Gas content from a volcano compared to the modern atmosphere.** These charts compare the gas content of an active volcano in Russia with the composition of the atmosphere. The most abundant gas erupted from volcanoes is water vapor. Oxygen is an important atmospheric gas not erupted by volcanoes.

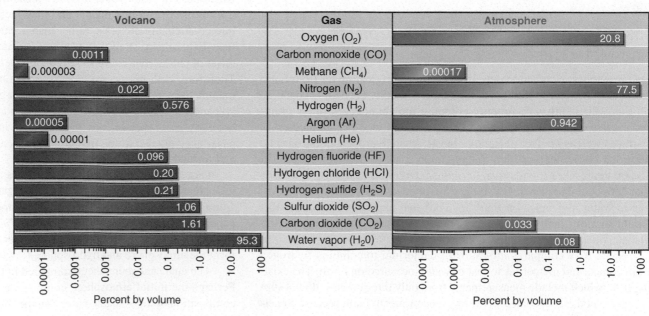

Volcano	Gas	Atmosphere
	Oxygen (O_2)	20.8
0.0011	Carbon monoxide (CO)	
0.000003	Methane (CH_4)	0.00017
0.022	Nitrogen (N_2)	77.5
0.576	Hydrogen (H_2)	
0.00005	Argon (Ar)	0.942
0.00001	Helium (He)	
0.096	Hydrogen fluoride (HF)	
0.20	Hydrogen chloride (HCl)	
0.21	Hydrogen sulfide (H_2S)	
1.06	Sulfur dioxide (SO_2)	
1.61	Carbon dioxide (CO_2)	0.033
95.3	Water vapor (H_2O)	0.08

Percent by volume

Percent by volume

9.5 How Do We Know . . . the Hydrosphere Came from the Geosphere?

Define the Problem

Did Early Earth Rocks Contain Oceans of Water? Does it seem hard to accept the hypothesis that all of the water you see on Earth, in addition to the water vapor in the atmosphere and ground water below your feet, originally escaped from the interior as volcanic vapor?

Unfortunately, geologists cannot gain much insight to this question by analyzing water contents of mantle peridotite or volcanic gases today. The reason this analysis is not fruitful is that water continually cycles between Earth's interior and exterior. Water exits the interior through volcanoes, but subduction of crust and sediment containing water-bearing minerals returns water to the mantle at convergent plate boundaries. Dehydration metamorphic reactions release the water from the subducted plate into the mantle. Some of this water resides in the mantle for a long time before joining partial melts of peridotite and returning to the surface through volcanoes. These observations mean that although water vapor is the most abundant gas emanating from modern erupting volcanoes (Figure 9.14), most of this water is recycled surface water, so it does not explain the origin of the first surface water.

State the Hypothesis

Could the Hydrosphere Originate from the Geosphere? Imagine a time near the beginning of Earth's history before volcano degassing substantially exported mantle water to the surface. Geologists hypothesize that in that primeval time nearly all of the water supply of the planet was in the geosphere, whereas today nearly all of it exists in the hydrosphere and atmosphere. Geologic processes, according to this hypothesis, redistributed the original water reservoir so that most of the water now exists in the atmosphere and hydrosphere rather than within rocks. Because most of the primordial water has left the mantle, how can geologists test the hypothesis that all of the water started out in the mantle?

A concise statement of the hypothesis is a good starting point for designing the test: The original mantle contained enough water to generate the modern hydrosphere by gas release through volcanoes.

Gather the Data

How Much Water Is on Earth, and How Much Was in the Original Mantle? Testing the stated hypothesis requires that we make some simple calculations to quantify how much water is enough water. To determine if the volume of the hydrosphere can come from the mantle, you need to estimate, and then compare, two values:

1. How much water is *currently* on Earth?
2. How much water was in the *original* mantle?

Figure 9.15 shows the estimated volumes of the major water reservoirs on Earth. The largest volume of water is in the ocean, and

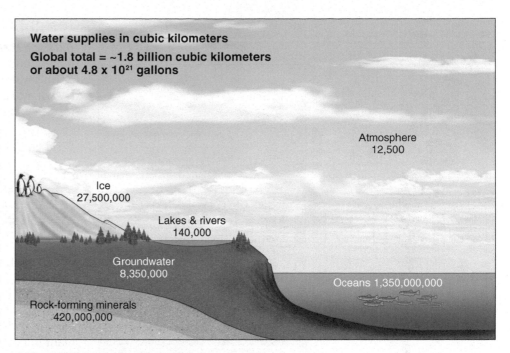

Water supplies in cubic kilometers

Global total = ~1.8 billion cubic kilometers or about 4.8×10^{21} gallons

Atmosphere 12,500

Ice 27,500,000

Lakes & rivers 140,000

Groundwater 8,350,000

Oceans 1,350,000,000

Rock-forming minerals 420,000,000

▲ Figure 9.15 Reservoirs of water on Earth.

this is a reliable value because the depth and area of Earth's oceans are well known. The sizes of most of the other reservoirs are almost as well known, and because they clearly are much smaller than the ocean volume, any uncertainties in these values are less important for the calculations. The amount of water retained by minerals in crust and mantle rocks is the least well-known value and is fairly large. Geochemical arguments suggest that the value shown in Figure 9.15 is reasonable and that the water in rocks cannot be more than twice the given value. The total estimated volume of the water on Earth is 1.8 billion cubic kilometers. This is a huge volume, equivalent to 4.8×10^{21} gallons. To put this number in perspective, think about the fact that if everyone currently living on Earth consumed 100 gallons of water each day, it would take 22 million years to use all this water.

How do geologists determine the amount of water in the original mantle? It is not helpful to look at the composition of *modern* mantle rocks for this value, because the hypothesis assumes that the mantle has already lost much of its water to create the present hydrosphere. You need, instead, to examine the water content of rock that might represent the *original* mantle. Geologists commonly turn to certain classes of meteorites to play this role. Because some meteorites are parts of small asteroids that never accreted in embryos, they are thought to represent the pristine starting materials for planet building. The measured water content of these meteorites, enclosed in mineral crystal structures, is highly variable and ranges from 0.5 percent to 9.0 percent, by mass.

Test the Hypothesis

Do the Numbers Add Up? **Figure 9.16** illustrates the calculations required to test the hypothesis that the original mantle contained enough water to produce the present-day hydrosphere. The calculated volume of water in the early mantle is 10 times larger than the volume currently

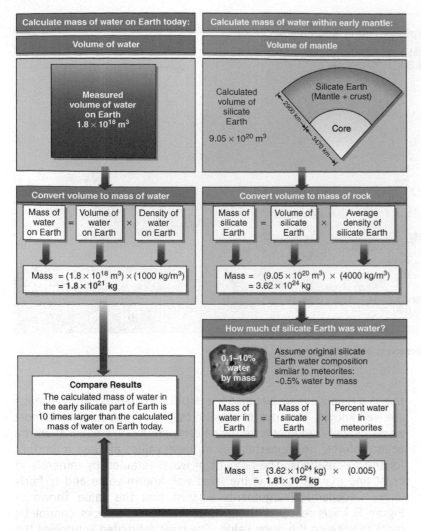

▲ **Figure 9.16 Calculations to test the hypothesis: The hydrosphere came from the geosphere.** This diagram illustrates the calculations that test the hypothesis that the original mantle, with a composition similar to meteorites, contained enough water to account for the water on and within Earth today. The masses of water present on Earth today and in the early silicate Earth are calculated separately and then compared.

present on, below, and above Earth's surface. The result, therefore, successfully tests the hypothesis so long as the assumption that certain meteorites resemble the composition of the early mantle is valid. The validity of that assumption is difficult to assess. This is why it is considered an assumption and not a separate hypothesis. Nonetheless, the calculation actually provides a relatively small number, rather than a suspiciously large value, for two reasons:

1. The calculations assume a conservative estimate of 0.5 percent water by mass from meteorites. Although some meteorites contain as much as 9 percent water, these types are not as abundant in meteorite collections as those with less water, so the calculation uses a smaller percentage.

2. The calculations ignore any water delivered from comets, which are as much as 80 percent water. If the calculations included comet water, then the volume of original water in the mantle could be even larger. Comets were left out of the calculation because there currently is no compelling evidence that comets were a major source of Earth's water (Figure 9.12), but they still may have been a minor player.

Insights

Did Earth Lose Some of Its Water? Assumptions, measurements, estimates, and calculations compose the test for the hypothesis that Earth's water originated in the geosphere. Some values used are well known, others poorly known, and some are fairly speculative, but the result is reasonable for two reasons:

1. The largest water reservoir on Earth, the oceans, is the best known, and uncertainties in the other reservoir volumes will not significantly affect the calculated volume of Earth's water.

2. Although the estimated volume of water in the early mantle rests on assumptions, it is also probably a minimum value, and it is still ten times greater than the amount required to successfully test the hypothesis.

Indeed, there may be a problem of missing water. The assumptions and calculations in this section permit the interpretation that Earth only has 10 percent, or less, of the water that it started with. Geologists think that it is very likely that Earth lost lots of water early in its history. Recall that Earth probably experienced countless giant impacts between 3.8 and 4.4 billion years ago when the Moon was heavily bombarded. The largest impacts may have vaporized the first-formed oceans and blown the early atmosphere into space. These losses of the early hydrosphere and atmosphere likely occurred several times.

Putting It Together—How Do We Know . . . the Hydrosphere Came from the Geosphere?

• Calculations of the amount of water available in the early mantle, using conservative estimates of water in meteorites, suggest that early Earth contained many more times the water than is currently found on or near the surface.

Where Are You and Where Are You Going?

A complete view of Earth's interior is emerging from the information supplied in this chapter and the previous chapters. Chapter 8 provided you with a fundamental understanding of how geoscientists determined the layered structure of the interior. This chapter showed how scientists derived hypotheses for the formation of the layers by differentiation, melting, and crystallization, and for the appearance of the hydrosphere and atmosphere. These hypotheses draw on observations from other solar system bodies, such as Earth's Moon and meteorites, in addition to data collected on Earth. The hypotheses also build on the framework of the planetary-origin theories that explain the origin of the diverse planets and minor bodies in our solar system.

Core formation resulted from planetary differentiation where dense metallic iron separated from the silicate mantle. Recent hypotheses suggest that core formation coincided with planetary formation and probably resulted from the sinking of dense metallic elements within a global magma ocean that was at least several hundred kilometers deep. The heat necessary for melting the planet to form the magma ocean was provided by giant impacts, as well as from heat produced by radioactive-isotope decay, which was much greater within the early Earth than it is at present. The giant-impactor hypothesis proposes that Earth collided with a large planetary embryo late in its formation history to produce a vapor and dust ring from which the Moon accreted in orbit around Earth.

Melting in the mantle produced less dense magmas that rose to the surface and solidified to form continental and oceanic crusts. Basaltic oceanic crust forms by partial melting of mantle peridotite. This process happens today, as well, and is observed at divergent plate boundaries and where plumes rise to the surface. The formation of more felsic continental crust requires multiple melting steps in the presence of water. Melting processes unique to the hot early Earth favored formation of large volumes of dioritic and tonalitic crust and help to account for the great antiquity of continental crust.

Gases escape from Earth's interior through volcanoes to provide components of the atmosphere and hydrosphere. This happens because partial melts of the mantle produce water-rich magmas; the water vapor escapes to the atmosphere through erupting volcanoes and then condenses to liquid water. With the exception of oxygen, volcanic gases can produce the components of Earth's atmosphere if reactions among the gases, reactions between gases and rocks, and biologic processes that consume and create atmospheric gases are taken into account. The advent of photosynthetic organisms, not geologic processes, best explains atmospheric oxygen content. Calculations successfully test the hypothesis that Earth's early mantle minerals contained more than enough water to account for Earth's current hydrosphere.

Two hypotheses exist to explain how water got into the mantle during planetary accretion. Long-held interpretations favor a late-stage delivery of water by comets and meteorites that formed much farther from the Sun than did Earth. New data support the contrasting hypothesis that water-rich planetesimals and embryos originally accreted to form the planet. Water is critical not only to the formation of the hydrosphere, but also as an essential ingredient in the recipes for making continental crust. Without water there would be no continents on Earth.

Understanding the internal makeup of the planet provides the knowledge you need to continue your quest for understanding how processes in the interior of the planet drive the external processes observed at the surface. The next step, then, is to understand the active processes within the mantle and core rather than viewing the mantle and core as static layers within the planet.

Extension Modules

Extension Module 9.1: A Geologic Tour of the Solar System. Learn the geologic basics of the planets, moons, asteroids, and comets of the solar system.

Extension Module 9.2: The Origin of the Moon. Learn about the different hypotheses for the origin of the Moon.

Confirm Your Knowledge

1. With respect to water, why is Earth unique compared to the other planets in our solar system?
2. List the main differences between the inner and outer planets.
3. Explain why the inner region of our solar system lost its lightweight gas molecules during planetary accretion.
4. How can we estimate the abundances of chemical elements in the early solar system?
5. Has the innermost part of Earth's core always been solid and constant in size?
6. What is the process of planetary differentiation? How did this process determine the composition of Earth?
7. What do meteorites tell us about Earth's origin?
8. What was the source(s) of the heat that made the early Earth much hotter than it is today?
9. What evidence tells us that early Earth contained radioactive isotopes that are no longer present?
10. Why is the oldest continental crust (4.4 billion years) so much older than the oldest oceanic crust (180 million years)?
11. Summarize the process of formation of oceanic crust.
12. How was the first continental crust formed?
13. What evidence supports the hypothesis that Earth's early atmosphere (before 2.3 billion years ago) did not contain oxygen?

Confirm Your Understanding

1. Write an answer for each question in the section headings.
2. Water is crucial to many geologic processes. Using the rock cycle as a guide, describe which geologic process would not happen if Earth were a waterless planet.
3. What is the evidence that the outer planets started to form before the inner planets?
4. Write a short paragraph in which you use the analogies of a bottle of salad dressing and hand clapping to explain planetary differentiation to a friend or family member who is unfamiliar with the concept.
5. How is the composition of Earth's present day mantle used to support the magma-ocean hypothesis?
6. What evidence from volcanic rocks indicates that Earth was much hotter in the past?
7. What evidence helps us determine when Earth received its water?
8. Compare and contrast the compositions of volcanic and atmospheric gases.

CHAPTER 10

Motion Inside Earth

Why Study Earth's Internal Motion?

One of the reasons we studied Earth's interior in Chapter 8 was to gain a better understanding of the internal forces that cause earthquakes and volcanoes. To complete this goal you need to know more about the internal workings of the planet.

This chapter focuses on the dynamic processes taking place throughout the Earth's concentric layers. Rather than viewing the interior as static, here you will explore the idea that Earth's mantle is in constant, albeit slow, motion. This concept may seem too incredible to believe—the solid rock of the mantle actually flows with a motion that is more familiar for liquid. Scientific advances commonly happen only because scientists are open-minded to follow where observations lead them without being restricted by preconceptions. You will find out how heat and gravity drive motion in the mantle and will begin to relate this motion to plate tectonics. You will also learn about the vigorous currents in the liquid outer core that generate the planet's magnetic field. Many Earth processes relate to its magnetic field; it permits a compass to work for navigation, protects Earth from harmful cosmic radiation, and even allows migrating animals to find their way. Motion inside Earth is fundamental to understanding processes at Earth's surface.

After Completing This Chapter, You Will Be Able to

- Explain the processes that cause motion within Earth.
- Describe where and at what speed the motion occurs.
- Explain how motion in the outer core generates the magnetic field.
- Distinguish heat transfer by conduction and convection and explain the factors controlling each process.

Pathway to Learning

10.1 How Does Convection Work?

10.2 What Does Mantle Convection Look Like?

EXTENSION MODULE 10.1
Is Mantle Convection Physically Possible?

Molten lava rises as fountains to the surface of a lava lake in Africa and then cools to a dark crust that sinks back to the depths along the hot, glowing cracks. Convection stirs this lava lake and similarly stirs Earth's interior.

10.4 How Do We Know . . . Earth's Core Is a Dynamo?

10.3 How Does Outer-Core Convection Generate the Magnetic Field?

IN THE LAB

A geophysicist visits your lab class and demonstrates a lava lamp, a popular mood setter in college dorm rooms in the 1960s, 1970s, and even today. Shortly after the lamp is turned on, blobs of orange liquid move up to the top, pause, and sink back down, as **Figure 10.1** shows. What causes this motion?

One thing you can infer right away is that heat must be essential to the movement of the orange liquid, because no motion occurs until heat is supplied at the base of the lamp. You may also guess right away that density differences between the orange and blue liquids could cause the motion, just as ice cubes rise through denser water, but you wonder how density can explain both up and down motion. Perhaps the density difference between the liquids varies with temperature. This suggestion makes sense—you recall that most compounds expand when heated, which means that the density of a substance is greater when it is cold and less when it is hot. When asked, the geophysicist confirms your hunch that heat and density are key factors to understanding motion in the lava lamp. The orange liquid is denser than the blue liquid at room temperature, but it expands more than the blue liquid when both are heated. This means that the orange liquid is less dense than the blue liquid when it heats up. As a result, the warmed orange liquid rises to the top of the lamp. When the orange blob reaches the top, it cools, becomes denser than the blue liquid again, and sinks to the bottom (Figure 10.1).

The motion of liquids in the lava lamp not only represents movement of mass, but also movement of heat. Heat moves with the orange liquid from the hot light at the bottom of the lamp up to the top, where it dissipates into the surrounding cooler environment. You may recall that motion that simultaneously moves matter and heat is convection, a process introduced in Section 1.6.

Watching the lava lamp is interesting, but what does this have to do with geology? The geophysicist indicates that the lava lamp is a simplistic analog for motion inside Earth. It is this internal motion that causes volcanoes to erupt, earthquakes to happen, mountains to rise, basins to sink, and a magnetic compass to function. Motion inside Earth is, therefore, fundamental to understanding geology at the surface.

Chapters 8 and 9 presented arguments stating that most of Earth's interior is solid and that density increases downward within Earth (Figure 8.20), so how is convective motion like that seen in the lava lamp possible inside the planet? To appreciate how convection occurs within Earth, you need to understand more about how heat and gravity cause convection.

▶ Figure 10.1 **Convection explains how a lava lamp works.**

ACTIVE ART

Convection in a Lava Lamp. *See how convection works inside a lava lamp.*

10.1 How Does Convection Work?

You can visualize how heat moves by a quick analysis of an everyday kitchen experience. **Figure 10.2** shows an all-metal pot filled with soup on a stove. The pot and its contents start out at room temperature (Figure 10.2a). Shortly after the stove is turned on, the end of the pot handle is too hot to touch without a hot pad, and the air temperature around the pot is slightly warmer than before. These observations show that heat conducts from the stove burner through the pot and out to the end of the handle, and then radiates into the surrounding air (Figure 10.2b). Conduction and radiation transport heat without moving matter.

After a few more minutes of heating, the soup begins to move. Soup wells up to the surface at the edge of the pot and then seems to sink near

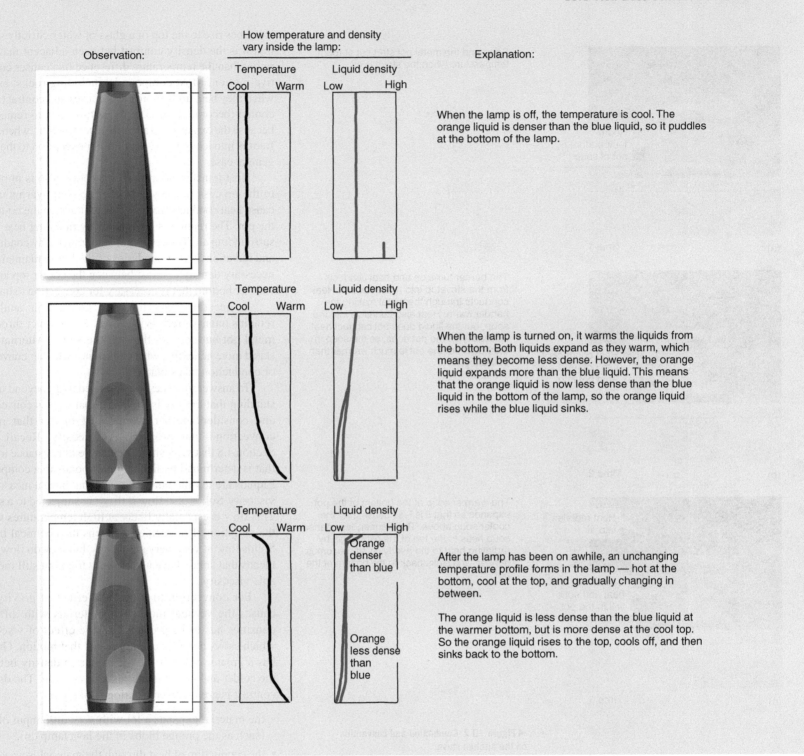

Observation:

How temperature and density vary inside the lamp:

Explanation:

Temperature — Cool / Warm   Liquid density — Low / High

When the lamp is off, the temperature is cool. The orange liquid is denser than the blue liquid, so it puddles at the bottom of the lamp.

Temperature — Cool / Warm   Liquid density — Low / High

When the lamp is turned on, it warms the liquids from the bottom. Both liquids expand as they warm, which means they become less dense. However, the orange liquid expands more than the blue liquid. This means that the orange liquid is now less dense than the blue liquid in the bottom of the lamp, so the orange liquid rises while the blue liquid sinks.

Temperature — Cool / Warm   Liquid density — Low / High

Orange denser than blue

Orange less dense than blue

After the lamp has been on awhile, an unchanging temperature profile forms in the lamp — hot at the bottom, cool at the top, and gradually changing in between.

The orange liquid is less dense than the blue liquid at the warmer bottom, but is more dense at the cool top. So the orange liquid rises to the top, cools off, and then sinks back to the bottom.

the center. This movement is convection—hot, expanded, less dense soup moves from the bottom of the pot to the top, while an equal volume of cooler, denser soup moves downward (Figure 10.2c).

Your first thought might be that this makes sense for the kitchen but not for Earth. In the kitchen, you see heat transfer by conduction in solids and by convection in liquids. Earth is mostly solid, so how is convection possible?

What Causes Convection Rather Than Conduction?

If you need a refresher on the concepts of conduction, radiation, and convection, then review Section 1.6 before proceeding. Conduction is the transfer *only* of heat from hot regions to cold regions. The rate of heat conduction varies for different materials because heat transfer relates to

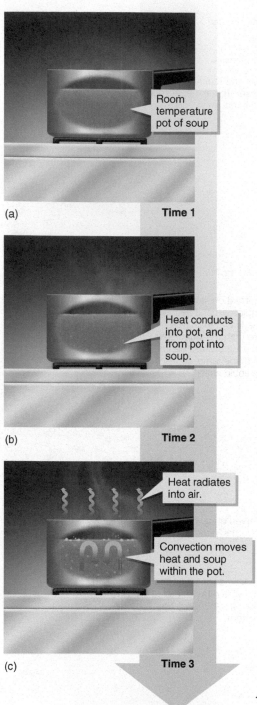

Soup and the metal pot start out at room temperature when the stove is off.

Room temperature pot of soup

(a)   **Time 1**

The burner turns on and heat conducts from the stovetop into the metal pot. Heat conducts through the metal making the handle warm. Heat also conducts into the soup, but the liquid does not conduct heat as readily as the pot does, so the soup in the bottom of the pot is much warmer than at the top.

Heat conducts into pot, and from pot into soup.

(b)   **Time 2**

The warmer soup at the bottom of the pot expands so that it is less dense than the cooler soup above. The warmer, less dense soup rises to the top of the pan, cools by radiating heat to the overlying air, becomes denser, and sinks back to the bottom of the pot.

Heat radiates into air.

Convection moves heat and soup within the pot.

(c)   **Time 3**

◄ **Figure 10.2 Conduction and convection on the kitchen stove.**

properties of atoms and mineral crystal structures. Convection is the simultaneous transfer of matter *and* heat by movement of material. Convection is commonly described as a process in which warm material, such as air or water, rises, while cold material sinks. While temperature is certainly part of convection, it is gravity rather than heat that directly causes convection.

To understand the role of gravity in convection, it is useful to consider the analogy of two objects of unequal density—such as water and ice cubes. The force of gravity exerted by the mass of Earth pulls downward more strongly on the denser water than on the less dense ice cubes. This is why

ice cubes rise to the top of a glass of water. Strictly speaking, it is the density contrast between adjacent materials rather than the temperature difference that causes convective motion. Most compounds, including rocks, expand when they heat up (become less dense) and contract when cooled (become denser). This is important to remember, because the expansion, rather than contraction, when water freezes into ice is a rare, but familiar exception to the more general case.

Heat transfer other than convection also is important to the process in the soup pot. The soup warms up because heat conducts from the burner through the bottom of the pot. The rising hot soup cools by radiating heat to the surrounding air. The continual heat transfer by conduction and radiation at the boundaries of the liquid maintains the necessary density contrast between the cooler top and the warmer bottom that is necessary for convection to happen.

A question regarding the convection analogies remains unanswered: Why does heat conduct through a metal pot but convect through the soup? Alternatively, stated more generally, what determines whether convection or conduction takes place?

To answer this question you need to go beyond understanding that gravity is the force that causes convection, and consider another property of matter that resists convection—that property is viscosity. Recall from Section 4.8 that viscosity is a measure of resistance to flow that is determined by the nature of bonds in a compound. Liquids are not the only materials that have a measurable viscosity. Solids also flow if they are subjected to a stress. This flow is especially likely at high temperatures where the vibration of molecules weakens the chemical bonds. Solids flow at very, very slow rates, but they do flow. This means that solids have a very, very high but still measurable viscosity.

For convection to occur, the effect of gravity that causes the vertical motion of materials with different densities has to be greater than the effect of viscosity, which slows down or even prevents that motion. Gravity has a greater effect if the difference in density between the colder and hotter materials is very large. The density contrast is great and convection is likely if

- the material expands a lot with very little input of heat (such as the orange blobs in the lava lamp do).
- the conduction of heat through the material is very slow. If heat conducts rapidly, then the whole object expands, and there is not enough density difference between top and bottom to drive convection. In such a case, heat transfers entirely by conduction.

We can now complete our understanding of why heat conducts through the pot and convects in the soup. Watery soup is a very slow conductor of heat, but water expands considerably when heated and has low viscosity; this combination of properties allows the soup to convect. In contrast, metals are excellent conductors of heat, but they do not expand much when heated, and they have high viscosity; these properties explain why the soup pot conducts heat instead of convecting.

How Solids Can Convect

Now you know what causes convection rather than conduction of heat, but perhaps you still have trouble picturing how convection occurs in a highly viscous solid, such as rock. Most silicate rocks, which dominate in the crust and mantle, exhibit very low heat conductivity and significant expansion when heated. These characteristics favor the transfer of heat by convection *if* the large viscosity force is overcome. The key factors that determine whether convection will occur in a solid are

- the masses of material with different densities that gravity tries to move; movement of large masses favor convection.
- the time permitted for gravity-driven motion to occur; long periods favor convection.
- the viscosity of the solid, which typically decreases with heating; low viscosity favors convection.

Another kitchen analogy depicted in **Figure 10.3** demonstrates the effects of these three factors. Picture three sticks of butter: one in the freezer, one at room temperature, and one heated on a sunny windowsill, which becomes soft but not melted. Imagine placing steel balls of equal mass on top of each butter stick. The density of the steel ball is five times greater than that of butter. Gravity, therefore, pulls down on the ball with more force than it pulls on the butter. If you considered the force of gravity alone, you

might predict that the steel ball would move down into the butter. Nonetheless, the ball on the frozen butter just sits there—no movement. Initially, the ball on the room-temperature butter stick does not seem to move either, but the next day you find a small dimple under the ball—the ball is slowly moving downward. The ball on the warmed butter by the windowsill immediately settles slightly downward and has sunk even farther by the next day. The gravity force, which relates to the mass of the steel ball, is the same in all three cases, but the viscosity resistance is different. The viscosity of the frozen butter is too great to permit the ball to sink. The viscosity of the sun-warmed butter is low compared to the gravity force and allows the ball to sink. The forces are more closely balanced for the butter at room temperature, although the gravity force is slightly stronger, permitting the ball to sink very slowly. If you use a bigger steel ball on top of the frozen butter, it will sink because the gravity force is greater for the bigger steel ball and eventually overcomes the viscous force of the frozen butter.

If a high-density solid is placed on top of a low-density solid (such as the steel ball on top of butter), then it is possible for the high-density solid to move downward, which displaces the low-density solid upward. The key conclusions, then, are that convection of solids

- is favored at higher temperature where solid viscosity is lower,
- occurs more readily if the masses of the solids are large,
- takes place very slowly.

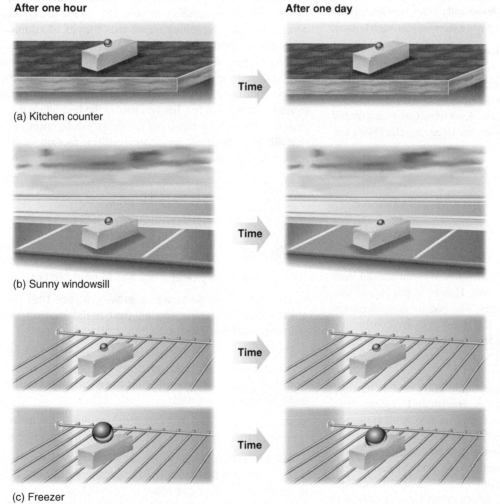

After one hour

After one day

Time

(a) Kitchen counter

Time

(b) Sunny windowsill

Time

Time

(c) Freezer

◄ Figure 10.3 **How gravity and viscous forces compete in solids.**

A steel ball sinks most readily in the butter that has warmed up on a window sill because the viscosity is lower at higher temperature.

The viscosity of frozen butter resists gravity, and a small steel ball does not sink noticeably. However, a larger ball weighs more, so the gravity force overrides the viscosity resistance, and the ball sinks into the butter.

Where Can Convection Occur Inside Earth?

Determining whether or not convection occurs in solids involves using simple calculations to compare the magnitude of the gravity and viscosity forces. The challenge is determining the values to plug into the calculations. We can proceed with reasonable confidence using inferences on the composition and temperature inside Earth developed in Chapter 8, along with measurements and theoretical values describing how well materials conduct heat and how much they expand when heated.

If we plug in those values and perform the calculations, they indicate that conduction dominates heat transfer within the lithosphere, except for localized convection in molten magma chambers. Convection is unlikely between the bottom and the top of the solid lithosphere because the viscosity in this rigid layer is extremely high.

Combining the lithosphere with the remainder of the mantle leads to a different result, however. Calculations show that convection in the mantle *as a whole* is to be expected because

- viscosity in the asthenosphere is relatively low because the rock is close to its melting temperature (Figure 8.22).
- some of the lithosphere is denser than the asthenosphere, which is why subduction takes place at convergent plate boundaries.
- the thickness and mass of the hotter, less dense and colder, denser regions of the mantle are huge (analogous to moving a big steel ball through butter).

This means that while heat does not cause motion *within* the lithosphere, the entire mantle, including the crust that is part of the lithosphere, can experience sinking and rising because of convection. This convective motion is slow, however, because viscosity is high. In the mantle, the average viscosity is 10^{20} times greater than cold molasses. As a result, the calculated velocity of mantle convection is very slow—about 10 centimeters per year.

EXTENSION MODULE 10.1

Is Mantle Convection Physically Possible? Learn the simple mathematical calculations that determine the likelihood of convection in Earth's mantle.

What about convection in Earth's core? The liquid outer core has an extremely low viscosity, about equal to that of water at room temperature and pressure. Low viscosity favors convection. However, the liquid iron of the core also conducts heat very efficiently. The conduction efficiency means that heat readily transfers through the outer core by conduction rather than convection. The combination of the high heat conductivity and high viscosity of solid iron (similar to those of the soup pot) makes convection even less probable in the inner core.

Nonetheless, this does not mean that the liquid outer core is motionless. Keep in mind that an imbalance between gravity and viscous force causes convection and that the gravity-driven motion depends on density contrasts. Geophysicists argue that compositional differences, rather than temperature differences, provide the necessary density contrasts for convection in the low-viscosity outer core. Seismic-wave data suggest that the inner core is nearly pure iron, whereas the outer core has nickel and light elements in addition to iron. The inner core slowly grows by crystallization

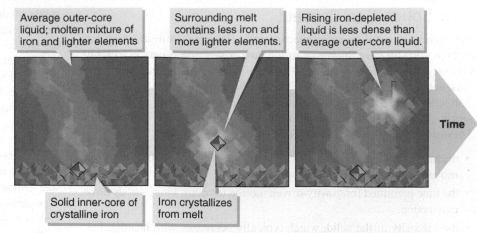

| Average outer-core liquid; molten mixture of iron and lighter elements | Surrounding melt contains less iron and more lighter elements. | Rising iron-depleted liquid is less dense than average outer-core liquid. |

Time

| Solid inner-core of crystalline iron | Iron crystallizes from melt |

▲ **Figure 10.4 How chemical convection works in the core.** Compositional differences, rather than temperature differences, produce density variations that cause fluid motion in the outer core.

of iron from the liquid outer core. When iron crystallizes, the nearby remaining liquid contains less iron than the average outer-core liquid.

Figure 10.4 illustrates how this compositional change resulting from iron crystallization causes convection in the outer core. The iron-depleted liquid left over from iron crystallization adjacent to the inner core is less dense than the average outer-core liquid. This means that the newly formed blobs of iron-depleted liquid move up, while denser, more iron-rich liquid moves down. The upward-moving fluid also transports heat away from the hotter inner-core boundary region. Due to the low viscosity of the outer core, convection velocities are on the order of 10 kilometers per year—in other words, about 100,000 times faster than in the mantle.

Two types of convection are now apparent. Fluid motion in the outer core is described as *chemical convection* (density differences caused by composition) and differs from the *thermal convection* (density difference caused by temperature) that you saw in the lava lamp, the soup pot, and the mantle. Although the causes of the density differences required for convection differ for these two types, the resulting motion moves heat along with matter in both cases.

Putting It Together—How Does Convection Work?

- Convection is the simultaneous movement (transfer) of matter and heat, whereas conduction transfers only heat from hot to cold regions.

- Convection happens when gravity forces that cause motion are greater than viscosity forces that resist motion; in other circumstances, heat transfers by conduction.

- Thermal convection originates from density contrasts between hot and cold materials. Higher density contrast, which favors convection, is present in materials that have low heat conductivity and large expansion when heated.

- Calculations show that convection should occur in the whole mantle and in the outer core, but not within the lithosphere or inner core.

- Convection in the outer core results from density variations caused by compositional differences, rather than by temperature differences that cause mantle convection.

10.2 What Does Mantle Convection Look Like?

The calculations summarized in the previous section include some speculation and inference about the properties of inaccessible materials deep within Earth, so you can rightly ask whether any evidence exists for convection in the mantle to support the results. You may also wonder whether convection can even occur when the picture of Earth's interior painted by earthquake data in Chapter 8 clearly suggests that density increases with depth (Figure 8.20). How is it possible to get rock that is high up in the mantle to sink if lower-density rock overlies higher-density rock? Even if convection occurs in the mantle, does it really resemble what happens in the lava lamp and soup pot?

What Heat-Flow Data Show

Recall that temperature increases downward into Earth. Temperature measurements in deeply drilled wells permit calculation of the total amount of heat escaping the surface of Earth—a quantity called **heat flow**.

Shortly after the discovery of radioactivity in the early 1900s, geologists hypothesized that heat flow through continents should be greater than through ocean floors. This conclusion arose from the observation that granite, which is common in continental crust, contains a greater abundance of radioactive, heat-producing elements than basalt, which composes oceanic crust (Figure 8.23). Nearly all heat-flow measurements at that time were from continents, so there were insufficient data to test the hypothesis.

By the mid-1950s, however, enough heat-flow data existed from areas of oceanic crust to disprove the hypothesis, as demonstrated in **Figure 10.5**. Heat-flow values vary considerably across both oceans and continents but are, on average, *higher* in oceans, not lower as hypothesized. This implies that most of the heat must be coming from below the crust because radioactive-heat production in basaltic oceanic crust should be small. The only way to account for so much heat flow from the mantle would be to multiply the huge volume of the mantle by its miniscule radioactive heat

production (Figure 8.23). While this adds up to much more heat than is generated by the thin crust, the vast majority of the heat is much too far below the crust to conduct effectively to the surface, given the extraordinarily low heat conductivity of silicate rock. The best way to account for the high heat flow from the mantle is to conclude that it convects to the base of the lithosphere from deep within the mantle. From the base of the lithosphere, which is generally thinner beneath oceans than continents, the heat conducts to the surface, where it is measured as heat flow.

What Causes the Density Contrasts in the Mantle Required for Convection?

Figure 10.6 shows how it is possible for the mantle to convect. Section 8.5 presented estimates of the temperature variation within Earth. The key to explaining mantle convection is that the temperature changes at the top and bottom of the mantle are much more abrupt than the temperature change through most of the mantle thickness. Heat conducted upward from the outer core causes the high temperature gradient at the base of the mantle, whereas cooling by conduction and radiation at Earth's surface causes the abrupt temperature decrease at the top of the mantle. Notice what happens if upward motion starts in the lower mantle, perhaps because of rock expansion caused by heating from the core. This rising parcel cools slightly, but not very much because silicate rocks do not conduct heat efficiently. Therefore, the rising parcel remains warmer, and therefore less dense, than the neighboring mantle rock, so it continues to rise. Likewise, if cold, dense lithosphere at the surface starts to sink at a subduction zone, it will warm up slightly but remain colder and denser than the neighboring mantle and continue to sink.

Solid Convection Generates Magma

Without convection, Earth would have no volcanoes. Convective rise of mantle rock accounts for one of the most important mechanisms for producing magma, explored in Chapter 4. Decompression melting takes place where hot rock rises from depth while cooling only slightly (Figures 4.16, 4.19, and 10.6). Eventually, the hot rock rises to where the pressure is

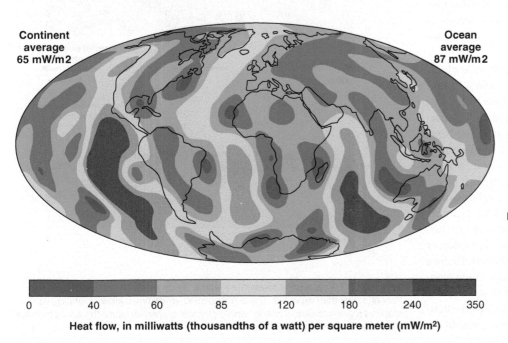

Continent average 65 mW/m2

Ocean average 87 mW/m2

0   40   60   85   120   180   240   350

Heat flow, in milliwatts (thousandths of a watt) per square meter (mW/m²)

100 watt bulb

2 million m/W/m3

Earth heat flow seems tiny compared to the output of a 100-watt light bulb, but when multiplied by the surface area of Earth, this heat energy is more than 3 times greater than annual human energy consumption on the entire planet.

◀ **Figure 10.5 Heat flow from Earth's surface.** Temperatures measured in deep drill holes permit calculation of the amount of heat energy emerging from within Earth. Geologists measure the quantity of heat in thousandths of a watt of energy per square meter of Earth's surface. The values range greatly from place to place but, on average, more heat moves through oceanic crust than through continental crust.

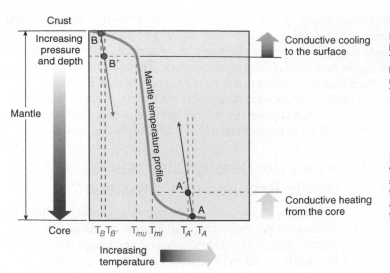

Crust

Increasing
pressure
and depth

Mantle

Core

Mantle temperature profile

Conductive cooling
to the surface

Conductive heating
from the core

$T_B T_{B'}$ $T_{mu} T_{ml}$ $T_{A'} T_A$

Increasing
temperature

If parcel B moves down, then its temperature increases from T_B to $T_{B'}$. It is still cooler, and hence denser, than surrounding upper mantle at temperature T_{mu} and continues to sink.

If parcel A moves up to A' it cools from temperature T_A to $T_{A'}$. It is still hotter, and hence less dense, than the surrounding lower mantle at temperature T_{ml} and continues to rise.

◀ **Figure 10.6 Why the mantle temperature profile permits convection.** The mantle temperature profile curves into a backward-S shape. This shape exists because conductive cooling through the lithosphere, at the top, and conductive heating from the core, at the bottom, produce steeper temperature gradients in these regions. Therefore, parcels of rising lower mantle are always warmer than adjacent rocks, and parcels of sinking upper mantle are always cooler than adjacent rocks, thus allowing convection to happen.

sufficiently low for the rock to melt, while surrounding, cooler rock remains completely solid. Geologists studying igneous rocks were proponents of convection in the mantle before other evidence for this process was known, basically because they knew how difficult it is to come up with a mechanism to partly melt peridotite to make basaltic magma without convection.

What Seismic Data Show

The preceding arguments suggest that if convection occurs in the mantle, then variation in density should exist at any level in the mantle. There should be regions of lower density representing upwardly mobile mantle and denser regions where mantle is sinking. If the density is different in adjacent regions, then we should expect seismic velocity also to differ because velocity partly depends on density. Calculations show that if a parcel of mantle is both less dense and hotter than its surroundings, then the P and S waves should move through that parcel more slowly.

However, the data in Chapter 8 did not indicate any such variations in seismic-wave velocity. The curves in Figure 8.20 imply that the only thing determining seismic velocity is the depth below the surface (or distance from the center). For example, the curves assume that the wave velocities at 500 kilometers below Los Angeles, California, are the same as at 500 kilometers below Beijing, China.

Actually, seismologists know from earthquake data that there are local variations in rock properties within Earth that cause wave velocities to vary as much as 2 percent at any particular level. Two percent seems like a very small difference, and ignoring that difference in Chapter 8 made comparing internal properties of Earth at various depths a simpler process. Now, however, we realize that these minor variations from place to place may play an important role. Assume for a moment that a 0.5 percent increase in seismic velocity equates to a 0.5 percent increase in density. Is it possible that such a small density increase permits this denser parcel to sink in viscous, solid mantle? Recall the steel balls and frozen butter example (Figure 10.3): If the high-density parcel is very small relative to its surroundings, then it may not move. If, on the contrary, the parcel of denser mantle rock is very large relative to its surroundings, this density difference may be adequate for convection to occur. Given the huge mass of the mantle, these small density differences are sufficient for convection.

Figure 10.7 shows how seismologists use seismic records from thousands of earthquakes recorded at thousands of seismic stations around the

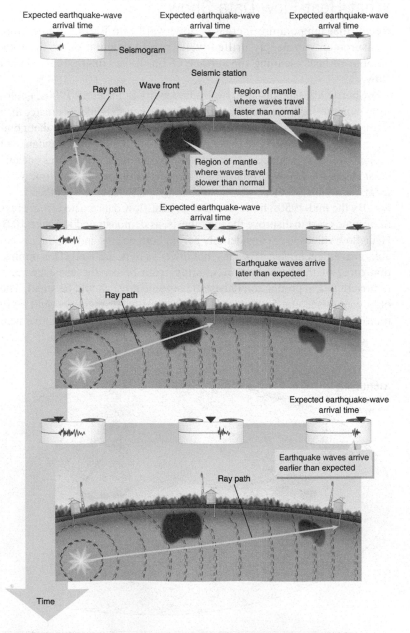

▲ **Figure 10.7 How seismic tomography works.** Earthquake waves sometimes arrive later or earlier than expected. These unexpected arrival times allow seismologists to map out regions of Earth's interior where physical properties cause the waves to move more slowly or faster than normal.

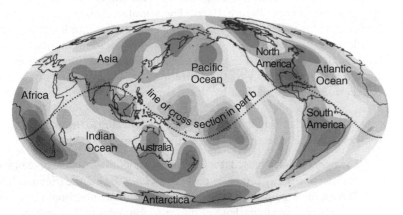

Seismic waves move slower than expected through the redder (probably warmer, less dense) regions and faster than expected through the bluer (probably cooler, denser) regions.

Percent difference in S-wave velocity compared to average velocity at 900–2880 km depth

Slower by 1% Hotter | Average | Faster by 1% Colder

(a)

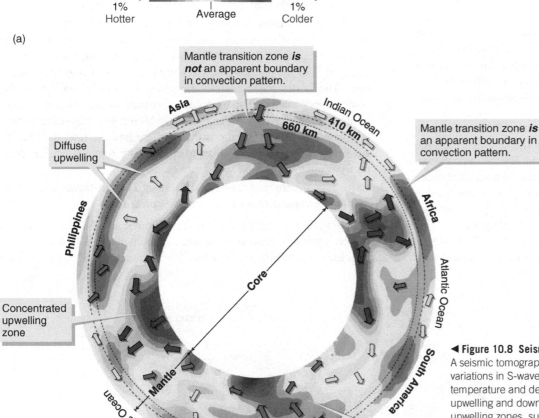

Mantle transition zone **is not** an apparent boundary in convection pattern.

Diffuse upwelling

Concentrated upwelling zone

Mantle transition zone **is** an apparent boundary in convection pattern.

This cross-section shows the variation in S-wave velocity within the mantle along the dotted line shown in the map (a).

• Red and yellow arrows show interpreted upwelling of slower, hotter mantle depicted in red and yellow shades.

• Blue arrows show interpreted downwelling of faster, cooler mantle depicted in blue shades.

Downwelling zone

◀ **Figure 10.8 Seismic-velocity variations map convection in the mantle.** A seismic tomographic map (a) and cross section (b) of the mantle show variations in S-wave velocity that most geophysicists explain by variations in temperature and density. Arrows show the interpreted pattern of convective upwelling and downwelling implied by the temperature variations. Deep upwelling zones, such as those beneath Africa and the Pacific Ocean, are round in map view and relatively narrow in cross section—their three-dimensional shape is like a column. The blue downwelling zones are irregularly shaped and elongate on the map and narrow in the cross section—their three-dimensional shape is like a rumpled sheet.

(b)

planet to map out regions of the mantle where seismic waves travel slightly faster or slightly slower than in adjacent rock at the same depth. This method of determining Earth's internal structure closely resembles the diagnostic medical tomography methods for imaging the human body—so the method applied to Earth is called **seismic tomography**.

Figure 10.8 shows a seismic-tomographic map and cross section of Earth based on variations in S-wave velocities. The illustrations reveal complicated patterns, and geophysicists actively debate what causes these patterns. Do the variations in seismic-wave velocity represent different rock types or different rock temperatures? Both factors likely apply, but

combining the tomographic data with other types of geophysical measurements suggests that temperature differences are more important than compositional differences. Therefore, the regions of seismically slow mantle are presumed to be warmer, less dense, and rising, compared to adjacent regions. Also, complementary regions of seismically fast mantle are interpreted to be colder, denser, and sinking. The cross section (Figure 10.8b) shows that some of the areas of interpreted upward or downward flow transect the entire thickness of the mantle. In other places, the pattern seems interrupted by the transition zone between 410 and 660 kilometers depth (Section 8.4). The map and cross section show that most cold areas of downward flow tend to be very elongate, so they look like rumpled sheets in three dimensions. Some warm areas have diffuse boundaries, whereas others are sharply defined columns rising through the mantle with semicircular shapes in map view. Another important consequence of this convection is apparent: In order to maintain a reasonably smooth surface at the top of the mantle, nearly horizontal flows must connect the upwelling and downwelling regions. Hot mantle rises to the surface, spreads sideways and cools, so that it sinks. This horizontal flow near Earth's surface relates to the motion of plates, as mentioned in the first chapter (Section 1.5). We can see an emerging connection between convection in the mantle and deformation at Earth's surface; Chapter 12 fully develops this connection.

What Mantle Convection Should Look Like

If the seismic-tomographic images in Figure 10.8 truly capture upward and downward movement in the mantle caused by convection, then the pattern is quite complicated and not very similar to the lava lamp and soup pot analogies. It is fair to ask, then, whether the pattern of mantle convection interpreted from the seismic data is physically plausible. To answer this question requires considering physics beyond the scope of this book. The results of certain experiments and computer simulations pertinent to the problem can be covered here, however.

Figure 10.9 portrays a simplified model of convection. In this case, heat enters at the bottom of a flat plate, like the light in the lava lamp and the stovetop burner under the soup pot. The heated material has the same viscosity everywhere from top to bottom. Convection produces symmetrical zones of hot upwelling and cold downwelling. Notice that the convective motion is not simply up and down—horizontal flow occurs at the top and bottom, and connects each upwelling zone with adjacent downwelling regions.

Does this convection resemble the motion implied by the seismic-tomographic images in Figure 10.8? Two features do seem similar—the

upwelling zones are narrow columns, whereas the downwelling regions are elongate sheets (Figure 10.9). The simulated convection pattern, however, is far more regular than the chaotic pattern implied for Earth. In addition, the distance between an upwelling region and the closest downwelling region in Figure 10.9 is almost exactly the same as the thickness from the top to the bottom of the convecting layer. The convection pattern implied by the seismic data does not show such a consistent pattern.

Is this simple form of convection really applicable to the mantle? No— for that to be the case, three complexities need to be added:

1. The mantle is also heated from within and not only from the core below. Radioactive decay produces heat *throughout* the mantle (Section 8.5). Current measurements and estimates suggest that only about 10 percent of the heat rising to the base of the crust originates in the core; the bulk of the heat is generated in the mantle. Computer simulations must, therefore, include heat distributed throughout the mantle.

2. Mantle rocks almost certainly do not have the same viscosity from top to bottom. Geophysicists estimate, for example, that the lower mantle viscosity is about 1000 times greater than the asthenosphere viscosity. For most materials, viscosity decreases at higher temperature. The hotter lower mantle, however, is more viscous than the colder upper mantle. This is because pressure causes closer packing of atoms in the lower-mantle minerals, which resists flow even at high temperature. The top of the asthenosphere has the lowest viscosity because the peridotite is close to its melting temperature, or even partly molten, so that it flows more readily.

3. Earth is a sphere, not a plate, so heat moves out through an ever-increasing volume of rock as it moves upward.

Plugging these complexities into the computer simulations produces a much more complex convection pattern, as shown in **Figure 10.10**. The regular spacing of upwelling and downwelling zones depicted in Figure 10.9

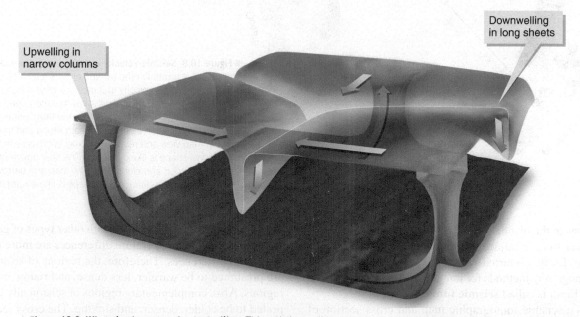

Upwelling in narrow columns

Downwelling in long sheets

▲ **Figure 10.9 What simple convection looks like.** This artistic rendition of convection is based on computer simulations. The simulated convection describes the motion of material that is of uniform viscosity and heated only at the base. Hot material rises in narrow columns, spreads out and cools near the surface, and descends in long, thin sheets. This pattern agrees closely with the photograph of the lava lake at the beginning of this chapter, in which lava rises along central columns, spreads out and cools, and then sinks in sheets of dense liquid.

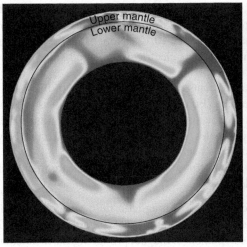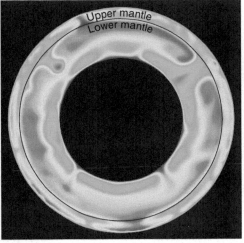

Red shades represent warmer, upwelling regions.

Blue shades represent cooler. downwelling regions.

Boundary between upper and lower mantle drawn at 660 km.

▲ **Figure 10.10 What complex convection looks like.** These two diagrams are from a computer simulation of convection under more Earth-like conditions than those shown in Figure 10.9. This model includes internal sources of heat in the mantle, viscosity that increases below the boundary between upper and lower mantle (660 km), and a spherical shape. The two illustrations are drawn from different times within the simulation and show that the convection pattern changes with time.

is now gone, and the distances between these zones tend to be greater than the thickness of the convecting mantle. In some places, the upwelling and downwelling zones pass through the entire simulated mantle. In other places, motion in the upper mantle seems disconnected from motion in the lower mantle, thereby emphasizing the importance of the viscosity change where mineral transformations occur in the transition zone. The down-welling zones seem sharply defined in narrow zones. In contrast, most upwelling zones have irregular shapes with diffuse outlines that reflect internal heating by radioactive elements in the mantle. Much of this diffuse upwelling occurs simply because sinking regions of dense, colder mantle displace adjacent regions of less dense, warmer mantle toward the surface. There are a few more intense, localized upwelling zones that seem to originate at the core-mantle boundary. The pattern in Figure 10.10 more closely resembles the pattern implied by the seismic-tomographic data illustrated in Figure 10.8, and adds confidence to the conclusion that the mantle does convect.

Putting It Together—What Does Mantle Convection Look Like?

• Mantle convection explains why heat flow is greater through the seafloor than through continents and why decompression occurs that melts rock to form magma.

• Mantle convection occurs because upward-moving rock expands more by decompression than it contracts by cooling, allowing it to remain hotter and less dense than its surroundings, so it continues to move up. Likewise, sinking mantle compresses more because of high pressure than it expands by warming, so that it remains cooler and denser than its surroundings, and it continues to move down.

• Seismic tomography shows regions of the mantle where seismic waves travel slightly slower or slightly faster than in adjacent rock. The slower regions are interpreted to represent warmer, upwelling

mantle, and the faster regions are interpreted to represent cooler, downward-flowing rock.

• Computer simulations of mantle convection produce patterns similar to those inferred from seismic-tomographic data.

10.3 How Does Outer-Core Convection Generate the Magnetic Field?

Motion in the outer core is so distant from the surface that it may, at first, seem too esoteric to worry about. At the beginning of this chapter you read that convection in this deeply remote part of Earth produces the magnetic field. The magnetic field allows humans to navigate with a compass, permits migrating animals to navigate using magnetic cell structures in their brains, and deflects hazardous cosmic radiation away from Earth. It is, therefore, important to understand why Earth has a magnetic field and how that field relates to convection.

What the Magnetic Field Looks Like

Before you can understand how Earth creates magnetic force, you first need to identify the phenomena that require explanation. For starters, consider what a compass reveals about Earth's magnetic field. A typical compass, shown in **Figure 10.11**, consists of an arrow-shaped magnetized needle. A magnetized object is one that consists of material, such as iron or the mineral magnetite, in which atoms align with the direction of magnetic force. Earth's magnetic force is oriented almost, although not exactly, along the axis through the planet around which the planet rotates. This means that the compass needle points, more or less, toward the north. In the most sophisticated compasses, the end of the needle opposite the point has a weight on it. This is because the magnetic force not only causes the compass needle to pivot toward a northerly point but also deflects the needle up or down (Figure 10.11). The weight keeps the compass needle freely spinning in a horizontal plane. Columbus and Magellan used compasses when they sailed to explore the Americas in the fifteenth and

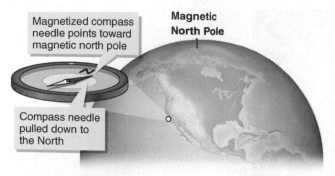

Magnetized compass needle points toward magnetic north pole

Magnetic North Pole

Compass needle pulled down to the North

▲ **Figure 10.11 How a compass reveals the magnetic field.** The magnetized arrow in a compass is pulled toward the magnetic North Pole and, in the northern hemisphere, it is also pulled downward.

sixteenth centuries, so the knowledge that Earth has a magnetic field has been utilized for a long time, even though the origin of the field has remained a mystery until recently.

You can map the orientation of the magnetic force by taking careful measurements with a freely suspended magnetized needle at many places on Earth and recording the direction that the needle points and the extent to which it deflects up or down compared to Earth's surface. Using these

ACTIVE ART

Understanding Inclination and Declination. See how inclination and declination change with location on Earth.

observations, it is then possible to draw lines to show the direction of the magnetic force at each location. These lines define the magnetic field, which is illustrated in **Figure 10.12**a. Notice that the field extends into space as well, where it deflects electrically charged solar particles and cosmic rays away from the planet.

Two measurements define the orientation of the field at any location on Earth's surface (Figure 10.12a). **Declination** is the horizontal angle made by lines connecting any measurement location to the magnetic and geographic north poles. **Inclination** is the vertical angle between the field-force line and Earth's surface.

Two key features now emerge when we examine the map of Earth's magnetic field:

1. The pattern of Earth's magnetic field very closely, but not exactly, resembles the field of a bar magnet (Figure 10.12b). The analogy works only if the hypothetical magnet producing Earth's field is located deep in the interior. This observation, made in the nineteenth century, led to the conclusion that Earth's magnetism originates deep within Earth and not at the surface or somewhere in space.

2. The north magnetic pole, where the force lines go into Earth (Figure 10.12a), and south magnetic pole, where the force lines emerge, do not coincide exactly with the geographic north and south poles. Currently, the angular distance between the magnetic and geographic poles is about 18 degrees.

The Magnetic Field Changes with Time

Careful measurements show that the magnetic field is not a static feature of Earth. Curiously, the location of the magnetic poles continually changes, as illustrated in **Figure 10.13**. This means that the field shifts position, which also affects declination and inclination angles everywhere on Earth.

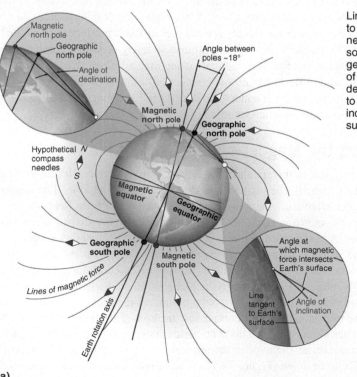

(a)

Lines of magnetic force pass through and encircle the planet to define Earth's magnetic field. Hypothetical compass needles show the orientation of the field. The north and south magnetic poles do not coincide exactly with the geographic poles along Earth's rotation axis. The orientation of the field at any point on the surface is described by the declination (the angle between the directions from the point to the geographic and magnetic north poles) and the inclination (the angle between the magnetic force and the surface at the point).

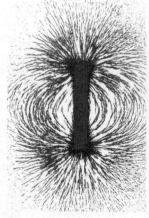

(b)

This photograph illustrates the magnetic field of a bar magnet. The red magnet is below a transparent sheet of plastic partly covered in iron filings. The magnetic iron fragments align themselves with the field of the magnet. The circular pattern resembles Earth's magnetic field depicted in (a) and suggests that most of the magnetic field originates deep within Earth and has a geometry similar to that of a bar magnet.

▲ **Figure 10.12 What Earth's magnetic field looks like.**

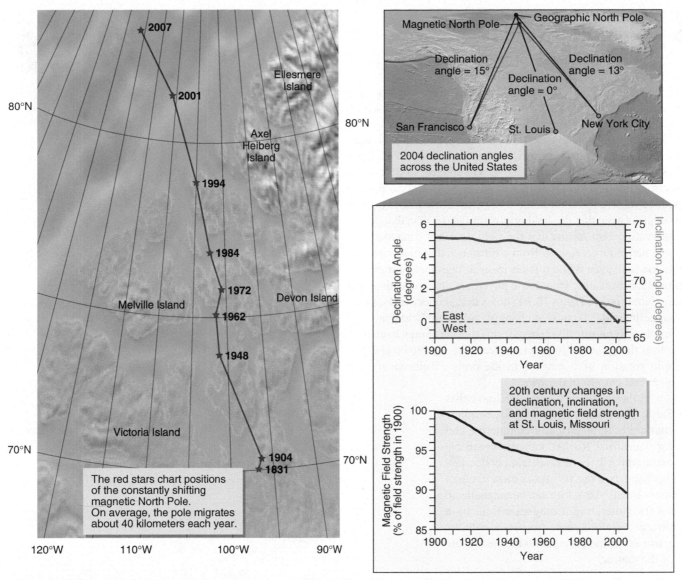

▲ **Figure 10.13 The magnetic poles are on the move.** The magnetic North Pole shifts about 40 kilometers each year and, if it stays on its current trajectory, will cross the Arctic into Siberia by 2050. Currently, there is almost no magnetic declination in St. Louis, but there is a 13° westerly declination at New York City and a 15° easterly declination at San Francisco. The declination angle at St. Louis changed by about 5 degrees during the twentieth century because of the moving magnetic pole. This means that an uncorrected compass that points north today in St. Louis would have pointed 5 degrees east of north in 1900. The strength of the magnetic field at St. Louis also decreased by more than 10 percent since 1900.

Navigation instruments must be adjusted for this shift to maintain their accuracy. Also, the strength of the field changes by as much as 0.01 percent per year in some parts of the world.

The magnetic minerals found in any rock preserve a record of the orientation of the magnetic field at the time the rock formed. Measurements of this orientation show great variations in inclination and declination, which is not surprising because such changes also occur in historical time (Figure 10.13). However, the measurements of ancient magnetic-field orientations also reveal **magnetic reversals**, when the magnetic poles swapped positions. During a period of reversed magnetism your compass would point south rather than north! Periods when the field is oriented as it is now are **normal polarity intervals**, and periods when the field is oriented in the opposite direction are called **reversed polarity intervals**. The field reversed several hundred times during the last 160 million years. Over the last 15 million years the intervals have ranged considerably in duration, although the average polarity interval has lasted about 200,000 years. The last reversal was 780,000 years ago.

The record of the ancient magnetic field in rocks indicates that reversals occur fairly abruptly, over intervals of 1000 to 6000 years. The intensity (strength) of the field also decreases significantly during reversals. When the magnetic field is weak organisms on the surface may be subjected to dangerous levels of cosmic radiation, which is less effectively repelled when the magnetic field is weak. Regardless of how the magnetic field forms, the process is capable of causing rapid and substantial changes in the orientation and strength of the field.

Why Must the Magnetic Field Form in the Core?

The pattern of the magnetic field implies that the field originates within Earth (Figure 10.12). So, where does it form within the planet?

Although the magnetic fields of Earth and of a bar magnet appear very similar, there cannot be a big mass of magnetized material deep within Earth that is a close analog to a bar magnet. Geologists know this because magnets function only at relatively low temperatures. The mineral magnetite, for example, loses its magnetism when heated above 580°C. Even pure iron loses its magnetization above 800°C. Given a reasonable near-surface geothermal gradient of 25–30°C/km, this means that no permanently magnetized minerals should be able to exist below roughly 25 or 30 kilometers depth. Given that the field originates deeper than that within Earth, this observation also means that the field does not emanate from permanently magnetized material but from a substance that must continually regenerate the field even while it fades away at high temperature.

A fundamental observation from basic physics is that magnetism and electricity are related forces. **Figure 10.14** shows that an electrical current generates a magnetic field. Likewise, a changing magnetic field generates an electric current. A **dynamo**, which transforms motion energy to electrical energy, exploits this principle to generate electricity. As shown in **Figure 10.15**, the rapid rotation of a magnet inside coils of electrical wire produces an electrical current in the wire.

Scientists hypothesize that a natural dynamo exists deep within Earth. The natural dynamo must be made of materials that are both electrically conductive and capable of being magnetized. Rocks dominated by silicate minerals are extraordinarily poor electrical conductors. Iron, on the other hand, is about 10^{16} times more conductive than granite or basalt. Also, iron can be magnetized. The conditions for generating a magnetic field by a natural dynamo are, therefore, more consistent with the iron-rich composition of the core than with the silicate composition of the mantle.

How the Magnetic Field Forms

The next piece of the puzzle is to determine just how this hypothesized dynamo works in the core. Geologists cannot directly examine the core, so the process for generating the magnetic field is only a hypothesis, but it is one that remains consistent with observed data and computer simulations.

A self-sustaining natural dynamo, sometimes called the **geodynamo**, is illustrated in **Figure 10.16**. The geodynamo requires an electrical conductor, such as iron, and motion caused by convection and Earth's rotation. Movement of the liquid-iron conductor within a magnetic field generates an electrical current. The electrical current, in turn, produces a magnetic field. The magnetic field then generates an electrical current in the convecting molten iron, which continually contributes to the magnetic field and so on, in a perpetually repeating loop. The continual generation of the magnetic field avoids the impossible requirement for permanently magnetized objects at high temperature deep within Earth.

▲ **Figure 10.14 Electrical current generates a magnetic field.** This picture shows a vertical electrical wire passing through a horizontal board that is partly covered with iron filings. The electrical current running through the wire produces a magnetic field that encircles the wire. The magnetic iron filings outline the resulting circular magnetic field.

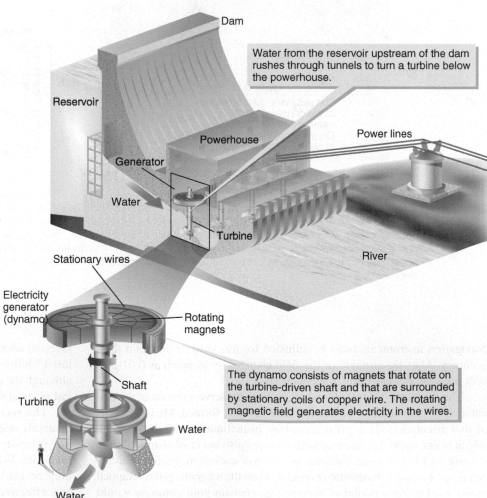

Water from the reservoir upstream of the dam rushes through tunnels to turn a turbine below the powerhouse.

The dynamo consists of magnets that rotate on the turbine-driven shaft and that are surrounded by stationary coils of copper wire. The rotating magnetic field generates electricity in the wires.

▲ **Figure 10.15 How an electromagnetic dynamo works.** Dams generate electricity when waterpower spins magnets surrounded by electrically conductive stationary wires. The changing magnetic field produced by the spinning magnets generates electricity in the wires.

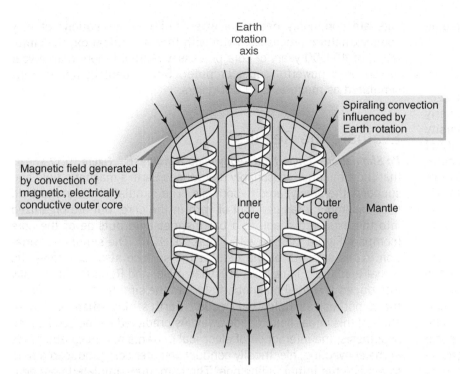

▲ Figure 10.16 Visualizing the geodynamo. The geodynamo results from convection in the outer core. The molten iron there conducts electricity and can be magnetized. Shifting electrical currents in the convecting core produce a magnetic field (Figure 10.14). This magnetization fades away at high core temperatures but not before convective motion generates electricity, like the electromagnetic dynamo in a dam (Figure 10.15). This electrical current regenerates the magnetic field. Convection, therefore, sustains magnetic and electric fields within the outer core. Earth's rotation likely influences the convection pattern and, therefore, the orientation of the continually regenerated magnetic field.

The geodynamo hypothesis also offers an explanation for why the magnetic field closely parallels Earth's rotation axis. Rotation may cause spiraling convection currents that closely parallel the rotation axis (Figure 10.16). Earth's rotation more likely affects convection in the outer core than in the mantle because outer-core viscosity is very low, close to that of water. In addition, convection in the extremely low-viscosity outer core is much faster than mantle convection, so outer-core motion more easily explains the measured rapid changes in the magnetic field than does sluggish movement of the mantle.

One important question remains: Where does the initial magnetic field come from that starts the geodynamo? The initial field does not have to be very strong and might have originated within the Sun at the time Earth formed. That hypothesis is somewhat problematic, though, because we do not know how the magnetic field first formed in the Sun. So, you can understand why this concept is still the subject of considerable research.

Why Does the Magnetic Field Change?

Scientists have a reasonable hypothesis for explaining the presence of Earth's magnetic field. However, can geologists also explain why the field changes orientations through time or completely reverses polarity? Geophysicists still wrestle with answers to these questions but many natural complications in Earth are likely to cause these phenomena.

For starters, convection in the outer core is probably much more complicated than the process shown in Figure 10.16. Although the inner core is too hot to retain a magnetic field for very long, the iron is very electrically conductive, so it should conduct the electrical currents generated by

the geodynamo in the outer core. This electrical energy in the inner core can then affect the magnetic field that forms in the outer core.

Another issue is that scientists have yet to understand how convection currents in the outer core likely interact with the mantle. Recall from Section 8.4 that there are scattered patches of low-seismic velocity in the lowermost mantle, which suggest partial melting, mixing of core and mantle ingredients, or both, at those locations. If these patches are hotter, then they will affect how heat conducts from the core into the mantle, and that process will affect the outer-core convection pattern. Any or all of these hypothesized complexities may account for variations in the strength and orientation of Earth's magnetic field over time as well as explain the polarity reversals.

Putting It Together—How Does Outer-Core Convection Generate the Magnetic Field?

• The shape of Earth's magnetic field resembles that produced by a laboratory bar magnet. Geologists know from the lines of magnetic force that the field originates deep within Earth, where temperatures are too high for the existence of permanently magnetized material.

• The magnetic field must be continually generated by the geodynamo—rapid convective motion of electrically conductive, iron-rich liquid in the outer core.

• Complexities in rapid outer-core convection caused by electrical conductance through the inner core and transfer of heat from the outer core to the mantle may account for rapid fluctuations in the strength and orientation of the magnetic field and even reversals of the field.

10.4 How Do We Know . . . Earth's Core Is a Dynamo?

Define the Problem

Why Create a Computer Simulation of Natural Processes? The geodynamo hypothesis for Earth's magnetic field remains speculative unless it can be tested. Clearly, geologists cannot go to the core to make the pertinent observations and measurements of convective motion, electrical currents, and magnetic fields. Designing an experimental model of the core that includes all of the materials, critical forces, and very high temperature and pressure also borders on impossible.

When necessary, computer models replace field observations and experiments to test hypotheses. Computer programs describe the natural processes by mathematical equations and set the values for material properties and rates of processes that are difficult or impossible to control in the laboratory. However, even computer simulations have limitations, not only because of uncertainties in values for deep-Earth materials and processes, but also because they require immense computing power to simulate what happens within the huge volume of Earth during long periods of geologic time.

Build the Simulation

What Goes Into a Computer Model of the Core? An important geodynamo simulation was constructed by Gary Glatzmaier, formerly of Los Alamos National Laboratory and now at the University of California at Santa Cruz, and Paul Roberts, of the University of California at Los Angeles. The model includes equations to represent the fundamental physics of magnetism and electricity, Earth's rotation, and convective motion. The model also includes values assigned to various Earth properties.

Some aspects of the real world are not incorporated in the simulation. The majority of the assigned values in the simulation, such as the size, mass, rotation rate, and most of the material properties in the core and mantle, are very Earth-like. However, to have the computer program reproduce the compositionally driven convection thought to occur in the outer core (Figure 10.4) would add considerably to the computational time. Rather than sacrifice other elements of the simulation, Glatzmaier and Roberts had the computer model simulate thermal convection in place of chemical convection. They accomplished this substitution by increasing the heat input across the boundary from inner core to outer core and increasing the viscosity of the liquid outer core. Such computational short cuts are commonly necessary when building simulations of very complex natural processes. Even with this simplification, the simulation of 45,000 years of core processes required more than a year to run on a powerful supercomputer. Subsequent computer runs simulated as much as 300,000 years.

Evaluate the Results

Do Simulations Support the Geodynamo Hypothesis? **Figure 10.17** shows that the simulation reasonably reproduces Earth's magnetic field and supports the geodynamo hypothesis. The simulated magnetic-field lines resemble a tangled ball of yarn in the core, but are organized into the expected shape of a bar magnet field outside of the core (compare Figure 10.17a with Figure 10.12b). The simulated variation in the strength of the field at the surface also resembles the actual field (see Figure 10.17b). Glatzmaier and Roberts numerically introduced an initial magnetic field into the simulation to represent the hypothesized field that seeded Earth's magnetism and then allowed this initial field to fade away. As predicted by the geodynamo hypothesis, the interaction of electrical currents and magnetic fields in the convecting, electrically conductive outer core produced a field to replace the initial fading one. The computer-simulated core generated a self-sustaining magnetic field that showed no indication of decaying away during the simulations.

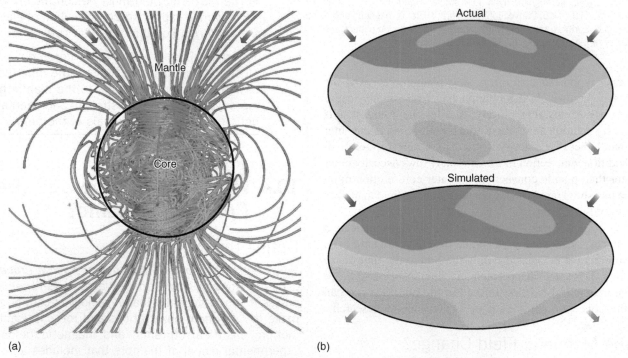

(a)

(b)

▲ **Figure 10.17 What the simulated magnetic field looks like.** In these two diagrams, blue shows the inward-directed part of the magnetic field and yellow shows the outward-directed part of the field.

a. This diagram shows simulated magnetic field lines that resemble a tangled ball of yarn in the core, but resemble a bar magnet farther out in the mantle, as is actually measured (compare to Figure 10.12b). Earth's surface is approximately located at the outer edge of the illustration.

b. This diagram compares the actual measured magnetic field with the computer-simulated field. Darker shading represents higher field intensity than lighter shading. The simulated field resembles the general patterns of the actual field intensity and direction. Although the simulated and actual patterns are not the same, both change rapidly through time, so only their general appearance is expected to be the same.

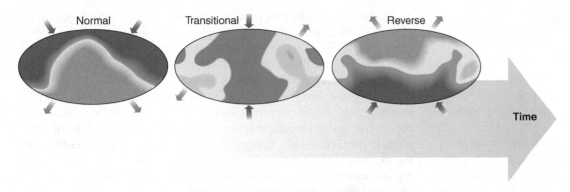

Normal   Transitional   Reverse

Time

▲ **Figure 10.18 A computer-simulated magnetic-field reversal.** The shading represents the intensity of the field. Blue areas indicate where the field is directed into the planet, and yellow areas indicate where the field is directed outward. The pale colors in the transitional period show that the magnetic field strength decreases during a reversal.

The Unexpected Results

Why Does Earth's Magnetic Field Reverse? The first Glatzmaier-Roberts simulation also produced an unexpected result even more noteworthy than successfully testing the physics behind the geodynamo hypothesis—the simulated magnetic field reversed on its own! **Figure 10.18** shows snapshot views of the simulated magnetic field reversing. In the simulation, about 1000 years elapsed in the transition from normal to reverse polarity. When Glatzmaier and Roberts combined the results from several simulations, they found that reversals occur about every 200,000 simulated years, which is about the average period between actual reversals (Section 10.3). The computer model calculates that the strength of the magnetic field weakens during reversals (Figure 10.18), which is consistent with measurements of magnetic intensity in rocks that record reversals during Earth history. None of these results were obvious outcomes of the way the computer simulation was designed, so magnetic-field reversal relates to fundamental physical processes that were not fully appreciated before the simulations were made.

Analysis of the computer-model output suggests that interactions between magnetic fields in the inner and outer cores cause polarity reversals. Rapid, chaotic convection currents cause a very complicated magnetic field in the outer core (Figure 10.16a) that attempts to change its overall orientation about every 100 years. The temporary magnetization induced in the hot, solid inner core is more stable in direction, counteracts the attempted reversal in the outer core, and maintains a stable orientation of the overall Earth field for tens of thousands to hundreds of thousands of years. Only once in many attempts did the convection-produced magnetic field in the outer core remain flipped long enough to induce a reversal in the inner-core part of the field. Once such a flip in both the inner and outer core takes place a full polarity reversal occurs.

Insights

Are Computer Models Sufficient to Test Hypotheses? The Glatzmaier-Roberts simulations reproduce observed natural phenomena while admittedly not realistically representing all of the natural variables. Does this mean that the simulations are insufficient tests of the geodynamo hypothesis? Similar criticisms arise whenever scientists apply computer models, with necessarily simplified input data and equations, to offer explanations of complex natural phenomena. With no hope of directly observing slow-acting processes deep within Earth, computer models are the best approaches to hypothesis testing.

Although the Glatzmaier-Roberts computer simulations do not include all of the complex processes likely to occur in the core, they add support to the geodynamo hypothesis to explain Earth's magnetic field. Not only do the computer models reproduce a self-sustaining, Earthlike magnetic field, but the physical processes captured by the mathematical formulas produce magnetic reversals and variations in the strength and orientation of the field just as geologists actually observe.

If, on the other hand, the simulations revealed something altogether different, then geologists would not know whether the hypothesis was wrong or the model was too unrealistic to simulate nature. One could argue, therefore, that although simplified models can lend support to a hypothesis, these models cannot falsify a hypothesis because the failure might be attributed to the model rather than the hypothesis.

Glatzmaier recognizes and states this problem clearly:

"We still have a long way to go before we can be confident that these results are robust—they may not be." Glatzmaier and his colleagues know what is necessary to make more realistic models but "the computational costs will be staggering by today's standards. . . . Significant breakthroughs that will fundamentally improve our understanding of the geodynamo will come from young people who are inspired to work on what others proclaim is impossible."

Those are the breakthroughs that continually advance science. For now, Glatzmaier and Roberts provided the breakthrough that strongly suggests that the geodynamo hypothesis is more correct than it is incorrect.

Putting It Together—How Do We Know . . . Earth's Core Is a Dynamo?

• Computer-simulation results, which include most but not all of the interpreted complexity of core processes, are consistent with the geodynamo hypothesis.

• The Glatzmaier-Roberts simulations not only reproduce the key features of the observed magnetic field at the surface but also generate simulated polarity reversals.

• The computer output suggests that magnetic-field-polarity reversals are caused by complex interactions between magnetism in the inner and outer cores.

Where Are You and Where Are You Going?

You have now completed a study of Earth's interior. You know that physical properties of Earth measured at the surface, such as seismic-wave velocities and the magnetic field, are consistent with dynamic motion inside the planet.

Convection transports matter and heat within Earth. Heat transfers by convection, rather than conduction, when the gravity force acting to move materials of different density is sufficiently greater than the viscosity force that resists this movement. Calculations show that conduction, rather than convection, is likely within the lithosphere and inner core, but that convection should occur in the mantle as a whole and separately in the outer core.

Mantle convection occurs because the silicate minerals composing mantle rocks are very poor conductors of heat, but they do expand when heated. Mathematical calculations indicate that when large masses of rock are heated at depth and expand, they rise as cooler, denser mantle sinks. Seismic-tomographic maps, depicting small variations in seismic velocity at different levels within Earth, reveal subtle density differences that outline upwelling zones and downwelling bands in the mantle. This pattern is consistent with computer simulations of convection when heat is provided within and at the base of the mantle and where mantle viscosity increases at greater depth.

Outer-core convection depends on density differences more likely caused by different liquid compositions than by temperature changes. Crystallization of iron at the margin of the inner core removes iron from the adjacent outer-core liquid. The relatively iron-poor liquid is less dense than average outer-core melt and rises as the denser liquid sinks.

A self-sustaining geodynamo in the outer core generates Earth's magnetic field. Rapid convective motion of low-viscosity and electrically conductive outer-core liquid causes electrical currents that, in turn, produce magnetic fields. Computer simulations indicate that changes in the intensity and orientation of the field measured at the surface likely result from two factors: (1) the complex convection in the outer core, and (2) interactions between the magnetic field in the outer core and the field temporarily induced in the inner core. These interactions also cause the magnetic field to reverse polarity at very irregular intervals that average about 200,000 years.

In the next group of chapters, you will embark on further studies of the consequences of dynamic motion within Earth. Convection provides an important mechanism for generating magma by decompressing hot, rising mantle rock, as well as a connection between volcanoes and internal processes. Convective motion describes the motion of lithospheric plates, which, in turn, explains the formation of mountains and the occurrence of devastating earthquakes. The evidence of this deformation at the surface is laid out in the next part of the book.

Active Art

Convection in a Lava Lamp. See how convection works inside a lava lamp.

Understanding Inclination and Declination. See how inclination and declination change with location on Earth.

Extension Module

Extension Module 10.1: Is Mantle Convection Physically Possible? Learn the simple mathematical calculations that determine the likelihood of convection in Earth's mantle.

Confirm Your Knowledge

1. How does convection differ from conduction? What causes convection?
2. Explain what chemical convection is and where it may be occurring within Earth.
3. Explain the factors that determine whether or not convection occurs.
4. Why is conduction the dominant form of heat transfer within the lithosphere?

5. How does evidence from seismic tomography support the hypothesis of mantle convection?
6. What is the difference between magnetic declination and magnetic inclination?
7. List the different ways Earth's magnetic field changes over time.
8. What causes Earth to have a magnetic field?

Confirm Your Understanding

1. Write an answer for each question in the section headings.
2. Describe the effects of thermal expansion, conduction of heat, and viscosity on the determination of whether heat transfers through material by conduction or convection.
3. Explain why there would be no volcanoes on Earth without convection.
4. Using what we know about thermal expansion, conduction of heat, and viscosity, should we expect thermal convection within Earth's core?
5. How do we know that most of the heat leaving Earth's interior is coming from below the crust?

6. How do heat-flow measurements support the hypothesis of mantle convection?
7. How do we know that Earth's magnetic field originates deep within Earth?
8. How do we know that Earth's magnetic field does not originate from permanently magnetized material?
9. Explain what will eventually happen to the magnetic field as Earth continues cooling.

CHAPTER 11

Deformation of Rocks

Why Study Rock Deformation?

Some rocks look like shattered glass, while others seem to contort like swirled taffy. Geologists describe broken and folded rocks as *deformed*.

Earthquakes result from rapid rock deformation that is accompanied by violent shaking, which can destroy property and kill people in seconds. Geologists strive to understand how and why rocks deform in hope of predicting earthquakes.

Deformation also can benefit society. Rock deformation helps to form accumulations of natural gas, petroleum, and important metal ores. Understanding how rocks deform helps us to understand how resources form and how to extract them.

Geologists collect outcrop information about deformed rocks to locate economic resources, to interpret the geologic history of an area, and to assess risks from earthquakes. Chapter 12 will explore the connection between convection inside Earth and the resulting deformation at the surface, but first, before we discuss that topic, you must understand how to recognize rock deformation and how something as solid as a rock can shatter, fold, and swirl.

After Completing This Chapter, You Will Be Able to

- Identify and describe the types of deformation visible at Earth's surface.

- Explain the relationship between deformation, called strain, and the forces, called stresses, that cause it.

- Apply the concepts of stress and strain to explain the causes and effects of earthquakes.

- Apply understanding of rock deformation to explain the origin of some Earth resources.

Pathway to Learning

11.1 What Do Deformed Rocks Look Like?

11.2 How Are Resources Related to Geologic Structures?

11.3 Why Do Rocks Deform?

11.4 How Do We Know . . . Why Some Rocks Break and Others Flow?

11.5 How Do Geologic Structures Relate to Stress, Strain, and Strength?

Erosion accentuates a surreal pattern of folded rocks in this aerial view of central Wyoming.

IN THE FIELD

ooking out across the rough, rocky landscape illustrated in **Figure 11.1**, you notice that the multicolored sedimentary rocks are not lying flat; they are tilted and folded. In the broad valley in front of you is an oil well. You wonder why the oil company decided to place the oil well there. How did people know that there was oil below ground at that exact location? You set out on a short hike around the area. Climbing across the outcrops, you identify layers of sedimentary rocks: pink mudstone, white limestone, interlayered red sandstone and mudstone, tan sandstone, and gray shale. You sketch a map in your field notebook that shows the areas where each rock type is found. Here and there you stop to take photos of what you see; some of these photographs appear in Figure 11:1 along with your sketch map.

You notice that the rock layers are inclined at angles (Photo C) and that some layers have shifted, as if the rock broke and moved along the break (Photo D). You add these observations to your sketch map. You clamber across a valley eroded in soft shale, past the oil well, and encounter a ridge of the same tan sandstone that you just left on the opposite side of the valley. But, here the sandstone bed is inclined in the opposite direction. As you hike back to your starting point and look back at where you walked, you notice that the rocks bend to form a pattern like the letter "S" on its side; this is best seen by tracing the pattern of the red sandstone and mudstone on your sketch map.

As you complete your sketch, you wonder what forces are necessary to break and bend rocks. How long does it take to deform rock? You may have seen images on television of rocks broken during earthquakes; these images attest to the fact that rocks can break quickly. Do all rocks respond similarly to deforming forces, or do some rocks fold or break more easily? Does the location of the oil well have something to do with the deformed rock?

Photo A

▲ Figure 11.1 The field site.

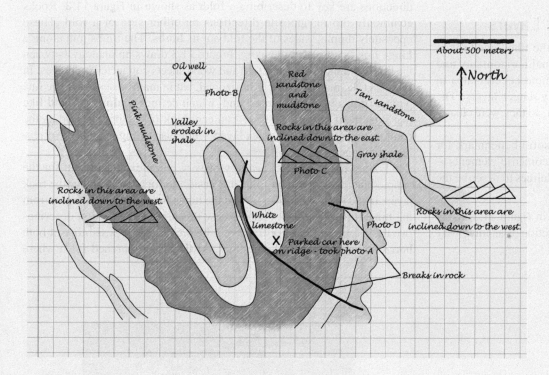

Photo B

Photo C

These are the same tan sandstone layers, but they do not line up.

All of the tilted layers are broken and displaced along this line

Photo D

Inclined rock layers

▲ Figure 11.1 (continued)

11.1 What Do Deformed Rocks Look Like?

Before considering the forces that deform rocks, you first need to understand what a deformed rock looks like. Field observations, such as those documented in Figure 11.1, record the shapes of folded rocks, the orientations of tilted rock layers, and the directions that rocks have moved relative to one another across breaks. Geologists use observations to develop systematic types of measurements and terms that describe the appearances of deformed rocks.

Describing the Orientation of Rock Layers

What is the best way to describe the orientation of the tilted and broken rocks in the field area? Recall the principles of original horizontality and superposition (Section 7.2), which state that sediment accumulates in nearly flat layers, with older layers located below younger layers. The sedimentary rocks in your field area are not horizontal, so they were clearly deformed sometime after deposition.

The tilted bedding planes in the rocks are a useful benchmark for describing the orientation of the deformed layers. Geometric elements—lines, planes, and angles—provide measurement guidelines to describe the orientation of the rock layers.

- The **strike** of a nonhorizontal bed, as illustrated in **Figure 11.2**, is the compass orientation of a line formed by the intersection of an imaginary horizontal plane with the inclined bedding plane.

- The **dip** of the inclined rock layer is the angle between the imaginary horizontal plane and the inclined rock layer.

Geologists describe the orientation of any planar geologic feature, not just bedding planes, by measuring strike and dip and recording them on a map with a standard geologic symbol (Figure 11.2).

Describing Folded Rocks

The folded rocks at the field site dip in different directions and these dip directions are key to describing a fold, as shown in **Figure 11.3**. Rocks commonly dip in opposite directions on either side of a fold. These oppositely dipping sides of the fold are its **limbs**. The limbs join along a **hinge line** (also called the axis). Figure 11.3 draws an analogy between the limbs and hinge line of a fold and wood sheets connected by hinges. In Figure 11.3a, the hinge line is horizontal, but in many cases the hinge line plunges downward into the ground, which defines a **plunging fold** like that illustrated in Figure 11.3b. You can think of a plunging fold as the combination of folding the rock layers while also tilting them in the direction of plunge.

Folded rocks exhibit a variety of very complex shapes, but for the simpler and most common folds it is easiest to focus on whether the rock layers in the limbs dip down toward the hinge line or dip down away from the hinge line. **Figure 11.4** shows how this simple distinction describes two basic types of folds, or four types by including the possibility of both horizontal and plunging hinge lines. **Anticlines** are arched folds where

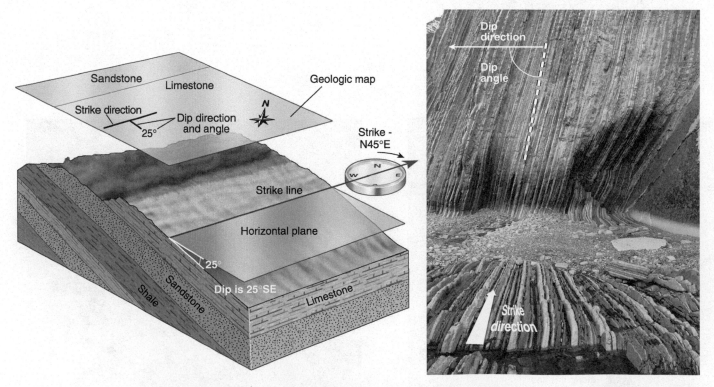

▲ **Figure 11.2 Visualizing strike and dip.** The diagram illustrates the geometry of strike and dip. Imagine a horizontal plane that intersects an inclined layer of rock. The strike line is the intersection of the inclined layer with the horizontal plane. The strike direction is the angle measured from the strike line to true north. The dip is the angle formed between the horizontal plane and the inclined bed. The strike-and-dip symbol is sketched on the simple geologic map above the block. Wave erosion of sedimentary rocks along a coastline, seen on the right, exposes cliffs of tilted sedimentary rock and reveals their strike and dip.

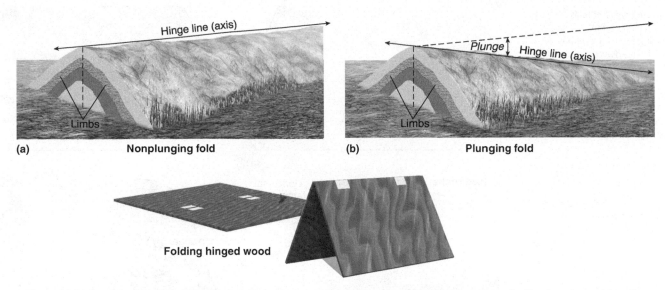

▲ **Figure 11.3 Visualizing parts of a fold.** Simple folds in layered rocks are defined by limbs that dip in opposite directions from a hinge line, which is also known as the fold axis. Two sheets of hinged wood provide an analogy for the hinge line and limbs of a fold. Plunging folds have an inclined hinge line.

limbs dip *away* from the hinge line, while **synclines** are trough-shaped folds where limbs dip *toward* the hinge line. Figure 11.4 also illustrates the surface patterns of eroded and folded rocks. Notice that the oldest rocks are exposed along the axis of an eroded anticline, whereas the youngest rocks are found along the axis of an eroded syncline. You can sometimes recognize folds within a single rock sample that you can hold in your hand. Larger folds, such as the ones seen at the field site (Figure 11.1), are hundreds of meters or even kilometers across, and their hinge lines can be traced across the landscape for tens of kilometers.

Figure 11.5 illustrates a rock that is so highly deformed it does not seem reasonable to identify anticlines and synclines. The folds illustrated in Figure 11.4 resemble a rumpled rug, but the outcrop in Figure 11.5 more closely resembles the swirling of fudge sauce and melted ice cream. These rocks have flowed rather than simply folded along a hinge line.

Describing Broken Rocks

Geologists also use special terms to describe broken rocks. Fractures of various lengths and shapes are common in almost every rock outcrop, just as you see cracks in building walls and ceilings and in pieces of lumber. Geologists make an important distinction between situations in which rocks simply crack and ones in which rocks slide past one another along a fracture.

- **Joints** are cracks where no displacement of the rock occurred, as pictured in **Figure 11.6**.
- **Faults** are fracture planes along which rocks moved.

The names used to describe the rocks adjacent to fault surfaces were originally assigned to these formations by miners and geologists who excavated tunnels to extract minerals. Faults are important to miners because

economically significant minerals precipitate from water within open spaces in the crushed and fractured rock found along faults. A critical observation that is useful for naming rocks along faults is that faults are rarely vertical. **Figure 11.7** depicts a miner walking on the rocks below a fault, called the **footwall**, and below the **hanging wall**, which is composed of rocks above the fault.

The next step in describing a fault is to determine how the footwall and the hanging-wall blocks moved past one another. As an analogy, place your palms together in front of you. Now, move your hands up and down past one another. This simulates movement parallel to the dip of the fault plane. Next, move one hand away from your body while moving the other one toward you. This motion is comparable to the movement along the strike of the fault plane. A third possibility is to move your hands up and down while simultaneously moving them toward and away from you. **Figure 11.8** applies these types of motion to classify faults. The faults illustrated in the photographs in Figure 11.8 exhibit displacements that are easily recognized at the scale of the picture. However, rocks are commonly displaced many kilometers across a fault, and a single fault can, in many cases, be traced across the landscape for 100 kilometers or even farther.

Movement of rock along the dip of the fault plane defines **dip-slip faults**. There are three kinds of dip-slip faults. A **normal fault** forms when the hanging-wall rock moves downward compared to the footwall. On the other hand, if the hanging wall moves up along a dip-slip fault compared to the footwall, then the fault is called a **reverse fault** if the fault dips at an angle steeper than 45 degrees, or a **thrust fault** if the angle is less.

Strike-slip faults form by horizontal movement along the strike direction of the fault plane (Figure 11.8). Comparing the relative movement of the rocks on either side of the fault distinguishes two types of strike-slip faults. If features appear shifted to the left from one side of the fault to the other, then a **left-lateral strike-slip fault** exists. To form a **right-lateral**

▶ **Figure 11.4 Visualizing types of folds.**
Folds form by horizontal forces that squeeze rocks. Plunging folds form where rocks both fold and tilt. The diagrams show the shapes of folded layers and how the layers look when folds erode, which exposes older layers along the axis of anticlines and younger layers along the axis of synclines. Geologists use special symbols to show the trace of fold axes on maps.

ACTIVE ART

Folding Rock. See how rocks fold.

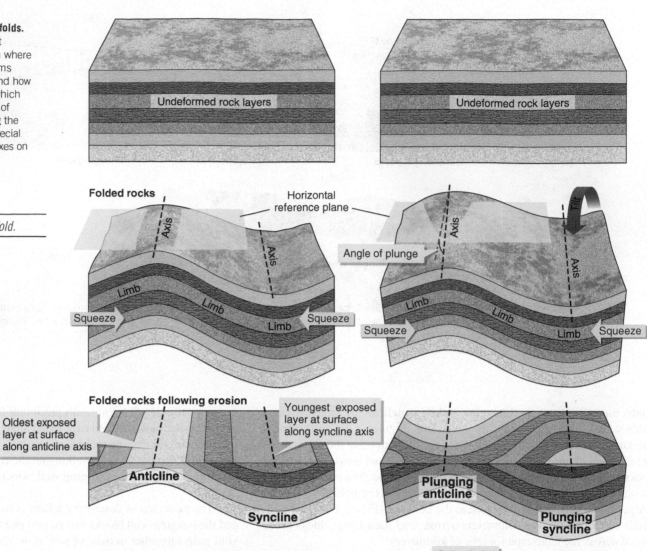

Undeformed rock layers

Undeformed rock layers

Folded rocks

Horizontal reference plane

Axis

Axis

Angle of plunge

Tilt

Axis

Axis

Limb

Limb

Limb

Squeeze

Squeeze

Limb

Limb

Limb

Squeeze

Squeeze

Folded rocks following erosion

Oldest exposed layer at surface along anticline axis

Youngest exposed layer at surface along syncline axis

Anticline

Syncline

Plunging anticline

Plunging syncline

Folded rocks depicted on a geologic map

Strike & dip

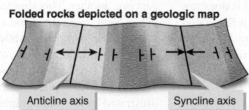

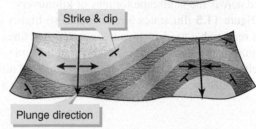

Anticline axis

Syncline axis

Plunge direction

Anticline

Syncline

Anticline and syncline exposed in cross-section (side view) in a roadcut, New Jersey

Axis of plunging anticline

Aerial view looking down on a plunging anticline, Sheep Mountain, Wyoming

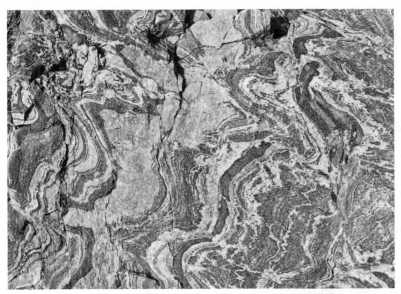

▲ **Figure 11.5 Highly contorted folds.** Foliation bands in this gneiss exhibit an extremely contorted folding pattern caused by rock flow at high temperature and pressure.

Checkerboard Mesa at Zion National Park, Utah, features vertical joints that do not displace the cross-beds.

Cross-beds

Joints

strike-slip fault, features shift to the right across the fault. You might think that right- and left-lateral motion depends on your perspective in looking at features displaced across the fault, but it does not matter. To check this, look at the strike-slip fault diagrams in Figure 11.8 by turning the book first to the left and then to the right, and you will see that displacement type is the same regardless of which side of the fault you are on.

Faults can show a combination of dip-slip and strike-slip movements. These are **oblique-slip faults**, in which diagonal motion occurs along the fault plane, both along the strike and dip as shown in Figure 11.8.

Parallel joints in sandstone appear to be etched into the landscape in this view in central Utah. Water preferentially accumulates in and flows along the joints, weathering and eroding the sandstone to accentuate the joints.

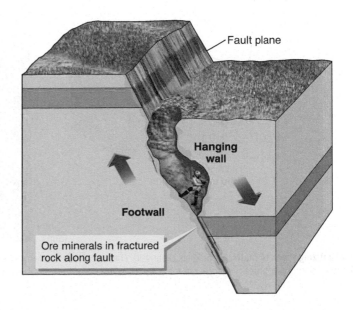

▶ **Figure 11.6 What joints look like.**

Making a Geologic Map and Cross Section

Equipped with these concepts for describing the orientation of deformed layers and the types of folds and faults, you can reexamine the sketch of the rocks at the field site (Figure 11.1). Your sketch is a roughly drawn **geologic map**, which shows the locations of the different rock types, their strike and dip, and locations of fold hinge lines and faults. **Figure 11.9** shows the actual geologic map for the field site. Color bands trace out where the different rock types are present. Strike-and-dip symbols show the orientations of the inclined sedimentary layers. The "S" pattern of deformed beds that is conspicuous in the field sketch (Figure 11.1) reveals a side-by-side plunging anticline and syncline on the geologic map.

Geologic cross sections are interpretations of the subsurface geology based on a geologist's measurements of the orientations of rock layers, folds, and faults at the surface. The schematic diagrams of folds and faults in Figures 11.4 and 11.8 include both map views and cross-section views

▶ **Figure 11.7 Describing faulted rocks.** Fault planes separate footwall and hanging-wall blocks of displaced rock. The terms for rocks on either side of a fault originated with miners who tunneled to extract ore minerals in fractured rock along the fault. Imagine walking in an open tunnel along a sloping fault plane. You would walk on the footwall, and the hanging wall would hang over your head.

Fault plane

Hanging wall

Footwall

Ore minerals in fractured rock along fault

of each structure. Figure 11.9 includes a cross section of the field site and shows how the rock layers are folded into an anticline and a syncline and offset by faults.

A cross section is made by drawing the subsurface continuation of the rock layers seen at the surface. The measured dip angles at the surface provide the necessary data for drawing the subsurface rock layers in the proper orientation. Faults mapped at the surface continue downward and offset layers in the subsurface. The cross section matches with the geologic map to complete a three-dimensional illustration of the deformed rocks.

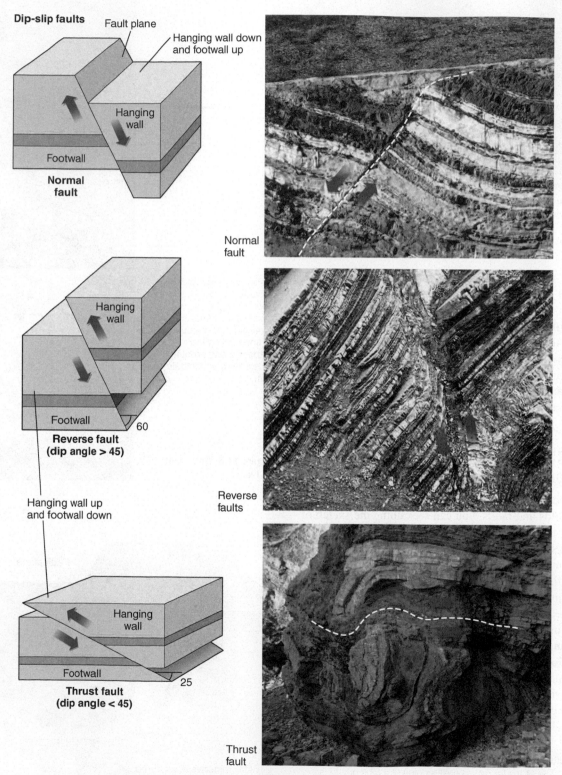

▲ **Figure 11.8 Visualizing types of faults.** Geologists distinguish between different fault types by (1) recognizing whether displacement parallels the dip, strike, or both strike and dip of the fault plane, and (2) noting the displacement of the hanging wall up or down compared to the footwall.

Strike-slip faults

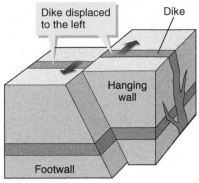

Left-lateral fault

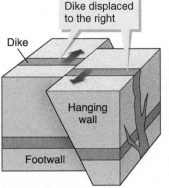

Right-lateral fault

Right-lateral strike-slip fault, Nevada

◀ **Figure 11.8** (*continued*)

Oblique-slip faults

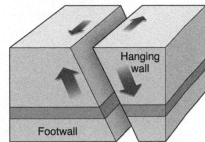

ACTIVE ART

Fault Motions. *See how motion along normal, reverse, and strike-slip faults displaces rocks.*

Putting It Together—What Do Deformed Rocks Look Like?

• Strike and dip describe the orientation of planar geologic features, such as inclined sedimentary layers and faults.

• Anticlines and synclines are distinguished by noting the dip directions of rocks forming the limbs of the fold.

• Breaks in rock include joints, where little or no displacement has taken place along fractures, and faults, where rocks are offset.

• Dip-slip faults include normal, reverse, and thrust faults. Strike-slip faults include left-lateral and right-lateral faults. A combination of movements along both the dip and strike of the fault plane produces oblique-slip faults.

• Geologic maps contain colors, patterns, and standard symbols that depict rock types, their location and orientation, and the nature and location of folds and faults that disrupt the rocks.

• Geologic cross sections are interpretations of rocks and structures below the ground surface; they are based on the measured orientations of rocks and structures observed on the surface.

11.2 How Are Resources Related to Geologic Structures?

Now you have enough background to consider why there is an oil well in your field area (Figure 11.1, Photo B). Geologists purposely located the oil well along the axis of the anticline.

Oil and Gas Traps

In Chapter 5 (Section 5.5), you learned that oil and natural gas are fossil fuels, which form from organic materials deposited with sediment. Oil and gas form from algae, bacteria, and tiny plants and animals (plankton) that live in the sea or in lakes. When the sediment and organic material are buried to depths where the temperature rises above 100°C, the organic compounds undergo changes comparable to mineral metamorphism. These changes convert the organic material into oil and gas.

Oil and gas, like ground water, move through pore spaces in rock. The three fluids have different densities, as shown in **Figure 11.10**. Oil is less dense than water, and natural gas is less dense than either oil or water. This

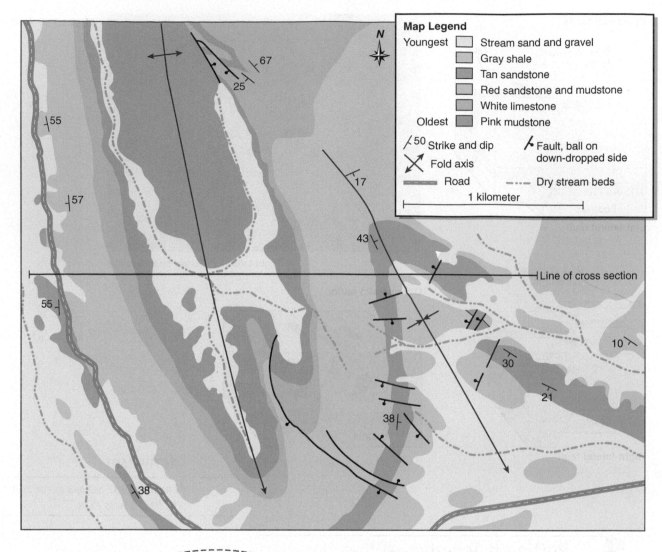

Map Legend

Youngest		Stream sand and gravel
		Gray shale
		Tan sandstone
		Red sandstone and mudstone
		White limestone
Oldest		Pink mudstone

Strike and dip   Fault, ball on down-dropped side

Fold axis

Road   Dry stream beds

1 kilometer

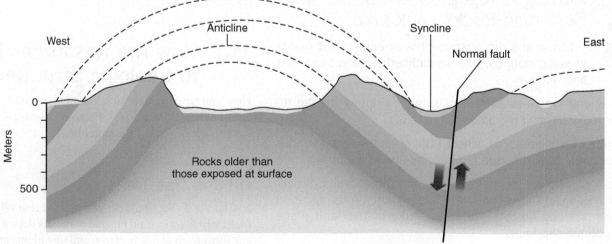

► **Figure 11.9 Geologic map and cross section of the field site.** The geologic map looks similar to the field sketch (Figure 11.1) but this map has more detailed geologic information, such as strike-and-dip symbols, fold axes, and displacement directions along faults. The cross section illustrates an interpretation of the geology below ground level based on the strikes and dips of rock layers that are measured in the field and labeled on the map.

means that oil and gas move upward through ground water in porous rock and eventually leak out to the surface unless they are trapped below less porous rock. For example, poorly cemented sandstone typically has large pores between sand grains, whereas shale has virtually no pore space. If a shale layer overlies a sandstone layer, then the shale forms a rock roof that traps oil and gas in the sandstone.

Anticlines commonly form oil and gas traps. **Figure 11.11**a shows the trapping process. If a low-porosity rock layer overlies a high-porosity layer, then the oil and gas migrate upward to the highest part of the porous layer and are trapped at the top of the anticline.

Thrust and reverse faults also trap oil and gas (Figure 11.11b). These traps form where movement along a fault displaces low-porosity rock on

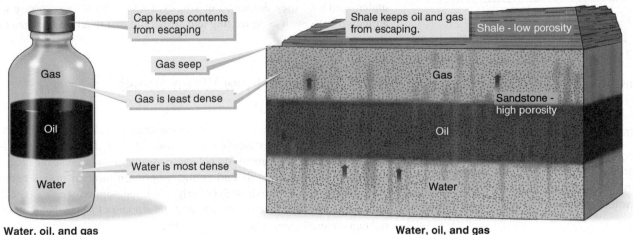

◀ **Figure 11.10 Density determines where oil and gas are found.** Gas, oil, and water separate according to their densities. The lower densities of oil and gas compared to water allow the fossil fuels to rise through water-filled pores in rocks. This upward movement stops only where low-porosity rock traps the oil and gas below the surface.

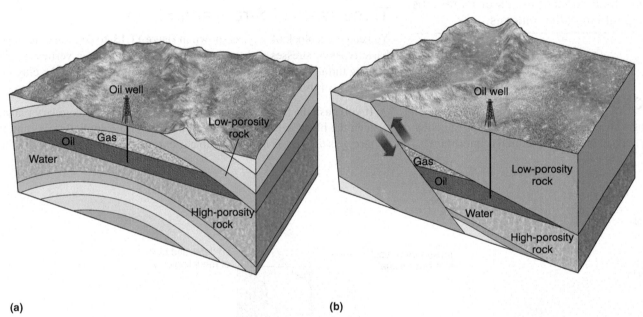

(a)   **(b)**

◀ **Figure 11.11 How geologic structures trap oil and gas.** (a) Oil and gas rise through water-saturated pores in the porous rock and are trapped beneath the low-porosity rock along the axis of the anticline. (b) Fault movements trap oil and gas by displacing high-porosity rocks next to low-porosity rocks.

top of high-porosity rock. When oil and gas move upward through the porous rock, the fossil fuels become trapped against the fault.

Metallic Mineral Resources

Economically important mineral deposits commonly coincide with faults. Most mines extracting metal ores are located in highly fractured rock. Underground shafts and tunnels in mines sometimes follow faults (Figure 11.7). **Figure 11.12** shows an example of a mineral vein along a fault. The minerals typically precipitate from hot water that originates near igneous intrusions (Section 4.10), as a result of metamorphism (Section 6.10), or that moves upward through deeply buried warm rocks within sedimentary basins. Hot water dissolves more metal ions than cold water, so metal ore minerals precipitate as the hot, metal-rich water moves toward the surface and cools. Faults and joints provide avenues for metal-rich fluid to move toward the surface, which explains why mineral deposits preferentially form along these fractures.

▲ **Figure 11.12 Minerals accumulate along faults and joints.** Many economically important mineral deposits form along fractures in rock. Shiny sulfide minerals form a vein that follows a fault cutting through volcanic rocks in this mine in Peru. The sulfide minerals include ores of copper, silver, lead, and zinc.

Putting It Together—How Are Resources Related to Geologic Structures?

- Gas is less dense than oil, and oil is less dense than water. As a result of this density contrast, oil and natural gas migrate upward through pores toward Earth's surface unless impeded by low-porosity layers, which trap the oil and gas.

- Anticlines and faults form common traps for oil and natural gas. The oil and gas migrate upward to the highest point below a nonporous layer along the hinge line of an anticline and are trapped beneath reverse and thrust faults where low-porosity layers are displaced on top of high-porosity layers.

- Metal-rich mineral resources commonly form along joints and faults where warm fluids with high concentrations of metal ions readily flow and cool to precipitate metallic minerals.

11.3 Why Do Rocks Deform?

Describing and mapping geologic structures are essential steps to understanding *how* rocks deform, but it is at least equally important to understand *why* they deform. Put simply, rocks deform in response to applied forces, such as squeezing and stretching that result in changes in size or shape.

Rocks Break or Flow When Stress Exceeds Strength

Stress is the force applied over a particular area of rock surface (also see Section 6.3), and **strain** describes the measurable rock deformation resulting from stress. Small stresses do not cause visible strain because the chemical bonds within minerals and contacts between mineral grains resist deformation. Rocks break or flow only when the applied stress exceeds the rock **strength**, which is a measure of the amount of stress that a material can endure before it permanently fails.

Three Types of Stress and Strain

You can use a block of clay, as shown in **Figure 11.13**, to demonstrate three basic types of stresses and the resulting strains. Imagine a vertical plane passing through the middle of the block with pins inserted in the clay on either side of this plane.

- If you push the block from opposite sides *toward* the imaginary plane, then the block gets *shorter* in the direction of the applied stress and it also gets *thicker*. The pins end up closer to one another and closer to the

▶ **Figure 11.13 Visualizing stress and strain.** Pushing, pulling, and sliding a clay block represent the stresses and strains that deform rocks. Pins placed near an imaginary reference plane help us to visualize the strain.

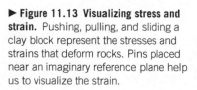

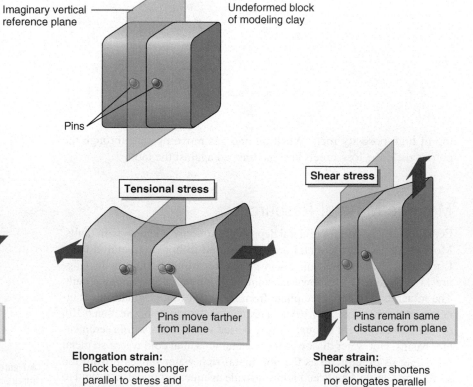

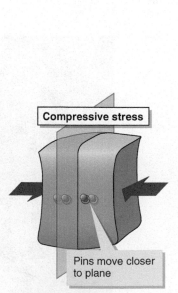

Imaginary vertical reference plane

Undeformed block of modeling clay

Pins

Compressive stress

Pins move closer to plane

Shortening strain: Block becomes shorter parallel to stress and thicker perpendicular to stress.

Tensional stress

Pins move farther from plane

Elongation strain: Block becomes longer parallel to stress and thinner perpendicular to stress.

Shear stress

Pins remain same distance from plane

Shear strain: Block neither shortens nor elongates parallel to stress.

imaginary plane. This pushing-together stress is **compression**, and the resulting strain is **shortening** (also called "contraction").

- If you pull the block from opposite sides *away* from the imaginary plane, then the block gets *longer* in the direction of the applied stress and it also gets *thinner*. The pins end up farther from one another and farther from the imaginary plane. This pulling-apart stress is **tension**, and the resulting strain is **elongation** (also called "extension").

- If you pull the block on one side and push it on the other parallel to the imaginary plane, then the block warps but the pins remain the same distance from the imaginary plane that they were prior to deformation. The term **shear** describes both the stress and the strain of sliding rocks in different directions without overall shortening or elongation.

Faulting, Folding, and Flowing Are Strain

Geologists design laboratory experiments to reproduce and explain the structures (such as folds and faults) of deformed rocks. Some informative experiments use materials that are weaker than real rocks so that structures form more easily in response to small stresses.

The relationship between stress, strain, and structures is illustrated by the sandbox experiment shown in **Figure 11.14**. Tension (pulling apart) in the left side of the sandbox causes the layers of loose sand and powder to elongate. Compression (pushing together) in the right side of the sandbox causes the layers to shorten. The strains in each part of the sandbox occur mostly along faults. If you look at the elongating side of the box in the bottom image, you will notice the normal faults form where the hanging-wall block slides off of the footwall block. If you look on the shortening side of the box, you will see that reverse faults form where the hanging-wall

block shoves over the footwall block. Small folds also form in response to the shortening.

It is important to notice that the faults are not perpendicular to the stresses created when the vertical board is cranked through the sandbox, because this reveals something consistent with what geologists know about faults. Sand and powder particles on opposite sides of a fault move past one another without getting closer to or farther from the fault plane. The way layers move past each other along the fault is, therefore, shear strain. At the same time, however, the movement of hanging-wall blocks relative to footwall blocks shows that normal faults lead to elongation where tension occurs and reverse faults result in shortening where compression occurs. Stated another way, in either compartment of the sandbox, the larger-scale deformation is elongation or shortening, and locally, or at the smaller scale, shear occurs along faults.

Figure 11.15 illustrates a classic nineteenth-century experiment where weak wax layers were used to represent sedimentary-rock beds. Wax layers shortened when squeezed from opposite sides. The layers initially folded into anticlines and synclines. More stress caused the folds to grow higher and narrower, and then faults appeared. The bottom wax layer flowed into the centers of the arching anticlines, rather than folding or breaking. The wax-model experiment shows that folds are a form of shortening caused by compression. In addition, notice how the total thickness of wax increased as the wax model shortened.

The stress required to deform real rocks is much greater than the stresses applied in the sandbox or wax experiments. So how do geologists determine the strength of real rocks and predict whether rocks will deform by faulting, folding, or flowing? In order to determine how rocks deform, you need to delve further into relationships between stress, strain, and strength.

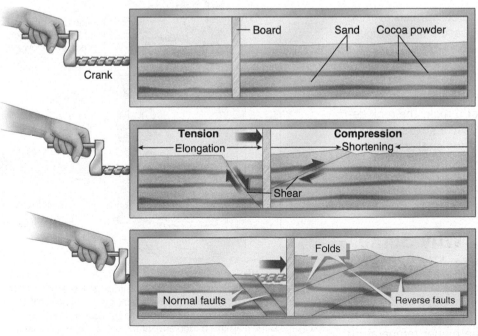

Layers of sand and cocoa powder fill the box. Turning the hand crank on the left side causes the vertical board in the center of the sandbox to shift to the right.

The board slowly moves to the right, causing elongation of the layers on the left side of the box and shortening on the right side.

Folds and reverse faults form in the shortening side of the sandbox and normal faults form in the elongated side. Normal faults form in response to tensional stress and reverse faults form by compression. However, the local motion along the faults is caused by shear stress because the fault planes are not perpendicular to the stresses.

▲ **Figure 11.14 Making faults and folds in a sandbox.** This sandbox experiment illustrates how different stresses form folds and faults.

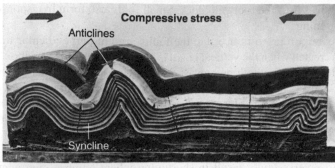

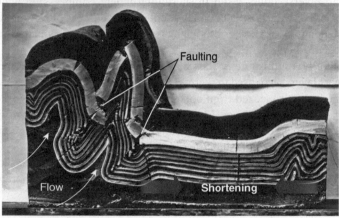

▲ **Figure 11.15 Visualizing folds and faults with wax.** These photos show cross sections through a progressively deformed wax model. Anticlines and synclines form as the wax layers fold during initial shortening. With increased shortening, the layers break along faults, and some wax flows into the growing anticlines.

Putting It Together—Why Do Rocks Deform?

• Stress is the magnitude of a force applied over an indicated area, and strain is the measure of the amount of deformation that results from the stress.

• Rock strength is the amount of stress that the material can endure before it strains.

• When applied stress exceeds the strength of a rock, the rock deforms by folding, faulting, or flowing. Compressive stress shortens rocks parallel to the stress direction, whereas tensional stress elongates rocks. Shear stress moves materials in opposite directions without shortening or elongation.

11.4 How Do We Know . . . Why Some Rocks Break and Others Flow?

Define the Problem

How Strong Are Rocks? Are all rocks equally strong? What factors determine whether rocks break or flow? Do the higher temperature and pressure below Earth's surface affect rock strength? The answers to these important questions help us understand how rocks deform

and explain the origins of structures observed and measured in the field. Geologists answer these questions by performing laboratory experiments.

The Experimental Setup

How Do Geologists Deform Rocks in the Lab? Every rock has a stress limit and breaks or flows if stress increases beyond that limit. This stress limit, known as the **yield strength**, can be determined for different rock types in a laboratory apparatus.

Figure 11.16 depicts a typical experimental setup. Rock samples are cut into small cylinders, and the lengths and diameters of the cylinders are measured. Then, a rock cylinder is placed between an anvil and a moveable piston. Hydraulic fluid is poured into a chamber encircling the rock and maintained at a chosen confining pressure, which represents surrounding rocks at a selected depth below Earth's surface. For example, by applying 250 bars of confining pressure, which is about 250 times greater than air pressure at sea level, a geologist can measure rock properties in conditions equivalent to one kilometer below the surface. Also, the whole apparatus can be heated to reproduce temperatures found below the surface. Pushing the piston down on the sample produces a compressional stress that is calculated from the amount of force exerted on the piston. Measured vertical shortening of the rock specimen records the resulting strain.

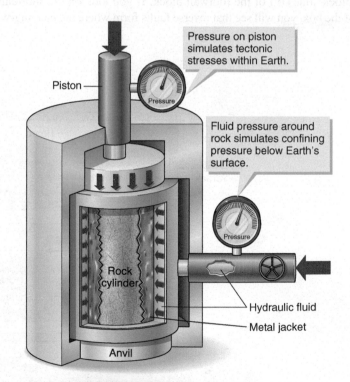

▲ **Figure 11.16 Experimental setup for measuring rock strength.** A small, cylindrical rock specimen is placed inside the metal jacket in the center of the sample chamber. Compressional stress is applied from the top by pressing down on a piston, which forces the sample against a fixed anvil at the bottom of the chamber. Confining pressure is controlled by pumping in hydraulic fluid to surround the jacket and keep it from buckling. Also, the chamber can be heated to investigate deformation at different temperatures.

First Results

How Do Rocks Strain? **Figure 11.17** illustrates results of rock-deformation experiments with cylinders of limestone subjected to compression. At low confining pressure (equivalent to shallow depths within Earth) and small stress, the cylinder does not change shape (Rock 1). With increasing stress the cylinder shortens by breaking and sliding along fractures that are like faults seen in the field (Rock 2). At higher confining pressure, equivalent to greater depth within Earth, the cylinders shorten by flow without visible fractures (Rock 3). Flow results from microscopic shifts within the crystal structures of the minerals composing the rock. These experimental results indicate that small stress produces no deformation and that rocks strain by either breaking or flowing when subjected to large stresses. We gain a clearer understanding of what actually happens during deformation if we compare the amount of strain that results from applying different amounts of stress.

Figure 11.18 illustrates the relationship between stress and strain revealed by experiments with three different rocks. The confining pressure is equal to Earth's surface for all three experiments. The rock cylinders exhibit different shortening strains at the same stress. Rock A deforms by breaking, whereas the other two flow (Figure 11.18a). These results translate into graph curves (Figure 11.18b).

The plots for all three rocks start with a sloping line that shows strain increasing as stress increases (Figure 11.18b). Within this straight-line part of each plot, if the stress is reduced, the strain also diminishes. In other words, the rock returns to its original dimensions after the stress is removed, just like an elastic rubber band returns to its original shape after you stretch it and then let go. Deformation under these low stress and low strain conditions are **elastic** (also see Section 8.2). In contrast, when stress increases to the yield strength of the rock, the resulting deformation is permanent and cannot be reversed by decreasing the stress. When stress exceeds the yield strength, the rock either breaks or flows.

Rock A in Figure 11.18 breaks when stress exceeds the yield strength. The term **brittle** describes the deformation of rock A because it breaks at a relatively small strain.

Instead of breaking, Rocks B and C permanently deform by flow when stress exceeds the yield strength. Geologists use the term **plastic** (or ductile) to describe rocks that flow without breaking.

How do you compare the strengths of the three rocks in Figure 11.18? Rock A is the strongest because it has the highest yield strength, which is the same as saying that it sustains more stress before exhibiting permanent strain. Rock B is described as

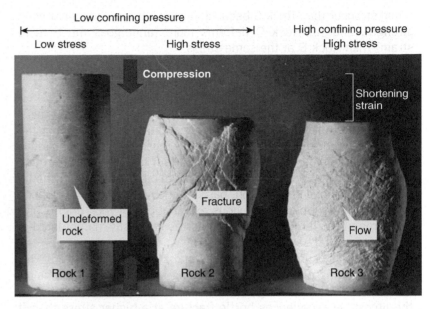

▲ **Figure 11.17 Rock cylinders deform in the lab.** These are examples of rock cylinders used to determine the strength and type of strain exhibited by limestone. Flow strain occurred at higher confining pressure than fracture strain. Notice that Rocks 2, 3, and 4 exhibit the same vertical shortening compared to the undeformed sample (Rock 1).

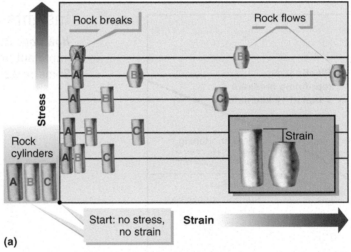

When rock cylinders initially compress, they shorten slightly in an elastic fashion. At higher stress, rock A breaks by fractures (like rock 2 in Figure 11.17). Rocks B and C deform by flow at yet higher stresses.

(a)

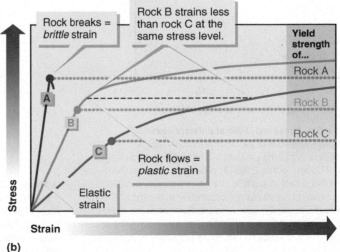

This graph summarizes the stress and strain relationships for the three rocks. Initial elastic strain makes straight lines on the graph up to the stress value that equates to the yield strength of the rock. Brittle fracture occurs at the end of the elastic strain line for rock A, and curves track the plastic flow of rocks B and C at higher stress.

(b)

◀ **Figure 11.18 Understanding stress-strain curves.**

being stronger than Rock C because (a) Rock B has the higher yield strength, and (b) Rock C deforms more (undergoes much more strain) than Rock B at the same stress.

Second Results

What Determines Rock Strength? The observation that different rocks possess different strengths at surface conditions indicates that the types of minerals found in rocks and how those mineral grains interconnect with one another must be important factors that determine rock strength. Temperature and pressure are other factors that should come to mind for consideration, based of the persistent reappearance of these two variables in previous explanations of rock origins and rock properties. Temperature and pressure are easily manipulated in the experimental setup.

Figure 11.19 compares the stress and strain of limestone at different confining pressures corresponding to different depths below the surface. At a confining pressure equivalent to about 1 kilometer, the limestone experiences brittle fracture at a higher stress than it does at surface pressure. At higher confining pressures, representing depths of 6 and 12 kilometers, the limestone exhibits plastic, rather than brittle, deformation. Two conclusions emerge from these experiments:

1. The strength of the rock increases with increasing confining pressure, which implies that rock becomes stronger where it is deeper below the surface. The explanation for the changing strength is that the high pressure compresses the chemical bonds within minerals and compresses the mineral grains against one another, both of which increase rock strength and make it increasingly difficult to deform the rocks.

2. When the stress exceeds rock strength, brittle fracture occurs close to the surface but plastic flow occurs at greater depths below the surface. The explanation for this difference in type of strain is that strain more likely occurs by dislocations of atoms within the rock-forming minerals at high pressure (causing the appearance of flow), whereas boundaries between minerals and chemical bonds within minerals physically break at low pressure.

This experiment with limestone does not account for the possible effect of increasing temperature below the surface. Now let's look at results of experiments carried out at different temperatures while keeping a constant confining pressure.

Figure 11.20 shows how increasing temperature of marble lowers the strength of the rock. The graph of stress and strain shows that the same stress causes much more plastic flow at 150°C than at room temperature. Molecules vibrate more rapidly at high temperature than at low temperature, and the vibration weakens the bonds that resist strain.

Insights

How Does Deformation Vary from Place to Place? The lab results have important implications for how deformation should vary within Earth. Some rocks are strong and others are weak; some are more prone

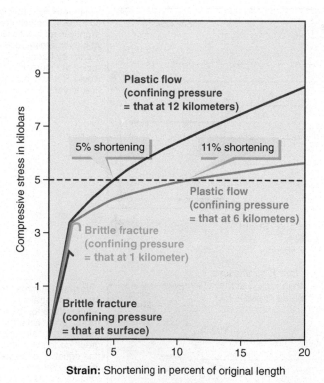

▲ Figure 11.19 **Experimental results—stress and strain at different confining pressures.** This graph summarizes the results of four compressional-stress experiments on limestone at different confining pressures. These different confining pressures correspond to different depths within Earth. At pressures between the surface and 1 kilometer, the limestone fails in a brittle manner, displaying greater strength at higher pressure. At confining pressures corresponding to depths of 6 and 12 kilometers, the limestone deforms by plastic flow. Less plastic shortening strain occurs at a particular stress (at 5 kilobars, for example) at a pressure equivalent to 12 kilometers than at 6 kilometers, which indicates that rocks are stronger at greater depth.

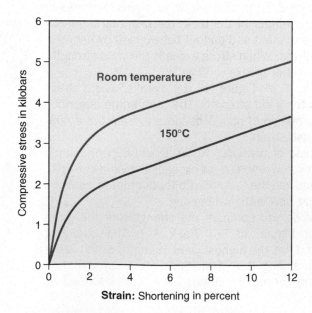

▲ Figure 11.20 **Experimental results—stress and strain at different temperature.** A graph of the experimental stress and strain data shows that marble is weaker at higher temperature than at low temperature. The same confining pressure, equivalent to about 30 kilometers deep in the crust, was applied in all the experiments. Increasing temperature from room temperature to 150°C lowers the strength of the rock as demonstrated by the greater strain at the same compressional stress.

to brittle fracture and others to plastic flow. At increasing depth within Earth, pressure and temperature work together, to decrease brittle fracture and increase plastic flow. Therefore, we can conclude that faults and joints, which are expressions of brittle deformation, should occur only near Earth's surface, whereas plastic flow is more likely at depth. The experimental results that describe and explain brittle and plastic deformation can now be combined with stress types so that we can fully understand rock deformation.

Putting It Together—How Do We Know . . . Why Some Rocks Break and Others Flow?

- At different combinations of stress and strain rocks exhibit elastic, brittle, or plastic deformation.

- Increasing confining pressure increases rock strength and leads to plastic flow rather than brittle fracture.

- Increasing temperature decreases rock strength and enhances plastic flow.

11.5 How Do Geologic Structures Relate to Stress, Strain, and Strength?

Sandboxes, wax models, and compression of rock cylinders all provide key pieces of understanding about how and why rocks deform. Geologists combine these laboratory insights with field observations to form a comprehensive picture of the deformed crust summarized in **Figure 11.21**. **Figure 11.22** illustrates examples of regional landscapes in the United States that relate to rock deformation.

It is also important to keep in mind the origin of stress. Up to this point you have seen the results of squeezing wax, cranking a board through a sandbox, and compressing rock cylinders between pistons and anvils. In nature, plate motions are the origin of most of the stresses that deform rocks. You will learn more about plate motions in the next chapter. Here, it is simpler to think of these plate-tectonic compressional, tensional, and shear stresses as caused by motions within the interior of Earth and expressed at the surface as rock deformation.

Deformation by Tension

Figure 11.21a illustrates the deformation that occurs in areas that experience tensional stress. Tension is typical of locations where upward moving and spreading mantle stretches the lithosphere, such as at divergent plate boundaries. The overall deformation pattern is one of

(a) Deformation associated with tensional stress

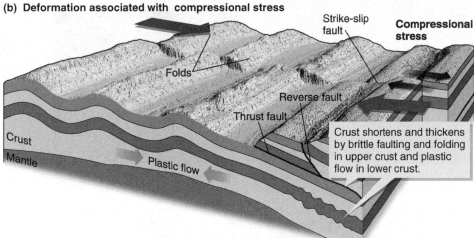

Strike-slip fault

Tensional stress

Horst

Graben

Crust

Graben

Normal faults

Mantle

Plastic flow

Crust stretches and thins by brittle normal faulting in upper crust and plastic flow in lower crust.

(b) Deformation associated with compressional stress

Strike-slip fault

Compressional stress

Folds

Reverse fault

Thrust fault

Crust

Mantle

Plastic flow

Crust shortens and thickens by brittle faulting and folding in upper crust and plastic flow in lower crust.

(c) Deformation associated with shear stress

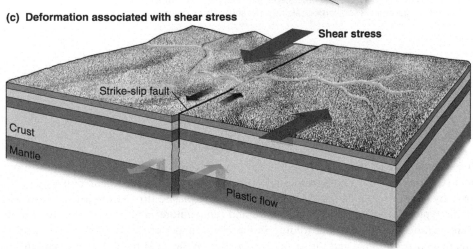

Shear stress

Strike-slip fault

Crust

Mantle

Plastic flow

◀ **Figure 11.21 What rock deformation looks like on a large scale.** Stresses deform rocks by brittle fracture and folding at shallow depths. Plastic flow occurs in the lower crust and mantle where pressure and temperature are much higher than in the upper crust. Tensional stress forms horst and graben blocks bounded by normal faults. Compressional stress forms thrust and reverse faults along with folds. Strike-slip faults form where faults are parallel to extensional and compressional stresses. Strike-slip faults also are the dominant structure where shear stress is the principal deforming stress.

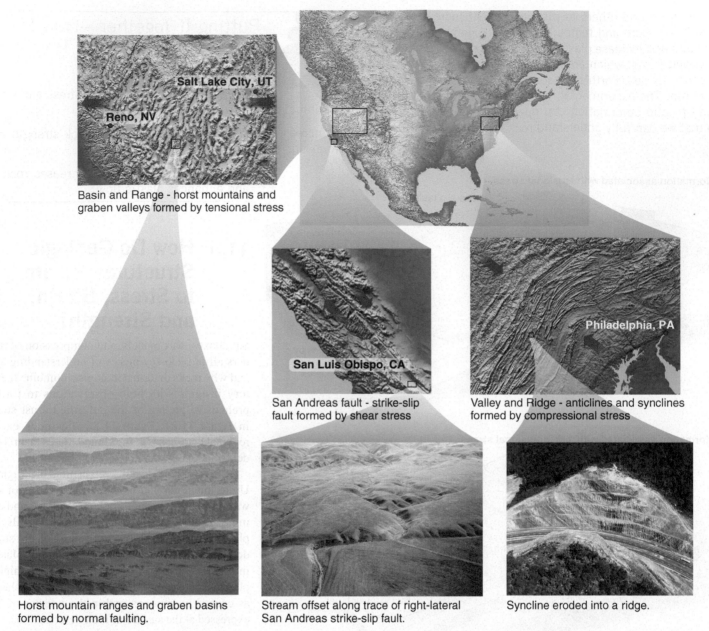

Basin and Range - horst mountains and graben valleys formed by tensional stress

San Andreas fault - strike-slip fault formed by shear stress

Valley and Ridge - anticlines and synclines formed by compressional stress

Horst mountain ranges and graben basins formed by normal faulting.

Stream offset along trace of right-lateral San Andreas strike-slip fault.

Syncline eroded into a ridge.

▲ **Figure 11.22 Visualizing landscapes of deformed rock.** Many large-scale landscape features of North America relate directly to rock deformation. The Basin and Range region of Nevada and adjacent states illustrates landscapes formed by normal faults. The Valley and Ridge region of the Appalachian Mountains formed by compressional stresses about 300 million years ago and produced a landscape that resembles a large rumpled carpet. Long, linear valleys are eroded along the San Andreas Fault in California, where large-scale regional shear stress deforms the crust.

elongation and thinning of the crust. Plastic flow in the lower crust resembles stretching taffy, whereas normal faults break the brittle upper crust and thin the crust as hanging-wall blocks slide down and away from footwall blocks. Where fractures closely parallel the stress direction, rocks slide past one another on strike-slip faults.

Blocks of crust jostled upward along normal faults are **horsts**, while those jostled downward are **grabens**. "Graben" originates from the German term *grabe*, meaning "ditch," and "horst" describes the nests built by large birds of prey, typically on a high cliff. Figure 11.22 illustrates the typical jumbled landscape resulting from tensional tectonic stress. Many parallel horsts and grabens form the Basin and Range region of Nevada and parts

of adjacent states; the valley basins are grabens and the mountain ranges are horsts.

Deformation by Compression

Figure 11.21b summarizes the deformation that happens in areas subjected to compressional stress, which usually are near convergent plate boundaries. Shortening occurs by plastic flow in the lower crust, where relatively high pressure and temperature inhibit brittle fracture, and by faulting and folding in the brittle upper crust. The faults are mostly reverse and thrust faults that shorten the crust by sliding the hanging-wall block up and over the footwall block. Where fractures are oriented parallel to the stress direction, the rocks slide past one another on strike-slip faults. Shortening also thickens crust and raises surface elevations in a fashion similar to that seen in the wax model (Figure 11.15). Compression formed the plunging anticline and syncline at the field site (Figure 11.9).

Deformation by Shear

Figure 11.21c illustrates structures formed in a region where shear stress dominates the deformation without significant shortening or elongating. Transform plate boundaries are an example of such a stress scenario. Strike-slip faults form in the upper crust and continue downward into a plastic zone where rocks flow past one another in opposite directions.

How Folds Form Near the Surface

Distinguishing brittle fracture from plastic flow does not fully explain formation of folds. Some folds do record plastic flow and these flow folds are dramatically illustrated in metamorphic rocks that form deep below the surface, where high temperature and pressure favor plastic deformation (Figure 11.5). However, we know that not all folding occurs by plastic flow because folds are most common in sedimentary rocks, such as those pictured in Figure 11.4 and at your virtual field site. Most folded sedimentary rocks could not have deformed by plastic flow because plastic flow requires temperatures and pressures where the rocks would be metamorphic rather than sedimentary.

One cause of folding at low temperature and pressure is seen where thin sedimentary layers bend significantly before they break. An analogy would be to bend a sheet of plywood considerably before it breaks, as **Figure 11.23** shows. Similarly, the top layer of wax in the model illustrated in Figure 11.15 folded to some extent before it started to break. Sedimentary layers commonly crack into joints and small faults as they bend, which further demonstrates that folding takes place under conditions of brittle strain. These small fractures are visible in Figure 11.4 and in the wax model in Figure 11.15.

Sedimentary rocks also fold because weak bedding planes allow rock layers to slip past one another without breaking. **Figure 11.24** draws an analogy to folding a stack of papers—the sheets bend and slip past one another when compressed. Distortion of writing on the edge of the stack of sheets demonstrates that the sheets slide past one another.

Some folds form near the surface in layered rocks that overlie faulted rocks. **Figure 11.25** illustrates this scenario where a fault in unlayered igneous rock passes upward into folded sedimentary rocks. The deforming sedimentary rocks slip along their bedding planes to produce a fold rather than breaking along the fault. These types of folds and related underlying faults are common features in the Rocky Mountains of the western United States.

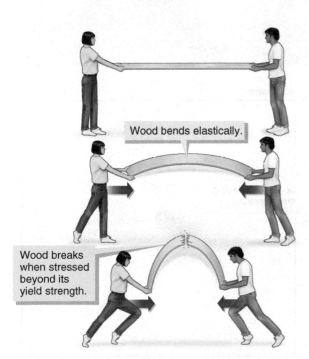

Wood bends elastically.

Wood breaks when stressed beyond its yield strength.

▲ **Figure 11.23 Using plywood as an analogy for rock bending.** When two people walk toward one another while grasping a sheet of plywood, the wood bends until the stress reaches the yield strength of the wood, at which point it breaks. Thin, wide sheets of plywood have similar relative dimensions to thin sedimentary beds that are continuous over large areas. In a similar fashion, rock layers fold before faulting when compressed.

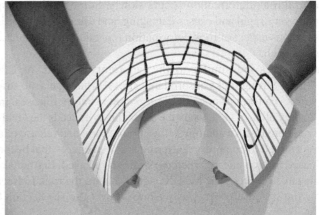

▲ **Figure 11.24 Folding a stack of paper is like folding sedimentary rocks.** When you fold a stack of paper, the sheets slide past one another, as indicated by the changing shape of the letters written on the edge of the stack. Similarly, sedimentary layers slide along bedding planes when compressive stress folds the rocks.

▶ **Figure 11.25 Folds may form above faults.** The diagram shows how movement along a fault causes overlying sedimentary rocks to fold over, or drape, the uplifted fault block. Although the fault displacement may later break to the surface, at this stage the folded sedimentary layers are the only hint of the presence of the buried fault. The photo shows a locality in northern Arizona where sedimentary rocks are folded above a concealed fault.

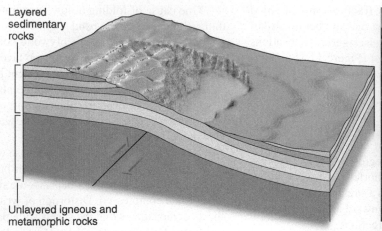

Layered sedimentary rocks

Unlayered igneous and metamorphic rocks

Putting It Together—How Do Geologic Structures Relate to Stress, Strain, and Strength?

• Brittle faults and joints form in rocks close to the surface whereas plastic flow occurs at deeper levels where high pressure inhibits fracturing and temperature lowers rock strength.

• Tension elongates and thins crust by movement on normal faults in shallow rocks and taffy-like stretching of plastic rocks at deeper levels.

• Compression shortens and thickens crust by producing folds, reverse faults, and thrust faults in shallow rock, and plastic flow in deeper rock.

• Folds form by plastic flow deep in the crust but also develop at and near the surface in sedimentary rocks where bending of thin layers and slippage along layers take place.

11.6 How Do Earthquakes Relate to Rock Deformation?

Although rock, sand, and wax deform rapidly in laboratory experiments, most natural rock deformation occurs slowly. Some motion along faults, however, is very rapid and causes damaging, and often deadly, earthquakes. This hazardous aspect of rock deformation represents a major area of geoscience research.

Earthquakes are similar to the laboratory fracture of Rock 2 in Figure 11.17. Strong rocks absorb large stresses during elastic strain. That stored stress is a form of potential energy that instantly releases, mostly as motion energy along a fault, when the stress finally exceeds the yield strength and the rock breaks. An analogy is the compression and release of a spring; the stress applied to compress the spring is stored energy that becomes rapidly released motion energy when the spring is released. Earthquakes occur when the stress exceeds the rock strength, causing the rock to fracture and releasing the stored-up energy as seismic waves (also see Section 8.2).

How Earthquakes Relate to Faults

Historic observations demonstrate that earthquakes are caused by movement along faults. **Figure 11.26**, for example, shows how earthquake foci and epicenters line up along recognizable faults (the focus is the location of

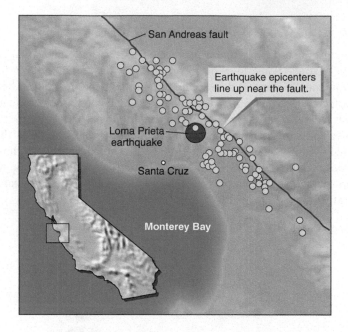

San Andreas fault

Earthquake epicenters line up near the fault.

Loma Prieta earthquake

Santa Cruz

Monterey Bay

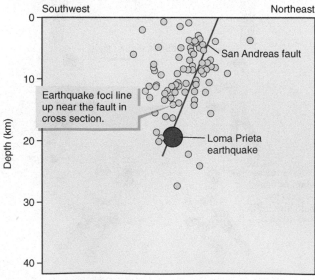

Southwest   Northeast

San Andreas fault

Earthquake foci line up near the fault in cross section.

Loma Prieta earthquake

Depth (km)

▲ **Figure 11.26 Earthquakes occur along faults.** The map and cross section show locations of earthquakes near the central California coast during October 1989. The larger circle marks the location of the Loma Prieta earthquake, which caused substantial damage in Santa Cruz and the San Francisco Bay area. Notice that most of the earthquakes coincide closely with the location of the San Andreas Fault.

1-meter offset of highway pavement in southern California occurred during movement on strike-slip fault in 1992.

3-meter high scarp in central Idaho occurred during movement on normal dip-slip fault in 1983.

◀ **Figure 11.27 Faults move during earthquakes.** These photographs, taken a short time after powerful earthquakes, document that blocks of crust move along faults during earthquakes.

the earthquake and the epicenter is the spot on Earth's surface directly above the earthquake; Figure 8.10). **Figure 11.27** illustrates observed displacement along faults during historic earthquakes. Movement along a dip-slip fault causes an abrupt vertical displacement of the ground surface called a **fault scarp**.

Millions of faults crisscross Earth's crust and mark planes of brittle rock failure. Some faults, such as the famous San Andreas Fault in California, coincide with tectonic plate boundaries. Most faults are relatively insignificant producers of modern earthquakes—they are only a few kilometers long, or have experienced very small amounts of slip, or do not show evidence of displacement in the recent geologic past. Not all faults are exposed at Earth's surface. Instead, recently deposited sediment or folded rock layers conceal many deeper faults until earthquakes reveal their locations.

Earthquakes result from brittle failure along faults, so you can predict that earthquake foci should occur at relatively shallow depths, where temperature and pressure are low. An example graph in **Figure 11.28** shows that earthquakes in continents overwhelmingly occur at depths less than 15 kilometers. Deformation at greater depth is typically plastic flow rather than brittle fracturing that produces faults.

This is not to say that earthquakes do not occur at greater depths. Earthquakes as deep as 640 kilometers have been recorded but they are restricted to narrow zones near convergent plate boundaries. Chapter 12 examines clues for the origin of these very deep earthquakes.

Unlike a single rock specimen that breaks once in the lab, rocks along faults can move many times. Movement along some faults has occurred several times during recorded history yielding total displacements of meters to tens of meters. Measurements of rock layers offset across the same faults also show displacements of many kilometers. These larger displacements indicate that many earthquakes occurred along these faults during long intervals of geologic time.

Repeated failure along faults indicates that they have strength, just as we describe a rock has strength. The tectonic stress must exceed the rock or fault strength in order for displacement to occur. Faults have strength because they are not perfectly smooth planes. Instead, irregular surfaces along the fractured rocks on either side of the fault fit like puzzle pieces. The irregular interlocking surfaces along the fault resist movement, which increases the strength of the fault. Even on smoother parts of the fault surface, friction impedes movement. Interlocked crushed rock fragments along the fault are, however, much weaker than unbroken rock a few meters away. Therefore, as stress builds up, the relatively weak locked-up fault fails first rather than a new fault forming in stronger, unbroken rock.

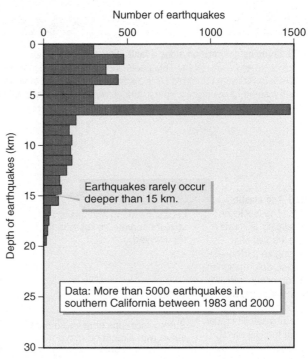

▲ **Figure 11.28 Earthquakes occur in the strong upper crust.** The graph shows the abundance of earthquake foci at different depths below the surface. Notice that earthquake foci are very rare below 15 kilometers and are almost nonexistent below 20 kilometers. Deeper than 20 kilometers the continental crust deforms by plastic flow without producing earthquakes.

Fault Movement Without Earthquakes

Not all faults have experienced historic earthquakes, yet the fact that these faults have moved in recent years is readily visible, as illustrated in **Figure 11.29**. Some faults are too weak to store up the energy of applied stress. As a result, rocks creep past one another on either side of the fault without generating earthquakes.

Elastic Rebound Theory Explains Building Stress Along Faults

The **elastic rebound theory**, illustrated in **Figure 11.30**, explains earthquakes as sudden brittle failure following the buildup of elastic strain. The

▲ **Figure 11.29 Creeping deformation along a fault.** Earthquakes do not accompany all movements along faults. This photo shows right-lateral strike-slip offset of a sidewalk in Hollister, California. This displacement did not occur during earthquakes but, instead, takes place by persistent, slow, creeping movement along the fault.

concept compares well with the representation of Rock A on the stress-strain graph in Figure 11.18, which experiences some degree of elastic strain occurs before sudden brittle fracture. Rocks on either side of a locked-up, nonmoving fault slowly bend as stress builds up. When the stress eventually exceeds the strength of the fault, the bent rocks suddenly snap back to their prestressed orientation, along with an abrupt displacement along the fault. The bending of the rock is nonpermanent elastic deformation that is recovered, or rebounds, after the fault breaks; hence the name "elastic rebound." The ability of the rock to bend and store energy is much like the illustrated bending of plywood prior to breaking (Figure 11.23), or the stretching of a rubber band until it breaks.

Careful surveying of locations on either side of faults before and after earthquakes provides evidence supporting the elastic rebound theory. Example data are mapped in Figure 11.30 and show elastic strain along the Hayward fault in California. Almost no displacement occurs near the locked strike-slip fault. Instead, the shear stress bends the rock on either side of the fault. The bending is elastic strain caused by stress along the fault that will eventually reach the yield strength of the fault and result in an earthquake. Unfortunately, it is difficult to determine the threshold of stress accumulation that is required for the fault to rupture, or else specific earthquake predictions might be possible.

Foreshocks, Mainshock, and Aftershocks

All the stored energy in the bent rocks rarely releases in a single, large earthquake. *Most* of the energy releases during a single **mainshock**, but smaller

▶ **Figure 11.30 The elastic rebound theory.** The blocks on the left schematically illustrate surveyed locations that shift before and during an earthquake because of elastic deformation in blocks of rock on either side of the fault. The map on the right uses colors to show surface motion detected by satellite radar on either side of the Hayward fault in California. The pattern of displacement indicates elastic strain across the fault.

Locations along an initially straight survey are repeatedly resurveyed.

Survey locations shift into a bent curve with almost no movement near the fault but increasing movement away from the fault.

Eventually, the stress reaches the yield strength of the fault and brittle failure occurs, which causes an earthquake.

Resurvey of the initial locations shows that they form two line segments, displaced by the fault.

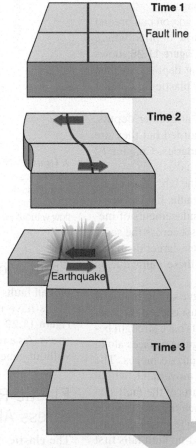

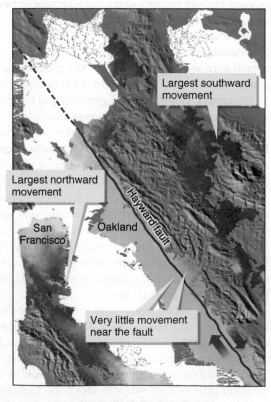

Detected motion, (mm/yr)

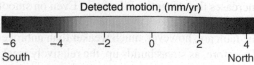

−6 South −4 −2 0 2 4 North

ACTIVE ART

Elastic Rebound. See how elastic strain deforms rocks along a fault before an earthquake and what happens during an earthquake.

foreshocks and **aftershocks** precede and follow the mainshock. The mainshock records breakage of most of the fault surface. Foreshocks represent small displacements of the weaker parts of the fault surface as stress rises, while aftershocks are breaks along small areas of the strongest part of the fault surface and fractures in nearby rocks. Returning to the stressed plywood analogy (Figure 11.23), we can say that cracking sounds heard before the board breaks are comparable to foreshocks. After the board breaks, representing the mainshock, there may be a few splinters that still connect the pieces until the board bends further. Breaking these last connecting splinters is analogous to some aftershocks. Additional aftershocks occur as some fracturing takes place in the rocks near the fault where not all rebound is purely elastic.

It is difficult to know whether a small earthquake is a foreshock preceding a major earthquake or simply a minor solo event. Foreshocks usually are recognized as such only after the mainshock occurs, so foreshocks are rarely useful in forecasting a major earthquake.

Hundreds of aftershocks commonly follow large earthquakes and may persist for weeks or even months. Most aftershocks are relatively small, but some are large enough to cause the collapse of buildings damaged by the mainshock. Aftershocks, therefore, greatly hamper relief and rescue efforts following earthquake disasters.

Figure 11.31 illustrates foreshocks and aftershocks related to a great earthquake in central Alaska in 2002. The modest Nenana Mountain earthquake and related aftershocks resulted from brittle failure along part of the Denali strike-slip fault. These earthquakes were apparently foreshocks to the larger Denali earthquake, which happened about a week after the Nenana Mountain events. The Denali fault earthquake started with failure at the focus along the Susitna Glacier thrust fault, and displacement continued eastward and southeastward for 90 seconds to cause 200 kilometers of rupture along two strike-slip faults. Hundreds of aftershocks occurred in the following weeks. Most of the aftershocks occurred along the same three faults that failed during the mainshock and represent small areas of the fault surfaces that did not fail during the larger earthquake. It is important to note, however, that not all aftershocks take place on the same fault plane where the mainshock occurred. Release of stress along the mainshock fault may transfer stress to adjacent faults, causing them to rupture and contribute to the aftershock pattern.

Putting It Together—How Do Earthquakes Relate to Rock Deformation?

• Earthquakes result from motion on faults. Earthquakes are caused by brittle failure of rock and hence occur in the upper crust where temperature and pressure are relatively low.

• Not all motion on faults produces earthquakes; rocks creep along faults when faults are too weak to store up the energy of prolonged stress.

• Elastic rebound theory explains deformation before and during earthquakes as brittle failure following accumulation of elastic strain. The bending of rocks on either side of a locked fault stores energy that is then released when stress exceeds the fault yield strength. The release of the stored energy causes the earthquake.

• Foreshocks and aftershocks are smaller earthquakes that occur before and after the mainshock.

11.7 How Are Earthquakes Measured?

News reports of earthquakes always mention the earthquake magnitude. Magnitude is a way of communicating the size of an earthquake and the amount of strain energy that was released. Early efforts to measure earthquake size relied on describing earthquake effects, but now geologists measure earthquake size based on measurement of earthquake waves recorded on seismometers (Figure 8.8).

Measuring Intensity

Intensity measures the violence of ground shaking during an earthquake in terms of the extent to which people felt the earthquake, the damage to structures, and the severity of secondary effects such as landslides. Giuseppi Mercalli developed the first intensity scale in 1902, and the updated version is the Modified Mercalli Intensity Scale. **Table 11.1** describes the scale, which assigns the typical observed effects of earthquakes to categories indicated by Roman numerals from I to XII. The effects of an earthquake vary from place to place, so the intensity of the earthquake also differs at different locations. This means that a single earthquake does not have a single intensity value.

Figure 11.32 illustrates Modified Mercalli Intensity maps for several large U.S. earthquakes. To make these maps, geologists examine the damage done by earthquakes in different locations and interview witnesses to determine how severe the shaking was at these locations. Historic accounts in newspapers and diaries provide the information to determine the effects for earthquakes that happened prior to the development of the intensity scale. The appropriate Roman numeral is placed on a map at each location to indicate the intensity of the earthquake at that point. Then, lines

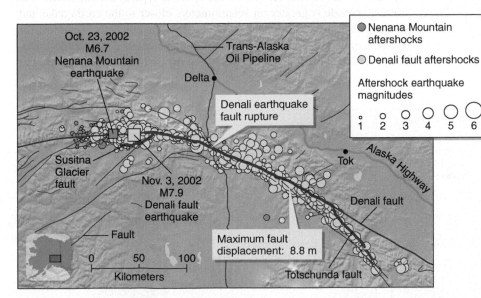

▲ **Figure 11.31 Foreshocks and aftershocks show rupture along faults.** This map plots earthquake epicenters in central Alaska in fall 2002. The Nenana Mountain earthquake and its aftershocks were possibly foreshocks for the larger Denali fault earthquake. Movement occurred on all three labeled faults during the Denali fault earthquake. Small areas of these fault planes that did not break during the Denali fault earthquake ruptured over the next few weeks, producing the plotted aftershocks. Some aftershocks also occurred on nearby faults where stress transferred after the big earthquake. The Denali fault rupture passed beneath the Trans-Alaska Oil Pipeline but did not break the pipeline.

Table 11.1 Modified Mercalli Intensity Scale

Intensity	Abbreviated Description of Effects
I	Not felt except by a very few under especially favorable conditions.
II	Felt only by a few persons at rest, especially on upper floors of buildings.
III	Felt quite noticeably by persons indoors, especially on upper floors of buildings. Many people do not recognize it as an earthquake. Standing motor cars may rock slightly. Vibrations similar to the passing of a truck. Duration estimated.
IV	Felt indoors by many, outdoors by few during the day. At night, some awakened. Dishes, windows, doors disturbed; walls make cracking sound. Sensation like heavy truck striking building. Standing motor cars rocked noticeably.
V	Felt by nearly everyone; many awakened. Some dishes, windows broken. Unstable objects overturned. Pendulum clocks may stop.
VI	Felt by all, many frightened. Some heavy furniture moved; a few instances of fallen plaster. Damage slight.
VII	Damage negligible in buildings of good design and construction; slight to moderate in well-built ordinary structures; considerable damage in poorly built or badly designed structures; some chimneys broken.
VIII	Damage slight in specially designed structures; considerable damage in ordinary substantial buildings with partial collapse. Damage great in poorly built structures. Fall of chimneys, factory stacks, columns, monuments, walls. Heavy furniture overturned.
IX	Damage considerable in specially designed structures; well-designed frame structures thrown out of plumb. Damage great in substantial buildings, with partial collapse. Buildings shifted off foundations.
X	Some well-built wooden structures destroyed; most masonry and frame structures destroyed. Rails bent.
XI	Few, if any (masonry) structures remain standing. Bridges destroyed. Rails bent greatly.
XII	Damage total. Lines of sight and level are distorted. Objects thrown into the air.

(*Source:* USGS—http://neic.usgs.gov/neis/general/mercalli.html)

are drawn to enclose areas of equal intensity. The maximum intensities, indicated by the highest Roman numerals, are located near the epicenters of the earthquakes depicted in Figure 11.32. Notice that the numbers generally decrease away from the epicenter. Before the development of seismographs, intensity maps were commonly used to approximately locate earthquake epicenters. The Modified Mercalli Intensity map of the New Madrid, Missouri, earthquakes shows how much damage can result at great distances from the epicenter. The largest New Madrid earthquake was felt 2000 kilometers away, and its maximum intensity reached Roman numeral XII. Damage was reported over a 500,000-square-kilometer area, which is larger than the area of California.

The Modified Mercalli Intensity Scale contains numerical values but it is still really a qualitative measure of an earthquake. It is challenging to apply this scale when earthquakes take place in unpopulated areas with few reports of the intensity of the shaking and a lack of buildings. Nonetheless, maps such as those shown in Figure 11.32 are useful to emergency management agencies and relief organizations because they provide an indication of the areas potentially affected by large earthquakes in the future.

Measuring Magnitude

Magnitude is a measure of earthquake size that is related to the energy released during the earthquake. It is not a directly measured physical quantity, such as temperature, but rather a value that is calculated from measurements made after an earthquake. Some magnitude calculations rely on measurement of the height (also known as the amplitude) of seismic waves on seismograms. Different magnitude scales use the amplitude of primary (P), secondary (S), or surface waves (see Section 8.2 for a refresher on types of earthquake waves).

Charles Richter developed the first magnitude scale in 1935, and **Figure 11.33** shows how Richter magnitude is calculated. To arrive at the Richter magnitude, seismologists measure the maximum amplitude (up-and-down motion) recorded on a specific type of seismometer. The wave amplitude is greater on seismometers closer to the earthquake and smaller when the seismometer is farther from the epicenter. To adjust for the different wave amplitudes at different distances, the magnitude calculation also uses the time elapsed between the arrival of P and S waves at

▶ **Figure 11.32 What earthquake intensity maps look like.** These maps illustrate the distribution of earthquake intensity effects for three U.S. earthquakes. Intensity is highest close to each earthquake epicenter and diminishes with distance away from the epicenter. Regions of equal intensity for the 1906 San Francisco earthquake and 1992 Landers earthquake are elongated northwest-southeast parallel to the fault ruptures during these events. Stronger earthquakes (San Francisco and New Madrid) are felt over larger areas than smaller earthquakes (Landers). The intensity map is incomplete west of the New Madrid epicenter because of a lack of population to report earthquake effects.

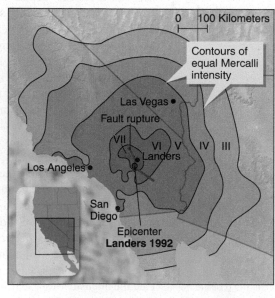

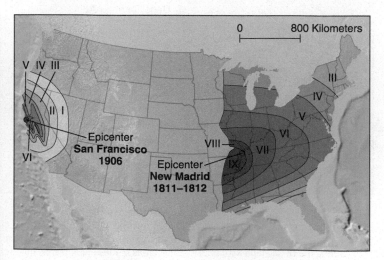

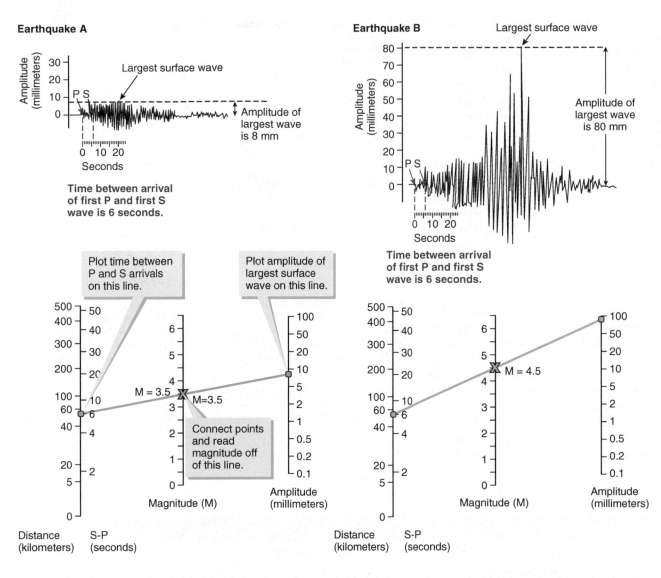

Earthquake A

Largest surface wave

Amplitude of largest wave is 8 mm

Time between arrival of first P and first S wave is 6 seconds.

Plot time between P and S arrivals on this line.

Plot amplitude of largest surface wave on this line.

M = 3.5   M=3.5

Connect points and read magnitude off of this line.

Distance (kilometers)   S-P (seconds)   Magnitude (M)   Amplitude (millimeters)

Earthquake B

Largest surface wave

Amplitude of largest wave is 80 mm

Time between arrival of first P and first S wave is 6 seconds.

M = 4.5

Distance (kilometers)   S-P (seconds)   Magnitude (M)   Amplitude (millimeters)

◄ **Figure 11.33 How to determine Richter magnitude.** Richter magnitude is determined by making two measurements on the seismogram: (1) the time that passes between arrival of the first P and the first S waves (which relates to the distance of the seismic station from focus), and (2) the amplitude, or height, of the largest recorded surface wave. If a line is drawn between plots of these two measurement points on a special graph, the line intersects the magnitude scale, providing the Richter magnitude. Notice that the distance from the seismic station to the focus is the same for the two earthquakes (as indicated by the time elapsed between arrival of the first P and first S waves). Earthquake B produced surface waves that were 10 times higher than those produced during earthquake A, but the earthquake magnitudes differ by only one unit (3.5 compared to 4.5).

the seismic station. Consider looking back at Section 8.2 and Figure 8.9 to refresh your memory of how this time elapsed on the seismogram relates to the distance between the earthquake focus and the seismic station. For every whole-number increase on the magnitude scale, the measured seismic-wave amplitude increases tenfold (Figure 11.33); therefore, a magnitude-5 (M5) earthquake has a wave amplitude 10 times larger than an M4, and 100 times larger than an M3 earthquake.

Unlike intensity, which varies from place to place for a single earthquake, there is only a single magnitude assigned to an earthquake. The news media commonly refer to this value as the "Richter magnitude," even though seismologists more frequently apply different, more recently developed methods to determine magnitude.

Magnitude scales based on seismograms assume that an earthquake occurs at one point and releases all of its energy in a single, very short instant of time. This approach is contrary to observations such as those recorded for the Denali fault earthquake in Figure 11.31, where the earthquake energy was released along nearly 300 kilometers of ruptured faults over more than 90 seconds. Richter's approach is analogous to the instantaneous breaking of a window when struck by a rock, whereas large earthquakes result from tearing of the upper crust over long distances over periods of many seconds to tens of minutes. An amazing example is provided by the horrific December 2004 earthquake near the

Indonesian island of Sumatra, which led to the deaths of about 230,000 people. The earthquake began at one end of a long thrust fault along a convergent plate boundary. The fault tore through the crust at a speed in excess of 8000 km/hr for nearly 10 minutes, rupturing more than 1500 km of the fault (roughly equal to the distance between San Francisco and Kansas City). But, the earthquake was still not over; slow motion continued along more of the fault for another 30 minutes until the final displacement, ranging in different places from 7 to 20 meters, was accomplished. Energy recorded as wavy squiggles on seismographs all over the world was released over a period of 40 minutes and originated from movement across more than 2,000,000 square kilometers of fault surface.

Because large earthquakes do not fit the assumptions of the Richter scale calculation, geologists calculate the "moment magnitude," which is a better measure of the actual energy released. Because an earthquake results from a rupture of a fault plane, not a line or a point, geologists need to measure the energy radiated from the entire plane on which there was movement over the time required for all of the rupture to take place. The moment magnitude allows them to do this. The **seismic moment** is calculated from measurements of the area of the fault plane that ruptured and moved, how far the rocks were displaced on either side of the fault, and rock strength. Field measurements reveal the area of the fault and the amount of displacement or slip, and the mainshock and aftershock foci reveal how deeply the fault ruptured. The seismic moment calculation is schematically shown in **Figure 11.34**. Analogous to the Richter scale, each whole-number increase in the moment magnitude represents a tenfold increase in the seismic moment. In contrast to the moment magnitude, magnitudes based on seismograms (like the Richter scale) underestimate the energy release because the energy is spread out over all of the waves generated during the duration of the earthquake and no single wave height really represents the energy of the whole event.

Table 11.2 lists some notable earthquakes within historic time. Earthquakes with magnitudes below M4 (moment magnitude 4) rarely cause significant damage, whereas magnitudes between M5 and M7 typically represent large earthquakes, and earthquakes with magnitudes above M8 are catastrophic if the epicenter is in a populated area. Although no magnitude scale has an upper limit, rocks are probably not strong enough to store up the elastic strain energy to result in earthquakes with moment magnitudes larger than M10.

The magnitude scales are numerical values that give an idea of the size of an earthquake, but how do you compare an M5 earthquake to an M7 earthquake? The M7 quake is 100 times larger in seismic moment than the M5 event, but how much more energetic is the M7 earthquake? For every one unit of magnitude increase, the energy increases 32-fold. This means that an M7 earthquake releases 32 times more energy than an M6 event and 1024 times more energy than an M5 earthquake (32 times 32 equals 1024). **Figure 11.35** graphically illustrates the energy released for earthquakes of different magnitudes and compares this to the energy of explosives and to events other than earthquakes. The M9.3 Sumatra earthquake released an amount of energy equivalent to all of the energy used in the United States in 6 months.

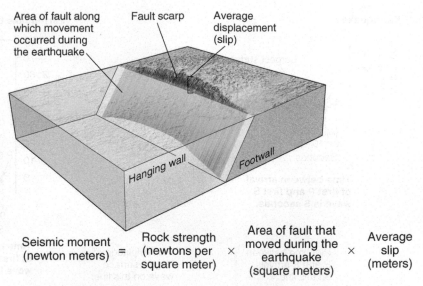

$$\begin{array}{c}\text{Seismic moment} \\ \text{(newton meters)}\end{array} = \begin{array}{c}\text{Rock strength} \\ \text{(newtons per} \\ \text{square meter)}\end{array} \times \begin{array}{c}\text{Area of fault that} \\ \text{moved during the} \\ \text{earthquake} \\ \text{(square meters)}\end{array} \times \begin{array}{c}\text{Average} \\ \text{slip} \\ \text{(meters)}\end{array}$$

▲ **Figure 11.34 Determining seismic moment.** The moment-magnitude scale uses the seismic moment to describe the size of an earthquake. This diagram shows how the seismic moment is calculated from measuring (1) the strength of the rocks offset along the fault, (2) the area of the fault plane along which movement occurred during the earthquake, and (3) the average amount of displacement that occurred on the fault.

EXTENSION MODULE 11.1

Calculating Magnitude and Energy Released from an Earthquake. Learn the basic mathematics behind the magnitude scales.

Table 11.2 Some Notable Earthquakes

Date	Place	Moment Magnitude	Fatalities
December 16, 1811	New Madrid, MO	8.1	
January 23, 1812	New Madrid, MO	7.8	
February 7, 1812	New Madrid, MO	8.0	
April 18, 1906	San Francisco, CA	7.8	>3000
November 4, 1952	Kamchatka, Russia	9.0	
May 22, 1960	Chile	9.5	5,700
March 28, 1964	Prince William Sound, AK	9.2	125
July 27, 1976	Tangshan, China	7.5	255,000–655,000
December 7, 1988	Apitak, Armenia	6.8	25,000
October 17, 1989	Loma Prieta, CA	6.9	63
June 28, 1992	Landers, CA	7.3	1
January 17, 1994	Northridge, CA	6.7	57
January 16, 1995	Kobe, Japan	6.9	5,502
August 17, 1999	Izmit, Turkey	7.6	15,637
September 21, 1999	Taicheng, Taiwan	7.8	2,100
December 26, 2003	Bam, Iran	6.6	>30,000
December 26, 2004	Sumatra, Indonesia	9.3	230,000
October 8, 2005	Kashmir, Pakistan-India	7.6	75,000
May 12, 2008	Sichuan, China	7.9	70,000

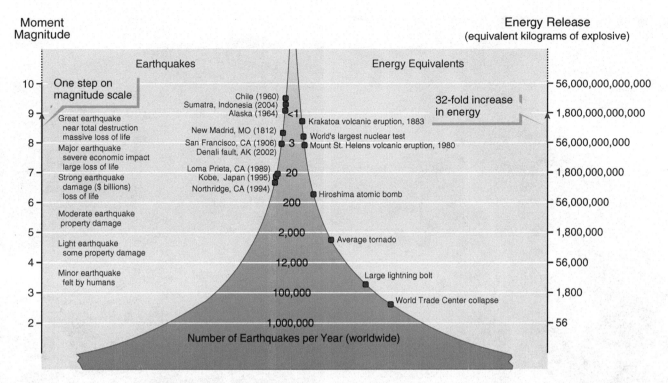

▲ **Figure 11.35 How much energy is released by an earthquake?** This diagram compares the energy released by an earthquake to the earthquake magnitude, and to the energy released by other natural and human-caused phenomena. The chart also indicates how many earthquakes of a given magnitude occur each year around the world. For instance, the 2002 Denali fault, Alaska, earthquake had a magnitude of 7.9; on average, three earthquakes of this magnitude occur each year. This earthquake released energy equivalent to detonating 56 billion kilograms of explosive and about 1000 times the energy released by the atomic bomb that destroyed Hiroshima, Japan, in 1945.

Putting It Together—How Are Earthquakes Measured?

• Earthquake intensity, or violence, is estimated from assessment of damage and eyewitness accounts. Many values of intensity are recorded for a single earthquake, and values are usually highest closest to the epicenter.

• Earthquake magnitude relates to the energy released by an earthquake. In general, each whole-number increase in magnitude is a 32-fold increase in energy released.

• An earthquake has only one magnitude value, on any given magnitude scale. Some magnitude scales rely on the response of seismometers to the passage of earthquake waves. The more recently used moment-magnitude scale is based on the area of the fault plane that ruptured, the distance that it moved, and the strength of the rock.

11.8 Why Are Earthquakes Destructive?

The destructive effects of earthquakes are not limited to the ground rupture along the fault. Energy released by large earthquakes has far-reaching effects that can lead to catastrophic losses in urban areas, as implied by Mercalli Intensity maps (Figure 11.33) and dramatically illustrated in **Figure 11.36**.

Ground Shaking

S-waves and surface waves radiating outward from the earthquake epicenter, or longer lengths of ruptured faults, cause the ground to shake up and down and side to side (Section 8.2). This ground motion exerts stress on buildings, bridges, dams, and other structures. The stress may exceed the strength of the beams or masonry, causing them to break and crumble. In some cases, pillars between floors of multistory buildings fail, causing the floors to crash down on one another (Figure 11.36c). Buildings sway back and forth and hit each other as the earthquake waves pass along the surface. Ground shaking can rupture natural-gas lines or knock over open-flame lanterns and cooking devices, triggering fires that commonly destroy more property than the ground shaking itself.

Ground shaking also triggers rapid movement of materials at Earth's surface. Figure 11.36b illustrates a landslide shaken loose by an M7.6 earthquake in El Salvador in 2001. Surface waves stressed the volcanic rocks on the hillside beyond their strength limit. The landslide buried 268 homes and killed 700 people.

Ground-shaking damage varies from locality to locality because the amplitude of seismic surface waves, which describes the shaking, depends not only on proximity to the epicenter, but also on the characteristics of the materials that the waves pass through. For example, seismic-wave amplitude is greater in loose sediment than in hard rock.

Figure 11.37 illustrates how geological materials partly determine the effects of an earthquake. The intensity map for the 1989 Loma Prieta, California, earthquake reveals that the highest intensity (IX) was

(a) Bridge collapse. Kobe, Japan, 1995. Moment magnitude 6.9.

(b) Earthquake-triggered landslide. Santa Tecla, El Salvador, 2001. Moment magnitude 7.6.

(c) Building collapse, Beichuang, China, 2008. Moment magnitude 7.9.

(d) Ground-shaking damage, California State University-Northridge bookstore, 1995. Moment magnitude 6.7.

▲ **Figure 11.36 How earthquakes cause damage.** Earthquake surface waves cause large vertical and horizontal shaking motions that topple buildings and bridges, trigger landslides, and throw objects from shelves.

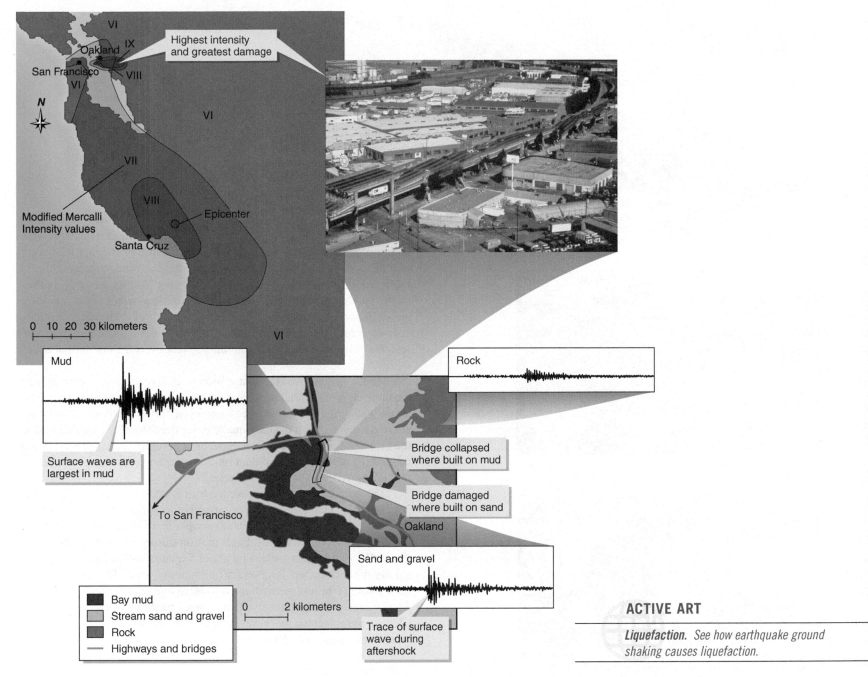

Figure 11.37 How geologic materials determine earthquake damage. The intensity map of the 1989 Loma Prieta, California, earthquake shows that the greatest ground shaking (intensity IX), where the deadliest damage involved a highway bridge collapse, occurred far from the epicenter. The geologic map below shows that severity of bridge damage coincided with different geologic materials; complete collapse occurred where the bridge was built on bay mud. Surface-wave heights measured at three nearby seismic stations during an aftershock show that ground shaking is most intense where waves pass through mud and least intense where waves pass through rock.

experienced far from the epicenter. This area of intense, deadly destruction included the collapse of a double-deck freeway bridge in Oakland, 100 kilometers from the epicenter. The segment of the bridge that collapsed was built on mud. Measurement of surface-wave amplitudes in mud, stream sand and gravel, and rock during aftershocks shows that the three materials responded very differently to the passing seismic waves. The higher-amplitude ground motion, where the surface waves passed through the mud, caused more intense shaking and failure of the bridge supports.

Where ground water is close to the surface, ground shaking causes loose sediment to behave like liquid. This process, called **liquefaction**, occurs in water-saturated sediment when grains settle during ground shaking, displacing the water upward in the intervening pore spaces such that the water pressure moves the grains apart and the whole sediment-water mixture loses strength. The wet ground turns to fluid slurry as the surface waves pass and then settles unevenly, resulting in severe damage to buildings and other structures. Liquefaction of wet mud contributed to the bridge failure illustrated in Figure 11.37.

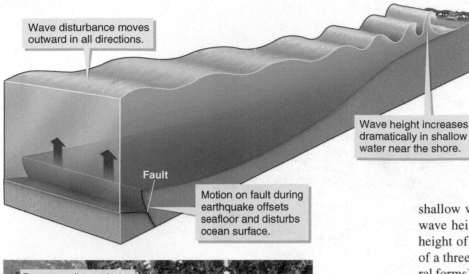

Wave disturbance moves outward in all directions.

Wave height increases dramatically in shallow water near the shore.

Fault

Motion on fault during earthquake offsets seafloor and disturbs ocean surface.

◄ Figure 11.38 **Why tsunami are destructive.**

Brown, sediment-laden water surges inland more than 500 meters.

Tsunami ravaged Indian Ocean shorelines following a magnitude 9.3 earthquake near Sumatra in December 2004. A satellite image (top) shows inundation of coastal Sri Lanka, more than 1600 km from the epicenter, about one hour after the first wave arrived. Along the Sumatra coast (bottom) the tsunami caused complete devastation.

Tsunami

Tsunami, or seismic sea waves, are caused by sudden vertical displacement of the seafloor and produce significant damage, as shown in **Figure 11.38**. These waves are sometimes erroneously called "tidal waves," although they are not caused by the gravitational pull of the Sun and Moon as tides are. Instead, fault displacement of the seafloor during an earthquake disturbs the ocean surface. Tsunami waves are usually less than a meter high and rarely noticed in the open sea, but the wave energy is confined into a progressively smaller space as the wave moves into

shallow water near shore. This confinement of wave energy causes the wave height to increase dramatically in shallow water near land—the height of some tsunami waves exceeds 10 meters, or roughly the height of a three-story building. "Tsunami" (the same in both singular and plural forms) comes from a Japanese term meaning "harbor wave" and refers to the sudden buildup of high waves in shallow water. Tsunami form not only during earthquakes, but also as a result of underwater volcanic eruptions and submarine landslides. Section 19.2 presents additional insights to tsunami dynamics and the ways they differ from normal ocean waves.

Unlike ground shaking that affects a relatively small region around the fault-rupture site, catastrophically damaging tsunami may strike coastal areas where the earthquake was not even felt, sometimes more than 1000 kilometers from the earthquake epicenter. The Hawaiian Islands, in the middle of the Pacific Ocean, have been severely damaged by tsunami originating from distant subduction-zone earthquakes along continental margins in Alaska and South America.

The deadliest tsunami in recorded history accompanied the December 26, 2004, M9.3 earthquake that occurred below the Indian Ocean offshore of Sumatra. Approximately 230,000 people in 14 countries rimming the Indian Ocean lost their lives. Fault motion during the earthquake raised the seafloor and ocean surface by about 5 meters within an area approximately 400 kilometers long and 50 kilometers wide. This motion lifted more than 100 billion metric tons of water above sea level over a matter of a few minutes. The resulting tsunami crossed the Indian Ocean in 10 hours. Close to the epicenter, the waves were as much as 30 meters high and penetrated inland as far as 50 kilometers.

Risk Assessment Tools

Geologists apply research on earthquakes and their effects to help save lives and property. Seismic-hazard mapping involves examining the geology of the area, mapping locations of the potentially active faults, and reviewing any past earthquake activity along the faults. The maps highlight the locations of geologic materials that are especially susceptible to failure by intense ground shaking or liquefaction. This information is used to map out the earthquake risk, which is the probable building damage and number of expected casualties based on the population in particular areas. Risk maps generally focus on the likely maximum intensity of ground shaking.

Figure 11.39 is a seismic-risk map of potential ground shaking for the continental United States. Notice that the higher risk zones are near active plate boundaries along the west coast of the United States (compare to the plate-boundary map in Figure 1.10). In addition, there are regions in

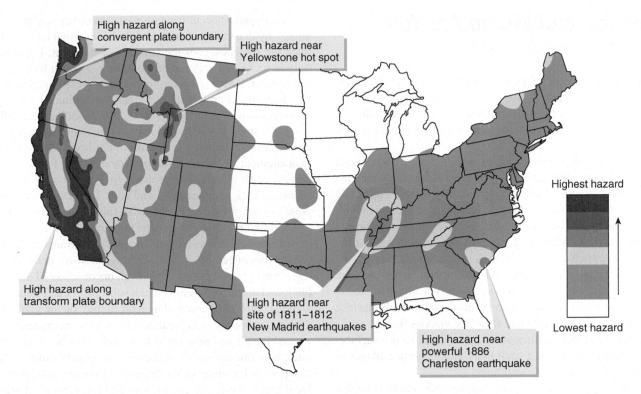

High hazard along convergent plate boundary

High hazard near Yellowstone hot spot

High hazard along transform plate boundary

High hazard near site of 1811–1812 New Madrid earthquakes

High hazard near powerful 1886 Charleston earthquake

Highest hazard

Lowest hazard

▲ **Figure 11.39 Seismic hazard map.** This seismic hazard map of the contiguous United States uses different colors to represent the expected hazard of damage by ground shaking during an earthquake.

the continental interior, such as the New Madrid, Missouri, seismic zone, where large earthquakes, although rare compared to events closer to plate boundaries, have occurred during historic time (Figure 11.32).

Geologists also routinely prepare risk maps at the scale of states, counties, and cities. Planners consult these maps when making land-use decisions. For example, they may decide to place golf courses and parks, rather than dense concentrations of buildings, along faults and in areas subject to intense ground shaking. The maps also guide the development of building codes and engineering designs, so that buildings, bridges, and dams are constructed to withstand expected ground shaking and liquefaction.

Earthquake Prediction

Can earthquakes be predicted? At present, seismologists cannot reliably predict the specific time, place, and magnitude of an earthquake. In general terms, seismologists anticipate a 67-percent chance that there will be a major (M > 6.7) earthquake in the San Francisco Bay area within the next 30 years, and a 60-percent chance of one in the Los Angeles area. These general forecasts are based on the frequency of earthquakes in the past and the measured buildup of strain in the rocks.

A general forecast of a big earthquake over a 30-year period may not seem very useful to you. More specific prediction approaches are being studied. Most of these studies focus on detecting phenomena that reliably occur before a large earthquake. The establishment of reliable precursors has been elusive, however, and different precursor phenomena are likely to be associated with different faults, so knowledge gained in one location may not apply in another. Nevertheless, seismologists keep investigating,

constantly seeking more information that could help develop models for prediction.

Putting It Together—Why Are Earthquakes Destructive?

• Earthquakes cause damage by fault rupture, ground shaking, landslides, and tsunami.

• Seismic surface waves cause ground shaking. The amplitude of the waves, and hence the damage that results, varies depending on the intensity of the earthquake and the nature of geologic materials that the waves travel through.

• Earthquake rupture of the seafloor generates tsunami that cause destruction both close to the epicenter and as far as thousands of kilometers away from it.

• Predictions for specific earthquakes are still not possible, but risk maps identify hazards for consideration in land-use planning, building codes, and engineering designs that diminish the destructive effects of earthquakes.

EXTENSION MODULE 11.2

Mitigating and Forecasting Earthquake Hazards. *Learn about ground shaking, liquefaction, tsunami, landslides, and fault-rupture hazards.*

Where Are You and Where Are You Going?

You now know how Earth's crust deforms and how to recognize the deformation. Deformation, or strain, results from stresses that break or otherwise change the shape of rock. Geologic structures, such as folds and faults, are examples of strain resulting from tectonic stresses that exceeded the rock strength.

Rock strength, in turn, depends on temperature, pressure, and mineral composition. Increasing pressure generally increases strength, but increasing temperature lowers strength. A rock may initially behave elastically when stressed, absorbing the applied stress as elastic strain energy and returning to its original shape and size if the stress releases. If stress increases instead of releasing, the rock undergoes permanent deformation when stress exceeds the yield strength. This deformation is brittle breakage or plastic flow. Brittle failure occurs near the surface at low temperature and pressure and produces faults and joints. Plastic deformation typically happens at higher-temperature and higher-pressure conditions deep below the surface and causes metamorphic rocks to fold and flow like viscous fluids.

Folding commonly affects sedimentary rocks close to the surface. Sedimentary rocks bend and slip past each other along bedding planes to a substantial extent before breaking along faults.

Deformation of Earth's crust produces both hazards and benefits for society. Hazards include destructive earthquakes as rock breaks and moves when stress exceeds the rock strength. Benefits include mineral and energy deposits. Structurally deformed areas commonly determine the formation and locations of economic deposits. Mineral deposits rich in metallic elements commonly form in fractured rocks where metal-rich fluids migrate into the fractures and precipitate ore minerals. Relatively low density oil and natural gas move upward through porous rocks until trapped below less porous layers. These traps commonly occur along the axes of anticlines and below faults.

Earthquakes occur because rocks on either side of a fault store strain energy by deforming near locked faults, and that energy abruptly releases once the stress exceeds the strength of the fault. Earthquakes can occur only in rocks that undergo brittle failure, so most foci are restricted to the upper crust, where temperature and pressure are relatively low.

Intensity of ground motion and the magnitude, or energy released, are two ways of measuring earthquakes. The Modified Mercalli Intensity Scale assigns values to different locations based on severity of damage and eyewitness accounts. Earthquake magnitude is calculated either from measurement of the response of seismographs to the passage of seismic waves, or from calculations involving the area of the fault plane that ruptured, rock displacement, and rock strength. Each whole-number increase in magnitude represents a 32-fold increase in the released energy.

Ground rupture, ground shaking, and tsunami are some of the effects of earthquakes. The severity of damage from ground shaking varies with different types of geologic materials. Seismic waves behave differently as they travel through different materials. Wet, unconsolidated deposits form less stable foundations for structures than solid bedrock. Tsunami are huge ocean waves, commonly generated by fault rupture on the seafloor, that can cause destruction locally or at great distances from earthquake epicenters.

Although you now know how rocks deform, you still need to determine why this deformation occurs. The observations of faults and folds, along with locations of earthquakes, provide indications of stresses in local areas. Now, you are prepared to examine the bigger regional and global scale of stress and strain encompassed in the discipline of tectonics. What are the origins of the compressional, extensional, and shear stresses that produce the wide variety of geologic structures visible at the surface? Ultimately, convection processes you learned about in Chapter 10 produce most of these stresses within Earth. In the next chapter, you will learn how to explain global tectonic processes in terms of convective motion within Earth. You will also learn how these processes produce the intense deformation that is concentrated along narrow zones on the planet—the plate boundaries.

Active Art

Folding Rock. See how rocks fold.

Fault Motions. See how motion along normal, reverse, and strike-slip faults displaces rocks.

Graphing Stress and Strain. See how to interpret graphs of stress and strain in rocks.

Elastic Rebound. See how elastic strain deforms rocks along a fault before an earthquake and what happens during an earthquake.

Liquefaction. See how earthquake ground shaking causes liquefaction.

Tsunami. See how an earthquake causes a tsunami.

Extension Modules

Extension Module 11.1: Calculating Magnitude and Energy Released from an Earthquake. Learn the basic mathematics behind the magnitude scales.

Extension Module 11.2: Mitigating and Forecasting Earthquake Hazards. Learn about ground shaking, liquefaction, tsunami, landslides, and fault rupture hazards.

Confirm Your Knowledge

1. Why is it important to study deformed rocks?
2. Describe the terms "strike" and "dip" in your own words.
3. Explain the difference between a plunging fold and a nonplunging fold.
4. Draw cross sections of an eroded anticline and an eroded syncline; label the youngest and the oldest rocks in each drawing. What is the difference between an anticline and a syncline?
5. Explain the difference between a fault and a joint.
6. Explain the difference between a footwall and a hanging wall.
7. Describe the types of dip-slip and strike-slip faults.
8. Explain the difference between stress and strain.
9. List and describe the three basic types of stress that can be applied to rocks and the resulting strains.
10. What type of deformation occurs when a rock undergoes stress less than its yield strength? What happens when stress exceeds the yield strength?
11. What factors determine rock strength?
12. List examples of geologic structures resulting from tension, compression, and shear.
13. For each of the following geologic structures decide whether the regional stress that caused the structure was compression, tension, or shear: Normal fault, reverse fault, thrust fault, strike-slip fault, anticline fold, syncline fold.
14. Define "fault scarp." What do scarps reveal about fault motion in the past?
15. Explain surveying data that support the elastic rebound theory.
16. How are earthquakes measured?
17. List and explain the hazards associated with earthquakes.

Confirm Your Understanding

1. Write an answer for each question in the section headings.
2. If you were exploring for natural resources, such as metallic minerals, oil, or gas, what geologic structures would you look for and why?
3. Explain the effects of composition, pressure, and temperature on rock strength.
4. For each of the localities described, determine whether the dominant deformation is due to tension, compression, or shear.
   - The fold and thrust belt in Wyoming
   - The normal-faulted basin and range region in Nevada
   - The right-lateral San Andreas Fault in California
5. How do we know that earthquakes are the result of movement along faults?

CHAPTER 12

Global Tectonics: Plates and Plumes

Why Study Tectonic Plates and Their Motions?

Earth moves beneath your feet. If you were born in North America about 20 years ago, then your birthplace is almost half a meter farther from Europe than it was on the day you were born. Perhaps that changing distance does not seem impressive; after all, it implies a speed similar to the rate of fingernail growth. When, however, you take into the account the astonishing weight of lithosphere plates, even this sluggish speed indicates that there are gargantuan forces acting at plate boundaries. These forces account for the catastrophic earthquakes and volcanic eruptions that grab media headlines, as well as for the patient construction of mountain ranges, the buildup of oil and mineral resources, and the formation of entire ocean basins over the vastness of geologic time.

This chapter will both broaden your understanding of plate tectonics theory and serve as a checkpoint to integrate the concepts introduced in previous chapters. Indeed, the power of plate tectonics lies in the ability of this all-encompassing theory to explain many geologic observations, most of which do not seem obviously linked at first glance.

Your learning approach will be a bit different in this chapter. In most of this book you have experienced an inductive approach to learning science—you have focused on observations first and then considered explanations for what happened. Here you will use deduction—if plate tectonics is a viable theory what does it predict and what evidence supports those predictions?

After Completing This Chapter, You Will Be Able to

- Evaluate the evidence for how geoscientists know that lithospheric plates move.
- Explain why plates move.
- Combine the observations of what happens at plate boundaries to explain why these processes occur.
- Extend the implications of modern plate motion for understanding Earth's history.
- Describe a hypothesis for the existence of volcanic hot spots.

Pathway to Learning

12.1 How Does Continental Drift Relate to Plate Tectonics?

12.2 What Is the Evidence That Plates Are Rigid?

12.3 What Is the Evidence That Plates Move Apart at Divergent Boundaries?

12.4 What Is the Evidence That Subduction Occurs at Convergent Plate Boundaries?

12.5 What Is the Evidence That Plates Slide Past One Another at Transform Plate Boundaries?

The earthquake that destroyed this city in Sichuan, China, in 2008 is evidence of the deformation caused by the motion of lithospheric plates.

IN THE LAB

You arrive in the geology lab to find the broken fragments of an intricately patterned dinner plate spread on the tabletop, as shown in **Figure 12.1**a. A geologist challenges you to assemble the pieces to reconstruct the plate. You fit fragments together by matching edges of similar shape. You also check that the ornamental pattern passes continuously from one fragment to the next. The result is a reconstructed plate without gaps between pieces and without overlapping pieces. What does this exercise have to do with geology?

To answer this question, the geologist gives you cardboard pieces cut in the shapes of Africa and South America (Figure 12.1b). The puzzle-piece look of eastern South America and western Africa tempt you to place the continents against one another, ignoring for the moment the presence of the South Atlantic Ocean. The notion that the two continents were once joined originated in the late sixteenth century, when explorers first mapped the coastlines of the continents. Is it just a coincidence that the outlines of the two continents match, or were they really joined in the past?

Geologic information ornaments the cardboard pieces in Figure 12.1b. You can match these geologic patterns between the cardboard continents, just as you matched patterns on the dinner plate fragments. Areas of different-age crust, rock deformation, and ancient glaciers all terminate abruptly at the edge of each continent (Figure 12.1b). However, when you match the outlines of the continents, a continuous distribution of rock and structure extends from one landmass to the other. Clearly, the match-up of the continents is not a coincidence. The visiting geologist tells you that German meteorologist Alfred Wegener did not think it a coincidence either. The matching coastlines and geology of Africa and South America were a small, but crucial, part of his writing about the continental drift hypothesis in the early twentieth century.

If South America and Africa were once joined and then separated from one another, then what does that mean for the origin of the South Atlantic Ocean? You consult a map (Figure 12.1c) and are impressed that the Mid-Atlantic Ridge, running like a spine down the center of the ocean, mimics the curves of the adjacent continent shorelines.

The information displayed in Figure 12.1 stimulates a number of questions. If Africa and South America moved away from one another, how did the oceanic lithosphere form between them? Why does the Mid-Atlantic Ridge have the same shape as the outline of the coast of Africa? How fast did Africa and South America move away from one another, and are they still moving? When did they separate? What forces cause the continents to move? The theory of plate tectonics provides answers to all these questions.

Scientific theories are not speculative ideas. Theories are rigorously tested concepts that provide the best explanations for observed natural phenomena known at the time and are widely accepted in the scientific community. When studying any theory, you should ask to see the supporting evidence. Throughout this chapter, you will explore the observed phenomena and tests that support plate tectonics as a valid explanation of a wide variety of geologic features and processes.

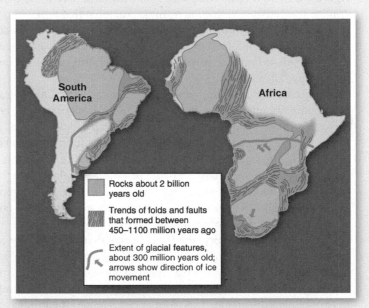

(a) The pieces of this broken plate can be fitted back together by matching shapes and patterns.

Rocks about 2 billion
years old

Trends of folds and faults
that formed between
450–1100 million years ago

Extent of glacial features,
about 300 million years old;
arrows show direction of ice
movement

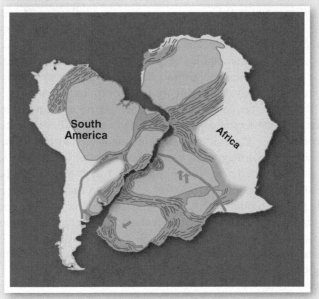

(b) Matching shapes and patterns of these cardboard geologic
models of South America and Africa implies that the two continents
were once joined and later drifted apart.

(c) This map shows the Atlantic Ocean between South America and Africa,
and the location of the Mid-Atlantic Ridge.

▲ Figure 12.1 Putting a puzzle together.

12.1 How Does Continental Drift Relate to Plate Tectonics?

Alfred Wegener advocated matching continents by shape and geological similarities in a series of books published between 1912 and 1928. **Figure 12.2** highlights Wegener's **continental drift** hypothesis: All continents were once joined in a single continent that broke into pieces to form the modern continents, which drifted to their current positions. He named the ancient supercontinent *Pangea* (derived from Greek words meaning "all of Earth"), and he called the seas surrounding it *Panthalassa* (meaning, "all ocean").

Wegener collected impressive data that support combining the modern continents into Pangea but had difficulty explaining how continents moved from the Pangea configuration to their present positions. He speculated that continental crust plowed through oceanic crust, like an icebreaker crashing through an icy sea. Most geologists rejected this concept because rocks are too strong for one type of crust to plow through another, and Wegener did not propose a driving force for the shifting continents. Wegener's inability to explain how continents moved was an obstacle to widespread acceptance of continental drift.

Plate tectonics theory, the blockbuster that dramatically advanced geologic science, developed quickly after a slow start. Cold War friction among countries after World War II led to the establishment of worldwide seismograph networks to monitor underground nuclear testing. These networks also documented the global occurrence of earthquakes. Large sums of money funded projects to survey the seafloor to find ways of safe navigation and detection of submarines. These surveys also measured the age, structure, and thickness of the virtually unknown oceanic crust. All of these data were assembled at oceanographic and geophysical institutes in the United States and United Kingdom and unexpectedly provided the basis for quickly developing the plate tectonics theory during the 1960s.

Therefore, you can consider continental drift as the unsuccessful forerunner of plate tectonics. Both plate tectonics and continental drift explain the abundant evidence of movement on Earth's surface and the assembly and disintegration of supercontinents. Plate tectonics, however, not only is armed with more supporting data than were available to Wegener, but also offers a different explanation for the drifting continents. In the plate tectonics theory, the continents are simply passengers on raft-like plates composed mostly of oceanic lithosphere. This means that continents and oceans within the same plate move together. This is different and more supportable than Wegener's idea that continents plow through oceans, and you will see why as we go through the chapter.

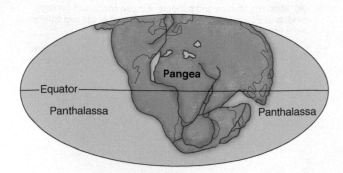

▲ **Figure 12.2 Fitting continents together.** Alfred Wegener hypothesized about the existence of a late Paleozoic supercontinent, which he named Pangea, surrounded by the seas of Panthalassa.

Putting It Together—How Does Continental Drift Relate to Plate Tectonics?

• The continental drift hypothesis, the forerunner to plate tectonics, proposed that all landmasses were once joined in a supercontinent, called Pangea, which later broke apart.

• The continental drift hypothesis was not universally accepted because it lacked a reasonable explanation for why continents moved. The plate tectonics theory is more supportable because it contends that continents are passengers on plates.

12.2 What Is the Evidence That Plates Are Rigid?

Plate tectonics theory states that the strong lithosphere (consisting of the crust and uppermost mantle) consists of pieces that move about on top of the weak, easily deformed asthenosphere. These rigid pieces are the tectonic plates depicted in **Figure 12.3**. Let's explore what is meant by "rigid plates" and then look at the evidence that supports this major premise of the plate tectonics theory. Deformation occurs at or near the edges of rigid plates where they interact with neighboring plates. The plate interiors are, by comparison, relatively undeformed as shown in **Figure 12.4**.

Shifting sea ice provides an analogy for rigid plate motion, as seen in **Figure 12.5**. Ice floats on the water and separates into slabs with rough edges that move toward, away from, or alongside one another, but the interiors of the slabs remain smooth and undeformed. Is this really how the lithosphere deforms, or does it more closely resemble the semirigid or weak boundary interactions described in Figure 12.4?

What Surface Deformation Reveals

One way to test for rigid plates is to see whether deformation occurs only at the edges of plates. **Figure 12.6** uses the global distribution of earthquakes between 1960 and 2000 to investigate where deformation happens. Earthquake epicenters are not uniformly spaced across Earth but are concentrated along and near plate boundaries (compare Figures 12.3 and 12.6).

Although earthquake epicenters trace narrow bands along the divergent and transform boundaries within ocean basins, they are much more scattered within continents (Figure 12.6). Some plate boundaries, especially on continents, seem to be broad, diffuse zones rather than narrow, sharp lines. This is particularly true near convergent boundaries. Some earthquakes also occur within plates, far from plate boundaries. Part of the explanation for more widespread deformation in continents is that laboratory experiments suggest that the rocks forming continental lithosphere are not as strong as oceanic lithosphere.

Taking all of the evidence from earthquakes, it seems that some plates are rigid, whereas others are only semirigid (Figure 12.4), so that some plate boundaries are actually zones hundreds of kilometers wide rather than rigid edges. Recent calculations, based on earthquake data from throughout the world, suggest that more than 80 percent of lithosphere deformation occurs at narrow plate boundaries. This means that it is reasonable to assume the plates are basically rigid while acknowledging semirigid behavior along some plate boundaries.

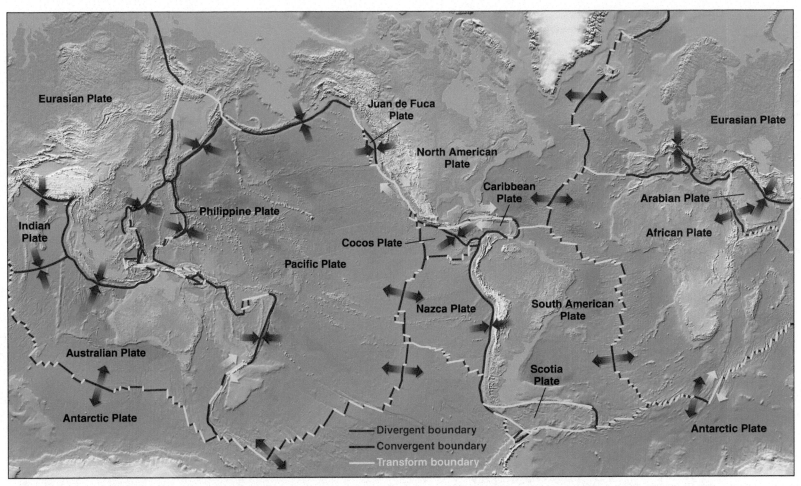

▲ **Figure 12.3 Where the plates are located.** Each plate is named for a prominent geographic feature on the plate. The boundaries between plates are described as
- divergent, where plates move away from one another, as shown by red arrows.
- convergent, where plates move toward one another, as shown by dark green arrows.
- transform, where plates slide past one another, as shown by yellow arrows.

ACTIVE ART

Motion at Plate Boundaries. See how plates move along their boundaries.

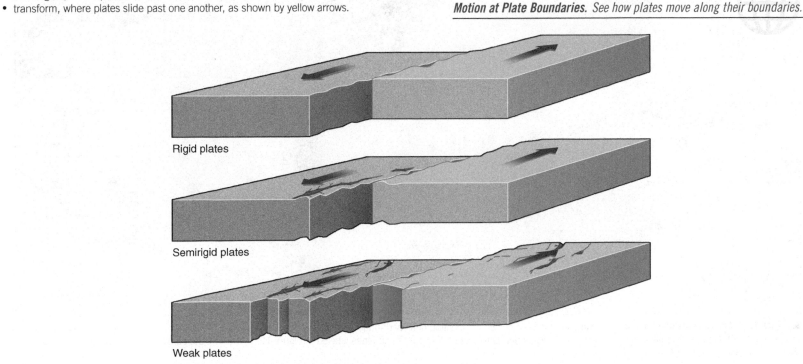

▲ **Figure 12.4 What is meant by a rigid plate?** Interacting rigid plates deform only at their edges, without any deformation in the interior. Weak plates deform throughout when they interact with neighbors and semi-rigid materials exhibit intermediate behavior.

(a)

(b)

◄ **Figure 12.5 Sea ice as an analogy for plate tectonics.** Some aspects of lithospheric plate motion resemble rigid slabs of ice moved about by waves and currents. (a) In frozen seas ice cracks and adjacent slabs move away from one another to expose open water. Notice how the outlines of the broken ice slabs can be matched together, similar to the outlines of Africa and South America (Figure 12.1). The interior part of each slab is unchanged by breaking at the margin. (b) Where slabs of sea ice converge toward one another, they buckle up into the air but the deformation occurs only in a narrow zone where the slabs collide.

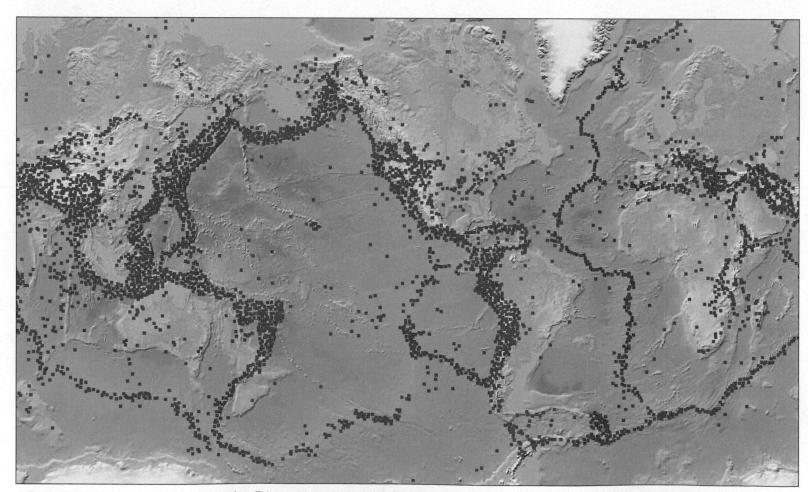

▲ **Figure 12.6 Earthquakes outline plate boundaries.** This map shows the location of all earthquake epicenters between 1960 and 2000, for earthquakes with magnitudes greater than 4.0. Most earthquakes occur in relatively narrow bands that mark the plate boundaries labeled in Figure 12.3. Within-plate earthquakes are most notable within North America and eastern Asia.

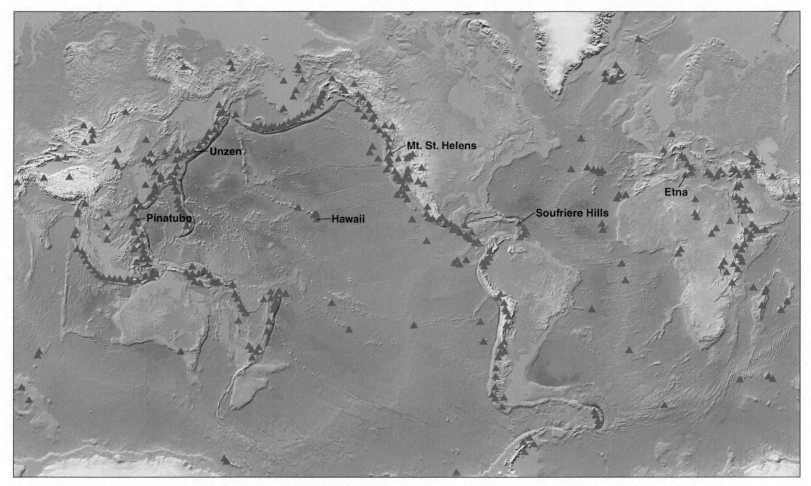

▲ **Figure 12.7 Volcanoes form at and near plate boundaries.** This map shows the location of all volcanoes that are known to have erupted within the last 10,000 years. Compare this map to Figures 12.3 and 12.6 and notice that most volcanoes erupt along the same narrow zones where earthquakes occur at plate boundaries. Some volcanoes, such as those in Hawaii, occur distant from plate boundaries and are not explained by plate interactions. The locations of some volcanoes discussed in Chapter 4 are indicated.

What the Locations of Volcanoes Reveal

Figure 12.7 is a plot of the volcanoes that erupted over the last 10,000 years. The distribution of recently active volcanoes is similar to that of earthquakes and, for the most part, defines narrow, linear, and arc-shaped belts, which are along or near plate boundaries. The low number of volcanoes mapped along divergent boundaries does not mean that there are few volcanoes along mid-ocean ridges. Geologists simply do not know the locations of most recently active, submerged seafloor volcanoes.

The presence of volcanoes near plate boundaries suggests a link between plate motion and locations where magma forms, where magma can reach the surface, or both. Recall from Section 4.6 how magma generation relates to plate tectonics. At divergent boundaries, the asthenosphere rises and melts by decompression. At convergent boundaries, metamorphic reactions in subducting crust release water into the overlying asthenosphere; the water decreases the mantle-rock melting temperature and triggers magma generation.

Not explained by plate tectonics are the large volcanoes scattered within plates, far from plate boundaries (Figure 12.7). Examples of these include the active shield volcanoes of Hawaii near the center of the Pacific plate, the many volcanoes scattered across Africa, and recently active Yellowstone National Park in the middle of the North American plate. These hot spots require a separate and complementary hypothesis (mentioned in Figure 1.13). They account for very few volcanoes and do not weaken the power of plate tectonics to explain most global volcanism.

What the Distribution of Mountain Belts Suggests

If folding and faulting occurs mostly at plate boundaries, then mountain-building processes should be concentrated at these boundaries. **Figure 12.8** is a map of active mountain belts, which are the areas displaying evidence of significant uplift and rock deformation within the last few million years. The outlines of these mountain belts are only approximate because not all areas of the globe are sufficiently well studied to know the exact age of mountain building. Nonetheless, young mountains largely coincide with the occurrences of earthquakes and active volcanoes (compare Figure 12.8 with 12.6 and 12.7). Except in some areas of western North America and southeastern Asia, the active mountain belts are close to or along plate

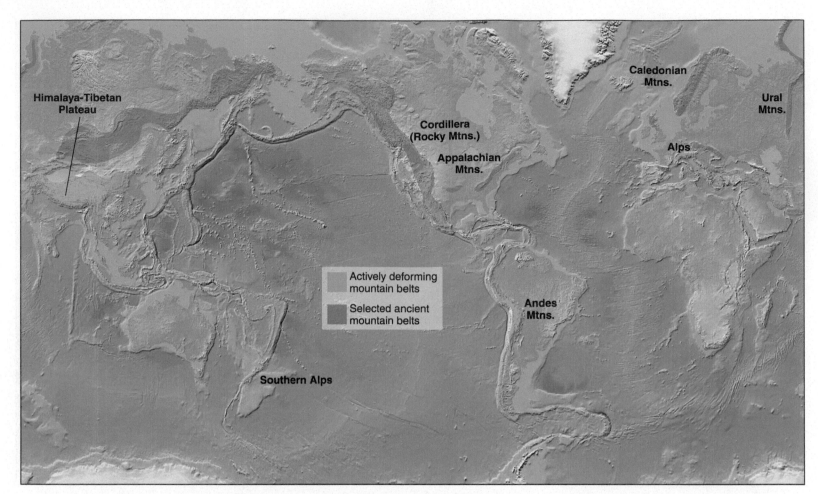

▲ **Figure 12.8 Mountains record deformation at and near plate boundaries.** Recently uplifted mountains on continents coincide with locations of earthquakes (Figure 12.6), which reveal that mountain building is ongoing at present. Most active mountain belts are located near plate boundaries. Some ancient mountain belts, the Appalachian Mountains for example, are far from plate boundaries and not associated with significant modern earthquake or volcanic activity. By application of uniformitarianism, these older mountain belts are interpreted to have formed near ancient plate boundaries.

ACTIVE ART

Correlating Processes at Plate Boundaries. See how volcanoes, earthquakes, and young mountain belts line up with plate boundaries.

boundaries. Wider mountain belts in continents are yet another indication of the relative weakness of continental lithosphere that leads to broader regions of deformation where plates interact with one another.

Some mountain belts of deformed rocks do not coincide with active volcanoes and only few, if any, earthquake epicenters. These mountains, such as the Appalachians in eastern North America, are inactive, ancient mountains and are generally lower in elevation and more rounded by erosion than the tall, actively rising mountains. Ancient mountain belts contain igneous rocks such as those found near modern convergent plate boundaries. You can apply uniformitarianism to hypothesize that in the geologic past, plate boundaries were in different places, near the ancient mountains. This means that plate tectonics theory not only explains active processes at present, but it also describes changing features on Earth through geologic time.

Putting It Together—What Is the Evidence That Plates Are Rigid?

• Plate tectonics theory states that lithospheric plates are rigid. Deformation concentrates at the narrow boundaries where plates interact with neighboring lithosphere and there is very little deformation in plate interiors.

• Most earthquakes and volcanoes concentrate in narrow zones at or near plate boundaries, because deformation is restricted mostly to the margins of rigid plates. Wider bands of earthquake epicenters, especially within continents, imply a semirigid behavior for some plates.

• Young, still-growing mountain belts mostly coincide with deforming plate boundaries. Ancient mountain belts mark the locations of former plate boundaries.

12.3 What Is the Evidence That Plates Move Apart at Divergent Boundaries?

The Mid-Atlantic Ridge forms a divergent boundary between Africa and South America (Figures 12.1 and 12.3). Matching the edges of the two continents into a single landmass requires two assumptions: (1) The lithosphere containing Africa and South America existed before the ocean that now separates the continents, and (2) the Atlantic Ocean grows wider through time. Accordingly, the plate tectonics theory predicts that oceanic lithosphere forms where continental lithosphere separates into two or more continents. Is the geology of the seafloor consistent with this prediction?

What Do Divergent Plate Boundaries Look Like?

Plate tectonics theory incorporates the concept of **seafloor spreading**, which emerged as a successfully tested hypothesis in the mid-1960s. Seafloor spreading describes the process of oceanic lithosphere spreading apart and also being created along mid-ocean ridges, as shown in **Figure 12.9**. Igneous rocks of the crust part of the new lithosphere crystallize from magma where asthenosphere peridotite decompresses and partially melts while rising into the gap between the spreading plates. The lower, mantle part of the new lithosphere contains the unmelted residue

▼ **Figure 12.9 What happens at a divergent plate boundary?** Most divergent plate boundaries coincide with mid-ocean ridges. The gap between separating lithospheric plates fills with upwelling asthenosphere, which partly melts to produce mafic magma that crystallizes to form new oceanic crust. The lithosphere cools as it moves away from the mid-ocean ridge by continued divergent motion. As the mantle cools, the critical temperature defining the boundary between strong lithosphere and weak asthenosphere occurs at greater depth, so the lithosphere thickens as it moves away from the divergent boundary.

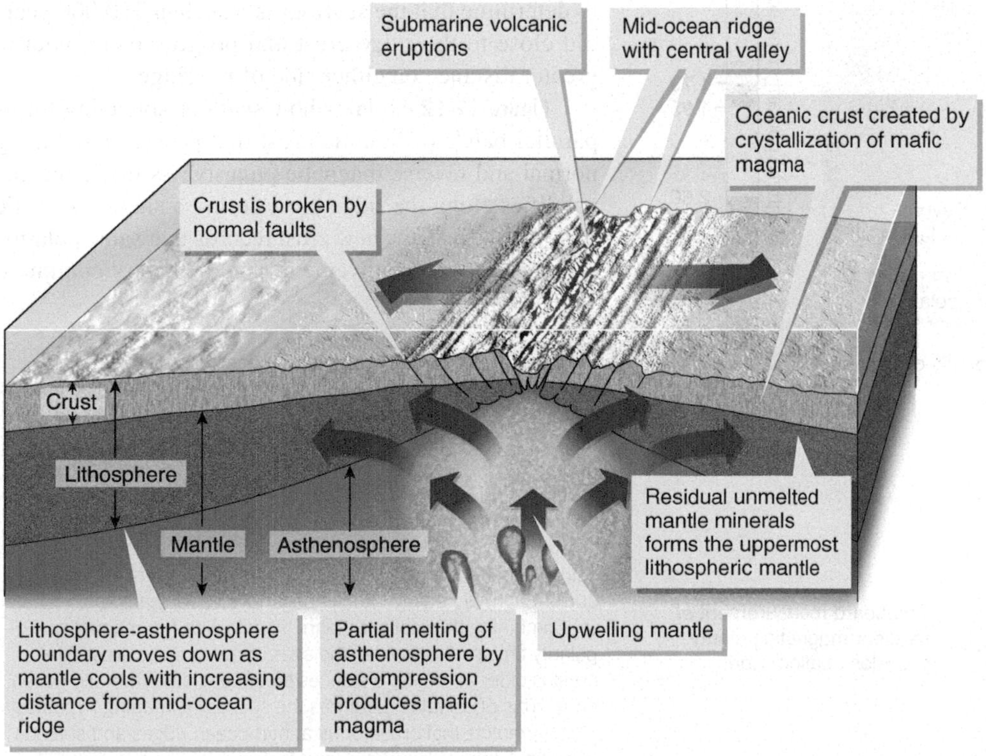

Submarine volcanic eruptions

Mid-ocean ridge with central valley

Oceanic crust created by crystallization of mafic magma

Crust is broken by normal faults

Crust

Lithosphere

Mantle   Asthenosphere

Residual unmelted mantle minerals forms the uppermost lithospheric mantle

Lithosphere-asthenosphere boundary moves down as mantle cools with increasing distance from mid-ocean ridge

Partial melting of asthenosphere by decompression produces mafic magma

Upwelling mantle

from the original asthenosphere peridotite. The seafloor spreading hypothesis is supported by many testable predictions.

What Oceanic-Crust Composition Reveals

Seafloor spreading explains the production of oceanic crust by igneous processes at mid-ocean ridges. Laboratory experiments (Section 4.6) show that partial melting of mantle peridotite produces mafic magma. Oceanic crust, therefore, should have mafic composition, and active volcanoes should occur along the mid-ocean ridges.

More than 1700 scientific drill holes into oceanic crust confirm the predicted mafic composition (Section 8.1). Approximately 18 cubic kilometers of basalt erupt each year along the 65,000-kilometer-long mid-ocean ridge system. The high rate of volcanic activity and related high heat flow from Earth's interior easily account for extensive hydrothermal metamorphism and submarine hot springs (described and pictured in Section 6.10).

The Significance of Mid-Ocean Earthquakes and Faults

If mid-ocean ridges mark locations of plate divergence, then they should coincide with earthquakes along normal faults formed by tensional stress where plates move away from one another. Data depicted in Figure 12.6 confirm a high level of earthquake activity, and **Figure 12.10** illustrates examples of the predicted normal faults. Faults are easily mapped at the surface in Iceland, where the Mid-Atlantic Ridge rises above sea level for easy viewing (Figure 12.10a). Geologists use small research submarines to examine normal faults along the more typical submerged parts of the ridge (Figure 12.10b).

What the Age and Magnetism of Oceanic Crust Reveal

If seafloor spreading happens, then the oceanic crust should be youngest along mid-ocean ridges and progressively older at greater distances from the ridge crest. A key way to test this is to measure the magnetic properties of ocean-floor basalt. Earth's magnetic field flips polarity at irregular intervals. The current condition, where a compass needle points toward north, is called "normal polarity." The reverse polarity condition, which is recorded in many ancient rocks, represents a flip in the magnetic field that would cause a compass to point southward. The details of the formation and changes of Earth's magnetic field are explained in Section 10.3).

Shipboard measurements of the magnetic polarity of seafloor rocks taken in the 1960s show parallel stripes of normal and reverse polarity that are symmetrically distributed on either side of mid-ocean ridges. **Figure 12.11** illustrates an example magnetic-polarity pattern in the Pacific Ocean and explains how the magnetic data reveal the age of the crust. The measured magnetic signal originates in the mineral magnetite within the seafloor basalt. Normal-polarity basalt erupts to form oceanic crust when Earth's

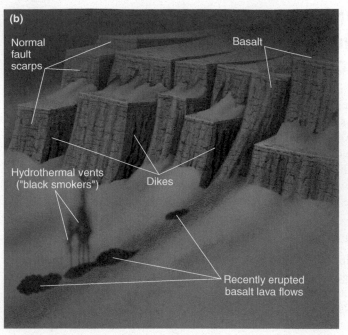

▲ **Figure 12.10 Normal faults exist along mid-ocean ridges.** (a) Iceland is a rare place where part of a mid-ocean ridge is exposed above sea level. Normal faults form prominent gashes through basalt lava flows, and careful surveying shows that the island is stretching 4 mm/year. Notice the highways for scale. (b) Geologists made this sketch after observing the seafloor along the Mid-Atlantic Ridge east of Florida from a small research submarine. The normal fault scarps depicted here are several hundred meters high.

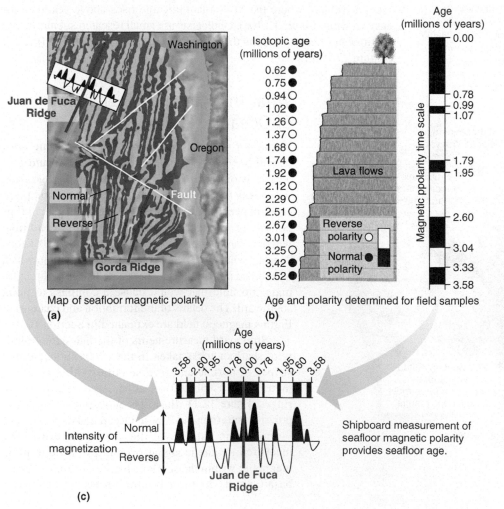

magnetic field is the same as it is at the present time. Reverse-polarity basalt dates to times in the past when Earth's magnetic field was reversed. The times of normal and reverse magnetic fields are determined by measuring polarities in more accessible lava flows on land and dating them with radioactive-isotope methods. Geophysicists combine the data from land and sea (Figure 12.11) to determine that the seafloor is less than 780,000 years old close to the ridge crest and progressively older at greater distance on either side of the ridge.

Figure 12.12 explains how seafloor spreading forms parallel bands of basaltic crust that possess alternating normal and reverse magnetic polarity. As mafic magma solidifies along the ridge crest during a single magnetic polarity interval, the new crust records that same polarity. As spreading at the divergent plate boundary continues,

◄ **Figure 12.11 How to determine seafloor age.** (a) Shipboard measurements of seafloor-crust magnetization show a symmetrical pattern of normal and reverse magnetic polarity on either side of mid-ocean ridges. Normal-polarity crust along the ridges formed during the current time of normal polarity. (b) Geologists studying volcanoes on land determine the eruption age and magnetic polarity for many lava flows. These data provide a magnetic polarity time scale, which depicts the times when Earth's magnetic field was either normal or reverse polarity. For example, the current normal polarity interval began 780,000 years ago. (c) Combining the seafloor magnetic data with the magnetic polarity time scale provides the ages of the oceanic crust. The crust is older at greater distances on either side of the mid-ocean ridge. This observation confirms the seafloor spreading hypothesis, which predicts that crust forms at mid-ocean ridges and spreads away from the ridge as new crust forms.

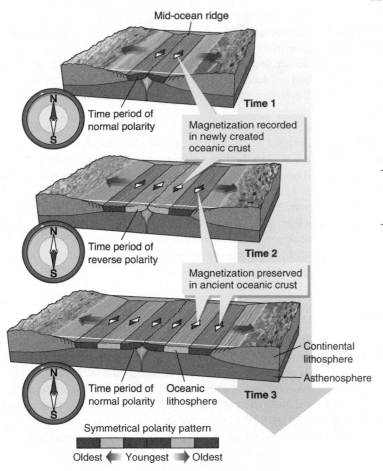

Mid-ocean ridge

Time 1

Time period of normal polarity

Magnetization recorded in newly created oceanic crust

Time 2

Time period of reverse polarity

Magnetization preserved in ancient oceanic crust

Continental lithosphere

Asthenosphere

Time 3

Time period of normal polarity

Oceanic lithosphere

Symmetrical polarity pattern

Oldest ← Youngest → Oldest

◄ **Figure 12.12 Seafloor spreading explains seafloor magnetization record.** Magnetite crystals in basalt record the orientation of Earth's magnetic field at the time lava erupts and cools at the mid-ocean spreading ridge. This record documents the flip-flopping reversals of the magnetic field during the time when the crust forms. As the two plates separate along the spreading ridge, new seafloor forms along the ridge axis while already formed crust moves aside. The spreading process creates vertical bands of crust on either side of the ridge that have different magnetic polarity.

ACTIVE ART

Seafloor Spreading and Rock Magnetism. *See how seafloor spreading at divergent boundaries produces bands of crust with alternating magnetic polarities.*

the original crust separates into two sections, one on either side of where still newer crust forms at the ridge crest. When the magnetic-field polarity reverses, horizontal transitions from normal-polarity crust to reverse-polarity crust can be seen on either side of the ridge crest.

Figure 12.13 portrays the age of oceanic crust based on measurements of seafloor magnetic polarity and the known polarity reversals back to about 180 million years ago. Mid-ocean ridges coincide everywhere with the youngest crust, and they form the centerline for symmetrical patterns of seafloor age.

▼ **Figure 12.13 Visualizing seafloor age.** This map of seafloor age results from shipboard measurements of magnetic polarity. Black lines outline the plates. Oceanic crust is progressively older at greater distances on either side of divergent plate boundaries along mid-ocean ridges.

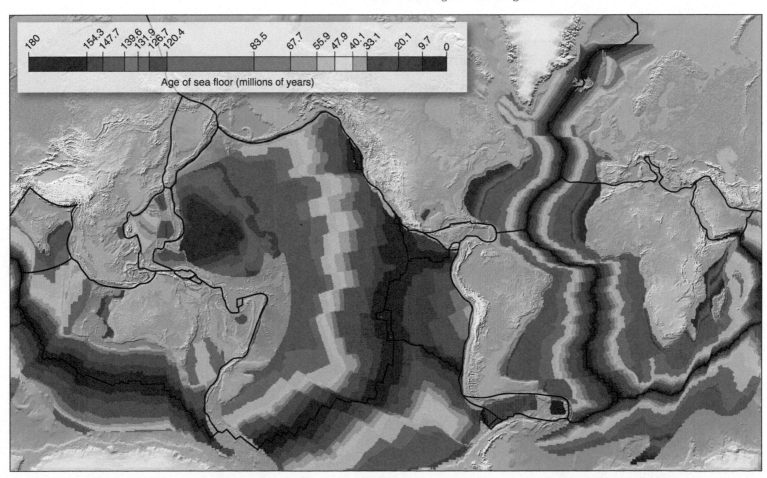

180 | 154.3 147.7 139.6 131.9 126.7 120.4 | 83.5 | 67.7 | 55.9 47.9 40.1 33.1 | 20.1 | 9.7 | 0

Age of sea floor (millions of years)

Examination of Figure 12.13 also answers the earlier questions about when Africa and South America separated and how fast they move apart. The oldest oceanic crust along the eastern margin of South America and western margin of Africa is 130 million years old. This was the first oceanic crust to form when the continents separated. Along the equator, the two continents are about 3900 kilometers apart. Divide the 3900 kilometers by the elapsed time of 130 million years and you calculate an average separation rate of 3 cm/yr. The symmetrical pattern of crust age on either side of the Mid-Atlantic Ridge (Figure 12.13) means that South America moves 1.5 cm/yr away from Africa, while Africa moves 1.5 cm/yr away from South America. Similar calculations across other divergent plate boundaries indicate that plates move between 1 and 10 cm/yr relative to their neighbors.

What Heat-Flow Data Show

Seafloor spreading predicts that divergent plate boundaries are unusually hot, because hot, partially melted asthenosphere rises close to Earth's surface to form magma that solidifies to make new crust (Figure 12.9). Data illustrated in **Figure 12.14** demonstrate that oceanic heat flow is indeed highest along a mid-ocean-ridge crest. The heat flow decreases with increasing distance from the ridge just as expected for conductive cooling of hot, newly formed lithosphere that moves away from the ridge. A comparison of global heat flow (Figure 10.6) and the location of divergent plate boundaries (Figure 12.3) shows that all divergent boundaries are areas of high heat flow.

The decreasing elevation of the seafloor away from mid-ocean ridges relates to the cooling of the lithosphere. Lithosphere contracts while it cools down, becomes denser, and sinks into the weak asthenosphere. **Figure 12.15** shows that the observed decrease in seafloor elevation almost exactly matches the calculated effect of simply cooling off the lithosphere as it moves away from spreading ridges.

The lithosphere also gets thicker as its ages and cools. The thickening, illustrated in Figures 12.9 and 12.15, occurs because rock strength is what defines the boundary between the strong lithosphere and the weak asthenosphere. Rocks are weaker at higher temperature (supporting data shown in Figure 11.20). This means that the critical temperature that distinguishes strong lithosphere from weak asthenosphere is located at deeper depths as lithosphere cools with increasing age and distance from the ridge. Hot, weak upper mantle beneath a mid-ocean ridge is part of the asthenosphere. As the upper mantle moves horizontally and cools, some of it becomes part of the stronger lithosphere.

What Seismic Tomography Reveals

Seafloor spreading predicts that asthenosphere moves upward below mid-ocean ridges (Figure 12.9). Therefore, the rising asthenosphere beneath the ridges should be hotter than its surroundings and even partly molten, so it should transmit seismic waves at slower velocities (see Section 10.2 for explanation and illustration of relationships between seismic velocity and mantle properties). **Figure 12.16** portrays seismically slower, and probably warmer, upper mantle below divergent plate boundaries, as predicted.

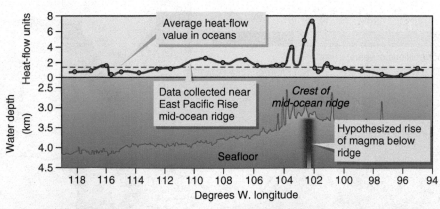

Heat flow measured by instruments towed behind research ships is unusually high over mid-ocean ridge crests. These data were collected over the divergent boundary between the Pacific and Nazca plates.

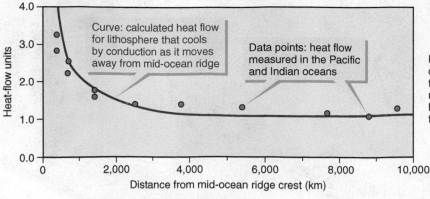

Measured heat flow in the oceans closely follows the heat flow predicted if the lithosphere is originally very hot at mid-ocean ridge crests, and then cools by conduction as it moves slowly away from the divergent plate boundary.

▲ **Figure 12.14 Visualizing high heat flow at mid-ocean ridges.**

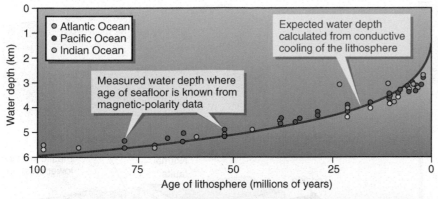

Water depth is shallowest at mid-ocean ridges and increases where the seafloor is older. The increase in water depth is predicted by cooling, contraction and sinking of the lithosphere as it moves away from the ridges.

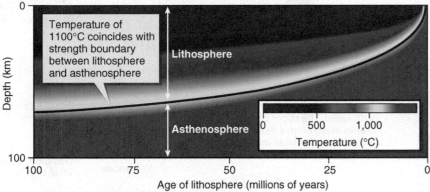

Heat-flow measurements allow calculation of temperatures within lithosphere as it cools and moves away from mid-ocean ridges. A temperature of 1100°C coincides with the transition from strong lithosphere, to weak asthenosphere. Lithosphere thickens rapidly to about 60 km at an age of 50 million years, and then more slowly to 75 km by 100 million years.

▲ Figure 12.15 Oceanic lithosphere cools, thickens, and subsides with age.

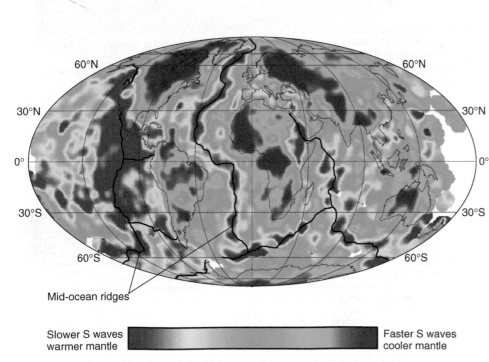

▲ Figure 12.16 How seismic-tomography data reflect divergent-boundary processes. This map shows variation in S-wave velocity between 175 and 250 kilometers below the surface. The color scale depicts how much the seismic velocity differs at each location from the average value. Notice that the upper mantle is seismically slower (red color) along almost all of the mid-ocean ridges, which include divergent plate boundaries. Unusually high temperatures are the most reasonable explanation for why the earthquake waves move more slowly in these regions of the mantle.

The Significance of Continental Rift Valleys

If ancient continents separated to make ocean basins, such as the one between Africa and South America, then it follows that divergent margins can originate within continents. What does a divergent plate boundary look like in the continental crust, rather than at a mid-ocean ridge? Do we see evidence today of such a beginning of a divergent boundary?

East Africa provides the best place to see the birth of a divergent plate boundary within a continent. The region, shown in **Figure 12.17**, is crossed by **rift valleys**, which are blocks of crust that drop along normal faults and are commonly associated with volcanoes. The East African rift valleys are as much as 4000 kilometers long and partly occupied by deep lakes. The rifts join with a divergent boundary that created the Red Sea between Africa and the Arabian Peninsula beginning about 5 million years ago. Ongoing tectonic tension may someday split Africa into two pieces, on separate plates.

Figure 12.18 combines geologic observations from rift valleys and the broken edges of separated continents, such as the east coast of North America, to illustrate the stages in breaking continents apart. The tensional stress that is first evident in the formation of a rift valley eventually stretches and separates the continental lithosphere into two fragments. New oceanic lithosphere forms to fill in the gap between the separating continents.

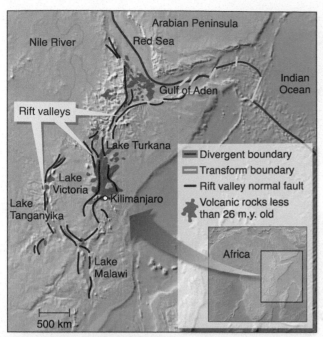

▲ **Figure 12.17 Divergence in East Africa.** Tensional stretching and thinning of continental lithosphere forms the rift valleys of East Africa. Graben blocks subside along normal faults to form elongate valleys, some of which partly fill with lakes. Upward-moving mantle melts and erupts at the surface to form volcanoes. The rifting connects northward with the divergent plate boundaries in the Red Sea and Gulf of Aden. If rifting continues, then the eastern horn of Africa may separate from the rest of the continent forming a new tectonic plate. The photo shows Lake Natron and the volcano Oldoinyo Lengai in Kenya.

▶ **Figure 12.18 Visualizing the origin of a divergent boundary within a continent.**

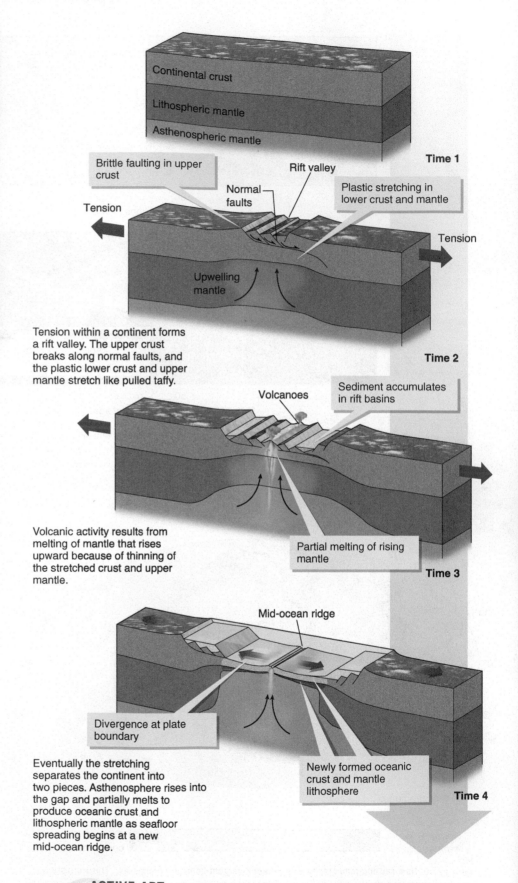

Tension within a continent forms a rift valley. The upper crust breaks along normal faults, and the plastic lower crust and upper mantle stretch like pulled taffy.

Volcanic activity results from melting of mantle that rises upward because of thinning of the stretched crust and upper mantle.

Eventually the stretching separates the continent into two pieces. Asthenosphere rises into the gap and partially melts to produce oceanic crust and lithospheric mantle as seafloor spreading begins at a new mid-ocean ridge.

ACTIVE ART

Forming a Divergent Boundary. See how a divergent boundary originates by rifting a continent.

Not all rift valleys pull apart sufficiently to form new ocean basins between fragmented continents. Some rift valleys, referred to as failed rifts, form long tectonic gashes across continents but then the tensional stress decreases or ceases. Geologists are not sure why some rift valleys become new continental margins, whereas others fail and remain as long troughs within continents. Besides East Africa, active rifts are found in other continents, forming low valleys followed by rivers, such as the Rhine in Germany and the Rio Grande in the southwestern United States. Ancient, failed rifts are recognized in many places on all continents. The Mississippi River flows along such an ancient rift valley and the within-plate earthquakes near New Madrid, Missouri (Figure 11.32), occur when stresses reactivate the old normal faults.

Putting It Together—What Is the Evidence That Plates Move Apart at Divergent Plate Boundaries?

• Plates move apart at divergent plate boundaries by the process of seafloor spreading, which creates new, mafic oceanic crust and mantle.

• Abundant earthquakes, basaltic volcanic activity, high heat flow, high seafloor elevation, and a seismically slow upper mantle confirm plate tectonics predictions for mid-ocean ridges.

• As predicted by the seafloor spreading hypothesis, magnetic polarity of seafloor crust, along with radioactive-isotope ages, demonstrates that oceanic crust is progressively older at greater distances from mid-ocean spreading ridges.

• Divergent plate boundaries can originate within continents to form rift valleys. Persistent extension along a rift valley causes the lithosphere to break into two continental fragments. Not all rift valleys extend sufficiently to form new continents and plate boundaries.

12.4 What Is the Evidence That Subduction Occurs at Convergent Plate Boundaries?

Convergent plate boundaries strongly contrast with divergent boundaries in three major ways.

1. Plates move toward one another at convergent boundaries and apart at divergent boundaries.
2. Subduction destroys lithosphere at convergent boundaries, whereas lithosphere is created at divergent boundaries.
3. Convergent boundaries are asymmetrical, with one plate angled downward beneath its neighbor. Divergent boundaries are vertical, symmetrical boundaries between plates.

What Does a Convergent Boundary Look Like?

Convergent plate boundaries coincide with some of the most spectacular landscapes and seascapes on Earth. A glance at southeastern Asia, shown in **Figure 12.19**, illustrates this point.

Deep-sea trenches, many kilometers deep, mark the curving line where one plate subducts beneath another in the ocean. Magma generated in the asthenosphere above the subducting plate feeds towering volcanoes on the overriding plate. The Indonesian volcanoes pictured in Figure 12.19 are just part of the great "Ring of Fire," a nearly continuous chain of volcanic

▼ **Figure 12.19 Convergent plate boundaries.** Plate convergence produces some of the most dramatic landscapes on Earth, as illustrated by southeastern Asia. Deep-sea trenches, where the ocean is commonly more than 5 kilometers deep, coincide with subduction of oceanic lithosphere. Volcanic island chains rise above the subducting plate, as is the case in Indonesia, where 76 volcanoes have erupted since 1800 in an area only about twice the size of Texas. Huge mountains, such as the Himalayas, form where continental lithospheric plates collide along convergent boundaries.

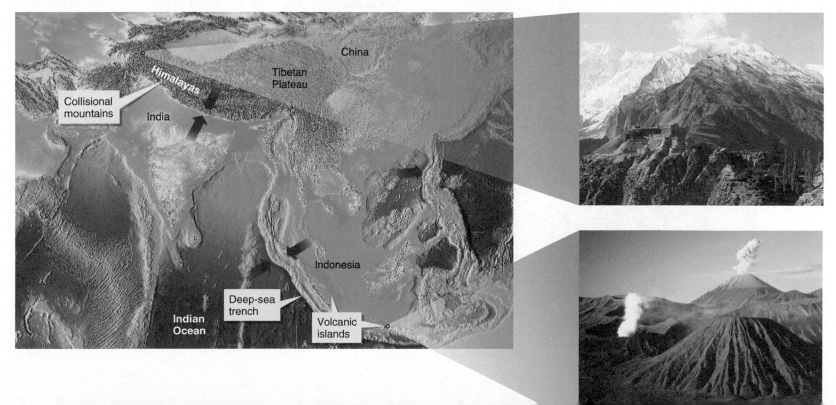

islands and continental volcanoes that stand above convergent plate boundaries encircling the Pacific Ocean (Figures 12.3 and 12.7). Some of these volcanoes erupt basalt, but most of them also erupt andesite, dacite, and occasionally rhyolite. The great diversity of magma composition suggests different processes of magma generation at convergent boundaries than occur at basaltic mid-ocean ridges.

Compressional stresses heave up great mountain ranges, especially where continents collide at convergent boundaries. The Himalayas are the most dramatic example of a modern-day convergent-margin mountain chain, but others are notable around the world (Figure 12.8). Compression forces blocks of crust tightly together along reverse and thrust faults. This tight squeezing tends to lock the faults so that huge elastic-strain energy builds up before the faults fracture. The result is large fault movements over long distances to produce dramatic and devastating earthquakes. The largest historic earthquake, an M9.5 event in Chile, in 1960 (Figure 11.35), occurred along a convergent boundary. The great Denali fault earthquake of 2002 (Figure 11.31) and the devastating 2004 earthquake offshore of Sumatra, Indonesia, are other examples of powerful earthquakes associated with convergent-boundary stress. Convergent plate boundaries are also the only locations where earthquakes routinely occur at depths greater than 50 kilometers. The deepest recorded earthquake occurred 640 kilometers beneath Bolivia, South America, in 1994.

What Happens at Convergent Boundaries?

Figure 12.20 illustrates how the plate tectonics theory explains all of the geologic features and phenomena that occur at convergent boundaries.

- Deep-sea trenches mark the actual plate boundary, where one plate bends and descends into the deeper asthenosphere.
- The plate collision causes compressional stress, which accounts for the folding, faulting, and associated earthquakes.
- Deep earthquakes occur within the subducted plate, which remains brittle at depth because it is colder than its deep mantle surroundings.
- Water-rich magma forms by partial melting above subducted lithosphere and undergoes fractional crystallization to produce compositionally diverse igneous rocks.

The actual plate boundary is the narrow trench, but the convergent-boundary processes are spread out over a wide zone that contrasts with narrow divergent boundaries. Earthquakes not only occur along the trench, but also in the deeply subducting plate, and in compressed rocks of the overriding plate, so that the zone of earthquake epicenters is very wide. Unlike divergent-boundary volcanoes, which form right at the plate boundary, convergent-margin magmas rise from an area in the asthenosphere above the subducted lithosphere and form volcanoes on the overriding plate many tens or hundreds of kilometers from the plate boundary.

The convergent-plate-boundary processes depicted in Figure 12.20 summarize decades of geologic mapping, geophysical measurements, geochemical studies of igneous and metamorphic rocks, and earthquake research. These data, in turn, allow us to test the explanations that plate tectonics theory provides for features observed around the Ring of Fire and at other convergent boundaries.

What Earthquakes Reveal

If deep earthquakes occur in the subducted plate bending down into Earth, as shown in Figure 12.20, then earthquake foci below the overriding plate should be deeper at greater distances from the trench. You can test this prediction by carefully examining the depth of earthquake foci. **Figure 12.21** shows the depth of earthquakes occurring below central South America in relation to the distance from the subduction-zone trench. Earthquakes east of the subduction zone and shallower than 50 kilometers record shortening strain in the South American crust caused by plate convergence. The deeper earthquakes occur at progressively greater depths east of the trench and define the outline of the inclined subducting plate. Earthquakes happen because of bending and downward-directed tension in the subducting plate, which occurs until the descending lithosphere becomes too hot for brittle fracture to take place. The inclined zone of earthquake foci characteristic of subduction zones is called the **Wadati-Benioff zone** in recognition of Japanese seismologist Kigoo Wadati and American seismologist Hugo Benioff who first described this feature.

What Seismic Tomography Data Reveal

Seismic tomography data, such as those shown in **Figure 12.22**, also detect subducting plates. The cooler and more rigid subducting lithosphere transmits seismic waves faster than the adjacent hot, plastic asthenosphere. Tomographic studies confirm the presence of subducting plates around the world.

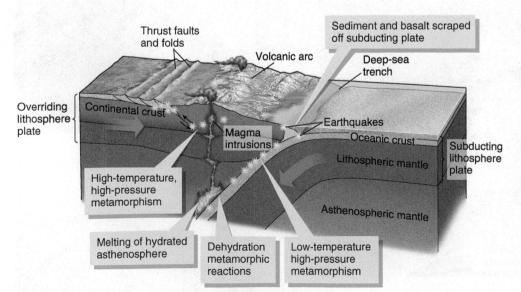

◀ **Figure 12.20 What happens at a convergent plate boundary?** Plates move together at convergent boundaries. A deep-sea trench marks the location of subduction, where convergence occurs within or at the margin of an ocean. Folding, faulting, volcanic activity, and metamorphism may extend 500 kilometers or more from the deep-sea trench because of the inclined orientation of the subducted plate and the large compressive stresses transferred to the overriding plate. Dehydrating metamorphic reactions release water into the asthenosphere, which promotes melting. Rocks within and adjacent to the cold, subducting plate experience low-temperature, high-pressure metamorphism. High-temperature, high-pressure metamorphism occurs where subduction-zone magmas rise into the overriding plate.

What Heat-Flow Data Reveal

Plate tectonics theory predicts a complicated pattern of heat flow from the mantle to the surface near convergent plate boundaries. On one hand, the subducted lithosphere is colder than the surrounding asthenosphere and draws in heat from the surrounding mantle by conduction. Therefore, the mantle close to the trench should be cooler than average. On the other hand, magma forms above the subducted plate and rises, so unusually high heat flow should occur in the overriding plate in the proximity of the rising magma.

Figure 12.23 tests these predictions with a heat-flow profile across the convergent plate boundary in the northwestern United States. As predicted by plate tectonics, heat flow is low where the cold lithosphere subducts and is high where magma migrates upward to feed the volcanic chain.

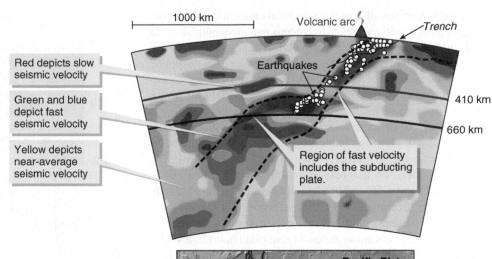

Red depicts slow seismic velocity

Green and blue depict fast seismic velocity

Yellow depicts near-average seismic velocity

Region of fast velocity includes the subducting plate.

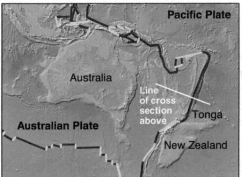

▲ **Figure 12.22 Seismic tomography detects subducted plates.** Seismic tomography data across the convergent plate boundary north of New Zealand reveals an inclined zone (green and blue) where earthquake waves move unusually fast. This fast zone is consistent with the location of relatively cold and rigid lithosphere, which is also partly outlined by earthquake foci.

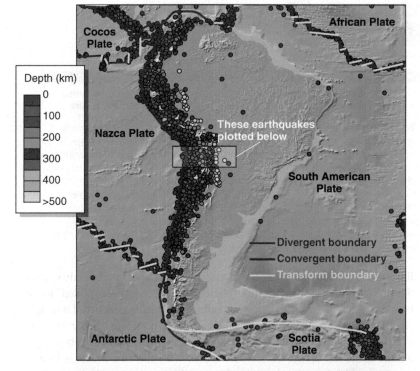

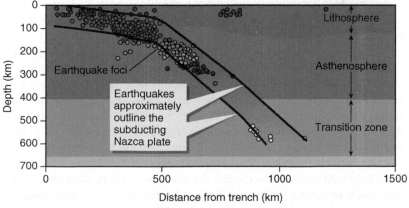

▲ **Figure 12.21 Earthquakes outline a subducting plate.** The graph shows that the deep earthquakes outline the Nazca plate subducting beneath South America, The subducting plate penetrates through the asthenosphere and into the transition zone. Compression also generates shallow earthquakes in the South American plate.

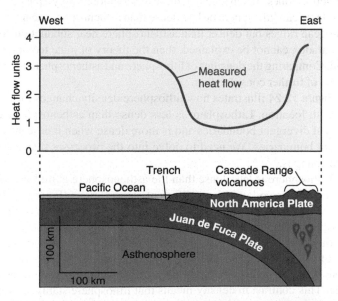

▲ **Figure 12.23 Heat-flow data are consistent with convergent-boundary processes.** A graph of surface heat flow across the subduction zone in the northwestern United States shows two important features: (1) Lower heat flow just east of the trench, which is consistent with the presence of relatively cold, subducted lithosphere at depth. (2) Higher heat flow in the vicinity of the Cascade Range volcanoes, which is consistent with the upward rise of hot magma generated by melting above the subducted plate.

A quick review of how these thermal characteristics of convergent plate boundaries relate to igneous and metamorphic rocks that form in these settings is worthwhile. Magma does not form at convergent margins because of heating in the asthenosphere; indeed, the insertion of a cold subducted plate refrigerates the surrounding asthenosphere much like dropping ice cubes in water. Instead, magma forms because dehydration metamorphism of the subducted plate releases fluid into the asthenosphere (Figure 12.20). This fluid reduces the melting temperature of the peridotite so that it partially melts (see Figure 4.19b). The resulting water-rich magmas erupt explosively (Section 4.9, Figures 4.25 and 4.26) and produce economically valuable mineral deposits (Section 4.10, Figures 4.29 and 4.31).

Paired metamorphic belts, described and explained in Section 6.10, are common at convergent plate boundaries. One member of the pair is a belt of high-pressure but relatively low-temperature metamorphic rocks, whose origin is readily explained by rocks moving downward at cold subduction zones (Figure 6.29). These rocks experience increasing pressure and metamorphic reactions as they descend. The high-temperature, low-pressure belt of metamorphism, the second member of the pair, coincides with where magma rises into the lithosphere above the subducted plate (Figure 12.21).

Why Does Subduction Occur?

Why does subduction happen? After all, when ice floes converge in the ocean the fractured ice rises into the air, rather than descending into the water (Figure 12.5). Ice cannot "subduct" into water because ice is less dense than water. By analogy, then, lithosphere must be denser than asthenosphere for subduction to happen. But wait—if this were generally true on Earth, then lithosphere should sink into asthenosphere everywhere, rather than only at subduction zones. Taken together, these observations imply a paradox. How can lithosphere be less dense than asthenosphere near mid-ocean ridges but denser than asthenosphere near subduction zones? If this paradox cannot be explained, then the theory of plate tectonics cannot stand. Comparing the densities of lithosphere and asthenosphere is, therefore, worthy of further consideration.

Figure 12.24 illustrates how lithosphere density changes through time and with location. Lithosphere is less dense than asthenosphere when it forms at divergent boundaries and is more dense when it descends at convergent boundaries. We need to delve into the processes that account for these changes.

Lithosphere is less dense than the asthenosphere at mid-ocean ridges because these two layers have different compositions (Figure 12.24). The asthenosphere peridotite partly melts and forms two parts of the young lithosphere—one is the solidified basalt and gabbro of the oceanic crust, and the other is a residue of unmelted peridotite minerals in the mantle part of the oceanic lithosphere. Both of these lithosphere layers are less dense than the underlying unmelted asthenosphere peridotite below the ridge. This contrast in density means that lithosphere stably sits on top of the asthenosphere like an ice cube in water.

However, lithosphere density increases over time because of changing temperature and composition. The lithosphere cools, contracts, and becomes denser with increasing age and distance from the mid-ocean ridge where it formed. The lithosphere also thickens with age (see Figure 12.15). The

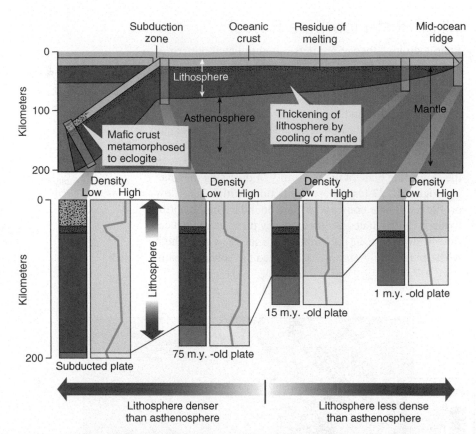

▲ **Figure 12.24 Why oceanic lithosphere subducts.** This diagram illustrates how the density of oceanic lithosphere changes while the lithosphere moves away from the mid-ocean ridge. When first formed (right side of diagram) the lithosphere is mostly low-density crust and the residues of melting to make that crust; so the lithosphere is significantly less dense than the asthenosphere. Farther from the mid-ocean ridge, mantle cooling causes the boundary between strong lithosphere and weak asthenosphere to move downward. The thickening lithospheric mantle becomes denser as it cools and contracts. Lithosphere that is older than 15 million years is denser than asthenosphere and can subduct into the asthenosphere. During subduction, mafic crust metamorphoses to very dense eclogite, which further increases the density of the subducting lithosphere and pulls the plate into the subduction zone.

lithosphere thickens at its base when asthenosphere cools sufficiently to have the strength characteristics of lithosphere. The asthenosphere "added" to the base of the cooling and thickening lithosphere has the high density of nonmelted upper mantle. So, two processes increase the lithosphere density: (1) contraction during cooling, and (2) the addition of high-density, non-melted upper mantle as the lithosphere thickens from below. By the time a plate is 15 million years old, it has gained enough dense peridotite along its thickening base and has cooled and contracted enough so that the overall lithosphere density is about equal to that of the underlying asthenosphere (Figure 12.24).

However, just because oceanic lithosphere older than 15 million years is denser than the asthenosphere, does not mean that this lithosphere immediately sinks into the deeper asthenospheric mantle. Three factors cause a delay in the beginning of subduction. First, the strong lithosphere resists the necessary downward bending. Second, the high viscosity of the asthenosphere also resists the sinking of the lithosphere. Last, any plate that is actively growing at a mid-ocean ridge contains some parts that are young, warm, thin, and less dense than asthenosphere, and these less dense parts help hold up the denser areas. An analogy would be your ability to float easily in water if you wear a buoyant life preserver; less dense young lithosphere likewise buoys up denser, older lithosphere.

In addition to having the right density contrast for subduction to take place, we need to have the plates moving toward one another. When plates converge, and the lithosphere is denser than the asthenosphere, then subduction begins with the densest plate sliding into the asthenosphere. At this point, a third factor, in addition to contractional cooling and addition of dense mantle, increases the density of the lithosphere to aid subduction. Where the subducting plate reaches a depth of about 80 to 100 kilometers, the basaltic crust metamorphoses to the very dense rock, eclogite. The metamorphism converts the crust from rock that is less dense than asthenosphere into rock that is denser than asthenosphere (Figure 12.24). The sinking edge of the dense subducted plate is now like a heavy anchor pulling a chain downward through water. As more lithosphere subducts, and the volume of the dense eclogite anchor increases, it becomes even more difficult for less dense parts of the plate to resist being pulled down the subduction zone. Indeed, as we will explore a little bit later, sinking dense lithosphere is what moves the plates.

The Types of Convergent Plate Boundaries

Features and processes at convergent margins vary from place to place depending on the types of plates that converge. There are two types of plate lithosphere (continental, oceanic), so there are three types of boundaries: (1) oceanic-oceanic, (2) continental-continental, and (3) continental-oceanic. **Figure 12.25** illustrates how these three types of subduction zones work.

The different characteristics of the three boundary types tie into the earlier consideration of how subduction takes place to begin with—the subducted plate must be denser not only than the asthenosphere, but also than the overriding plate. Continental crust is much less dense than mantle peridotite. As an outcome, when oceanic and continental plates converge, the oceanic plate always subducts (Figure 12.25).

The different "subductability" of continental and oceanic crust also explains another mystery left over from Chapters 8 and 9—the antiquity of continental crust compared to the youthfulness of oceanic crust. Plates with thin, dense oceanic crust can subduct at convergent plate boundaries, but thick, low-density continental crust cannot subduct to any great extent. Continental crust is virtually permanent, except to the extent that it gradually weathers and erodes away as sediment. By contrast, oceanic crust is continuously destroyed by subduction at the same rate it is created at mid-ocean ridges. As a result, crust older than 1 billion years dominates continents, but there is no crust present in the ocean basins older than about 180 million years. Faults, folds, and igneous plutons resulting from billions of years of

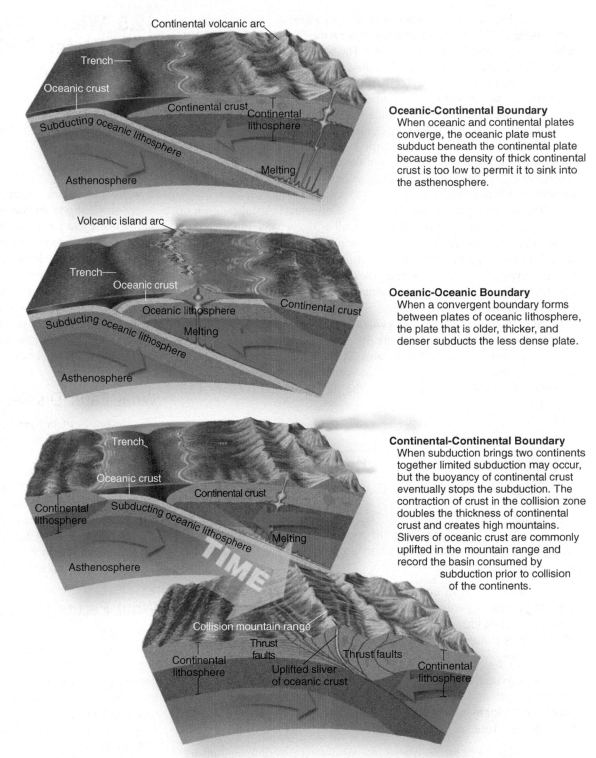

Oceanic-Continental Boundary
When oceanic and continental plates converge, the oceanic plate must subduct beneath the continental plate because the density of thick continental crust is too low to permit it to sink into the asthenosphere.

Oceanic-Oceanic Boundary
When a convergent boundary forms between plates of oceanic lithosphere, the plate that is older, thicker, and denser subducts the less dense plate.

Continental-Continental Boundary
When subduction brings two continents together limited subduction may occur, but the buoyancy of continental crust eventually stops the subduction. The contraction of crust in the collision zone doubles the thickness of continental crust and creates high mountains. Slivers of oceanic crust are commonly uplifted in the mountain range and record the basin consumed by subduction prior to collision of the continents.

▲ Figure 12.25 Three types of convergent plate boundaries.

plate tectonics scar the ancient continental crust, and these faults are weaknesses that account for some within-plate earthquakes.

As depicted in Figure 12.25, continent-continent convergent boundaries are extremely complex and cannot persist for long intervals of geologic time because of the buoyant nature of continental crust compared to asthenospheric mantle. Crust on both sides of the plate boundary rise because continental crust, like the converging ice floes in Figure 12.5,

cannot subduct. The best modern example of continent-continent convergence is the collision of India with Southeast Asia (Figures 12.3 and 12.19),
which began about 50 million years ago. The collision doubled up the crust,
uplifted the Himalayas as the highest mountains on Earth, and raised the
Tibetan Plateau (an area roughly half the size of the contiguous United
States) to a mean elevation of 5 kilometers above sea level (which is higher than any location in the contiguous U.S.). The collision is comparable
to wrecking a house with a bulldozer, shattering the building and shoving
it off its foundation. As a result, eastern Asia shattered along ancient faults
and continues to slides eastward out of the way toward the Pacific Ocean.
These far-flung effects of the Indian collision account for earthquakes
throughout China (Figures 12.6 and 12.8).

Oceanic lithosphere subducts beneath continental lithosphere prior to
continent-continent collision. Eventually, all of the oceanic lithosphere is
consumed, which brings another continental block into the trench beneath
the edge of the overriding continental lithosphere (Figure 12.25). Thrust-
faulted slivers of oceanic crust and lithospheric mantle slide up onto land
during convergence and mark the former presence of an ancient ocean that
once existed between the continents.

All modern continents show evidence of former continent-continent
collision zones in the form of highly deformed metamorphic rocks in the
eroded remnants of once-tall mountains. The Appalachian Mountains, for
example, mark a zone of continent-continent convergence when Pangea
formed from colliding continents in the late Paleozoic (Figures 12.2
and 12.8).

Putting It Together—What Is the Evidence that Subduction Occurs at Convergent Plate Boundaries?

• The Wadati-Benioff zone of earthquake foci outlines
subducted plates descending at an angle from deep-sea trenches
into the asthenosphere.

• Seismic tomography shows subducted plates as inclined zones of
unusually fast seismic velocity in the asthenosphere. Subduction
zones also coincide with areas of low heat flow. These data reveal the
cold, dense plates sinking into the asthenosphere at subduction
zones.

• Metamorphic fluids released from the subducted plate cause melting in the surrounding asthenosphere. Magma intruding into the overriding lithosphere accounts for volcanic activity, high heat flow, and
high-temperature metamorphism.

• Subduction is possible where old lithosphere is denser than asthenosphere. Metamorphism of mafic crust to eclogite further increases the density of the subducting plate and pulls it downward in
the asthenosphere like an anchor.

• Where continental plates collide, high mountain ranges form because thick, low-density continental lithosphere cannot subduct.

• Oceanic lithosphere is continuously consumed at convergent plate
boundaries so there is no ocean floor older than 180 million years.
Continental crust cannot subduct and is mostly more than 1 billion
years old.

12.5 What Is the Evidence that Plates Slide Past One Another at Transform Plate Boundaries?

According to the theory of plate tectonics, not all plate boundaries are sites
of lithosphere creation or destruction. The lithosphere is conserved along
transform boundaries because the plates move alongside one another.

What a Transform Boundary Looks Like

Transform boundaries show up on the plate-boundary map (Figure 12.3)
where strike-slip faults join parts of other plate boundaries. **Figure 12.26**
shows that these particular strike-slip faults are called "transforms" because plate boundary motion transforms from one boundary to another
along these faults. Most transform plate boundaries are short and
connect spreading-ridge segments along divergent plate boundaries (see
Figure 12.3). Only a few transforms, including the San Andreas Fault in
California (Figure 12.26), form long plate boundaries. The San Andreas
transform connects a divergent boundary, to the south, with a convergent
boundary, to the north, while also forming the boundary between the
Pacific and North American plates (Figure 12.3).

What Fault Displacements Show

Plate tectonics theory predicts that strike-slip faults should be found at
all interpreted transform boundaries. This prediction is easily tested for
continental transforms by examining features displaced across the fault.
Figure 12.27 shows such a test for the San Andreas transform boundary
in California. Not only do the field relationships confirm the predicted
strike-slip motion, but the rate of motion along the transform, about
5.6 cm/yr, is comparable to plate velocities at mid-ocean ridges. Notice,
as well, in Figure 12.27 that this plate boundary in California is a zone, with
movement across many faults, rather than a simple line on the map. In
stronger oceanic lithosphere, most transform boundaries are single faults.

What Earthquakes on Oceanic Transforms Show

Figure 12.28 illustrates transform faults where they connect segments of mid-
ocean ridge. In this example, it *looks as if* two mid-ocean-ridge segments are
displaced by a left-lateral strike-slip fault. However, first appearances can be
deceiving. According to plate tectonics theory the ridge is not an old hill
that is later displaced by a younger left-lateral fault. Instead, the ridge segments and transform fault are simultaneously active parts of the boundary
between plates A and B. Two hypotheses from plate tectonics led to this
interpretation.

First hypothesis—the displacement across the example transform fault
should be right lateral, not left lateral, because the plates move away from
one another along the segmented divergent plate boundary. To see why this
is the case, notice in Figure 12.28 that the lithosphere of Plate B moves
east away from the mid-ocean ridges, whereas lithosphere of Plate A moves
west. These opposite directions of motion require right-lateral displacement along the transform boundary.

Second hypothesis—the transform fault is only actively moving between the ridge segments that it connects. East and west of the connected
ridge segments the continuation of the fault line separates lithosphere belonging to the same plate. There should not be movement across these

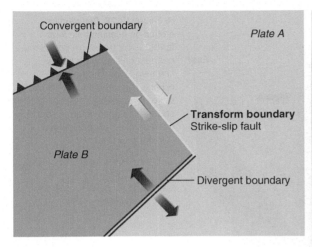

▲ **Figure 12.26 What a transform boundary looks like.** Plates slide past one another along strike-slip faults at transform boundaries. Plates A and B in the diagram move in opposite directions and share a transform boundary that connects a divergent boundary with a convergent boundary. The San Andreas Fault in California is part of a transform plate boundary zone separating the North America and Pacific plates.

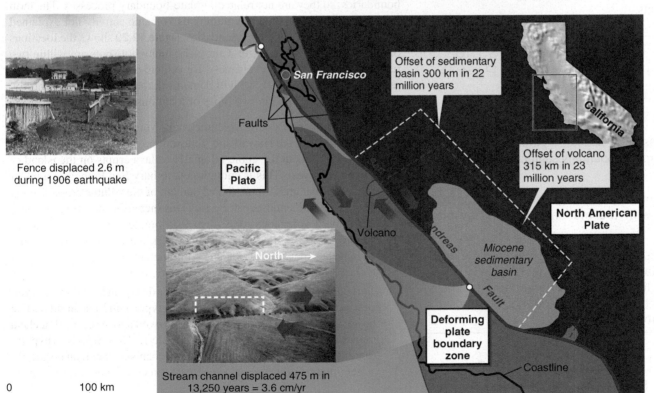

◄ **Figure 12.27 Calculating plate motion along the San Andreas transform boundary.** The Pacific plate slides northwestward past the North American plate along a transform boundary. The San Andreas Fault is the longest of several nearly parallel faults that form the plate boundary zone. Offset of geologic features that match across the fault provides estimates of the speed of plate motion, which currently is 3.6 centimeters per year. Adding this speed to calculations for other faults in the boundary zone reveals a total movement of 5.6 centimeters per year between the Pacific and North American plates.

continuations of the fault line because the lithosphere on either side moves in the *same* direction and at the *same* velocity.

Earthquake seismographs successfully tested these two hypotheses. The first critical piece of evidence is that earthquakes only occur along the transform between the ridge segments and along the divergent boundary

(Figure 12.28). The *apparent* fault traces that *seem* to extend farther beyond the mid-ocean ridge segments are not faults but simply **fracture zones** that separate lithosphere of different age and thickness within the same plate. A second important item of evidence comes from sophisticated analyses of seismograms that allow geologists to determine the direction of movement

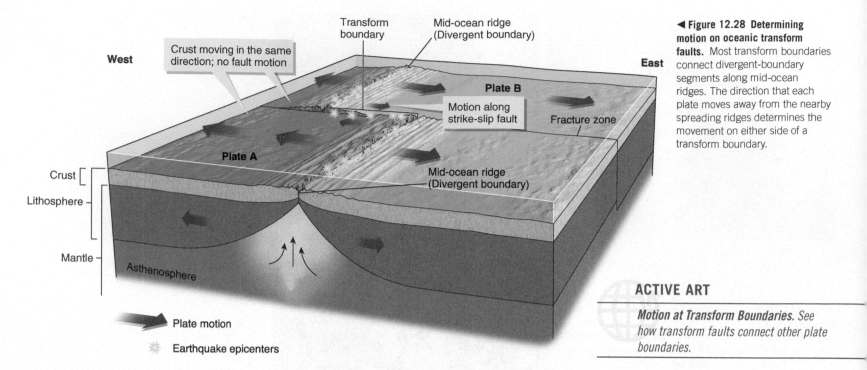

West

Crust moving in the same
direction; no fault motion

Transform
boundary

Mid-ocean ridge
(Divergent boundary)

Plate B

East

Motion along
strike-slip fault

Fracture zone

Plate A

Crust

Lithosphere

Mid-ocean ridge
(Divergent boundary)

Mantle

Asthenosphere

Plate motion

Earthquake epicenters

◀ **Figure 12.28 Determining
motion on oceanic transform
faults.** Most transform boundaries
connect divergent-boundary
segments along mid-ocean
ridges. The direction that each
plate moves away from the nearby
spreading ridges determines the
movement on either side of a
transform boundary.

ACTIVE ART

Motion at Transform Boundaries. See
how transform faults connect other plate
boundaries.

along faults during earthquakes. In the Figure 12.28 example, the seismograms reveal that transform-boundary earthquakes are caused by right-lateral strike-slip motion consistent with the plate motion inferred from the divergent-boundary segments.

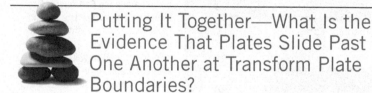

Putting It Together—What Is the Evidence That Plates Slide Past One Another at Transform Plate Boundaries?

• Transform plate boundaries are a special type of strike-slip fault along which plate motion from one boundary segment is transformed to the next segment.

• The direction and velocity of strike-slip displacement on transform boundaries are consistent with the motions predicted by plate tectonics.

• Most transform boundaries are short faults between mid-ocean-ridge segments. Transform faults line up with tectonically inactive fracture zones that separate lithosphere of different age, thickness, and elevation but belong to the same plate and move in the same direction at the same velocity.

12.6 What Does the Mantle-Plume Hypothesis Explain that Plate Tectonics Cannot Explain?

Plate tectonics theory explains processes at, or close to, plate boundaries but it does not explain active volcanoes and deformation *within* plates. The causes of these phenomena are less well established than the plate tectonics theory. This section explains the mantle plume hypothesis, an incompletely tested companion to plate tectonics theory that offers the potential to round out a global view of tectonic processes.

The Problem of Hot Spots

Figure 12.7 reveals many volcanoes at locations that are distant from plate boundaries, so they are not related to plate-boundary processes. The term **hot spot** describes an area of voluminous volcanic activity not explained by melting processes at plate boundaries. **Figure 12.29** shows the locations of about 40 hot spots on Earth. Curiously, many hot spots are within two regions of the world where seismic-wave velocities are unusually slow (the red areas in Figure 12.29); we will come back to this observation shortly.

Hot spots are found in two types of locations.

1. Large volumes of young volcanic rocks very far from plate boundaries. Examples are Hawaii and Yellowstone, pictured in **Figure 12.30**. On average, about 0.1 cubic kilometer of lava erupts on the island of Hawaii each year. This is enough lava to bury San Francisco a meter deep each year and is more than 5 percent of the volume erupted along all of the world's mid-ocean ridges. This incredible eruption rate built huge shield volcanoes more than 9 kilometers above the seafloor (Figure 12.30). Yellowstone National Park marks a similarly prolific volcanic hot spot where 6000 cubic kilometers of magma erupted over the last 2 million years, mostly as rhyolitic tuff.

2. Unusually prolific volcanism along or near a divergent boundary. The volcanic activity is so excessive compared to typical mid-ocean ridges that the hot spot volcanoes commonly build up well above sea level. Iceland is an example of this type of hot spot (Figure 12.10a). Scientists hypothesize that there must certainly be some mechanism other than normal divergent-margin processes to generate the excessive amounts of magma.

The Mantle-Plume Hypothesis

The mantle-plume hypothesis emerged to explain hot spots in the late 1960s at the same time that plate tectonics theory was established. The hypothesis, illustrated by **Figure 12.31**, has two fundamental components:

1. Hot spots form where narrow columns of unusually hot mantle, called **plumes**, rise to the surface from the core-mantle boundary.

2. The plume locations are stationary in the mantle.

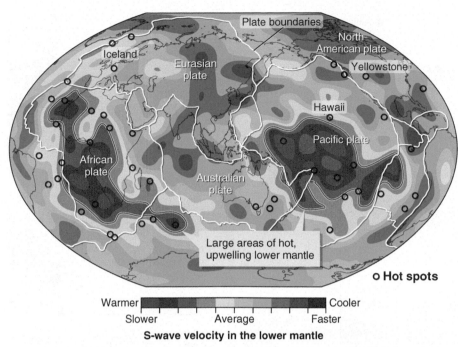

Warmer [scale] Cooler
Slower Average Faster
S-wave velocity in the lower mantle

▲ **Figure 12.29 Where hot spots are located.** Each circle on this map is the location of a hot spot—an area of extraordinary, high-volume volcanic activity that usually is not at a plate boundary. The map also shows seismic-tomography results from the lower mantle, just above the outer core. Many hot spots coincide with seismically slower, warmer, lower mantle "superplumes" centered beneath Africa and the South Pacific.

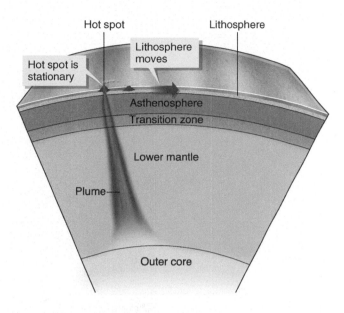

▲ **Figure 12.31 The mantle-plume hypothesis.** The hypothesis proposes that hot spots are the surface expression of plumes, which are stationary, unusually hot columns of the mantle that rise because of convection from the core-mantle boundary. Plumes are about 200–500 km in diameter and are about 200–300°C warmer than the surrounding mantle.

▼ **Figure 12.30 What hot spots look like.** The "Big Island" of Hawaii (left) marks the location of a hot spot that has built volcanoes more than 9 kilometers above the seafloor in the middle of the Pacific plate. The black "fingers" visible in this Space Shuttle photograph are historically erupted basalt lava flows. The famous geysers and other thermal features of Yellowstone National Park (right) relate to a volcanic hot spot within the North American plate. The last major volcanic eruption at Yellowstone, about 640,000 years ago, blanketed most of North America in ash.

The first component of the hypothesis explains the peculiar composition of hot spot-island basalts. Rising asthenosphere decompresses and partially melts to form basaltic magma. According to the hypothesis, the melting asthenosphere originates deep below the plates at hot spots, so the resulting basaltic magma has a different composition from the magma produced by melting near the base of the lithosphere, which forms mid-ocean-ridge basalt at divergent plate boundaries (see Figure 4.19a for a refresher on magma production at hot spots and mid-ocean ridges). Indeed, the chemical composition of basaltic lava flows in places such as Hawaii and Iceland is subtly different from basalt erupted along the mid-ocean ridges. Although these differences involve elements that are only present in trace amounts, these differences are consistent with part of the magma at hot spots originating deeper in the mantle than the magma erupted at mid-ocean ridges.

The second component of the hypothesis explains the ages of extinct volcanoes that trail away from hot spots. For example, **Figure 12.32**a shows lines of extinct volcanic islands and submarine volcanoes, called **seamounts**, that continue for thousands of kilometers across the Pacific Ocean until ending at an active hot spot volcano. Volcanic rocks are progressively older along

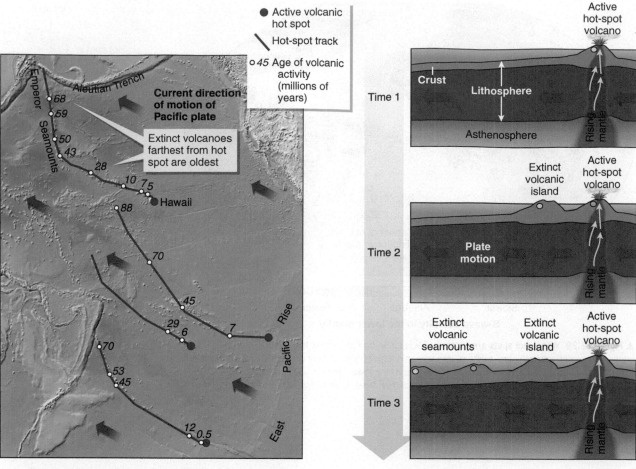

● Active volcanic hot spot

— Hot-spot track

○45 Age of volcanic activity (millions of years)

Current direction of motion of Pacific plate

Extinct volcanoes farthest from hot spot are oldest

Time 1

Crust

Lithosphere

Asthenosphere

Active hot-spot volcano

Rising mantle

Time 2

Extinct volcanic island

Active hot-spot volcano

Plate motion

Rising mantle

Time 3

Extinct volcanic seamounts

Extinct volcanic island

Active hot-spot volcano

Rising mantle

ACTIVE ART

Hot Spots and Plumes. See how an island chain forms by plate motion across a hot spot, which might connect to a hypothesized mantle plume.

(a) Active hot-spot volcanoes in the Pacific Ocean are present at one end of long chains of volcanic islands and submerged volcanic seamounts. Radioactive-isotope ages of volcanic rocks show that the volcanoes are progressively older at greater distances from the active hot spot.

(b) Hot-spot tracks form where lithosphere moves slowly across a nearly stationary column of rising and melting mantle. Volcanoes form at the hot spot but become extinct because plate motion carries them away from the hot spot. New volcanoes form behind the extinct volcanoes. The colored dots show shifting positions of volcanoes through time as plate motion carries the lithosphere over the hot spot.

▶ **Figure 12.32 Hot spots leave tracks on moving plates.**

each line, away from the active hot spot. Figure 12.32b shows how these data suggest a stationary plume of rising hot asthenosphere that melts to form a line of volcanoes as the lithospheric plate passes overhead. *If* this hypothesis is correct, then **Figure 12.33** shows how to use the locations and ages of volcanic rocks along a hot spot track to determine the speed and

direction of plate motion. Of course, this plate "speedometer" only works if the plume is stationary. It is plausible that the hypothetical plume moves, which negates the plate-velocity calculation unless the plume velocity is also known. Whether hot spots move or not remains unclear. Most data suggest that hot spots do move but usually at rates slower than plates.

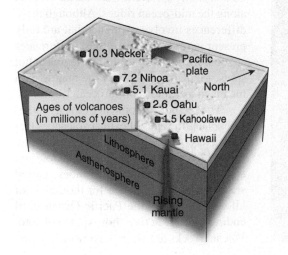

10.3 Necker

Pacific plate

North

7.2 Nihoa

5.1 Kauai

2.6 Oahu

1.5 Kahoolawe

Ages of volcanoes (in millions of years)

Hawaii

Lithosphere

Asthenosphere

Rising mantle

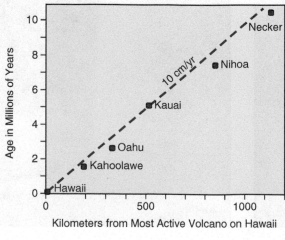

◀ **Figure 12.33 Using hot spots to estimate plate velocity.** Pacific plate movement produced a hot spot track of progressively older volcanoes along a line northwestward from the active volcanoes on Hawaii. *If* the hot spot beneath Hawaii is stationary, then the age and distance of the older volcanoes from Hawaii indicate how fast the plate moves. On a graph of volcano age plotted against distance from Hawaii, the slope of the line is velocity. The ages and distances of the volcanoes suggest a plate velocity of about 10 centimeters per year.

Testing the Plume Hypothesis

If seismic tomography data revealed the presence of narrow cylinders of lower-velocity (warmer) mantle extending from the core-mantle boundary up to hot spot volcanoes, then the plume hypothesis would be positively tested. Figure 12.29 indicates the existence of seismically slow, lower mantle around the locations of some hot spots, but not all of them. Research suggests that some of the slowing of seismic waves relates to compositional differences compared to the surrounding lower mantle. However, composition alone cannot easily explain the large differences in velocity, which implies that these regions of lower mantle are also hotter, and hence rising, compared to the neighboring rocks. These large areas of seismically slow lower mantle rising upward beneath Africa and the southern Pacific are called "superplumes." While superplumes may document convective upwelling from deep in the lower mantle they are not, as the hypothesis proposes, narrow plume columns connecting to distinct hot spots such as Hawaii.

Seismic tomography alone may not be adequate to test the plume hypothesis, because these data have limitations. Seismic-tomography data reveal many details about the upper mantle, but they still provide only a hazy view of the deepest mantle because fewer measured seismic waves pass through the lower mantle than the upper mantle. In addition, the standard seismic-tomography methods detect only large features, and plume columns less than 500 kilometers wide will not show up in images such as the ones shown in Figures 10.8 and 12.16. Recent advances in seismic tomography have increased the sensitivity of the method to detect narrow plumes rising from the base of the mantle. The results remain ambiguous in most places but do suggest deeply rooted plumes below at least seven hot spots, including Hawaii.

The plume hypothesis for explaining hot spots is not yet adequately tested and is very controversial. Alternative ideas explain the locations of hot spot volcanoes by tears in the interiors of nonrigid plates that permit local decompression melting in the upper mantle. This hypothesis is consistent with tomographic data that suggest some hot spots originate within 200 kilometers of the surface. It also explains why age progressions, such as those illustrated in Figure 12.32a, are not present at all within-plate volcanic regions. In this alternative hypothesis, the unusual chemistry of erupted basalt is attributed to variations in the composition of the upper mantle, rather than requiring a deeper source. It is possible that not all hot spots have the same origin. Some hot spots may be deeply rooted plumes, whereas others may result from other processes. **Figure 12.34** illustrates a modified plume hypothesis where only some plumes are narrow vertical columns extending from the base of the mantle to Earth's surface. Perhaps the variable viscosity of the mantle and some physical boundaries, such as those defining the transition zone, also influence the shape of plumes. Some narrow plumes might rise from the transition zone

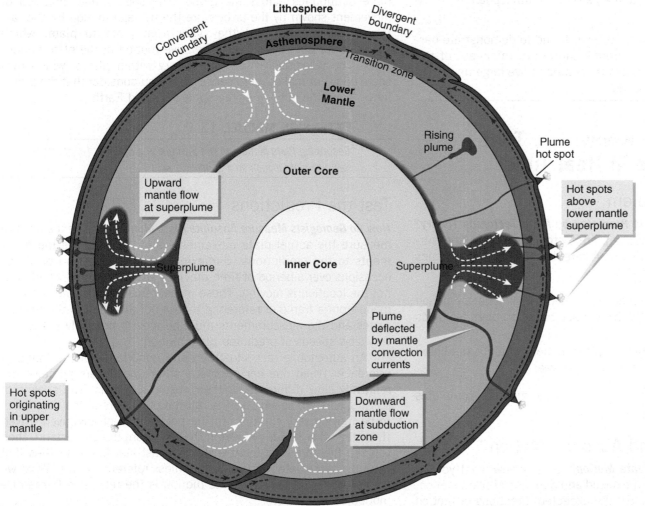

◄ **Figure 12.34 Varieties of plumes and hot spots.** Current data suggest that some hot spots are really plumes that rise from the core-mantle boundary. Other hot spots may originate near the transition zone and possibly form above lower-mantle superplumes that are centered on opposite sides of Earth beneath the Pacific Ocean and Africa (Figure 12.29). Other hot spots may originate directly below the lithosphere in the upper asthenosphere. In some cases, mantle convection currents may deflect proposed plumes to form hot spots at the surface far from where they originate in the lower mantle.

above lower mantle superplumes rather than directly from the top of the core, which would explain the occurrence of many hot spots above the superplumes (Figure 12.29).

Tracking narrow plumes downward through the lower mantle is challenging not only because of the limitations of seismic-tomography data, but also because of mantle motion. Convection currents may deflect a vertically rising plume so that its surface expression is 1000 kilometers away from where it is hypothetically rooted in the lower mantle (Figure 12.34). This likely mantle motion also calls into question the validity of the stationary plume concept.

Putting It Together—What Does the Mantle-Plume Hypothesis Explain that Plate Tectonics Cannot Explain?

• Hot spots are areas of prolific generation of igneous rocks not explained by plate tectonics. Some hot spots are present within plates, whereas others are areas of unusually intense volcanic activity at or near divergent plate boundaries.

• Extinct volcanoes are progressively older at greater distance from many active hot spot volcanoes.

• The plume hypothesis suggests that hot spots mark places where hot mantle convectively rises to the base of the lithosphere from the core-mantle boundary.

• Conclusive evidence has not yet been found to demonstrate narrow plumes of mantle rising through the entire thickness of the mantle. However, most hot spots are located above large upwelling superplumes in the lower mantle.

12.7 How Do We Know . . . That Plates Move in Real Time?

Understand the Problem

Does the Lithosphere Move as Predicted by Plate Tectonics Theory? Consider the possibility of testing plate tectonics theory by measuring actual lithosphere movement. A direct test of plate tectonics requires two pieces of information:

1. Theoretical calculations of predicted speed and direction of motion at different locations on Earth's surface based on plate tectonics theory.

2. A method to measure very slow plate motion, which is predicted to be only about 1–10 centimeters per year.

Comparison of predicted and measured motions completes the test.

Visualize Relative and Absolute Motion

How Do Geologists Predict Plate Motion? Let's review what you do know about data that predict the speed and direction of plate motion. The arrows in Figure 12.3 show the expected directions of motion

between plates where they share a boundary, based on whether that boundary is divergent, convergent, or transform. Maps of seafloor age (Figure 12.13) show the rate of plate separation by seafloor spreading along divergent boundaries. Similarly, data such as those shown in Figure 12.27 provide the rate of motion between two plates along a transform boundary.

These uses of plate-boundary information reveal *relative velocities* between plates along the boundary. These data do not provide *absolute velocities*, or how fast, and in what direction, each plate moves independently of its neighbors. **Figure 12.35** illustrates the difference between relative and absolute velocities. If you catch up with and pass a car on the highway, then your speed is faster relative to the speed of the other car, but this relative speed does not specify the absolute speed of either vehicle. The absolute speeds are known from reading the speedometers in each car. Notice in Figure 12.35 that cars, and plates, can move in the same direction in an absolute sense but still be moving toward, away from, or past one another in a relative sense. Geologic data reveal the relative speeds at which two plates move toward, away from, or past one another at a shared boundary, but these data do not specify the absolute velocity of any plate, and plates do not have speedometers.

Figure 12.36 illustrates the predicted absolute plate motions that correspond to what is known about relative plate motions. These predictions are made by calculating the relative velocities across all plate boundaries across the globe. The speeds and direction of movement shown by the arrows are the average motion for the last three million years and they vary from plate to plate, which accounts for directly measured relative motion at the plate boundaries. The velocities also seem to vary within plates, which may seem odd, but makes more sense when you consider that the plates move along the outside of the sphere-shaped Earth.

EXTENSION MODULE 12.1

Describing Plate Motion on the Surface of a Sphere. Learn how to describe plate motion on a sphere.

Test the Predictions

How Do Geologists Measure Absolute Plate Motion? The next step is to measure the actual plate movements and compare the measurements to the predictions. Surveying the same location on many occasions over a period of time provides an indication of whether or not the location is moving. These measurements must, however, be made from a frame of reference that is not attached to the moving plates and the measurements must be able to precisely detect the very slow speeds of predicted plate motion.

An automobile race illustrates the importance of the frame of reference. When one car passes another, the drivers perceive the relatively slow difference in their absolute speeds. The speeds of passing cars are so similar, that all the other cars in the race seem to be moving very slowly from the perspective of any given driver. If, on the other hand, you watch the race from the grandstand, then you see the cars whiz by at dizzyingly fast absolute velocities that you fully appreciate from your motionless reference frame. What we need in order to measure plate motion is the reference frame of a motionless spectator.

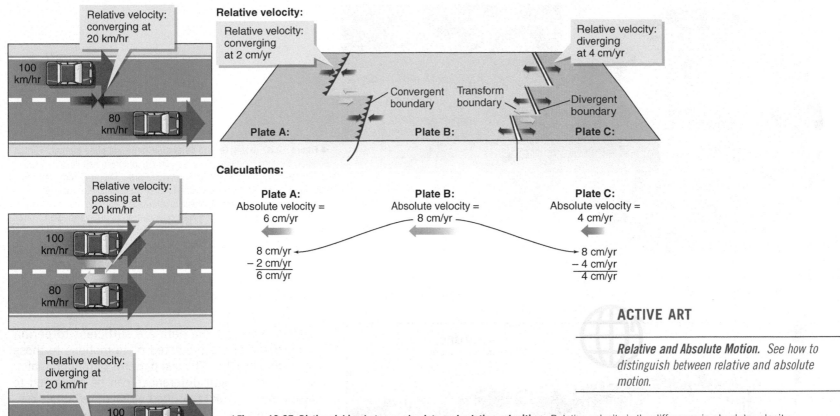

ACTIVE ART

Relative and Absolute Motion. See how to distinguish between relative and absolute motion.

◀ **Figure 12.35 Distinguishing between absolute and relative velocities.** Relative velocity is the difference in absolute velocity between two moving objects. Two automobiles moving in the same direction at different absolute velocities have relative velocities that show the vehicles converging, passing, and diverging. Likewise, relative velocities of plates determined by spreading rates at divergent boundaries, slip rates along transform boundaries, and convergence rates at convergent boundaries do not reveal absolute velocities. In this example, three plates have absolute velocities that move them in the same direction but at different speeds. As with the passing automobiles, this means that their relative velocities are such that they have divergent, convergent, or transform boundaries.

The Global Positioning System, or GPS, provides the ability to make the necessary measurements. GPS surveying, illustrated in **Figure 12.37**, uses radio signals from special satellites orbiting Earth. The positions of the satellites in space are known to an extremely high degree of precision and, importantly, are not attached to Earth, so they are not moving with the plates. The time it takes for a signal to travel from a satellite to a GPS receiver on the ground determines the distance between the receiver and the satellite. The position of the receiver on Earth is accurately determined by simultaneously determining the distances from the receiver to four or more satellites.

You may be familiar with small, inexpensive, handheld GPS receivers that confidently locate hikers and boaters, or even provide navigation aids within some automobiles. Geologists use more expensive, research-grade GPS equipment to locate a point on Earth to within a few millimeters (about half the diameter of a U.S. dime). To determine plate motions, each survey station is permanently marked by a concrete monument, and surveyed repeatedly over a number of years. **Figure 12.38** shows how the

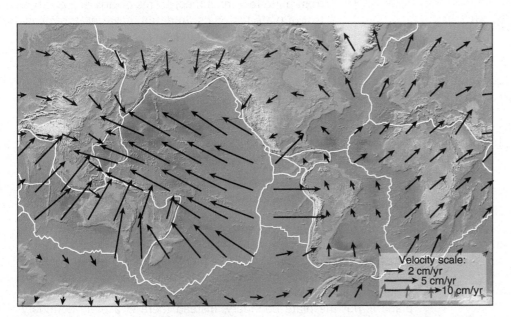

▲ **Figure 12.36 Predicting plate motion.** These arrows show a predicted pattern of global plate motions that is consistent with the relative motions observed at all plate boundaries.

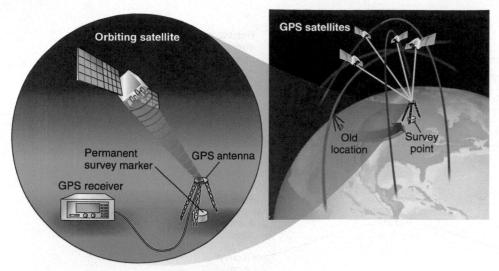

◀ Figure 12.37 **Using GPS to determine real-time plate motion.** Global Position System receivers detect radio signals from satellites located at very precisely known positions above Earth. The time required for the signal to travel from a satellite to a GPS receiver depends on the distance between them. Geologists calculate a very precise location of the receiver by analyzing the signals simultaneously received from many satellites. If repeat measurements for the permanent survey markers at the same station over a period of years yield different locations, then the station is moving.

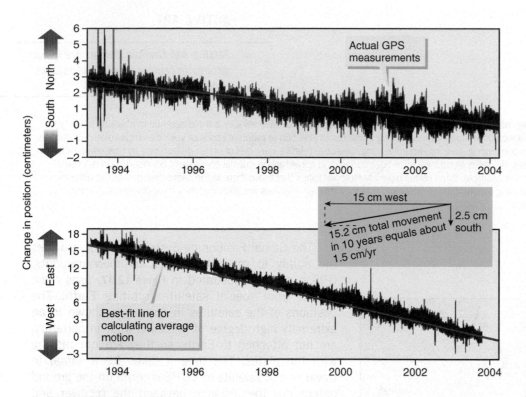

▲ Figure 12.38 **How fast does North America move?** These graphs plot GPS data collected at a location in eastern Wisconsin. The line drawn through the data averages the variability in the measurements. Between 1994 and 2004, this location moved southward by 2.5 cm, and westward by about 15 cm. This translates into plate motion of 1.5 cm/yr in a direction slightly south of due west.

survey data demonstrate the speed and direction of a GPS monument permanently attached to a moving plate. Geologists routinely monitor real-time plate motion using data from more than one hundred GPS stations from around the world.

Evaluate the Results

Do Plates Move as Predicted? **Figure 12.39** shows the current ongoing lithosphere motions indicated by GPS measurements. GPS stations are only on land, so they are not widely distributed over all of the

plates. However, the data are sufficient to permit comparison of the predicted plate motions to those measured by GPS. The two patterns of motion match very closely. Slight differences can be explained to result from the fact that the predicted velocities represent averages over the last 3 million years, whereas the GPS data depict average motion over 10 years; velocities may not be steady at the same rate over long time periods.

Insights

How Do the Results Test Plate Tectonics Theory? Real-time measurements of absolute plate motion match the velocities and relative plate motions predicted from plate tectonics theory. This clearly is a positive test of plate tectonics. In detail, however, the measurements show that plates do not move in a perfectly rigid fashion, as predicted by the theory when it was first proposed in the late 1960s.

Figure 12.40 illustrates data that demonstrate this nonrigid behavior in western North America. The map was made using GPS data to calculate the movement of GPS stations relative to movement of the North American plate. If the plates are perfectly rigid without internal deformation (Figure 12.4), then all locations on the North American plate would appear stationary on this map, and only those locations on the adjacent Pacific plate, in western California, should move. Instead, there *is* movement of some stations in western North America, and this movement increases toward the San Andreas transform fault. The data confirm, therefore, that the San Andreas Fault does not represent all of the movement between the North American and Pacific plates (Figure 12.27). This single-fault plane is not the plate boundary, instead there is a plate-boundary zone that is several hundred kilometers wide so that large areas of the western United States are being dragged northward alongside the Pacific plate.

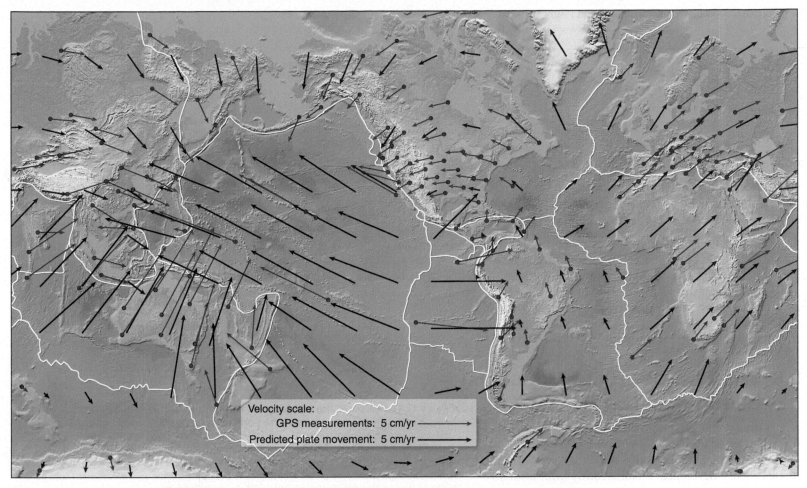

▲ **Figure 12.39 Map of real-time plate motion.** The red arrows on this map depict the actual measured direction and speed of surface movement at GPS stations located around the world. These measurements correspond very closely to the predicted plate motions, which are shown by black arrows.

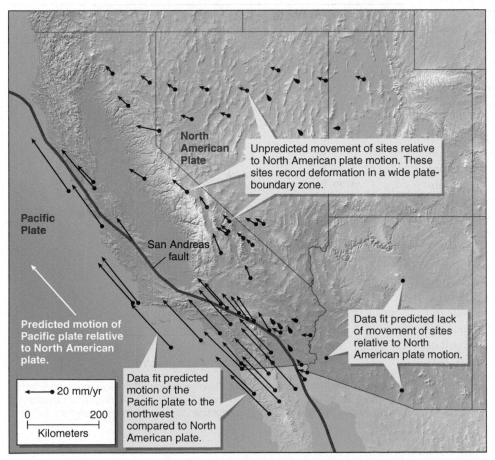

North American Plate

Unpredicted movement of sites relative to North American plate motion. These sites record deformation in a wide plate-boundary zone.

Pacific Plate

San Andreas fault

Predicted motion of Pacific plate relative to North American plate.

Data fit predicted lack of movement of sites relative to North American plate motion.

Data fit predicted motion of the Pacific plate to the northwest compared to North American plate.

◄—● 20 mm/yr

0   200
Kilometers

◄ **Figure 12.40 Western North America is a plate boundary zone.** Arrows on this map depict measured motion of GPS stations compared to the center of North America: Stations in western California move northwestward on the Pacific plate. Other stations in the western United States also move slightly northwest compared to the center of the continent. These results show that the North American plate is not perfectly rigid but deforms near its boundary with the Pacific plate.

Putting It Together—How Do We Know . . . That Plates Move in Real Time?

• Relative motion describes the direction and speed of one plate compared to another, whereas absolute motion describes the actual velocity of a plate.

• Global Positioning System (GPS) satellites provide a real-time test of plate motion by repeatedly surveying locations on Earth's surface to see whether the locations move.

• GPS data show that land sites move with speeds and directions that are generally compatible with plate tectonics theory but also show that the nonrigid characteristics of plates can produce wide boundary zones rather than narrow, discrete boundaries as originally proposed by the theory.

12.8 What Causes Plate Motion and Plumes?

When Alfred Wegener proposed continental drift, many geologists rejected his hypothesis because it lacked a mechanism to explain large-scale motion on Earth's surface. Plate tectonics theory developed simultaneously with efforts to document mantle convection (Chapter 10). Linking convection with plate tectonics explains plate motion, as shown in **Figure 12.41**. The cold, downwelling part of convection occurs as subduction at convergent plate boundaries, and the hot, upwelling part of convection occurs along divergent boundaries and at hypothesized mantle plumes.

Downwelling at Convergent Boundaries

During convection, the upper part of the convecting system cools and sinks in long, linear downwelling zones (Figure 10.8). In plate tectonics, lithosphere cools conductively as it moves away from a mid-ocean ridge (Figure 12.15) and becomes denser than its surroundings (Figure 12.24). The strength of the lithosphere does not permit downwelling to occur as readily as it would for a viscous fluid, but where plates move toward one another, the downwelling part of the convection system forms long subduction zones (Figures 12.34 and 12.41). Some subducting plates penetrate into the lower mantle, as illustrated in Figure 12.41, but the high viscosity of the lower mantle causes other sinking plates to pause or even stop in the transition zone.

Convection downwelling at subduction zones determines the direction and speed of plate motion. Geologists reached this conclusion by examining data shown in **Figure 12.42**, which demonstrates that the fastest plates also have the longest subduction zones along their margins. Gravity pulls the dense subducted slab downward, which drags the rest of the plate into the trench (Figure 12.24). Careful calculations show that the downward pull of subducting slabs, called the slab-pull force, accounts for more than 90 percent of the total force required to explain plate motion. Nearly all of the remaining force comes from slab suction, which describes the flow in the asthenosphere caused by the downward movement of the subducted plate that also draws the overriding plate toward the trench (Figure 12.41).

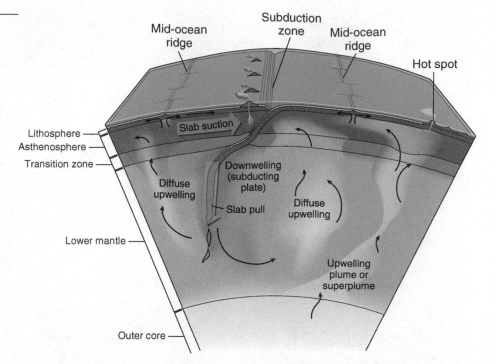

▲ **Figure 12.41 How convection explains global tectonics.** Subduction is the downwelling part of convective circulation. The downward motion of the subducted slab exerts a slab-pull force. Subduction also causes slow currents in the asthenosphere that draw overriding plates toward the trenches; this is the slab-suction force. These two forces drive plate motion. Plumes and superplumes may be caused by the convective upwelling that originates in the lowermost mantle. Some upwelling currents are probably restricted to the upper mantle where downward subduction displaces deeper, warmer mantle upward in diffuse upwelling zones.

ACTIVE ART

Convection and Tectonics. *See how convection motion relates to plate tectonics and plumes.*

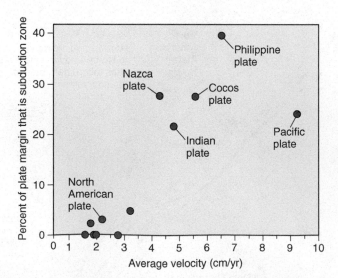

▲ **Figure 12.42 Slab pull moves plates.** Plates with long subduction zones move faster than those with short or no subduction-zone boundaries. This relationship supports the hypothesis that dense, sinking, subducting slabs provide the primary force for plate motion.

Upwelling at Plumes, Superplumes, and Divergent Boundaries

The connection between plate tectonics and convective upwelling is more complex. The upwelling, as delineated by zones of relatively slow seismic waves, tends to be more broadly distributed in the mantle compared to the narrow downwelling zones (Figure 12.41). This more diffuse nature of upwelling is consistent, however, with convection in the mantle, where viscosity varies considerably and heat is provided internally as well as from below (Figure 10.9). Also, much of this diffuse, upward mantle motion can be simply viewed as regions of the asthenosphere that are displaced upward by downward injection of subducted plates, just as water level rises in a glass if you stick your finger into the water.

Divergent plate boundaries represent only part of the upwelling expected by mantle convection. Certainly, asthenosphere moves upward at divergent boundaries, but this upward motion occurs only because mantle peridotite rises to fill the spaces created where the lithosphere stretches apart by the stresses originating as slab pull and slab suction at subduction zones. This upwelling mantle beneath most mid-ocean ridges only occupies shallow levels of the asthenosphere rather than being a part of convection rising through the whole thickness of the mantle.

Mantle convection should also include deep upwelling from near the core-mantle boundary, because at least some subduction downwelling seems to persist to this great depth (Figure 12.41). Plumes, if they exist, would be the deep upwelling component of convection. Seismically slow zones implying large masses of rising mantle from near the core-mantle boundary do appear in the seismic tomography data as the superplumes (Figures 12.29 and 12.34), which enclose most hot spots. In some cases, upwelling superplume mantle may migrate toward divergent boundaries (Figure 12.41), partly linking mid-ocean ridges to convective upwelling, and also accounting for the tendency for many hot spots to coincide with divergent plate boundaries.

Putting It Together—What Forces Cause Plate Motion and Plumes?

- Plate tectonics, mantle plumes, and superplumes are expressions of convection.

- Subducting lithosphere represents the downwelling process of mantle convection.

- Subducting plates provide nearly all of the force required to cause plate motion by slab pull and slab suction.

- Convective upwelling is diffuse and occurs in the upper mantle at divergent plate boundaries to fill the void left by subducting plates, and in the lower mantle as superplumes, which probably connect to at least some hot spots at the surface.

12.9 What Are the Consequences of Plate Motion over Geologic Time?

Current plate boundaries and plate motions are just a snapshot in a long-running movie of Earth history. Plates move so slowly that there are no discernible changes in continental positions and sizes of ocean basins during the relatively short time of human history.

Nonetheless, it is clear that plate motions caused profound changes on the surface during the long history of Earth (Section 12.1). Although the distance a plate moves over centuries and millennia seems trivial it is really quite substantial over the dimensions of geologic times. Movement of 5 cm/yr for 10 million years, for example, adds up to 500 kilometers of travel. How has plate tectonics changed the appearance of Earth through geologic time?

Reconstructing Past Plate Motions

A variety of geologic data help us to decipher long-term plate-motion history. The distributions of seafloor crust of different ages show where spreading ridges existed in the past and permit calculations of the direction and rate of relative motion away from those ridges. However, because oceanic lithosphere recycles into the asthenosphere, the record of oceanic crust is helpful only as far back as the age of crust in the modern ocean basins—about 180 million years.

How can we reconstruct plate motions prior to 180 million years ago? The archive of geologic history preserved on more permanent continental crust yields information that helps with this. Examples of data that geologists use to reconstruct past positions of continents as passengers on lithospheric plates include the following:

- Ancient igneous rocks have compositions consistent with divergent or convergent boundaries, or with hot spots. Thus, their compositions tell us about the tectonic settings for magma generation. Radioactive-isotope dates on these rocks indicate when these tectonic environments existed.

- Regional metamorphism closely relates to convergent plate boundaries (Section 6.10), and it is most intense where continents collide and the partially subducted crust experiences unusually high pressure (Figure 12.25). Isotope-dating methods applied to metamorphic rocks reveal when these collisions occurred.

- Sedimentary rocks record erosion from mountains and deposition in basins shaped by plate-boundary processes. Fossils within the sedimentary strata indicate when these tectonic elements existed in the landscape and unconformities record when tectonic forces deformed the sedimentary basins.

EXTENSION MODULE 12.2

Using Paleomagnetism to Reconstruct Past Continental Positions. Learn how paleomagnetism is used to determine past continental positions as passengers on moving plates.

Paleozoic Plate Tectonics

Figure 12.43 is your visual aid for fast time travel through reconstructions of Earth's past. As the illustrations show, plate motions over geologic time account for prominent geologic features in North America and elsewhere in the world. Reconstructions for older times are more speculative than those for recent times. Older rocks are rare; they get buried under younger sedimentary and volcanic rocks and metamorphosed near convergent boundaries.

Figure 12.43a–c depicts global geography at three snapshots during the Paleozoic Era (see Figure 7.9 if you need to refresh your knowledge of the geologic time scale). The continents do not have familiar outlines because current continental margins only existed after Wegener's Pangea supercontinent broke apart later in the Mesozoic. Most of the land composing modern North America was near the equator and rotated almost 90 degrees clockwise from its present orientation; what is now

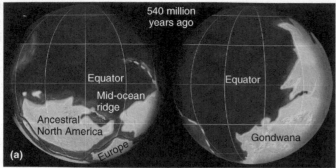

(a)

At the beginning of the Phanerozoic Eon, the continent most closely resembling modern North America was near the equator. The Gondwana continent included crust that is now found in Africa, South America, Antarctica, Australia, and India.

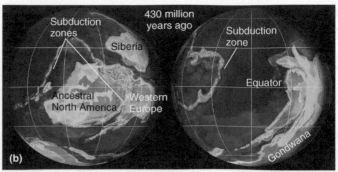

(b)

Collision of ancestral North America with continental blocks now found in Western Europe created high mountain ranges about 430 million years ago (the Caledonian Mountains shown in Figure 12.8). Sea level was high, and continents were partly submerged below shallow seas.

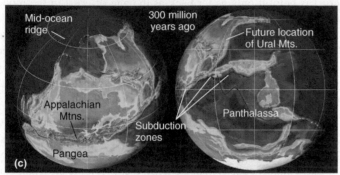

(c)

The Appalachian Mountains and other mountain belts (e.g., Ural Mountains, Figure 12.8) formed by continent-continent collisions at convergent plate boundaries when the Pangea supercontinent assembled during the late Paleozoic Era.

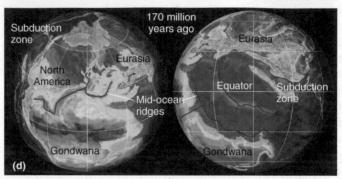

(d)

Pangea began splitting apart about 180 million years ago, as indicated by the oldest oceanic crust in the North Atlantic Ocean and elsewhere.

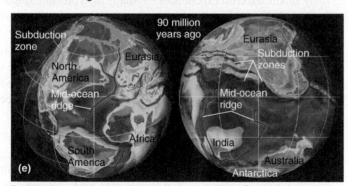

(e)

By 90 million years ago, Africa had separated from South America, and India was moving northward toward its eventual collision with Eurasia. A subduction zone was present along western North America where magma formed the batholiths of the Sierra Nevada in California, and folding and thrust faulting formed mountains throughout western North America.

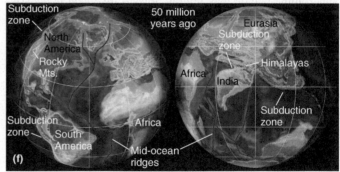

(f)

India made first contact with southern Asia about 50 million years ago, initiating uplift of the Himalayan Mountains, which continues today. The Rocky Mountains formed because of convergent-margin processes off the west coast of North America. A mid-ocean ridge was also located close to this coastline.

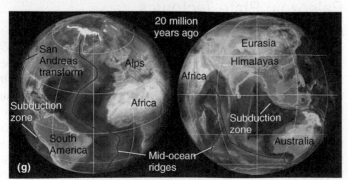

(g)

By 20 million years ago, continent-continent collision formed a nearly continuous belt of high mountain ranges from southeast Asia, through the Middle East, to the Alps of southern Europe. The earlier mid-ocean ridge off the west coast of North America began subducting beneath the continent, which changed relative plate motions and initiated the San Andreas transform boundary.

ACTIVE ART

Plate Motions through Time. *See how continents changed positions through geologic time as a result of plate motions.*

◀ **Figure 12.43 Visualizing plate motions through geologic time.**
Geology and art combine to render images of how Earth's surface changed over geologic time because of plate tectonics. On these images, only a few plate boundaries are indicated for simplicity.

northern Canada was on the east side of the ancient continent, and what is now the eastern United States bordered a southern ocean. The Atlantic Ocean did not start forming until the Mesozoic (Figure 12.13), so it does not appear in this Paleozoic view. Ancient oceans, whose lithosphere has long since subducted out of view, surrounded North America. The oceans to the south and east became progressively smaller during the Paleozoic as oceanic crust subducted and North America collided with neighboring continents.

The Appalachian Mountains (Figure 12.8 and partly illustrated in Figure 11.21) formed near Paleozoic subduction zones. Mountain building began in the Ordovician and was punctuated by two large continent-continent collisions. The first, in the Silurian and Devonian Periods (362 to 439 million years ago), involved collisions with continental crust now part of Western Europe and Siberia in northeastern Asia (Figure 12.43b). The collisions created Himalaya-style mountains in what are now the northeastern United States, northern Canada, Greenland, the British Isles, and Scandinavia. Collision of ancestral North America with Gondwana, a continent that included modern Africa and South America, occurred during the Pennsylvanian Period as Pangea formed (Figure 12.43c). The resulting mountains stretched throughout the eastern United States as the Appalachian Mountains, over to the present-day Ozark and Ouachita Mountains in Oklahoma and Arkansas, and are still visible today as subtle, mostly buried features across the whole width of Texas.

Mesozoic Plate Tectonics

Plate motions during the Mesozoic, illustrated in Figure 12.43d–e, featured the creation of the modern ocean basins and more-familiar-looking continents. Pangea began splitting apart in the Triassic, and new mid-ocean ridges were generating oceanic crust by the Jurassic Period (Figure 12.43d). Crust of an equatorial ocean subducted beneath southern Eurasia, bringing some fragments of old Gondwana, including India and Africa, closer to Eurasia.

Continental rifting in eastern North America during the Triassic and early Jurassic produced rift valleys from what is now Massachusetts to North Carolina. These valleys resembled the modern East Africa rift valleys. Volcanic activity accompanied rifting, and numerous gabbro sills are prominent in the northeastern United States, where they form the Palisades of the Hudson River near West Point, the prominent strategic ridges of the Civil War battlefield at Gettysburg, Pennsylvania, and innumerable quarried rock resources from Connecticut to New Jersey.

A convergent plate boundary lurked off the coast of western North America from the middle part of the Paleozoic until the early Cenozoic Era. Many features of western North America formed above this subduction zone. Compression caused uplift, folding, and the formation of reverse and thrust faults from the west coast to the Rocky Mountains (Figure 11.25). San Francisco is built on a chaotic mess of sedimentary rocks and chunks of ocean-floor basalt that were scraped off of this subducting plate. Subduction-zone magmas intruded western North America and Mesozoic and early Cenozoic andesitic volcanic rocks are common in that region. The Sierra Nevada batholith, including the rocks at picturesque Yosemite National Park (Figure 4.1c), is the uplifted and exposed intrusions related to this subduction zone. The igneous rocks generated above the Mesozoic and Cenozoic subduction zones contain most of the rich mineral deposits mined in western North America.

Cenozoic Plate Tectonics

Subduction beneath southern Europe and Asia led to Tertiary continent-continent collisions, depicted in Figure 12.43f–g. The resulting mountain chains include the Himalayas, the mountainous terrain of central Asia and the Middle East, and the Alps and related mountain ranges of southern Europe.

Cenozoic tectonic events in the western United States are very complicated. A mid-ocean ridge was located off the west coast during the early Tertiary (see Figure 12.43f). The westward motion of the North American plate was, however, faster than the rate of oceanic-crust production at this mid-ocean ridge. As a result, North America ran over most of the mid-ocean ridge, beginning about 26 million years ago. The mid-ocean ridge is largely subducted out of view, but a small part remains off the Pacific Northwest coast, where the tiny Juan de Fuca plate is created only to subduct a short distance away beneath Oregon, Washington, and northern California (Figure 12.3). Subduction of the spreading ridge placed the North American plate in contact with the Pacific plate in western California. Both plates move in westerly directions, and the Pacific plate is moving much faster than the North American plate (see Figure 12.39). The San Andreas transform boundary, therefore, replaced the convergent plate boundary that existed along the west side of North America since the middle Paleozoic.

The transition from convergence to a mostly transform boundary along the western edge of the North American plate coincided with a change from compressional to tensional stress throughout most of the western United States. The exact cause of this change in stress is the subject of vigorous ongoing research, but it at least partly relates to changing forces at the changing plate boundary. Tensional stress causes normal faults, which are responsible for uplift of mountain ranges and down dropping of valley floors in many western states (see Figure 11.22).

Putting It Together—What Are the Consequences of Plate Motion over Geologic Time?

• Plate tectonics explains the history of continents and ocean basins by linking rocks and structures with plate-boundary processes.

• Paleozoic mountain ranges, such as the Appalachian Mountains, formed along convergent plate boundaries. Continent-continent collisions along plate boundaries assembled the Pangea supercontinent by the end of the Paleozoic Era.

• Pangea rifted apart during the Mesozoic, and the modern continents dispersed to their present positions since that time. Rift valleys formed in eastern North America prior to the formation of the oldest Atlantic Ocean crust when the continents separated.

• A convergent plate boundary along the west coast of North America from the middle Paleozoic until the middle Cenozoic caused faulting and folding, intrusion and crystallization of batholiths, erupting volcanoes, and formation of valuable mineral deposits in the western United States.

• Relative plate motions changed along the western boundary of the North American plate beginning about 26 million years ago. Subduction continues offshore of southwestern Alaska, Washington, Oregon, and northern California, but transform boundaries formed elsewhere along the western edge of the plate.

Where Are You and Where Are You Going?

You now know far more about plate tectonics than the bits and pieces that were revealed in Chapters 1–11. Most importantly, you know the evidence that supports plate tectonics as a viable explanation for a wide range of geological processes. Global tectonic processes, in turn, are reliably linked to convection within Earth. The theory also explains the ancient tectonic history of Earth, including familiar landscapes of North America.

Relative motions of plates at divergent, convergent, and transform plate boundaries are consistent with all available geologic and geophysical data. These data include the following:

- Locations of earthquakes, volcanoes, and active mountain belts, which mostly coincide with the narrow interactive boundaries between rigid plates
- Locations of normal, reverse and thrust, and strike-slip faults, which are consistent with the stresses associated with each plate boundary
- Heat-flow and seismic-tomography data, which delineate where hot asthenosphere wells up beneath mid-ocean ridges, and where cold lithosphere sinks to the base of the mantle
- The antiquity of continental crust, which cannot be readily subducted, and the youth of oceanic crust, which is continuously created at divergent boundaries and consumed at convergent boundaries.

Hot spots are areas of especially prolific volcanic activity not explained by plate-boundary processes. Many hot spots occur within plates, and some are found along or near divergent plate boundaries. Active hot spot volcanoes reside at one end of a long line of extinct volcanoes that are progressively older at greater distance from the hot spot. These hot spot tracks, and the odd composition of hot spot basalt, imply that this volcanism results from mantle processes originating below the plates. The mantle-plume hypothesis explains hot spots as the surface expression of hot, upwelling mantle that rises all the way from the core-mantle boundary. Proving the hypothesis is difficult to do, but the coincidence of most hot spots with two broad superplume regions where hot mantle is rising from the lower mantle suggests that the hypothesis is at least partly correct.

Plate motion and mantle plumes are expressions of mantle convection. Subducting lithosphere at convergent plate boundaries records the convective downwelling of cold material. The downward pull of subducting plates provides nearly all of the driving force for global plate motion. Shallow upwelling in the asthenosphere is recorded at divergent plate boundaries. Deeper upwelling is indicated in seismic tomography by the superplumes, which may be connected with the narrower plumes hypothesized to cause most hot spots.

The next chapter completes your large-scale, global view of tectonics. You will learn why continents are high-standing areas on Earth's surface, whereas oceanic crust is so much lower that it is submerged beneath the sea. You will see evidence that continents grow over time and learn how the plate tectonics theory explains this growth.

Active Art

Motion at Plate Boundaries. See how plates move along their boundaries.

Correlating Processes at Plate Boundaries. See how volcanoes, earthquakes, and young mountain belts line up with plate boundaries.

Seafloor Spreading and Rock Magnetism. See how seafloor spreading at divergent boundaries produces bands of crust with alternating magnetic polarities.

Forming a Divergent Boundary. See how a divergent boundary originates by rifting a continent.

Motion at Transform Boundaries. See how transform faults connect other plate boundaries.

Hot Spots and Plumes. See how an island chain forms by plate motion across a hot spot, which might connect to a hypothesized mantle plume.

Relative and Absolute Motion. See the difference between relative and absolute motion.

Convection and Tectonics. See how convection motion relates to plate tectonics and plumes.

Plate Motions through Time. See how continents changed positions through geologic time as a result of plate motions.

Extension Modules

Extension Module 12.1: Describing Plate Motion on the Surface of a Sphere. Learn how to describe plate motion on a sphere.

Extension Module 12.2: Using Paleomagnetism to Reconstruct Past Continental Positions. Learn how paleomagnetism is used to determine past continental positions as passengers on moving plates.

Confirm Your Knowledge

1. Define the continental drift hypothesis. How does it differ from plate tectonics theory?
2. Contrast the geologic processes at plate boundaries with those at plate interiors. Explain how these differences relate to the concept of rigid plates.
3. How does the distribution of volcanoes and mountains support the theory of plate tectonics? What aspects of the distributions of these features are not obviously explained by plate tectonics?
4. List and explain, in your own words, the evidence for the existence of divergent plate boundaries?
5. List and explain, in your own words, the evidence for the existence of convergent plate boundaries?
6. Why are there mountain belts on both the east and western margins of North America if only the western margin is a plate boundary?
7. Which is stronger, the lithosphere or asthenosphere?
8. On average, how fast do the plates move?
9. Where on Earth is a likely place for a new ocean basin to form in the future? Explain why you picked this location.
10. How does the oceanic lithosphere become denser with increasing age and distance from a mid-ocean ridge?
11. Explain the relationship between oceanic lithosphere density and the process of subduction.
12. List and explain, in your own words, the evidence for transform faults.
13. What are the similarities and differences among the Hawaiian, Icelandic, and Yellowstone hot spots?
14. What is the difference between relative and absolute velocities?
15. In your own words, explain how GPS data are used to test the plate tectonics theory.
16. What forces cause plates to move?

Confirm Your Understanding

1. Write an answer for each question in the section headings.
2. What does the word "theory" mean when used to define plate tectonic theory or the theory of evolution?
3. Although he collected impressive evidence to support his continental drift hypothesis, Alfred Wegener did not receive widespread acceptance of his idea. What obstacle prevented many scientists from accepting continental drift?
4. Use Figure 12.43 to locate the approximate location where you were born on each map. Describe where your birth location was on the globe at 540, 430, 300, 170, 90, 50, and 20 million years ago.
5. How old is the oldest crust in the oceans? Why is the oldest crust located where it is?
6. What are the dominant geologic events in North America during the Paleozoic, Mesozoic, and Cenozoic eras, and how do these events relate to plate tectonics?
7. A satellite returns images of the surface of a distant planet. Conical volcanoes are clearly visible on the images. How can you determine whether this planet experiences plate tectonics, plumes, or both based on the distribution of visible volcanoes?
8. The evidence for the geologic history of plate motions on Earth prior to about 180 million years ago comes entirely from evidence found on continents and not from rocks found in the oceans. Explain why this evidence is restricted to continents.

CHAPTER 13

Tectonics and Surface Relief

Why Study Surface Relief?

Earth has a highly irregular surface that ranges from dramatic, towering mountains to seemingly bottomless deep-sea trenches. Beyond providing impressive scenery, Earth's undulating surface has played important roles in world history. Mountains blocked invading armies and formed isolating barriers to economic and cultural exchange until modern transportation surmounted these obstacles. Agriculture started and flourishes in low-lying places where nourishing sediment washes from eroding highlands to produce fertile soil.

What causes the varied topography of Earth? Surface elevation and shape, which are the elements of topography, result from the interplay of internal and external forces. Motion driven by heat inside Earth deforms the surface, causing some areas to rise while others sink. External processes of weathering and erosion sculpt the crust into a myriad of landforms.

This chapter focuses on pursuing knowledge of the relevant internal processes as a prelude to our study of the external, landscape-sculpting forces. We will start with helpful analogies that provide general explanations but leave many natural details to be filled in. Complexity gradually increases until you have a satisfactory understanding of the whole picture.

After Completing This Chapter, You Will Be Able to

- Explain processes in the crust and mantle that determine the varying elevations of the surface.

- Relate mountain building processes to the formation and growth of continents.

Pathway to Learning

13.1 Why Are Continents High and Oceans Low?

13.2 How Do We Know . . . That Mountains Have Roots?

13.3 How Does Isostasy Relate to Active Geologic Processes?

13.4 Why Does Sea Level Change?

Mountains rise abruptly more than 1500 m above the sea at Glacier Bay National Park, Alaska.

In the Field

Imagine yourself vacationing across the rugged terrain of the western United States. You reach the lowest elevation in the western hemisphere—Death Valley, California, portrayed in **Figure 13.1**. You stand at a point that is 85 meters *below* sea level. The high point in the adjacent mountains is 3346 meters *above* sea level. The difference in elevation between two locations is the **relief**; you can describe the relief between the bottom of Death Valley and the top of the nearby mountains as being equal to 3431 meters.

On your virtual road trip onward to the Pacific Ocean, you keep track of mapped elevations along the way in your field book. You use these data and maps to make a profile of the varying surface elevations across the landscape (Figure 13.1). From Death Valley, it is only 200 kilometers to the top of Mount Whitney, the highest elevation in the contiguous 48 states at 4418 meters high (Figure 13.1). You are amazed to find such large differences in surface elevation over relatively short distances. What processes account for these differences? When you reach the shore of the Pacific Ocean, you look out across the gently rolling surface of the sea, and for a moment you are impressed by how smooth and gentle the ocean surface seems compared to the rough and rugged land surface. The next moment you realize that this is not a fair comparison—the seafloor is probably not as smooth as the sea surface and may well be just as rugged as the land you have traveled across.

You stop at a library to find some maps that depict the ocean depths. The maps allow you to continue profiling seafloor elevations west of California into the Pacific Ocean (Figure 13.1). Clearly, the seafloor also displays substantial relief.

The total relief on Earth, the difference in elevation between Mount Everest (8850 meters *above* sea level) and the Mariana Trench in the western Pacific Ocean (11,040 meters *below* sea level), is 19,890 meters, nearly 20 kilometers. This seems like a huge variance—compared to your height, for instance—until you remember that our planet is more than 12,740 kilometers in diameter. The relief on Earth's surface seems almost trivial when stacked up against the vast size of the entire planet. All the same, to the inhabitants of the planet, Earth's surface is very rough and uneven. On the largest scale is it worthwhile to ask why are continental areas overwhelmingly elevated above sea level, whereas other regions are much lower and submerged beneath the seas? The range in elevations of Earth's surface is certainly one of the most basic features of Earth that geologists should be able to explain.

▶ Figure 13.1 **Visualizing elevation relief in California.**

13.1 Why Are Continents High and Oceans Low?

The phrases *above sea level* and *below sea level* probably sound familiar to you. Sea level is a convenient reference point for describing elevations of Earth's surface features, so elevations commonly are reported relative to sea level. Describing surface elevation as either above or below sea level also usually distinguishes land from ocean. What determines whether regions are above or below sea level?

Elevation Relates to Crust Type

We need to consider the total range of elevations on Earth in order to gain insights to this question. **Figure 13.2** depicts the percentages of Earth's surface area that are at various elevations and illustrates two important features:

1. About 29 percent of Earth's surface is land, because it is above sea level, and 71 percent is ocean.

2. Two elevations, at about 0.5 kilometer above sea level and 4.5 kilometers below sea level, are the most common elevations.

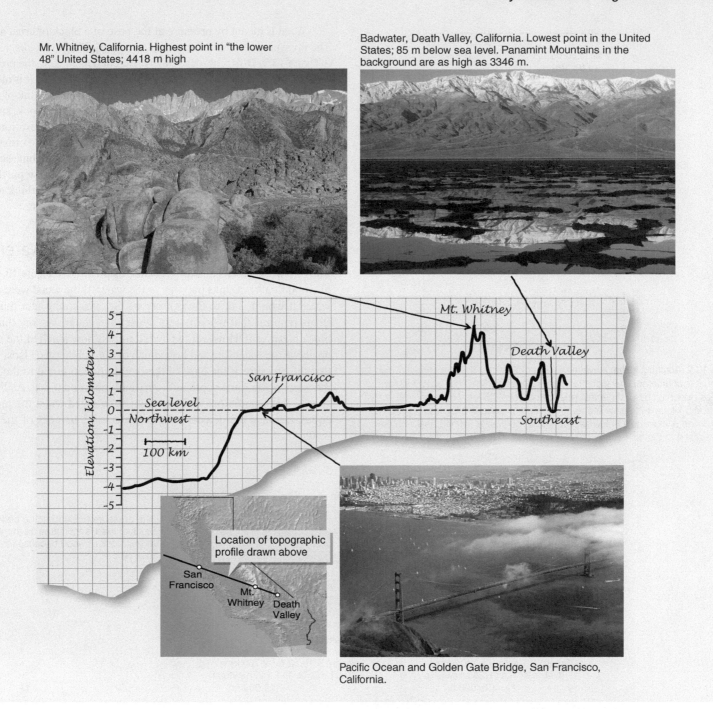

Mr. Whitney, California. Highest point in "the lower 48" United States; 4418 m high

Badwater, Death Valley, California. Lowest point in the United States; 85 m below sea level. Panamint Mountains in the background are as high as 3346 m.

Location of topographic profile drawn above

San Francisco Mt. Whitney Death Valley

Pacific Ocean and Golden Gate Bridge, San Francisco, California.

Judging from Figure 13.2, the main factor that determines the elevation seems to be the type of crust present at a location. Continental crust underlies most of the areas represented by the graph peak at 0.5-kilometer elevation and at higher elevations. Oceanic crust underlies the lower elevations. As you can see in Figure 13.2 however, while sea level conveniently divides land and sea, it does not exactly distinguish areas of continental and oceanic crust because shallow water submerges some areas of continental crust (Figure 13.2). Next, let's find out how the type of crust determines surface elevation and why some parts of continents are above sea level whereas other parts are submerged.

Isostasy Determines Elevation

The **principle of isostasy** explains variations in surface elevations in terms of the type, thickness, and density of the crust. According to this principle, low-density crust rests on denser, underlying mantle so that the pressure is the same at the base of any given imaginary blocks of rock that consists of both the crust and upper mantle. The word "isostasy" derives from Greek roots *iso*, "equal," and *stasis*, "standing." These word roots emphasize that adjacent blocks of crust are equal in terms of the pressure beneath them even if they are at different elevations.

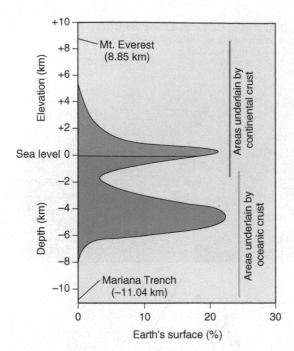

▲ **Figure 13.2 Graphing global elevation.** This graph shows the percentage of Earth's surface that is at different elevations (above sea level) and depths (below sea level). Most areas underlain by continental crust are between 1.5 kilometers above sea level and about 0.5 kilometers below sea level. Most areas underlain by oceanic crust are between 3 and 6 kilometers below sea level. The two clearly defined peaks on the graph suggest that the type of crust is primarily what determines surface elevation.

What is meant by pressure at the base of a block of crust and mantle? The pressure is equal to the weight of material in the blocks, as illustrated in **Figure 13.3.** Different rocks have different densities, so the pressure at the base of a short block of high-density rock is the same as it is at the base of a tall block of low-density rock. Therefore, it helps to think of high elevations as being associated with regions of lower-density rock, such as granite, that is found in continental crust. Conversely low elevations are made up of denser rock, such as basalt, that composes oceanic crust. However, the principle of isostasy also relates to how the crust interacts with the mantle. The weight measurements in Figure 13.3 only partly describe isostasy. To understand isostasy completely, we need to think about buoyancy, the ability of something to float, as well as pressure.

Demonstrating Isostasy in a Tub of Water

Imagine wood blocks in a tub of water, as pictured in **Figure 13.4**a. A block of wood floats but sinks partway in the water. If the wood were denser than water it would sink to the bottom of the tub, so the fact that the wood floats requires that wood is less dense than water. We say that the wood block is buoyant. However, what determines how much of the wood sticks up above the water line and how much of it is submerged? Isostasy requires that the pressure at the base of the wood block is equal to the pressure in the water at the same depth as the base of the wood block. As Figure 13.4a shows, this wood block is 75 percent submerged when the pressures are equal, because the density of the wood is 75 percent of the density of water.

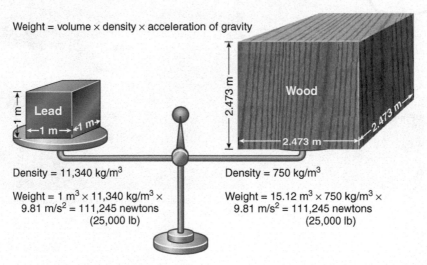

Weight = volume × density × acceleration of gravity

Lead

Density = 11,340 kg/m³

Weight = 1 m³ × 11,340 kg/m³ × 9.81 m/s² = 111,245 newtons (25,000 lb)

Wood

Density = 750 kg/m³

Weight = 15.12 m³ × 750 kg/m³ × 9.81 m/s² = 111,245 newtons (25,000 lb)

The pressure (weight) at the base of a small block of lead is the same as the weight of a larger block of wood because lead is denser than wood.

► **Figure 13.3 Weight is pressure.** The pressure at the base of a block of any material is equal to the weight of the material. Weight is equal to the volume of the block, multiplied by the density of the material and the acceleration of gravity.

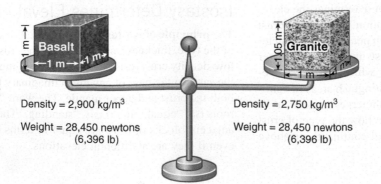

Basalt

Density = 2,900 kg/m³

Weight = 28,450 newtons (6,396 lb)

Granite

Density = 2,750 kg/m³

Weight = 28,450 newtons (6,396 lb)

Equal-weight blocks of common crustal rocks, basalt and granite, are different sizes, because basalt is denser than granite.

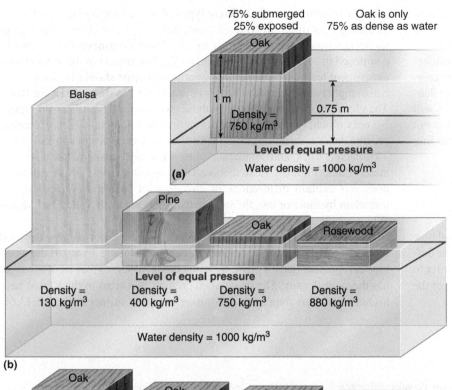

Isostasy determines the thickness of wood above and below the water line. The pressure at the base of the block must be the same as the pressure in the water at the same depth.

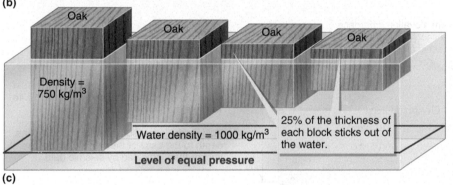

Wood blocks of different densities and thickness, cut to thicknesses so that they all sink to the same level in the water. Less dense blocks project higher above the water than denser blocks.

Wood blocks of the same density, but different thicknesses. The same proportion of each block is exposed as it is submerged, so thick blocks have higher surface elevations than thin blocks. The thick blocks also sink lower in the water.

◀ **Figure 13.4 Demonstrating the principle of isostasy.** The top-surface elevations of wood blocks floating in water are determined by the densities of the wood and the thicknesses of the blocks.

Wood blocks of different densities and sizes are shown in Figure 13.4b. The blocks are cut so that the submerged bases of the blocks are at the same level in the water. The high-density blocks almost completely submerge, whereas very little of the low-density wood sinks below the water line. The principle of isostasy explains why blocks with lower densities have higher surface elevations and blocks with higher densities have lower surface elevations.

This result from isostasy also explains why most of an iceberg is submerged. Ice is only slightly less dense than water, as **Figure 13.5** shows. For the pressure to be the same at the base of the iceberg as it is below the open-ocean surface, most (about 90 percent) of the iceberg must be submerged.

Figure 13.4c illustrates a second result from isostasy. In this bathtub experiment, all of the blocks are cut from the same wood, so they all have

Density of ice: ~900 kg/m³

Density of water: ~1000 kg/m³

About 90% of the mass of an iceberg is submerged

▶ **Figure 13.5 Isostasy explains icebergs.** Ice floats in water because ice is less dense than water. The density difference is small, so 90 percent of the ice is submerged, as predicted by the principle of isostasy. All sailors navigating icy waters know not to travel close to an iceberg. Even if an iceberg seems small above water, most of its size is submerged and hidden from view.

the same density. The blocks are, however, cut to different thicknesses. The fraction of each block exposed above the water line must be the same for all of the blocks, because all of them have the same density. However, because the blocks have different thicknesses isostasy causes the thicker blocks to project farther above the water than the thin blocks. In this case, the principle of isostasy tells us that when considering blocks of equal density, the thicker blocks will have higher elevations and thinner blocks will have lower elevations. In addition, notice that thicker blocks stick down into the water more than thin blocks. This means that higher-standing blocks also have thicker "roots" below them.

How Isostasy Explains Elevations of Continents and Oceans

How do wood blocks floating in water explain elevations on Earth and how does this all relate to the imaginary blocks of crust and mantle in our definition of isostasy? The key is to apply the principle of isostasy in order to draw conclusions about the elevations of Earth's surface by thinking of the wood as blocks of crust and the water as the mantle.

The experiment with different types of wood blocks (Figure 13.4b) is an analogy for different types of crust: denser oceanic crust composed of mafic igneous rocks, in contrast to less dense continental crust mostly composed of intermediate and felsic rocks. The results of the wood-block experiment lead us to predict that continental crust should be thicker and rise to higher elevations than denser oceanic crust, as shown in **Figure 13.6**a. The elevation difference between these two types of crust is as predicted, but only if both types of crust are more than 80 kilometers thick. However, seismic data summarized in Chapter 8 show that oceanic crust is only about 7 kilometers thick, and continental crust is usually between 25 and 50 kilometers thick (Figure 8.7), so this is an unacceptable result. This model also does not explain differences in elevation of locations on continents underlain by more or less the same crust, such as Death Valley and Mount Whitney (Figure 13.1).

Let's try another approach. Think back to the experiment with wood blocks of equal density but different thicknesses (Figure 13.4c). If the wood blocks represent Earth's crust, then this model assumes that all of the crust has the same density. The model then predicts that areas of thick crust have higher elevations than areas of thinner crust, as shown in Figure 13.6b.

▶ **Figure 13.6 Using isostasy to explain elevation.** Each of these models for Earth elevation follows the analogy of the wood-block experiments (Figure 13.4) combined with data revealing a 5-kilometer average elevation difference between continents and oceans (Figure 13.2). Model A and Model B do not, however, fit with other geologic data. Only Model C, which combines aspects of the other two models, is consistent with what is known about the thickness and composition of continental and oceanic crusts.

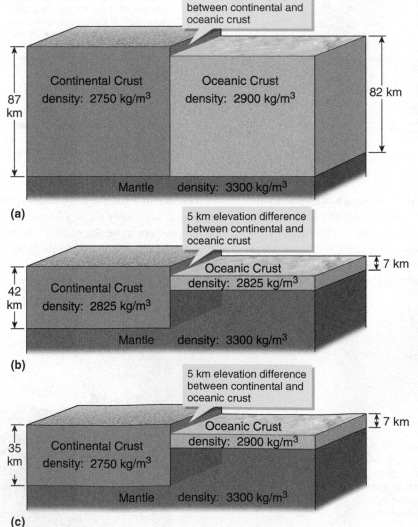

Model A. Oceanic crust is lower than continental crust *only* because oceanic crust is denser. This is analogous to Figure 13.4b where blocks have different density.

This model is invalid because it requires the depth to the base of the crust to be much thicker than indicated by seismic data.

Model B. Oceanic crust is lower than continental crust *only* because oceanic crust is thinner, and because both crust types are assumed to have the same density. This is analogous to Figure 13.4c where wood blocks have the same density.

This model is invalid because continental and oceanic crusts actually have different densities.

Model C. Continents are higher than areas underlain by oceanic crust because continental crust is both thicker *and* less dense than oceanic crust.

This model, combining models A and B, provides reasonable estimates of crustal thickness and crustal densities.

Most of the thickness difference between these blocks of continental and oceanic crust appears as a deep continental root in the mantle, just like most of the thickness of a thick wood block, or an iceberg, is submerged in the water. The problem with this model is that it assigns the same density to continental and oceanic crust. Yet, we know from rock samples and seismic data that oceanic crust must consist of denser rock than continental crust (look back to Table 8.1 and Figure 8.2 for data about differences between continental and oceanic crust). Model B might explain the highly varied relief on continents where the crust is of similar density, but it is unsatisfactory for explaining differences in elevation of oceanic crust versus continental crust.

Figure 13.6c shows that the best explanation for the different elevations of continental and oceanic crust combines knowledge from both experiments. Continental and oceanic crusts differ both in density *and* thickness. Continental crust is both less dense and thicker than oceanic crust, so areas of continental crust tend to stand at higher elevations above the mantle than do areas of oceanic crust as we see represented in Model C.

Putting It Together—Why Are Continents High and Oceans Low?

• The principle of isostasy states that the weight at the base of any block of crust and mantle must be the same as below every other block.

• Isostasy explains variations in surface elevation as a result of variations in thickness *and* density of crust.

• Areas of continental crust are higher than areas of oceanic crust because continental crust is thicker *and* less dense than oceanic crust.

13.2 How Do We Know . . . That Mountains Have Roots?

Define the Problem

Do Mountains Actually Have Roots? Analogies between wood blocks in water and crustal blocks in the mantle offer an explanation for different surface elevations, but these analogies do not prove that the required density and thickness differences actually exist below Earth's surface. What is the evidence that the principle of isostasy really explains surface relief? Isostasy provides a test. High, mountainous regions should have thicker crust than lowland areas of continents, like the wood blocks illustrated in Figure 13.4c. In other words, mountains should have roots.

Examine Two Historical Hypotheses

How Did the Himalayas Cause Survey Errors in India? Scientists developed the principle of isostasy in the nineteenth century to explain puzzling surveying measurements near the highest points on Earth—the Himalayas in Asia. George Everest, the Surveyor General of India and the namesake for Earth's highest mountain, supervised an ambitious survey of India between 1840 and 1859. During this work,

errors appeared in the measured locations of some survey stations in valleys in northern India, next to the towering Himalayas. Everest dismissed the errors as simple inaccuracies in the surveying method.

However, British mathematician John Pratt proposed a different explanation for the erroneous measurements. Pratt considered how some of the surveying was done using astronomical methods illustrated in **Figure 13.7**. A table is set up such that its level surface is perpendicular to a line directed toward Earth's center. Surveyors then sight a telescope on the North Star, which is almost directly above Earth's rotation axis. A plumb weight suspended beneath the table establishes the line toward Earth's center (Figure 13.7b). The angle between the line sighted to the star and a line drawn directly toward Earth's center permits surveyors to calculate latitude.

Pratt realized that the plumb line only points to Earth's center, the assumed center of gravitational attraction, *if* mass is uniformly distributed within Earth and the surface is smooth. Pratt suspected that the surveying errors only occurred in the plains of northern India because of proximity to the mountainous Himalayas. He hypothesized that the great mass of rock at elevations much higher than the survey stations exerts a gravitational attraction that deflects the plumb line away from the center of Earth and causes an erroneous measurement of latitude (Figure 13.7c). The incorrect measurement of latitude caused the errors in location. Pratt calculated the deflection expected to result from the presence of the mountains but was surprised that his calculation was nearly three times larger than the actual errors in Everest's survey data.

Pratt modified his hypothesis to account for this discrepancy. He suggested that the deflection caused by the mass of the high mountains was partly offset by the presence of less mass than expected below the mountains. If, for example, the crust beneath the mountain is less dense than below the survey location, then the mountainous region does not exert as much gravitational attraction as it would if all the crust had equal density. The Pratt model of isostasy accounts for different elevations with crust of different density, following Model A in Figure 13.6.

British Royal Astronomer George Airy almost immediately proposed an alternative to Pratt's explanation. Airy agreed with Pratt that the plumb-line deflections were probably caused by the presence of less mass than expected below the mountains. However, he proposed that the lower mass did not result from unusually low-density crust beneath the mountains. Instead, Airy hypothesized that similar crust existed everywhere but that the crust is thicker below the mountains than below the lowlands. The Airy model of isostasy accounts for different elevations by variations in crustal thickness, similar to Model B in Figure 13.6.

Evaluate the Hypotheses

What Matters More, Crust Density or Thickness? Which hypothesis is correct, or is neither Pratt nor Airy correct? Pratt explained elevations and gravitational pull by changing crustal density from place to place, whereas Airy explained these observations by different thickness of crust of the same density. Both hypotheses explained the errors in the survey equally well. The merits of each hypothesis were not tested until the last half of the twentieth century, when abundant seismic data provided the necessary knowledge of the density and thickness characteristics of crust.

► **Figure 13.7 Explaining surveying errors in India.** Surveyors miscalculated latitudes in India because plumb lines on the surveyors' tables did not point directly to Earth's center as expected. This diagram explains the surveying technique and Pratt's hypothesis accounting for the plumb-line deflection.

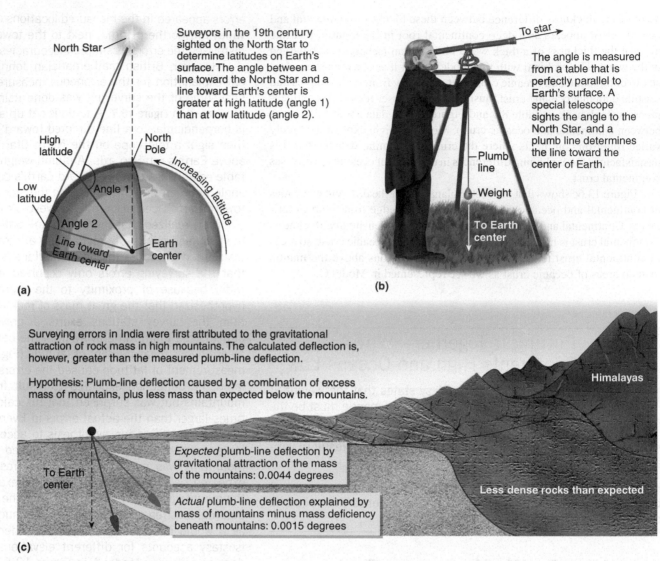

Suveyors in the 19th century sighted on the North Star to determine latitudes on Earth's surface. The angle between a line toward the North Star and a line toward Earth's center is greater at high latitude (angle 1) than at low latitude (angle 2).

North Star

High latitude

Low latitude

Angle 1

Angle 2

North Pole

Line toward Earth center

Earth center

Increasing latitude

(a)

The angle is measured from a table that is perfectly parallel to Earth's surface. A special telescope sights the angle to the North Star, and a plumb line determines the line toward the center of Earth.

To star

Plumb line

Weight

To Earth center

(b)

Surveying errors in India were first attributed to the gravitational attraction of rock mass in high mountains. The calculated deflection is, however, greater than the measured plumb-line deflection.

Hypothesis: Plumb-line deflection caused by a combination of excess mass of mountains, plus less mass than expected below the mountains.

Himalayas

To Earth center

Expected plumb-line deflection by gravitational attraction of the mass of the mountains: 0.0044 degrees

Actual plumb-line deflection explained by mass of mountains minus mass deficiency beneath mountains: 0.0015 degrees

Less dense rocks than expected

(c)

When distinguishing between two hypotheses, it is important to focus on testing predictions that only support one hypothesis and not the other. Airy's hypothesis predicts that the base of the crust is deepest beneath areas of highest elevation; in other words, mountains should have roots in the mantle. In contrast, Pratt's hypothesis calls for no change in the depth to the base of the crust between lowlands and mountains.

Figure 13.8 depicts the thickness of continental crust determined from seismic data below a line from the plains of northern India, over the Himalayas, and across the high Tibetan Plateau (illustrated

in Figure 12.19). Earthquake waves move faster through mantle rocks than through the crust. By comparing the arrival times of earthquake waves at different locations, it is possible to determine the velocity of the waves at different depths, which permits identification of the depth of the crust-mantle boundary. These data show a crustal root protruding deeper into the mantle below the highest surface elevations, which supports Airy's hypothesis. In the matter of the erroneous Everest survey, variations in crust thickness are to blame for the deflected plumb line and provide an isostatic explanation for surface elevation.

► **Figure 13.8 Mountains do have roots.** This diagram shows surface elevation and crust thickness between the low plains of India, the high Himalayas, and the Tibetan Plateau. The depth to the base of the crust is calculated from the velocity of seismic waves at different depths below the surface. The base of the crust is deeper under the higher elevations, showing that mountains have roots protruding down into the mantle.

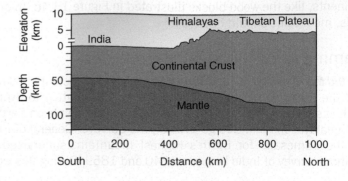

Himalayas   Tibetan Plateau

India

Continental Crust

Mantle

Elevation (km)

Depth (km)

South   Distance (km)   North

Insights

Does Airy Isostasy Work Everywhere? Before accepting that Airy was entirely correct and Pratt entirely wrong, take another look at Figure 13.6. Airy's hypothesis, which compares well with Model B in Figure 13.6, does not explain the larger-scale differences in elevation between continents and oceans. When considering elevations over the whole globe, geologists must combine Pratt and Airy isostasy hypotheses, as we do in Model C of Figure 13.6.

Here is an important lesson about scientific method: when testing competing hypotheses, it is always important to determine whether it is impossible for both ideas to be correct. Nothing about either Pratt's or Airy's hypotheses requires that the alternative view is incorrect. Indeed, it is common in science to combine hypotheses or to see that one applies best for one set of circumstances (Pratt's model is essential to explain elevation differences between oceans and continents), while the other applies best in other situations (Airy's hypothesis best explains elevation differences within areas of continental crust).

Before we move ahead, think back to the field trip through California. Is the crust beneath Mount Whitney thicker than beneath Death Valley, as predicted by Airy isostasy? Seismic data suggest that the crust is only 30 kilometers thick beneath Death Valley compared to 45 kilometers thick beneath Mount Whitney. There is a problem, however. The thickness of the crust beneath Mount Whitney does not seem thick enough to explain the very high elevation. In other words, Mount Whitney has a root but it is not deep enough to support the high elevation. This means that additional hypotheses must be added to Airy and Pratt isostasy to complete your understanding of Earth's relief. The content in the following section will help us do just that.

Putting It Together—How Do We Know . . . That Mountains Have Roots?

• The Pratt and Airy isostasy hypotheses both successfully explain differences in surface elevation. Although it is not possible for both hypotheses to be independently correct for all situations, combinations of each are satisfactory in most, though not all, cases.

• Seismic data demonstrate the presence of thick roots of crust projecting downward into the mantle beneath mountains as predicted by Airy's model. However, Pratt's model is essential to explain elevation differences between oceans and continents.

13.3 How Does Isostasy Relate to Active Geologic Processes?

Erosion and rock deformation cause changes in surface elevation so these processes must be related to isostatic principles. If the thickness or mass in an isostatically balanced block of crust changes, then the weight at the base of that block becomes greater or less than that of neighboring blocks. **Figure 13.9** illustrates this idea with another simple experiment in a water

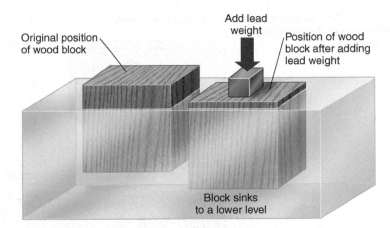

▲ **Figure 13.9 What happens when you add weight to a block?** A wood block is initially in a stable isostatic condition. Then a lead weight is placed on the block, pushing it down in the water to a new isostatically stable position. If the lead weight was removed, the wood block would rise back to its original position. By analogy, changes in the distribution of weight in Earth's crust should cause some areas to subside and others to rise.

tub. The conclusion is that when geologic processes change the thickness or mass, or both, of areas of crust, then the crust isostatically adjusts to a new stable position.

The Effects of Tectonic Shortening and Stretching

Compressional stress shortens and thickens the crust by plastic flow in the lower crust and while forming reverse and thrust faults in the upper crust (Section 11.5 explains the relationship between stress and deformation). **Figure 13.10** shows that the principle of isostasy requires the elevation of a shortened region to increase because the thickness of crust increases. To understand why this happens, compare the change in land elevation to the elevations of thin and thick wood blocks in Figure 13.4c; the thicker block has a higher elevation.

Tensional stress thins and stretches the lower crust by plastic flow, while normal faults form in the upper crust. In Figure 13.10, you can see how elevation decreases where the crust becomes thinner. Looking back to Figure 13.4c, note how a thinning of the continental crust produces a block that is lower in elevation and whose root shrinks. Isostasy explains why low-elevation rift valleys form where continents lengthen by tension (Figure 12.18).

Isostasy also explains why continents have wide, submerged continental shelves. Continents separate from one another where divergent plate boundaries form (Figure 12.18). When tension stretches the crust at these divergent boundaries, the crust is also thinner, forming continental shelves with lower elevation than the interior of the continent (where the nonstretched crust remains thicker). The elevation of thinned crust is lower than sea level, causing submergence of part of the continental crust. This application of isostasy explains why a small area of Earth that is underlain by continental crust is also low enough to be covered by seawater (Figure 13.2).

The Effects of Erosion and Deposition

Erosion and deposition of sediment redistribute mass and change the thickness of the crust. One more look at the wood blocks in the tub shows how

▶ **Figure 13.10 How shortening and lengthening the crust change elevations.**

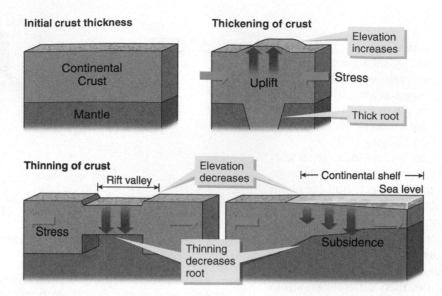

Initial crust thickness

Continental Crust

Mantle

Thickening of crust

Elevation increases

Uplift

Stress

Thick root

Compressed continental crust shortens and thickens. As a result, elevation increases above a thick root that projects into the mantle.

Thinning of crust

Rift valley

Elevation decreases

Stress

Thinning decreases root

◄— Continental shelf —▶

Sea level

Subsidence

Extended continental crust lengthens and thins. The thinned crust subsides because of isostasy to form a rift valley. If rifting separates the continent into fragments, then the elevation of the thinned continental margin is lower than normal-thickness crust in the continental interior, forming a submerged continental shelf.

this works. The two wood blocks shown in **Figure 13.11** are analogous to blocks of continental crust with the same thickness. As shown in the drawing, Block A is "eroded" with a wood planer, and the shavings accumulate as "sediment" on Block B.

Isostasy requires adjustments to the blocks in Figure 13.11 because the weights and thicknesses of the blocks changed. Block A now weighs less so it rises. However, this uplift only partly offsets the erosional loss in height of the block, so the final surface elevation is somewhat lower than before because it is now also thinner. In contrast, Block B subsides because of the added weight of the wood shavings. However, this subsidence only partly offsets the depositional gain in height (thickness), so the final surface elevation of Block B increases overall. You can conclude from these experiments that isostasy results in uplift where erosion occurs, even though overall elevation decreases, and results in subsidence where deposition occurs, even though overall elevation increases. Now, let's apply the wood-block analogy to real geology.

We can use isostasy to explain why mountains continue to rise long after tectonic forces cease to thicken crust and drive uplift along faults.

This is because mountains result from *both* tectonic deformation and isostatic response to erosion. **Figure 13.12** illustrates how isostasy combines with erosion to cause rock uplift, even as surface elevations gradually decrease. As long as the crust beneath a mountain range is thicker than in adjacent areas, the mountain range stands higher than the adjacent lowlands. This observation explains why ancient mountains, such as the 300-million-year-old Appalachian Mountains, persist as high topographic features long after they form. Isostasy also offers an explanation for how metamorphic rocks that form deep beneath the surface are later exposed at the surface (Figure 13.12).

Now, let's turn from erosion to the isostatic consequences of sediment deposition. When geologists measure the total basin subsidence in a rift valley, they find that the subsidence is greater than what crustal thinning can explain. The "extra" subsidence occurs because, according to isostasy, the weight of accumulating sediment pushes down the crust. The thinned, submerged continental shelves along rifted continental margins also subside under the weight of accumulating sediment delivered by rivers and also in response to the weight of overlying seawater.

▶ **Figure 13.11 How erosion and sedimentation change elevations.** These two wood blocks start out with the same density and thickness, and they float at the same level in the water. Block A is "eroded" with a wood planer, and the shavings are "deposited" on top of Block B. Eroded Block A is now thinner and it moves up as it erodes, although its elevation is lower than before because it has lost some height in the erosion process. Block B subsides under the added weight of the shavings, although the added thickness of the deposit causes its surface elevation to increase slightly.

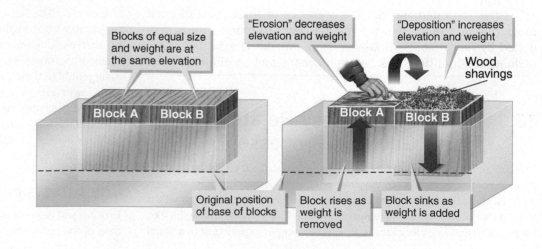

Blocks of equal size and weight are at the same elevation

Block A   Block B

"Erosion" decreases elevation and weight

"Deposition" increases elevation and weight

Wood shavings

Block A   Block B

Original position of base of blocks

Block rises as weight is removed

Block sinks as weight is added

▶ **Figure 13.12 Isostasy and erosion cause rock uplift.** Tracking the reference point in these diagrams indicates that isostatic adjustment results in rock uplift even as mountains erode. Erosion decreases the thickness of the crust, which means that the thickness of the root projecting into the mantle must also decrease. Although the surface elevation of the mountains gradually decreases through time, the underlying rocks persistently rise. Erosion eventually exposes metamorphic and plutonic-igneous rocks that originally formed deep in the crust.

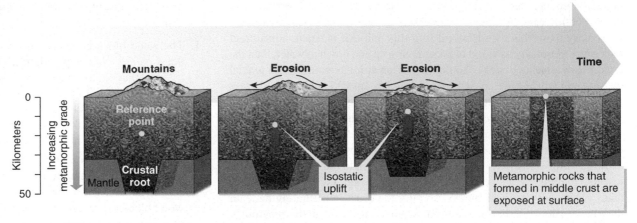

What Elevation Changes Related to Ice-Age Glaciers Reveal About Isostasy

According to isostasy, adding weight to Earth's surface causes subsidence and removing it causes uplift. Ice-age glacial ice was as much as 5 kilometers thick across large areas of northern North America and Eurasia about 21,000 years ago (the evidence for this appears in Chapter 18). That ice is nearly all gone now, except in Greenland. The addition of ice-age glaciers should have caused the underlying crust to subside. Then, when the ice melted there should have been uplift of the land surface. Is this what happened?

For relevant data to answer this question, consider **Figure 13.13**a, which illustrates old shorelines that are now visible well above sea level along Hudson Bay, Canada, where the ice-age glaciers were thickest. The highest-elevation shoreline is about 150 meters above sea level and formed about 8000 years ago, based on isotope dates on seashells left behind on the abandoned beach. It is unlikely that these ancient shorelines record falling sea level because global sea level instead has risen over the last 21,000 years as glaciers melted and added water to the oceans. The only way to explain the old, elevated beaches is that the land is rising faster than sea level is rising. This seems consistent with our expectations from the principle of isostasy because the Hudson Bay region would have been depressed under the weight of the ice-age glaciers and then raised more recently after the ice melted.

Additional data in Figure 13.13b demonstrate that uplift is still taking place in Canada, and nearby regions not affected by glaciation are subsiding. Current rates of uplift and subsidence in eastern North America detected by Global Positioning System measurements (Section 12.7 explains these types

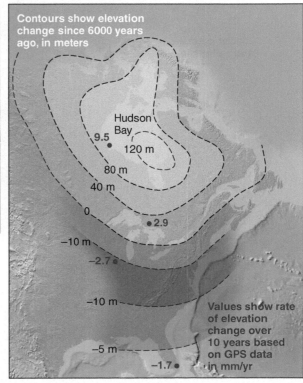

▶ **Figure 13.13 Elevation changes document isostasy.** The photograph on the left shows old raised beaches along Hudson Bay. The most recent exposed beaches to emerge above sea level are light-colored ribbons of sand. Dark-green trees cover older beach ridges that stand higher than the intervening light-green swampy areas. The map on the right shows the changes in elevation of the land surface in eastern North America over the last 6000 years. Most of the data used to draw the map come from uplifted or submerged shoreline features along seacoasts and lake margins. Active rates of uplift and subsidence obtained from Global Positioning System data are shown in red. The area of modern isostatic uplift corresponds to the area depressed by the weight of the former ice-age glaciers. The area of modern isostatic subsidence represents the area that bulged up when the flexible crust deformed under the weight of the ice.

of measurements) match the longer-term pattern recorded by shorelines over the last 6000 years. However, these observations are not obviously related to isostasy as we have understood it to this point. If isostatic adjustment took place as quickly as wood blocks moving up and down in water, then the crust below glaciated areas should have risen simultaneously with glacial melting and would have ceased rising long ago. Instead, all glaciated areas in the northern hemisphere are still rising, and adjacent areas are sinking.

Figure 13.14 illustrates two explanations for the isostatic response to weight shifting on the crust when ice-age glaciers formed and then melted away. In the combined Pratt-Airy isostasy view (Figure 13.14a) the addition of the thick mass of ice also added weight to the continental crust, just like placing a lead weight on the block of wood. Isostatic adjustment takes place only on local blocks, so only the crust immediately beneath the glacier subsided. In this scenario, when the ice melted, the previously weighed down block of crust would have instantaneously popped back up to its original position.

However, the wood-block analogies of Earth-scale isostasy have two limitations:

1. The wood-block models assume a very weak crust that cannot resist deformation when any weight is place on top of it. The models also assume that faults penetrate completely through the crust so that adjacent blocks of crust readily slide up and down past one another while adjusting to changes in crustal thickness. In reality, the upper crust is strong, and the entire crust is not penetrated by faults because plastic flow occurs in the hot lower crust (Section 11.6).

2. The mantle is almost entirely made up of solid rock, not liquid. Although the hot mantle flows like a very high viscosity fluid, this flow is much slower than the adjustment of water level to bobbing wood blocks. The crust can only move as quickly as the mantle very, very slowly flows from one place to another to equalize the pressure below regions of changing crustal weight.

ACTIVE ART

Glacial Isostasy. See how the crust isostatically responds to the growth and shrinkage of glaciers.

▼ **Figure 13.14 Ice-age glaciers provide a test of isostasy.** These diagrams compare two scenarios for isostatic adjustments resulting from the formation and melting of a thick glacier. Flexural isostasy, illustrated on the right, most satisfactorily explains observations, such as those shown in Figure 13.13.

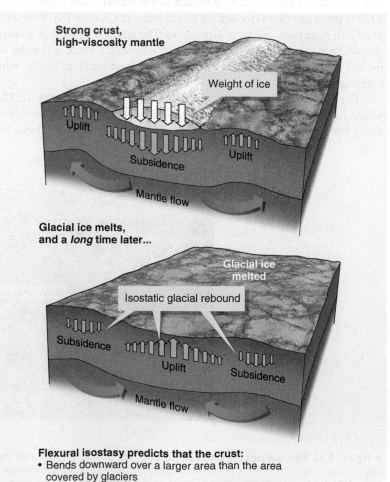

Weak crust, low-viscosity mantle

Weight of ice

Subsidence · Continental crust · Mantle

Glacial ice melts, and a *short* time later...

Glacial ice melted

Strong crust, high-viscosity mantle

Weight of ice · Uplift · Subsidence · Uplift · Mantle flow

Glacial ice melts, and a *long* time later...

Glacial ice melted · Isostatic glacial rebound · Subsidence · Uplift · Subsidence · Mantle flow

Pratt-Airy isostasy predicts that the crust:
• Subsides only directly below the ice
• Returns to original elevation after the glacial ice melts

(a)

Flexural isostasy predicts that the crust:
• Bends downward over a larger area than the area covered by glaciers
• Bulges upward slightly beyond the area weighed down by the glaciers
• Adjusts elevation very slowly after removal of the ice, because of the very slow flow of the viscous mantle

(b)

Figure 13.14b illustrates an alternative view where the crust flexes down over a larger area than simply beneath the glacier, because the strong crust supports the weight over a large area. Mantle displaced below the subsiding area slowly flows to adjacent areas, causing a small upward bulge. After the ice melts, the crust everywhere very slowly returns to its previous elevations. The adjustments are slow because highly viscous mantle does not flow fast enough to instantaneously even out pressure differences caused by the rapid redistribution of weight when the ice melts. This view of isostasy, called *flexural isostasy* because of the slow flexing of the strong crust, is more consistent with the data portrayed in Figure 13.13. The eastern North American landscape is still responding to the disappearance of the ice-age glaciers. The area that flexed downward beneath the glaciers is still rising. Areas that bulged upward beyond the glacier margins during the ice age are still slowly sinking back to where they were before. The ongoing isostatic adjustment of land elevations to the melted ice is called **glacial rebound**. Geologists calculate that an additional 330 meters of uplift will occur at Hudson Bay before isostatic stability is restored. At the current rate of rebound, this deformation will continue for another 30,000 years.

The illustration of flexural isostasy by glacial rebound suggests that sitting down on a waterbed is better than blocks in water as an analogy to how isostasy works. Water redistributes when your weight is added to the bed. The top of the waterbed sags beneath your body and slightly bulges up next to you. When you stand up, the surface rebounds back to its original shape, showing that the deformation was elastic. In real life, unlike on a waterbed, the elastic sheet of crust rests on very high viscosity mantle rock. The mantle must flow away from areas of the crust where the weight increases and rise where the weight decreases. The crust can only flex down or up as fast as the mantle can flow, which given its viscosity is not that fast. This means that real isostatic adjustments are long term, not instantaneous, responses to changing distributions of mass at or near the surface.

It is important to understand, however, that the Pratt and Airy style isostasy, modeled by wood blocks in water, is not really incorrect so much as it is incomplete. The Pratt and Airy approaches focus on differences in crustal thickness, density, or both, that adequately explain the different elevations of continental crust and oceanic crust and the broad variations in continental elevation from place to place. Flexural isostasy includes these same variations in crustal density and thickness but the flexure approach (a) more accurately describes the boundaries of areas that sink and rise isostatically, (b) more accurately estimates the amount of subsidence and uplift, and (c) more realistically calculates the time required for isostatic adjustments to take place.

Isostatic Adjustment Causes Earthquakes

Movement of crust because of isostasy must stress the rocks. What, for example, are the effects of these stresses in areas undergoing glacial rebound? Most rocks are strong enough not to be affected by the small calculated stresses, but the stresses are large enough to reactivate old faults where broken rock has low strength. Scattered earthquakes, some with moment magnitudes as high as 6.0, occur in northeastern North America despite the fact that this region is very distant from active plate boundaries. Glacial rebound, rather than plate motion, explains most of these earthquakes. Glacial-rebound stress shifts rocks along faults that formed long ago when plate boundaries existed near this region (Figure 12.43). Some geologists think that the great New Madrid, Missouri, earthquakes of

1811–1812 (Figure 11.32 and Table 11.2) may partly relate to glacial-rebound stresses, because there is no geologic evidence of Cenozoic fault movement in Missouri before the ice age. The New Madrid area isostatically rose adjacent to the ice-age glaciers and is now actively sinking back to its former elevation.

The take-home point is that relatively small stresses can cause earthquakes along old, weak faults and, with isostatic adjustment, it does not take much weight to cause earthquakes. For example, the weight of water filling reservoirs behind new dams commonly triggers hundreds of small earthquakes, and a few of these cause minor damage.

Putting It Together—How Does Isostasy Relate to Active Geologic Processes?

• Isostatic adjustments to changes in crustal weight and thickness cause vertical motion of the crust.

• Compressional stress that shortens the crust also thickens it, causing isostatic uplift of mountains. Tensional stress thins crust, causing isostatic subsidence of sedimentary basins.

• Isostasy causes crust beneath eroding mountains to rise slowly even though the overall elevations decrease, because the weight of the crust decreases as the mountains erode. The eroded sediment accumulates in sedimentary basins, causing subsidence.

• Slow flexural isostatic adjustment to changing weight is well documented by the patterns of recent and ongoing uplift and subsidence in eastern North America, which are attributed to glacial rebound.

• Isostatic movement of the crust causes movement along old faults, thereby accounting for earthquakes that occur far from plate boundaries.

13.4 Why Does Sea Level Change?

Sea level is a convenient reference position for comparing elevations, but sea level has not always been at the *same* level. We know that the edges of continents submerged today were dry land at times in the past. On the other hand, widespread marine sedimentary rocks in continental interiors, far from present shorelines, indicate times in the geologic past when whole continents were largely submerged. The maps in Figure 12.43 illustrate varying levels of continental submergence through geologic time. Of course the submergence of part of a continent beneath seawater could result from either subsidence of the land, rise in sea level, or a combination of both. It is very difficult, therefore, to clearly determine how sea level has changed independently of the uplift and subsidence of continents. **Figure 13.15** illustrates two different estimates of real sea-level changes for the last 140 million years. The differences between the two interpreted sea-level curves result from the difficulty of separating out the effects of land uplift and subsidence. However, the general patterns of these two interpretations are similar and suggest that between 50 and 100 million years ago global sea level was about 200 meters higher than it is now.

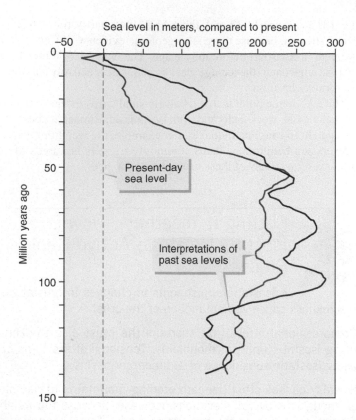

Sea level in meters, compared to present

Million years ago

Present-day sea level

Interpretations of past sea levels

◄ **Figure 13.15 Sea-level changes through time.** The graph shows two interpretations of long-term variations in sea level, compared to present sea level, during the last 140 million years. Geologists use the distribution of marine sedimentary deposits on continents to estimate ancient sea levels.

fluctuations caused by the growth and melting of glaciers. At the peak of the last ice age (21,000 years ago), sea level was at least 100 meters lower than it is today simply because large volumes of water were stored on land as glacial ice. When the glaciers melted, the water flowed off into the oceans and sea level rose to its present position. If the remaining glaciers melt (mostly in Greenland and Antarctica), then global sea level will rise an additional 80 meters.

Changing the Size of the Container

Another explanation for fluctuating sea levels states that the size of ocean basins changes through time. For an analogy, if you pour water from a large bowl into a small bowl, the water will spill over the side because the small bowl cannot hold all of the water. Figure 13.16b shows a similar situation for oceans. If ocean-basin depths decrease, similar to making a smaller bowl, then the water spreads out onto previous dry land, therefore raising sea level without changing the volume of water in the oceans. This hypothesis of changing the size of ocean basins is necessary because there are times in Earth history when shoreline sedimentary deposits show changes in sea level without geologic evidence for the presence of glaciers. In addition, in a completely ice-free world we would expect sea level to only be about 80 m higher than it is now, but the estimates of sea levels in the past are about three times greater (Figure 3.15).

Plate tectonics potentially causes the changes in the size of ocean basins that are implied by the sea level curves. To link plate tectonics with sea level you need to make use of the fact that the elevation of seafloor relates to its age (look back to Figure 12.15 to see a graph of this relationship). Let's look closely at what causes greater water depth above old lithosphere and shallower water depth above young lithosphere.

Why does sea level rise and fall? If you think of the world ocean as one large container of water, then you can change the water level in the container two ways. You can either (1) change the amount of water within the container, or (2) change the size of the container that holds the water.

Changing the Volume of Water in the Container

One easy explanation for changing sea level is that the amount of water in the oceans changes through time. **Figure 13.16**a explains sea-level

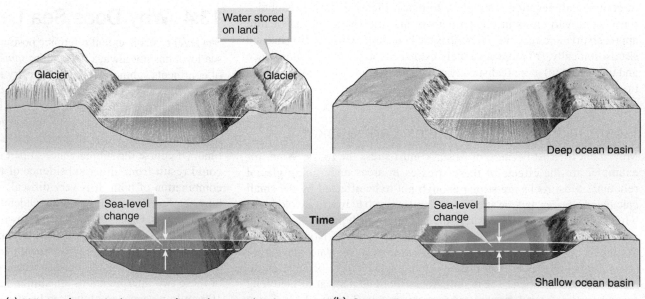

Water stored on land

Glacier   Glacier

Deep ocean basin

Sea-level change

Time

Sea-level change

Shallow ocean basin

▶ **Figure 13.16 Why sea level changes.**

(a) Volume of seawater increases after an ice age when ice melts and adds water to oceans

(b) Same volume of water covers a larger surface area if ocean basins become shallower

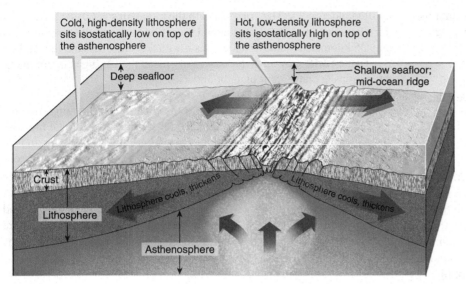

Cold, high-density lithosphere sits isostatically low on top of the asthenosphere

Hot, low-density lithosphere sits isostatically high on top of the asthenosphere

Deep seafloor

Shallow seafloor; mid-ocean ridge

Crust

Lithosphere

Lithosphere cools, thickens

Lithosphere cools, thickens

Asthenosphere

◀ **Figure 13.17 Isostasy explains water depth in the ocean.** This diagram shows that water depth increases away from a mid-ocean ridge because the lithosphere cools, thickens, and becomes denser over time. The older, colder, denser lithosphere subsides because of isostasy in comparison to the younger, hotter, less-dense lithosphere underlying the mid-ocean ridge.

How Isostasy Explains Water Depth in Oceans

Figure 13.17 illustrates how you can apply the principle of isostasy to explain seafloor elevations. When the oceanic lithosphere forms at a mid-ocean ridge it is hot, which also means that it is expanded and has a relatively low density. Then, the lithosphere cools, contracts, thickens, and becomes denser as seafloor spreading moves it away from the ridge (Section 12.3). Older seafloor at greater distance from the ridge is underlain by progressively thicker, older, cooler, and denser lithospheric mantle. When applying the principle of isostasy to these changes in thickness and density, it turns out the surface elevation of the lithosphere is affected much more by the increasing density than by the increasing thickness. This means that the lithosphere isostatically sinks into the underlying asthenosphere more when it is old, cold, and dense, than when it is young, warm, and less dense. There is room for a greater depth of seawater above the lower old lithosphere than above the higher young lithosphere. Therefore, ocean depth increases as the age of the underlying lithosphere increases and the depths are shallowest over the very young lithosphere at mid-ocean ridges.

Changes in lithosphere age, therefore, change ocean depth. If all the oceanic lithosphere is mostly young and warm, then isostasy keeps the lithosphere riding high in the asthenosphere, oceans are shallow, and sea level rises onto continents. If oceanic lithosphere is mostly old and cold, then it rides lower in the asthenosphere, oceans are deep, and continents stand higher above sea level (and sea level appears low). A change from deep oceans to shallow oceans is a change in the size of the ocean container; there is less room for water in the oceans and sea level rises onto the continents (Figure 13.16).

Sea Level Changes when Average Lithosphere Age Changes

Some geologists apply isostasy to this likely connection between lithosphere age and water depth to form a hypothesis for the fluctuating sea levels during Earth history. The average age of all oceanic lithosphere on Earth at a particular time depends on how much new lithosphere is created at divergent plate boundaries and the age of the lithosphere simultaneously destroyed at convergent boundaries. On average the world oceans are shallower (higher sea level) if the global average lithosphere age decreases. This might happen if new mid-ocean ridges form, for instance. In that case,

all of the seawater will not all fit into the shallower oceans and some of it will "spill" onto the continents. If the average oceanic lithosphere age later increases, then the lithosphere isostatically sinks lower in the asthenosphere, ocean basins become deeper; the deeper container is large enough to hold all of the seawater that had flooded onto the continents so sea level falls.

Geologists use maps of seafloor age (Figure 12.13) to test the relationship between age and sea level. Maps can be used to calculate the average age of oceanic lithosphere at any one time over the last 180 million years. The average age of oceanic crust today, for example, is about 65 million years old. Back when sea level was very high, between 50 and 100 million years old, the average age of oceanic crust was only about 40 million years old. Using average age for particular time intervals permits a calculation of seafloor depth, using isostasy, and the size of the global ocean container from which the elevation of sea level can then be estimated. These calculations have large uncertainties associated with them, but within the range of these uncertainties they provide estimates of sea level that are comparable to the estimates based on the geologic evidence in Figure 13.15. The similarity of these two approaches to estimating sea level in the past supports the isostatic hypothesis of sea-level change.

Putting It Together—Why Does Sea Level Change?

• Isostasy explains why ocean depth correlates with seafloor age: The older the lithosphere, the greater its density, and the lower its surface elevation.

• Sea-level changes are explained either by changing the volume of seawater or by changing the size of ocean basins that contain the water.

• Plate tectonics affects global sea level through changes in oceanic lithosphere age. When the global average lithosphere age decreases, the average seafloor elevation increases. This causes sea level to rise so that seawater floods onto continents.

13.5 How and Where Do Mountains Form?

We have demonstrated that vertical movements of Earth's surface depend on isostasy. However, the major features of Earth's surface are related to plate tectonics, and plates move horizontally. So how do these vertical and horizontal motions relate to one another? A key place to answer this question is in Earth's long, high mountain ranges. Most mountain chains relate to processes at convergent plate boundaries. To explain mountains, geologists link plate-boundary processes with vertical isostatic adjustments.

Mountain Belts Near Continental Margins

Uplift to make mountains near convergent plate boundaries can be related to isostasy. Compressional shortening thickens the crust (Figure 11.13). Thickening, in turn, leads to uplift (Figure 13.10). However, the relationships between processes at convergent margins and surface topography are more complex.

The Andes in western South America, portrayed in **Figure 13.18**, are a useful example of mountains formed near an oceanic-continental convergent boundary. The highest mountains did not form right at the trench that marks the plate boundary. Instead, the high mountains are found 200–400 kilometers away on the overriding plate. Some of the mountains are volcanoes that grow upward by accumulation of erupted volcanic materials, but the volcanoes themselves are built on a foundation of highly

uplifted older rocks. The crust beneath the mountains is 60–70 kilometers thick compared to typical continental crust east of the mountains, which is only 35 kilometers thick. What causes the thicker crust beneath the Andes?

Figure 13.19 shows two processes of crustal thickening near continental margins at convergent plate boundaries.

1. Volume and thickness are added to the crust by crystallizing magma. Plutonic rocks solidify from igneous intrusions within and at the base of the crust (see Section 4.10 for explanations of why most magma solidifies below Earth's surface). The volume of plutonic rocks is probably at least 10 times greater than the volume of volcanic rocks that also add an upper veneer to the total crustal thickness.

2. Crust thickens during shortening by plastic flow of lower crust caused by compression at the plate boundary. Plastic flow and thickening are greatest where magmas pass through the crust because this area is hotter and hot rocks flow more easily.

▶ **Figure 13.18 The Andes are an active mountain belt.** The shaded-relief map of South America shows that the high Andes Mountains, like other mountains near oceanic-continental convergent boundaries, are located several hundred kilometers from the plate boundary. The photo illustrates the highest mountain in South America; 6914-meter-high Aconcagua, in Argentina.

Andes Mountains located 200–400 km from plate boundary

Deep-sea trench marks subduction at plate boundary

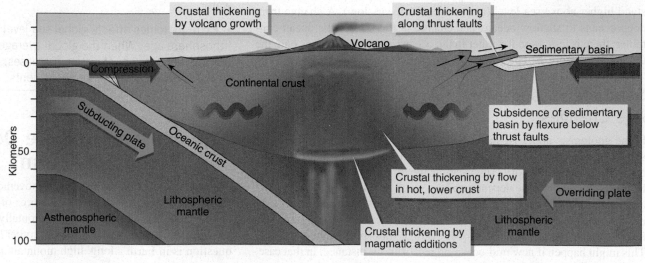

▶ **Figure 13.19 How continental-margin mountain belts form.** Thick crust underlies high mountains. Most crustal thickening occurs because of (1) addition of magma from below that solidifies as plutonic and volcanic rocks and (2) plastic flow of weak, hot compressed lower crust. Less thickening occurs where thrust faults cause crustal shortening. The added weight of the overthrust rocks flexes the crust to form a basin that accumulates sediment eroded from the mountain belt.

Crustal thickening by volcano growth

Crustal thickening along thrust faults

Volcano

Sedimentary basin

Compression

Continental crust

Subducting plate

Oceanic crust

Subsidence of sedimentary basin by flexure below thrust faults

Overriding plate

Crustal thickening by flow in hot, lower crust

Asthenospheric mantle

Lithospheric mantle

Lithospheric mantle

Crustal thickening by magmatic additions

Kilometers — 0, 50, 100

The process of crustal thickening by magmatic additions and plastic flow, and the resulting isostatic adjustments, explains why high mountain belts rise skyward well within the overriding plate rather than adjacent to the trench at convergent plate boundaries.

Compression also forms thrust faults in the brittle upper crust that shove rocks away from the volcanic chain toward the center of the continent (see right side of Figure 13.19). These thrust faults thicken the crust only slightly, so the resulting uplifts are not as high as those closer to the continental margin. As the thrust faults stack up rocks on the continent, the added weight causes the crust to flex downward over a wide region, as predicted by flexural isostasy. Part of the resulting depression fills with sediment eroded from the thrust-uplifted rocks and the more distant higher mountains. The weight of the accumulating sediment causes additional subsidence of this sedimentary basin.

Mesozoic and early Cenozoic thrust faults in western North America formed mountains and deep basins, visible in **Figure 13.20**. The weight of the thrust-faulted rocks formed adjacent sedimentary depressions (Figure 13.20b) that contain more than 6 kilometers of sedimentary rock in some places. The sedimentary rocks in these basins, and in parts of the thrust belt, are the source for prolific supplies of oil, natural gas, and coal throughout the Rocky Mountain region of the United States and Canada. Similar sedimentary basins adjacent to the Paleozoic Appalachian Mountains also host huge coal and natural gas resources.

Mountain Belts Where Continents Collide

The mountain building where continents collide at convergent plate boundaries (Figure 12.25) is slightly different. **Figure 13.21** summarizes the

◀ **Figure 13.20 Thrust faults make mountains and basins.** (a) This view of Banff National Park in Alberta, Canada, is typical of the Rocky Mountains of Montana and western Canada, where sedimentary rocks were shoved eastward along thrust faults during the Mesozoic, when North America was near a convergent plate boundary. (b) This view in western Montana shows the plains east of the Rocky Mountains, which form the distant skyline. Sedimentary rocks are as much as 6 kilometers thick beneath the plains and accumulated in the basin that flexed down under the weight of the overthrusted rocks. These sedimentary rocks contain rich resources of oil, natural gas, and coal.

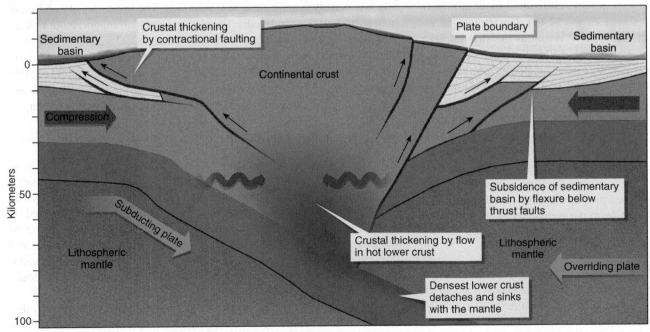

◀ **Figure 13.21 Characteristics of a collisional mountain belt.** The highest mountains form where continental plates collide, because the crust becomes very thick. Most of the crust on the subducting plate is too buoyant to subduct and is shoved up on faults and thickens by plastic flow at depth. Some of the densest lower crust and lithospheric mantle detach from the subducting continent and sink into the mantle. Thrust faults shorten and thicken the upper crust on both sides of the collision zone, forming deep sedimentary basins.

► **Figure 13.22 Some mountains do not have roots.** Seismic data across the United States do not show the thicker crust beneath the highest elevations as expected from isostasy. The high, mountainous region west of the Rocky Mountains does not have a thick crustal root, and the crust beneath the Rockies is insufficient to explain all of its high elevation. However, seismic data also reveal contrasting properties of the upper mantle beneath the eastern and western United States. Seismically slow buoyant mantle with the properties of the asthenosphere or unusually hot or low-density lithospheric mantle is present beneath most of the western United States. This buoyant mantle is what holds up the high elevation of the region.

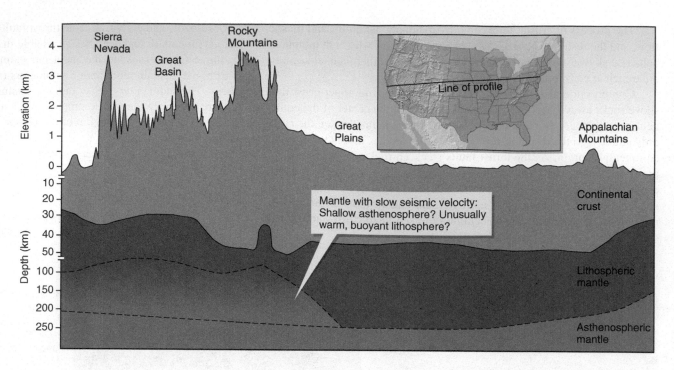

features of collisional mountain belts. Crustal thickening by igneous-rock additions and shortening by plastic flow are similar to the case of continental-margin mountain belts. In collisional mountain belts, however, the shortening takes place in both plates, because the low-density continental crust does not readily subduct as oceanic crust does. As a result, the mountain belt has a more symmetrical cross section than we see in the case of oceanic-continental convergence (compare Figures 13.19 and 13.21). Huge thrust faults shove slivers of crust many kilometers thick outward from the collision zone. These thrust faults cut deeply through the crust and effectively double up the thickness of continental crust. The overthrusted rocks add considerable weight to crust on both sides of the mountain belt, flexurally depressing basins that fill with 10 kilometers or more of sediment.

The extraordinary thickening of crust in collisional mountain belts forms the highest mountains (Figures 1.2a and 12.19). The Himalayas are the highest mountains on Earth and are still rising as the slow-motion collision between India and Asia, which started 50 million years ago, continues. The Alps of southern Europe formed when small continental blocks collided with the larger European continent beginning about 40 million years ago.

According to the principle of isostasy, high mountains have thick roots (Figure 13.8). The crust in the deep roots experiences extremely high temperature and pressure, which causes high-grade metamorphism and even some melting. Melting of continental crust forms granitic magma that rises to solidify in the middle crust. In a continental collision zone, the high-grade metamorphic rocks in the lowermost crust of the subducting plate may be sufficiently dense to detach from the remainder of the crust and sink along with the underlying mantle lithosphere (Figure 13.21).

The Role of the Mantle in Surface Topography

The high elevations of the Rocky Mountains, Sierra Nevada, and many other mountain ranges define the Western Cordillera of North America (outlined in Figure 12.8). Mountain uplift in western North America occurred during the Mesozoic and early Cenozoic when there was a convergent plate boundary along the entire west coast (described in Section 12.9).

Data illustrated in **Figure 13.22** do not show, however, a systematic relationship between surface elevation and crust thickness in the western United States. The crust below the high elevations of the Sierra Nevada (including Mount Whitney, Figure 13.1) is actually thinner than the crust of the western Great Plains, which is 3 kilometers lower in elevation. The mountains do not have the predicted thick crustal root, a problem that we first noted at the end of Section 13.2. How can this be?

The mantle is the key for explaining the high elevation of the western United States. Seismic data indicate a large region of mantle below the western United States where earthquake waves move unexpectedly slowly. Calculations suggest that this area of slow seismic waves consists of unusually low-density mantle close to the surface. The lower-density and more buoyant mantle rises and pushes the surface upward to support mountainous elevations even where the crust is not very thick. Geologists and geophysicists do not yet understand whether the buoyant mantle is unusually low-density lithospheric mantle or if, instead, the lithosphere is uncommonly thin and the seismically slow mantle is part of the hotter asthenosphere. Either way, isostasy and buoyant mantle explain high elevations in this region.

How Fast Do Mountains Rise?

All of this discussion of mountain-building processes leads to another popular question: How fast does mountain uplift happen? Plates move horizontally at about 1 to 10 centimeters per year, but how fast does Earth's surface move vertically? As one example, nearly 15 meters of uplift occurred on a thrust fault along the convergent boundary in Alaska over a matter of minutes during a single earthquake in 1899. However, many decades and even centuries of almost no uplift intervene between such great earthquakes, so these single spasms are not a good measure of long-term rates of mountain uplift.

To measure uplift rates, geologists must know a vertical uplift distance and the time during which that uplift took place. One site they have studied is a coral reef near a convergent boundary in the southwest Pacific Ocean that is now 400 meters above sea level. Radioactive-isotope ages measured from the calcite in the coral indicate that the reefs were thriving below sea level 120,000 years ago. These two pieces of data indicate that the old seafloor rose at least 400 meters in no more than 120,000 years; this means that the average uplift rate was at least 3.3 millimeters per year. Another example comes from the Andes of South America, where metamorphic rocks exposed at the surface contain index minerals indicating a temperature and pressure of metamorphism equivalent to about 10 kilometers depth. Radioactive-isotope ages on these minerals show that they were at this depth only 2 million years ago, which leads to calculation of an average rock-uplift rate of 5 millimeters per year, and an equal erosion rate to remove the 10 kilometers of overlying rock.

Analyses of this sort in active tectonic settings around the globe show that the crust beneath mountains rises at rates of about 3 to 10 millimeters per year, when averaged over long time periods. Therefore, on average, these figures tell us that plate motion moves Earth's surface horizontally at speeds about 10 times faster than isostasy causes vertical surface displacement.

EXTENSION MODULE 13.1

Measuring Uplift Rates. Learn how geologists measure mountain-uplift rates.

Putting It Together—How and Where Do Mountains Form?

• Most mountain belts form near convergent plate boundaries where crust thickens by compressional shortening and by intrusion and crystallization of magma.

• The high elevations in the western United States do not have a thick crustal root but are held up, instead, by unusually low-density mantle.

• Even as crustal thickening causes uplift of mountains, thrust faults weigh down adjacent lowlands to form deep basins. These basins fill with sediment eroded from the mountains and contain rich resources of oil, natural gas, and coal.

• Mountains rise vertically at rates of 3–10 millimeters per year, or only about one-tenth the speed of horizontal plate motion.

13.6 How Does Mountain Building Relate to Continent Growth?

Mountain building is essential to making continents because mountain building thickens the crust. The thickness of continental crust, not just its low density, explains why continents are higher than oceans (Figure 13.6, Model C). After all, we know that where continental crust is thin, surface elevations are below sea level (Figure 13.10). The crust must be thicker than about 30 kilometers in order to stand higher than current sea level. Crust achieves this thickness by mountain building, even in areas where those mountains have long since eroded away.

Another critical observation, illustrated in **Figure 13.23**, is that sedimentary rocks record the coastline edge of western North America at the

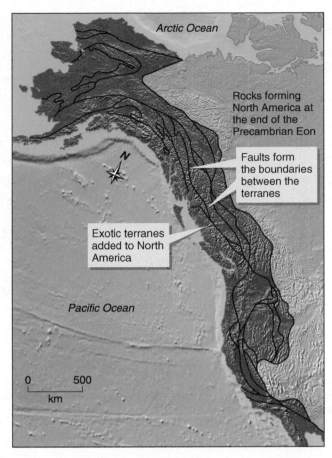

▲ **Figure 13.23 Most of western North America is exotic.** Crust that originated somewhere other than North America underlies the entire red region in the map. Tectonic collisions at convergent plate boundaries added the exotic crust to the continent. The added-on crust includes many crustal blocks, outlined by faults, which have different geologic histories than neighboring blocks or North America.

end of the Precambrian (about 540 million years ago) that was more than 500 kilometers east of where it is today. The mountains of western North America occupy this area where North America grew westward to the modern shoreline. These two observations suggest an important role of mountain building in the growth of continents.

Assembly at Subduction Zones

Geologic maps show that far-western North America consists of distinct crustal blocks, each one separated from its neighbors by faults (Figure 13.23). The oldest rocks exposed in each block do not resemble rocks of the same age in adjacent blocks or in the rest of North America. Each block of crust instead appears to originate hundreds or thousands of kilometers from North America and was later added onto the continent. These blocks are called **exotic terranes** or **accreted terranes**. These terms refer to regions where the crust is exotic to North America and accreted (added on) to the continent.

Crust accretes to a continental edge along a subduction zone because some of the crust entering the zone cannot subduct. Continental crust resists subduction because it has a low density (Section 12.4). Crust formed at volcanic arcs near convergent plate boundaries also will not readily subduct because it is commonly greater than 20 kilometers thick and contains relatively low-density intermediate to felsic igneous rocks. Unusually thick oceanic crust, typical of locations experiencing excessive

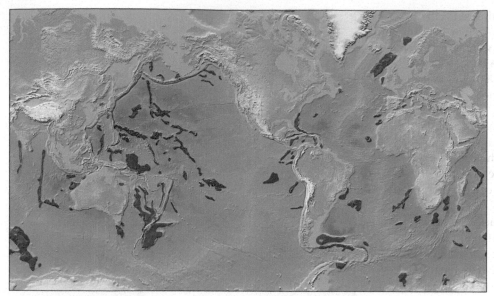

◀ **Figure 13.24 Exotic terranes of the future?** The map shows locations of crust within modern ocean basins that is too thick to subduct. Each red area is a potential exotic terrane of the future; if it enters a subduction zone and accretes onto a continent.

hot spot volcanism, is also too thick and buoyant to subduct at convergent plate boundaries. **Figure 13.24** shows large, mostly submerged areas between the present-day continents that consist of crust that is too thick to subduct. If any of these areas reach convergent plate boundaries adjoining continents, then they will be accreted onto the continent.

Figure 13.25 illustrates two scenarios for how accretion and continental growth happen.

1. Where any piece of "nonsubductable" continental crust, unusually thick oceanic crust, or volcanic-arc crust enters a subduction zone at the edge of a continent, it is thrust onto the edge of the continent (Figure 13.25a).

2. If a continent follows subducting oceanic lithosphere into a trench, the buoyant continent cannot subduct and instead collides with the volcanic arc on the overriding plate (Figure 13.25b). Then, a new subduction zone forms that faces in the opposite direction. The arc transfers from one plate to another and becomes part of the continent.

▼ **Figure 13.25 How exotic terranes accrete to a continent.** These diagrams show how continents grow along convergent plate boundaries by the addition of crustal blocks that originated in far distant places.

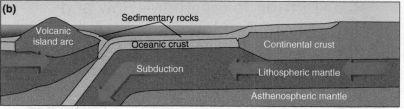

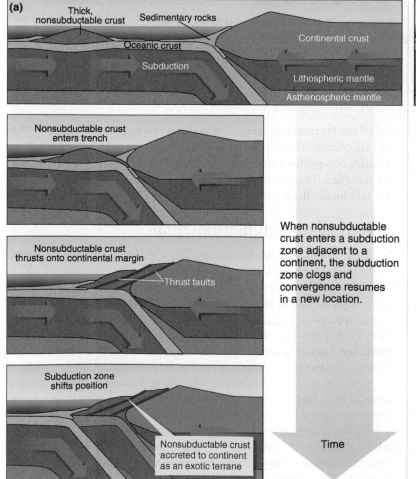

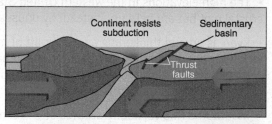

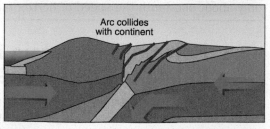

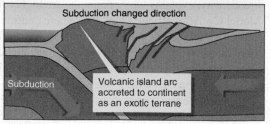

The Coast Range of Oregon and Washington consists of seamount and island volcanoes, similar to modern Hawaii, which accreted to North America about 45 million years ago because the volcanic crust was too thick to subduct beneath the continent.

(a)

The crust underlying this area of the Appalachian Mountains in central Massachusetts formed as a volcanic island arc that was shoved onto North America about 450 million years ago.

(b)

▲ Figure 13.26 Exotic crust in North America.

Phanerozoic sedimentary rocks that rest on older Precambrian metamorphic and plutonic rocks. Landscapes of the North American craton are shown in **Figure 13.27**. A map of North America, presented as **Figure 13.28**, outlines the North American craton and shows that the Precambrian crust is divided into discrete blocks, called **provinces**, each of which is composed of igneous and metamorphic rocks of different ages from its neighbors.

The Precambrian crust in the North American craton, ranging in age from about one to four billion years, is metamorphic and plutonic rock (Figure 13.27). This observation implies two things:

1. The crust probably formed by mountain building near convergent plate margins, where magma intrusion and metamorphism of thickened crust are common.

2. Considerable uplift and erosion, including isostatic adjustments, happened in order to expose the metamorphic and plutonic rocks at the surface before they became buried beneath much younger sedimentary rocks (Figure 13.12).

Superbly exposed rocks in northwestern Canada are highly metamorphosed Precambrian rocks typical of the craton. The low topographic relief reveals the current tectonic stability of the continental interior, although the metamorphism attests to intense deformation during Precambrian time.

(a)

As you might imagine, the collision of buoyant crustal blocks imparts unusually large compressional stress across the convergent-boundary continental margin. The large stresses cause considerable crustal thickening. Accretion of exotic terranes is, therefore, an important part of mountain building. All mountain belts contain accreted terranes, which illustrate the growth of continents by the assembly of pieces of nonsubductable crust at convergent plate boundaries. **Figure 13.26** illustrates examples of exotic-terrane landscape within North American mountain ranges.

Making the Ancient Continental Centers

To complete the connection of mountain building to the growth of continents, consider the origin of continental crust in the low-lying interiors of continents, far from recent, or even recognizable, older mountains. These interior regions of continents, called **cratons**, have been tectonically stable compared to continental margins for more than 500 million years. These low-elevation areas are typically 0.5 kilometer or less above sea level. Rocks exposed at the surface are either very ancient Precambrian rocks, or relatively thin coverings of mostly undeformed

Deep erosion in the Grand Canyon, Arizona, exposes the unconformity between Precambrian metamorphic and igneous rocks, recording mountain-building events that constructed North America, and much younger horizontal sedimentary rock. The undeformed sedimentary rocks are typical of the tectonically stable craton. Deep erosion exposes the underlying Precambrian rocks, whereas in most of the United States the nature of older rocks is known primarily from deeply drilled oil-exploration wells.

▶ Figure 13.27 Tectonically stable landscapes of the North American craton.

Sedimentary rocks

Unconformity

Metamorphic and igneous rocks

(b)

► **Figure 13.28 North American craton consists of Precambrian provinces.** The North American craton is the region where ancient Precambrian rocks are exposed at the surface (mostly in the Canadian Shield, outlined in red) or only thinly buried by sedimentary rocks. The craton is divided into provinces, each of which has rocks that metamorphosed during different times of mountain building that enlarged the continent. The craton has experienced only very minor tectonic deformation during the last 1 billion years, except in the Rocky Mountain region. The craton is partly rimmed by the Western Cordillera and the Appalachian Mountains, areas of highly deformed, in some cases metamorphosed, rocks affected by mountain building over the last 450 million years.

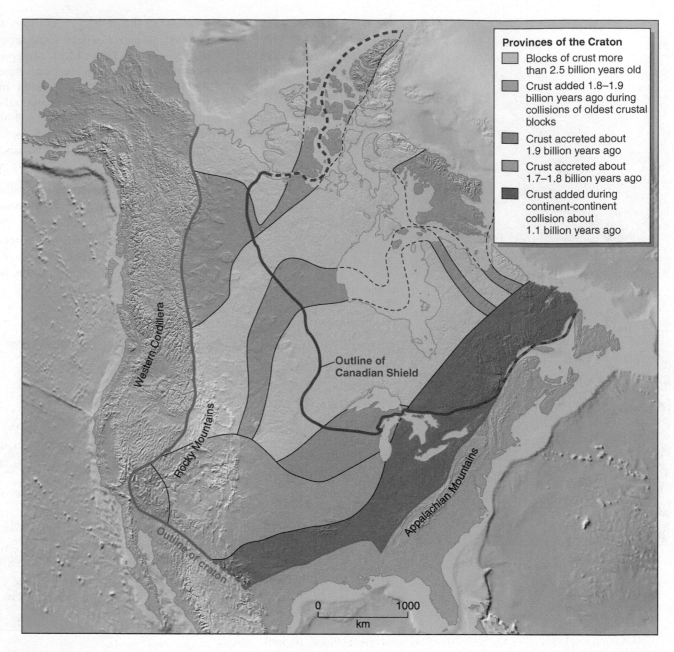

Provinces of the Craton

- Blocks of crust more than 2.5 billion years old
- Crust added 1.8–1.9 billion years ago during collisions of oldest crustal blocks
- Crust accreted about 1.9 billion years ago
- Crust accreted about 1.7–1.8 billion years ago
- Crust added during continent-continent collision about 1.1 billion years ago

Western Cordillera

Outline of Canadian Shield

Rocky Mountains

Appalachian Mountains

Outline of craton

0   1000
km

Figure 13.29 integrates these two conclusions to show that the exposed Precambrian rocks probably formed in the middle to lower crust during thickening of the crust by mountain building. Erosion of the high mountains, with accompanying isostatic uplift, eventually removed the now missing rocks that formed the Precambrian upper crust. The mountains eventually eroded to low-relief topography (Figure 13.27a) with sufficient remaining crustal thickness to maintain the surface elevation at, or slightly above, sea level. At times when sea level was unusually high, thin layers of marine sedimentary rocks accumulated on top of the Precambrian rocks.

Closer scrutiny of the geology of the Precambrian rocks of the craton and the map pattern in Figure 13.28 reveals how North America formed by mountain building along convergent plate boundaries. **Figure 13.30**

summarizes the story. Each province that consists of rock older than 2.5 billion years is made up of jammed-together, convergent-boundary volcanic arcs. These small, continental embryos of amalgamated arc crust then collided with each other, mostly between 1.8 and 2.0 billion years ago. Metamorphism of this age in Canada and the northern United States defines mountain belts that formed between the older provinces (Figure 13.28). Later, large masses of crust were added along the southern and eastern margins of North America by major collisions with continents and volcanic island arcs between 1.0 and 1.7 billion years ago.

What all of this tells us is that continents are produced by mountain building at ancient convergent plate boundaries. Where mountains stand high above the surrounding landscape, mountain-building processes are easily recognized. However, even the crust below the low-lying interior of

Precambrian mountain building, metamorphism, and crustal thickening

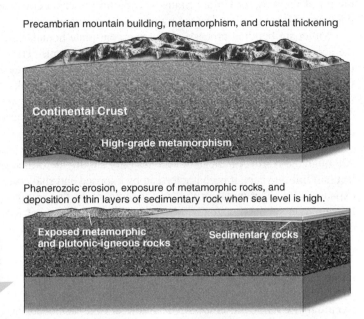

Continental Crust

High-grade metamorphism

Phanerozoic erosion, exposure of metamorphic rocks, and deposition of thin layers of sedimentary rock when sea level is high.

Exposed metamorphic and plutonic-igneous rocks

Sedimentary rocks

Time

▲ **Figure 13.29 Cratons started as mountains.** Precambrian metamorphic and plutonic rocks now found in the craton formed by mountain building and crustal thickening near ancient convergent plate boundaries (top). High mountains gradually wore down by erosion, combined with isostatic uplift (Figure 13.12), until low-relief regions remained (bottom). The crustal thickness of the low regions is sufficient to keep the surface above sea level, except when sea level is exceptionally high.

North America is the product of mountain building that occurred as long as 4 billion years ago. The now-stable craton stands above sea level as an isostatic consequence of the development of relatively thick, low-density crust in mountain belts.

Putting It Together—How Does Mountain Building Relate to the Growth of Continents?

• Continents grow through time by the collision and accretion of crustal fragments that cannot be subducted along convergent plate boundaries.

• The low-elevation regions of continents are areas of long-term tectonic stability, called cratons, where Precambrian metamorphic and plutonic rocks are present at or near the surface.

• Craton crust formed during mountain-building events between 1 and 4 billion years ago. These mountain-building events featured collisions of thick blocks of low-density, mostly igneous crust. North American crust is a collage of these crustal blocks.

• Although the Precambrian mountains were long ago eroded down and partly buried, the elevation of the continental interior of North America above sea level is a result of the crustal thickening that occurred during ancient periods of mountain building.

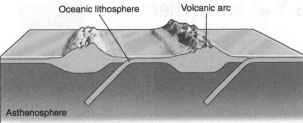

Oceanic lithosphere   Volcanic arc

Asthenosphere

Precambrian rocks more than 2.5 billion years old represent highly deformed and metamorphosed volcanic arcs that formed at convergent boundaries.

Island arcs collide at subduction zones to form early continents

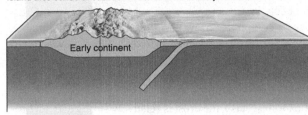

Early continent

Volcanic arcs collided in mountain-building events that formed the oldest provinces of the craton

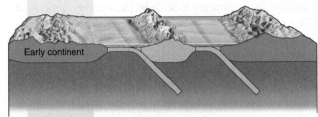

Early continent

Collisional zone of volcanic arcs between early continental blocks

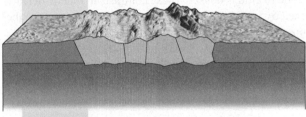

The small, early Precambrian continents collided with one another, and with intervening volcanic arcs, between 2.0 and 1.8 billion years ago to form larger continents.

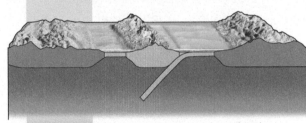

Continent-continent collision and accretion of arc and oceanic rocks

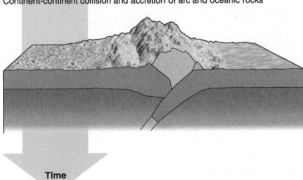

Since 1.8 billion years ago, continents grew by accretion of exotic blocks of crust, many of which originated as volcanic arcs.

Time

▲ **Figure 13.30 Continents are made at convergent plate boundaries.** These highly schematic diagrams show how continents have grown through time because of igneous and collisional processes at convergent plate boundaries.

Where Are You and Where Are You Going?

Isostasy explains the rough relief of Earth's surface. Areas of thick crust, or low-density crust, or both, exhibit a higher elevation than areas of thinner or denser crust. High-standing mountains, therefore, typically have a thick root of low-density rocks. Continental crust is both thicker and less dense than oceanic crust, which explains why continental areas are largely above sea level and oceanic areas are submerged beneath the sea. Crust thickens in response to compression, forming high elevation mountains, whereas thinning occurs during tension and produces low-elevation rift valleys and submerged continental shelves. Isostatic adjustments to changes in the distribution of mass within the crust commonly cause earthquakes within plates, far from plate boundaries.

According to flexural isostasy, the crust is relatively strong and not broken all of the way through into discrete blocks, and the viscous, solid mantle flows very slowly. As a result, the crust bends under the added weight from deposited sediment or glacial ice and slowly rebounds when weight is removed by erosion or melting of ice. The rebound of crust that flexed under the weight of the great ice-age glaciers, which melted away thousands of years ago, explains the slow rising and sinking of eastern North America.

The density of Earth's mantle also plays a role in surface elevation. Where the mantle is relatively less dense, surface elevations are higher than in areas where the mantle is denser. These contrasts in mantle density are mostly determined by temperature, which in turn determines where the mantle has the properties of cold, strong lithosphere in contrast to weak, hot asthenosphere. The relatively high elevations of mid-ocean ridges, compared to surrounding seafloor, are explained by the very thin nature of the lithosphere along divergent plate boundaries. Most of the mountainous region of the western United States surprisingly lacks a corresponding thick crust; it is underlain instead by unusually buoyant mantle.

Mountain-building processes at convergent-plate boundaries include the accretion of blocks of crust to the edges of continents. These exotic blocks of crust, which originated far from the continent where they are now found, are too thick to subduct completely. When these blocks reach a subduction zone, they are shoved onto, or jammed beneath, the continent. Continents grow through time by the accretion of these exotic terranes at convergent boundaries.

The craton is the geologically most stable, interior part of a continent. Although the North American craton has only slightly deformed over the last half billion years, it is characterized by exposed and shallowly buried expanses of ancient metamorphic and igneous rocks that originated near Precambrian subduction zones. Most of the continental crust seen today originated more than a billion years ago as a result of convergent-boundary magmatism and collisional accretion of nonsubductable blocks of crust. The old Precambrian rocks of the cratons were once deep below the surface in the thick crustal roots of ancient mountains. Over the course of geologic time, the mountains eroded as isostasy kept raising the crust. Eventually, continental crust of the stable cratons achieved the normal thickness and an average elevation slightly higher than long-term average sea level.

This chapter completes your study of deformation within and at the surface of Earth. However, we have still not explained all of the characteristics of Earth's surface. Isostasy and, especially, tectonic forces driven by motion within the planet determine the general form of surface features, but water, wind, ice, and even living organisms sculpt the varied landscapes of Earth. To complete your understanding of how Earth works, you now embark on studies of Earth surface processes—where the forces of the geosphere interact with the hydrosphere, atmosphere, and biosphere.

Active Art

Glacial Isostasy. See how the crust isostatically responds to the growth and shrinkage of glaciers.

Extension Module

Extension Module 13.1: Measuring Uplift Rates. Learn how geologists measure mountain-uplift rates.

Confirm Your Knowledge

1. Define "relief." What is the total relief exhibited on Earth's surface? What is the relief within the state where you live?
2. What is the principle of isostasy?
3. How does isostasy determine surface elevation?
4. Explain how icebergs are used as analogies for surface elevations.
5. How does isostasy relate erosion and deformation to changes in surface elevation?
6. How were the elevations of the Himalayas surveyed in the mid-nineteenth century? How were the errors in Everest's survey data interpreted by John Pratt and George Airy?

7. Are either the Pratt or Airy models correct in understanding isostasy? What are the limitations to each model?
8. What is flexural isostasy? How is flexural isostasy different from the Pratt and Airy models? Use flexural isostasy to explain glacial rebound.
9. How do changes in glacier volume on land affect sea level?
10. How does the age of the seafloor affect sea level?
11. How are mountains formed at convergent boundaries?
12. Why are the highest mountains 200 to 400 kilometers inland of oceanic-continental convergent plate boundaries rather than right at the boundary?

13. Cite two examples of mountain belts formed by continental collision.
14. Explain why the mountains of the Western Cordillera of North America exhibit high elevations but do not have deep crustal roots.
15. What is the observed range of mountain uplift rates? How does this range in values compare with the speed of plate motion?
16. Continents tend to have the oldest rocks in the center with younger rocks flanking the center. Explain this observation, with reference to cratons and exotic terranes.

Confirm Your Understanding

1. Write an answer for each question in the section headings.
2. What information do you need to calculate the pressure (weight) of a column of rock? Calculate the pressure at the base of oceanic and continental crust for columns of rock that are one square meter in area at the surface.
3. If the mantle has a density of 3000 kg/m^3 and a section of crust has a density of 2000 kg/m^3, what percent of the crust would stick up above the top of the mantle and what percent would stick down into the mantle?
4. Explain two lines of evidence for how we know that mountains have roots.
5. Why does a mountain develop a root as well as height?

6. What are some of the problems with an isostatic model of crustal elevations that relies on wood blocks floating in water as an analogy?
7. What evidence did the most recent ice age provide us to evaluate the merits of the Pratt-Airy isostasy and flexural isostasy models?
8. The Appalachian Mountains formed by continent-continent collision 300 million years ago. Explain why they persist as a mountain range today despite 300 million years of erosion.
9. Explain why the two elevations, around 0.5 kilometers above and 4.5 kilometers below sea level, are the most common elevations on Earth?
10. Describe the "life cycle" of a mountain in terms of geologic processes.

CHAPTER 14

Soil Formation and Landscape Stability

Why Study Soils?

Earth-surface processes, such as soil formation, take place at the interface of the geosphere with the hydrosphere, atmosphere, and biosphere. Soils are the weathering products of rocks that remain in place at Earth's surface. All other weathering products erode or dissolve away, usually to be recast elsewhere as sediment. The many soil varieties are fascinating mixes to dig into, to build on, to farm, or to use to grow a garden.

Soil is essential to our survival. Agricultural economies depend on fertile soils for success, and they fail when the nutrients are exhausted or the soil erodes away. Engineers must be aware of the strength of soil, to ensure that the natural soil foundation will support the weight and function of any structure being built. Degradation of soil is a universal problem, from contamination to soil erosion. Soil is a nonrenewable resource, and, as such, the study of soil and its conservation will always be important.

Geologists study soils as guides to the history of landscapes. Soil properties slowly mature through time. Erosion or sediment deposition changes landscapes faster than the soil matures. Mature soils, therefore, cover only the landscapes that neither erode nor accumulate sediment.

You may take soil for granted. Language offers quite a few disparaging words about "dirt." Soil, however, is not just "dirt." It is a critical factor in human existence and a key to understanding the history of Earth's landscapes.

After Completing This Chapter, You Will Be Able to

- Describe soil characteristics, including soil horizons.
- Explain the factors controlling soil formation.
- Relate soils to landscape processes and human activities.

Pathway to Learning

14.1 What Is Soil?

14.2 What Distinguishes Soil Horizons?

14.3 How Do Soils Form?

14.4 What Factors Determine Soil Characteristics?

This view in eastern Washington emphasizes that soil covers most of Earth's land surface and is the most essential ingredient to successful agriculture.

14.5 What Are the Types of Soils?

14.7 How Do Human Activities Affect Soils?

14.6 How Do We Know . . . That Soils Include Atmospheric Additions?

IN THE FIELD

While on a virtual drive in Ohio, you stop to look at rocks exposed in an old quarry. You notice that hard rock does not extend all the way to the top of the quarry wall. Instead, a loose mixture of sand and pebble-size rock fragments, along with fine-grained clay and decomposing plant material, sits just below the surface, as shown in **Figure 14.1**a. This loose mixture is what you commonly refer to as "dirt," but you take the time to notice several interesting features. The dirt bears distinct color bands, with gray shades near the top, and orange to red colors below. Trees and shrubs grow in the dirt, and accumulations of dead leaves, twigs, and small roots seem to account for the gray color close to the surface. When you poke around with the tip of your pocketknife, the dirt falls apart in clods—loosely held together clumps of mineral grains and organic material. Where the loose dirt is moist, it is sticky, and you can roll it between your palms into balls. The sticky, clumping characteristics suggest the presence of clay minerals.

This unconsolidated material is "soil." It is the stuff that you enjoyed digging in as a child, that is stirred up in the preparation of planting a garden or a farm field, and that is piled up during excavations for building foundations or highways. You see soil almost everywhere at the ground surface, but you may not have given much thought to how it came to be there.

On another occasion, you examine soil in a sand pit in the New Mexico desert, which is illustrated in Figure 14.1b.

This soil overlies loose, stream-deposited gravel and sand, with distinct bedding that curiously disappears upward. There is some reddish color in the nonbedded soil zone, similar to but not as red as the red band at the Ohio quarry, but the desert soil lacks the gray band containing abundant organic matter at the top (Figure 14.1b). Hard, white, mineral nodules are a conspicuous feature of the desert soil (Figure 14.1b). You can easily scratch the white nodules with your knife but not with your fingernail. The hardness of the mineral is appropriate for calcite. When you place a drop of weak acid on a nodule, it fizzes, confirming that it is calcite.

Your observations lead to a number of questions. Soils link to weathering processes, but how does weathering account for the observed features of the soils? These features include the color bands seen in the soils in both locations, the tendency of both soils to fall apart in clods, and the presence of calcite in the desert soil. How does soil form, and what does it form from? How do soils relate to minerals and rocks and to the landscapes the soils cover? Why do the two soils illustrated in Figure 14.1 look different? If soils differ in physical and mineral properties, do these properties relate to their suitability to support growing crops? How is soil significant to understanding geological processes? Why is soil, rather than bare rock, the most common material found right at the land surface?

▶ Figure 14.1 What soil looks like.

14.1 What Is Soil?

Soil means different things to different people. Engineers define soil to include any surface material that is not solid rock. In this usage, soil includes recently deposited sediment that has not yet consolidated into rock, as well as loose material resulting from rock weathering. Soil scientists evaluate soil usefulness for different agricultural functions, so they define soil as the medium for plant growth. The term "topsoil" refers to just that upper part of the soil disturbed by crop cultivation or shoveling in a backyard garden. Clearly, there is no single, everyday definition of soil.

A Geologic Definition of Soil

Your field observations (Figure 14.1) suggest that soils contain loose mineral and organic materials and have subtle, colored layers and are somehow related to the weathering of sediment or rock. These conclusions are consistent with how geologists define **soil** as a layered mixture of loose mineral and organic constituents that have different physical or compositional properties (or both) than the original, nonweathered material. Chemical and physical weathering processes convert preexisting materials into soil. Plants that take root in the soil and animals that live within it also influence

Red soil on top of loose rock rubble

Gray band

Orange-red band

About 0.5 m

(a) Soil exposed in a rock quarry in Ohio

No bedding visible in reddish soil

Bedding visible in sand and gravel

About 0.5 m

White calcite coating on pebbles and sand grains

(b) Soil exposed in a sand pit in New Mexico

the chemical and physical weathering processes. Soil formation, therefore, occurs close to the surface and links to the climate variables of moisture and temperature, which determine both chemical-weathering reactions and the type of vegetation living on the surface.

The subtle color bands visible in the field examples (Figure 14.1), are an essential component of the soil definition. These **soil horizons** are distinguished from one another by different particle sizes and mineral compositions. Your field observations suggest that horizons are not like sedimentary beds that accumulate one on top of the other. Instead, soil

horizons form in place during weathering and reveal different physical and chemical processes at different depths below the ground surface.

Plants and soils are linked. Plants grow up from the surface, and their roots penetrate down into the soil. Fallen leaves and dead stems accumulate on the soil surface and are mixed into the soil by burrowing insects and other animals, whose remains also accumulate on the surface and at shallow depth in the soil. Air and water mixed into the soil aid in plant growth.

Biologic activities affect the entire thickness of the soil, even if decaying plant organic matter is mostly just at the top. Every anthill you see is a reminder that animals living in soil, especially insects and worms, persistently move soil particles. The growth of plant roots also moves soil. The disruption of the original arrangement of sediment grains by developing plant roots and burrowing animals explains the lack of bedding where soil formed in the sedimentary deposit illustrated in Figure 14.1b.

Soil Forms from Parent Material

What is soil made from? At the quarry, soil seemingly formed in broken rock, whereas the desert soil formed in loose sediment (Figure 14.1). In general, then, geologic materials at Earth's surface are either solid rock or loose unconsolidated material that overlies solid rock. **Bedrock** describes large, continuous occurrences of solid rock; this term is distinguished from the term "rock," which by itself could just as easily describe a small fragment that you pick up from the ground. Loose fragments of weathered bedrock, rather than bedrock itself, cover most of Earth's surface. All unconsolidated deposits overlying bedrock comprise **regolith**, a term derived from the Greek *rhegos*, meaning "blanket," and *lithos*, meaning "stone." Some regolith is simply fragments from underlying or nearby outcrops of rock that were dislodged by physical weathering but have not moved, or at least have not moved far; this type of regolith overlies the quarry bedrock in Figure 14.1a. Other regolith is sediment, the remains of weathered rock transported from their place of origin, as seen in the excavation in Figure 14.1b. The regolith from which a soil forms is called the **parent material** for the soil.

We can restrict our definition of soil to that part of the regolith that contains distinct soil horizons. **Figure 14.2** illustrates the differences among bedrock, regolith, and soil. The upper boundary of a soil is the atmosphere, but the lower boundary is sometimes indistinct. In most cases, the downward decrease in intensity of weathering and plant activity is gradual. Soil

▲ **Figure 14.2 Visualizing soil and parent material.** This labeled photograph shows the components of the near-surface environment that are important to understanding soil formation. Regolith is unconsolidated rock and organic material that overlies bedrock. Soil is that part of the regolith that is modified by chemical weathering and biological activity. The parent material is what weathers to produce the soil. Soil commonly exhibits horizons of different physical and compositional properties, which usually appear as color differences, too.

scientists commonly assign a maximum thickness of two meters to a soil, and this assigned lower boundary usually encloses most plant roots and the most active weathering processes.

Putting It Together—What Is Soil?

• Soil is naturally occurring horizons of mostly loose mineral and organic constituents formed by weathering and biologic processes.

• The soil parent material is usually regolith rather than hard bedrock. Regolith is either fragments generated by physical weathering of underlying rock, or loose sediment transported from elsewhere. Soil is the part of the regolith that has horizons.

• Soil horizons differ from one another and from parent material by contrasting grain size, mineral composition, or both.

14.2 What Distinguishes Soil Horizons?

Horizons are essential to the definition of soil, so it is important to understand why horizons exist and how they relate to soil formation. The first step is to determine the characteristics that distinguish each soil horizon from its neighbors. Explanations for how the horizons form must be consistent with these characteristics.

Even brief field examination of soil (Figure 14.1) suggests at least three horizons:

A. A top horizon that is sometimes gray because it contains organic matter

B. A middle horizon that can be reddish in color and contain clay

C. A bottom horizon that is not very different, if at all, from the parent material.

These are the three most common horizons, simply lettered A, B, and C, from top to bottom. Some soils contain additional horizons, labeled O and E. **Figure 14.3** illustrates and summarizes the key features of the five soil horizons, and **Figure 14.4** identifies horizons in soils that formed in contrasting forest and desert environments, such as those at the Ohio and New Mexico field sites.

Horizons Close to the Surface: A, O, and E

Organic matter is most abundant near the top of the soil where biologic activity is highest and dead leaves and branches accumulate beneath the plants that grow in the soil. Organic matter produces the distinctive "earthy" odor of moist soil. The presence of organic matter mixed with mineral and rock fragments defines the **A horizon**. Compared to the parent material, A horizons also lack the minerals that are most susceptible to dissolution and transformation to new minerals by chemical weathering (such as the iron and magnesium silicates, micas, and soluble minerals such as calcite). Where plant growth is dense, as in a forest, for example, there may be so much litter of decaying leaves, conifer needles, and wood on the ground that the surface horizon consists only of organic matter without minerals and thus forms an **O horizon** ("O" for "organic").

In some cases a distinctly paler, perhaps even white, horizon with few or no colored minerals, clay, or organic matter exists just below the A horizon (Figures 14.3 and 14.4). The absence of organic matter, easily weathered minerals, and weathering products such as clay, oxide, and hydroxide

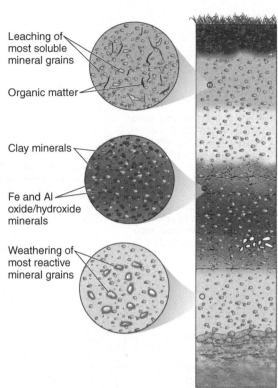

O horizon
Surface horizon of organic residues from dead plants and animals in varying stages of decomposition. Color is dark in shades of gray and brown.

A horizon
The uppermost horizon composed of minerals; can be the surface horizon. Contains decomposed organic matter, especially in wetter climates, which darkens this horizon compared to lower horizons. Readily dissolved minerals are less abundant than in lower horizons.

E horizon
Light-colored horizon lacking clay, organic matter, and easily weathered minerals.

B horizon
The horizon of maximum accumulation of clay minerals, and iron and aluminum oxide and hydroxide minerals. In dry climates, calcite may accumulate in this horizon, too. Compaction and shrinking and swelling of clay minerals leads to aggregates of soil minerals into clods. Iron and aluminum oxides and hydroxides typically impart shades of yellow, orange, or red, depending on mineral type and abundance. Calcite forms white nodules and layers.

C horizon
The regolith below A and B horizons that exhibits little if any evidence of weathering; commonly the parent material for the soil. The color of this horizon is determined by the colors of the original minerals; oxidation may produce pale yellow and orange coloration.

Bedrock

▲ **Figure 14.3 Describing soil horizons.**

▲ **Figure 14.4 What soil horizons look like.** These photographs contrast the horizons of a forest soil (left) with a desert soil (right). The forest soil has an O horizon of decaying leaves that overlies the dark A horizon. There is also a light-colored, highly weathered E horizon. The B horizon is shades of orange because of the accumulation of iron-hydroxide minerals and clay. The desert soil has a thin, light-colored A horizon without very much organic matter. The B horizon is orange-brown at the top, because of accumulation of iron-hydroxide minerals and clay, and has a lower, light-colored zone containing white calcite.

minerals are key features of this horizon. This is the **E horizon**, which receives its label from the word "eluviation," derived from the Latin word roots *e* and *lavere*, which mean "to wash out." Some E horizons contain nothing but quartz, which is the most weathering-resistant, common mineral in rocks.

The Middle Horizon of Mineral Accumulation: B

The next lowest horizon in the field examples (Figures 14.1 and 14.4) is reddish in color, sometimes contains calcite, and is sticky when wet. Stickiness suggests the presence of clay minerals that attach to one another because of stray electrical charges around the grain boundaries. The coloration and the abundance of clay and calcite (when present) are not characteristics of the parent material. This is the **B horizon**, which is most distinctive for colors and textures indicating accumulation of minerals that are not present in the parent material.

The accumulation of minerals in the middle part of the soil defines the B horizon (Figure 14.3). The presence of hydroxide and oxide minerals containing iron, aluminum, or both, accounts for the distinctive red, orange, and yellow hues of B horizons. Clay minerals are abundant in many B horizons, even where clay is absent from the parent material. Although soluble minerals such as calcite are rare or absent in the A horizon, they can be more abundant in the B horizon than in the underlying parent material, but only where the climate is dry (Figure 14.4). In microscopic view (Figure 14.3), all of the added minerals—oxides, hydroxides, clays, and sometimes calcite—form coatings on the other mineral and rock fragments in the soil.

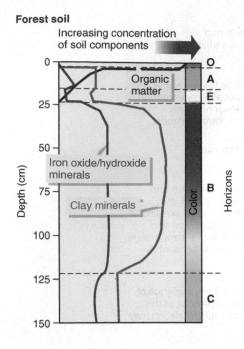

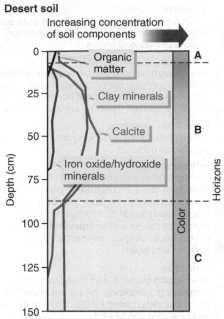

Organic matter concentrates in the surface horizons, including an **O** horizon in the forest soil.

A and **E** horizons have lower mineral abundances than the **C** horizon, indicating subtraction of components from the upper horizons.

The **B** horizon contains the greatest abundances of iron oxides and hydroxides, clay minerals, and calcite (in the desert soil).

◀ **Figure 14.5 Comparing the composition of soil horizons.** This diagram compares and contrasts the distribution of various components in a humid-region forest soil and an arid-region desert soil. Using the parent material in the C horizon as the starting composition, it is apparent that mineral components are subtracted from the A and E horizons and added in the B horizon.

The Bottom Horizon of the Soil: C

The bottom horizon of the soil is the least modified horizon and commonly is the parent material for soil formation. This is the **C horizon**, which is regolith that lacks the organic matter found in the A horizon and contains little if any of the red colors, clay content, or calcite occurrences that define the B horizon.

Soil Horizons Result from Mineral Additions, Subtractions, and Transformations

Figure 14.5 graphically summarizes data collected from different soil horizons. Assuming that the composition of the C horizon approximates the composition of the parent material, then comparisons of the O through B horizons to the C horizon reveal the soil-forming processes that add to, subtract from, or transform the parent material. *Added* components or those *transformed* from the parent material by chemical weathering reactions are more abundant in the upper horizons compared to the C horizon. *Subtracted* components are less abundant in the upper horizons compared to the C horizon. Formation of soil horizons, therefore, relates to chemical and biologic processes that remove components from some levels in the soil and add components or transform them into new minerals at other levels. The defining characteristics of the soil horizons, therefore, provide important insights into soil formation:

- Some horizon colors imply addition of constituents not present in the parent material. The gray color of the O and A horizons, the reddish colors lower down, and the white calcite nodules in the desert soil are characteristics of the soil that do not appear in the parent material.
- Organic matter is added to make O and A horizons.
- The most readily weathered minerals are subtracted from A and E horizons.
- Oxide, hydroxide, clay minerals and, in dry regions, soluble minerals such as calcite appear in B horizons by addition by transformation during weathering reactions, or both.

Putting It Together—What Distinguishes Soil Horizons?

- Soil horizons are distinct because of differences in organic matter and mineral content that result from additions, subtractions, and transformations of minerals resulting from chemical weathering, and addition of organic matter.

- A and O horizons contain added organic matter. A horizons also show evidence of mineral subtraction.

- The E horizon, where present, is a nearly white horizon defined by an absence of organic matter, clay, colorful oxide and hydroxide minerals, or easily weathered minerals present in the parent material.

- The B horizon is notable for additions of colorful oxide and hydroxide minerals and clays. Calcite is present in arid-region B horizons.

- The C horizon is weakly weathered, unconsolidated regolith below the zone of accumulated minerals defining the B horizon.

14.3 How Do Soils Form?

The material additions, subtractions, and transformations that explain soil horizon characteristics imply active physical and chemical processes that convert parent material to soil. **Figure 14.6** summarizes these processes.

Explaining Mineral Subtractions from A and E Horizons

The A and E horizons contain partly dissolved minerals and lack the most easily dissolved minerals altogether. This is the result of chemical weathering, which primarily involves dissolution, oxidation, and hydrolysis reactions. (Section 5.1 explains these reactions.)

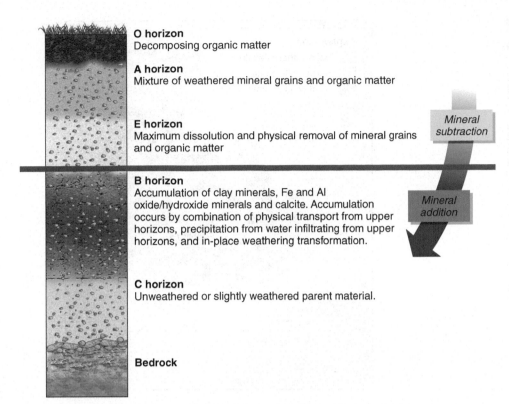

O horizon
Decomposing organic matter

A horizon
Mixture of weathered mineral grains and organic matter

E horizon
Maximum dissolution and physical removal of mineral grains and organic matter

Mineral subtraction

B horizon
Accumulation of clay minerals, Fe and Al oxide/hydroxide minerals and calcite. Accumulation occurs by combination of physical transport from upper horizons, precipitation from water infiltrating from upper horizons, and in-place weathering transformation.

Mineral addition

C horizon
Unweathered or slightly weathered parent material.

Bedrock

▲ **Figure 14.6 Explaining soil horizons.** Weathering is most intense near the surface because of moisture availability and natural acidity. Minerals are subtracted from the A and E horizons by dissolution and physical movement of small particles by downward-infiltrating water. Minerals are added to the B horizon by both chemical and physical processes.

Why do minerals dissolve away near the surface? Natural acids and organic compounds offer an explanation. Carbon dioxide in the atmosphere, or respired into the soil through plant roots, makes a weak acid when mixed with water infiltrating from rain and snowmelt. Although organic matter is always present in the A horizon, it also easily dissolves as quickly as it accumulates. You witness the solubility of organic compounds whenever you make tea or coffee; the hot water dissolves some of the organic solids to color and flavor the water. Decaying and dissolving organic matter also generates additional acids (such as vinegar and citric acid) and other organic compounds that enhance mineral dissolution. Minerals containing calcium, sodium, magnesium, and potassium held in crystal structures by relatively weak ionic bonds are most susceptible to dissolution by these acids, so these minerals are the first to be removed from the A horizon.

The evidence of mineral dissolution in surface horizons but the low abundances of weathering products in these same horizons mean that two things happen:

1. Weathering reactions chemically remove soluble chemical components from soil minerals, and infiltrating water then flushes the dissolved ions downward.

2. Solid weathering products of oxidation and hydrolysis reactions (e.g., clays and iron oxides) physically wash downward with infiltrating water, so that only the most weathering-resistant minerals and persistently accumulating organic matter remain.

E horizons represent the most intense flushing out of reactive components and weathering products. Water-soluble organic compounds in the E

horizon leach aluminum and iron from crystal structures, which even breaks down relatively resistant clay minerals. As a result, E horizons are clay poor and have less aluminum and iron than the parent material (see Figure 14.5). Even organic matter dissolves almost completely from E horizons, and only the most strongly weathering-resistant minerals, such as quartz, remain. The intensity of chemical and physical removal observed in E horizons requires considerable throughput of water and lots of reactive organic acids. These requirements are usually only met in humid forests, which explains why E horizons only exist in some forest soils (Figure 14.3).

Explaining Soil Fertility and Infertility

The dissolving of minerals in the A and E horizons also plays a critical role in soil fertility. Natural soil nutrients such as potassium, phosphorus, iron, and trace metals are released from minerals during weathering. Plants absorb these nutrients where they dissolve in water or bond to organic molecules. Plants rarely are able to draw the elements directly from mineral crystals, so the parent material must weather in order to be fertile. Too much weathering, however, completely removes the nutrient elements from the soil, unless the plant tissues that absorb nutrients return these components to the soil through their own decay.

Soils become infertile if the mineral nutrients extracted by weathering and plant growth are not returned as recycled organic matter. Infertility results from long periods of intense weathering that not only remove all of the nutrients from the minerals, but also oxidize and dissolve all of the nutritious organic matter. Intense agricultural activity hastens extraction of nourishing elements from the soil. In addition, most of the crops are removed from the land, so that the nutrients are not returned to the soil by plant decay. As a result, even the richest soils gradually lose their fertility and require addition of fertilizer to sustain agricultural productivity.

Explaining Mineral Additions and Transformations in the B Horizon

Although water flushes the most soluble mineral components from the whole thickness of a soil, the B horizon is primarily the site of mineral addition and mineral transformation, rather than subtraction (Figures 14.5 and 14.6). Water infiltrating down to the B horizon brings large quantities of dissolved ions and tiny clay particles resulting from mineral weathering in the A and E horizons. However, there are limits to the ion-carrying capacity of the water, so minerals begin to precipitate from solution in the B horizon, especially when the water evaporates. Additional weathering in moist B horizons transforms parent-material minerals into weathering products that remain where they form.

Three processes, therefore, are observed to account for accumulation of colorful iron oxide and hydroxide minerals and clay minerals in B horizons:

1. Minerals precipitate in the B horizon from water that contains large concentrations of the ions that dissolved from minerals in the A and E horizons.

2. Hydrolysis and oxidation of feldspars and iron- and-magnesium-bearing silicate minerals in the B horizon form clay, oxide, and

▲ **Figure 14.7 Soil particles combine to form aggregates.** Soil falls apart into clods that are loosely held together by fine roots and the cohesion between clay minerals. Open cracks separate the soil aggregates.

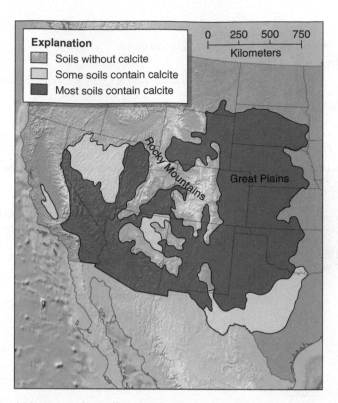

▲ **Figure 14.8 Where soil contains calcite.** Soils where calcite accumulates in the B horizons are found in the western Great Plains and in lower-elevation areas of the southwestern United States. Soil calcite only accumulates where the climate is both dry and seasonally hot. Infiltrating moisture leaches calcite from the soil where precipitation is abundant or where cool temperatures diminish evaporation even where rainfall is relatively sparse.

hydroxide minerals in the moist, oxidizing environment. Infiltrating water carries away only the most soluble ions.

3.   Tiny clay grains drain downward with infiltrating water from the overlying horizons.

The abundance of clay minerals in B horizons explains particle aggregates, such as the soil "clods" that you observed in the field. **Figure 14.7** illustrates cracks in the B horizon that allow the soil to be pried apart into aggregates, several centimeters across. The soil aggregates are a mild form of consolidation (discussed in Section 5.5) caused by the electrical attraction of clay minerals, which commonly have stray electrical charges around the periphery of individual crystals. When clay minerals press close to one another, they tend to adhere and stick together.

The particle aggregates and the cracks between them are important features of B horizons. The cracks between soil aggregates are important pathways for downward transport of water and clay particles from above. Washed-in clay particles coat the surfaces of the cracks, and roots preferentially follow the cracks.

Explaining Calcite in Desert Soil

Calcite is a common ingredient of arid-region B horizons (see Figures 14.4 and 14.5), and **Figure 14.8** shows how common calcite is in the dry areas of the western United States. The presence of calcite indicates that downward-percolating soil water cannot flush out all of the soluble ions. Calcite dissolves when naturally acidic water moves through the upper soil horizons and only precipitates again as mineral crystals when ion concentrations in the water are very high. These high ion concentrations only occur because of evaporation, or nearly complete withdrawal of soil moisture by plants.

These observations explain why calcite most commonly accumulates in B horizons that form in dry climates. In the contrasting humid regions there is almost always sufficient water draining through the soil to prohibit precipitation of soluble calcite, and the calcium and carbonate ions are

▲ **Figure 14.9 Calcite cements soil into rock.** Over time, enough calcite accumulates in desert soils to cement the soil particles into solid rock, which is popularly called "caliche." Calcite cemented most of the B horizon in this soil in southwest Texas.

flushed into deeper ground water. This is why calcite is absent from the B horizons of the example moist forest soils in Figures 14.1 and 14.4.

In some cases, there is enough precipitation of calcite to cement the soil particles into hard rock, as illustrated in **Figure 14.9**. These soils are not easily excavated for foundation construction, plants have difficulty extending deep roots, and water cannot readily percolate downward. Some geologists consider this white, hard, rocky part of the soil a different horizon, called a "calcic horizon," but soil scientists do not distinguish a separate horizon. The term "caliche" is popularly used to refer to well-cemented calcic B horizons, but this term is not formally used by geologists or soil scientists. Calcite is simply another example of mineral accumulation in B horizons, but soil calcite is restricted in occurrence by the availability of moisture.

Putting It Together—How Do Soils Form?

• Chemical-weathering processes form soils, as indicated by the mineral additions, subtractions, and transformations that define soil horizons.

• Mineral dissolution and downward flushing of dissolved ions and fine-grained weathering products are characteristic of A and E horizons, where water enters the soil from rain or snowmelt and biologic activity forms acids and organic compounds that enhance mineral weathering.

• The B horizon contains colorful oxide and hydroxide minerals and clays that physically wash down from overlying horizons, precipitate from ions dissolved in the overlying horizons, or form in place by weathering reactions. Calcite precipitates in arid-region B horizons when water evaporates.

14.4 What Factors Determine Soil Characteristics?

What factors account for the overall differences noted in the forest and desert soils illustrated in Figure 14.1? The grayer color of the A horizon above the quarry, compared to the desert soil, suggests a greater accumulation of plant organic matter because of greater plant abundance in the more humid climate. The presence of calcite in the desert soil, but not in the forest soil, is consistent with accumulation of soil calcite in arid regions but not in humid regions. Vegetation and climate, then, are two factors that affect soil characteristics. Geologists and soil scientists identify additional factors, including the mineral content and texture of the parent material, location on irregular landscapes, and the length of time that the soil has been forming. **Figure 14.10** summarizes the relationships between these soil-forming factors and soil characteristics, which are explored in the following paragraphs.

The Role of Parent Material

Soils result from weathering, so the weathering of different rocks should produce different soil characteristics. Minerals have varying susceptibilities to chemical weathering and they break down into different weathering products. For these reasons, the parent material from which soil forms must affect the composition of soil.

For example, clay minerals form relatively quickly in soils whose parent material contains abundant feldspar, mica, or iron- and magnesium-bearing silicate minerals that typically weather to clay (see Table 5.1). Clay is already abundant in fine-grained sedimentary deposits or shale that forms the parent material for other soils. On the other hand, parent material composed mostly of quartz sand will not weather to produce a thick, clay-rich B horizon.

Another example of the significance of parent material for soil characteristics is volcanic ash, which covers large areas near active volcanoes. The glassy ash particles dissolve in water to provide essential elements for vegetation growth. Plants extract these nutrients from water that passes through the soil and weathers the ash. This explains why agriculture in tropical, volcanically active regions, such as Central and South America, Indonesia, and the Philippines, focuses close to the volcanoes in those areas.

The thickest fertile soils develop on sedimentary deposits. In contrast, by the time weathering forms thick regolith on bedrock, most of the mineral nutrients are already leached out of the soil. Fertile soils of the upper Midwest and Northeast United States developed in widespread sedimentary deposits formed during the last ice age, which ended about 10,000 years ago. The glaciers left behind a blanket of crushed rock that was also widely redistributed by rivers and wind. These deposits have not yet weathered sufficiently to remove mineral nutrients.

The Role of Climate

Rainfall and temperature are key variables in soil formation. Moisture is required for weathering reactions to take place, and the volume of water

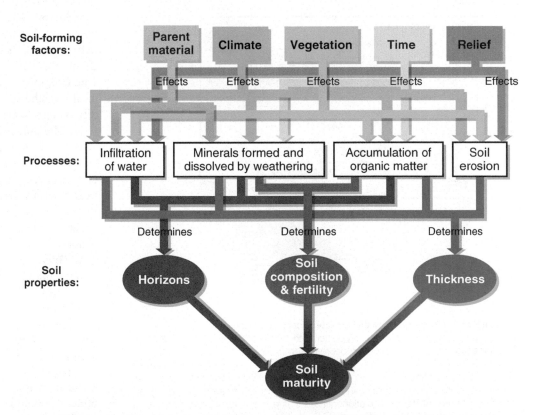

▲ **Figure 14.10 Links between soil-forming factors, processes, and soil properties.** Follow the arrows to see how the soil-forming factors affect the processes determine soil properties.

moving down through the soil determines the depths to the boundaries between horizons. Chemical-weathering reactions happen faster in the high temperatures of the hot tropics, whereas the freeze-thaw process in colder regions enhance physical weathering to break rock into more weatherable fragments.

If all other variables affecting soil formation are the same in humid and arid regions, then you can predict important differences in the soils formed in these regions. These differences are seen at the field sites (Figure 14.1) and portrayed in Figure 14.5. In the humid region there is more mineral dissolution in near-surface horizons and greater accumulation of clay, oxide, and hydroxide minerals in the B horizon. As a result, the humid-region soil has a redder and more clay-rich B horizon, and a more heavily weathered and leached A horizon when compared to the desert soil. The aridity of the desert soil causes the formation of a calcite-rich zone in the B horizon that is not present in the humid-region soil (Figure 14.8).

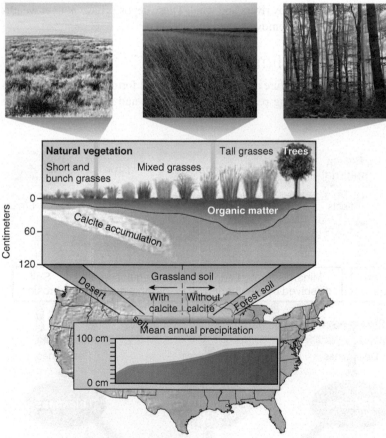

▲ Figure 14.11 **How vegetation and climate affect soil horizons.** Soil characteristics change with observed changes in vegetation and precipitation along a line from Wyoming to Wisconsin. Precipitation decreases from east to west, so vegetation changes from forest in Wisconsin, to grassland in the plains states, to desert scrub in northeastern Wyoming. Organic-rich A horizons are thickest below the tall-grass prairie of Minnesota. Calcite accumulates in the B horizon only in the western, dryer region. The depth to the calcite accumulation increases as precipitation increases, because the dissolved calcite is carried deeper into the soil where there is more water moving through the soil.

The Role of Vegetation

The effects of climate and vegetation on soil formation are strongly interconnected. This relationship exists because climate determines the type and abundance of vegetation. For example, **Figure 14.11** illustrates how soil-horizon properties change between the upper Great Lakes and the east slope of the Rocky Mountains. Rainfall decreases westward across this region and, as a result, forests in the east give way to grasslands and desert shrubs farther west. The different vegetation types determine the thickness and richness of organic matter in the A horizon. Dense mats of fine roots generate thick, organic-rich A horizons in grassland soils. Forest soils have O horizons of leaf and needle litter. There is very little organic matter in most desert soils, simply because vegetation is scarce.

The Role of Time

Up to this point, we have examined and interpreted soil characteristics at single snapshots in time, associated with particular characteristics of parent material, climate, and vegetation. However, even when these soil-forming factors remain the same, chemical weathering progressively alters the parent material, so we expect soil properties to change. Geologists describe this by saying that soils mature through time.

Most natural changes in soil properties are not recognizable on human time scales. It is challenging, therefore, to understand the role of time in determining soil properties. To understand how soils mature, geologists compare properties of soils that started forming at different times and are, therefore, of different ages. For example, soil does not form on sedimentary deposits until after sediment deposition, so comparison of soils formed on deposits of different ages reveals how soils change over time.

Soils of different ages differ mostly in the thickness and mineral content of horizons because these properties relate to the extent of weathering. **Figure 14.12** illustrates changing soil characteristics over time in grassland, forest, and desert environments. Very young, immature soils exhibit very little chemical weathering and contain only a thin B horizon, if any; these soils may simply contain an A horizon resting on a C horizon. Old, mature soils have well-developed horizons, with removal of all easily weathered minerals from upper horizons, and extensive mineral additions to a thick B horizon. Soil thickness generally increases through time. The longer that water moves through the regolith, the deeper the effects of chemical weathering.

Weathering effects add up. The abundance of easily weathered minerals decreases in progressively older soils, and the clay content increases in B horizons. Increasing clay content usually slows water infiltration so that water remains in the soil longer to benefit plants that require abundant water. Calcite accumulation, in dryer climates, gradually cements the lower B horizon into rock. Water does not easily infiltrate this rock-hard B horizon, which means that rainfall more likely runs off to streams rather than soaking into the ground.

How long does it take to form soil? There is no straightforward answer to this commonly asked question. The rates of soil formation strongly link to parent material, climate, and vegetation. This means that soils with very similar physical and mineral properties may be of different ages in different places. The formation of an A horizon on recently exposed regolith may require only a few decades. Soils containing B horizons more than a meter thick may represent tens of thousands of years of weathering and mineral additions. On average, however, mature soils require at least several thousand years to form. This means that in terms of human lifetimes, soils that erode away are nonrenewable resources.

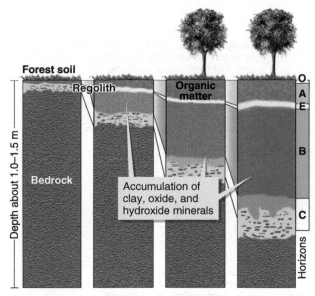

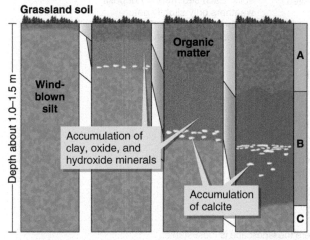

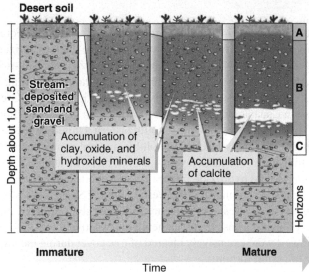

Forest soil

Depth about 1.0–1.5 m

Regolith

Bedrock

Organic matter

Accumulation of clay, oxide, and hydroxide minerals

O A E B C — Horizons

An immature forest soil consists of an **A** horizon developed on unweathered regolith. Over time, the **B** horizon thickens and an **E** horizon develops where organic acids infiltrated from thick **O** and **A** horizons. Weathering converts more and more of the original rock to regolith, which is further modified into soil.

Grassland soil

Depth about 1.0–1.5 m

Wind-blown silt

Organic matter

Accumulation of clay, oxide, and hydroxide minerals

Accumulation of calcite

A B C

Grassland soils gradually develop a thick, organic-rich **A** horizon. In dry regions, calcite accumulates in the **B** horizon, and the depth to calcite accumulation increases through time.

Desert soil

Depth about 1.0–1.5 m

Stream-deposited sand and gravel

Accumulation of clay, oxide, and hydroxide minerals

Accumulation of calcite

A B C — Horizons

Desert soil **A** horizons are thin and low in organic content throughout their development. Calcite increases in abundance through time and eventually cements part of the soil into hard rock.

Immature ————————————→ Mature

Time

▲ **Figure 14.12 How soils mature.** The diagrams show changes in soil thickness and properties through time for three different climatic and vegetation situations. Different starting materials are also depicted to show how soil formation modifies parent material. Soil maturity is most quickly recognized by increased soil thickness and increased reddening of the B horizon over time.

The Role of Landscape Relief

The irregular topography of Earth's surface also plays a role in soil development. Mature soils require adequate time for soil processes to generate thick B horizons and extensively weathered minerals. Mature soils, therefore, require land surfaces that are undisturbed, neither eroding nor being buried under new sediment layers, for many thousands of years.

Figure 14.13 shows how landscape relief affects soil development. Thick soil cannot form on steep slopes, because erosion removes regolith before chemical weathering substantially modifies it into soil. The material eroded from the hillside accumulates at the base of the slope. This means that regolith is thicker at the base of the hill than along the hillside, but the persistent deposition of material at the bottom of the slope continually buries the land surface and interrupts soil formation. Figure 14.13 shows that the most mature soil forms on flat surfaces where neither erosion nor deposition takes place.

Flat areas near rivers tend to lack mature soils, however, because new increments of sediment bury the soil each time the river floods (Figure 14.13). Soil formation starts over at the upper surface of the newly deposited sediment, only to stop and start over when the next flood occurs. On a positive note, the newly deposited sediment replenishes the surface with the soluble nutrient elements that provide high soil fertility.

Geologists use soil maturity as a gauge of landscape disturbance. Where the land surface is underlain by a mature soil, the landscape is stable because there has been neither erosion nor deposition for a long time. Where soil is thin or immature, a geologist infers that the landscape has changed at some point in the recent past, because the current land surface has not existed long enough for mature soil to form. This observation suggests high rates of erosion or deposition that may make a location unsuitable for buildings and highways.

Putting It Together—What Factors Determine Soil Characteristics?

• The factors determining soil characteristics are parent material, climate, vegetation, time, and landscape relief and location.

• Parent material determines which minerals dissolve and which precipitate during chemical weathering.

• Climate and vegetation are closely linked variables in soil formation. Vegetation is denser in moister climates and the combination of greater moisture and greater development of organic acids from decaying vegetation leads to soils with the most distinctive horizons.

• Soils mature through time, because the effects of chemical weathering add up. Soils generally become thicker and contain more clay or calcite or both in their B horizons with increasing age.

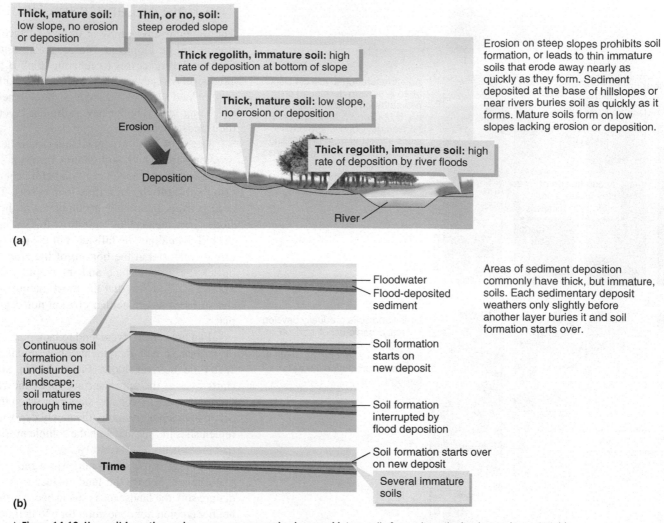

Thick, mature soil: low slope, no erosion or deposition

Thin, or no, soil: steep eroded slope

Thick regolith, immature soil: high rate of deposition at bottom of slope

Thick, mature soil: low slope, no erosion or deposition

Erosion

Deposition

Thick regolith, immature soil: high rate of deposition by river floods

River

Erosion on steep slopes prohibits soil formation, or leads to thin immature soils that erode away nearly as quickly as they form. Sediment deposited at the base of hillslopes or near rivers buries soil as quickly as it forms. Mature soils form on low slopes lacking erosion or deposition.

(a)

Floodwater
Flood-deposited sediment

Continuous soil formation on undisturbed landscape; soil matures through time

Soil formation starts on new deposit

Soil formation interrupted by flood deposition

Time

Soil formation starts over on new deposit

Several immature soils

Areas of sediment deposition commonly have thick, but immature, soils. Each sedimentary deposit weathers only slightly before another layer buries it and soil formation starts over.

(b)

▲ **Figure 14.13 How soil formation varies across an uneven landscape.** Mature soils form where the landscape is most stable.

• Unchanging or very slowly changing landscapes have mature soils. Unstable landscapes, where rates of erosion or deposition are faster than rates of soil formation, are marked by immature soils or no soil.

14.5 What Are the Types of Soils?

Variations in soil-forming factors, including parent material, climate, vegetation, relief, and time, create different soil types. The properties of different soils determine agricultural uses and productivity. Most countries where agriculture forms a major part of the economy use classification schemes based on characteristics that are affected by the soil-forming factors. The classifications vary from country to country because of regional differences in climate and parent materials. **Table 14.1** presents a simple, informal classification of soil, along with the formal U.S. Department of Agriculture terms. **Figure 14.14** illustrates the locations of these soil types in North America. Brief consideration of soil types allows you to integrate the knowledge of soil horizons and soil-forming factors in previous sections.

Forest Soils

Soils in temperate-zone forests, such as those found in eastern North America and parts of the mountain West (Figure 14.14), have several common ingredients (Table 14.1 and Figures 14.1, 14.4, and 14.5). Forest soils commonly have thin A horizons; instead they have thick O horizons composed of decaying leaves and conifer needles. B horizons are clay-rich.

Figure 14.15 illustrates how horizon development varies according to soil moisture. Very moist forest soils in New England, the upper Great Lakes, and eastern Canada form in cool, wet climates, so they have E horizons. Forest soils in the warm, rainy southeastern United States are highly oxidizing and strongly leached of soluble mineral components. As a result, these soils have very red, clay-rich B horizons with abundant iron and aluminum oxide and hydroxide minerals. Forest soils in the Midwest, Mississippi Valley, and mountains of the western United States form where precipitation is not as great, so the soils are less strongly weathered than those in wetter climates. These less weathered and relatively fertile forest soils in the Midwest were extensively cleared of tree cover during the nineteenth century to permit farming.

Rainforest Soils

The most intense chemical weathering occurs in hot, wet, tropical rainforests. In the United States, these conditions occur only in small areas of Hawaii and Puerto Rico, but rainforest soils are common in South America and Africa, as shown in **Figure 14.16**. The large volume of water that passes through the soil, along with the high acidity resulting from the decaying plant matter, strongly weathers almost all minerals, even including quartz. In the most

Table 14.1 Characteristics and Fertility of Major Soil Types

Soil Type	Characteristics	Fertility	U.S. Department of Agriculture Soil Order Names
Forest soils	Usually include an O horizon, a thin A horizon, and a clay-rich B horizon. An E horizon is present where the weathering by organic acid is intense.	Fertility depends on the extent of weathering to dissolve and remove mineral nutrients. Soils that form under the wettest, warmest, and most acidic conditions are the least fertile.	Alfisols—moderately leached, fertile soil; forms in humid climate. Ultisols—highly leached, low-fertility soil; forms in humid to tropical climates. Spodosols—highly leached, low fertility soil with E horizon; forms in cool, humid forests.
Rainforest soils	Highly weathered soils that contain very little organic matter, and consist largely or entirely of iron and aluminum oxides and hydroxides with clay minerals.	Soils are infertile because mineral nutrients are dissolved and removed and organic matter is not retained in the soil.	Oxisols
Grassland soils	Thick, dark, organic-rich, A horizons are typical. B horizons vary considerably and may or may not contain calcite.	Soils are very fertile primarily because of retention of organic-matter nutrients.	Mollisols
Desert soils	Light-colored, organic-poor A horizons are typical. B horizons vary considerably in clay content and color, but typically contain calcite. Degree of mineral weathering is less than for most other soil types.	Soils are fertile when irrigated but otherwise are found in areas where the climate is typically too dry for productive agriculture. Organic content is low but the soils contain mineral nutrients that are not readily dissolved in the dry conditions.	Aridisols
Wetland soils	Very dark, organic-rich soils with more than 20% plant organic matter. The soil is very lightweight when dry because of the abundance of low-density organic particles.	Soils are very nutrient rich but may be difficult to farm because of water-logged conditions.	Histosols
Soils with weakly developed horizons	Weak horizon development because of limited weathering, either because soil is very young or forms in a cold climate where weathering rates are very slow. Some soils lack horizons because soil constituents are mixed so that horizon boundaries are destroyed.	Soils are commonly fertile and support productive agriculture with adequate moisture and growing season because mineral nutrients have not been removed by weathering.	Entisols—very slightly weathered to unweathered soil in very recently deposited or rapidly eroded parent material. Inceptisols—slightly weathered soil with formation of a slightly reddish, clay poor B horizon. Andisols—weakly weathered young soil formed on recently deposited volcanic ash. Gelisols—weakly weathered soil in very cold climates where soil is frozen at least part of the year. Vertisols—soils formed in clay-rich parent material that shrinks and swells by wetting and drying; shrink and swell physically mixes horizons as organic material falls deep into the soil shrinkage cracks.

extreme cases, only iron and aluminum oxides and hydroxides remain, with varying proportions of clay minerals. These oxide and hydroxide minerals may be sufficiently abundant to cement the soil into rock.

Rainforest soils are very infertile, because the intense dissolution and oxidation destroy most of the organic matter along with mineral nutrients. Agriculture in the tropics commonly involves clearing a region that is only suitable for farming for a few years. This necessitates moving on to clear new sections of rainforest at frequent intervals. This agricultural practice plays an important role in deforestation, especially in South America.

Rainforest soils can form economically important mineral deposits. Soils rich in aluminum hydroxide minerals are the primary sources of aluminum, a versatile, lightweight, flexible, corrosion-resistant metal. Aluminum ore is called **bauxite**, a multicolored mixture of aluminum hydroxide minerals (Figure 14.16). Bauxite mining occurs in the modern tropics in South America, Africa, and Jamaica, and also in areas covered by rainforest in the recent geologic past. Both the world's largest bauxite deposits, in northern Australia, and smaller deposits in Arkansas, are ancient rainforest soils. Some iron-oxide-rich tropical soils are mined as iron ore.

Grassland Soils

Grassland soils exist throughout the center of the United States (Figure 14.14) and are especially fertile for growing wheat, corn, and soybeans. Grasslands form in climate zones that are drier than forested regions, so grassland soils are not strongly leached of soluble mineral components compared to forest soils. The closely spaced network of thin roots, along with fast-decaying small grass blades, leads to the formation of thick, dark A horizons as seen in **Figure 14.17** (also see Figure 14.12). The abundant

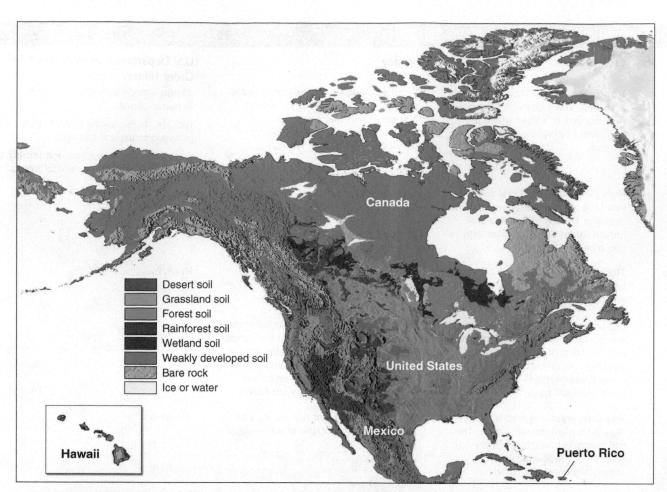

▲ **Figure 14.14 Distribution of soil types in North America.** Soil types in North America closely relate to climate and vegetation, and to the maturity of soil development. Immature soils form in cold areas (where weathering is very slow), on steep mountainous terrain (where erosion removes regolith before it is deeply weathered), and along rivers (where soil formation is interrupted by deposition during floods).

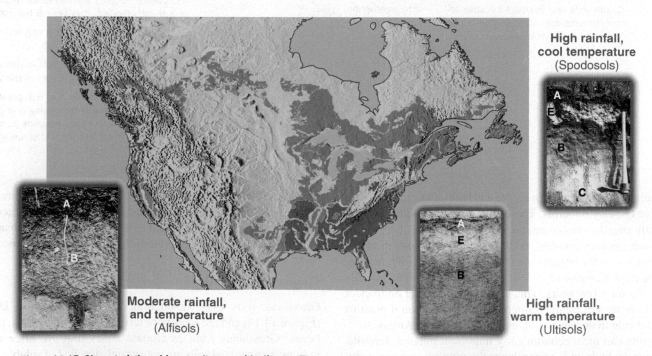

▲ **Figure 14.15 Characteristics of forest soils vary with climate.** This map shows the distribution of three forest soil types in North America. In areas where soil moisture is highest, soils have the best developed E horizons and reddest B horizons, which indicate the most intense weathering of minerals near the surface and accumulation of minerals at depth. Warmer temperature promotes more development of red iron oxide minerals.

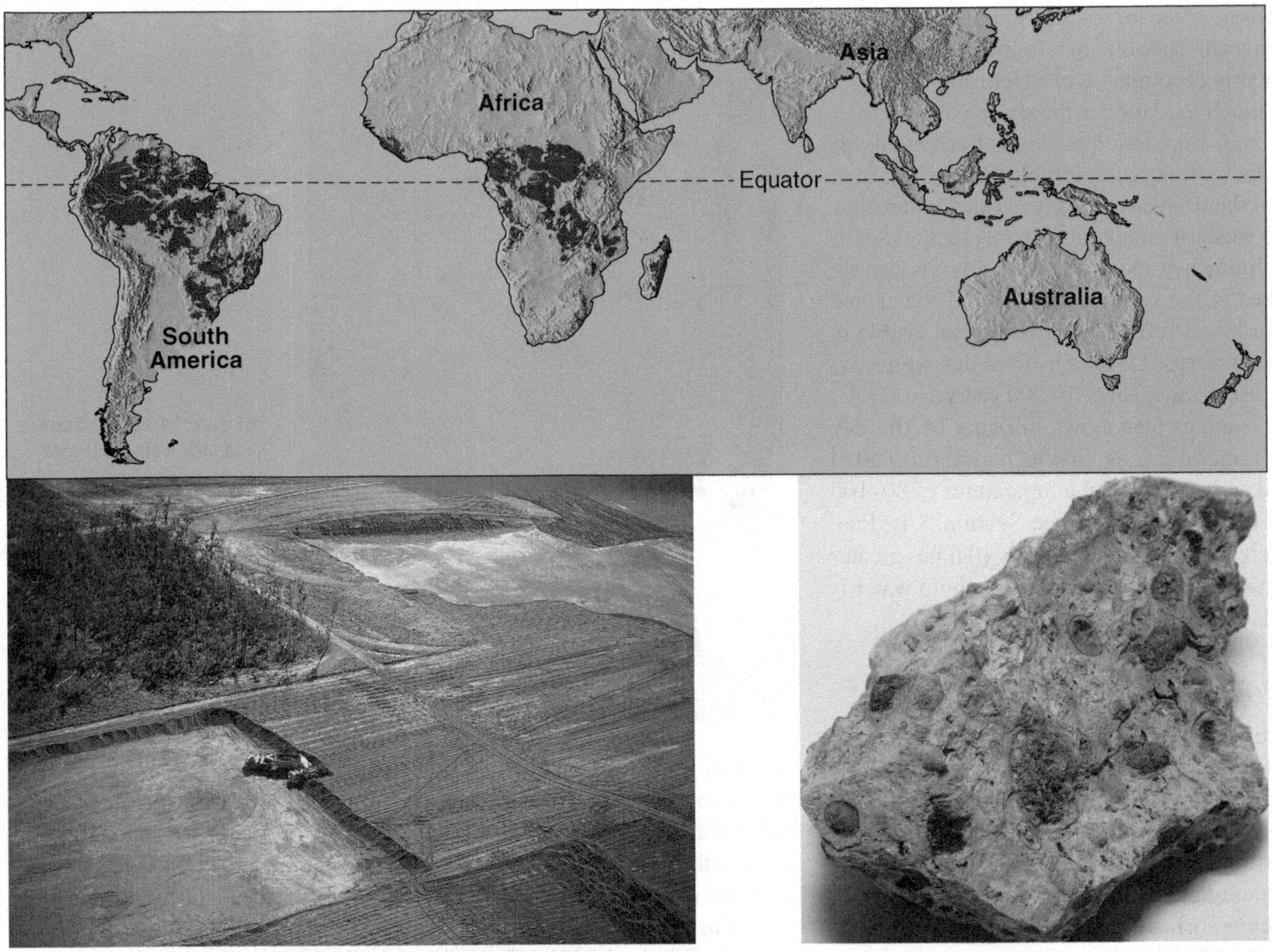

◄ **Figure 14.16 Rainforest soils are the most weathered soils.** The map shows the distribution of the most strongly weathered soils on Earth. These soils mostly form in the near-equatorial rainforests of Africa and South America where rainfall is heavy and temperatures are very warm year-round. Almost all of the parent-material minerals and organic matter are dissolved away, producing a soil that is very rich in brightly colored iron and aluminum hydroxide and oxide minerals. Modern and ancient rainforest soils are mined for the aluminum ore bauxite, composed primarily of aluminum hydroxide minerals. The photos show a bauxite mine in Australia and a sample of bauxite.

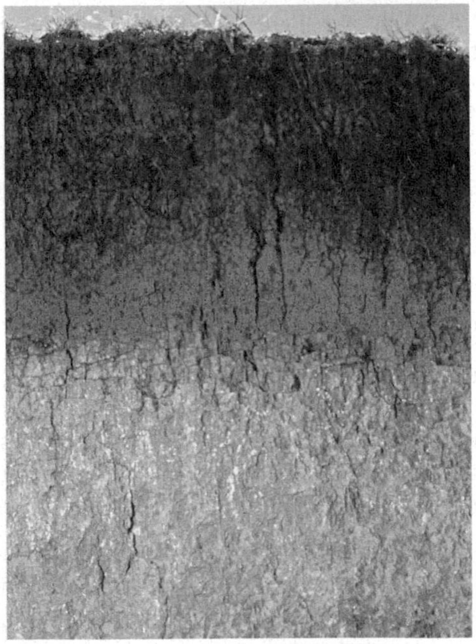

▲ **Figure 14.17 Grassland soils have dark and light horizons.** This excavated soil from Kansas shows the dark, organic-rich A horizon that commonly forms grasslands, where dense networks of fine roots and abundant decaying organic material contribute substantially to the composition and texture of the soil. Limited weathering in a relatively dry climate produces a light-colored B horizon that includes streaks of white calcite and only minor accumulation of iron oxides or hydroxides.

organic matter provides much of the nutrient base for successful agriculture. Grassland soils in the United States straddle the boundary between climate regimes that are appropriate for, or inappropriate for, the formation of calcite-rich B horizons, so some grassland soils contain calcite and others do not (see Figures 14.8 and 14.11).

Desert Soils

Desert soils are common in the western United States (Figure 14.14), where they form in a dry climate with very sparse vegetation. As a result, the soils (a) contain very little organic matter in the A horizon, (b) have very thin and clay-poor B horizons unless the soil is very old or formed by weathering clay-rich parent material, and (c) commonly contain calcite in the B horizon. Figures 14.1, 14.4, 14.5, 14.9, and 14.12 illustrate these features. Limited moisture means that mineral nutrients are not strongly leached out of desert soils, but nutritious organic content is very low. Agricultural production depends on irrigation, because of aridity, and fertilizers, because of low soil fertility. Although some of the most productive farmland in the world is irrigated desert soil, the soil quality degrades by the accumulation of evaporite minerals, including halite, from evaporation of irrigation water in the hot, dry climate.

Wetland Soils

Marshy wetlands are at the opposite climate and vegetation extreme from desert soils. Wetland soils commonly consist entirely of O and A horizons.

The soil is water saturated so there is no place for soil water and dissolved ions to percolate toward to form B horizons. The soil contains more than 20 percent organic matter, primarily as poorly decomposed plant remains. The abundance of organic carbon prevents much oxidation in the water. Not only does this mean that there are no oxidizing weathering reactions to create oxide and hydroxide minerals, but the lack of oxygen prohibits further decay of plant residues. North America's wetland soils are mostly in Canada and New England, and in small areas of the western mountains, where precipitation is abundant and cool temperatures diminish evaporation. Wetland soils also are present locally as coastal marshes in the southeastern United States and alongside streams and lakes. The largest areas of wetland soils visible in Figure 14.14 stretch discontinuously across Canada in areas that were covered by lakes near the end of the last ice age, about 10,000 years ago.

The black, organic-rich soil, such as that shown in **Figure 14.18,** may contain more plant residue than mineral grains, in which case it is called **peat** (or peat moss). When buried and subjected to temperatures of 80–100 degrees centigrade, peat transforms to coal (also see Section 5.4). Peat holds as much as four times its weight in moisture, which is 10 times greater than the wettest mineral soils. This extraordinary capacity to hold water is why peat is added to potting soils for houseplants.

Soils with Weakly Developed Horizons

Large areas of North America are covered by soils with weakly developed horizons, regardless of climate and vegetation type (Figure 14.14). Weakly formed horizons, such as seen in **Figure 14.19**, exhibit only subtle color contrasts, little or no evidence of mineral additions and subtractions, and easily weathered minerals still exist in the A horizon. In most cases these immature soils are not old enough to exhibit strongly developed horizons because they exist on landscape surfaces that are frequently disturbed by erosion or deposition (Figure 14.13).

Regions experiencing frequent ash deposition downwind of active volcanoes also have soils without distinctive horizons, because soil formation starts over every time a new layer of thick ash accumulates. These soils are

◄ **Figure 14.19 Immature soils lack well-developed soil horizons.** This excavation exposes an immature soil where weathering has not modified the parent material. Organic material forms a thin, dark A horizon at the surface, but the soil lacks horizons with different mineral contents.

common in Hawaii and in the northwest United States near the active volcanoes of the Cascade Range. Volcanic soils are typically fertile because nutritious elements that are essential for plant growth are released more rapidly by weathering glassy volcanic ash than crystalline minerals. Volcanic soils support robust agriculture all around the Pacific Ocean "Ring of Fire."

The widespread presence of immature soil in Alaska and northern Canada (Figure 14.14) reflects the very slow rates of mineral weathering and organic matter decay in very cold climates. In many places, water is permanently frozen in these soils, which greatly decreases the rates of weathering reactions.

◄ **Figure 14.18 Wetland soils are mostly organic matter.** Large amounts of plant material accumulate in marshy wetlands and swamps along with only minor, if any, mineral sediment. The soil is saturated in water and contains very little oxygen, so the organic matter does not readily decay. The abundant organic matter accounts for the very dark color of this illustrated wetland soil. Notice the water ponded in the bottom of the excavation, which demonstrates that the soil is saturated like a sponge.

Putting It Together—What Are the Types of Soils?

• Soil types relate to the different soil-forming factors in different locations.

• Forest soils usually contain less organic matter but more clay-enriched B horizons than do grassland soils. Desert soils contain the least organic matter of all soils, and clay accumulates slowly in the B horizon because of limited chemical weathering. Desert soils and grassland soils in relatively dry areas contain calcite in the B horizon.

• Rainforest soils are very strongly weathered and usually infertile because of a nearly complete dissolution of organic matter and mineral nutrients. Highly weathered tropical soils are the source of bauxite, which is mined to produce aluminum.

• Wetland soils are composed primarily of organic matter.

• Immature soils, with only faintly developed horizons, form where there has been insufficient time for significant weathering. Most immature soils indicate unstable landscapes or areas too cold for weathering reactions to occur effectively.

14.6 How Do We Know . . . That Soils Include Atmospheric Additions?

Picture the Problem

How Can Geologists Explain Calcite-Rich Soil? A problem emerges when examining a thick zone of calcite in a desert soil such as that seen in Figure 14.9. For each molecule of calcite in the soil, there must have been a molecule of calcium liberated by weathering within the parent material. The calcium dissolves from the parent material and is then incorporated in the precipitated calcite. In many cases, however, tests show that the parent material contains very little calcium, so it is difficult to account for the large amount of calcite present in the soil.

Figure 14.20 illustrates the problem of a calcium-rich soil that formed in calcium-poor, igneous-rock regolith. Weathering would need to dissolve as much as 100 meters of rock to account for the mass of calcite in the soil. Other observations summarized in Figure 14.20 argue against such extensive chemical weathering, however. For instance, the presence of vesicular, broken up rock typical of the top of a lava flow rules out the possibility that a great thickness of the lava flow weathered away. In addition, although calcium dissolves from silicate minerals in the rock, the much more abundant silicon and other elements should remain in relatively dry desert soil. If 100 meters of rock weathered to release the required amount of calcium, then there should be an immense, thick residue of insoluble minerals at the surface, and this is not the case. These observations indicate that the composition of the parent material cannot closely match the composition of the regolith in the C horizon or the underlying bedrock. What, then, weathered to form the soil? Calcite-rich B horizons are common over large areas of the western United States that lack calcium-rich rock (Figure 14.8), so this is an important problem to solve.

State the Hypothesis

Is Calcium Delivered in Dust and Rainfall? Any visitor to an arid region is immediately impressed with, and possibly distressed by, the abundance of windblown dust. Deserts are dusty because the surface is dry, and there is very little vegetation to protect soil particles from blowing away or to blunt the force of strong wind blowing across the surface.

Of importance to our problem of calcite in soil, scientists hypothesized that accumulating dust provides continuous additions

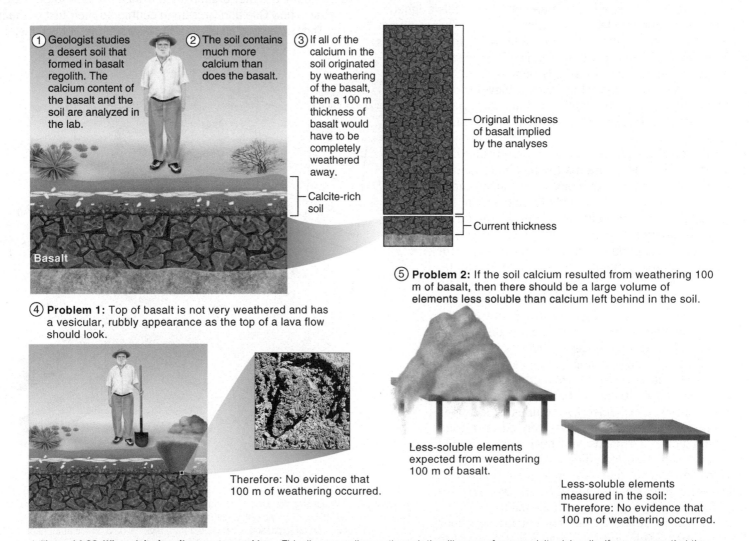

① Geologist studies a desert soil that formed in basalt regolith. The calcium content of the basalt and the soil are analyzed in the lab.

② The soil contains much more calcium than does the basalt.

③ If all of the calcium in the soil originated by weathering of the basalt, then a 100 m thickness of basalt would have to be completely weathered away.

Calcite-rich soil

Basalt

Original thickness of basalt implied by the analyses

Current thickness

④ **Problem 1:** Top of basalt is not very weathered and has a vesicular, rubbly appearance as the top of a lava flow should look.

Therefore: No evidence that 100 m of weathering occurred.

⑤ **Problem 2:** If the soil calcium resulted from weathering 100 m of basalt, then there should be a large volume of elements less soluble than calcium left behind in the soil.

Less-soluble elements expected from weathering 100 m of basalt.

Less-soluble elements measured in the soil: Therefore: No evidence that 100 m of weathering occurred.

▲ **Figure 14.20 Why calcite in soil presents a problem.** This diagram walks you through the dilemma of many calcite-rich soils, if you assume that the soil formed by the weathering of only the underlying rock. To account for the amount of calcite in the soil requires an unrealistic amount of rock weathering and would leave behind in the soil a far greater abundance of relatively insoluble elements than actually exist.

of calcium, and other ions, to the surface soil horizons. The wind-blown dust may originate far from where the soil develops and can, therefore, contain minerals different from those of the underlying parent material. There was also speculation that rainfall delivers dissolved calcium ions. Infiltrating water then dissolves calcium delivered by dust and rainfall from the surface A horizon and carries it downward to precipitate as calcite in the B horizon. The parent material composition, in other words, is not constant but receives new components from the atmosphere while the soil forms.

Collect the Data

How Much Calcium Is Present in Dust and Rainfall? To test the hypothesis, Leland Gile and Robert Grossman, of the U.S. Soil Conservation Service (now the U.S. Natural Resources Conservation Service), conducted dust-sampling experiments over a 10-year period in the 1960s. Eight trays, called dust traps, were set up more than 1 meter above the ground surface in the southern New Mexico desert to collect dust, which was then analyzed for calcium content. Gile and Grossman showed that between 10 and 125 grams of dust fall on each square meter of the desert surface each year after taking into account the size of the pans in the traps and the period when dust was collected.

The geology of the experiment area consists of felsic igneous rocks and sediment eroded from those rocks. These parent materials contain very little calcium, and what little exists is found in low-solubility silicate minerals. This means that dust with calcium did not simply blow up into the traps from the nearby surface, but instead blew in from some distant location where regolith contains dust-sized grains of water-soluble calcium-bearing minerals, such as calcite or gypsum.

All of the dust samples contained calcium that dissolves in water; **Figure 14.21** graphs the results of laboratory analyses of calcium abundance. To allow easier comparison to the amounts of calcite present in the nearby soils, the calcium contents were calculated as the equivalent amount of calcite that could precipitate from the calcium supplied in the dust. Six of the eight traps yielded similar amounts of equivalent calcite, between 0.35 and 0.55 g/m^2/year (grams per square meter of ground surface per year). The other two dust traps yielded much higher values, near 1.3 g/m^2/year.

The two very high values are clear outliers on the data plot (Figure 14.21). Outlier values are common in scientific data collection, and it is important to know whether they represent measurement error or are naturally explained. Gile and Grossman determined that the data were correct but were explained by the unusual locations of these two dust traps. One trap was very close to sand dunes, which probably explains why it accumulated up to ten times as much dust, including calcium, as the other seven traps. The other outlier data point came from a trap located where eroded soil containing calcite was exposed at the surface, so this trap had a potential calcium source that the other traps did not. Given the unusual locations for these two traps, Gile and Grossman felt it best not to include these large calcite contents in their data analysis. Averaging the values from the other six traps they

concluded that about 0.5 grams of calcite accumulates per square meter of desert surface each year.

Next, Gile and Grossman considered how much calcium, as a building block for making calcite, might be delivered by rain. Rainwater contains ions dissolved from dust in the atmosphere (the primary source of calcium), volcanic gas, and air pollution. Analyses of rainwater in southern New Mexico indicate that each liter of water contains about three thousandths of a gram of dissolved calcium ions. This is an extremely small concentration, about equal to dissolving one teaspoon of calcium in 400 gallons of water. Nonetheless, even in this desert region there is sufficient rainfall during an average year to deliver about 1.5 grams of calcite to each square meter of the soil. Adding this amount to the dust-trap calcium abundances, and excluding the two data outliers, indicates that the total equivalent calcite added by atmospheric dust and rainfall is about 2.0 g/m^2/year (see Figure 14.21).

Use the Data

Can Atmospheric Calcium Additions Account for Soil Calcite? The data show that wind and rain supply calcium to soils as hypothesized, but is the modern atmospheric calcium addition sufficient to account for the amount of calcite seen in the local desert soils? **Figure 14.22** illustrates how Gile and Grossman continued their test of the hypothesis. They excavated all of the soil below a 1-meter-by-1-meter square

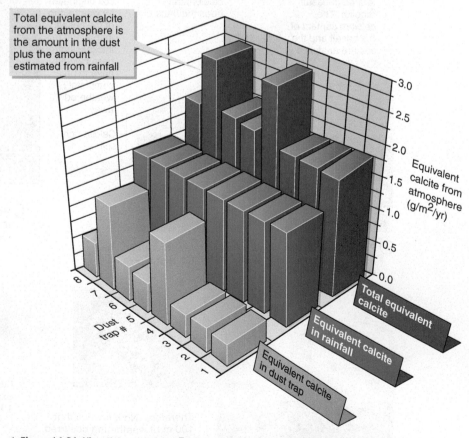

▲ **Figure 14.21 Visualizing the data.** Each dust trap accumulated calcite-equivalent calcium at different rates, as indicated by the yellow columns in the graph. The contribution of calcite-equivalent calcium from precipitation is calculated from regional data on the amount of calcium dissolved in rainfall, and is shown by the blue columns representing the same value at each location. The total calcite contribution from atmospheric sources, shown by the purple columns, is the sum of the calcium in the dust and rainfall.

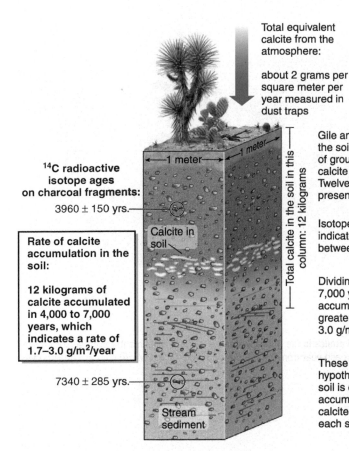

Total equivalent calcite from the atmosphere:

about 2 grams per square meter per year measured in dust traps

14C radioactive isotope ages on charcoal fragments:

3960 ± 150 yrs.

Rate of calcite accumulation in the soil:

12 kilograms of calcite accumulated in 4,000 to 7,000 years, which indicates a rate of 1.7–3.0 g/m²/year

7340 ± 285 yrs.

Calcite in soil

Stream sediment

Total calcite in the soil in this column: 12 kilograms

Gile and Grossman excavated all of the soil below a one square meter plot of ground and measured the total calcite content by laboratory analyses. Twelve kilograms of calcite are present in the soil column.

Isotope ages of charcoal fragments indicate that the soil started forming between 4,000 and 7,000 years ago.

Dividing 12 kilograms by 4,000 and 7,000 years indicates that calcite accumulated at an average rate greater than 1.7 g/m²/yr and less than 3.0 g/m²/yr.

These values are consistent with the hypothesis that all of the calcite in the soil is explained by the present-day accumulation of about 2 grams of calcite-equivalent dust that falls on each square meter each year.

▲ Figure 14.22 **A simple calculation tests the hypothesis.**

on the surface. The calcite content was measured in the laboratory and found to be about 12 kilograms. If the hypothesis is correct, then nearly all of the 12 kilograms of calcite should result from atmospheric addition of calcium ions over the history of soil formation.

To determine the duration of soil formation, the soil scientists separated charcoal fragments from the soil and from underlying stream sediment. The ¹⁴C radioactive-isotope dating method provided ages of about 4000 years and 7000 years for these charcoal samples (Figure 14.22). The older charcoal was deposited in the stream sediment *before* the soil formed, and the younger charcoal may have been mixed into the soil *while* the soil formed. This suggests that the soil formed over at least 4000 years but no more than 7000 years. To account for the 12 kilograms of calcite over these times requires addition of somewhere between 1.7 and 3.0 grams per year, on average, into the square meter of surface where the soil was excavated.

The dust trap and rainfall data match well with the observed calcite content of the soil. The measured atmospheric addition of 2.0 g/m²/year compares closely to the rate of calcite accumulation in the soil, ranging between 1.7 and 3.0 g/m²/year.

Insights

Do Soils Form at a Steady Rate? The dust-trap data match up reasonably well with the amount of calcite in the soils to support the hypothesis that explains calcite accumulation where parent material appears to be calcium deficient. You can ask, however, whether it is valid to compare 10 years of dust-trap data with the amount of calcite that accumulates in a desert soil over thousands of years. This problem frequently arises in geologic studies where measurements cannot be

made over the same time spans as geologic processes lasting thousands or even millions of years.

Is the rate of calcite accumulation and, therefore, the rate of soil formation constant over thousands of years? If not, then do the dust-trap data adequately test the hypothesis? There is no reason to expect that calcite accumulates in southern New Mexico soils at a constant rate of 2 grams over each square meter each and every year for thousands of years. Climate changes over short intervals likely change the rate of calcium addition to the soil. The amount of accumulating dust and calcium is probably greater during dry times. During wetter times, the amount of calcium delivered in dust probably decreases, but more is delivered by rainwater, which contains much more calcium than does the dust. On the other hand, not all of the calcium in the rainwater soaks into the soil, because some of the water runs off the ground surface into streams.

The important thing to remember in a study of this type is not that the numbers add up exactly, because they do not. What is important is that the amount of calcium in dust and rainwater is simply "in the ballpark" of that required to account for the amount of calcite in the soil. Inasmuch as the parent material contains virtually no calcite, the hypothesis that the necessary calcium arrives from the atmosphere is reasonably supported by the data collected over a time interval that is necessarily much shorter than the natural process.

Putting It Together—How Do We Know . . . That Soils Include Atmospheric Additions?

• Soil parent material is not simply rock and regolith of unchanging composition. Windblown dust and rainfall deliver new chemical components to the surface, where they are chemically moved into lower soil horizons.

• Measurements of calcium in dust and rainfall support the hypothesis that calcite in most desert soil forms from atmospheric additions rather than the original parent material.

14.7 How Do Human Activities Affect Soils?

Soil fertility and soil conservation are essential considerations for successfully raising crops. Soils, however, form within natural ecosystems, rather than managed pastures and fields of specialized crops. Agricultural activities also disturb soil horizons by plowing and cultivating, and, in some cases, reshape the land surface, which changes patterns of water runoff and infiltration into the soil.

Soil erosion is the greatest threat to food production. Scientists estimate that 8000 km³ of soil has eroded away around the world during human history; enough sediment to cover Earth's entire surface to a depth of 6 centimeters. Each year about 2 billion metric tons of soil erodes away in the United States. **Figure 14.23** shows that this erosion affects most of the country, with some areas losing more than 2 millimeters of soil each year. Flowing water causes about two-thirds of this erosion, and the remainder results from blowing wind. The problem for agriculture is that more than half of the soil erosion in the United States is from croplands. Even where some soil remains, the upper horizons containing the nutrient-rich organic matter are commonly lost.

To what extent is soil erosion in agricultural regions the natural process of regolith erosion, and to what extent is the erosion enhanced by human activity? Landscape stability is important for soil formation (see Sections 14.3 and 14.4). Soil erosion results from landscape disturbance and the removal of weathered regolith from where it accumulated. To evaluate soil erosion, it is important to understand how landscapes of soil formation become landscapes of soil destruction.

Soil Erosion by Flowing Water

Agricultural activities disturb landscapes by increasing water runoff and erosion potential. Erosion in natural forests and grasslands is small compared to soil loss in the same regions after clearing of natural vegetation and conversion to crop production. Rainfall infiltrates efficiently into natural soils covered in thick vegetation and decaying organic matter, because plants and dead plant matter slow down

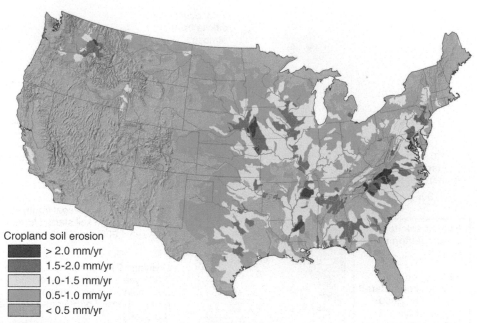

Cropland soil erosion

- ▓ > 2.0 mm/yr
- ▓ 1.5-2.0 mm/yr
- ░ 1.0-1.5 mm/yr
- ▓ 0.5-1.0 mm/yr
- ▓ < 0.5 mm/yr

▲ **Figure 14.23 How much soil erodes in the United States.** This map illustrates the average soil erosion that takes place in the country each year from cropland, which is farmland where soils are disturbed to grow crops.

◀ **Figure 14.24 Soil erodes from bare fields.** (a) Muddy water rushes from a farm field during a heavy rainstorm. (b) Shallow gullies formed in this field because of heavy rainfall prior to crop planting in the spring. After crops are well established, the surface-water runoff moves more slowly and is not as erosive. (c) Runoff from heavy rainfall on this overgrazed pasture eroded deep gullies. So much soil was removed that tree roots were undermined, causing them to topple.

water flowing across the surface and increase the opportunities for it to soak into the soil. In contrast, many farm fields lie bare and unused for parts of a year, or are only sparsely covered by plants sown in rows and then tilled to remove weeds. In addition, heavy farm implements moving through a field compact the soil and decrease its porosity. Infiltration is lower on bare, compacted ground than on loose vegetated soil covered in plant debris.

Lower infiltration means greater runoff from rainfall and snowmelt, and flowing water erodes the soil. The water may flow in shallow continuous sheets across the fields, or it may focus into small gullies where deeper flow enhances erosion. **Figure 14.24** shows how the flow of water through gullies removes nutrient-rich upper soil horizons from fields.

Soil Erosion by Blowing Wind

Wind erosion is most common in dry regions where agricultural production may already be marginal. Wind erodes sand- and silt-size particles

▲ **Figure 14.25 Soil erosion during the Dust Bowl of the 1930s.** (a) Vegetation died over large areas of the Great Plains during the 1930s drought, turning the region into a "Dust Bowl" of blowing soil. (b) During the largest dust storms, dark dust clouds towered more than 100 meters over the surface and obscured sunlight to cause total darkness.

▲ **Figure 14.26 Evidence of soil erosion.** The soil scientist points to where ground surface was when the grasses started to grow. Later, wind erosion of soil lowered the land surface to the level at his feet. To the left of the soil scientist is a light-colored sand dune. Strong winds frequently move the sand so that no vegetation is reestablished and there is no stable land surface for new soil development.

from the soil where the soil is dry and vegetation is sparse. Vegetation decreases the effectiveness of wind erosion by binding soil with roots and by providing obstacles that substantially decrease wind velocity. Wind erosion of soil generally happens during periods of drought when natural or crop vegetation dies off over large areas. The decreased soil moisture further enhances the likelihood of wind erosion because dry soil particles do not clump as well as moist ones.

Droughts are natural events, but agricultural activities may enhance wind erosion. In some locations, water-demanding crops replace native drought tolerant grasses. Parts of the landscape covered by living native grasses erode less strongly than cultivated areas where crops die for lack of moisture. In other cases, native grasses die or are removed by livestock grazing, which decreases plant cover during drought even among drought-tolerant plants.

Figure 14.25 shows dramatic examples of soil erosion by wind. These photographs show dust storms in the western plains of the central United States during a protracted period of drought in the 1930s. The combination of low rainfall and soil loss by wind erosion turned the region into what came to be known as the Dust Bowl. One sky-darkening windstorm stripped 350 million metric tons of soil (enough to fill more than 4 million railroad cars) from the Great Plains and carried it eastward where it fell like fine snow for two days along the Atlantic coast.

Figure 14.26 shows how to easily recognize soil loss by wind erosion. Plant roots bind soil particles and protect soil from erosion. In a sparsely vegetated landscape, therefore, the soil preferentially blows away in areas between plants, leaving the plants on noneroded pedestals.

Preventing Soil Erosion

Slope is an important factor in soil erosion. Water flows faster down steep slopes and causes the most erosion on the steepest slopes. Where possible, it is best to leave steep slopes undisturbed. Where necessary to farm in steep terrain, it is best to modify the slopes by making terraces, as shown in **Figure 14.27**. The flat terrace surfaces not only slow the water and decrease erosion, but also allow the water to soak into the soil to support plant growth rather than flowing downhill to streams.

Soil-erosion damage occurs even on gentle slopes of rolling hills. **Figure 14.28** shows how crop-planting practices can diminish this erosion. If crop rows run downhill, the water follows the rows and causes erosion. Corn rows shown in Figure 14.28 follow along the contours of the slopes rather than down the hillside. This planting practice allows the crops to interrupt downslope water

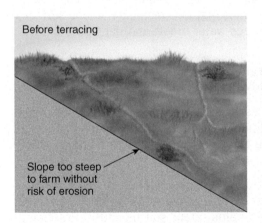

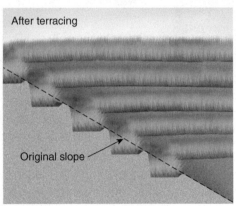

Before terracing

Slope too steep to farm without risk of erosion

After terracing

Original slope

◀ **Figure 14.27 Terraces diminish soil erosion on steep slopes.** The diagram shows how disturbance of vegetation and soil on a steep slope increases the risk of erosion by surface water runoff. A solution, shown on the right, is to cut terraces on the slope. Flat terrace surfaces slow down the runoff so that water infiltrates to support crop growth, rather than rushing downslope and eroding the soil. The photo shows a California vineyard established on a terraced hillside.

Crops planted
in strips along
contours of hillside

Corn

Grass planted
along drainage

Hay

◀ **Figure 14.28 Planting crops to decrease water erosion.** The pattern of crop planting on this hilly Iowa farm diminishes soil erosion. Crop rows are parallel to the contours of the hillsides, rather than running straight downhill where they would convey rapid and erosive water runoff. The alternation of corn, planted in spaced-out rows, with hay, which forms a dense cover of grass, also diminishes erosion because water flows very slowly through the hay. Grass planted along steeper drainage channels keeps water from flowing fast enough to erode gullies.

▲ **Figure 14.29 Windbreaks diminish wind erosion.** Rows of trees form windbreaks between these North Dakota wheat fields. The potential for water erosion is minimal, because the land is nearly flat. However, the soil is easily eroded by wind when the fields are bare between harvest and new growth. The windbreaks slow down the wind blowing along the ground surface and deflect the wind upward away from the surface; both of these effects decrease erosion of fine soil particles.

flow, which simultaneously decreases erosion and increases infiltration. The farmers also alternate the planting of corn with hay or grains, which, like natural grass, is not grown in wide rows but forms a continuous cover of vegetation. Water runoff is generally slow and non-erosive through the close-spaced grasses. Erosion in the row-crop part of the field can be further diminished during the winter months by leaving the unutilized crop residues in the field rather than plowing them under. This "no-till" farming practice decreases erosion because the plant remains slow down runoff rather than leaving bare soil exposed to be washed or blown away. The decaying plant remains also add nutrients back to the soil.

Windbreaks, like the rows of trees shown in **Figure 14.29**, diminish the erosive force of wind. The trees form obstacles to wind that both slow

down the wind speed and force the blowing air up and away from the land surface. By keeping the wind from blowing at high velocity across large distances of dusty fields, there is very little wind erosion of the soil, and soil particles are not transported out of the field.

Putting It Together—How Do Human Activities Affect Soils?

• Plowing and cultivating disturb soil horizons. Agricultural activities may also reshape the land surface, which increases water runoff and decreases infiltration into the soil.

• Bare areas, rows of sown crops, and compacted land surfaces leave significant parts of the field exposed to water runoff and soil erosion.

• The loss of vegetation, especially during drought, promotes soil erosion by wind. Dry conditions and the lack of vegetation decrease soil moisture and enhance the ability of wind to pick up and move soil particles.

• Terracing steep slopes and planting parallel to contours diminishes water runoff. Planting continuous covers of vegetation and trees serves as a windbreak and helps retain soil moisture. Practicing "no-till" methods also decreases erosion by inhibiting water runoff and restores nutrients to the soil.

Where Are You and Where Are You Going?

Soil is not just dirt—it is a mixture of organic matter and that part of surface regolith that has weathered but not been transported away as sediment. Soil not only is important for growing productive crops, but also is the source of aluminum ore and peat, the forerunner of coal. In addition, soil characteristics allow geologists to estimate landscape stability.

Soil formation is a complex series of processes that act on the starting parent material.

- *Mineral changes:* Chemical reactions, enhanced by the breakdown of organic matter, dissolve some minerals and provide the chemical building blocks to form new minerals.
- *Movement of soil components:* Infiltrating water and, to a lesser extent, burrowing organisms and plant roots move materials from one horizon down to another, usually from near-surface A, O, or E horizons into the B horizon.
- *Gain of materials:* Organic matter from decaying plants, dust from the atmosphere, and ions dissolved in rainwater are added into the soil as it forms.
- *Loss of materials:* Leaching of the most soluble components into ground water, and surface erosion by water and wind remove material from the soil.

The variations in the physical and compositional characteristics of soils from place to place are linked to critical soil-forming factors. These factors are

- the parent material that weathers to form each soil.
- the climate and vegetation that determine the chemical-weathering reactions that form the soil.
- the duration of the soil-forming processes that determines maturity and strength of horizon development.
- the relief and stability of the landscape, which determine how long a soil forms before burial or erosion. Mature soils are associated with stable landscapes, because well-developed soils will not form on landscapes that are actively eroding or being buried beneath sediment.

Agricultural activities may destabilize the landscape and cause soil erosion, which also removes nutrients. Planting agricultural crops in place of native vegetation, along with soil compaction by farm implements, commonly leads to greater water runoff, which erodes soil. Wind also removes the finer particles of soil, especially in dry seasons or during droughts in arid regions.

In the coming chapters you will learn about different processes that sculpt Earth's surface and reflect interactions among the geosphere, hydrosphere, atmosphere, and biosphere. These processes include the failure of hillsides by mass movements, the transport of water and sediment in rivers, the dynamics of glaciers, the action of waves on shorelines, and the movement of surface sediment by wind. In all of these situations, soil development plays an important role in understanding the landscape history, because soils record the stability of the landscape and the environmental conditions that have existed during the evolution of the landscape.

Confirm Your Knowledge

1. What is the geologic definition of "soil"? How does it differ from other definitions?
2. What is the difference between the formation of soil horizons and sedimentary layers?
3. What are the principle soil horizons? What processes dominate in each horizon?
4. Why is soil development critical to the ability of a soil to support plant growth?
5. Identify the factors that affect the overall characteristics of a soil.
6. Which soil horizon contains an abundance of soil aggregates? How do they form? Why are they not as abundant in the other soil horizons? Why are they important?
7. Why is it that rocks containing calcite exist all over the United States, but soils containing calcite are restricted to the southwestern region?
8. How would you distinguish a mature soil from an immature soil?

9. Figure 14.12 illustrates how soils mature. What other factors that affect the overall characteristics of a soil are illustrated in the figure?
10. How does location in the landscape affect soil development at a particular site?
11. Describe the main types of soils and the processes and features that distinguish them from one another?
12. Describe how soil formation can produce an economically significant mineral deposit.
13. Why are large areas of North America covered with immature soils characterized by weakly developed horizons?
14. How can calcium-rich B horizons form in soils on top of calcium-poor rock in hot and arid regions?
15. Explain why erosion is a major threat to agriculture.
16. Why did the 1930s Dust Bowl happen?
17. Explain some ways to reduce soil erosion.

Confirm Your Understanding

1. Write an answer for each question in the section headings.
2. Find a soil profile exposed somewhere in your community (perhaps in a road cut, quarry, or building excavation). Identify, describe, and measure the soil horizons present in your profile. (Note: If circumstances prevent you from visiting an actual soil profile, work from a photograph and omit the measurements if a scale is not provided.)
3. Why do E horizons not occur everywhere?
4. You discover an ancient reddish soil deposit buried beneath younger sediment and the modern soil. What horizon does the reddish deposit likely represent? What would you look for in this ancient soil to determine the ancient climate?

5. Plants are important factors in producing soils. How do plants at the surface affect the B horizon?
6. If you lived in the tropics, what environmental factors would you look for in choosing a site for farming?
7. Explain the apparent discrepancy that lush tropical rainforests characteristically grow on very infertile soils.
8. Examine the full-page photograph at the beginning of this chapter. The aerial view in eastern Washington shows natural hillslopes and farmland. The yellow areas are growing wheat. The remaining areas are covered in native grasses. Explain how the farming practices visible in the photograph are designed to minimize soil erosion.

CHAPTER 15

Mass Movements: Landscapes in Motion

Why Study Landslides?

Falling, sliding, and flowing masses of rock and regolith are found almost everywhere on Earth, even underwater. This chapter focuses on how rock and regolith move downslope only because of the downward pull of gravity, without flowing water, blowing wind, or creeping ice.

"Landslide" is a commonly used term for the gravity-driven, downslope movement of rock and regolith, and the facing page illustrates a dramatic example of this movement. Geologists, however, prefer to use terms that describe and define the processes taking place rather than just referring to "land that slides." You are about to embark on an investigation of these processes.

Catastrophic landslides are infrequent so people in communities at risk can have a false sense of security. Added up, however, small and large landslides cause staggering numbers of casualties and substantial economic losses—close to 2 billion dollars annually in the United States. Urban growth into areas prone to slope failure, deforestation, and the local increase in precipitation caused by changing climate patterns all contribute to increasing losses. By understanding how landslides happen and why they occur, geologists help detect, predict, and mitigate potential hazards.

After Completing This Chapter, You Will Be Able to

- Explain the causes of gravity-driven, downslope movement of rock and regolith.
- Relate these processes to features that are visible in natural landscapes.
- Apply your knowledge to recognize areas where potential hazards exist.

A massive landslide in 2006 buried an entire village in the Philippines, killing 1100 people.

15.5 How Do We Know . . . How to Map Mass-Movement Hazards?

15.4 When Do Mass Movements Occur?

15.6 How Do Mass Movements Sculpt the Landscape?

IN THE FIELD

What do landslides look like? To answer that question, you will imagine witnessing four events and examine the different processes and results. Notice which processes the events have in common, how they are different, and what they reveal about how and why rock and regolith move downslope.

Your first stop is Yosemite National Park, California, on the evening of July 10, 1996. It is the height of the tourist season; the campgrounds are full, and hikers crowd the trails. At 6:52 P.M., without warning, a giant block of granite separates from a cliff high above Yosemite Valley. The result is shown in **Figure 15.1**a. The granite block has a volume of about 30,000 cubic meters (equivalent to filling 250 eighteen-wheeler semi-trailers with rock). The rock slides down the slope in contact with the cliff for 167 meters and then free-falls 550 meters to the valley below. The impact of the rock on the valley floor shakes the ground strong enough to show up on seismometers 200 kilometers away. The air displaced from beneath the falling rock produces an air blast that wipes out 1000 trees in the forest covering an area equivalent to 24 football fields. The combination of falling rock and blasting air damages a nature center and several bridges, destroys a snack bar, kills one person, and seriously injures several others. The entire event is over in a few seconds.

The next field-trip stop is Madison Canyon, Montana, on August 17, 1959. Near midnight, a magnitude-7.3 earthquake shakes southwestern Montana. Hundreds of vacationers are violently awakened in their tents and camp trailers along the Madison River, near Yellowstone National Park. The ground heaves beneath campers' sleeping bags, and their tents shudder. At the peak of the ground shaking, a huge mass of rock and loose regolith rips from the mountainside and slides down the canyon wall. A wave of broken rock, soil, and trees descends through the darkness. It moves so fast that it moves across the river and travels partway up the opposite side of the steep canyon. Campers, trailers, and cars are buried within seconds—28 people die.

The earthquake shaking dislodged 30 million cubic meters of rock and regolith on the steep canyon slope. This landslide, illustrated in Figure 15.1b, was 1000 times larger in mass than the block of rock that later fell in Yosemite Valley. The Madison Canyon landslide destroyed part of a campground, buried the only highway in the valley, and dammed the Madison River to form Earthquake Lake.

The next field excursion is a real-time trip to Slumgullion Creek near Lake City in southwest Colorado. Figure 15.1c shows a slope in the San Juan Mountains where regolith has slowly and continuously crept downhill for 2000 years. Along this 6.8-kilometer-long hillside more than 170 million cubic meters of regolith and forest have moved hundreds of meters. The outline of the moving regolith in Figure 15.1c resembles an oozing, pasty flow of earthen debris. The surface of the shifting debris is cracked and heaved in different directions. About 700 years ago, the slowly moving mass dammed the Lake Fork of the Gunnison River to form Lake San Cristobal. Most landslide events happen quickly and are studied after the fact, but the Slumgullion slide moves so slowly that it can be studied in real time. You can go out and actually observe the movement in action at rates between 10 centimeters and 6 meters per year.

Your last field trip is to the San Bernardino Mountains, east of Los Angeles, California. Wildfires burned more than 350 square kilometers of steep, forested mountainsides in October 2003, removing vegetation and leaving the ground covered with ash and charred wood. Imagine that you are there on Christmas Day of that year, when storms strike and drop as much as 20 centimeters of rain in a 24-hour period, the heaviest rainfall in 20 years. Steep hillsides, covered in burned debris and loose soil, seem to melt into flowing masses of regolith and water. You hear a sound like thundering freight trains crashing through narrow mountain canyons. The saturated debris resembles a flow of wet concrete more than four meters thick that plows through a church camp at 50 kilometers per hour. Buildings are knocked from their foundations, walls collapse, and half of a group of 30 people gathered for the holiday perish. Figure 15.1d shows the devastation.

(a) Yosemite National Park, July 1996. A rock climber took this photograph (right) as a slab of granite crashed to the valley floor in a huge cloud of dust. Air displaced by the falling rock blasted down trees in the forest. The overhead view from a helicopter (far right) shows the results.

Dust cloud

Rock slab detached from here

Rock slab detatched from here

Shattered granite at base of cliff

Trees blown down by air blast

Damaged building

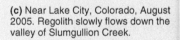

Scar left by landslide

Slide debris

Landslide dammed the river to form a lake

(b) Madison Canyon, Montana, August 1959. An earthquake-triggered landslide fills the river canyon (above). Effects of the slide are still obvious 41 years later (right). No vegetation has grown on the slide scar and drowned trees rise as snags in Earthquake Lake.

(c) Near Lake City, Colorado, August 2005. Regolith slowly flows down the valley of Slumgullion Creek.

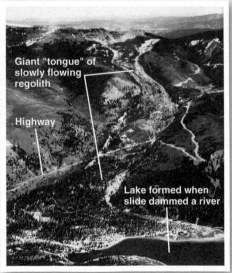

Giant "tongue" of slowly flowing regolith

Highway

Lake formed when slide dammed a river

(d) San Bernadino Mountains, California, December 2003. Boulders, sand, mud, water, and trees flowed rapidly through this canyon, demolishing buildings and burying victims.

▲ Figure 15.1 What mass movements look like.

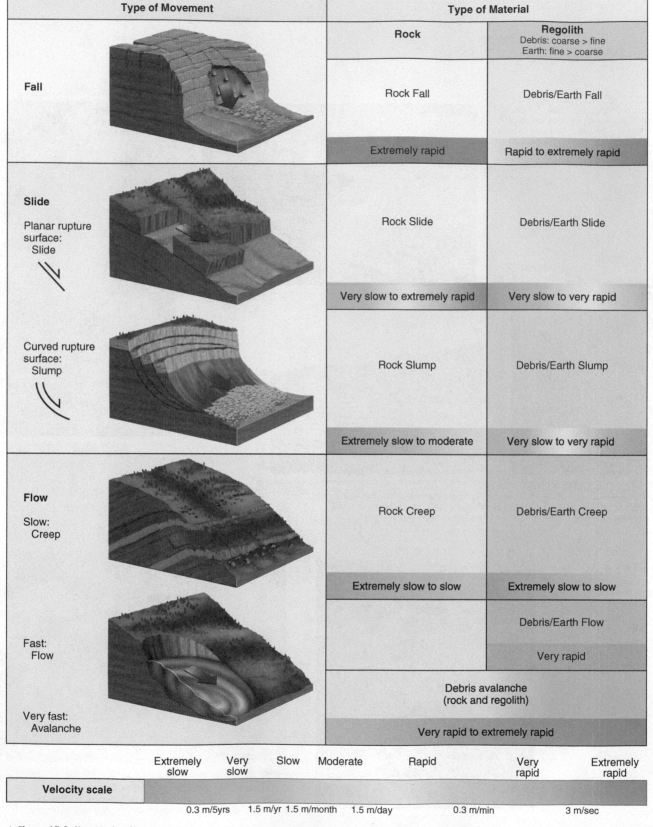

Type of Movement		Type of Material	
		Rock	**Regolith** Debris: coarse > fine Earth: fine > coarse
Fall		Rock Fall	Debris/Earth Fall
		Extremely rapid	Rapid to extremely rapid
Slide Planar rupture surface: Slide		Rock Slide	Debris/Earth Slide
		Very slow to extremely rapid	Very slow to very rapid
Curved rupture surface: Slump		Rock Slump	Debris/Earth Slump
		Extremely slow to moderate	Very slow to very rapid
Flow Slow: Creep		Rock Creep	Debris/Earth Creep
		Extremely slow to slow	Extremely slow to slow
			Debris/Earth Flow
Fast: Flow			Very rapid
		Debris avalanche (rock and regolith)	
Very fast: Avalanche		Very rapid to extremely rapid	

Velocity scale	Extremely slow	Very slow	Slow	Moderate	Rapid	Very rapid	Extremely rapid
	0.3 m/5yrs	1.5 m/yr 1.5 m/month	1.5 m/day		0.3 m/min		3 m/sec

▲ **Figure 15.2 How to classify mass movements.** Mass-movement classification relies mostly on identifying the type of material that moves and the type of movement. The velocity of the movement is distinctive of some processes but ranges considerably for others.

ACTIVE ART

Mass Movements. See how mass movements work.

15.1 What Are the Characteristics of Mass Movements?

The real events described above are examples of what geologists call **mass movement**—gravity-driven downslope motion of rock and regolith. Although water and ice commonly participate in mass movements of rock and regolith, the motion occurs because of the force of gravity pulling on the solids and not because of flowing water.

Criteria for a Meaningful Classification

How would you describe the processes involved in the four events that you "witnessed" in the previous section? All of these events are mass movements, but what distinguishes each from the others? Answers to these questions provide insights into how mass movements work, and form the basis for a useful, process-based classification.

The transported materials differ among the events—solid rock at Yosemite, mostly regolith at Slumgullion, and a mixture of rock and regolith in Madison Canyon and San Bernardino Mountains. Water played an important role in the San Bernardino event but seems to have been of negligible, if any, consequence in the others.

The material also moved or moves in different ways. The Yosemite event mostly featured free-falling rock. Rock and regolith at Madison Canyon primarily moved with a sliding motion. The masses moving at Slumgullion and that moved in southern California, although consisting mostly of solid objects, flow like a highly viscous fluid.

Speed is another distinguishing characteristic of the different mass movements. The flow of regolith along Slumgullion Creek is slow and ongoing, whereas at the other three localities the events happened quickly and ended abruptly.

Based on these observations, it is easy to see why three factors—(1) the nature of the mixture of solids (rock or regolith), (2) the type of motion (sliding, falling, flowing), and (3) the velocity of motion (fast or slow)—describe the variety of mass-movement processes. These characteristics are the criteria for the classification scheme of mass movements illustrated in **Figure 15.2**.

Distinguishing the Materials in Motion

Rock and regolith are the two types of moving material. In the classification, regolith only refers to the weathered loose blanket of debris on the original hillslope. The key factor in this part of the classification is noting what the material was prior to its descent, because originally solid rock breaks up during movement and may resemble regolith when the event is over. Debris and earth are categories of regolith. "Debris" is coarse grained; 20 to 80 percent of the fragments are larger than 2 millimeters in size. In the category "earth," 80 percent or more of the particles are less than 2 millimeters in size. The terms "rock," "debris," and "earth" completely describe the range of material that moves downslope and allow for specific classification. For example, rock fall and debris fall share the same process of movement but involve the transport of different types of material.

Mass Movement by Free Fall

During a **fall**, material detaches from a steep slope and then free falls through the air, or bounces and rolls downslope (Figure 15.2). The key feature is that the moving mass loses contact with the surface. Rock falls are common where the rock is highly jointed and on a steep slope, such as shown in **Figure 15.3**, and describe the observed event at Yosemite Valley (Figure 15.1a).

Falls produce piles of loose rock and debris at the base of steep slopes, as seen in **Figure 15.4**. These accumulations are called **talus**. The lack of vegetation or soil on talus indicates that material is actively accumulating.

Before

After

▲ Figure 15.3 Old Man is a victim of rock fall. The New Hampshire state symbol, the Old Man of the Mountain, was a jointed, overhanging granite cliff with an outline that resembled the silhouette of a man's face. Despite efforts to strengthen the crumbling cliff, the Old Man collapsed in a rock fall in May 2003.

Rock fall talus

► Figure 15.4 Talus slopes, indicators of falls. Rocks falling from these steep sedimentary-rock cliffs in the Canadian Rockies produce towering cone-shaped slopes of talus. The lack of soil or vegetation on the talus indicates that rock falls occur frequently and that the slopes are very unstable.

Mass Movement by Sliding Along a Surface

Slides move downslope in contact with a **surface of rupture**, which separates moving material from stationary material (Figure 15.2). In some cases, the material moves as a single large block along the surface. In other cases, the moving rock and regolith fragments jostle around but do not lose contact with neighboring fragments or with the rupture surface. Recognizing whether the mass moves along a planar or curved surface allows distinction of two types of slides.

Planar slides, such as the one illustrated in **Figure 15.5**a, typically occur along bedding planes, foliation, or joint planes oriented parallel to the slope (Figure 15.2). These relatively smooth, sloping planes provide a weak rupture surface for sliding. At Madison Canyon, rock slid along a surface of rupture defined by foliation in the metamorphic rocks underlying the steep slopes above the river.

If the rupture surface curves, then the slide mass rotates as it moves downslope, causing rock layers and surface features to tilt (Figure 15.2). The

This highway clings to an unstable slope south of San Francisco, California. Steeply dipping sedimentary rocks slide along bedding planes into the ocean, leaving a barren, unvegetated slope. Rock slides frequently close the road.

Dipping sedimentary beds

(a)

This earth slump along the Genesee River in New York shows the curving, scoop-shaped rupture scarp typical of a slump. Notice that the broken pavement and trees tilt back into the slumping hillside, which indicates rotational movement along a curved rupture surface.

(b)

◄ Figure 15.5 What slides and slumps look like.

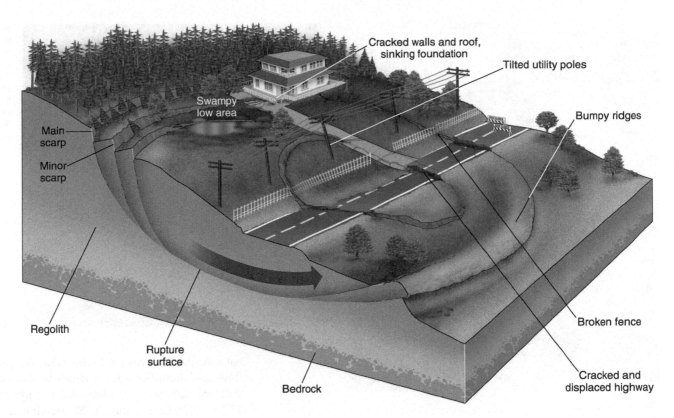

▶ **Figure 15.6 Recognizing a slump.** This illustration shows the distinctive features produced by an active or recently formed slump (compare to Figure 15.5b).

Labels in figure: Cracked walls and roof, sinking foundation; Tilted utility poles; Bumpy ridges; Swampy low area; Main scarp; Minor scarp; Regolith; Rupture surface; Bedrock; Broken fence; Cracked and displaced highway

curved rotational slide surfaces are scoop shaped, and they usually form in regolith or poorly consolidated or weak rock where bedding or joints do not influence failure. These rotational slides on curved surfaces are called **slumps**, and an example is shown in Figure 15.5b.

Figure 15.6 shows the landscape features to look for where a debris slump recently occurred or is actively moving. The surface of rupture looks like a fault, and the downslope sliding and sinking of the regolith produces scarps at the top of the slide. If the slump moved recently or is actively moving, then the scarp may be almost vertical and devoid of vegetation. Older scarps erode and are less obvious. Sliding regolith conceals the rest of the surface of rupture, but its curved shape is revealed by the rotation of the slumping mass, which tilts the land surface back into the hillside (see Figures 15.5b and 15.6). This tilting may disrupt downslope water runoff to form swampy areas or ponds along the hillslope. Open cracks form in the land surface and are especially evident where road pavement breaks and tilts. Where the rupture surface curves back up at the bottom of the slump, the ground bulges unevenly into very irregular and bumpy topography.

Mass Movement by Flow Resembling a Liquid

Flows are the continuous movement of rock, regolith, or both that behaves like a high-viscosity liquid (Figure 15.2). **Figure 15.7** illustrates the liquid appearance of moving flows and the nature of the deposits.

You were a virtual witness to two flows at the beginning of the chapter. Based on fragment size, the Slumgullion example is an earth flow (Figure 15.1c), whereas the San Bernardino Mountains event was a debris flow (Figure 15.1d). The fluid-like characteristics of debris flows partly result from water included in the flows,

(a)

A moving debris flow in the Philippines has the typical flowing-concrete appearance. Although water-saturated and moving like a fluid, the debris flow consists mostly of sand and gravel. The wave in the foreground is about 1 meter high.

(b)

This coarse, bouldery debris flow damaged homes near Durango, Colorado. Notice that the deposit is a very poorly sorted mixture of boulders, sand, mud, and broken trees.

◀ **Figure 15.7 What debris flows look like.**

These nearly vertical
sedimentary rock layers
bend downslope, to the
right, because of rock creep.

Bent over tree trunks are an
indication of earth creep on
this steep hillside. The
bending results from
simultaneous vertical growth
of the tree and downslope
tilting caused by creep of
near-surface regolith above
the main root zone of the
tree.

◀ **Figure 15.8 Hillsides that creep.**

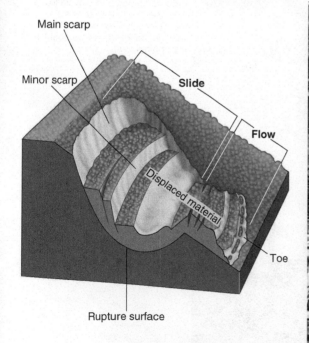

but solid particles constitute more than three-quarters of the mass. Debris flows and earth flows leave deposits of chaotically mixed fragments of different sizes, with very irregular, bumpy upper surfaces (Figures 15.1c, 15.1d, 15.7). Contrasting the Slumgullion and San Bernardino Mountain events also illustrates a wide range in flow velocity.

Figure 15.8 shows how very slow flows, called **creep**, are detected only by dislocation or bending of features at the surface. The bending of the rock layers in Figure 15.8a indicates that the surface regolith and part of the rock are flowing downslope without evidence of a rupture surface that would define a slide. The tree roots anchor the trees in stable regolith at depth while surface regolith slowly moves downslope. The movement tilts the trees but they adjust their growth back to a vertical position, causing a downslope bend in the tree trunks that becomes more obvious over time (see Figure 15.8b).

The term "avalanche" also describes some flows. Most familiar might be snow and ice avalanches, which are mass movements that sometimes bury skiers and mountain towns. **Debris avalanches** are very rapid flows of rock, regolith, vegetation, and sometimes ice. Notice that in this definition of a debris avalanche, the moving material is not restricted to regolith of debris size, despite the name. Some debris avalanches are gigantic in volume and involve the failure of significant parts of whole mountains, as depicted in the photograph at the beginning of this chapter. Despite very rapid flowing motion, debris avalanches contain very little water.

Single Events—Multiple Processes

Many mass-movement events involve combinations of processes. **Figure 15.9** shows how movement sometimes originates at the upslope end as a slide, but then the material breaks up and flows beyond the surface of rupture at the downslope end. The displaced rock mass at Yosemite slid along a rupture surface before falling to the

▲ **Figure 15.9 Single event, multiple processes.** The drawing illustrates the parts of a complex mass movement that is both a slide and a flow. These same features are visible in the 1995 photograph of La Conchita, California. Another flow at this location in January 2005 destroyed thirteen homes and killed 10 people. The scarp is the exposed part of the rupture surface for the slide. The toe is the farthest traveled material. The mass movement is a slump type of slide near the top, as indicated by the curved surface of rupture. The irregular bumpy topography at the base indicates the part of the mass movement that flowed beyond the rupture surface.

valley floor (Figure 15.1a). The steep scars at the top of the Slumgullion earth flow (Figure 15.1c) and the debris avalanche illustrated in chapter opening photograph are evidence of a slide at the upslope end of the mass movement. The detached regolith and rock then continue downslope as a flow.

Putting It Together—What Are the Characteristics of Mass Movements?

• Mass movements are downslope movement of rock, regolith, or both, caused by the downward pull of the force of gravity.

• The three characteristics used to describe mass movements are (1) the material that moves (rock, debris, or earth), (2) the movement process (fall, slide, or flow), and (3) the velocity of movement (fast, slow).

• Falls are material falling free from a very steep slope. Slides move along planar or curved surfaces of rupture. Flows move like viscous liquids despite being composed primarily of solid particles.

• Landscape features such as scarps, surface cracks, tilted or bent trees, bumpy topography, and talus provide clues to the occurrence and type of mass movement affecting a slope.

• Some mass-movement events involve more than one process. Rock and regolith initially separate along a rupture surface, but the slide-displaced material continues downslope as a fall or flow.

lower angle of tilt (Figure 15.10b). What can you conclude? That smoother surfaces favor motion compared to rough surfaces.

Finally, wet the sanded board with a film of water and repeat the experiment. This time the brick begins to move at an even lower tilt angle (Figure 15.10c). What can you conclude? That a wet surface favors motion on that surface.

This simple series of experiments suggests that slope, surface roughness, and water are factors affecting the balance between driving and resisting forces.

Gravity Is the Driving Force

The gravity force can be described as the weight of an object pulling straight down toward the center of Earth. A simple illustration, **Figure 15.11,** shows that the slope angle determines the size of the gravity force. Gravity holds a boulder stationary against a horizontal surface. In contrast, along a vertical surface, gravity pulls the boulder downward. On a slope that is in between horizontal and vertical, there is a component of the gravity pull that is parallel to the sloping surface, and this pull increases as the slope angle increases (Figure 15.11a). The driving force for mass movement, therefore, is greater on steep slopes than on gentle slopes. This explains why the brick moved down the board when the board was tilted to a steeper angle.

Friction and Cohesion Are the Resisting Forces

Forces acting between the boulder and the underlying material resist movement of the boulder down the slope. The resisting forces are friction and cohesion, which combine as the *resisting strength* of the material. With

15.2 What Causes Mass Movements?

Field observations provide descriptions useful for classifying the types of mass movements, but what causes these movements? A fundamental law of physics is that all movement requires applying a force to get objects to move. Gravity exerts a force that pulls rock and regolith downhill. However, not all hillslopes fall, slide, or flow away. This suggests that there are also forces that resist movement. Understanding what causes mass movement requires that we examine both the driving forces that pull material downslope, and the resisting forces that tend to keep material where it is. Mass movement occurs when the driving forces exceed the resisting forces.

Using an Analogy of Mass Movement

To help you understand the factors that enhance or impede mass movements, consider the simple experiment conducted with a brick on an inclined board that is illustrated in **Figure 15.10**. Place the brick on a horizontal board and then gradually raise one end of the board. Eventually the board rises to a critical slope angle where the brick moves. What can you conclude? That increasing slope favors motion.

Next, sand the board to make it smoother, and repeat the experiment. This time, the brick begins to move at a

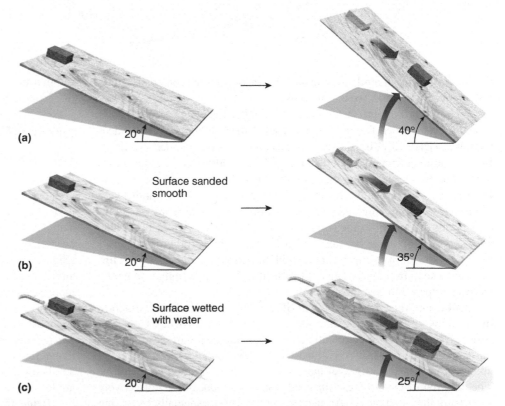

▲ **Figure 15.10 Experiments to understand causes of mass movement.** (a) A brick slides down the inclined slope of a wood board when the board tilts past a critically steep angle. (b) The critical angle for the onset of movement decreases if the board is first sanded smooth. (c) The angle decreases even further if the board is wetted with water.

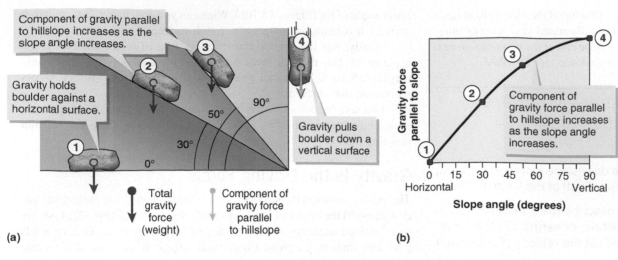

Component of gravity parallel to hillslope increases as the slope angle increases.

Gravity holds boulder against a horizontal surface.

Gravity pulls boulder down a vertical surface

90°

50°

30°

0°

Total gravity force (weight)

Component of gravity force parallel to hillslope

(a)

Component of gravity force parallel to hillslope increases as the slope angle increases.

Gravity force parallel to slope

0 15 30 45 60 75 90
Horizontal   Vertical

Slope angle (degrees)

(b)

◄ **Figure 15.11 Visualizing gravity forces acting on a boulder.** (a) On a flat surface (position 1), the total gravity force of the boulder pulls the boulder down against the surface, so it does not move. The gravity force also pulls downward parallel to a vertical surface (position 4), which likely means that the boulder falls. On an inclined surface (such as positions 2 and 3), a component of the gravity force is directed parallel to the surface (follow the yellow arrows).
(b) The graph shows how the gravity force parallel to the ground surface increases as the slope angle increases. The labeled positions on the curve correspond to the four positions of the boulder illustrated in panel (a).

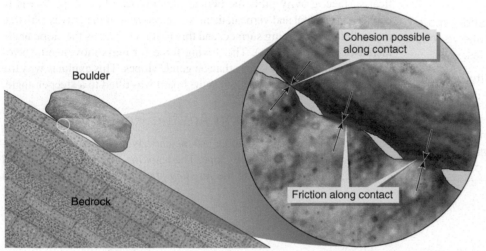

Cohesion possible along contact

Boulder

Bedrock

Friction along contact

◄ **Figure 15.12 Resisting strength is friction plus cohesion.** Friction relates to the smoothness of the hillslope surface and the smoothness of the surface of the boulder. The friction force also depends on the slope, because at lower slope the weight of the boulder more effectively presses the boulder into contact against the underlying rock. Cohesion is the attraction of particles to each other at the atomic level.

this definition of strength we can say that stronger materials resist mass movements better than weak materials.

Friction is the force that opposes motion between two objects that are touching one another. The friction increases as the roughness of the surfaces increases. **Figure 15.12** illustrates how friction affects the boulder on a slope. In this example, friction depends both on the smoothness of the surface on which the boulder rests and the smoothness of the surface of the boulder. Friction is lower on a smooth surface than a rough one. This explains why the brick started moving at a lower slope angle when the board was sanded smooth compared to the rough surface (Figure 15.10). Friction also varies with slope angle because on lower slopes the weight of fragments pulls them down in stronger contact with underlying material. This is important to keep in mind because it means that friction is higher on gentle slopes compared to steep slopes.

Cohesion, the other component of resisting strength, is the attraction of particles to each other at the atomic level (Figure 15.12). Cohesion results from opposite electrostatic charges on adjacent particles, and it is particularly important in clay minerals (Section 5.4 and 14.2). Materials such as loose sand and gravel have very little cohesion between grains. The addition of clay to these materials raises the cohesion by effectively increasing the "stickiness." An increase in cohesion, especially by adding clay particles, helps material stay intact and resist mass movement. We can now see a connection between slope and surface roughness with our brick and board experiments and the driving and resisting forces. We will soon

see that the presence of water also relates to these forces, as implied by the experiment.

Driving Force Greater than Resisting Force Equals Movement

For motion to happen, the gravity force parallel to the surface must be greater than the resisting strength. If the resisting strength of the material is small, then objects move on a low slope, such as when a pencil rolls off a nonlevel tabletop. If the resisting strength is large, then objects only move on a steep slope, as shown by the brick resting on a rough, dry board (Figure 15.10a). The graph in **Figure 15.13** shows the importance of slope angle to determine whether motion occurs, because the driving force increases with increasing slope angle, whereas resisting strength decreases with increasing slope angle. For every case, therefore, there is a critical angle below which there is no movement and above which movement takes place. This angle depends on the cohesive and frictional characteristics that define the resisting strength of the materials. The rock and regolith on a hillslope is stable below the critical angle and unstable above the critical angle.

Figure 15.14 and **Figure 15.15** extend the simple case of a boulder sitting on a hillside (Figures 15.11–15.13) to other mass-movement scenarios. For a fractured rock outcrop (Figure 15.14), the friction and cohesion adjacent blocks of rock are touching determine the resisting strength.

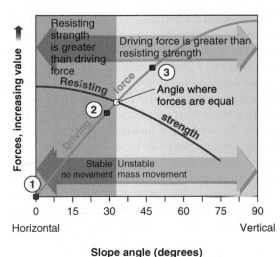

Resisting strength is greater than driving force

Driving force is greater than resisting strength

Resisting strength

Driving force

Angle where forces are equal

Stable no movement

Unstable mass movement

Forces, increasing value

0   15   30   45   60   75   90
Horizontal   Vertical

Slope angle (degrees)

(a)

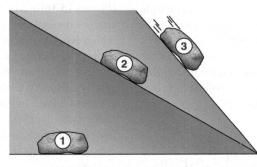

(b)

◄ **Figure 15.13 The onset of mass movement.** Mass movement occurs when the driving force of gravity exceeds the resisting strength of the rock or regolith. Boulders resting on slopes of different angles, labeled positions 1, 2, and 3, experience forces of different values, as shown in the graph. Driving force increases and the resisting strength decreases as the slope angle increases. For the illustrated examples, boulders in positions 1 and 2 are stable whereas the boulder in position 3 moves because it is unstable.

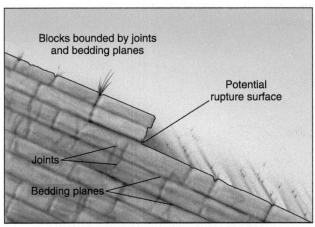

Blocks bounded by joints and bedding planes

Potential rupture surface

Joints

Bedding planes

Resisting force greater than driving force: No movement

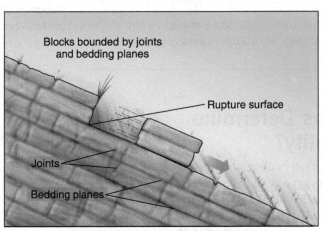

Blocks bounded by joints and bedding planes

Rupture surface

Joints

Bedding planes

Resisting force less than driving force: Movement

◄ **Figure 15.14 Resisting strength in fractured rock.** So-called "solid rock" is actually broken along joints and sedimentary bedding planes or metamorphic foliation. Mass movement, as a slide or fall, occurs by loss of frictional and cohesive resistance along these preexisting breaks in the rock.

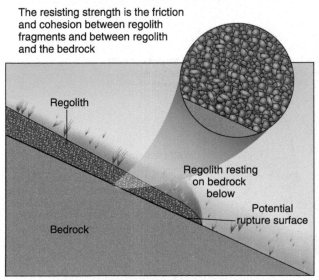

The resisting strength is the friction and cohesion between regolith fragments and between regolith and the bedrock

Regolith

Regolith resting on bedrock below

Potential rupture surface

Bedrock

Resisting force greater than driving force: No movement

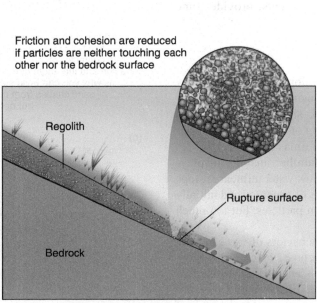

Friction and cohesion are reduced if particles are neither touching each other nor the bedrock surface

Regolith

Rupture surface

Bedrock

Resisting force less than driving force: Movement

◄ **Figure 15.15 Resisting strength in regolith.**

The frictional resistance to sliding along a bedding plane, as a possible rupture surface, is greater if the weight of the rock above the surface is large and if the surface is rough. This situation compares to the brick on the rough, unsanded board in Figure 15.10. The cohesion is mostly determined by the minerals in the rock, and it is usually low between blocks of rock. For regolith (Figure 15.15), the resisting strength includes friction and cohesion between regolith particles and between the regolith and the underlying bedrock. Cohesion may be more important in the case of regolith compared to rock because regolith commonly contains clay minerals.

Putting It Together—What Causes Mass Movements?

• Mass movements happen if the gravitational force is greater than the resisting strength of the rock or regolith.

• The resisting strength is the sum of the frictional resistance to movement and the cohesion between minerals in the rock or regolith.

• The size of the gravitational force directed parallel to the slope is greater at steeper slope angles, so steep slopes are more likely unstable than gentle slopes.

15.3 What Factors Determine Slope Stability?

Rock and regolith are not continually falling, sliding, and flowing down hillsides, which suggests that mass movement happens when something increases the size of the driving force, decreases the resisting strength, or both. What processes change the slope, friction, or cohesion of a mass of rock or regolith so that movement occurs? The field observations at the beginning of the chapter, along with the brick and board experiments, provide some insights.

The Role of Water

Water commonly contributes to slope instability. You saw how significant water is for causing mass movement by noticing how easily the brick slides on a wet board (Figure 15.10c), and by thinking of the debris flows triggered by heavy rainfall in California (Figure 15.1d).

Figure 15.16 provides a familiar analogy of playground or beach sand to show how water influences slope stability. Adding small amounts of water to sand initially increases the cohesion between particles, but too much water leads to failure.

To understand these two behaviors of wet sand, we need to consider the **angle of repose**, which is the steepest angle of a stable slope. This angle is determined by friction and cohesion. Figure 15.16a shows that loose dry sand pours into a pile whose sides are equal to the angle of repose. Adding a small amount of water to loose particles

produces thin films of water between the particles. The thin water films increase the cohesion between particles and permit a higher angle of repose (Figure 15.16b). The increased cohesion is caused by the weak electrical charges on each end of the water molecule (Section 2.4 and Figure 2.15), which weakly attaches to adjacent grains. However, when you add more water, the weight of the water exerts pressure on the sand grains and moves them apart. Water fills the pore spaces between grains and completely surrounds each grain. The sand grains no longer touch one another (Figure 15.16c). In this condition, the friction is reduced because the water partly supports the weight of the sand grains and the grains are not touching one another. Abundant water, therefore, exerts a pore pressure that reduces resisting strength and causes slope instability. Furthermore, the added water adds to the weight of the sand-and-water mixture, which increases the gravity driving force.

The Role of Geologic Materials

Field and lab studies show that the composition and grain size of the slope material also influence stability. Cohesion depends on the minerals that form the rock or regolith. Clay minerals have high cohesive properties. Cohesion also increases where mineral cements, such as calcite, are present between fragments.

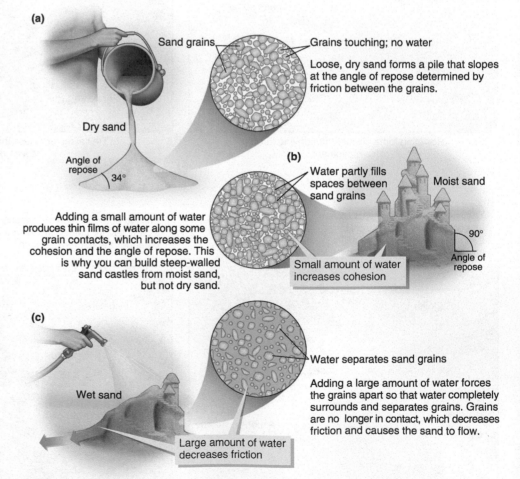

▲ Figure 15.16 Visualizing the effect of water on the angle of repose of loose sand.

The Role of Slope

The importance of slope steepness to cause mass movements is clear from the brick and board experiments illustrated in Figure 15.10 and from the analysis of the gravity force increasing with steeper slope (Figure 15.11). **Figure 15.17** illustrates natural and artificial processes that increase slope

angles and, therefore, enhance mass movement. Slopes increase along stream channels because many streams erode valleys downward into the landscape. As a stream cuts down, the adjacent slopes become steeper (Figure 15.17a) and may reach the critical angle where driving force exceeds resisting strength. Stream bank erosion caused the slump illustrated in Figure 15.5b. Excavation and artificial filling on hillsides also steepens slopes, which increases the likelihood of mass movement (Figure 15.17b).

The Role of Geologic Structures

Joints, bedding planes, foliation, and faults are planes that interrupt cohesion otherwise provided by interlocking mineral grains (Figure 15.14). These planes may be smooth so that there is also little frictional resistance to movement. If smooth bedding or foliation planes are parallel to the surface slope, then movement is more likely to occur than if the planes dip into the slope, as illustrated in **Figure 15.18**. Joints, bedding planes, foliation, and faults also provide avenues for water to move downward from the surface. Water along the planes further reduces friction and cohesion to favor mass movement.

The Role of Weathering

Chemical and physical weathering produce regolith and determine its resisting strength. Freezing and thawing of water in cracks or shrinking and swelling of clay minerals by wetting and drying decreases friction and cohesiveness, which promotes slope failure. The rock-fall destruction of the Old Man of the Mountain in New Hampshire, shown in Figure 15.3, illustrates these effects. Physical weathering weakened the jointed rock. Following several days with temperature fluctuations that caused freezing and thawing, the Old Man fell from the steep cliff.

The Role of Vegetation

Vegetation is a very important factor in slope stability. Roots penetrate and bind together regolith and absorb water from precipitation. When vegetation is removed and deep roots rot away, the added cohesion and frictional resistance contributed

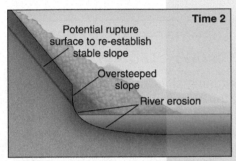
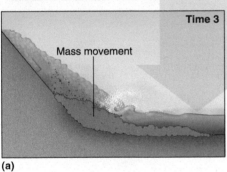
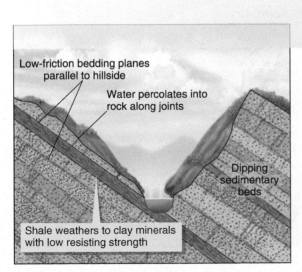
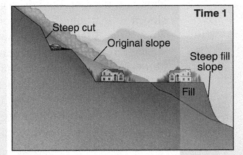

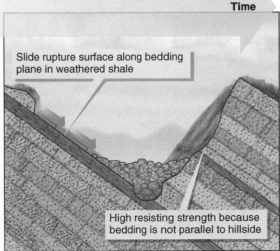

◀ Figure 15.17 Changing slope influences slope stability. Natural processes and human activities make slope angles steeper, which increases the likelihood of mass movements. (a) Vertical and horizontal river erosion forms steep unstable slopes. (b) Hillside construction can make slopes more susceptible to mass movement, as does using unconsolidated fill with steep slopes as a building foundation.

◀ Figure 15.18 Local geology influences slope stability. The bedding planes of the rock on the left side of the valley dip parallel to the hillslope. The addition of water along joints adds weight and causes loss of cohesion in the clay-rich shale layer. The decrease in friction and cohesion causes a rock slide. By comparison, the situation on the steeper right side of the valley shows the bedding planes oriented in the opposite direction of the slope. Although rock falls may occur on this steep slope, the orientation of the bedding planes means that slides are highly unlikely.

▲ **Figure 15.19 Mass movements follow wildfires.** This photo from western Colorado shows debris flows related to the destruction of regolith-stabilizing vegetation by wildfire. Heavy rain causes debris slides and earth flows that carry loose regolith, ash, and charred wood into canyon bottoms, where the debris mixes with stream water to form debris flows.

by plants disappears. Examples of vegetation removal from hillslopes include tree logging, clearing land for agricultural and urban expansion, livestock overgrazing, and wildfire. **Figure 15.19** shows debris flows following rainfall on wildfire-denuded slopes. Large, fire-related debris flows caused the Christmas 2003 disaster in southern California (Figure 15.1d).

Figure 15.20 illustrates the role of vegetation, along with climate, to determine the thickness of regolith on hillslopes. Where plants are abundant, the roots bind the regolith that forms by weathering on the hillslopes so that runoff erodes only a small volume of the potential sediment. In arid regions with low vegetation abundance and slow rates of weathering, surface-water runoff removes most regolith particles almost as quickly as they loosen from bedrock. Hillslopes in the relatively humid eastern United States and moist regions farther west, have thick blankets of regolith that mostly fail in debris or earth flows, slides, slumps, and creep. In arid parts of the western United States, however, regolith is thin or absent and slopes mostly fail by rock falls and slides.

Putting It Together—What Factors Determine Slope Stability?

- Factors determining slope stability are the abundance of water, the composition and texture of material, presence and orientation of planar features in the rock that may form rupture surfaces, the steepness of the slope, the amount of weathering, and vegetation.

▶ **Figure 15.20 Climate and vegetation affect regolith thickness.** In humid regions, dense vegetation stops erosive runoff and roots bind together regolith fragments to form rounded hillslopes with thick regolith. Mass movements primarily happen in regolith, usually as creep, slumps, and flows. In contrast, arid regions without abundant vegetation lack thick regolith because weathered fragments wash away soon after they separate from the rock. Mass movements are typically rock falls and rock slides.

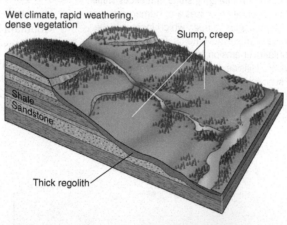

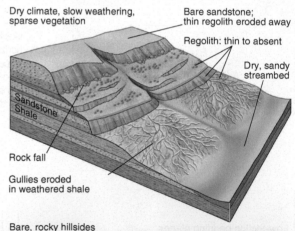

Great Smoky Mountains National Park, Tennessee

Arches National Park, Utah

15.4 When Do Mass Movements Occur?

We have identified the factors that change the balance of driving and resisting forces to permit mass movement, but what determines *when* the movement occurs? The changing magnitudes of the driving and resisting forces must reach a point where cohesion and friction are insufficient to offset the downslope pull of gravity. In some situations, the changing force magnitudes are gradual and unsuspected, leading to a sudden and unexpected slope failure, such as the rock falls at Yosemite (Figure 15.1a) and the Old Man of the Mountain (Figure 15.3). In other cases, there is a sudden stimulus that causes a rapid change in the slope, resisting strength, or both that cross the threshold for movement on what was previously a stable slope. To the extent that these stimuli, called "triggers," can be anticipated in terms of either location or timing, or both, it is possible for geologists to assess the potential hazard from mass movements.

Rainfall and Snowmelt Triggers

Addition of water to hillslope materials increases the likelihood of mass movement, so heavy rainfall can trigger failures. This was the case in the San Bernardino Mountains field example, where extremely heavy rainfall in a short period of time caused damaging and deadly debris flows (Figure 15.1d). Another case is illustrated in the photo at the beginning of the chapter. After several days of unusually heavy rain in February 2006, the village of Guinsaugon, Philippines, completely disappeared when a debris avalanche swallowed more than 350 houses and a school.

Melting snow not only adds lubricating moisture, but also adds weight. The weight of snow on a hillslope along with the weight of infiltrating meltwater increases the magnitude of the gravity driving force for slope failure.

These weather-related triggers of mass movement happen on steep hillsides across the United States and are linked to extreme weather conditions associated with unusually heavy rain, hurricanes, and rapid spring warming. If weather forecasts suggest that these conditions may occur, then warnings are issued for areas having steep slopes that may be most susceptible to slope failure. Special attention is given to hillsides recently stripped of vegetation by wildfire or human activity.

Earthquake Triggers

Strong earthquake ground shaking can trigger mass movements. The most common effects are rock and debris falls that happen because the seismic surface waves literally lift rock fragments off the ground and away from neighboring fragments. As the shaking particles lose contact with one another the frictional resistance to movement decreases and the fragments bounce and roll downhill. Dust clouds from innumerable rock falls fill the air after large earthquakes.

Vibrations from a magnitude-7.9 earthquake triggered a debris avalanche on Nevado Huascaran, Peru, in May of 1970 that killed 18,000 people. The avalanche buried the entire town of Yungay and part of another. Damage from this devastating avalanche is shown in **Figure 15.21**. The Madison Canyon rock slide (Figure 15.1b) is a smaller-volume example of mass movement triggered by an earthquake.

You learned in Chapter 11 about liquefaction, the process that causes mass movement of water-saturated regolith when it is shaken by strong earthquake waves (see Section 11.9). Earthquake-wave vibrations cause grains in the regolith to compact closer together in the same way that you shake a box to settle its contents, as illustrated in **Figure 15.22**a. However, when there is water between the grains, the compacting grains displace water and that forces other grains apart, which decreases cohesion and friction, as illustrated in Figure 15.22b. The water-rich layer of particles behaves like a fluid and flows even on gentle slopes. Damage from an earthquake-induced liquefaction is shown in Figure 15.22c.

Volcanic Eruption Triggers

The largest debris flows, rock slides, and debris avalanches in history have originated on the slopes of volcanoes. Mass-movements are expected near steep volcanoes at the beginning of any eruption, so threatened communities are usually evacuated.

A combination of circumstances causes volcanic debris flows, also known as lahars (see Section 4.9). First, explosive volcanic eruptions deposit thick accumulations of loose volcanic ash and pumice lapilli on steep mountain slopes. Second, eruptions typically destroy large areas of slope-stabilizing vegetation. Third, snowmelt generated by lava flows and pyroclastic flows, or heavy rainfall following eruptions, causes rapid erosion

▲ **Figure 15.21 Earthquake-triggered debris avalanche destroyed Peruvian village.** Prior to 1970, the tall peaks of Nevado Huascaran loomed above the village of Yungay, Peru. A 1970 earthquake shook loose rock, regolith, and ice near the top of the mountain to form a gigantic debris avalanche that buried Yungay and 18,000 of its residents.

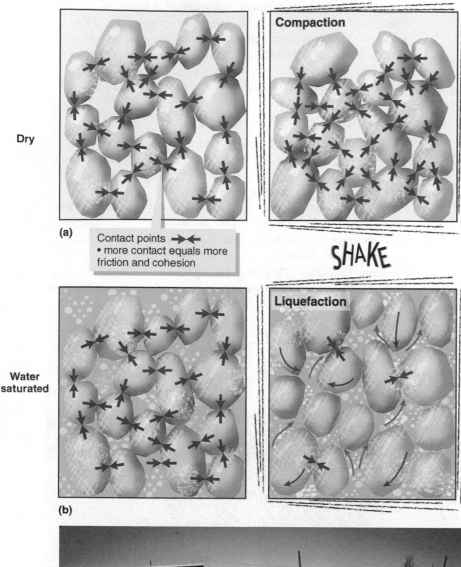

Dry

Contact points ➤◄
• more contact equals more friction and cohesion

(a)

Compaction

SHAKE

Water saturated

Liquefaction

(b)

and sliding of the loose ash, which mixes with water in stream channels to make huge debris flows. In short, almost all of the ingredients for mass movement are commonly present during or after a volcanic eruption, and may affect communities as much as 100 km away for many years.

For example, heavy rains triggered many debris flows during and for years following the June 1991 eruption of Mount Pinatubo in the Philippines (see the beginning of Chapter 4). Over the next three years about two cubic kilometers of volcanic ash, pumice, and rock washed and slid from steep hillsides and debris-choked valleys and then was transported mostly as lahars that flowed as far as 50 kilometers from the volcano. To get an idea of the volume of these lahar deposits, imagine debris covering all of Washington, D.C., up to three stories high. The lahars tore out bridges, buried entire cities, and filled in river valleys, causing streams to change course and destroy villages and farmland. **Figure 15.23** shows some of the damage caused by these events. Similarly, tragic lahars generated from melting snow by an eruption of Nevada del Ruiz destroyed the city of Armero, Colombia, in 1985 (see Section 4.9) and killed at least 23,000 people.

Many volcanoes literally fall apart in huge rock slides and debris avalanches. An excellent example is the largest mass movement ever witnessed by humans. When Mount St. Helens, Washington, erupted in 1980, the intruding magma swelled the flank of the volcano. The swelling steepened the

(c)

▲ **Figure 15.22 Earthquake-induced liquefaction causes mass movements.** (a) During earthquake ground shaking, dry silt and sand grains compact, and the number of contact points between grains increases. (b) In water-saturated silt and sand, the grains attempt to compact during ground shaking, but the water cannot compact and instead is forced out of the pore spaces, which causes the grains to move apart. Friction decreases as the number of grain contacts decreases, and the mass behaves like a fluid. (c) Liquefaction during the great Alaskan earthquake of 1964 caused mass movements that destroyed these homes in Anchorage.

▲ **Figure 15.23 Debris flows following a volcanic eruption.** This view of a city street in the Philippines shows the buildings buried almost to the top of the first story by debris-flow deposits. The debris flows, also called lahars, were triggered by heavy rainfall on ash and pumice that were deposited on the steep slope of Pinatubo volcano by an explosive eruption in 1991.

volcano slope and stressed the rock to the point where it began to crack. These changes simultaneously reduced the resisting strength of the rock and increased the slope. The threshold for failure was reached, possibly enhanced by ground shaking from an earthquake caused by magma movement. The result, illustrated in **Figure 15.24**, was a gigantic, 2.8-cubic-kilometer rock slide and debris avalanche that quickly descended the north slope of the volcano and traveled 22 kilometers down an adjacent river valley.

Figure 15.25 summarizes and integrates the processes and triggers that cause hillslopes to fail by mass movements. If something happens to increase slope angle, decrease cohesion, or decrease frictional resistance, then slope stability decreases.

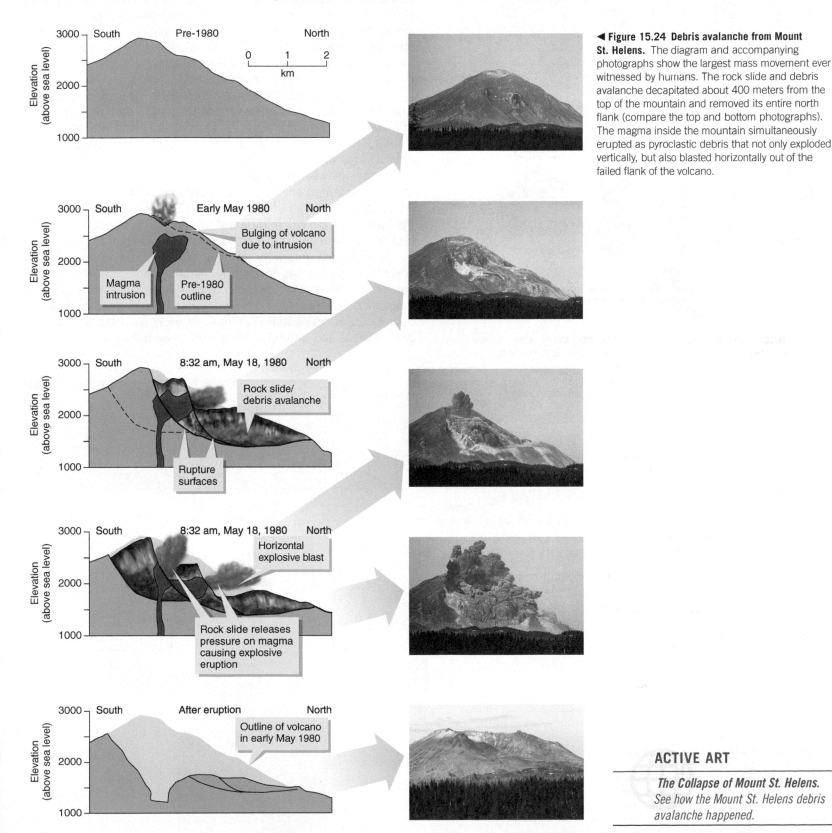

◄ **Figure 15.24 Debris avalanche from Mount St. Helens.** The diagram and accompanying photographs show the largest mass movement ever witnessed by humans. The rock slide and debris avalanche decapitated about 400 meters from the top of the mountain and removed its entire north flank (compare the top and bottom photographs). The magma inside the mountain simultaneously erupted as pyroclastic debris that not only exploded vertically, but also blasted horizontally out of the failed flank of the volcano.

ACTIVE ART

The Collapse of Mount St. Helens.
See how the Mount St. Helens debris avalanche happened.

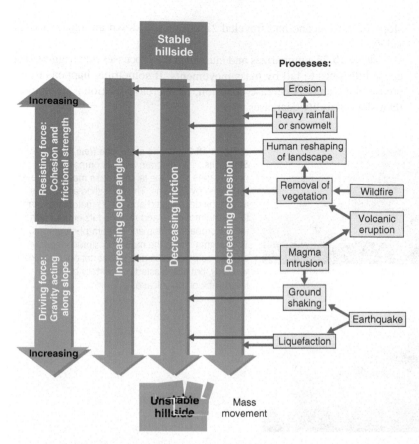

▲ **Figure 15.25 Linking triggers with driving and resisting forces.** This diagram schematically links natural and unnatural processes, listed on the right, to changes in slope, cohesion, or friction. Increasing slope, or decreasing cohesion or friction, changes the balance between driving and resisting forces that determines whether a hillside is stable and stationary or unstable and prone to mass movement.

Putting It Together—When Do Mass Movements Occur?

• Triggers are stimuli that abruptly imbalance driving and resisting forces governing the occurrence of mass movements.

• The most common triggers are addition of water by rainfall or snowmelt, ground shaking during earthquakes, and slope failures during volcanic eruptions.

• Mass movements do not require a specific trigger but instead occur when a threshold is gradually reached between forces that favor or resist slope failure.

▶ **Figure 15.26 Flowchart for making a hazard map.** Slope angles, rock types, and locations of past mass movements form three data layers in the geographic information system (GIS) for making a hazard map. Slope angles are calculated from digital versions of topographic-elevation maps, and past slides are inventoried from aerial photographs. A geologist compares and analyzes the data layers to produce the bottom map that provides a visual guide depicting areas where future hazards are greatest.

15.5 How Do We Know . . . How to Map Mass-Movement Hazards?

Picture the Goal

How Do Geologists Map Areas That Are Prone to Future Mass Movements?
Mass movements are clearly damaging and deadly events, so it is desirable to recognize hazardous areas when making land-use and construction decisions. The goal is to identify where a risk of mass movement exists and to outline the hazardous areas on maps that are consulted by planners and developers.

The hazard assessment must include a detailed inventory of slope-instability factors across the landscape. Based on your study of mass movement, these are the factors that most influence slope stability—slope angle, type of rock or regolith, orientation of bedding and foliation planes, extent of breakage by joints and faults, and extent of weathering of rock and regolith. The likelihood of mass movement at any particular location is, however, a complicated combination of all of these factors. To make a map that shows the level of hazard risk in different locations, it is necessary to combine all of these geologic data together in a meaningful way.

To show how this work is done, we will examine how a slide-susceptibility map was created by the U.S. Geological Survey for a part of San Mateo County south of San Francisco, California. The motivation for the study is rapid urban growth that pushes the boundaries of residential development away from flat valleys and coastal areas and into rural areas with steep hillslopes.

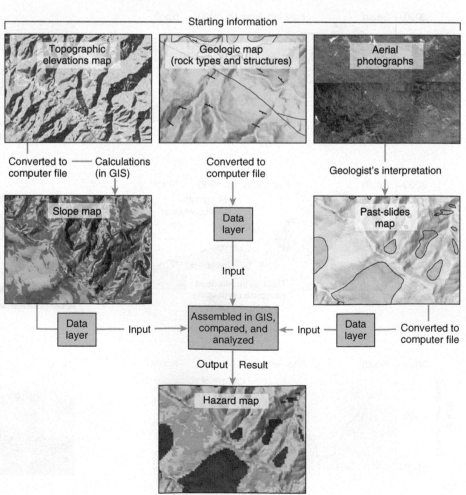

Assembling the Data

What Types of Maps Are Needed? **Figure 15.26** summarizes the data used to develop a hazard map. Judging from the factors that cause unstable hillslopes, the most basic information includes steepness of slopes and types of rocks and regolith. Knowing where mass movements happened in the past is also useful to identify where unstable hillslopes exist.

It would be quite confusing to illustrate all of this information on a single map, so separate maps, illustrated in **Figure 15.27**, are made for each data type. The topographic-elevation map is used to calculate the hillslope angles so that areas of very steep, and potentially unstable slopes, are recognized. A part of the resulting slope map is shown in Figure 15.27a. The geologic map, illustrated in Figure 15.27b, shows the distribution of rock types and also reveals locations of rocks with relatively low resisting strength—in this case siltstone, mudstone, and weathered basalt. Topographic and geologic maps are routinely prepared and are available for many parts of the United States, so these data are typically already on hand.

Another necessary map shows the locations of past mass movements because these definitively indicate the locations of unstable slopes. The map showing locations of past mass movements has to be prepared for the specific study area. The money and time expenditures of having a geologist walk over all of a large study area are prohibitive for most hazard-assessment projects. Instead, most of the effort to recognize past mass movements

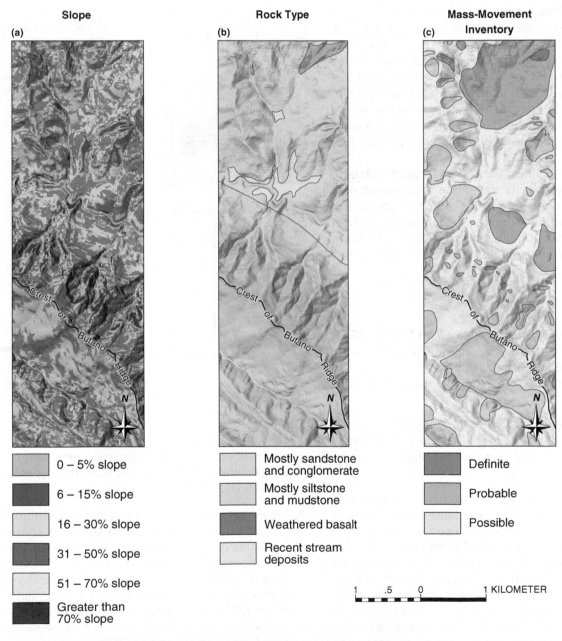

Slope	Rock Type	Mass-Movement Inventory
(a)	**(b)**	**(c)**

Slope legend:
- 0 – 5% slope
- 6 – 15% slope
- 16 – 30% slope
- 31 – 50% slope
- 51 – 70% slope
- Greater than 70% slope

Rock Type legend:
- Mostly sandstone and conglomerate
- Mostly siltstone and mudstone
- Weathered basalt
- Recent stream deposits

Mass-Movement Inventory legend:
- Definite
- Probable
- Possible

1 .5 0 1 KILOMETER

▲ **Figure 15.27 GIS data layers used to make a hazard susceptibility map.** Here are parts of the slope, rock-type, and mass-movement-inventory maps that form the data layers for GIS analysis of hazards in part of San Mateo County, California.

focuses on examining aerial photographs. Slide scarps, talus slopes, irregular ground surfaces underlain by material that has moved downslope, stream valleys blocked with slide debris, and areas of vegetation disturbed by movement on hillsides are visible on the photographs. Geologists transfer the locations of past movements from the photographs to maps. These characteristics of landscapes affected by mass movement (see Figure 15.6) are most obvious for events that happened recently. Older mass movements are more difficult to recognize after the deposits and scars erode and overgrow with vegetation. As a result, there is some uncertainty in the recognition of past mass movements from aerial photographs, so geologists classify and inventory the past slides as definite, probable, or possible, with decreasing confidence of recognition, as shown in Figure 15.27c.

Applying the Tool

How Do Geologists Bring the Data Together? Geologists must integrate the information on all of these maps along with interpretation of risk in order to determine the mass-movement hazard at any one location. Hazard maps are usually built by using **geographic information systems** (GIS). A GIS is a powerful set of computer-software tools that collects, stores, retrieves, analyzes, and displays data that pertain to particular locations. The GIS manipulates each component of the database in distinct data layers. For instance, each of the three maps of San Mateo County represents a single data layer in the GIS (see Figure 15.26).

GIS allows comparison of the information in each database layer at each location. If geologists determine a relationship of mass-movement hazard to slope and rock type, then analysis of the slope and rock-type map layers reveals slide susceptibility at every location. The slide-susceptibility determination forms a new layer in the GIS and is printed out as a separate hazard map (Figure 15.27).

Picture the Results

What Locations Are Most Susceptible to Future Slides? **Figure 15.28** illustrates the interpreted slide-hazard susceptibility for the same area in California that is depicted by data in Figure 15.27. The geologists used three interpretations to determine the varying levels of susceptibility:

1. Areas of possible, probable, or definite mass movement in the past are unstable areas where future movement is most likely. These locations are assigned the highest susceptibility (compare Figures 15.27c and 15.28).

2. Hillslopes underlain by some rock types are more susceptible to failure than other slopes associated with other rocks. The data layer depicting past slides was superimposed on the geologic data layer within the GIS to see whether some rock types were more likely than others to be associated with mass movement. By surveying all of rural San Mateo County, not just the areas shown in Figure 15.28, about half of the hillsides underlain by mostly clay-rich siltstone and mudstone or weathered basalt have possibly, probably, or definitely slid in the past. In contrast, only about one-third of the area underlain by mostly sandstone and

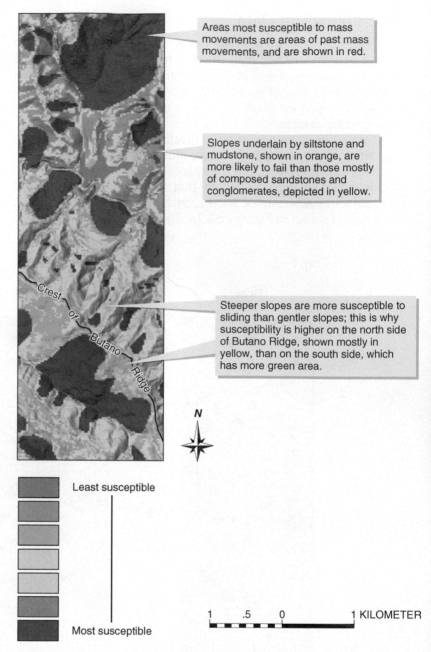

Areas most susceptible to mass movements are areas of past mass movements, and are shown in red.

Slopes underlain by siltstone and mudstone, shown in orange, are more likely to fail than those mostly of composed sandstones and conglomerates, depicted in yellow.

Steeper slopes are more susceptible to sliding than gentler slopes; this is why susceptibility is higher on the north side of Butano Ridge, shown mostly in yellow, than on the south side, which has more green area.

Least susceptible

Most susceptible

1 .5 0 1 KILOMETER

▲ **Figure 15.28 Rock and debris-slide susceptibility map.** The hazard map results from analysis of the three data-layer maps shown in Figure 15.28.

conglomerate is affected by past slides. Therefore, interpreted susceptibility is lower for the areas underlain by sandstone and conglomerate compared to the other rock types (compare Figures 15.27b and 15.28).

3. Steeper slopes are more likely to fail in future slides than are gentler slopes. The slope data layer was superimposed on the past-slide data layer. Not surprisingly, slides are more common on steeper slopes. As a result, within a particular rock type, higher slide susceptibility occurs on steeper slopes than on lower slopes. Contrasting the susceptibility of the steep north side of Butano Ridge with the less steep south side best shows this interpretation, even though both are underlain by the same rock type (compare Figures 15.27 and 15.28).

Insights

How Can These Results Be Understood Scientifically? The susceptibility map illustrates correlations between phenomena and locations, but scientists also seek to understand why these correlations exist. Comparison of the GIS data layers indicates that hillside slope steepness and type of underlying rock correlate with the number of past slides and the areas impacted by past slides.

Why do more slides happen on the steeper slopes? The relation between steeper slopes and a larger driving force for mass movement readily explains this correlation, as illustrated and developed in Figures 15.11 through 15.14.

Why are some rock types associated with more slides? The siltstone and mudstone contain abundant clay minerals. Shrinking and swelling of clay minerals during wet and dry periods cause these rocks to deteriorate rapidly by physical and chemical weathering. This results in low rock strength and the formation of thick, clayey regolith that readily loses frictional and cohesive strength when wet. The basalt weathers to form similar weak, clay-rich regolith. The sandstone and conglomerate beds do not weather as quickly, and the resulting regolith does not contain as much clay as is found on the siltstone, mudstone, and basalt hillslopes.

Putting It Together—How Do We Know . . . How to Map Mass-Movement Hazards?

• Hazard maps indicate areas susceptible to future mass movements. The hazard maps combine data from other maps that depict relevant geologic information and topographic information.

• GIS is computer software that is capable of collecting, storing, retrieving, and manipulating a large variety of data sets.

• GIS can be used to construct hazard maps by comparing and analyzing the assembled geologic and topographic data.

15.6 How Do Mass Movements Sculpt the Landscape?

We have explored why mass movements occur and why they are hazardous, but how important are these processes for determining the appearance of landscapes? Mass movements seem infrequent, at least based on how rarely they are reported in the news media. In reality, however, mass movements occur widely and frequently on Earth's surface, even though most events are not reported in the news. Mass movements account for most of the rock and regolith removed from mountainous hillslopes, so these movements play an important role in landscape development.

Mass Movements and Mountain Building Go Together

Of all the factors determining slope stability, the most important is slope steepness. Steep slopes are most common in mountainous landscapes, and slopes are steepest where mountains actively rise in response to ongoing tectonic activity.

Steep mountain slopes result from simultaneous uplift and erosion, illustrated in **Figure 15.29**. Mountain uplift increases land-surface elevation. Streams flow down the steep slopes between mountains and lowlands and erode deep canyons into the uplifting rock (canyon carving is further explained in Chapter 16). Stream erosion lowers elevation along valleys, which produces high relief between mountain peaks and valley bottoms. Canyon walls are very steep where rivers erode in hard rock, producing unstable slopes that are prone to mass movement. Slides, falls, and flows lower the relief, but the continued river incision into uplifting rock renews the steep slopes, which leads to more mass movement.

In addition to steep slopes, other factors make actively rising mountain hillslopes unstable:

• Tectonic deformation that drives mountain building pervasively fractures the rock, producing many planes of potential rupture.

• High-magnitude earthquakes that are common in areas of active mountain building and whose shaking triggers huge slides and debris avalanches, such as the one illustrated in Figure 15.21.

• Mountain building that brings metamorphic rocks to the surface, with closely spaced foliation planes that serve as rupture surfaces. Many metamorphic minerals weather to clay minerals that further weaken the rock.

• The heavy rain and snowfall that occur in most high mountain regions. Abundant moisture contributes to slope instability by lubricating regolith fragments and rupture surfaces and causes more weathering to weaken rock and generate still more loose, unstable regolith.

• The cold temperatures that occur in very high mountains and are necessary for the freeze and thaw that causes rock falls.

Mass Movements Move Mountains

Recent studies reveal that mass movements are the primary agents of regolith movement in mountains. As implied in Figure 15.29, the amount of rock and regolith removed by the incision of a valley by persistent flow of water in a narrow stream channel is small compared to the volume of rock and regolith that episodic falls, slides, and flows bring down the steep valley slopes.

A study in the tectonically active Southern Alps of New Zealand drives this point home. **Figure 15.30** shows the result of a recent rockslide and debris avalanche that lowered the elevation of the country's highest peak by about 10 meters. Comparison of aerial photographs taken 60 years apart showed that more than 7000 mass movements occurred in an area of about 5000 square kilometers within the Southern Alps. For comparison, this area is a little smaller than the state of Delaware. These slope failures affected areas ranging in size from as little as 100 square meters to about 1 square kilometer. Estimated volumes of rock and regolith moved during these events are equal to an average lowering of elevation by more than 5 millimeters per year over the entire area. This may seem like a small erosion rate, but if this rate continues for one million years, which is a short time for mountain building, then mass movement will remove a 5-kilometer thickness of rising crust!

There is little doubt that mass movements determine the steepness of slopes and the maximum heights of mountain peaks in regions of tectonic uplift. As tectonic forces push rocks high above the surrounding Earth surface, gravity works to pull the rocks back down in mass movements.

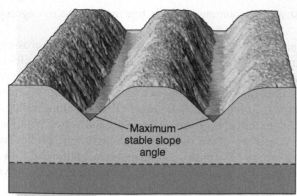

Maximum stable slope angle

Time 1

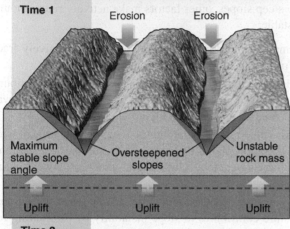

Erosion   Erosion

Maximum stable slope angle   Oversteepened slopes   Unstable rock mass

Uplift   Uplift   Uplift

Time 2

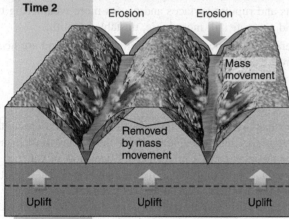

Erosion   Erosion

Mass movement

Removed by mass movement

Uplift   Uplift   Uplift

Time 3

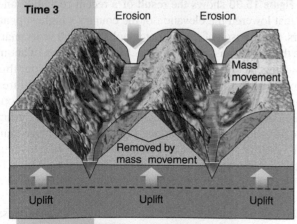

Erosion   Erosion

Mass movement

Removed by mass movement

Uplift   Uplift   Uplift

Time 4

◄ Figure 15.29 Uplift and mass movement work together. Uplift of mountains causes rivers to erode down into ever-deeper valleys. This erosion produces steep, unstable slopes that promote mass movements. The landslides return the slopes to stable angles and the rivers transport the mass-movement debris to distant lowlands. The river responds to the continuous uplift by more erosion, followed by more mass movements. Both mass movement and river erosion contribute to the changing face of the uplifting mountain landscape, but mass movements lower the elevations more effectively than river erosion.

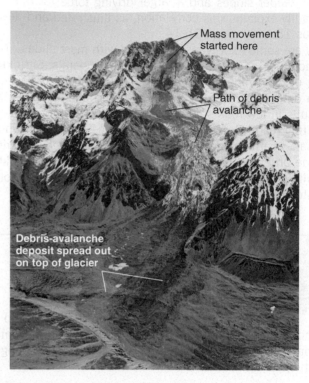

Mass movement started here

Path of debris avalanche

Debris-avalanche deposit spread out on top of glacier

▲ Figure 15.30 Mass movement moves mountain. Just after midnight on December 14, 1991, part of New Zealand's highest mountain, Mount Cook (also known as Aoraki), collapsed. The resulting debris avalanche traveled nearly 7.5 kilometers at speeds reaching 200 kilometers per hour. The mountain was shortened by 10 meters.

Putting It Together—How Do Mass Movements Sculpt the Landscape?

• Mass movements are the primary process of rock and regolith movement in mountainous regions and, therefore, in landscape development.

• Active mountain uplift increases land-surface elevation and produces high relief. Stream erosion lowers elevation along valleys, promoting very steep and unstable slopes prone to mass movement.

• In addition to steep slopes, rock fractures, earthquakes, foliation of metamorphic rocks, heavy rainfall, and freeze-thaw weathering all promote mass movements in mountains.

ACTIVE ART

Uplift and Mass Movement. See how tectonic uplift, river erosion, and mass movements combine to shape mountain landscapes.

Where Are You and Where Are You Going?

Mass movements are the gravity-driven downslope transport of rock and regolith. They occur widely over Earth's entire surface and shape landscapes. Mass movements are a very costly natural hazard, and geologists diminish their damaging effects by understanding the factors that cause mass movement and recognizing where those factors exist in the landscape.

Description and classification of mass movements relies on easily observed features of active mass movements, or their resulting landforms and deposits, or both. The classification emphasizes the nature of the moving material (rock or regolith) and the type of movement (e.g., flow, slide, fall). The speed of the movement is useful for describing mass movements; speed varies from fractions of a meter per year to tens of meters per second.

Mass movement happens when the driving force for motion exceeds the resisting strength of the material. Gravity is the driving force that pulls rock and regolith downslope. The magnitude of this driving force increases with increasing slope angle. Opposing gravity is a combination of friction and cohesion that defines the resisting strength, which holds material in place. Friction and cohesion are properties of the can material but change in response to human or natural causes so as to change the likelihood of slope failure. Friction, for example, is less on steeper slopes than on gentler ones, and friction diminishes if water exerts pressure along potential surfaces of movement. Variations in composition and textures of rocks, structural features of rocks (e.g., bedding and foliation planes, faults, and joints), the abundance of water and vegetation (and therefore climate), and slope angles contribute to the stability of slope materials by affecting the magnitudes of the gravity driving force and the frictional and cohesive resisting strength.

Mass movements are the primary process of rock and regolith movement in steep, mountainous terrains. A complex interplay of tectonic uplift of rock, stream erosion of canyons into the rock, and mass movements that reduce the slope angles between mountaintops and canyon bottoms produces the elevation and relief of mountains.

Mass movements are only one of many processes that modify and sculpt the surface of Earth. In the upcoming chapters, you will explore erosional processes, especially processes related to movement of water in both liquid and frozen states. You learned that mass movements tear down mountain sides, but you need to understand the role of flowing streams to generate the relief exploited by slides and falls in mountains, how the streams carry away the mass-movement debris, and where the debris ends up going. In the next chapter you will learn about the significance of streams as sculptors of the landscape, how human activities influence the work of rivers, and how and why rivers affect human lives.

Active Art

Mass Movements. See how mass movements work.

Liquefaction. See how earthquake ground shaking causes liquefaction.

The Collapse of Mount St. Helens. See how the Mount St. Helens debris avalanche happened.

Uplift and Mass Movement. See how tectonic uplift, river erosion, and mass movements combine to shape mountain landscapes.

Confirm Your Knowledge

1. Define "mass movement."
2. What three factors are used to describe and classify the variety of mass-movement processes?
3. What is the difference between a slide and a slump? What are the characteristic features they leave on the landscape?
4. What is the difference between a rock fall and a rock slide? How can you recognize a slump in the field?
5. Use the concepts of driving and resisting forces to define what is meant by a stable hillslope in contrast to an unstable hillslope.
6. What factors affect slope stability? Explain how each factor relates to slope stability.
7. What is the relationship between slope stability and resisting forces?
8. What is the "angle of repose"?
9. List and describe four triggers of mass movements.
10. What combination of circumstances is needed to cause a volcanic debris flow (also called a "lahar")?
11. What information is needed to create a mass-movement hazard map? How do you construct a hazard map?
12. What factors make actively rising mountain hillslopes unstable?

Confirm Your Understanding

1. Write an answer for each question in the section headings.
2. What factors affect the balance between the driving and resisting forces for mass movements? How do these factors favor, or reduce, the likelihood of mass movement?
3. A mass-movement hazard map has some uncertainty associated with it. For each type of information needed to create a mass-movement hazard map identify the cause of the uncertainty.
4. What human activities associated with building a house increase the chances of mass movement?
5. A developer hires you to determine the suitability of a steep hillside for the construction of apartment buildings. What observations would you make to determine mass-movement risk? Explain how each observation applies to your objective.
6. Section 3.5 introduced the concept of descriptive versus genetic classifications. Think about the many classifications that you have encountered so far in this text: the three rock groups, types of rocks within each rock group, types of plate boundaries, and types of mass movements. Analyze how each of these classifications is descriptive, genetic, or some combination of both.

CHAPTER 16

Streams: Flowing Water Shapes the Landscape

Why Study Streams?

Streams provide water for many uses, deposit sediment that becomes fertile soil, and are pathways for commerce. Streams erode curving valleys and sculpt scenic landscapes, such as the Grand Canyon and Niagara Falls. However, streams also flood with costly losses in lives, property, and crops.

Americans withdraw one trillion liters of fresh water each day from rivers, streams, and lakes. Most of this water is used to generate electricity and for irrigating crops. Only 10 percent of withdrawn surface water is used for drinking, but it accounts for 58 percent of all public and domestic water supplies (the remainder comes from ground water). Clearly, it is important to understand how much water flows in streams in order to allocate it for these many uses.

Farmland and cities occupy flood-prone lowlands adjacent to river channels. Floods are the costliest natural hazard in the United States, causing an average of 140 fatalities and $5 billion in damages each year. As a result, people want to understand why rivers flood so that they can predict floods and minimize their effects.

After Completing This Chapter, You Will Be Able to

- Explain how water and sediment move in channels and across floodplains.
- Relate patterns and dimensions of stream channels to water and sediment transport processes.
- Recognize and explain landscape features produced by river erosion and deposition.
- Analyze the causes of floods.
- Explain how streams change through time in response to natural processes and human activities.
- Explain how streams become naturally obstructed to produce lakes.

Pathway to Learning

16.1 Where Does the Water Come From?

16.2 Where Does Sediment Come From?

16.4 How Do Streams Transport Sediment?

16.6 Why Does a Stream Change along Its Course?

EXTENSION MODULE 16.1 *How a Stream Gage Works*

16.3 How Do Streams Pick Up Sediment?

16.5 Why Do Streams Deposit Sediment?

16.7 What Factors Determine Channel Pattern?

The Rio Grande in New Mexico carries water and sediment from highlands, such as the stream-carved mountains in the background, to the Gulf of Mexico and along the way provides irrigation water to support farming on fertile floodplain soils.

IN THE FIELD

Imagine yourself in the scene depicted in **Figure 16.1**a, alongside a roaring river in the wilds of Alaska. You walk along the riverbank on round, smoothly polished cobbles and boulders and gaze out over fast-moving water. The opposite bank is more than a hundred meters away, although small islands of bare gravel interrupt the flow in many places. There is no soil or vegetation on these midstream gravel accumulations, so these must be temporary parts of the river landscape that submerge when the river is deeper. Why does the river have more water at some times than at others?

The stream carries more than water. You hear what sounds like loudly grinding teeth and realize that these are the noisy collisions of rolling and bouncing of cobbles below the water surface. The rounded cobbles along the bank are evidence that these collisions grind and smooth sediment grains. You look into the river to watch the moving cobbles, but all you see is muddy, murky water. Small sediment particles suspended in the water obscure the rolling cobbles. What determines the sizes of sediment carried by the stream, and how does the stream pick up the sediment to begin with?

Natural benches rise along the margin of the valley and provide vantage points (Figure 16.1a). You observe that each bench consists of rounded gravel despite being tens of meters above the flowing stream. Thick soil and plants cover the bench, indicating that it is far above where the river now flows, even during floods. This bench must, therefore, record the higher elevation of the riverbed at an earlier time in its history. Why do rivers erode downward into the landscape? Do all rivers only cut deeply into Earth's surface, or are some riverbeds at stationary elevations or even rising as sediment fills up the channels?

On another occasion, you fly above the Mississippi River in Louisiana, taking in the views seen in Figure 16.1b. What a different scene from the river in Alaska. The Mississippi is nearly a kilometer across with no islands sticking up within the channel. Judging from the size of tugboats and barges navigating the river, the Mississippi channel seems deeper than the Alaskan stream. Although the river banks are far below you, the sediment exposed along the banks is much finer grained than that seen along the Alaskan river and is likely all sand and mud. Why are the dimensions and sediment characteristics of the two river channels so different?

From your vantage point above the Mississippi River you admire the long ribbon of water that bends back and forth in wide, horseshoe-shaped curves. Why is the channel not straighter, like the river in Alaska? Cultivated farmland extends along the river as far as you can see, interrupted here and there by towns and cities. Ridges of dirt and gravel follow alongside the curving course of the river channel. These are artificial levees built to protect the farmland and towns from floods. This powerful river provides commercial shipping and fertile farmland, but its value may be offset by the threat of floods, which you occasionally learn about in the local and national news. How do river processes affect human activities, and how do human activities affect rivers?

(a) Alongside a river in Alaska. The muddy water in the channel divides and rejoins around islands of gravel. Forested benches composed of river gravel overlook the valley in the photo on the right.

(b) Above the Mississippi River in Louisiana. The river is wide and deep enough for tugboats and barges to navigate the channel. Artificial levees protect communities and farms from floods.

Levee

▲ Figure 16.1 Field visits to see dynamic rivers.

16.1 Where Does the Water Come From?

The field observations demonstrate that **streams** are both flowing water in channels that drain water from land and agents of sediment transport. Studies by geologists and engineers show that streams are complex features governed by the physics of flowing water, the ease of erosion of geologic materials forming the banks, and tectonic forces that deform the landscape along the stream course. Worldwide, streams deliver about 700,000 cubic meters of water to the ocean each second. For comparison, this volume of water would fill 374 Olympic-size swimming pools. The most important place to begin learning about streams is to understand where the water comes from and what factors determine how much water flows through the channel.

The Hydrologic Cycle

Water falls from the atmosphere as rain or snow and then evaporates back into the atmosphere. Careful observation shows, however, that water follows many paths on and below Earth's surface rather than just shuffling back and forth between ground and sky.

Figure 16.2 illustrates the **hydrologic cycle**, a concept for describing the movement of liquid water and water vapor through all parts of the Earth system. Stream water starts out as precipitation (rainfall and snow). Precipitation results from the condensation of water vapor in the atmosphere into water drops and ice crystals. One important cause of condensation and precipitation is high continental elevations that force moist air to rise. The rising air cools, which causes the water vapor to condense. This process explains why precipitation is generally greatest in mountainous regions and along boundaries between lowlands and higher elevations.

Not all of the precipitation that reaches the land surface drains into streams. Some of the water soaks into the ground and follows different routes in the hydrologic cycle. Plants use part of this infiltrated water. Some soil moisture returns to the atmosphere by evaporation or **transpiration**, the process in which plant leaves and stems release water vapor. The remaining infiltrated water becomes **ground water**, which is the water present below Earth's surface in pore spaces and fractures within regolith and rock. The **water table** is the undulating boundary between unsaturated regolith or rock, above, and saturated regolith and rock, below. In the saturated zone, all pores and fractures are filled with water, whereas in the unsaturated zone, these open spaces are partly filled with water and partly filled with air.

Ground-water flow, which will be covered in Chapter 17, and stream flow are connected (Figure 16.2). Most ground water eventually comes back to the surface, especially in low-elevation valleys. This means that the total amount of water in a stream is the surface runoff of rainfall and snowmelt plus whatever infiltrated ground water seeps into the channel. If streams contained only surface runoff, then they would only flow when there was rainfall or melting snow to feed them. While this is true of many small streams, all large rivers and other smaller streams flow almost all of the time. Persistently flowing streams receive an influx of ground water during dry times, and this influx happens only if the bottom of the stream, called the bed, is lower than the water table (Figure 16.2).

The Drainage Basin: The Area Where the Water Comes From

How can you describe the size of a stream? There are several possibilities. The length of the river from its **headwaters**, where it begins, to its **mouth**, where it ends, is one measure of size. The measure of the width, depth, or

▶ **Figure 16.2 Visualizing the hydrologic cycle.** Water moves through the Earth system along pathways in the hydrologic cycle. Evaporation and transpiration (or combined as evapotranspiration) transport water vapor into the atmosphere from Earth's surface. Water vapor condenses into liquid or frozen water that falls as precipitation. Precipitation either runs off the surface or infiltrates into the ground. Some ground water seeps back to the surface. Streams transport surface water within the hydrologic cycle.

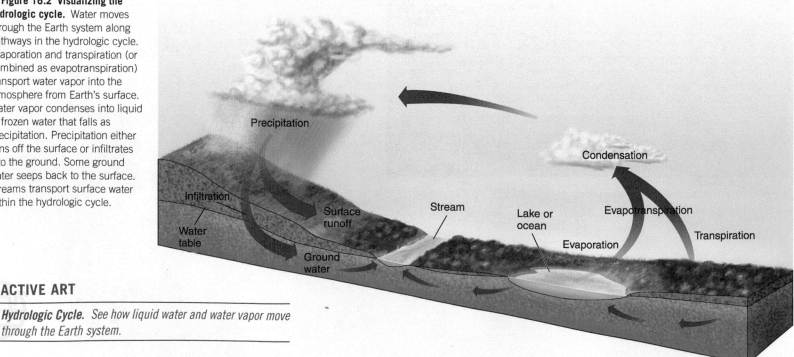

ACTIVE ART

Hydrologic Cycle. See how liquid water and water vapor move through the Earth system.

both at different locations along the stream can also give an indication of size. River length, along with channel width and depth, figures into two other ways to measure streams—drainage basin area and discharge.

The **drainage basin**, schematically illustrated in **Figure 16.3**, describes the area from which a stream gathers water. The drainage basin of one stream is separated from adjacent drainage basins by ridges called **divides**. The headwaters are at high elevations close to a divide, whereas the mouth is the lowest elevation in the drainage basin where the stream enters another stream, a lake, or the ocean. Smaller **tributary** streams flow into larger streams.

Most rivers ultimately transport water and sediment to the ocean. Fewer streams flow into low areas that are completely surrounded by divides, in which case the river terminates in a lake, rather than at the ocean.

Figure 16.4 shows the major divides and drainage basins in North America. The Mississippi River has the largest drainage basin, with an area of 3.3 million square kilometers encompassing parts or all of 33 states. Several drainage basins in the western United States do not reach the ocean but end at lake in the continental interior. The Great Basin, centered in the state of Nevada, is the largest area of dead-end drainage that does not reach the ocean (Figure 16.4). The continental divide, labeled in Figure 16.4, is a particularly notable drainage divide that separates surface water flowing to the Atlantic and Arctic Oceans from that flowing to the Pacific Ocean.

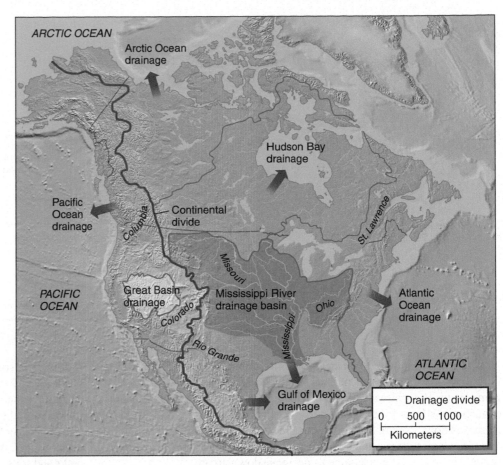

▲ **Figure 16.4 Outlining the drainage basins of North America.**
The red lines on this map are the large-scale drainage divides that separate streams draining to different oceans surrounding North America. The Continental Divide separates streams flowing westward toward the Pacific Ocean from those flowing eastward and northeastward to the Atlantic and Arctic Oceans. The Great Basin is an area of internal drainage, where runoff does not drain to the ocean but instead accumulates and evaporates in shallow lakes.

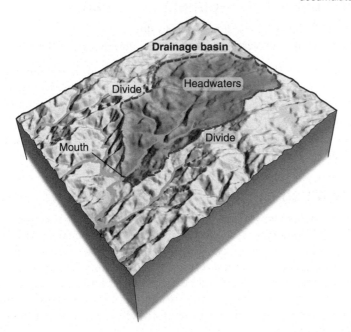

▲ **Figure 16.3 Visualizing drainage basins.** A drainage basin is outlined in red. Ridges outline the divides that separate neighboring drainage basins. All surface runoff from precipitation that falls within one drainage basin flows into the stream with the labeled headwaters and mouth.

Discharge: The Amount of Water Flowing in a Stream

Another way to describe stream size is to measure how much water flows through the channel. Think back to the field observations and imagine holding a rope across the channel and then measuring the volume of water passing under that rope in each second. You are measuring the volume of water per interval of time (e.g., cubic meters per second) that passes through that part of the channel. This measurement is called the **discharge**.

Figure 16.5 shows how to calculate the discharge by multiplying measurements of the flow velocity and the cross-section area of the flowing water. It is impractical to measure discharge everywhere along a stream. Instead, discharge is calculated from measurements of water depth at **stream gages** located along parts of the stream where the channel cross section is carefully surveyed. Data collected over many years permit estimation of discharge simply by knowing how deep the water is in the channel.

EXTENSION MODULE 16.1

How a Stream Gage Works. *Learn how a stream gage is constructed and how hydrologists use the gage data to determine discharge.*

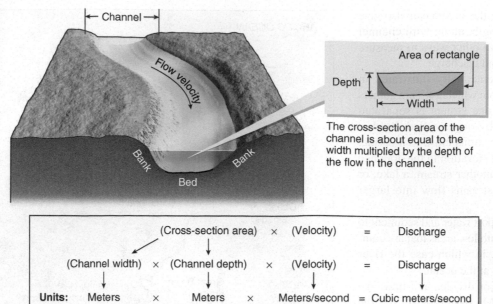

Geologist measuring water depth and flow velocity

The cross-section area of the channel is about equal to the width multiplied by the depth of the flow in the channel.

(Cross-section area)	×	(Velocity)	=	Discharge
(Channel width) × (Channel depth)	×	(Velocity)	=	Discharge
Units: Meters × Meters	×	Meters/second	=	Cubic meters/second

◄ **Figure 16.5 Visualizing the discharge of a stream.** Discharge is the amount of water flowing in a stream. It is defined as the volume of water passing through a particular cross section of channel in an interval of time. Discharge is calculated by multiplying the cross-section area of the flow and the average flow velocity. The cross-section area of flow is the width multiplied by the depth for a rectangular channel. Real channels do not have perfectly rectangular cross sections, so careful surveys of the channel bed and banks are needed to calculate the area. The average velocity is obtained by making many measurements of velocity across the channel, because the velocity varies from place to place.

Discharge changes downstream within a drainage basin. There is very little discharge in a stream near its headwaters, but the stream gradually gains water downstream by runoff from increasingly larger areas and contributions from tributaries. Therefore, the discharge of a river usually increases downstream. Exceptions to this general rule are streams in dry regions, where stream water soaks through the streambed to become ground water. This causes the discharge to decrease downstream.

The discharge at one location along a stream varies with time. Discharge increases when runoff is high because of heavy rain or rapidly melting snow. Most of the time, however, the water does not fill a channel to the top of its banks. Stream-gage data show that the discharge necessary to fill the channel usually occurs about once every two years. A **flood** occurs when the banks of the stream can no longer contain the discharge.

Putting It Together—Where Does the Water Come From?

• Stream water is surface runoff from rainfall and snowmelt plus infiltrated ground water that reemerges at the surface where the water table intersects stream channels.

• The drainage basin is the area from which water flows to a stream and divides separate adjacent drainage basins.

• Discharge is the volume of water that passes through a cross section of a stream channel during an interval of time.

16.2 Where Does Sediment Come From?

The rivers you examined in the virtual field (Figure 16.1) are as notable for the sediment they carry as for the amount of water flowing in the channel. It is worthwhile, therefore, to understand the origin of this sediment. You know from Chapters 5 and 14 that sediment results from

rock weathering, which also produces dissolved ions. Both sediment particles and dissolved ions are transported by streams. The particles carried by the stream are its **sediment load** and the invisible ions form the **dissolved load**.

When geologists examine the sediment load, they find that, as you might expect, the composition of the sediment corresponds to the types of rocks and soils observed in the drainage basin. Streams erode, transport, and deposit sediment liberated from rocks by weathering and mass movement. By these processes, high mountains gradually wear down and the resulting sediment accumulates in river lowlands or ocean basins.

The Sources of Sediment in Streams

Figure 16.6 shows how sediment gets to streams. See whether you can observe any of these processes in action the next time you visit a flowing stream.

• Surface runoff after rainfall or snowmelt picks up loose sediment and washes it into the stream (also see Figure 14.24).

• Mass movements deliver regolith and rock from hillslopes directly to the stream (also see Figure 15.30).

• Streams pick up sediment where they erode horizontally into the banks or erode vertically downward through their beds.

How Much Sediment Do Streams Carry?

Measurements show that each year, the world's rivers transport about 10 cubic kilometers of weathered material to the oceans and lakes as either sediment or dissolved load. About five-sixths of these river loads are sediment grains and the rest are dissolved ions. The annual global sediment load would fill a freight train that encircles Earth 24 times at the equator. If spread equally over Earth's land surfaces, this volume of sediment represents 7 millimeters of elevation–lowering of the continents every century.

Streams with large drainage basins and discharges, such as the Mississippi River, carry hundreds of millions of metric tons of sediment to their mouths each year. Not all of the sediment eroded within the drainage

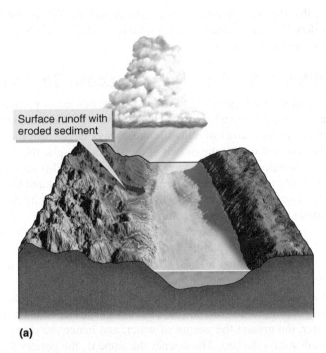

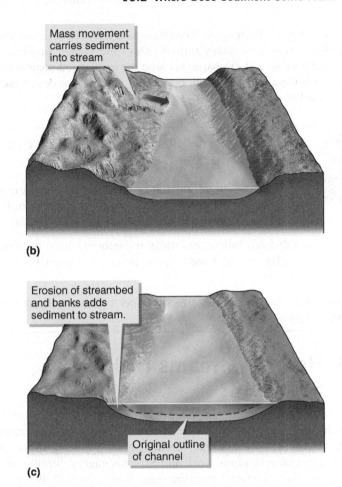

(a)

(b)

(c)

▲ Figure 16.6 Three ways to get sediment into flowing stream water.

Surface runoff with eroded sediment

Mass movement carries sediment into stream

Erosion of streambed and banks adds sediment to stream.

Original outline of channel

basin immediately flushes out to the river mouth. This is because streams not only erode and transport sediment, but they also store it to form the gravel bars and floodplains observed in the field (Figure 16.1).

To determine whether some rivers transport more sediment than other rivers of similar size, it is convenient to divide the total mass of sediment transported by the area of the drainage basin where the sediment comes from. On this basis, the sediment load of major rivers graphed in **Figure 16.7** varies from only about 10 metric tons per square kilometer per

year to more than 1700 metric tons per square kilometer per year. How can geologists explain these huge differences? Most of the extremely sediment-laden rivers are in Southeast Asia and drain the steep slopes of the Himalayas. Are these rivers carrying more sediment because steep

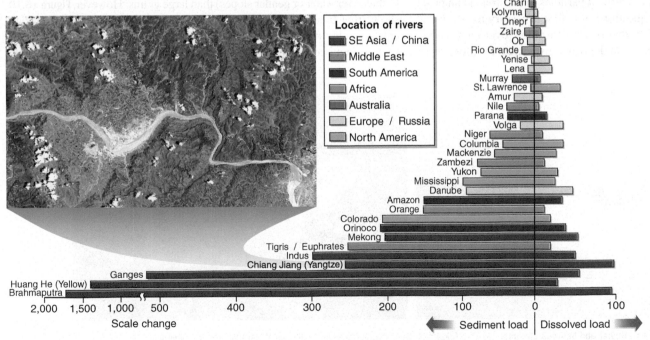

Location of rivers

- SE Asia / China
- Middle East
- South America
- Africa
- Australia
- Europe / Russia
- North America

Chari
Kolyma
Dnepr
Zaire
Ob
Rio Grande
Yenise
Lena
Murray
St. Lawrence
Amur
Nile
Parana
Volga
Niger
Columbia
Mackenzie
Zambezi
Yukon
Mississippi
Danube
Amazon
Orange
Colorado
Orinoco
Mekong
Tigris / Euphrates
Indus
Chiang Jiang (Yangtze)
Ganges
Huang He (Yellow)
Brahmaputra

2,000 1,500 1,000 500 400 300 200 100 0 100

Scale change

◀ Sediment load | Dissolved load ▶

Metric tons per square kilometer of drainage basin per year

◀ Figure 16.7 Rivers transport weathering products. This bar graph illustrates the annual sediment and dissolved loads for the world's major rivers. Dividing the total mass of sediment particles (sediment load) and dissolved ions (dissolved load) by the drainage basin area helps to compare rivers with different-sized drainage basins. The largest masses of sediment carried in rivers in Southeast Asia and China. The satellite image shows the muddy Chiang Jiang River in China. Dissolved load is smaller than sediment load for most rivers, with notable exceptions such as the St. Lawrence River, where weathering of limestone produces large masses of dissolved ions in comparison to sediment load.

slopes erode faster? Then again, Southeast Asia has a wet climate with a profound wet season of heavy rainfall. Does the greater precipitation increase weathering and discharge to cause the greater sediment delivery to the ocean? Geologic data show both relief and climate determine the sediment load of streams.

Putting It Together—Where Does Sediment Come From?

• Streams transport the products of rock weathering, which include particles comprising the sediment load, and dissolved ions constituting the dissolved load.

• Surface runoff from hillslopes, mass movements from hillslopes, and erosion of channel bed and banks deliver sediment to flowing streams.

• Streams with larger drainage basins and higher discharges usually carry more sediment.

16.3 How Do Streams Pick Up Sediment?

A stream must pick up sediment grains before carrying them downstream. Stream channels are evidence of the erosive power of water to carve through regolith or bedrock. Why do streams erode sediment? Picking up and moving sediment grains requires work and expends energy. Stream energy mostly results from the fact that stream water flows downhill. When water enters a stream channel, it possesses the potential energy of that elevation in the landscape (Section 1.6 and Figure 1.16 explain potential and motion energy). The potential energy converts to motion energy as the water flows downslope. The motion energy is sufficient for the water to do work—it picks up and moves sediment.

Streams have energy to move sediment but how do streams actually erode their beds? How big are the sediment particles that a stream can pick up and move? To answer these questions, it is first necessary to consider, using the examples in **Figure 16.8**, that rivers either flow over loose sediment or solid rock. The loose sediment deposited by a stream is **alluvium**,

and rivers that flow in alluvium are **alluvial streams**. In contrast, **bedrock streams** flow through channels cut into solid rock. Different processes are required to erode alluvium and bedrock.

How Alluvial Streams Pick Up Loose Sediment

It might seem logical to think that faster water carries larger particles, so a stationary particle eventually moves if the flow velocity increases enough. However, the particle sizes of alluvium are always larger near the headwaters than downstream at the mouth of the stream, despite the fact that measured flow velocity typically *increases* downstream. This implies that faster flows only transports finer rather than coarse sediment and leaves us looking for another explanation for why streams erode. Velocity does not really explain how streams erode and transport sediment.

Force, not velocity, must be exerted to move any object. Flowing water exerts a downstream force parallel to the streambed. This force is a shear stress that works similar to the shear stress applied parallel to a fault plane (see Section 11.3 for more about shear stress).

The shear stress that flowing water exerts on a streambed depends on the weight of the water and the steepness of the streambed. The deeper the water, the greater the weight of water, and hence the shear stress that bears down on the bed. The steeper the slope is, the greater the pull of gravity on the water running down it. Increasing the water depth (water weight), the streambed slope, or both increases the shear stress.

Figure 16.9 illustrates how shear stress moves sediment. The concept is similar to comparing the driving and resisting forces of mass movements in Chapter 15. Shear stress is a measure of the gravity driving force. The forces resisting movement are the weight of the particle plus friction and cohesion with its neighbors. To pick up sediment grains, the shear stress of the flowing water must be larger than the weight of the grains and the friction and cohesion between them. Particles at rest on the stream bed usually start to move when shear stress increases because of increasing water depth that occurs when discharge increases (Figure 16.9b).

Smaller sediment grains weigh less than larger grains, so you can initially assume that small grains erode at lower shear stresses (which equates to shallower water or gentler slopes) than large grains. However, **Figure 16.10** shows that this is only partly true. Silt- and clay-size particles include cohesive clay minerals (see Sections 5.4 and 15.2), so additional shear stress is required to overcome the cohesion beyond the weight of the grains.

▲ Figure 16.8 **Contrasting alluvial and bedrock streams.** (a) The Arkansas River in Oklahoma is an alluvial stream flowing in a valley of stream-deposited alluvium. (b) The Potomac River, at Great Falls, Virginia, is a bedrock stream, flowing directly on rock.

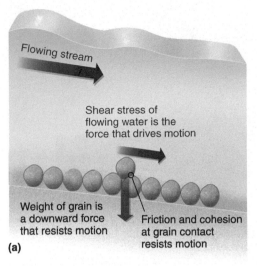

Shear stress of flowing water is the force that drives motion

Flowing stream

Weight of grain is a downward force that resists motion

Friction and cohesion at grain contact resists motion

(a)

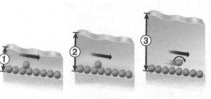

(b) Increasing water depth = increasing shear stress

(c) Increasing slope = increasing shear stress

▲ **Figure 16.9** What it takes to move alluvial sediment grains.

Weight, which is the downward pull of gravity, holds a sediment grain on the streambed. Flowing water, also driven by gravity, exerts a shear stress on the grain surface. For motion to occur the shear stress must exceed the weight of the grain, plus friction and cohesion at grain contacts.

Shear stress increases when water depth increases, and eventually the stress is sufficient to move the grain.

Shear stress increases with increasing channel slope, so sediment grains move most easily on steep slopes.

Once these small grains erode, however, they are so small and lightweight that streams easily transport them.

The overall conclusion reached from these observations is that loose sediment grains erode when the shear stress of the flowing water is sufficient to cause erosion. For example, gravel first moves at a higher shear stress than is required to move sand. For a stream to move gravel it must be deeper, or steeper, or both, than a stream that moves sand but does not move gravel. Cohesive clay and silt grains require relatively higher shear stresses to get picked up despite their small size.

How Bedrock Streams Break Off Pieces of Rock

Some streams flow in the bottoms of bedrock canyons that are more than 1000 meters deep. The Grand Canyon, pictured in Figure 1.2b, is a good example. Mass movements from the steep canyon walls widen the valley and bring particles to the stream, which then moves the particles as alluvium. Mass movements, however, do not deepen stream channels. In order to cut a deep canyon, the stream must break solid rock into moveable pieces.

Figures 16.11 and **16.12** illustrate two observable processes that dislodge rock fragments at the bottom of a bedrock stream:

1. Abrasion removes rock fragments from a solid rock surface by the hammerlike impact of sediment grains carried in the water.

2. Plucking gradually pries loose blocks of rock along preexisting joints and bedding planes.

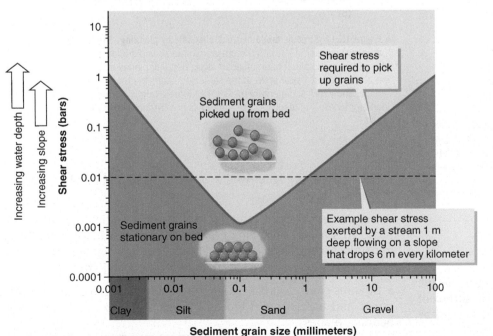

◄ **Figure 16.10** Different grain sizes move at different shear stresses. More shear stress is required to pick up gravel than sand because gravel is larger. Silt and clay particles are small but they require large shear stresses for movement because the grains are cohesive and stick to one another. The dashed line drawn on the graph shows that the shear stress exerted by a 1-meter-deep stream flowing on a slope of 6/1000 is sufficient to erode grains between 0.02 mm and 1 mm across (the grain-size values are on the horizontal axis where the dashed line crosses the graph).

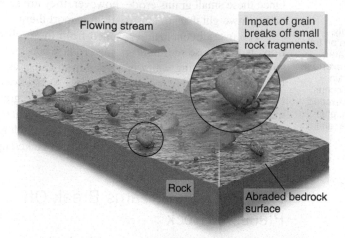

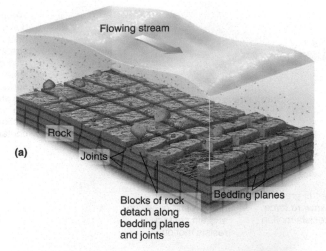

▲ **Figure 16.11 Streams erode bedrock channels by abrasion.** Sediment grains carried in the water act like sandpaper to break off and grind down the underlying bedrock. The photo shows scoured and polished bedrock exposed along the Indus River in Pakistan. This outcrop is submerged and abraded when the river flows with higher discharges during floods.

▲ **Figure 16.12 Streams erode bedrock channels by plucking.** The shear stress of the flowing water along with the pressure of water and sediment grains forced into cracks pluck bedrock apart along bedding planes and joints. The photos, taken 1 year apart at the same spot along a stream in Ontario, Canada, show how sedimentary rocks break loose along joints and bedding planes. The resulting rock fragments are carried away by the stream.

Putting It Together—How Do Streams Pick Up Sediment?

• Flowing water picks up loose sediment when the shear stress exerted by the flowing water is sufficient to overcome the resistance to movement caused by the weight of the grains, friction, and cohesion.

• Shear stress is higher where water is deeper, streambed slope is steeper, or both.

• Larger shear stress is required to erode large sediment grains compared to small grains. Fine-grained sediment with abundant clay minerals is very cohesive, however, and requires higher shear stresses to erode than otherwise is needed for the small grain size.

• Streams erode bedrock by abrading the bed with transported sediment and plucking out blocks of rock along existing joints and bedding planes.

16.4 How Do Streams Transport Sediment?

We now know how streams pick up sediment. Next, we need to examine two more questions: How do the grains move with the flowing water? Is there a limit to how much sediment can be transported by a stream?

Sediment Moves as Bedload and Suspended Load

From the banks of the virtual Alaskan river in Figure 16.1a, you heard cobbles banging into one another beneath the water surface but the moving gravel was invisible because the water was too murky to see through. A better way to study sediment movement in flowing water is to make laboratory observations of sediment and water moving through transparent glass-walled channels. **Figure 16.13** summarizes the observations from laboratory and natural channels. Moving sediment occupies two positions in the stream—(1) on or near the bed, and (2) suspended in the water.

Large grains roll, slide, and bounce along the bed and spend most of their time touching the bed. These particles comprise the **bedload** of the stream. The flow may mold bedload particles into submerged dunes and ripples, whose movement along the riverbed causes cross-bedding (look back to Section 5.6 and Figure 5.17 for more about dunes, ripples, and cross-bedding).

Most of the time only part of the sediment moves and the rest remains stationary. Movement of the largest grains requires shear stresses that occur only during unusually deep flows. Some sediment **bars** may be exposed above the water level and this sediment moves only when the flow is deeper. The unvegetated islands viewed in the Alaska field site are bars (Figure 16.1a) and are visible in the opening picture for this chapter and in Figure 16.8a. The fact that bars lack vegetation and soil indicates that they are not permanent features, but instead are submerged and moved frequently by high-discharge flows.

In contrast to the bedload is the **suspended load**, which consists of small sediment grains that mix with the flowing water and are transported above the bed, which they rarely touch. Suspended clay and silt grains give a muddy appearance to many streams and may obscure the view of bedload transport along the streambed.

Stream Power Limits Sediment Transport

Just because flowing water exerts enough shear stress to move sediment of a particular size does not necessarily mean that the particles travel very far. The stream has to do work and expend energy to erode and transport sediment. We can describe the work done by a stream by combining the shear stress, which picks up sediment, and the average flow velocity, which transports sediment. Multiplying these two measurements together provides a third value, called **stream power**, which describes the ability of the stream to do work.

The larger the volume of sediment to transport, the more stream power is required to do the job. Decades of measurements show that this concept of stream power explains where streams erode their beds and where they deposit sediment. If the stream power along a stretch of the stream is just sufficient to transport the sediment carried in from upstream, then all of the power is used up to continue moving that sediment, and no additional sediment erodes from the streambed. If, however, there is more stream power than is needed to transport the sediment from upstream, then more sediment erodes and is transported. If the stream power at a location is insufficient to carry what has come from upstream, then some sediment is deposited until the sediment load matches the power available to move it.

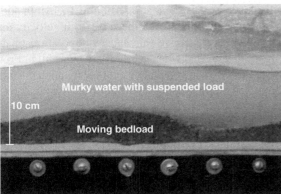

▲ **Figure 16.13 Visualizing sediment moving as bedload and suspended load.** Coarse grains roll, slide, and bounce along the bed as bedload. Finer grains remain suspended by eddies in the stream current and completely mix with the flowing water. The photo shows how geologists study bedload and suspended load transport by observing artificial streams through a glass-walled channel.

ACTIVE ART

How Streams Move Sediment. See how flowing water moves the sediment load.

Putting It Together—How Do Streams Transport Sediment?

• Sediment that rolls, slides, and bounces along the bed of the stream is bedload, whereas finer sediment intimately mixed with the water flowing above the bed is suspended load.

• Bars are mounds of sediment that are stationary and exposed at low discharge but are submerged and transported at high discharge.

• Stream power, which is the product of shear stress and flow velocity, describes the ability of a stream to do work. If the stream power is just what is necessary to transport the sediment delivered from upstream, then there will be no erosion or deposition.

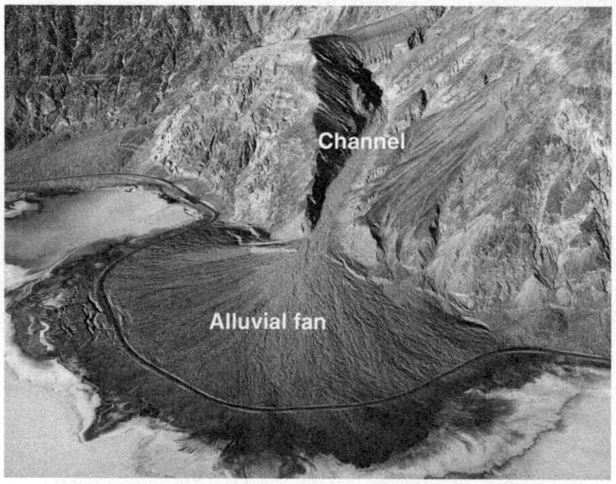

Deep water flow in confined channel

Sediment transport: high shear stress and stream power

Shallow water flow on unconfined alluvial fan

Sediment deposition: low shear stress and stream power

16.5 Why Do Streams Deposit Sediment?

What conditions determine where a stream deposits sediment? A stream must lose its ability, or power, to move sediment when it switches from erosion and transport of sediment to depositing sediment. Stream power is the product of shear stress and velocity; so for stream power to decrease, the shear stress, velocity, or both also must decrease. **Figure 16.14** shows how you can predict locations of sediment deposition from knowing the factors that change shear stress and velocity and, therefore, change stream power.

▲ **Figure 16.15 How alluvial fans form.** Alluvial fans form where water flowing in a confined channel abruptly spreads out on an unconfined valley floor. The abrupt increase in the width of flowing water causes an equally sudden decrease in the flow depth. The resulting decreases in shear stress and stream power cause deposition. The photograph shows an alluvial fan at the mouth of a bedrock canyon in Death Valley National Park, California.

Channel

Alluvial fan

Deposition Happens when Discharge Decreases

Consider the example of a brief increase and then decrease in stream flow following a heavy rain. When the discharge decreases, the velocity, width, and depth of flow decrease. Decrease in water depth also decreases shear stress (Figure 16.14). In addition, stream power decreases because both velocity and shear stress decrease, so this causes deposition of the sediment that eroded when the discharge was high. From this example we can conclude that erosion and deposition alternate over time along a stream because discharge changes.

Deposition Happens Where Water Depth Decreases

Figure 16.15 shows how streams flow through narrow bedrock channels in a mountain range, and then form fan-shaped masses of alluvium called **alluvial fans** where they enter a broad valley where there are no hard-rock valley walls to confine the flow.

▼ Figure 16.14 **Tracking the results of changing streamflow characteristics.**

Stream power = Shear Stress × Velocity

Increase water depth	Increases → Shear stress	Increases → Stream power	Results in →	Erosion
Decrease water depth	Decreases → Shear stress	Decreases → Stream power	Results in →	Deposition
Increase channel slope	Increases → Shear stress	Increases → Stream power	Results in →	Erosion
Decrease channel slope	Decreases → Shear stress	Decreases → Stream power	Results in →	Deposition
Increase flow velocity	Increases → Velocity	Increases → Stream power	Results in →	Erosion
Decrease flow velocity	Decreases → Velocity	Decreases → Stream power	Results in →	Deposition

Can we use our understanding of stream power to explain why the stream deposits the sediment that forms the alluvial fan? In this case, the discharge does not change, but the flow is deeper in the upstream channel than it is where the water spreads out downstream as a shallow sheet. The shallower flow exerts less shear stress, and this means the stream power drops, so sediment abruptly deposits to form the fan.

Deposition Happens Where Slope Decreases

Sediment grain size is coarser at the headwaters than at the mouth of a stream. We can explain this observation by recalling that shear stress is very sensitive to slope (Figures 16.9 and 16.14). Therefore, if the channel slope decreases without a compensating increase in water depth, then the shear stress decreases. Most importantly, the slope of most stream channels decreases downstream while depth increases, as will be examined more closely in Section 16.6. The slope decrease is more substantial than the depth increase, so shear stress usually decreases downstream. The downstream decrease in shear stress means that it is increasingly difficult for the stream to erode the largest particles on the bed, so they are left behind while finer sediment continues downstream.

Deposition Happens Where Velocity Decreases

What happens where streams flow into large, relatively still bodies of water such as the ocean, lakes, or artificial reservoirs? Where a river enters still water, the current velocity drops to almost nothing, and this causes water to back up at the mouth of the channel. If flow velocity decreases, then stream power also decreases, which diminishes the sediment-carrying capacity of a stream. The decrease in stream velocity where the stream approaches and enters a standing body of water should cause deposition of sediment.

A **delta** is the landform produced by deposition of sediment where a stream enters a lake, reservoir, or sea. The presence of deltas is consistent with our deduction that sediment is deposited where streams slow down. The largest delta in North America forms where the Mississippi River enters the Gulf of Mexico. This delta, shown in **Figure 16.16**, covers about 28,600 square kilometers, which by comparison is a little larger than the area of Maryland. Deposition of the coarsest bedload takes place first and clogs the channel. The resulting bars and vegetated islands of river-deposited sediment cause the channel to split into many smaller **distributary channels** (Figure 6.16). The deposited bedload accumulates along the coastline and is redistributed by waves and ocean currents. The suspended load continues out to sea, where it gradually settles to the seafloor.

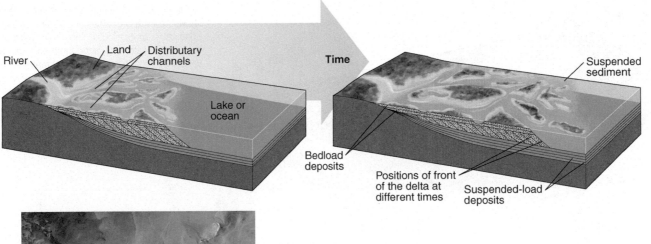

▲ **Figure 16.16 How a delta forms.** Deltas form where streams enter the ocean or a lake. Flow velocity decreases where the stream approaches still water, which also decreases the stream power and causes sediment deposition. Bedload accumulates inclined layers that build the delta up to sea level and also causes the shoreline to migrate outward from land. Bedload deposition also obstructs the channel, causing it to divide around swampy islands into many distributary channels. Suspended sediment settles out far from shore as muddy layers. The satellite photo shows the Mississippi River delta in Louisiana.

Putting It Together—Why Do Streams Deposit Sediment?

• Streams deposit sediment where the stream power decreases because the discharge decreases, the water depth decreases, the slope decreases, or the velocity decreases.

• Alluvial fans form where sediment deposition results from abrupt change from a narrow, deep, confined channel to a wide, shallow, unconfined sheet.

• Deltas result from deposition of sediment because of a decrease in flow velocity where a stream enters still water.

16.6 Why Does a Stream Change Along Its Course?

Figure 16.17 graphically summarizes hundreds of measurements of stream characteristics moving downstream from headwaters to mouth. The scientific challenge is to explain these measurements.

Some of the changes along the stream course are easy to explain. Clearly, elevation must decrease because water flows downhill. It also makes sense that discharge and sediment load increase toward the mouth because tributaries add more water and sediment.

Other downstream changes require more thought. For example, discharge increase could happen simply by increasing just width, depth, *or* velocity, but all three increase downstream. Why is this the case? Also, why do the slope angle and grain size decrease? And, what causes shear stress to decrease downstream? The graph in Figure 16.17 holds the clues to understanding many dynamic aspects of water and sediment transport in streams.

Streams Adjust to Carry Available Water and Sediment

Observations show that the width, depth, and slope of a channel at any location naturally adjust to carry the amounts of water and sediment delivered from immediately upstream. Any change in the amount of water or in the amount or grain size of sediment moving through the channel causes adjustments in channel dimensions or slope to match to the new conditions. These adjustments result in deepening or widening of the channel and erosion or deposition of sediment.

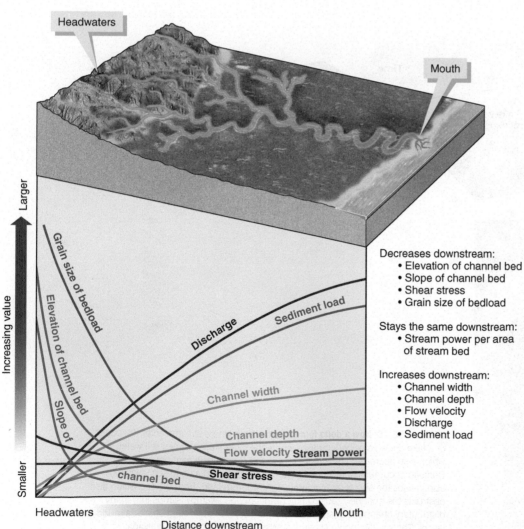

◄ **Figure 16.17 Graphing downstream changes.** This graph shows how the properties of a stream change from its headwaters to its mouth. The curves represent the changes seen in most streams, although there are many natural variations in each of these characteristics.

Decreases downstream:
• Elevation of channel bed
• Slope of channel bed
• Shear stress
• Grain size of bedload

Stays the same downstream:
• Stream power per area of stream bed

Increases downstream:
• Channel width
• Channel depth
• Flow velocity
• Discharge
• Sediment load

Streams Become Wider, Deeper, and Faster Downstream

Discharge increases downstream, so either the cross-sectional area of the stream, the velocity of flow, or both must also increase because discharge is the cross-sectional area multiplied by velocity. Surface runoff and tributary flow to the main stream not only add water to the channel, but also add sediment. This means that the river fine-tunes the channel width, depth, and velocity to carry the increasing amounts of both sediment and water. One important adjustment is widening of the channel so that there is more streambed for moving more sediment, just like a wide conveyor belt in a factory moves more material than a narrow one.

However, there is a limit to how much the stream can widen. For a particular discharge a wider channel is also shallower, and the flow is slower because of greater friction of water against the wider bed. If depth and velocity decrease as the channel widens, then the stream loses stream power and is unable to carry the increasing amount of sediment. This means that not only does the channel widen downstream, but it also becomes deeper, and the flow velocity increases slightly (Figure 16.1). In other words, all three components of discharge (width, depth, and velocity) increase downstream.

Slope Decreases Along the Base Level of Erosion

The downstream decrease in slope connects to the observation that the stream power, calculated as the product of shear stress and velocity, does not change along the length of a stream (Figure 16.17). Because velocity increases downstream, the shear stress must decrease downstream in order for the stream power to remain unchanged.

Figure 16.18 contrasts the slopes of a well-adjusted alluvial river and a bedrock river with waterfalls. Where the channel slope decreases downstream, a graph of elevation along the alluvial channel makes a smooth, concave-up curve that illustrates the profile of the stream channel. Some bedrock streams, however, lack smooth profiles. Relatively soft rocks erode easily to form a concave-up elevation curve. However, where hard, resistant rocks are present, the elevation curve has stretches of relatively low slope, alternating with steep slopes and waterfalls (also see Figure 16.8b).

Study of channel-slope profiles leads to the conclusion that a river has the ability to erode its bed down to a specific elevation, called **base level**, everywhere along its course. If the stream slope is adjusted to carry the water discharge and sediment load, then the base level along each point in the stream is the elevation along the expected concave-up elevation profile of the stream bed shown in Figure 16.18a. If the stream enters the ocean, its ultimate lowest elevation, or ultimate base level, is sea level.

Geologists assume that any deviations from a smooth concave-up profile reveal locations where the river has not eroded to its base level. For example, the bedrock stream in Figure 16.18b has not eroded to its potential base level because resistant rock perturbs the elevation profile at the waterfall.

When the elevation of a stream is at base level, the stream neither erodes nor deposits sediment. The stream simply transports the sediment that washes in from slopes or tributaries or enters the channel through mass movements. If changes in land-surface slope, water discharge, or sediment supply knock the stream out of adjustment, then it erodes or deposits sediment, or both in different places, to achieve a new base level.

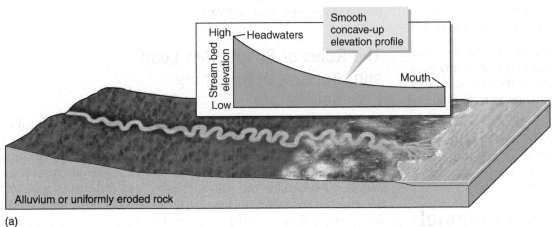

◄ Figure 16.18 Comparing river profiles. (a) The elevation profile of a stream commonly forms a concave-up curve, with steeper slopes near the headwaters and gentler slopes near the mouth. This profile indicates that the stream slope is well adjusted to the sediment load and discharge. Such a profile is typical of streams flowing over alluvium or uniformly eroded rock. (b) The smooth elevation profile is interrupted where a stream flows over rock that is resistant to erosion. A waterfall may form where the resistant rock forms the streambed. Further bedrock erosion takes place as the stream attempts to establish the smooth potential profile.

(a)

(b)

Sediment Size Decreases Downstream

Imagine measuring sediment grain size at many locations along the Mississippi River. Your data would show that gravel makes up 30 percent of the bedload in southern Illinois, but only the finest sand, silt, and clay make it to the mouth of the river. Why does grain size decrease downstream?

Stream-transported particles are rounded, so it is reasonable to suggest that large grains grind down into small grains. After all, sediment fragments loosened from rock by weathering are angular. With increasing downstream transport, however, they become more and more rounded (also see Figure 5.10). Rounding happens as the sharp edges and corners are knocked off by grain collisions, which also make the grains smaller.

However, careful observations indicate sudden decreases in grain size over short distances along streams that grain rounding and abrasion cannot explain. This means that in addition to the effects of abrasion, streams selectively transport smaller particles downstream and leave larger ones behind (Figure 16.17). The selective transport happens because even though water depth gradually increases downstream, the slope angle drops off dramatically so that the overall effect is to decrease shear stress. Large fragments only move when shear stress is high, so larger grains are left behind in upstream locations as shear stress decreases downstream.

Putting It Together—Why Does a Stream Change along Its Course?

• The width, depth, and slope of a channel adjust to changing discharge or sediment load in order to maintain the appropriate stream power to transport the available water and sediment.

• Discharge and sediment load increase downstream, causing channel width, channel depth, and flow velocity to increase downstream, too.

• Slope decreases downstream along the base level of erosion, which is the elevation to which the stream can erode its bed at any location along its course. If a stream is everywhere at its base level, then a graph of elevation of the streambed from headwaters to mouth is a smooth, concave-up curve.

• Sediment grain size decreases downstream. This happens mostly because decreasing slope causes diminished shear stress and partly because of abrasion of grains during transport.

16.7 What Factors Determine Channel Pattern?

Channel width, depth, slope, and flow velocity are not obvious when standing on a stream bank and require actual measurements. The outline of the channel itself is visible, however, and can be highly variable (see Figure 16.1). Think back to the field observations in Figure 16.1: Why are most channels very curvy, such as those observed for the Mississippi River, instead of straight? Why do some rivers have sediment bars within the channel, such as the Alaskan river, while others do not?

Meandering and Braided Channels

The Mississippi River channel is so curvy, it looks like a sinuous snake. This is normal for streams and rivers; in fact, it is extremely rare to find a natural stream channel that is perfectly straight. Most sinuous channels also shift across valleys by a process called **meandering**.

Figure 16.19 uses velocity measurements and shear stress to explain channel meandering. The flow is faster and deeper around the outside of a channel bend, causing erosion of a **cutbank**. At the same time, flow is slower and shallower on the inside of the bend, causing sediment deposition to form a **point bar**. Simultaneous erosion and deposition on opposite stream banks cause the channel to meander by shifting toward the cutbank.

Flow in the less sinuous, gravelly Alaskan river (Figure 16.1a) separates and rejoins around bars. This channel shape is described as **braided**, because the pattern of water flow around the bars resembles braided rope or hair.

Many streams are both meandering and braided, whereas others are described as primarily either braided or meandering. There is a complete gradation between, and combinations of, the attributes of these two channel patterns. Geologists combined many field observations to determine why meandering and braided patterns form. **Figure 16.20** summarizes the results.

The Roles of Bedload and Discharge in Determining Channel Pattern

Braiding is more common where streams carry large volumes of bedload. We can explain this observation by concluding that the channel must be wide to transport all of the sediment, but when the discharge does not completely fill the channel, some of the bedload is immobile and exposed as bars. The more variable the discharge, the longer the times that the stream is unable to transport all of its bedload. This circumstance favors the formation of bars in a braided stream.

The Roles of Suspended Load and Slope in Determining Channel Pattern

Meandering is more common where streams carry mostly suspended sediment and slopes are low. We can explain these observations by considering that fine-grained suspended sediment moves at low stream power. The stream flows on a low slope so that the stream power remains just adequate to transport mostly suspended load with minimal erosion of the bed. **Figure 16.21** demonstrates that a sinuous channel has a lower slope than a straight channel flowing in the same valley, which explains why streams carrying abundant suspended load are typically sinuous. Combine sinuous channel shape with the erosion and deposition on channel curves (Figure 16.18) and we see why channels with low slopes also meander.

The Role of Banks in Determining Channel Pattern

We should expect that bank erodibility influences channel pattern, because a channel can widen to enhance its sediment-transport capability only if the banks easily erode. Where banks easily erode, the channel tends to be wide and shallow, leading to a braided pattern. Where banks are stable because of cohesive clay or plant roots, then the channel tends to be narrow, deep, and sinuous.

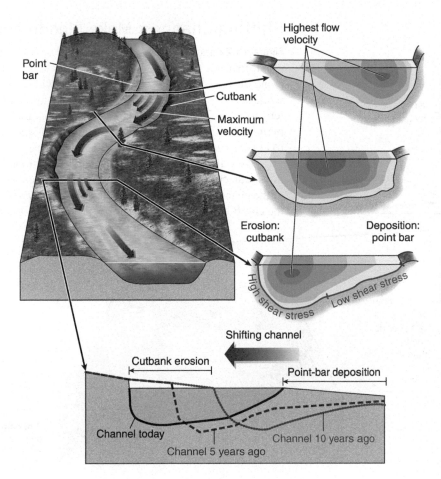

Flow velocity is slower close to banks and bed because of friction. When water flows around a curve in a channel, the flow is faster near the outside of the bend than it is near the inside of the curve. This velocity pattern causes erosion of the cutbank on the outside of the curve, and deposition of sediment to form a point bar on the inside.

Erosion maintains a deep channel on the outside of the bend, whereas point-bar deposition causes shallow water on the inside of the curve.

Simultaneous erosion and deposition on opposite banks cause the channel to shift position through time.

◀ **Figure 16.19 Why a sinuous stream meanders.**

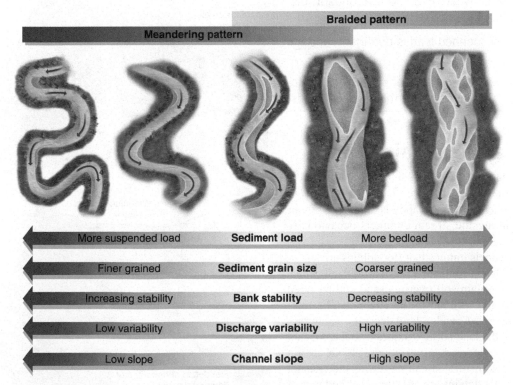

▲ **Figure 16.20 Factors determining channel pattern.** Most streams exhibit a meandering or braided channel pattern, or some combination of both patterns. The channel pattern relates to variations in the characteristics listed in the diagram.

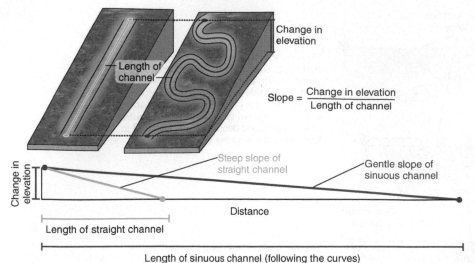

$$\text{Slope} = \frac{\text{Change in elevation}}{\text{Length of channel}}$$

▲ **Figure 16.21 A straight channel is steeper than a sinuous channel.** The channel slope equals the elevation change along the length of the channel. The length of channel is the actual distance that water flows in the channel. A straight channel between two points is shorter than a curving channel between the same two points. Therefore, the slope of a long, curvy, sinuous channel is less than the slope of a short, straight channel with the same downstream drop in elevation.

Putting It Together—What Factors Determine Channel Pattern?

• Channels braid where bedload is abundant, discharge is variable, and the banks erode easily.

• Channels meander where suspended load is dominant, and the banks are cohesive and difficult to erode.

• Highly sinuous channels meander because of simultaneous erosion and deposition on opposite banks where the channel bends.

16.8 How Does a Floodplain Form?

Streams are almost always flanked by flat areas that, together with the channel, form a valley bottom. This is true even for many streams in hilly and mountains regions where flat topography is otherwise rare or nonexistent. The flat areas alongside stream channels occasionally flood, so they are commonly called floodplains.

Their association with channels implies that floodplains are somehow constructed by streams. This is confirmed by shallow excavations showing that fine-grained floodplain soil overlies sand and rounded gravel deposits that resemble the bedload in the channel. The fine-grained soil consists of weathered suspended-load deposits. Therefore, we can define a **floodplain** as the land surface adjacent to the channel that is made by the river and is inundated during floods. Floods are not rare events, as discharge in most natural streams exceeds the channel depth about every two years. Floodplains are also the inhabited part of the river-formed landscape, so it is important to understand how floodplains form and change through time.

Shifting Channels Make Floodplains

Figure 16.22 shows how rivers widen their valley to form a floodplain. Erosion along one bank is compensated for by deposition along the opposite bank so that the channel width remains the same while shifting position. Channels typically shift from 0.5 to 10 meters per year. The former channel deposits that underlie the floodplain are commonly excavated as sources of sand and gravel for road construction and a variety of industrial uses.

Figure 16.23 shows that if the meandering stream is highly sinuous, adjacent cutbanks may erode toward one another until part of the channel is cutoff from the flow. Ground water may seep into the cutoff channel segment to form a lake on the floodplain. These

A stream channel may start out in a narrow valley, without a floodplain

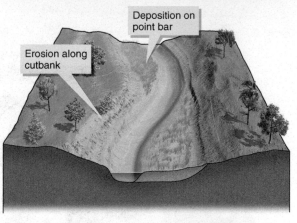

Erosion along cutbank

Deposition on point bar

Cutbank erosion and point-bar deposition cause the stream to shift horizontally.

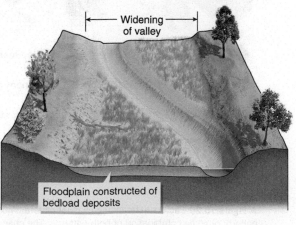

Widening of valley

Floodplain constructed of bedload deposits

Channel shifting, back and forth, gradually produces a wide, valley-floor floodplain that is underlain by channel-deposited sand and gravel

Time

▲ **Figure 16.22 Shifting channels form floodplains.**

crescent-shaped **oxbow lakes** host ecologically diverse wetland habitats on floodplains.

Overbank Flooding Builds Up Floodplains

Deposits close to the floodplain surface consist of suspended sediment that settles from flood waters that frequently inundate floodplains. **Figure 16.24** illustrates how deposited silt and clay raise the elevation by a few centimeters to as much as a meter during each flood. These periodic overbank inundations interrupt soil development and deposit new, less weathered sediment and organic matter that renew the nutrient content of floodplain soils (see Figure 14.13). The renewal of soil fertility by flooding has long drawn agricultural activity to the edges of rivers.

When river water and suspended sediment spill out of the channel and onto the floodplain, the flow spreads out as a thin sheet. The abrupt decrease in water depth leads to sharp decreases in shear stress and flow velocity that cause deposition of most of the transported sediment. More sediment accumulates next to the channel than farther away, and over time the depositional processes build up a ridge alongside and parallel to the channel, called a **natural levee** (see Figure 16.24). The gradual upbuilding of the natural levee raises the elevation of the stream banks. Natural levees are different from artificial levees, which are walls or earthen

ACTIVE ART

Meandering Stream Processes. *See how meandering streams form floodplains and oxbow lakes.*

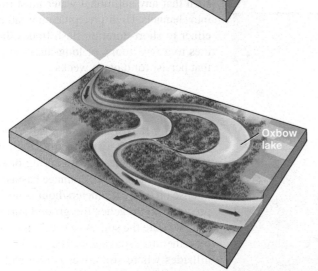

Cutbank erosion causes curves in the stream to erode toward one another.

Eventually the channel erodes through the narrow neck of land separating the curves. The cutoff meander loop may fill with water to form an oxbow lake on the floodplain.

Future meander cutoff will form an oxbow lake here.

Oxbow lakes caused by recent meander cutoffs.

The photo shows recently formed, and about-to-form, oxbow lakes along a meandering stream.

◄ Figure 16.23 How an oxbow lake forms.

Bedload moves at the bottom of the channel. Suspended load is distributed throughout the depth of the flowing water.

During a flood, the bedload remains at the bottom of the channel. Some suspended load moves onto the floodplain with the flooding water. Sediment quickly deposits on the floodplain, with the coarser grains accumulating closer to the stream.

Silt and clay deposited during floods builds up the floodplain, with greatest deposition next to the channel to form natural levees that are higher than the rest of the floodplain.

Cleaning up mud deposited from floodwater suspension is a common scene following a flood.

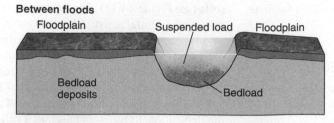

Between floods

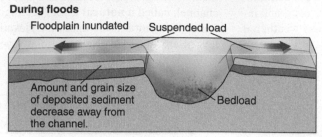

During floods

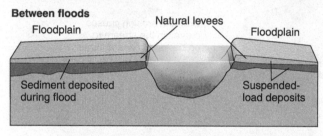

Between floods

► Figure 16.24 Suspended-load deposits build up floodplains.

ACTIVE ART

Flooding and the Formation of Natural Levees. See how floods deposit sediment and form natural levees.

Putting It Together— How Does a Floodplain Form?

• Floods occur when the discharge is too great for the flow to be contained between the stream banks.

• Floodplains are the low-relief areas adjacent to streams that inundate during floods.

• Rivers construct floodplains by horizontal shifting of the channel over time, by slow incremental deposition of suspended-load sediment during floods, or both.

16.9 Why Do Streams Flood?

What causes flood discharges that exceed the carrying capacity of the channel? Recalling that stream water originates as precipitation (Section 16.1), unusually high discharge must then relate to greater than normal precipitation. Floods occur when a drainage basin is incapable of soaking up all of the water from rainfall or snowmelt, such that any additional water must run off the surface into channels. High-precipitation weather conditions lead either to short-duration flash floods that last only minutes to a few hours, or long-duration prolonged floods that persist for days to weeks.

Flash Floods

Damaging flash floods, such as the ones illustrated in **Figure 16.25**, usually result from high-intensity rainfall, commonly associated with spring or summer thunderstorms, in small, steep, drainage basins. The rainfall is so intense (10–20 centimeters/hour is not unusual) that the precipitation reaches the ground surface faster than it can infiltrate the soil. As a result, most of the water runs off rather than soaking in. The problem is worse on steep hillsides where soil cover is thin and water infiltrates very slowly into underlying rock. Rapid runoff from steep saturated slopes overwhelms channels very quickly. Some of the most deadly flash floods happen where the heavy runoff completely fills reservoirs behind dams, causing the dams to fail and unleash even more water downstream.

Flash floods are deepest where a narrow bedrock canyon confines the channel. Where a canyon exits the mountains onto an alluvial fan, the flow expands over the wide fan surface as a relatively thin sheet of water. Figure 16.15 shows that although the water is shallower on the alluvial fan than in the canyon, the area of inundation can be quite large, so that the entire fan surface is, in effect, a floodplain (see Figure 16.25b). Many large western cities sit at the feet of mountain ranges, with large areas of the cities constructed on alluvial-fan surfaces. Examples of such flood-prone urban areas include much of greater Los Angeles, Las Vegas, and Salt Lake City.

embankments that humans build along river channels to raise the banks so that the channel holds more water at high discharges. Where artificial levees are not in place along rivers, overbank flooding occurs more frequently and damages property.

▲ **Figure 16.25 Flash flood!** (a) A flash flood did this damage in Fort Collins, Colorado, in 1997. (b) This city in Venezuela was devastated in 1999 by flood water and debris flows that spread out over a large alluvial fan after exiting the canyon visible in the upper left.

Flash floods are hard to predict, although weather forecasts can assess the likelihood of heavy thunderstorms that may cause these floods. People can then monitor water levels in streams and avoid travel along stream valleys. Street flooding in cities may occur with only modest rainfall because there is no infiltration where rain falls on roofs and pavement. Once a flood begins, the floodwaters move so rapidly that it is difficult to send effective warnings downstream. As a result, flash floods can be deadly.

Prolonged Floods

In contrast to flash floods are the cases in which water levels rise gradually along a stream over several days or weeks, and progressively submerge larger and larger areas of the floodplain. The worst such recent example in the United States is the 1993 flood in the upper Mississippi River drainage basin, illustrated in **Figure 16.26**, which submerged 26,000 square kilometers (which roughly compares to the state of Maryland). The upper Mississippi River drainage basin also experienced prolonged flooding in 2008. Prolonged floods dominate the news media for weeks with stories about the destruction of entire towns, as exemplified by **Figure 16.27**, and heroic efforts by emergency workers and volunteers to fortify the tops of levees with sandbags.

Prolonged floods occur along rivers in large drainage basins. Unusual weather conditions lead to exceptional water runoff. Heavy rain over days to weeks is the usual cause. Although infiltration occurs early in the rainy period, the ground eventually saturates so that additional rainfall runs off to channels. The higher-than-average runoff in streams with small drainage basins may not cause severe problems. However, each small stream is a tributary to a larger stream, and each successively larger stream accumulates all of the unusually high discharges from its overloaded tributaries. These discharges add up until the larger river is overwhelmed to a degree that is seen only every few decades, or perhaps only once a century.

Prolonged floods along rivers are extremely costly because these floods affect large urban and agricultural areas. The number of fatalities is small as a percentage of the number of people affected by the flood; unlike flash floods, prolonged floods rarely occur as a complete surprise, and water levels usually rise slowly enough to allow people to evacuate. Regional weather patterns leading to floods are commonly forecast several days to a week or more in advance.

◄ **Figure 16.26 The Great Flood of 1993.** Compare these two satellite images, which show the region where the Illinois and Missouri rivers join the Mississippi River near St. Louis, Missouri. The usual channel widths of the rivers are visible as dark ribbons in the image on the left. The image on the right, obtained near the peak of the 1993 flood, shows large areas of submerged floodplain.

▲ **Figure 16.27 Prolonged floods inundate large areas.** Flooding along the Iowa River in June 2008 completely submerged the business district of Cedar Rapids, Iowa, causing nearly $800 million in damage. Notice how far the flood waters, muddy with suspended sediment, extend beyond the labeled stream channel.

Putting It Together—Why Do Streams Flood?

• Floods occur because of unusually heavy rainfall or snowmelt that generates water far in excess of the volume that readily infiltrates into the ground. The resulting excess surface runoff and ground water flow to channels causes stream discharges that are too large to be confined between the stream banks.

• Flash floods start and end abruptly and usually result from heavy thunderstorm rainfall in steep, rocky, drainage basins where infiltration is low and surface runoff is rapid.

• Prolonged floods with slowly rising and falling discharges persist for days or weeks and result from unusually rainy conditions over days to months.

16.10 How Do We Know . . . the Extent of the "100-Year Flood"?

Picture the Problem

How Often Do Different Parts of a Floodplain Flood? There are substantial investments and economic incentives for expansion of communities and agriculture along nearly flat, fertile, floodplains close to river channels used for navigation and irrigation. Floodplains naturally flood, but not all areas of a floodplain are at equal risk of inundation. Areas of the floodplain that are at the lowest elevations, and usually close to the stream channel, are the areas that flood most often and most deeply.

When hearing about big floods on television or reading about them in the newspaper, you frequently learn that a particular flood is called the 50-year flood, the 100-year flood, or the 500-year flood. What do these numbers mean and how are they determined?

When evaluating acceptable risk of habitation on floodplains, the United States National Flood Insurance Program determines the chances that different parts of a floodplain flood each year. It was decided that the area with a 1 in 100, or greater, chance of flooding each year represents a significant risk. This chance, or probability, of 1 in 100 (also written as 0.01) is described as a **recurrence interval** of 100 years, which means that on average there is an expectation of such a flood occurring once during every 100 years.

Design the Analysis

What Are the Specific Questions to Answer? Federal law established the area inundated by a flood with a 100-year recurrence interval as the standard for planning and insuring. Most new developments are largely excluded from the 100-year floodplain, and existing properties within this area are insured with federal subsidies. The law requires maps to show the extent of the 100-year flood.

Geologists must answer two questions to construct these maps:

1. How big is the flood that has a 0.01 probability of occurring each year?
2. To what elevation will the water rise during such a flood?

Analyze the Discharge Data

How Big Is the 100-Year Flood? The size of the 100-year flood is determined by examining discharge data collected by stream gages over periods of several decades to perhaps as long as a century. **Figure 16.28** illustrates a 50-year flood record and shows how to do the analysis.

In the example, a measured discharge greater than or equal to 1500 cubic meters per second shows up in the stream-gage record five times over a period of 50 years (Figure 16.28b). You can reasonably expect, therefore, that a discharge of at least 1500 cubic meters per second will occur another 5 times during the next 50 years. This means that, on average the 1500-cubic-meters-per-second discharge occurs once every 10 years (50 years divided by 5 events). The recurrence interval for this discharge 10 years, and the probability of a flood at least this large happening during any particular year is 1 in 10, or 0.1. By comparison, when flipping a coin there is a one–in–two probability of the coin landing heads up, meaning a probability of 0.5 and a recurrence of the results one time in every two coin flips.

You can further analyze the discharge data with the same method and determine the discharge of the 25-year flood, the 50-year flood, and so forth. If, however, there is only 50 years of record, how can you determine the discharge of the 100-year flood?

Figure 16.28c shows how to use the graph of the 50-year stream-gage record to estimate the discharge for the 100-year flood. Simply draw a line through the data points, extend the line out to a recurrence interval of 100 years, and the estimated discharge of 3500 cubic meters per second is read from the graph (see Figure 16.28c). This process of extrapolating from known data to a value outside the data range has uncertainties that must also be considered, so a range of values for the 100-year discharge is possible.

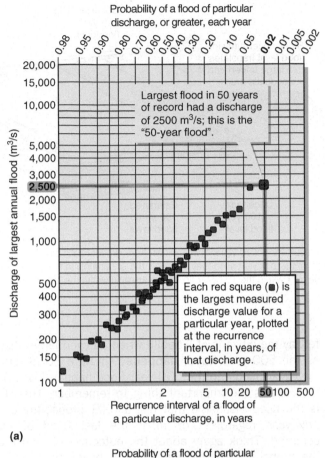

(a)

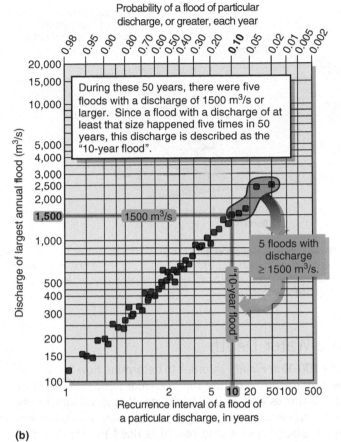

(b)

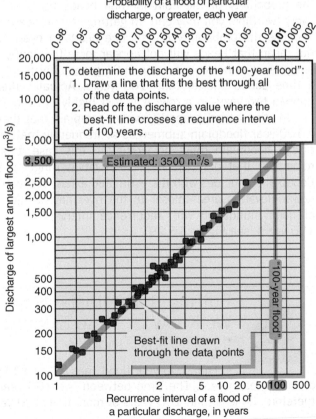

(c)

◀ **Figure 16.28 Graphing the 100-year flood.** These graphs show how to use stream-gage data to determine the probabilities of stream flows with different discharges. Discharges and recurrence intervals are plotted on special graphs with nonlinear axes that cause the data points to fall close to a straight line.

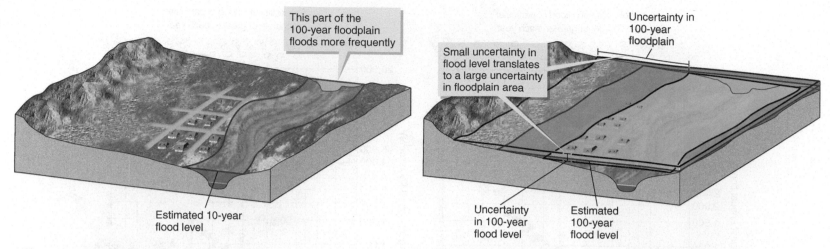

▲ **Figure 16.29 Uncertainty of floodplain extent.** The area affected by a frequent flood with a recurrence of 10 years is well known from experience. The 100-year floodplain, however, is estimated from computer calculations. Although the uncertainty in the water depth of the 100-year flood may be small, the uncertainty in flooded area is very large area on the map because the slope of the floodplain is very low. Properties located close to the river within the 100-year floodplain are also within the 10-year floodplain, so they have a higher flood probability than 0.01.

Analyze the Water Depth

How Deep Is the 100-Year Flood? In order to determine how far up the floodplain the water will rise during the 100-year flood, you need to know the water depth as well as the discharge. It is important to estimate this flood level everywhere along the floodplain, not just where a stream gage is placed and whose record is used to estimate the 100-year flood discharge.

Scientists and engineers use sophisticated computer models to calculate how rapidly the flood discharge moves downstream and to estimate discharges at locations between the stream gages. Combining discharge estimates with surveyed land elevations and channel-depth measurements provides estimates of the water depth of the 100-year flood discharge at every location along the valley.

However, the flood-depth estimates are uncertain because many factors influence the velocity of the flood water and the accuracy of the computer models. **Figure 16.29** illustrates how the uncertainty in flood depth affects the outline area inundated by the flood. As important as it is to delineate the 100-year floodplain for land management and insurance policies, it is impossible to mark its boundaries precisely. Just a minor uncertainty in the depth of the 100-year flood translates into a very large uncertainty in the flooded area.

Uncertainty is inherent to any scientific calculation or forecast. Everyone desires exact answers, but the very nature of scientific inquiry does not always permit high accuracy. Scientists acknowledge where the uncertainties are and how large they are so that planners recognize the uncertainties of the forecast.

Insights

What Does the "100-Year Flood" Really Mean? Some people mistakenly believe that if the 100-year flood happened last year, then such a flood will not occur again until 100 years from now. Consider this actual case. During the Great Flood of 1993, the town of Wapello, Iowa, experienced the worst flooding in its history. An analysis such as that done in Figure 16.28 equated the peak discharge on the Iowa River to the 200-year flood. Imagine, therefore, the frustration felt by the citizens of Wapello when an even larger disaster, ranked as the 500-year flood, happened in June 2008. How could the 200-year flood and the 500-year flood happen just 15 years apart?

Here is the important thing to remember: The 100-year flood is the flood discharge that has a 0.01 probability of occurring in *any* year, regardless of when the last flood of that size last occurred. Think again about the outcomes of coin flips. Although the probability of the coin landing heads up is 0.5, or once in every two flips, you may end up getting several heads in a row. Just because the probability is only one heads in every two flips does not mean that heads will not appear more frequently. By the same analysis, it is possible for the 0.01-probability flood to recur with a time period shorter than 100 years. Actually, this flood could happen two years in a row; it is just highly unlikely to do so.

In other cases, property owners complain that their property on a 100-year floodplain submerges so frequently that the delineation of the floodplain must be wrong. In this case it is essential to remember that only the highest-elevation fringe of the 100-year floodplain coincides with the estimated extent of the 100-year flood. Lower-elevation areas closer to the river have a higher probability of flood inundation with recurrence intervals much shorter than 100 years (see Figure 16.29).

Putting It Together—How Do We Know . . . the Extent of the "100-Year Flood"?

• The "100-year flood" has a 1-in-100 probability of occurring each year. The time between two 100-year floods may, therefore, be less than 100 years or more than 100 years.

• Where stream gage records are less than 100 years long, the 100-year-flood discharge is estimated by extrapolating the discharges measured for floods with shorter recurrence intervals.

- Computer programs combine stream-gage data, the physics of flow in stream channels, and the surveyed topography along the stream to estimate the area inundated by the 100-year flood.

- Like most scientific calculations, the discharge and depth of the 100-year flood are uncertain, which must be considered when determining the risk from the 100-year flood.

EXTENSION MODULE 16.2

How to Determine Recurrence Intervals of Floods. *Learn how to use stream-gage data to calculate the recurrence times of different discharges.*

EXTENSION MODULE 16.3

How to Reduce Flood Hazards. *Learn how dams, levees, and floodplain management diminish the destructive effects of floods.*

16.11 How Do Human Activities Affect Streams?

Human changes of drainage-basin landscapes and stream channels can throw streams out of natural balance. Some people view modifications of streams as detrimental to the environment, whereas others emphasize the resulting increased economic productivity or public safety.

These modifications are also unplanned scientific experiments. This is because the alterations of the stream channel, or changes in the supplies of water and sediment to the channel, provide tests and illustrations of how streams adjust to provide a balance among channel dimensions, slope, discharge, and sediment load (these variables were explored in Section 16.6). Let's apply our understanding of how streams work to explain stream responses to human changes of natural landscapes.

How Dams Affect Rivers

Dams and reservoirs interrupt nearly all large rivers and many small streams, and serve a variety of purposes. They control floods by holding back high discharges and gradually releasing water downstream to keep the flow below the stream banks. A dam stores some water during a rainy season or when spring snowmelt flows from mountains, and then releases it as needed during drier seasons when the river may not naturally transport enough water for drinking and farming. At some dams, water passes through tunnels from a high reservoir side to a low downstream side to turn turbines that generate electricity. Hydroelectricity accounts for 12 percent of electrical power produced in the United States. How do these interruptions in sediment and water transport affect streams? **Figure 16.30** summarizes the observed effects.

Dams trap 90–100 percent of the sediment loads of rivers within their reservoirs. A reservoir is a still body of water, and the water-surface elevation forms a new, higher base level for the stream. As a result, the river deposits sediment to reduce the slope as flow approaches the reservoir (Figure 16.30). The decrease in river velocity where flow enters the reservoir causes all of the bedload, and most or all of the suspended load, to

deposit as the stream power drops to zero. Over many decades or centuries, the persistent sediment deposition eventually fills in the reservoir and makes the dam obsolete.

Streambeds and banks almost always erode downstream of a dam. For example, after completion of Hoover Dam on the Colorado River near Las Vegas, Nevada, the riverbed eroded downward as much as 7 meters along more than 100 kilometers of channel during a 14-year period. You can explain this process by considering that the water released into the channel downstream of a dam contains almost no sediment, although the stream does have power to carry sediment. This means that there is excess stream power for erosion and the river erodes down to a new base level.

Other, predictable changes occur in channel pattern and dimensions downstream of a dam. Large discharges no longer happen because the dam impounds high flows in the reservoir. This means that the old channel below the dam is larger than required to transport the available water, so the channel fills in along its edges and becomes narrower (see Figure 16.30). Vegetation encroaches closer to the channel when there are no floods and the plant roots decrease the erodibility of the banks, which also contributes to the narrowing of the channel. The decrease in discharge ranges, the decrease in coarse sediment load is now trapped in the reservoir, and the increase in bank stability because of new vegetation growth, all combine to make channel meandering more likely than braiding (see Figure 16.19).

How Changing Land Use Affects Rivers

Human land use usually changes the natural movement of water and sediment to a stream channel. Some scientists estimate that global sediment

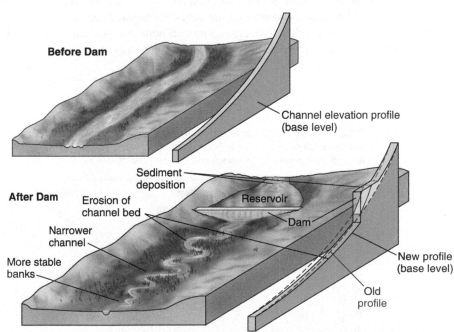

▲ **Figure 16.30 Dams change stream channels.** Upstream of a dam the water elevation in a reservoir establishes a new, higher base level, which causes sediment deposition in the channel and on the floor of the reservoir. Downstream of a dam, the stream is not carrying sediment, and excess stream power causes channel erosion to a deeper base level. Post-dam discharge is lower and varies less than pre-dam discharge, so the stream channel becomes narrower, and plants commonly stabilize the banks.

load in rivers has increased tenfold because of human activities. Consider these observations:

- Vegetation affects surface runoff to channels and also how much soil erosion or mass movement happens to deliver sediment to channels. As pointed out in Section 14.7, replacement of natural vegetation by cropland and pasture typically increases surface runoff and causes soil erosion, especially where slopes are steep.

- Logging of forests increases mass movement on steep slopes when the binding strength of tree roots to hold regolith in place is lost (see Section 15.3).

- Mining and large-scale construction projects produce large piles of loose rock and soil that erode easily to introduce large volumes of sediment into streams.

- Rainfall in cities mostly falls on pavement and roofs where there is no infiltration. **Figure 16.31** shows the result—surface runoff increases substantially when humans convert natural or rural landscapes to urban and suburban development.

How streams respond to land-use changes depends on the relative importance of increases in water to do work and increases in sediment to transport. If the addition of water is greater than the addition of sediment, then stream power exceeds what is required to transport the sediment. So, with this excess power the stream erodes its bed and banks to gain sediment to transport. The erosion increases the width and depth of the channel. This is one reason why city engineers commonly line urban stream channels with concrete, to reduce stream erosion that damages property and to keep the stream from shifting position into developed areas. If, on the other hand, sediment supply now exceeds the ability of streams to transport it, then the channels partly or even completely fill in. The channel adjusts to carry its newly acquired sediment load by increasing channel width so that there is more room on the bed to carry the sediment. This causes erosion of banks and increases the braided character of the stream. Channel depth also decreases as sediment is deposited on the bed, which increases the risk of overbank flooding at even modest discharges.

Putting It Together—How Do Human Activities Affect Streams?

- Human activities such as construction of dams and modification of naturally vegetated landscapes cause predictable adjustments in stream behavior.

- Changes in channel dimensions and shape result from modifications of the water discharge, sediment load, or both, of the stream.

- Stream adjustments may include detrimental erosion or deposition that requires lining channels with concrete to control channel location and the nature of water and sediment transport.

16.12 How Do Stream-Formed Landscapes Change Through Geologic Time?

Base-level and channel-shape change over years or decades because responses to human landscape modifications are analogs for how streams respond to natural variations in stream power and sediment load over longer intervals of geologic time. Not all rivers are adjusted to their base level of erosion. Some streams are currently cutting deep canyons to reach lower elevations, whereas other channels are filling with sediment to reach higher base-level elevations. These changes between stream erosion and deposition are important aspects of changing landscapes on Earth's surface.

Terraces—Evidence of Downcutting and Filling of Valleys

Figure 16.32 shows **terraces**, which are step-and-bench landforms alongside and above a river channel. The benches alongside the Alaskan river (Figure 16.1a) are terraces. Excavation of the flat top of a terrace reveals well-rounded gravel or sandy bedload deposits. The apparent river alluvium is, however, high above the current streambed and floodplain.

▼ **Figure 16.31 Visualizing the effect of urbanization on discharge.** These graphs summarize many observations of changes in stream discharge caused when urban development replaces natural landscapes. Infiltration is diminished after development, so more water runs off to streams and the runoff reaches the streams faster. As a result, stream discharges in urban areas rise rapidly to flood stage.

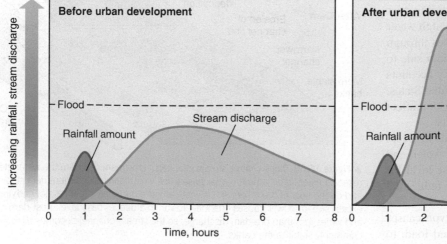
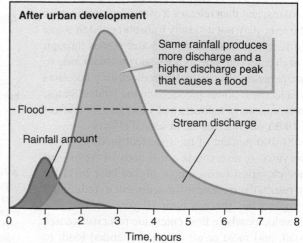

▲ **Figure 16.32 What stream terraces look like.** The stair-step-like benches alongside this river are terraces that record former positions of the stream channel and floodplain.

How do terrace landforms and their deposits originate? The presence of bedload deposits high above the present channel records an earlier time when the river occupied a higher position in the landscape. Each terrace level records a former position of the valley bottom as illustrated in **Figure 16.33**.

Terraces are evidence of the downcutting or filling of valleys over long times. River downcutting and deposition indicate adjustment of the stream to offset imbalances between stream power and sediment load. Climate change, tectonic processes, and fluctuations in sea level drive these imbalances that create terraces.

How Climate Changes River Characteristics

Precipitation and temperature vary over time at any location. These climate variations are known from historic records for human time frames and by fossils and sediment characteristics over geologic time frames.

Rivers are very sensitive to changes in precipitation for two reasons:

1. Stream flow originates from precipitation, so changes in precipitation cause changes in discharge (Section 16.1).

2. Sediment load varies depending on the amount and intensity of rainfall and the amount of vegetation cover of the land surface (Section 16.2).

Figure 16.34 describes two scenarios of climate change that cause a base-level change. Alternations between times of erosion and deposition produce terraces such as those illustrated in Figures 16.32 and 16.33. Keep in mind that the base level of a stream is a balance of available stream power to erode and transport sediment versus the amount of sediment available to move. If one of these two variables increases while the other decreases, or if both increase or decrease but by different amounts, then the stream is out of adjustment and erodes or deposits sediment to adjust the bed to a new, balanced base level.

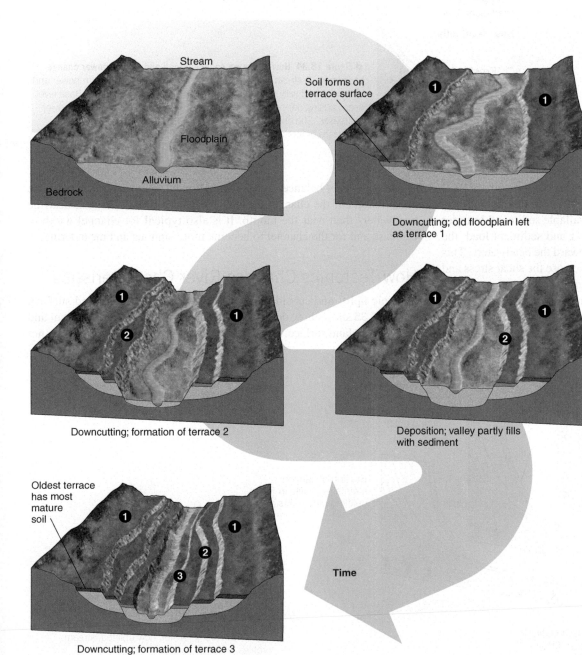

Downcutting; old floodplain left as terrace 1

Downcutting; formation of terrace 2

Deposition; valley partly fills with sediment

Oldest terrace has most mature soil

Time

Downcutting; formation of terrace 3

ACTIVE ART

Forming Stream Terraces. See how stream terraces form.

◄ **Figure 16.33 How stream terraces form.**
When a stream channel erodes downward, part of the former floodplain remains as a nearly flat terrace surface above the active channel. Soil forms on the terrace surface because it is too high above the channel for further deposition during floods. Time periods of downcutting alternate with times of deposition to form multiple terrace levels. The oldest terrace has the highest elevation, and the most mature (reddest) soil, because it has existed for the longest time.

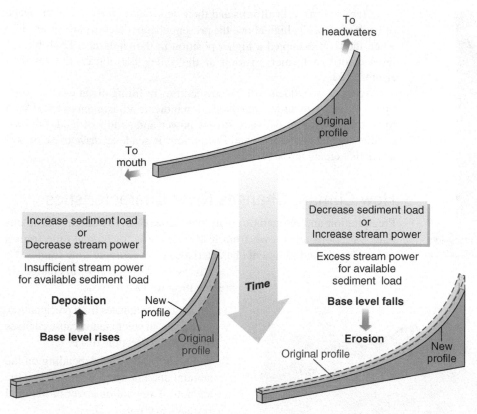

◄ **Figure 16.34 How changing sediment load and stream power change base level.** A stream-channel profile that is adjusted to stream power and sediment load will change if either of these variables changes. Base level rises causing deposition if the sediment load increases or stream power decreases. The channel erodes to a lower base level if sediment load decreases or stream power increases.

If sediment load increases or stream power decreases, then base level rises to a higher elevation (Figure 16.34). Deposition occurs because the stream lacks the power to transport all of the available sediment. In order to achieve a new balance between stream power and sediment load, the elevation of the streambed may increase more toward the headwaters. This causes an increase in slope that is necessary to increase the shear stress and velocity in order to move the additional sediment.

If sediment load decreases or stream power increases, then base level falls to a lower elevation (Figure 16.34). The stream has excess stream power for the available sediment, so it erodes down into its bed. In order

to achieve a new balance between stream power and sediment load, the overall slope of the channel decreases by having more incision near the headwaters than near the mouth. It is also typical for channel width to decrease and for the channel to become more sinuous and meandering.

How Tectonics Changes River Characteristics

Tectonic uplift and subsidence warp the elevations of the land surface. **Figure 16.35** shows how deformation changes the slope of the stream and the adjacent land surface. If the stream power and sediment supply are not

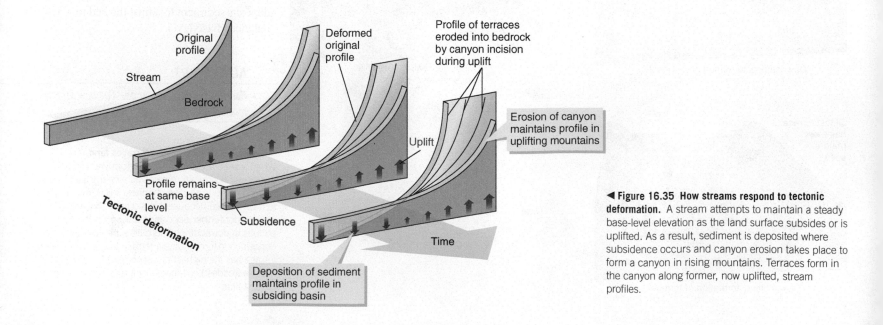

◄ **Figure 16.35 How streams respond to tectonic deformation.** A stream attempts to maintain a steady base-level elevation as the land surface subsides or is uplifted. As a result, sediment is deposited where subsidence occurs and canyon erosion takes place to form a canyon in rising mountains. Terraces form in the canyon along former, now uplifted, stream profiles.

changing, however, then the base level of the stream also does not change. This means that the channel adjusts for the warping of the surface to maintain a constant base level of erosion. The adjustments cause stream incision where uplift occurs and sediment deposition where the land subsides (Figure 16.35). Terraces form along the uplifting stretch of the river and mark former locations of the river floodplain that were abandoned as the river cut downward.

Incised Rivers—Tectonics or Climate?

Many rivers have cut deep, scenic canyons, as seen in **Figure 16.36**, and contain wild rafting rapids and waterfalls. These rivers actively erode bedrock, even where a meandering pattern suggests that the stream was once an alluvial river before cutting into the rock, as explained in **Figure 16.37**. Until recently, most geologists interpreted incised rivers to represent areas of active tectonic uplift.

An alternative possibility is that streams are responding to climate change that increased stream power and caused a downward adjustment of base level. This explanation is appealing to explain why river canyons recently formed in areas that are not tectonically active, such as the location illustrated in Figure 16.36a.

Which process causes river incision—uplift or climate change? Geologists wrestle with this question because it is very difficult to distinguish between these two causes of river downcutting. The two processes are likely intertwined and act together to at least some extent. Recall from Section 13.3 (see Figure 13.12) that erosion causes isostatic adjustment of the crust that results in uplift. The isostatic uplift increases relief, which encourages more erosion by rivers, which results in more isostatic uplift, and so on, in a perpetual loop between stream erosion and isostatic adjustment of elevation, even if the stream erosion was started by changing climate rather than by tectonic uplift.

How Sea-Level Variation Changes River Characteristics

Sea level is the lowest base-level elevation that a stream draining to the ocean can achieve. A change in sea level should change the streambed elevations that determine the slope of the channel upstream of the coastline. Geologists see terraces along coastal streams that document these predicted effects.

Figure 16.38 shows how sea-level fluctuation causes downcutting or sediment deposition in channels. The elevation profile of the stream defines the slope at any location that is necessary for transporting the available sediment. The slope is always very low near the mouth of the stream where it enters the ocean.

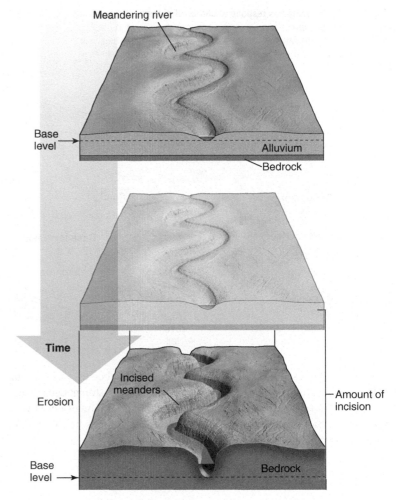

▲ Figure 16.37 **How incised meanders form.** A meandering stream pattern forms in alluvium. If the stream channel incises down through the alluvium into bedrock, then the sinuous channel form is etched into the underlying rock.

If sea level falls, then base level also falls, and the bed elevation along the stream adjusts so that the channel meets the new, lower shoreline. This adjustment causes incision of the lower part of the stream valley.

If sea level rises, then base level also rises, and the bed elevation along the stream adjusts to a higher shoreline position. The adjustment causes sediment deposition in the lower part of the stream valley. The effect is very similar to the base level adjustment where a reservoir fills behind a dam (compare Figure 16.38 with Figure 16.30).

▶ Figure 16.36 **Streams incise canyons.** The New River in West Virginia, on the left, and the Green River in Utah, on the right, eroded deep canyons into bedrock. The streams have a sinuous channel pattern, suggesting that they started out as alluvial channels at, or above, the elevations of the surrounding flat plateaus.

▼ **Figure 16.38 How streams respond to sea-level change.** Base level rises when sea level rises, causing sediment deposition in the lower part of a stream valley near the ocean. Base level falls when sea level falls, which causes erosion in the lower part of a stream valley.

If sea level rises, then base level also rises, and the bed elevation along the stream adjusts to a higher shoreline position. The adjustment causes sediment deposition in the lower part of the stream valley. The effect is very similar to the base level adjustment where a reservoir fills behind a dam (compare Figure 16.38 with Figure 16.30).

How Bedrock Geology Affects

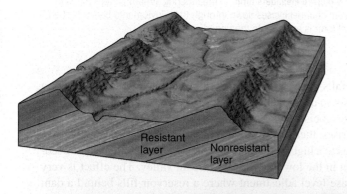

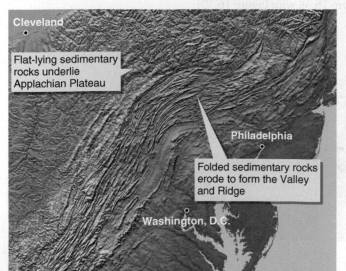

channels and valleys more easily erode in some rocks than in others. Second, the locations of rock with variable resistance to weathering and erosion determine where streams will erode most easily.

Figure 16.39 shows how dipping rock layers with different resistance to weathering and erosion determine the location of stream valleys. Big rivers with large stream power may cut across dipping layers, but smaller streams erode valleys in the most readily weathered and eroded rock types. The more resistant, inclined rock layers remain as higher ridges and drainage divides. Over long erosional time periods, stream erosion etches the pattern of the geologic structure into the landscape. This effect of erosion to accentuate inclined rock layers as ridges or valleys is also evident in many photographs in Chapter 11 (see Figures 11.4, 11.8, and 11.22).

Putting It Together— How Do Stream-Formed Landscapes Change through Geologic Time?

• Stream channels erode their beds or deposit sediment over long geologic intervals because of changes in stream power and sediment load.

• Terraces are benches alongside and higher than modern floodplains and channels. Terraces are former positions of the valley bottom and record the history of downcutting and sediment filling of stream channels.

• Changing streambed elevations result from (a) tectonic uplift and subsidence of the land surface, (b) variations in climate that determine discharge and sediment load, and (c) fluctuations in sea level that determine ultimate base level.

• Streams preferentially erode deeper and wider valleys where rock is most easily eroded, whereas more resistant rock forms ridges or cliffs.

ACTIVE ART

Erosion of Dipping Rocks. See how streams erode through dipping rock layers.

◄ **Figure 16.39 How streams erode dipping rock layers.** Streams preferentially erode valleys in rocks that weather into easily eroded regolith, whereas more resistant rock types remain as higher ridges. Parallel ridges and valleys form where dipping layers strike in the same direction. The photo to the upper right illustrates such a landscape that forms Comb Ridge in Utah, where knife-edge ridges of cemented sandstone rise above valleys eroded in mudstone. The image at the bottom left, made from satellite radar scans of Earth's surface, highlights folded sedimentary rocks of the Appalachian Mountains in the Valley and Ridge region. The patterns of the folded rocks are accentuated where valleys erode into less resistant rocks and more resistant rocks stand up as ridges.

16.13 How Do Lakes Form?

Not all stream water flows uninterrupted to the ocean. Some streams terminate in lakes. A lake may have a downslope outlet to another stream, or the water simply accumulates in a lake without an outlet and evaporates.

In simple terms, a lake accumulates water in a low part of the landscape that resembles a bowl, as shown in **Figure 16.40**. Surface runoff flows downslope into the bowl from all sides. Ground water seeps into the lake if the bottom of the bowl is lower than the water table. If the bowl completely fills, then water spills over the lowest elevation along the edge of the bowl and continues downslope in a stream channel. If water evaporates from the lake faster than it flows in, then it will not spill out. Figure 16.4 shows the outline of the Great Basin, an area where streams terminate in the continental interior rather than reaching the ocean. Stream flow in these drainage basins ends up in drying lakes that leave behind dusty salt flats called **playas** (Figure 16.40).

Some lakes, such as the Great Lakes, owe their origin to scouring of the landscape by glacial ice and will be considered further in Chapter 18. This section describes lakes related to surface processes that disturb the courses of rivers.

How to Naturally Dam a River

Many lakes form by natural blockage of a through-flowing stream, just like an artificial dam blocks a stream to form a reservoir. **Figure 16.41** illustrates the most common processes that naturally dam streams, where lava flows and mass movements (also see Figures 15.1b and 15.1c) block a stream and water pools on the upstream side of the obstacle. Eventually, the water rises to the elevation of the top of the natural dam and pours out through a newly formed outlet channel. Bedload and most suspended load sediment are deposited below the still water of the lake, just as in the case of an artificial reservoir (see Figure 16.30) so that the lake eventually fills in with sediment.

How Tectonic Activity Forms Lakes

Tectonic activity forms long-lasting lakes by dropping blocks of crust along faults so that surface runoff flows toward a central depression surrounded by higher elevations, as illustrated in **Figure 16.42**. If the down-dropped fault block subsides faster than the stream deposits sediment, then an enclosed depression collects all of the surface runoff and usually some ground-water seepage.

▼ **Figure 16.40 Lakes—open and closed.** Lakes form in low topographic bowls that collect runoff from both adjacent hillsides and ground water. Open-basin lakes, such as Lake Tahoe along the California-Nevada border at left, spill over into an outlet stream. However, if evaporation keeps the lake from filling up, then it is closed without an outlet, as illustrated on the right by Deep Springs Lake, California.

(Surface runoff) + (Ground-water flow) > Evaporation

(Surface runoff) + (Ground-water flow) < Evaporation

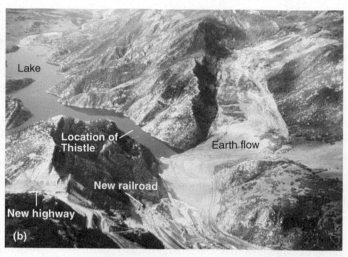

◄ **Figure 16.41 Naturally dammed streams.** (a) A lava flow erupted in 1650 from a volcano named Cinder Cone, in northeastern California, filled parts of two stream valleys, causing lakes to form on the upstream side of the lava. (b) When an earth flow blocked the Spanish Fork River in Utah in 1983, a lake formed upstream of the blockage and completely submerged the town of Thistle along with a major highway and railroad. The cost of relocating the highway and railroad, along with property losses in Thistle, totaled $400 million, making this the most costly mass-movement event in U.S. history.

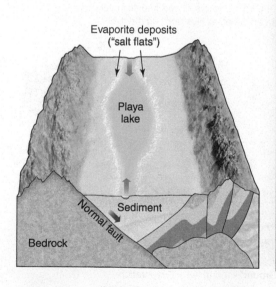

▲ **Figure 16.42 Lakes form in fault-block valleys.** Blocks of crust drop down along normal faults to form valleys between uplifted mountains. Runoff from the mountains forms lakes in the valleys. In dry regions, such as the Great Basin, evaporation leaves behind salty playa lakes within enclosed valleys without stream outlets, as seen in the Space Shuttle view looking southwestward across Nevada toward eastern California.

If more water enters the lake than is lost to evaporation, then the lake eventually fills to the lowest spot on its rim and spills over into an adjacent river valley. The lakes of the East African Rift Valleys (Figure 12.19) and Lake Baikal in Russia, the deepest lake in the world (1620 meters deep), are examples of lakes in tectonic basins with outlets.

In arid regions, such as the Great Basin, evaporation may keep the depression from filling with water. The resulting shallow, salty playa lakes may dry up completely during the hottest, driest times of the year (Figure 16.42). The valley floors surrounding and underlying the lake bed encrust with evaporite minerals such as halite, gypsum, and less common but commercially important minerals such as borax, which is used in the manufacture of detergents, glass, and ceramics. Great Salt Lake in Utah is the remains of a much larger lake that evaporated to leave surrounding salt flats. These chemical sedimentary deposits represent the dissolved load of streams that would flush out to the ocean if evaporation of the inland lakes did not occur.

Putting It Together—How Do Lakes Form?

• Lakes form naturally where lava flows and mass-movement deposits block a stream channel or where faults shift crustal blocks downward to produce a central depression that collects water.

• If lake water evaporates faster than water enters the lake from streams or ground water, then the lake dries up, at least in part, and precipitates evaporite minerals.

Where Are You and Where Are You Going?

Streams are an integral part of the hydrologic cycle. Stream water is the direct surface runoff or ground water that initially reached Earth's surface as rain or snow. Some surface precipitation also infiltrates deeply to form ground-water resources, and evaporation and plant transpiration carry water back to the atmosphere. Atmospheric water vapor condenses into water droplets or ice crystals and eventually returns to the surface as precipitation, thus completing the cycle.

Streams are complicated, dynamic features on Earth's surface. Stream erosion creates irregular relief on the surface by carving valleys separated by intervening drainage divides. Streams transport sediment, most of which erodes from areas of steep relief and areas where runoff from rainfall exceeds the landscape-stabilizing effect of vegetation. Shear stress of the flowing water, which depends on channel slope and water depth, determines the ability of streams to pick up sediment particles. Stream power, which relates equally to the shear stress and the velocity of the flowing water, determines the ability of streams to transport the picked-up sediment. Streams deposit sediment where the stream power is insufficient for continued sediment transport.

The width, depth, and slope of a stream channel adjust to establish the stream power required to transport the available sediment. Changes in discharge or sediment load cause changes in these channel characteristics; these changes in turn lead to erosion or deposition to establish new, well-adjusted channels. Stream adjustments over time create terraces that show how climate change, tectonic activity, sea-level change, or human activity affected stream processes. Responses of streams to human activities that change discharge, sediment load, or modify channel characteristics are predictable because geologists understand the physical processes that determine these responses.

Stream-channel patterns include mixtures of braided and meandering behavior. High coarse-grained-bedload sediment supply, steep slope, fluctuating discharge, and easily eroded banks favor braided streams. Flowing water diverts around exposed bars in braided streams. Large suspended load, low slope, consistent discharge, and less erodible banks favor laterally shifting, sinuous, meandering streams.

Streams form floodplains, which pose both hazards and benefits. Floodplain deposits accumulate from laterally shifting channels and by the settling out of suspended sediment from overbank floodwaters. Agricultural productivity depends on fertile floodplain soils renewed by floods and the proximity to irrigation water. Large rivers also support urban growth centered on shipping commerce in deep, wide channels. The concentration of populations close to rivers, however, increases human vulnerability to flood hazards. Geologists estimate the probability of floods inundating particular floodplain areas by combining analysis of stream-gage data, detailed knowledge of floodplain topography, and knowledge of the physics of river flow. The public learns these probabilities as recurrence intervals, although a flood with a particular recurrence interval may occur more or less frequently than the interval implies.

Natural lakes form where stream channel blockages form depressed areas where water accumulates. Processes that form lakes include eruption of lava flows or mass-movement events that block channels and tectonic subsidence of crust to form enclosed depressions. Where inflow to the lake from runoff and ground water is greater than evaporation, the lake fills up to the lowest spot on its edge and spills out into a stream channel. Where evaporation exceeds water inflow, the lake dries up, at least partly, and leaves behind evaporite deposits of possible economic value.

In the next chapter, you will consider the water that is present beneath Earth's surface. Ground water and surface water are interconnected, and both originate by precipitation at the surface. Ground water accounts for 42 percent of the public water supply in the United States, so it is very important to understand the geology of this important resource. Ground water flows through minute pores and cracks in rock and sediment and does not form underground rivers or stagnant lakes. Underground flow of water is very different from the flow observed in stream channels but it is important to understand in order to evaluate the extent of ground-water supplies in a region and how poor-quality water might move into drinking-water supplies. Although ground water is hidden beneath your feet, it still plays a role in the development of surface landscapes. You will soon learn how ground water makes spectacular caves, creates steep canyons, and sometimes forms holes that swallow whole city blocks.

Active Art

Hydrologic Cycle. See how liquid water and water vapor move through the Earth system.

How Streams Move Sediment. See how flowing water moves the sediment load.

Meandering-Stream Processes. See how meandering streams form floodplains and oxbow lakes.

Flooding and the Formation of Natural Levees. See how floods deposit sediment and form natural levees.

Forming Stream Terraces. See how stream terraces form over time.

Erosion of Dipping Rocks. See how streams erode through dipping rock layers.

Extension Modules

Extension Module 16.1: How a Stream Gage Works. Learn how a stream gage is constructed and how hydrologists use the gage data to determine discharge.

Extension Module 16.2: How to Determine Recurrence Intervals of Floods. Learn how to use stream-gage data to calculate the recurrence times of different discharges.

Extension Module 16.3: How to Reduce Flood Hazards. Learn how dams, levees, and floodplain management diminish the destructive effects of floods.

Confirm Your Knowledge

1. Define "stream." What role do streams play in distributing sediment on Earth?
2. Define and describe or sketch the hydrologic cycle.
3. What are the two paths that rainfall can take after it reaches the land surface?
4. Streams are important natural resources. How do humans use the water withdrawn from streams?
5. What is a drainage basin? Describe how water moves through the drainage basin.
6. About how often does a natural stream overflow its banks?
7. Define "discharge." How is discharge measured?
8. How does sediment get into a stream?
9. How is erosion by water flowing in an alluvial stream different from that in a bedrock stream?
10. Streams are important agents of transport for sediments. Define "sediment load" and explain the link between sediment load and drainage-basin climate.

11. What is the difference between bedload and suspended load? What is the dissolved load?
12. Define "stream power." What is the link between stream power and sediment deposition?
13. Define and describe the formation of an alluvial fan. How does it differ from a delta?
14. Define "base level." What processes cause base level to change?
15. What is a floodplain? How does it form?
16. Why does flooding occur?
17. What is the difference between a flash flood and a prolonged flood?
18. How could a 100-year flood happen two years in a row?
19. List the positive and negative consequences of building a dam on a river.
20. What are some of the ways that human activities affect rivers?
21. What are terraces and what do they tell us about the past?
22. What processes cause rivers to erode incised valleys and canyons?
23. Describe the processes that naturally form lakes.

Confirm Your Understanding

1. Write an answer for each question in the section headings.
2. What is the difference between stream discharge and stream velocity? What are the units used to measure discharge and velocity? Why does stream discharge change along the length of the stream? Why does stream velocity change along the length of a stream?

3. What would you suspect to be the cause of dramatic differences in the sediment load per square kilometer between two streams with very similar climates? What would you suspect to be the cause of dramatic differences in the dissolved load per square kilometer between two streams with very similar climates?

4. What would you suspect to happen to the load of a stream if its channel shape changes from being narrow and deep to being wide and shallow (assume the area of the channel and average velocity remain constant)?

5. Using Figure 16.17, list the stream properties that increase with downstream distance and those that decrease. Then, explain which properties are most responsible for the increase in sediment load with downstream distance and the decrease in the grain size of the sediment load with downstream distance.

6. Consider a planet with little precipitation. What would the landscape look like compared to Earth?

7. Using the map in Figure 16.4, determine where the streams in your region of the country flow. Where is the local divide?

8. Examine the photo at the beginning of this chapter. Describe the natural and human-created features visible in the photo and relate them to stream processes.

9. Go to http://water.usgs.gov/waterwatch/ and click on a stream gage location near your home or school. Examine the discharge and gage-height graphs. Use your knowledge of how the stream may be regulated by dams, along with recent weather, to explain variations in these data over the last week and how these values compare to long-term average values.

CHAPTER 17

Water Flowing Underground

Why Study Ground Water?

Americans pump nearly 314 billion liters of water out of the ground every day. This resource provides drinking water to approximately half of the population. Almost two-thirds of the ground water pumped in the United States irrigates farmland and supports livestock.

Ground water usually is a reliable water source. It is available year-round, whereas some streams stop flowing during dry seasons, just when crops need water the most. In locations where surface water is not immediately available, ground water is a valued resource. Ground water is less quickly polluted than surface water by sewage, industrial wastewater, and runoff from agricultural fields that might include pesticides, herbicides, or harmful microbes.

Earth materials dissolve in ground water. These compounds precipitate minerals that clog pipes and cement sediment grains into sedimentary rock, but that also form economically valuable mineral deposits.

Underground water helps to form surface landscapes, so the study of ground water relates to your understanding of landscape development and potential geologic hazards. Ground water reacts with subterranean minerals in the rock, regolith, and soil. These reactions dissolve some rocks to form large caverns. When these cavern roofs collapse, sinkholes form at the surface and occasionally "swallow" buildings.

After Completing This Chapter, You Will Be Able to

- Explain where ground water exists below the surface and the fundamentals of how ground water flows.

- Relate human activities to changes in ground-water flow and water quality.

- Apply an understanding of ground-water chemistry to water quality, cementation of sedimentary rocks, and formation of economic ore deposits.

- Relate ground-water flow to changes in surface landscapes.

Pathway to Learning

17.1 What Is Ground Water and Where Is It Found?

17.2 Why and How Does Ground Water Flow?

EXTENSION MODULE 17.1
Anatomy of a Water Well

EXTENSION MODULE 17.2
Darcy's Law

Ground water spills from springs that emerge from rocks high on a cliff above the Snake River in Idaho.

IN THE FIELD

Imagine taking a hike in the forests of southern Oklahoma. The trail climbs alongside Travertine Creek. To your surprise, the stream starts at a natural **spring**—a place where ground water emerges onto the surface. The stream suddenly appears from the subterranean world through cracks in rock, as seen in **Figure 17.1**a. A white mineral crust covers the banks alongside the spring. At a nearby visitor's center you see a display with a soft-drink bottle that is completely encased in the same mineral crust (Figure 17.1a). The display identifies the mineral as calcite that forms a type of limestone, called travertine, for which the creek is named. Your field observation indicates that the calcite precipitates from the spring water.

On another virtual trip, you vacation in Florida and come on a scene illustrated in Figure 17.1b. A hole in the ground was less than a meter across when it was first noticed, but in less than a day it gradually expanded to more than 75 meters in diameter. This natural hole swallowed a house, several vehicles, and part of a community swimming pool and severely damaged two businesses. A newspaper story tallies the damage at more than $2 million. The story calls the feature a sinkhole and explains that sinkholes form where underground water dissolves cavities in limestone.

Perhaps you have never given much thought to underground water. At the Oklahoma spring, however, you see underground water emerging onto the surface, and it makes you wonder. How much water in streams comes from underground rather than originating as surface runoff from rain and snow? You know that water wells pump underground water to the surface for drinking, cleaning, crop irrigation, and industrial processes. How much water really is present underground? How fast does it move and how does it move through rock? Are there underground rivers? Why does underground water emerge at the surface at springs? How do you know the water is safe to drink and not contaminated? What is the origin of the dissolved minerals that precipitate around the spring? Why does limestone precipitate along Travertine Creek but dissolve in Florida? What landscape features other than springs and sinkholes also relate to water flowing beneath the surface?

(a) Field observations of a spring and Travertine Creek, Oklahoma. The stream begins as a spring where water emerges from underground. Calcite precipitates from the water to form a crust along the stream bank (above) and encases a bottle that was discarded in the stream (right).

(b) Aerial (above) and ground (right) views of a sinkhole that suddenly opened in the town of Winter Park, Florida. The sinkhole is about 75 meters across. Notice the damage to buildings and the vehicles that fell into the hole.

▲ Figure 17.1 Evidence for ground water is visible at the surface.

17.1 What Is Ground Water and Where Is It Found?

In the hydrologic cycle, described in Section 16.1, underground water originates as rainfall and snowmelt that soaks downward into regolith and rock below Earth's surface. Some of this water moves short distances through the shallow soil horizons and reemerges to join surface runoff in streams. The remainder of the infiltrated water percolates deeper below the surface through open spaces and fractures. **Ground water** is the term that refers to this subterranean water.

How Ground Water Is Stored in Rock and Regolith

A simple lab experiment, illustrated in **Figure 17.2**, provides insights into where ground water resides. Water added to a container of loose solid particles fills in the spaces between the grains. Ground water similarly fills open spaces between particles in regolith and some sedimentary rocks, and it also fills fractures that cut through all types of rocks. These open spaces are **pores**. **Porosity** is the percentage of the total volume of the regolith or rock that consists of pores. **Table 17.1** lists typical porosity values for rock and sediment, ranging from just a few percent to more than 50 percent. In other words, although some rocks are very solid, others actually have more empty space than mineral grains. Materials with greater porosity have the potential to store greater amounts of ground water.

Only very rarely are pores more than a few centimeters across, and most are smaller than a millimeter. This means that it is extraordinarily rare to encounter ground water flowing in underground rivers.

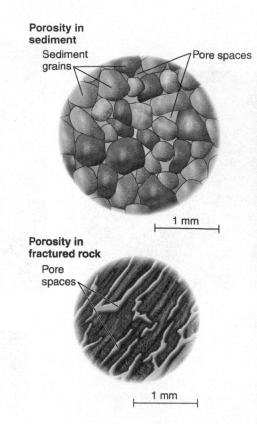

Porosity in sediment

Sediment grains · Pore spaces

1 mm

Porosity in fractured rock

Pore spaces

1 mm

Porous sediment

Water

Approximately 50% porosity

▶ **Figure 17.2 What porosity looks like.** Porosity is the percentage of pore spaces found between sediment grains and in fractures. The glass containers at right start out with equal volumes of sediment and water. Water poured into the container of sand fills the pores, which in this case represent about 50 percent of the volume of loose sand. The greater the porosity, the more water can be stored in sediment and rock.

TABLE 17.1 Example Porosity Ranges for Different Earth Materials

Material	Typical Porosity Range
Unconsolidated sediment	
Well-sorted sand and gravel	25–50%
Mixed sand and gravel	20–35%
Silt and clay	30–60%
Rock	
Shale	0–10%
Sandstone	3–30%
Limestone and dolostone	1–30%
Plutonic and metamorphic rocks	0–5%
Volcanic rocks	1–50%

Locating Ground Water

How do geologists know about ground water concealed from view beneath our feet? **Figure 17.3** shows how wells are drilled to determine where ground water exists below the surface. At some point while drilling downward, ground water starts to seep into the well. After drilling several wells in an area, one can see a pattern of how the depth to ground water below the land surface varies from place to place.

The **water table** is the surface that marks the top of the ground water. The wells first encountered water when they reached the water table. Notice in Figure 17.3 that the water table is not flat, despite what the word "table" might imply. Instead, the water table rises and falls in elevation in a pattern that reflects variations in the surface topography. Where the land surface is high, the water table is also relatively high. Where the land surface is low, the water table is also relatively low.

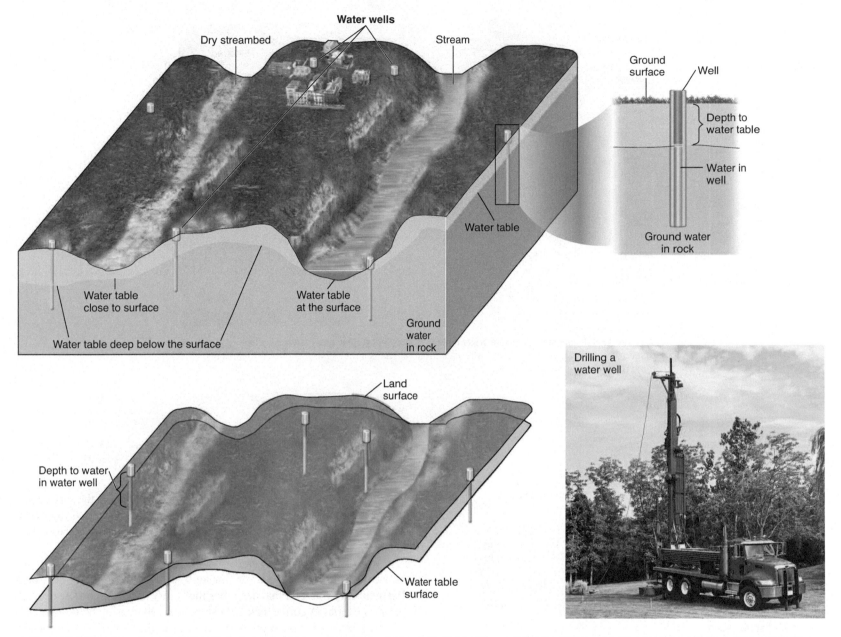

▲ Figure 17.3 Visualizing the water table. Hydrologists locate the water table by drilling wells. Most wells are drilled to withdraw ground water for use at the surface, although sometimes wells are drilled simply to find and monitor the elevation of the water table. The water table surface is not flat and mimics somewhat the ground surface; this means that the water table is usually low under valleys and high under hills and ridges. The relief of the water-table surface is typically less than the relief of the ground surface so that the water table is closer to the surface under valleys than under hills and ridges. Some valley elevations are lower than the ground surface so that ground water emerges to form flowing streams.

The water table intersects the ground surface along a deep stream valley illustrated in Figure 17.3. At this intersection some ground water seeps through the streambed and flows away as surface water. The flow of ground water into streams explains why many streams flow every day, year-round, even when it has not rained or snowed for many weeks. Water runoff to the channel from the adjacent hillsides ceases a few days after it rains. Ground water, however, seeps into the channel and sustains the persistent flow of streams with beds eroded to or below the water table. Where there is not a channel to drain the water away, natural and artificial lakes may occupy surface depressions where the water table intersects the surface; this case is shown in **Figure 17.4**.

Notice the dry streambed in Figure 17.3. This stream flows only when runoff enters the channel during and shortly after rainfall. The channel is not eroded deeply enough to intersect the water table, so it does not receive ongoing additions from seeping ground water.

Notice, too, in Figure 17.3, how the water-table elevation varies over short distances. This observation, along with knowledge that the water-table surface mimics the land-surface topography, means that once the depth to the ground water is known from several wells, it is possible to predict the depth to which a new well needs to be drilled.

EXTENSION MODULE 17.1

Anatomy of a Water Well. Learn how a well is drilled and how ground water is withdrawn from a well.

Ground surface

Flooded
rock quarry

Pond

Groundwater
in rock

Water table

▲ Figure 17.4 **Ponds form where the water table intersects topographic depressions.** Surface depressions that are lower than the water table fill with water. Natural ponds form this way, and rock quarries and mines commonly fill with ground water.

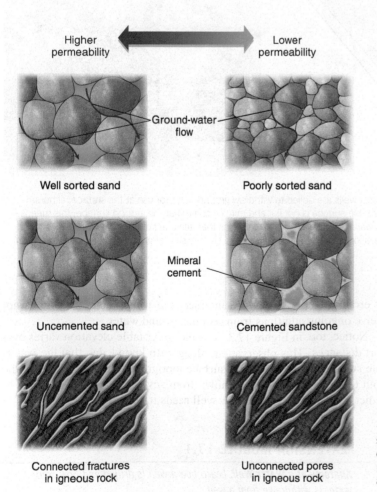

Higher
permeability

Lower
permeability

Ground-water
flow

Well sorted sand

Poorly sorted sand

Uncemented sand

Mineral
cement

Cemented sandstone

Connected fractures
in igneous rock

Unconnected pores
in igneous rock

Ground Water Flows

A pump lowered into a well extracts and brings the water to the surface, but the bottom part of the well keeps filling with water. This means that water flows through the pores in rock or regolith and into the well. How does ground water move?

Water flows from one pore space to another through narrow gaps between sediment grains, or through cracks in more solid rock. **Permeability** is a term that describes the ability of a fluid to flow through porous material. A coffee maker illustrates the relationship between porosity and permeability. After pouring water into a filter full of coffee grounds, the water flows into the pot. The water flowed through the pore spaces between the coffee grounds and then through tiny pores in the paper filter. The water flows into the pot almost as quickly as it is poured into the filter, which means that the coffee grounds and the filter paper are very permeable. For geologic materials, pores or fractures connect with one another so that water flows through the open spaces. Materials with high permeability have well-connected pores. Water flows faster through more permeable materials than through less permeable ones. **Figure 17.5** illustrates how differences in sediment grain size or sorting, or variations in the abundance and spacing of rock fractures, influence both the porosity and permeability of Earth materials.

◀ Figure 17.5 **Visualizing the factors that affect porosity and permeability.** Variations in sediment grain size and sorting, cementation, or abundance and spacing of fractures cause different porosity and permeability within rock and regolith. Porous rocks can still have low permeability, as shown by the examples on the right, because the pores are not well connected and water cannot easily flow from pore to pore.

The Water Table Separates Saturated and Unsaturated Zones

If you dig down to the water table, then you will discover that water completely fills pores in the **saturated zone** below the water table, whereas air partly fills pores in the overlying **unsaturated zone**. Therefore, the water table forms the boundary between the unsaturated and saturated zones, as shown in **Figure 17.6**. Infiltrating water moves down through the unsaturated zone to join ground water in the saturated zone. Importantly, water flows into wells only if the wells are drilled into the saturated zone.

Water in the unsaturated zone is the primary water source for most plant roots. Soil feels moist when you dig below the surface because there are thin films of water on the particles. Electrical charges at the molecular scale attract water molecules to adhere on mineral surfaces in the soil and regolith in this zone.

Why do separate saturated and unsaturated zones exist? *If* Earth materials were uniformly highly porous and permeable, as portrayed in **Figure 17.7**a, then gravity would consistently pull water down toward the center of the planet, and it would seemingly never saturate the pores close to the surface. However, Earth materials are *not* uniformly well suited for water storage or flow; see Figures 17.2, 17.5, and Table 17.1 for comparisons.

Measurements in wells show that porosity and permeability decrease as one moves farther down into Earth. Why is this? Compared to the situation in shallow sediment layers, grains in deeply buried sedimentary rocks are more closely compacted together, and pore spaces more likely are filled with cementing minerals. Porosity in igneous and metamorphic rocks that compose most of the crust is typically limited to fractures, which only form in the brittle upper crust. This means that there is effectively a downward limit to which water can readily move, so the pore spaces in the near-surface materials fill up to produce the saturated zone.

Why, then, does the saturated zone not rise to the surface with the addition of more water, as depicted in Figure 17.7b? In a very few places, this does happen, such as where ground water seeps into streams (Figures 17.1 and 17.3) or lakes (Figure 17.4). More commonly, however, ground water does not flood out onto the surface. Earth's surface is not flat, so when a rising water table reaches the ground surface at the low spots, the ground water flows out onto the surface. Figure 17.7c shows that removal of ground water at the low elevations in the landscape keeps the water table from rising to the surface everywhere.

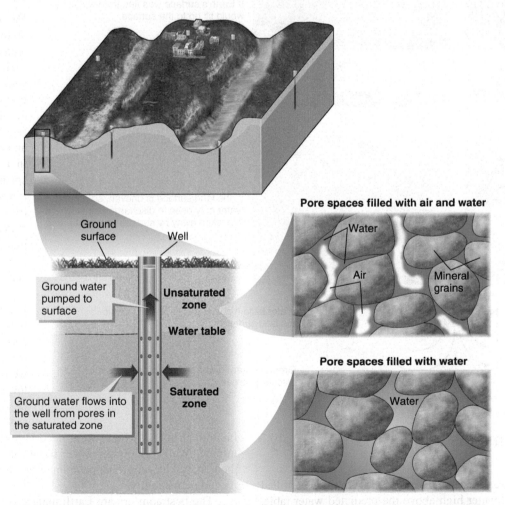

Pore spaces filled with air and water

Water

Air

Mineral grains

Pore spaces filled with water

Water

Ground surface

Well

Ground water pumped to surface

Unsaturated zone

Water table

Saturated zone

Ground water flows into the well from pores in the saturated zone

▲ **Figure 17.6 What happens above and below the water table.** The water table separates the unsaturated and saturated zones. In the unsaturated zone, air fills part of the pore spaces, and water simply adheres to the solids at the margins of the pores. In the saturated zone, water fills in all of the pore spaces. Ground water flows through the pores of the saturated zone and into a well through holes that puncture the outer part of the well pipe. Then, a pump brings the water to the surface.

If Earth materials were uniformly porous and permeable, then water would infiltrate deep into Earth and there would be no ground water near the surface **But...**

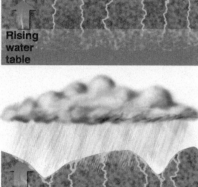

Rising water table

... porosity and permeability decrease downward, so infiltration is limited and pores fill with water.

If Earth's surface was flat, then water would fill up to the surface. **But...**

Rising water table

... the land surface is uneven, so ground water only rises to discharge points where it is taken away by surface flow in streams.

Water table highs and lows usually mimic the land surface.

Recharge   Discharge

Ground water flows from recharge areas to discharge areas.

▲ Figure 17.7 Understanding why the water table is not flat.

Why Water Sometimes Is Found at Surprisingly Shallow Depths

Figure 17.8 shows the same area as seen in Figure 17.3, with a newly drilled well that encounters ground water high above the predicted water table. Clearly, this is advantageous to the landowner, because the drilling cost was lower than expected, and it will also be less expensive to pump the water a shorter distance to the surface. However, what causes this unexpectedly shallow ground-water resource?

An important clue appears on a nearby hillside, where ground water emerges at a spring far above the valley floor. Close examination of the rocks exposed near the spring shows that the water emerges where a porous and permeable sand layer rests on top of hard, impermeable shale. The impermeable shale stops the downward-moving water and causes local pooling or, in other words, a local saturated horizon within the otherwise unsaturated zone. This is an example of **perched ground water**, where low permeability in specific spots creates a shallow saturated horizon perched above the water table.

Characteristics of Aquifers

Variations in porosity and permeability clearly affect where ground water is stored and how readily it flows. This means that not all geologic materials provide equally good supplies of ground water. An **aquifer** is a body of rock or regolith with sufficient porosity and permeability to provide water in useful quantities to wells or springs. **Confining beds** (sometimes also called "aquitards") are the contrasting low-permeability materials within the saturated zone that restrict the movement of ground water into or out of adjacent aquifers.

The term "useful" in the definition of aquifer may sound nebulous, but this is because not all parts of the saturated zone provide the same quantities of water to wells. A rock layer with relatively low porosity and permeability may be a sufficient aquifer for a few scattered wells supplying water to individual homes where water need is perhaps less than 100 liters per minute. Wells supplying water for a city, however, need to produce thousands of liters of water per minute and require aquifers with high porosity and permeability.

The best aquifers are Earth materials that have both high porosity and high permeability, as shown in **Figure 17.9**. This combination of characteristics means that there are large volumes of water stored in the aquifer (because of high porosity) and the water readily flows in large volumes to pumping wells (because of high permeability). The most productive aquifers, accounting for 80 percent of the United States ground-water resource, are unconsolidated sandy or gravelly sediment. Many of these aquifers are located in geologically recent deposits of streams and glaciers.

Some rocks also form good aquifers (see Figure 17.9). Sandstone and conglomerate that are only slightly cemented into rock can have high porosity and permeability. Ground water partially dissolves limestone (by processes explored in Section 17.4), which enlarges pores to the point of forming large caverns that fill with water and make superb aquifers. Open space along sedimentary-rock bedding planes allows rapid movement of ground water between rock layers. Volcanic rocks can be very porous and permeable, as illustrated by the photo at the beginning of this chapter, which shows springs emerging from basaltic lava flows in Idaho. Water wells tapping this lava-flow aquifer yield as much as 450 liters of water per second. This well yield would fill an Olympic-size pool in less than an hour. Lava flows have high porosity and permeability because they have many fractures formed by breakage of the congealing lava as it flowed or during contraction when the rock cooled. Any rock with closely spaced fractures can be sufficiently permeable to form a productive aquifer.

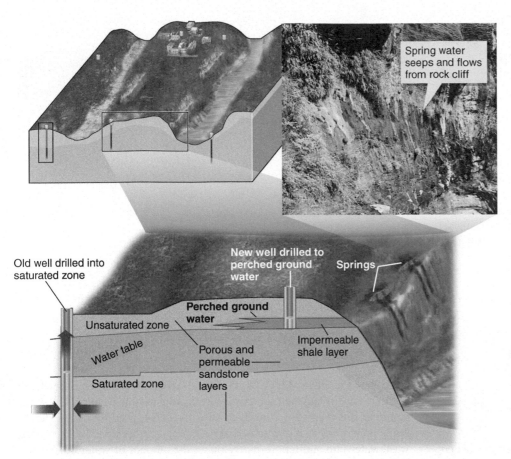

▲ **Figure 17.8 Visualizing perched ground water.** Impermeable layers in the unsaturated zone interrupt the infiltration of water to perch some ground water above the water table. These local areas of saturation within the otherwise unsaturated zone provide water to shallow wells. Perched ground water may also flow to the surface to produce springs on valley walls.

The Water-Table Level Changes Through Time

The inflow and outflow of ground water from an aquifer determines the elevation of the water table. Water exits the saturated zone where the water table reaches the land surface or where wells extract water. If more water does not enter the saturated zone to replace the water that leaves, then the volume of water in the saturated zone diminishes, so the elevation of the water table declines. Water removed from the saturated zone is discharge, and water added to the saturated zone is **recharge**.

Natural recharge occurs wherever surface water infiltrates completely through the unsaturated zone. The recharging water is rainfall and snowmelt that soak into the soil or stream water that soaks through the bottom of a streambed located at a higher elevation than the water table. Notice in **Figure 17.10** that some streams gain flow from ground water, whereas others lose flow to ground water.

▼ **Figure 17.9 Aquifer types and properties.** Aquifer materials vary across the United States depending on the local geology because permeability and porosity vary for different rock types. Materials with high porosity and high permeability form the most productive aquifers.

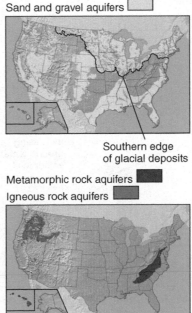

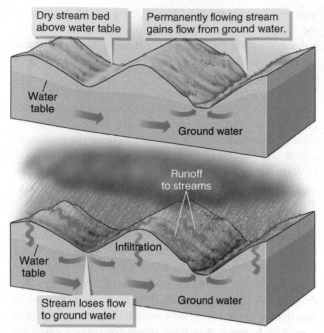

Dry stream bed above water table

Permanently flowing stream gains flow from ground water.

Water table

Ground water

Runoff to streams

Water table

Infiltration

Stream loses flow to ground water

Ground water

▲ **Figure 17.10 Gaining and losing streams.** Permanently flowing streams intersect the water table and gain discharge from ground-water flow. Streams not in contact with the water table are dry during periods of limited rainfall. When streams above the water table receive runoff, some of the flow is lost to ground water by infiltration through the streambed.

This is why some streams flow year-round, regardless of when the last precipitation fell on the surface, whereas others dry up shortly after each rainfall.

You may have noticed that people redistribute ground water on the surface or inject it into the aquifer, causing artificial discharge or recharge,

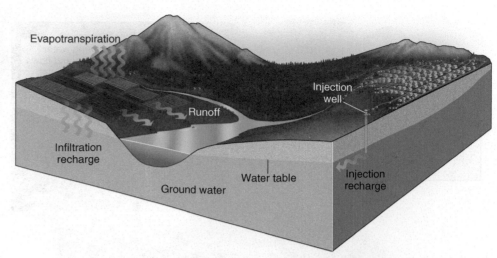

Evapotranspiration

Injection well

Runoff

Infiltration recharge

Water table

Injection recharge

Ground water

▲ **Figure 17.11 Artificial recharge.** Humans redistribute surface water and inject treated wastewater into aquifers. Irrigation and injection wells are sources of artificial recharge. Water not evaporated or used by plants during irrigation of cropland or pastures produces runoff to streams and infiltration to the saturated zone.

as depicted in **Figure 17.11**. Agricultural irrigation removes water from streams or wells and then spreads it across cropland and pasture. The water that does not evaporate, run off to streams, or get used by plants slowly seeps downward to the saturated zone. Some communities inject treated wastewater into aquifers to partly offset the amount of water discharged through wells.

Where the water table is close to the surface, its elevation may fluctuate seasonally or even daily, as shown in **Figure 17.12**. Where ground water

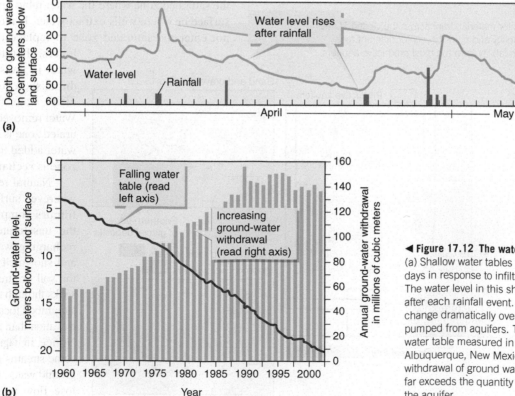

Water level rises after rainfall

Water level

Rainfall

April

May

(a)

Falling water table (read left axis)

Increasing ground-water withdrawal (read right axis)

1960 1965 1970 1975 1980 1985 1990 1995 2000

Year

(b)

◄ **Figure 17.12 The water table changes over time.**
(a) Shallow water tables may fluctuate over hours or days in response to infiltration following precipitation. The water level in this shallow well in Nevada rises after each rainfall event. (b) Water-table elevations change dramatically over long times when water is pumped from aquifers. This graph shows the falling water table measured in a non-pumping well in Albuquerque, New Mexico. The decline is caused by withdrawal of ground water throughout the city which far exceeds the quantity of water that naturally enters the aquifer.

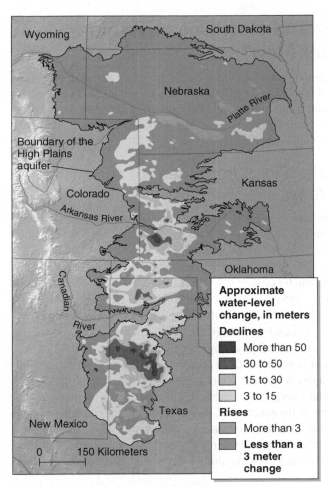

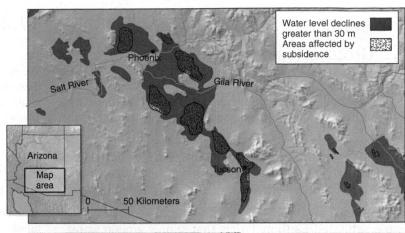

▲ Figure 17.13 Historical changes in the High Plains aquifer.
Withdrawal of ground water for irrigation has caused widespread
decline in the water table over large areas of the High Plains since
the late 1800s. Just within the Texas Panhandle and adjacent New
Mexico more than 170,000 wells pump out a total of 6 million
cubic meters of water each year, so that half of the aquifer
thickness has been drained. The water table rose during this same
period in a few small areas where stream water is used for irrigation
and then artificially recharges the aquifer.

is heavily used for irrigation, industrial uses, or to supply cities, the
discharge usually exceeds the recharge. This causes a decline in the
water table over large areas, a condition pictured in **Figure 17.13**. This
situation may require the drilling of new wells or deepening of
existing wells to "chase" the falling water level in the aquifer.

What do you think happens to the land surface when so much
ground water is extracted that the water table drops? **Figure 17.14**
illustrates that pumping large quantities of ground water causes the
land surface to sink, if the aquifer consists of uncemented sedimentary deposits. Pore water in sediment layers partly supports the
weight of the sediment grains. When the water table drops in elevation, the sediment grains compact more closely together where
the water disappears from the pores. The compaction decreases the
porosity of the aquifer and also decreases the volume of the sedimentary layers so that the land surface sinks (subsides). The subsidence
may cause large cracks to open across roads and below building
foundations.

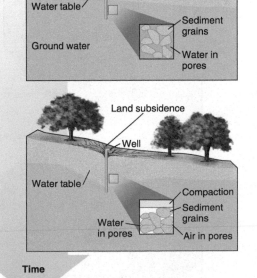

◄ Figure 17.14 Subsidence
caused by ground water
withdrawal. When ground
water is pumped out of pore
spaces in loosely consolidated
sediment, the sediment grains
compact more closely together.
The compaction causes the
land surface to subside. For
example, ground-water
withdrawal for irrigation and
drinking water has caused
both large declines in water
levels and large areas of
subsidence in southern
Arizona. The photographs
show open fissures caused by
subsidence.

Putting It Together—What Is Ground Water and Where Is It Found?

• Ground water occupies pore spaces in regolith and rock. Porosity is the percentage of the rock or regolith occupied by pores.

• The water table separates the unsaturated zone (where air and water fill pores) from the saturated zone (where water completely fills pores). Ground-water wells are drilled below the water table.

• Aquifers are saturated Earth materials that supply ground water. Productive aquifers have high porosity and high permeability.

• Water-table elevation remains the same only if the amount of water recharging the saturated zone equals the amount of water discharging to the surface at springs, streambeds, lakes, or wells.

• In most places, discharge through wells exceeds recharge, so water levels are falling. Falling water level in unconsolidated sediment may cause the sediment grains to compact more closely together, which leads to land-surface subsidence.

17.2 Why and How Does Ground Water Flow?

We previously noted that when wells pump ground water to the surface, more water flows into the well to replace it. This simple observation demonstrates that ground water moves and is not simply sitting still within pore spaces. Motion requires force and energy. What forces and sources of energy cause ground-water flow?

Ground Water Flows from Areas of High Energy to Areas of Low Energy

To understand the energy in ground-water flow, let's first consider the more familiar case of surface-water flow in a stream. Why does water always flow downhill? Gravity is the force that pulls the water molecules downward, and potential energy drives the motion. The potential energy is greater for objects at high elevation than for objects at low elevation. Objects move from areas where they possess high energy to areas where they possess low energy (see Section 1.6 for a refresher on potential energy). This is why water flows through a stream channel from high elevation to low elevation. Motion energy accounts for most of the change in energy along the flow path.

Ground water flows away from areas where the water-table elevation is high toward areas where the water table is low, because of the differences in potential energy between high and low elevation parts of the aquifer. The slope of the water table, therefore, determines the direction of ground-water flow. This does not mean, however, that ground water simply moves along the top of the water table from high elevation to low elevation like surface water flowing along a streambed. Ground water flows

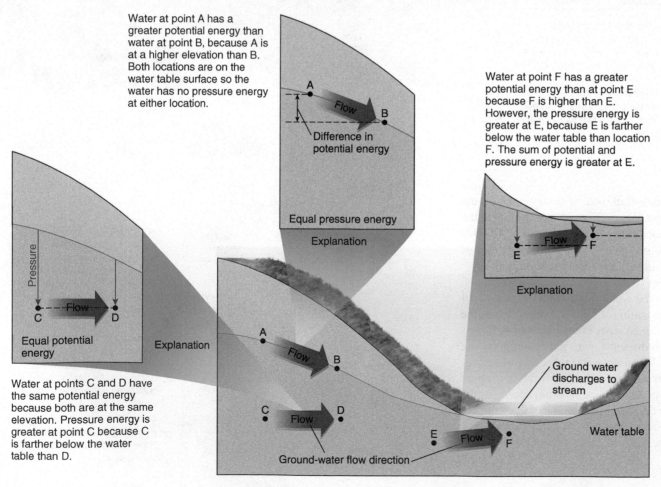

Water at point A has a greater potential energy than water at point B, because A is at a higher elevation than B. Both locations are on the water table surface so the water has no pressure energy at either location.

A
Flow
B
Difference in potential energy

Equal pressure energy

Explanation

Water at point F has a greater potential energy than at point E because F is higher than E. However, the pressure energy is greater at E, because E is farther below the water table than location F. The sum of potential and pressure energy is greater at E.

E
Flow
F

Explanation

Pressure

C — Flow → D

Equal potential energy

Explanation

Water at points C and D have the same potential energy because both are at the same elevation. Pressure energy is greater at point C because C is farther below the water table than D.

A
Flow
B

C
Flow
D

E
Flow
F

Ground water discharges to stream

Water table

Ground-water flow direction

◀ **Figure 17.15 Understanding the direction of ground-water flow.** The direction of ground-water flow between any two locations in an aquifer is determined by comparing the total energy of the water at each location. The total energy is the sum of the potential energy, determined by elevation, and the pressure energy, which relates to the depth below the water table, because the pressure energy is determined by the weight of overlying water in the aquifer. In this example, the overall flow pattern is downward and horizontal below the hill and upward toward the stream.

through pores *everywhere* in the saturated zone, which requires also taking into account energy generated by pressure.

To understand the role of pressure, consider the analogy of pressing down on a saturated sponge resting on a countertop. The water cannot pass downward into the impermeable countertop so some of it flows out the sides and top of the sponge as you exert pressure.

Aquifers, like the sponge, effectively have a bottom, like the countertop, where porosity and permeability are so low that only a tiny amount of water is present and it is barely able to move (see Figure 17.7c). The deeper the water is in the aquifer, however, the greater the pressure exerted on it by the weight of water in the pore spaces above. The pressure cannot simply push the water deeper because the permeability is too low at great depth. Therefore, as with the sponge analogy, pressure causes ground water to move horizontally or possibly even upward.

The energy that drives ground-water motion is the sum of the potential energy, related to elevation, and the pressure energy, determined by the weight of the overlying pore water. Potential energy is greater at the surface than at lower elevations below ground, but pressure energy increases with increasing depth below the water table, so ground-water flow from high energy to low energy is much more complex than water flowing in a stream. **Figure 17.15** explains ground-water flow by applying the combined effects of potential energy and pressure. Ground-water flow is not always downward but is commonly close to horizontal, and is actually directed upward at many locations.

The pressure is lowest at discharge locations, including wells, because withdrawal of water from the aquifer decreases the weight of the remaining water. Where ground water discharges into low-lying valleys, the potential energy is low because the elevation is low, and the pressure energy is low because there is no overlying ground water. Therefore, ground water always flows toward discharge locations (Figure 17.15).

EXTENSION MODULE 17.2

Darcy's Law. Learn how Henry Darcy's experiment defined a simple mathematical formula for describing ground-water flow.

Where Ground Water Flows

Figure 17.16 shows that ground water follows curving paths through the saturated zone from relatively high areas on the water-table surface toward discharge locations. The curving flow paths result from the combined effects of gravity pulling the water down, pressure forcing the water laterally, and the sum of the two being lowest where water discharges. Where the distance between points of recharge and discharge is short, the curving flow path remains close to the water table. Where the distance between recharge and discharge locations is long, then the curving flow path extends deeply down into the saturated zone.

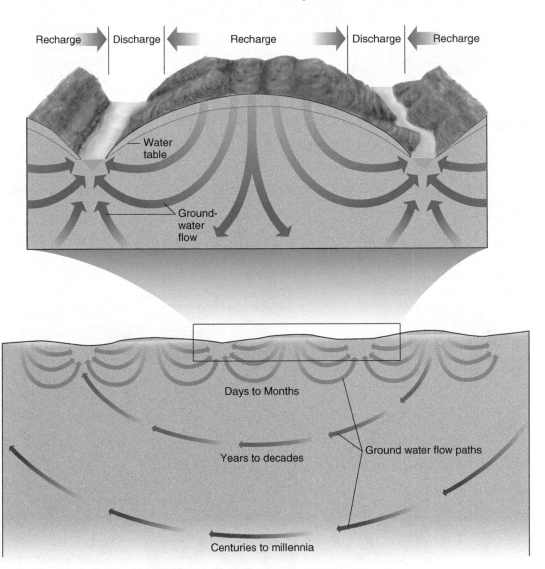

▲ **Figure 17.16 Ground water moves from recharge areas to discharge areas.** Ground water follows curving paths down and away from recharge areas and up and toward discharge areas. The time that the water spends within the aquifer depends partly on porosity and permeability, partly on the elevation difference of the water table between recharge and discharge areas, and mostly on the length of the flow path. Water following short flow paths close to the surface may reside in the aquifer for only days to years. Water following routes deep below the surface may not reach discharge locations for centuries or millennia.

Ground-water flow is a little more complicated in sedimentary aquifers consisting of both permeable and impermeable layers. **Figure 17.17** illustrates ground-water flow in a deep, permeable sand layer between impermeable clay confining beds. This is an example of a **confined aquifer** where impermeable layers separate, or confine, the permeable aquifer layers. The confining beds keep the shallow and deeper ground water from mixing.

Deeply Flowing Ground Water Forms Hot Springs

Ground-water flow paths help us understand the origin of warm springs or hot springs (distinguished only by the relative warmth of the discharging ground water) that have long been popular for relaxing and purportedly therapeutic soaking. Deep, far-traveling ground water passes through rocks that are much warmer than materials close to the surface. Heat conducts from rock into water more readily than heat conducts from water into rock.

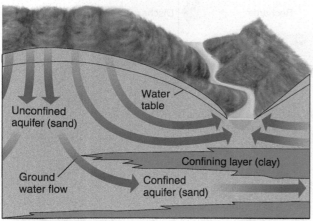

▲ **Figure 17.17 Ground-water flow in a confined aquifer.** Confined aquifers form where impermeable or low-permeability confining layers separate more permeable aquifer materials. Where the deeper flow is located between the confining layers, it remains mostly or entirely separate from the ground water in the shallower, unconfined aquifer.

This means that ground water acquires the temperature of the warmest rock that it passes through and then cools only slightly while moving toward discharge to the surface or a well. **Figure 17.18** illustrates that the discharging ground water is warm because it followed a subsurface path through warm rocks.

Ground water near shallow magma chambers in volcanically active areas may encounter rocks heated to more than 100°C at depths less than 1 kilometer below the surface (see Figure 17.18). This means that even shallowly circulating ground water can discharge as boiling hot springs at the surface. Well-known examples attract tourists and scientists to Yellowstone National Park.

Deeply circulating ground water causes warm springs in locations remote from volcanoes, such as Warm Springs, Georgia, and Hot Springs, Arkansas (see Figure 17.18). With typical geothermal gradients, ground

▶ **Figure 17.18 How ground water gets warmed up.** Hot springs and warm springs form where ground water circulates through hot rocks and then discharges at the surface. The hottest springs form where ground water heats up close to molten magma or igneous intrusions that have recently solidified and remain very hot. Many hot springs in the western United States relate to these igneous processes, including those in Yellowstone National Park, the Cascade Range in Oregon and Washington, and The Geysers north of San Francisco, California. Cooler warm springs form where deeply circulating ground water rises to the surface, usually along a boundary between permeable and impermeable rocks. The water heats up by conduction where it passes through warm rock at depth and then carries the heat to the surface.

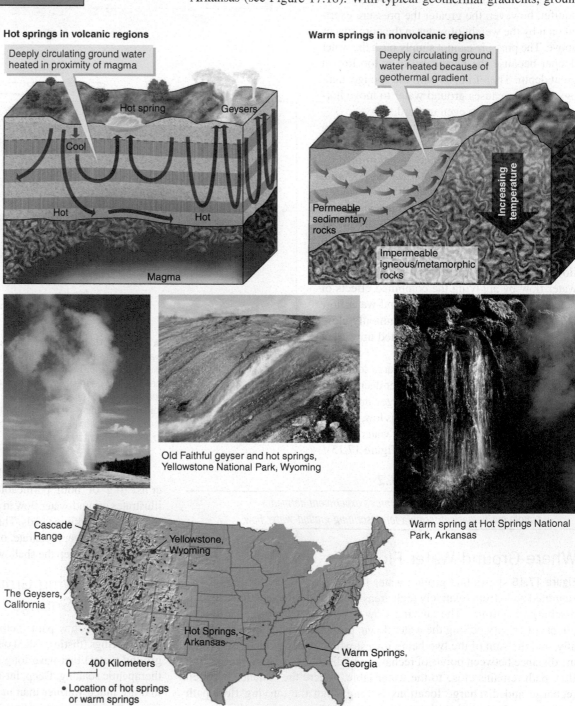

Hot springs in volcanic regions

Deeply circulating ground water heated in proximity of magma

Hot spring   Geysers

Cool

Hot   Hot

Magma

Warm springs in nonvolcanic regions

Deeply circulating ground water heated because of geothermal gradient

Increasing temperature

Permeable sedimentary rocks

Impermeable igneous/metamorphic rocks

Old Faithful geyser and hot springs, Yellowstone National Park, Wyoming

Warm spring at Hot Springs National Park, Arkansas

Cascade Range

Yellowstone, Wyoming

The Geysers, California

Hot Springs, Arkansas

Warm Springs, Georgia

0   400 Kilometers

• Location of hot springs or warm springs

water that circulates to a depth of 2 kilometers is heated to about 60°–75°C, making it comfortably warm when it later rises and discharges at the surface. Warm springs in the eastern United States, and other nonvolcanic regions of the world, are always associated with upward flow and discharge of deep ground water, usually where flow in thick sedimentary-rock aquifers is diverted to the surface along steep contacts with less permeable, commonly metamorphic or igneous, rocks.

Why the Water Table Is Lower Near a Pumping Well

Figure 17.19 shows how the shape of the water table changes near a pumping well. Monitoring of water levels in closely spaced wells shows that the water-table elevation declines as water is withdrawn, and the greatest decline occurs at the pumping well. The reshaped water table resembles a cone-shaped funnel that is centered on the well (Figure 17.19). Hydrologists call this local lowering of the water-table elevation the **cone of depression**.

Let's use our understanding of ground-water flow to explain the cone of depression. Water-table elevation declines when discharge exceeds recharge (Figure 17.14). It seems reasonable, therefore, to attribute the cone of depression to pumping water from the aquifer faster than water can flow toward the well to replace what is pumped out. Surprisingly, though, after a while the cone of depression no longer enlarges even though water is still pumped from the well at the same rate. To understand why this happens, remember that the well is a point of ground-water discharge, so ground water flows through pores toward the well. The direction of ground-water flow parallels the slope of the water table, so the cone of depression merely establishes the water-table slope required in order for the water to flow to the discharging well. As the cone gets larger, the well draws in ground water from a larger and larger region of the aquifer. Eventually, the increased flow toward the well balances the discharge of water pumped from the well, and the water-table elevations stabilize.

The shape of the cone of depression, therefore, reflects the balance between the pumped water and ground-water flow toward the well. If the discharge increases, then the cone of depression enlarges and then stabilizes again. If the discharge decreases, then the cone of depression shrinks to a new stable configuration. As long as water is pumped from the well, there will always be a cone of depression because ground water flows toward the well. The depth of the cone of depression will never fall below the bottom of the pumping well, because if the well does not discharge water, there is no flow toward the well and there is no cone of depression. Other nearby wells that are pumped infrequently or with very small discharges compared to a neighboring heavily pumped well may, however, be adversely affected by the cone of depression that forms around a heavily pumped well (Figure 17.19).

Why Ground Water Sometimes Flows Higher than the Ground Surface

Figure 17.20 shows another well drilled in the original study area but to a greater depth than the other wells. Unexpectedly, the water flows from the top of the well on its own without using a pump. Apparently, the water

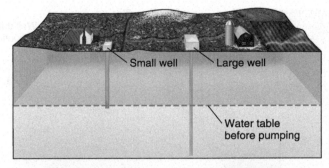

Small well   Large well

Water table before pumping

The water table slopes inward toward a pumping well because ground water always flows parallel to the slope of the water table to a discharge point.

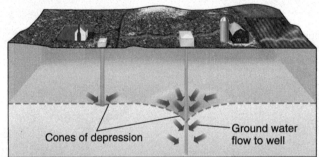

Cones of depression

Ground water flow to well

The size of the cone of depression is determined by the amount of water withdrawn.

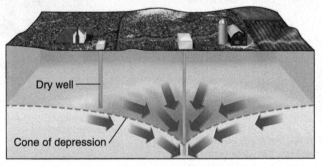

Dry well

Cone of depression

The cone of depression stabilizes when the amount of water flowing to the well matches the amount of water pumped from the well. The large cone of depression around a well that pumps a large amount of water may cause adjacent low-discharge wells to dry up.

▲ **Figure 17.19** How a cone of depression forms.

ACTIVE ART

Forming a Cone of Depression. See how a cone of depression forms.

pressure at the bottom of the well is sufficiently high to force the water up to an elevation higher than the ground surface. When water rises in a well or at a spring above the elevation of the water table, it is referred to as **artesian** discharge. Flowing wells and springs are the special case of artesian flows that rise on their own all of the way to the land surface. Although not common, flowing artesian wells have the advantage of providing water without the cost of pumping water from the well.

Figure 17.21 shows the conditions that form an artesian well. Artesian wells are not simply drilled into the top of the saturated zone. Instead, they penetrate deeper into the aquifer through one or more confining layers. So, our first important observation is that artesian wells are drilled into confined aquifers. Second, the confined aquifer recharges at a higher elevation than where the well is drilled. The confining impermeable layers force the ground water to flow between them rather than following curved paths back to surface-discharge locations. The upper confining layer is like a lid on this part of the aquifer and keeps the water at high pressure. Think of what would happen if you punched a hole in a running garden hose—the effect is similar to drilling a well into a confined aquifer. When a well is drilled into the confined aquifer, the high-pressure water rapidly discharges from the well.

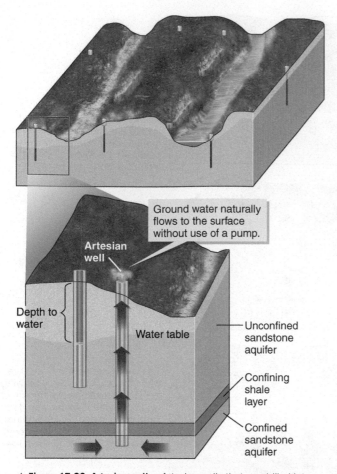

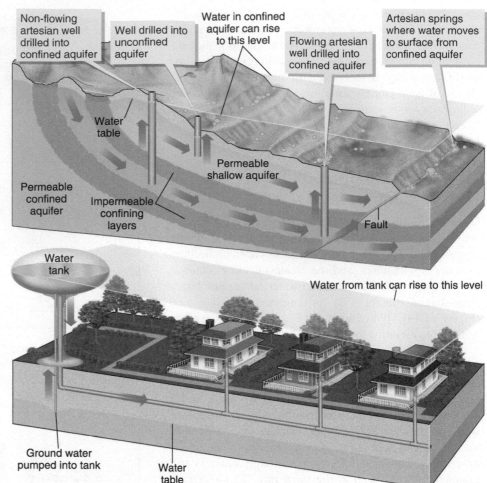

▲ **Figure 17.20 Artesian wells.** Artesian wells that are drilled into a confined aquifer below a confining layer, like an impermeable shale bed. The water in the confined aquifer is under artesian pressure, which causes it to flow up the well without being pumped. In contrast, ground water that enters a well drilled into the unconfined aquifer immediately below the water table must be pumped to the surface.

▲ **Figure 17.21 How an artesian aquifer works.** The combination of pressure and potential energy in a confined aquifer cause water to rise higher than the water table and even higher than the land surface where wells or faults penetrate the upper confining layer. The rise of water in wells can happen where the recharge area is much higher than where wells are drilled. The effect is very similar to how public water supplies are delivered from high water tanks. Water is pumped into the water tank and then flows to users. The total potential and pressure energy allows the water to rise in the plumbing nearly as high as the water level in the tank. The water-level elevation decreases with greater distance from the recharge area, for an artesian aquifer, and from the water tank, in a public water system, because energy also converts to heat by the friction of water flowing through pores spaces and pipes.

The elevation of the recharge area and the distance between the recharge area and the discharge point determine how high the water rises from the confined aquifer. The effect is similar to dropping a ball from a measured height above the floor. The ball starts out with the potential energy associated with the height above the floor. When the ball hits the floor, it bounces back up to nearly the same elevation as where it started. Energy cannot be gained or lost, but energy can be converted to different forms of energy, such as heat. The ball has to have the same energy after it bounced as before it was dropped, except for whatever energy was expended by frictional heating when the ball struck the floor. The ball rises less and less high on each successive bounce as more of the potential energy converts to heat by the friction of the bounce. In a similar fashion, the potential energy of the water in the confined aquifer is the energy associated with the high elevation where it entered the aquifer. The water would rise to the same elevation within the artesian well if it were not for energy expended as heat by the friction of the water flowing through the pores in the aquifer.

Water-supply systems use the same physical principles to distribute water through a community, as illustrated in the lower part of Figure 17.21. Water pumped into a high water tower has sufficient potential energy to flow through pipes in all buildings that are at a lower elevation than the tower.

Pollution Also Moves in Ground Water

We can also use our understanding of ground-water flow to determine the risks of pollution to aquifers. Human activities can introduce harmful chemicals or unhealthy microbes into ground water. Most pollutants enter the ground water along with recharge, move with the ground water, and then exit in discharge areas. Pollutants threaten the drinking-water supply where they enter wells. Where pollutants discharge to streams, they affect stream ecology and contaminate drinking water obtained from streams.

Figure 17.22 illustrates some sources of pollution. Chemicals spilled on the surface or applied as herbicides, pesticides, and fertilizer to farmland

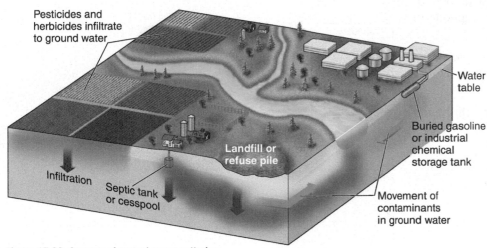

Pesticides and herbicides infiltrate to ground water

Water table

Buried gasoline or industrial chemical storage tank

Infiltration

Septic tank or cesspool

Landfill or refuse pile

Movement of contaminants in ground water

▲ Figure 17.22 Sources of ground-water pollution.

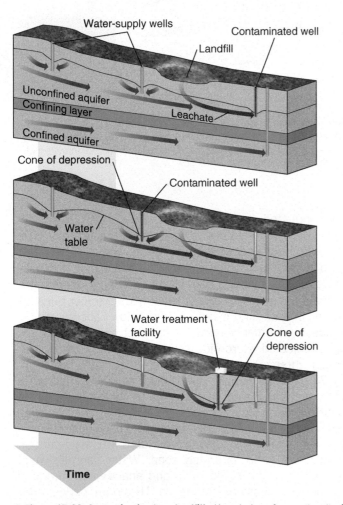

Water-supply wells

Contaminated well

Landfill

Unconfined aquifer
Confining layer
Confined aquifer

Leachate

Cone of depression

Contaminated well

Water table

Water treatment facility

Cone of depression

Time

Liquid leachate from landfill infiltrates to the ground water and flows into the water-supply well. Wells upslope of the landfill are not affected because leachate does not flow toward those wells. The well drilled into the confined aquifer is unaffected because the confining layer restricts contamination to the unconfined aquifer.

Increased pumping of the initially safe water-supply well enlarges the cone of depression, which draws the leachate into the well and contaminates it.

A new well is drilled to remove and treat the polluted ground water. The well is located on the down-flow side of the landfill and is pumped heavily to create a cone of depression that draws in contaminated water and does not allow the leachate to travel further through the aquifer.

▲ Figure 17.23 Contamination by a landfill. Knowledge of ground-water flow not only predicts which wells will be affected by the pollution, but also aids in cleaning up the contamination.

seep down to the saturated zone and flow in ground water. Water soaking through landfills dissolves some hazardous materials or mixes with disposed harmful fluids that then enter the ground water. Septic systems may leak waste into ground water. Underground gasoline, fuel oil, or chemical storage tanks also may leak.

Geologists apply their understanding of the direction and path of ground-water flow to determine where pollution travels. **Figure 17.23** shows an example of ground-water contamination by water percolating through a landfill. Pollution initially threatens only one of the local ground-water supply wells, because ground water does not flow toward the other wells. However, one of these unaffected wells pumps at an increasing rate, so that the enlarging cone of depression in the water table causes polluted ground water to flow to the well.

Figure 17.23 also shows how to use knowledge of ground-water flow to halt the spread of the contamination and possibly even clean up the water. A new well is drilled within the contaminated part of the aquifer to purposely create a cone of depression that draws the polluted water to the well. The polluted water not only stops flowing toward the at-risk drinking-water well, but the contaminated water also can be pumped from the new well, treated to remove the contaminants, and then either used at the surface or injected back into the aquifer.

Another thing to notice in Figure 17.23 is that water pumped from a deeper confined aquifer is unaffected by the landfill pollution. Many cities purposely drill to deeper confined aquifers, if they exist, to avoid near-surface pollution, even though the cost is greater to drill a deep well compared to a shallow one.

Putting It Together— Why and How Does Ground Water Flow?

• Ground water flows from points of high potential energy and high pressure energy to points of low energy, and from high elevations on the water-table surface toward low elevations.

• Ground water flows along curving paths from recharge areas at the water table toward discharge points into surface water or wells. The farthest traveled water follows the deepest flow path through the aquifer.

• Water flowing along deep flow paths heats up because rocks are hotter at greater depth below the surface. Warm springs and hot springs form where this heated water returns to the surface.

• Cones of depression in the water-table surface form where wells discharge ground water. The water-table surface slopes in all directions toward the well so that the amount of ground water flowing to the well is equal to the amount pumped from the well.

• Artesian pressure develops where confined aquifers are recharged at significantly higher elevations than where they discharge, so that the water may rise above the land surface.

• The occurrence and travel direction of ground-water pollution is predicted by understanding how ground water flows.

17.3 How Do We Know . . . How Fast Ground Water Moves?

Picture the Problem

How Fast Does Ground Water Flow? Sections 17.1 and 17.2 established the physical processes and material characteristics that determine how and where water moves in an aquifer. Not yet addressed, however, is how fast ground water moves. It is easy to think of situations in which it would be useful to know the flow velocity. Consider this scenario: Water levels declined in wells during a period of drought, but recently there have been several consecutive wet years. How long will it take the newly added recharge to reach the wells and potentially raise the water levels? Consider, too, the possible discovery of contamination in ground water just a few kilometers from a city water well. How long will it take the pollution to reach the water well, and how might this duration influence decisions on how to clean up the pollution or the need to drill a new water well elsewhere?

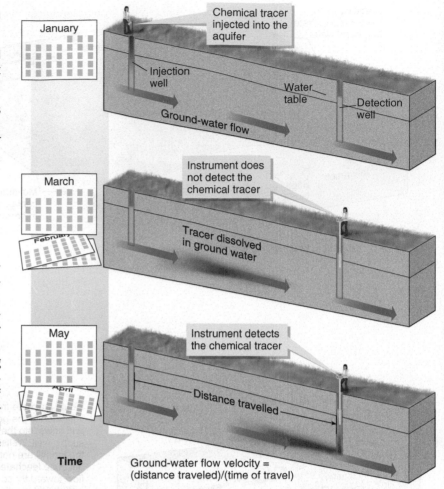

▲ **Figure 17.24 Using a tracer to measure ground-water flow velocity.** Ground-water flow velocity can be calculated by injecting into the aquifer a nontoxic chemical compound that dissolves in water. The tracer is then tested for in detection wells that are located along the direction of ground-water flow.

Design the Experiment

How Can Ground-Water Velocity Be Measured? Measuring the velocity of ground water is more difficult than measuring flow in a stream channel. Geologists easily know stream flow by placing meters into the water that measure how fast the water is moving (see Figure 16.5). Clearly, it is not easy to design a meter to insert into the tiny pore spaces of an aquifer to measure the velocity of the flowing water.

There is also a good reason to expect that ground water moves more slowly than surface water. Only a small part of the volume of water flowing in a stream channel is slowed down by friction with the solid banks and bed. For ground water, however, most of the water in a tiny pore or fracture is in contact with surrounding solids, so friction has a greater effect in slowing the flow of ground water than of surface water. Taken together, these observations mean that in order to measure the velocity of ground water geologists need a method that detects very slow velocities and that does not require placing a device into the underground flow.

The most effective way to measure ground-water flow velocity is illustrated in **Figure 17.24** and involves injecting an easily dissolved chemical substance, called a tracer, into a well. Water is then repeatedly sampled and analyzed for the presence of the tracer from other detection wells located along the ground-water-flow path. The time elapsed for the tracer to travel from the injection well to a detection well reveals the average ground-water flow velocity.

Research conducted in the 1980s by the U.S. Geological Survey at Cape Cod, Massachusetts, provides a case study. The aquifer at Cape Cod consists of loose sand and gravel. The water table is only about 5 meters below the ground surface, so it is easy and inexpensive to drill injection and detection wells into the saturated zone. Bromide ion was chosen as the injected tracer because it easily dissolves in water and is not harmful in low concentrations. In fact, bromide is commonly used to disinfect swimming pools. The hydrologists injected a bromide solution into the aquifer over 17 hours through three closely spaced wells.

Figure 17.25 shows the Cape Cod project layout. Data from a few preliminary wells revealed that the water table slopes down toward the south. Ground water flows in the direction of decreasing water-table elevation, so the detection wells were only located south of the injection wells (Figure 17.25). Six hundred fifty-six detection wells were drilled into the saturated zone. Water samples were extracted from some or all of the wells and analyzed for bromide concentration on 16 occasions during an 18-month period after injecting the tracer.

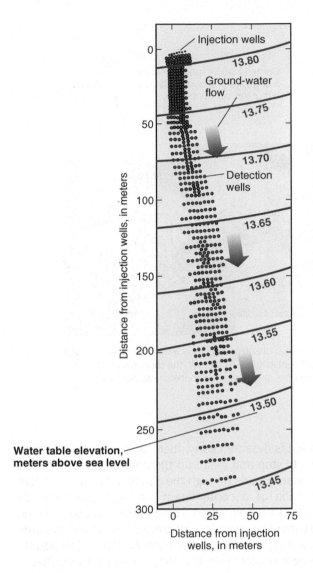

Water table elevation, meters above sea level

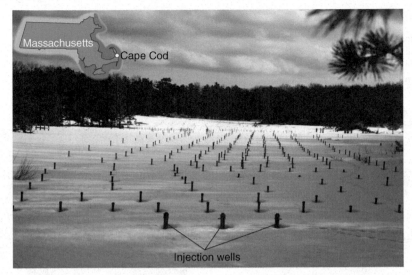

◀ **Figure 17.25 Designing a research project to measure ground-water flow.** The map shows the outline of the Cape Cod project area. Six hundred fifty-six monitoring wells were drilled in an area extending southeast from the injection wells. In the photograph, the injection and detection wells stick above the snow at the research site. The monitoring wells were drilled southeast of the injection wells because the measured water-table levels, which are contoured on the map, are high in the northwest and low in the southeast, which indicates ground-water flow to the southeast.

Analyze the Data

***How Fast Did the Ground Water Flow?* Figure 17.26** maps the movement of bromide ion through the Cape Cod aquifer. Thirty-three days after injection, the highest bromide concentration was detected 17 meters from the injection wells. Samples collected 461 days after injection revealed bromide in many detection wells, with the center of the area of wells containing the tracer located 198 meters from the injection wells. Using these distances and the elapsed times, the average velocity of the ground water is about 0.43 *meters per day*, or a distance equal to about one and a half times the height of this book page in a day. By comparison, water in a fast-flowing stream moves faster than 0.43 *meters per second*. Keep in mind, too, that permeability is high in gravel and sand so this is a rapid ground-water flow velocity. As expected, ground water flows very slowly!

The data depicted in Figure 17.26 also show other important features. First, the ground water does not flow at a uniform velocity. Four hundred sixty-one days into the experiment, some of the wells containing bromide are only 150 meters from the injection wells, whereas others are 235 meters from the injection wells. This means that some of the bromide moved in ground water at an

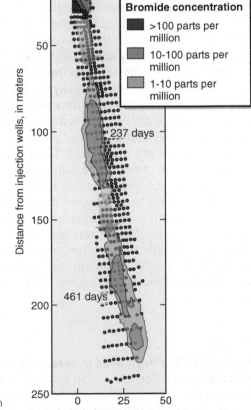

► **Figure 17.26 Tracking the tracer.** The colors in this map represent measured bromide concentrations in the detection wells at three times during the experiment.

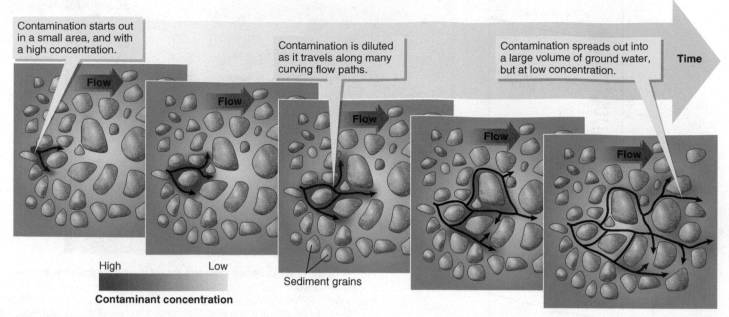

Contamination starts out in a small area, and with a high concentration.

Contamination is diluted as it travels along many curving flow paths.

Contamination spreads out into a large volume of ground water, but at low concentration.

Time

Flow

Flow

Flow

Flow

Flow

High   Low

Sediment grains

Contaminant concentration

▲ **Figure 17.27 Contamination spreads in an aquifer.** This series of diagrams shows the path of contamination dissolved in ground water as it moves through an aquifer. The flowing water divides around sediment grains and flows along many curving paths of different lengths. As a result, the contaminant spreads out in the aquifer both parallel to and perpendicular to the overall direction of ground-water flow. This process also dilutes the contaminant, because it gradually spreads out through a progressively larger volume of water.

average of only 0.33 meters per day, whereas other water flowed at an average velocity of 0.51 meters each day. Second, the bromide concentration started out at 640 parts per million at the injection wells, but after 461 days the highest detected concentration was only 39 parts per million. The data show that the change in concentration resulted from the bromide spreading out from the small injection site to occupy about 7400 cubic meters.

Figure 17.27 illustrates how hydrologists explain (a) why the velocity is not uniform and (b) why the bromide spread out through the ground water. Ground water carrying the bromide tracer ions follows circuitous curving paths around the sand grains separating the pores in the aquifer. Some paths are fairly straight and the straightest paths are the fastest. Ground water following the most roundabout path around the sand grains takes longer to cover the same straight-line distance from the injection point. Every place the bromide-bearing water divides around a sand grain, the tracer spreads out over a larger area, so the concentration at any one location is less than it was at the injection point. In addition, each path through the aquifer experiences variations in permeability resulting from different sediment sizes. This means that ground water flows fastest along paths with the highest average permeability.

Insights

How Is This Knowledge Used to Assess the Movement of Contaminants in Ground Water? One motivation for understanding ground-water flow velocity is to understand how quickly polluted ground water flows toward streams or drinking-water wells. The very slow velocities revealed by the Cape Cod experiment suggest that unless a stream or well is located very close to a contamination source, and as long

as the contamination is detected early, there should be enough time to undertake steps to stop and clean up the pollution.

The way the tracer spreads through the aquifer also demonstrates that the concentration of a contaminant may naturally diminish below harmful levels at long distances from the pollution site. On the other hand, the farther the contaminant travels, the larger the area that is affected, because the contaminant spreads out in the aquifer rather than traveling a straight line. If a contaminant is hazardous at very low concentrations, then it is also important to remember that the small amount of the contaminant that follows the shortest, fastest path through the aquifer moves faster than the average velocity and reaches wells or streams long before the more concentrated pollution that moves an average distance at an average velocity.

Putting It Together—How Do We Know . . . How Fast Ground Water Moves?

• Injecting a soluble tracer ion into an aquifer and then monitoring water composition along the flow path reveals how fast ground water flows.

• Ground water flows very slowly, compared to streams, because ground water flows in tortuous paths through pore spaces and is slowed by friction along the edges of the pores.

• Not all water molecules follow the same circuitous path through aquifer pore space. Chemicals dissolved in water, therefore, spread out through the aquifer so that the concentration of the chemicals diminishes with increasing travel distance.

17.4 What Is the Composition of Ground Water?

Precipitation of minerals around the spring you observed in the field (Figure 17.1a) shows that ground water contains dissolved ions that can bond to form minerals. Where do these naturally occurring dissolved ions come from and how are they important to understanding ground-water processes? Do compounds introduced into aquifers by human activity behave similarly to natural components in ground water?

Water-Mineral Reactions in Aquifers

Chemical reactions take place wherever water and minerals are in contact for extended periods. Mineral grains are in constant contact with water in the saturated zone, and the water moves very slowly (as we learned in Section 17.3), so there is plenty of time and opportunity for these reactions to occur. **Figure 17.28** shows that the chemical reactions in the aquifer dissolve some minerals, which also increases the concentration of dissolved ions in the ground water. Changing composition of the water may then cause bonding of concentrated ions to form minerals that fill in the pore spaces in the aquifer. The chemical reactions, therefore, may either increase (when minerals dissolve) or decrease (when minerals precipitate) aquifer porosity.

The farther the water flows through the aquifer, the greater the time it reacts with minerals. Ground water following short paths near the water table to discharge in wells or streams may not have sufficient reaction time with minerals to cause a significant compositional change. On the other hand, water following long flow paths deep into the saturated zone typically has very high concentrations of dissolved ions because the water is in contact with reactive mineral grains for thousands or tens of thousands of years. Elevated temperature at depth also enhances chemical reactions in the deep saturated zone.

Cementation of sedimentary rocks occurs where dissolved ions in ground water precipitate in pores as minerals (Figure 17.28). Cementation decreases porosity and permeability, which slows ground-water flow. When the water slows, the water molecules spend even more time in contact with mineral grains, which promotes further dissolution of some original minerals in the rock while new cement minerals form in other places.

Some elements dissolved in ground water are unhealthy to humans. The element arsenic has attracted considerable recent attention from hydrologists, health scientists, and government decision makers. Arsenic causes a variety of health problems, ranging from skin diseases to increased risk of the development of some cancers. **Figure 17.29** shows variable arsenic levels in ground water in the United States. Although some arsenic results from human pollution, the arsenic concentrations are highest where the water flows through volcanic rocks and some sedimentary rocks that contain arsenic in minerals such as pyrite. In most places, therefore, the harmful arsenic results from natural interactions of ground water with minerals. Prior to 2006, the acceptable level of arsenic in drinking water was 50 parts of arsenic per 1 billion parts of water, but now a new standard of only 10 parts per billion has to be met.

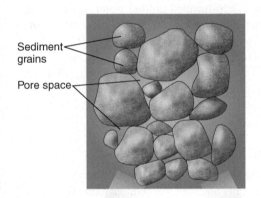

Sediment grains

Pore space

Water dissolves grains: Porosity and permeability increase

Minerals precipitate from water: Porosity and permeability decrease

Dissolved ions

Time

Time

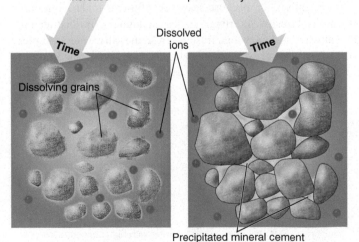

Dissolving grains

Precipitated mineral cement

▲ **Figure 17.28 How chemical reactions in ground water make or destroy porosity.** If chemical reactions dissolve the minerals in the aquifer, then the porosity and permeability increase as the pore spaces get larger. If the ground water contains large concentrations of dissolved ions, then new mineral crystals may precipitate in the pore spaces. This mineral cement fills in the pores and closes off the connections between them so that porosity and permeability decrease.

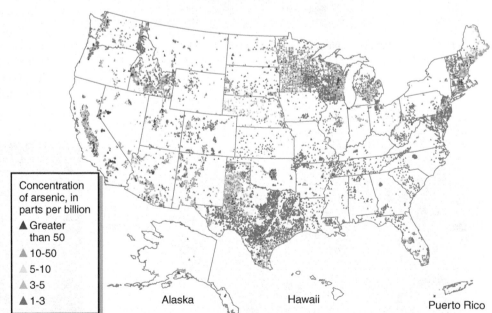

Concentration of arsenic, in parts per billion

▲ Greater than 50

▲ 10-50

▲ 5-10

▲ 3-5

▲ 1-3

Alaska

Hawaii

Puerto Rico

▲ **Figure 17.29 Arsenic threatens American ground water.** The maximum arsenic concentration permitted in United States drinking water is 10 parts per billion. Ground water used for drinking in many parts of the country exceeds this requirement and requires expensive treatment.

Hard and Soft Water

Minerals that precipitate from ground water not only fill in aquifer pore spaces, they also clog water-supply pipes, as illustrated in **Figure 17.30**. **Hard water** describes water with high concentrations of dissolved ions. Usually these are calcium and magnesium ions formed by dissolving calcite and dolomite within the aquifer. These ions commonly reach concentrations in the water where they precipitate and clog pipes with rings of calcite, dolomite, or other minerals. Hard water also diminishes the effectiveness of soaps, because the calcium and magnesium ions react with soap and laundry detergent to draw "soap scum" compounds out of the water so that the soap does not stay in the water as a cleanser. Hard-water ions also precipitate as minerals when water evaporates, which is why you may need to periodically clean coffee makers and showerheads with weak acid to dissolve clogging mineral precipitates.

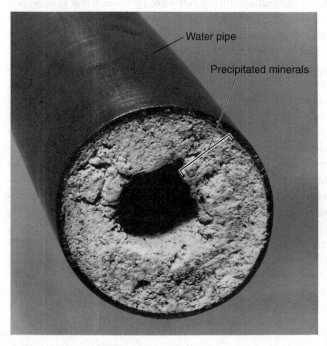

▲ **Figure 17.30 Hard water clogs pipes with minerals.** Minerals precipitated from hard ground water filled in most of this water pipe.

Soft water, in contrast, has low concentrations of dissolved ions. Most soft water is artificially produced from hard water by using water softeners, which are special filtration devices that chemically remove the calcium and magnesium ions. Soft water causes problems of its own, however, because water that lacks many dissolved ions is more reactive than water that is already carrying many dissolved constituents. As a result, soft water dissolves and corrodes pipes, especially in hot-water lines, which can add unhealthy levels of lead, copper, and other metals to drinking water.

Ground Water Forms Economic Mineral Deposits

We can extend the analogy of hard-water deposits in pipes to explain important ore deposits of zinc, lead, copper, and other metals that form in sedimentary rocks. Although the metal contents of rock are usually very low, many metallic elements do dissolve easily in warm water. The warm ground water that follows deep flow paths passes through a large volume of rock over long periods of time (see Figure 17.18) and dissolves the metals.

When the ground water eventually moves toward the surface, it cools slightly, mixes with shallow water that has a different chemistry, and it also reacts with near-surface rocks. All of these changing conditions also change the composition of the original, deep-traveled ground water in ways that decrease the solubility of the metal ions. As a result, metallic minerals, usually sulfide minerals such as the lead ore galena, precipitate in the pore spaces of sedimentary rocks to form ore deposits.

Figure 17.31 shows that rich ore deposits of lead and zinc minerals originating by this process are scattered across the central United States. In the mid-twentieth century, nearly half of the world's supply of lead and zinc came from mines in this region.

Ground-Water Chemistry Near Coastlines

Some water wells near coastlines yield good, freshwater for many years, but then the water abruptly becomes salty and unusable without expensive treatment. We can apply our understanding of ground-water flow and cones of depression to explain why this happens. Ground water beneath the land surface originates as freshwater precipitation. However, ground water beneath the seafloor is typically salty, because seawater infiltrates underlying pore spaces. Saltwater and freshwater mingle in coastal aquifers, as illustrated in **Figure 17.32**. Saltwater is denser than freshwater, so the salty water occupies the lower part of the aquifer while freshwater discharges onto the seafloor above the saltwater part of the aquifer. This density-driven flow explains the

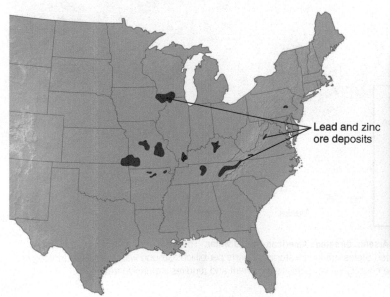

◀ **Figure 17.31 Lead and zinc deposits formed from ground water.** Deeply circulating, warm ground water with high concentrations of dissolved metal ions forms valuable mineral deposits where the water cools close to the surface and undergoes chemical reactions with near-surface ground water and rocks. Lead and zinc ore deposits in the central and eastern United States formed by this process.

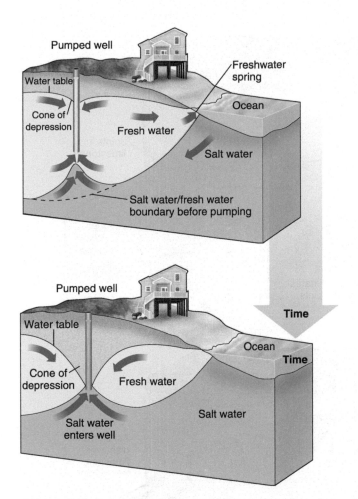

▲ **Figure 17.32 Freshwater and saltwater mix in coastal aquifers.** Dense, salty seawater moves landward beneath less dense freshwater. Freshwater moves seaward over the saltwater in the aquifer and may discharge onto the seafloor at freshwater springs. Pumping freshwater from the aquifer eventually draws the undrinkable saltwater upward into the well.

presence of freshwater springs on the seafloor near continents and islands.

Now let's see what happens when we start pumping freshwater from the upper part of a coastal aquifer. Ground water flows toward the well because it is a discharge site. Figure 17.32 shows how the ground-water flow not only forms a cone of depression in the low-density, freshwater part of the aquifer, but also how the high-density saltwater part of the aquifer also flows up toward the well because of the decreased aquifer pressure.

There is widespread loss of drinking water in coastal communities when increased pumping of freshwater causes saltwater to enter the wells. Water wells must be located farther from the shoreline, with greater expense to transport water to customers, or the salt has to be removed from the water. Either option is costly.

Unnatural Additions: Ground-Water Pollution

Pollution dissolved in ground water moves in the same way as tracers and naturally dissolved ions. The concentration of a pollutant is very high where it is first introduced into the ground water, but the pollutant spreads out so that the concentration decreases with distance traveled (see Figure 17.27).

Some contaminants precipitate as solids just as cement minerals precipitate in the aquifer. Therefore, a contaminant might be present at unhealthy levels close to the source, but be barely detectable at a greater distance because of spreading or precipitation of the contaminant as a solid.

Some harmful liquid contaminants do not mix in water but move instead as separate liquids, as shown in **Figure 17.33**. Oil and vinegar in salad dressing provide a familiar analogy for this separation behavior of some liquids. Gasoline is a common ground-water contaminant that does not mix with water. Gasoline is stored in underground storage tanks at fuel stations. If the tanks leak, then the fuel escapes into ground water. Gasoline is less dense than water, so it remains close to the water table rather than moving along deep flow paths into the aquifer. Technicians remove gasoline from ground water simply by pumping ground water from wells drilled to the water table at a lower elevation than the layer of contaminant. Unfortunately, a carcinogenic additive in the gasoline dissolves in ground water and is more difficult to remove. MTBE (methyl tertiary-butyl ether) was added to gasoline to reduce air pollution from vehicle exhaust but has now become a troublesome ground-water pollutant instead. Other harmful chemicals not only do not mix with water, but they also are denser than water. These chemicals sink deep into the aquifer regardless of the ground-water flow paths, and removal is costly. Solvents used in manufacturing industries and by dry cleaners are examples of these dense, insoluble liquids.

Most human-derived pollution is close to the surface, so deeper water-supply wells typically avoid most contamination. Deeper wells are even less susceptible to contamination where they tap into a confined aquifer that is separated from the shallow, polluted aquifer (see Figure 17.23). Poor water quality at greater depth, however, may result from increased dissolved-ion concentrations along the deep ground-water flow paths. Many variables, both natural and human, determine where to find the best-quality ground water.

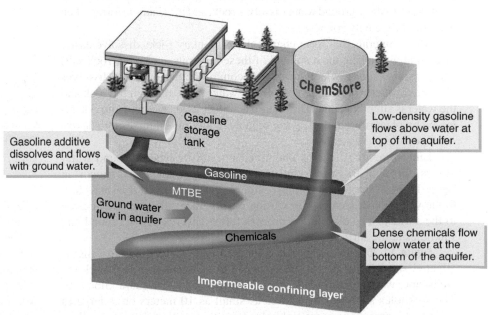

▲ **Figure 17.33 How liquid density determines contaminant flow.** Some liquid contaminants do not dissolve in water so their location in the aquifer depends on comparing their density to the density of water. Dense chemicals sink to the bottom of the aquifer. Low-density liquids, such as gasoline, float at the top of the aquifer. MTBE, a carcinogenic gasoline additive, dissolves in, and moves with, the ground water.

Putting It Together—What Is the Composition of Ground Water?

• Slow-moving ground water reacts with the minerals in the aquifer materials. Ions release into the water where minerals dissolve.

• Changes in ground-water chemistry, temperature, or both cause mineral precipitation that cements sedimentary rocks, clogs water pipes, and forms some economically important mineral deposits of lead and zinc.

• Coastal aquifers contain less dense freshwater above denser salt-water. An inverted cone of depression around a well may draw in the salty water.

• Not all contaminant liquids mix in water. Gasoline is a common pollutant that is less dense than water and floats near the water table. Some solvents, however, are denser than water and sink to the bottom of the aquifer.

17.5 How Does Ground Water Shape the Landscape?

Ground water is mostly out of sight below the surface. However, it still plays an important role in forming surface landscapes.

Landscapes Molded by Ground-Water Dissolution of Rock

The property destruction you saw in the field (Figure 17.1b) resulted from a collapsing sinkhole. A **sinkhole** is a depression on Earth's surface caused by the collapse of surface rock and regolith into a large underground cavity. The cavity forms where ground water dissolves rock, and the sinkhole forms when the roof of the cavity collapses.

Most minerals that form chemical sedimentary rocks dissolve easily in ground water. The most soluble of these minerals are halite (rock salt) and gypsum. Calcite, which forms very common limestone, also dissolves in ground water where large volumes of water move through the rock and the water is slightly acidic. The natural acid forms when carbon dioxide released by plant roots and decaying organic matter in soil mixes into downward percolating water.

Limestone dissolves by chemical weathering where ground water moves through fractures and along bedding planes, as shown in **Figure 17.34**. Water is usually most acidic in the unsaturated zone and close to the water table, so most dissolution occurs under the ground but close to the surface. The dissolution process forms caverns below ground and sinkholes at the surface.

Figure 17.35 shows stages in the development of a landscape that is affected strongly by ground-water dissolution. Sinkholes pockmark the landscape, with as many as 2500 such depressions per square kilometer. The sinkholes start out as features as small as 10 meters or as large as 1 kilometer across. As more sinkholes form they coalesce to form valleys that result from collapse of caverns rather than by stream erosion. Some streams disappear into sinkholes. In regions where the water table is declining, rock dissolution also increases downward. The collapse of

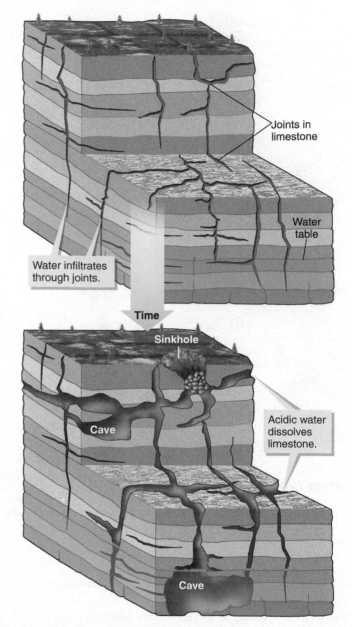

▲ Figure 17.34 **Where ground water dissolves limestone.** Infiltrating water mixes with carbon dioxide and organic acids in soil to form a weak acid. The acidic water preferentially moves down along joints in the rock. Limestone dissolves in contact with the weak acid, which enlarges joints and forms caves. Sinkholes form where rock and regolith collapse into near-surface cavities.

enlarging caverns eventually produces a highly irregular landscape of high-standing rock towers with intervening depressions and valleys.

Karst topography is the term geologists use to describe a terrain, such as that illustrated in Figure 17.35, with distinctive landforms and irregular drainage patterns caused by rock dissolution. The German word "karst" originates from Kr̆s; region of Slovenia, at the head of the Adriatic Sea, where this type of topography is dramatically represented. **Figure 17.36** shows the distribution of karst topography in the United States.

Limestone caverns, commonly just called caves, form where ground water dissolves rock, but they also contain many spectacular features

Sinkholes in Kentucky

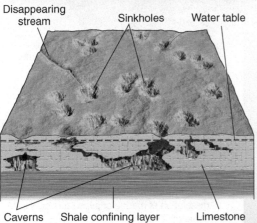
Disappearing stream   Sinkholes   Water table

Caverns   Shale confining layer   Limestone

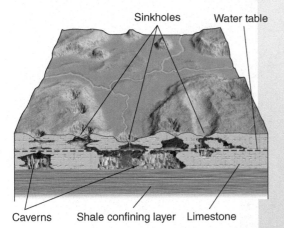
Sinkholes   Water table

Caverns   Shale confining layer   Limestone

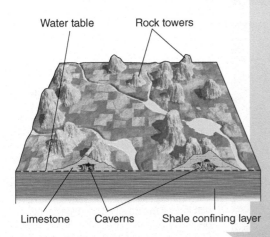
Water table   Rock towers

Limestone   Caverns   Shale confining layer

Time

Disappearing stream in Texas

Cave in Texas

Rock towers in China

◄ **Figure 17.35 How ground water forms karst landscapes.** Dissolution of limestone by ground water forms caverns that collapse to form sinkholes. The pockmarked surface of collapsed sinkholes is highly irregular, and surface streams disappear into solution-enlarged joints and sinkholes. After long periods most of the limestone may dissolve away leaving behind towering remnants of undissolved rock.

produced by mineral precipitation. **Figure 17.37** shows how ground water causes both mineral dissolution and precipitation. Carbon dioxide gas dissolved in water produces the acid that dissolves the calcite, which produces high concentrations of calcium and carbonate ions in the ground water. Where the ground water drips from the roof of a cave in the unsaturated zone, the carbon dioxide escapes from the water and passes as a gas into the cave atmosphere. This resulting decrease in the carbon dioxide content of the water also decreases the acidity, which causes calcite to precipitate. Columns, pillars, and tapestry-like sheets of precipitated calcite form the common scenic highlights of famous caverns. In a similar fashion, escape of carbon dioxide gas from emerging spring water caused the travertine precipitation illustrated in Figure 17.1a.

EXTENSION MODULE 17.3

The Geology of Caves. *Learn the origin of caves and the origins of the exotic rock formations commonly found in caves.*

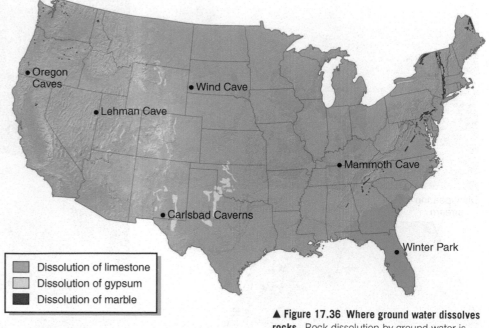

Dissolution of limestone
Dissolution of gypsum
Dissolution of marble

▲ **Figure 17.36 Where ground water dissolves rocks.** Rock dissolution by ground water is recognized over large areas of the United States. Most of these areas include caves and karst topography. Limestone dissolution accounts for most of these features, but dissolution of marble or gypsum occurs in some regions.

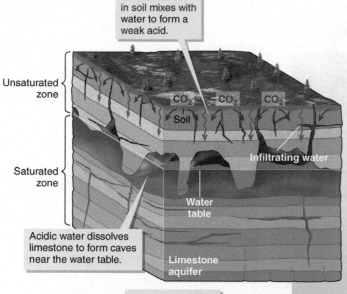

Carbon dioxide in soil mixes with water to form a weak acid.

Unsaturated zone

Saturated zone

Soil

CO$_2$ CO$_2$ CO$_2$

Infiltrating water

Water table

Acidic water dissolves limestone to form caves near the water table.

Limestone aquifer

Acidic water rich in ions from dissolved limestone and in dissolved CO$_2$.

Time

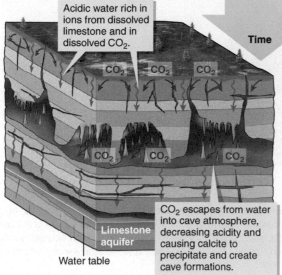

CO$_2$ CO$_2$ CO$_2$

CO$_2$ CO$_2$ CO$_2$

CO$_2$ escapes from water into cave atmosphere, decreasing acidity and causing calcite to precipitate and create cave formations.

Limestone aquifer

Water table

◄ **Figure 17.37 How ground water forms caves.** Limestone dissolves because the infiltrating water is weakly acidic from carbon dioxide (CO$_2$). Caverns typically form near the water table and are partly submerged in ground water. When the water table falls, the cave is entirely in the unsaturated zone. CO$_2$ escapes into the cave from the infiltrating ground water, which is rich in ions that resulted from limestone dissolution closer to the surface. Release of CO$_2$ into the cave atmosphere decreases water acidity so calcite precipitates to produce spectacular cave formations.

Plants grow around springs at base of overhanging alcoves.

Canyon width does not change downstream.

Small channel enters a much wider canyon.

Narrow stream channel in a very wide canyon bottom

◀ **Figure 17.38 Ground water helps form scenic canyons.** The labels on this photograph of stream canyons in southern Utah point out features that suggest the influence of ground water on canyon formation. Perched ground water exits at springs and wet seeps near the canyon bottom. Weathering of rock around the springs forms overhanging cliffs, which collapse as rock falls that produce sediment carried away by streams. The end result is large amphitheater-headed canyons that are too wide to be eroded by the surface streams without the assistance of ground-water weathering.

Landscapes Molded by Seeping Ground Water

Figure 17.38 illustrates a scenic canyon eroded in sandstone. Although a stream runs through the bottom of the canyon, several visible features are uncommon for stream-eroded valleys:

- The canyon floor is very wide compared to the size of the small stream at the bottom.
- The width of the valley does not change downstream, whereas stream-eroded valleys generally are wider in the downstream direction.
- The main canyon and side canyons terminate upslope in vertical to overhanging alcove walls that form natural amphitheaters. Stream channels either are missing altogether above these amphitheaters or are much narrower than the width of the alcove.
- Pockets of green vegetation mark locations of springs near the base of the canyon.

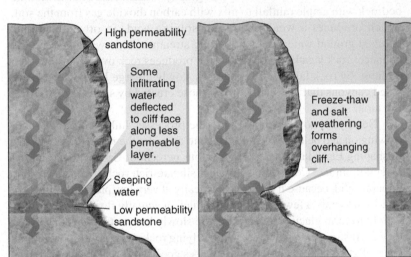

High permeability sandstone

Some infiltrating water deflected to cliff face along less permeable layer.

Seeping water

Low permeability sandstone

Freeze-thaw and salt weathering forms overhanging cliff.

Time

Unstable cliff fails in rock fall.

These observations are consistent with ground water playing a role in the landscape history. **Figure 17.39** illustrates how perched ground water enhances rock weathering where it emerges on a canyon wall. Wetting and drying, crystallization of salt from drying ground water, and freezing and thawing are processes that weaken and dislodge rock fragments where the water seeps out onto the outcrop. As the weathered particles fall, roll, and wash away, an overhang is produced above the ground-water discharge area. The overhanging cliff is unstable and collapses in a rock fall. The rock-fall debris weathers further into smaller fragments that are carried away by flowing surface water. In this fashion, ground water plays a major role in forming wide, steep-sided canyons.

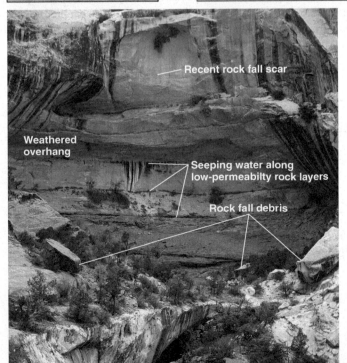

Recent rock fall scar

Weathered overhang

Seeping water along low-permeabilty rock layers

Rock fall debris

◀ **Figure 17.39 How perched ground water contributes to mass movement.** Infiltrating water moves to exposed cliff faces along low-permeability rock layers. The seeping water attracts plant growth and enhances weathering by freeze and thaw, salt precipitation, and plant roots. Weathering produces an overhanging alcove that eventually fails by rock fall. The photograph shows evidence of these processes in Utah.

Putting It Together—How Does Ground Water Shape the Landscape?

• Karst topography forms where ground water dissolves minerals in limestone and evaporite. Dissolution opens up spaces beneath the surface, including caverns, and causes the collapse of depressions, called sinkholes, that may capture surface drainage and make it flow underground.

• Perched ground water exiting on canyon walls enhances rock weathering and mass movement. These processes form deep, wide, steep-walled canyons that have distinctly different shapes from those carved by streams alone.

Where Are You and Where Are You Going?

Ground water flows through rock and regolith beneath Earth's surface and represents an essential water resource. Wells drilled below the water table, where pore spaces are completely filled with water, extract ground water for drinking, irrigating crops, and industrial use. The water-table elevation changes through time if there is an imbalance between the amount of ground water discharging to the surface at streams, lakes, or wells compared to the amount of water recharged from the surface.

Ground water flows from areas of high energy to areas of low energy where water discharges naturally to streams and springs or into artificial wells. Ground water follows curving subsurface paths so that the water that flows the greatest distance also travels deeper below the surface. Ground water flows much more slowly than surface water in streams, because water follows long, circuitous paths through small underground pores and is slowed by friction along pore edges.

Aquifers are Earth materials with high porosity and high permeability that provide sufficient quantities of water to wells to support the intended purpose and scale of ground-water withdrawal. Aquifers may be present immediately below the water table, perched above the water table along a local impermeable horizon, or confined between impermeable layers deep in the saturated zone.

Natural processes and human activities influence ground-water composition. Ground water reacts with minerals. These reactions dissolve some minerals and add dissolved ions to the water. The ions may bond to form new minerals that fill pore spaces, cement loose sediment into sedimentary rock, form valuable ore deposits, or even clog water pipes. Relatively dense salty seawater underlies less dense freshwater in coastal aquifers and encroaches into heavily pumped wells. Pollution enters ground water from many sources, including leaking fuel or chemical storage tanks; faulty septic systems; infiltration of water containing herbicides, pesticides, and fertilizers applied to fields; and contaminated streams. If contaminants dissolve in the water, then knowledge of how ground water flows also predicts the contaminant movement. However, some contaminant fluids do not mix with water; they float near the water table, if less dense than water, or sink through the aquifer, if denser than water. Pollutants typically spread out in the direction of ground-water flow, which means that the concentration decreases with increasing flow distance, although the affected area becomes larger.

Although hidden from view, ground water plays important roles in the formation of surface landscapes. Highly irregular karst topography, recognized by sinkhole depressions above underground caverns and disappearing streams, forms when ground-water dissolves rocks composed of soluble minerals. Most karst topography forms in areas of limestone bedrock with ample rainfall to mix with carbon dioxide gas from the soil, which produces a weak, calcite-dissolving acid. In other locations, perched ground water seeps out along stream-canyon walls and greatly enhances rock weathering. Weathering produces rock overhangs that then collapse by rock fall. These processes slowly enlarge long, wide, steep-sided canyons that are distinct from those created by stream erosion acting alone.

Next, you move from studying the processes linked to liquid water flowing on and below the land surface to consider the role of frozen water in shaping Earth landscapes. Eighty-four percent of Earth's freshwater is locked up in ice. Compared to typical silicate-rich rocks, ice is a curious mineral solid, because it deforms plastically at very low pressure. When ice builds up to only a few tens of meters thick, it actually flows under its own weight to form glaciers. Glaciers move slowly but are thick and heavy, so they exert strong erosive force on underlying rock and regolith. The importance of glaciers to sculpt Earth landscapes goes far beyond the mere 10 percent of the surface that they occupy today, because they occupied much larger areas during prehistoric ice ages.

Active Art

Why Is There a Water Table? See why a water table exists.

Forming a Cone of Depression. See how a cone of depression forms.

Extension Modules

Extension Module 17.1: Anatomy of a Water Well. Learn how a well is drilled and how ground water is withdrawn from a well.

Extension Module 17.2: Darcy's Law. Learn how Henry Darcy's experiment defined a simple mathematical formula for describing ground-water flow.

Extension Module 17.3: The Geology of Caves. Learn the origin of caves and the origins of the exotic rock formations commonly found in caves.

Confirm Your Knowledge

1. Define "ground water." Where is the ground water "stored"?
2. How do you find ground water?
3. What is the difference between the saturated zone and the unsaturated zone? What is the name of the undulating surface that forms the top of the saturated zone?
4. What is permeability? How does it relate to porosity?
5. How do porosity and permeability change with depth below the surface? Which Earth materials have the highest porosity and permeability? The lowest?
6. What are the characteristics of an aquifer?
7. What Earth materials form the best aquifers? Why are these materials the best aquifers?
8. How does a confined aquifer differ from an unconfined one?
9. Define "perched ground water."
10. How does pumping water affect the water table?
11. What is natural recharge? What is artificial recharge?
12. What factors control the flow of ground water? Can it flow upward, despite downward gravitational pull?
13. What is an artesian well? Explain how an artesian well works.
14. How do chemicals get into and pollute ground water? Give some examples.
15. Why does ground water flow more slowly than water in streams?
16. Does ground water flow at a uniform velocity through an aquifer? If not, why not?
17. What is the difference between hard water and soft water? What problems could each type cause in your home?
18. How do hot springs and warm springs form?
19. What is a sinkhole? How does it form? How do sinkholes relate to caves?

Confirm Your Understanding

1. Write an answer for each question in the section headings.
2. What is the major difference between an intermittent stream that is dry part of the year and a stream that flows year-long, even when it has not rained or snowed for several weeks?
3. Many brands of bottled water promote their product as artesian water. Is artesian water necessarily better than non-artesian water?
4. In the ground water study illustrated in Figures 17.25 and 17.26 a chemical tracer was injected at 640 parts per million. After 461 days the highest concentration measured was only 39 parts per million. How did the chemical tracer get so diluted?
5. If you were planning to purchase a house with a well and septic system, what would you look for in regards to water availability, water quality, and sinkhole risk? How would you determine where to locate your well relative to the location of your septic system?
6. Do you think there is a water table beneath the sea? Beneath lakes?
7. Water is a very reactive chemical compound. Explain how chemical reactions between ground water and minerals affect drinking-water quality, rock formation, mineral-resource formation, and development of landscapes. In this short essay, explain why ground-water chemistry tends to be very different from surface-water chemistry.

CHAPTER 18

Glaciers: Cold-Climate Sculptors of Continents

Why Study Glaciers?

Glaciers—flowing masses of snow and ice that persist from year to year—currently cover 10 percent of Earth's surface. Glacial ice holds 84 percent of Earth's fresh water. Most of this ice is in Antarctica and Greenland, with the remainder forming small glaciers at high latitudes or in high mountains scattered across Earth.

During geologically recent "ice ages," glaciers covered twice as much area as they do now. Therefore, glacial erosion and deposition during past ice ages shaped landscapes over large areas where glaciers are no longer seen. The recognition of ice ages in the recent geologic past indicates that Earth's climate shifts between cold and warm extremes. The most recent glacial advance reached its peak approximately 21,000 years ago, and glaciers retreated to their present positions by about 6000 years ago.

Glaciers account for more than scenic alpine landscapes. Glacial deposits form fertile soils in the northern United States and southern Canada and many ground-water aquifers. Glaciers sculpted the Great Lakes, which are important to the economy and history of central North America. The history of glaciation on land is tied to changes in global sea level that are affecting coastlines today.

After Completing This Chapter, You Will Be Able to

- Explain why glaciers form and how they move.
- Explain how glacial erosional and depositional landforms originate.
- Describe when and why glacial climates happen.

Pathway to Learning

18.1 What Is a Glacier?

18.2 How Does Glacial Ice Form?

18.3 How Does Ice Flow?

18.4 How Do Glaciers Erode and Transport Sediment?

18.5 How Do Glaciers Deposit Sediment?

18.6 What Happens when Glaciers Reach the Ocean?

18.7 How Do Valley Glaciers Modify the Landscape?

A backcountry camper enjoys an early evening view of Mendenhall Glacier in southeastern Alaska.

IN THE FIELD

Consider a virtual summer vacation to the Canadian Rocky Mountains. Your itinerary includes a visit to the Athabasca Glacier, in Jasper National Park, Alberta. **Figure 18.1**a shows a view of the park from a distance. The glacier looks like a huge icy tongue, part of which is snow white and part of which is dirty, resembling winter slush in a city street. The glacier fills in the bottom of a wide, steep-sided valley that looks as if it were carved out with a giant ice-cream scoop. You go closer and see streams emerging from beneath the glacier (Figure 18.1b); the water is clouded with suspended sediment. You encounter bare rock that is remarkably smooth but in places exhibits long, parallel scratches that look like somebody slid a giant knife across the outcrop (Figure 18.1c).

▶ Figure 18.1 **Visit to the Athabasca Glacier.**

(a) Distant view of the Athabasca

(b) The Athabasca River flows from beneath the glacier, muddy with suspended sediment

(c) Scratched and polished rock in front of glacier

(d) Moraine ridge in front of glacier. Sign points out where the front of the glacier was in 1992

(e) Top of the glacier is hard ice with a bumpy, cracked surface

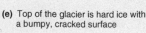

(f) A stream of water on the glacier surface disappears into a deep crack.

You also notice lumpy piles and low ridges of unsorted boulders, pebbles, and sand that are scattered around the margin of the glacier (Figure 18.1d). You first wonder whether these are bulldozed remains of a construction project; but a sign explains that the deposits are **moraines**, a French word that describes natural heaps of stony debris. Some of the moraine ridges coincide with posted signs that mark locations of the front of the glacier at various times over the last century.

You record your observations along with insights from exhibits in the visitors' center in your notebook, as seen in **Figure 18.2**. The moraines and scratched rock surfaces were beneath or alongside the glacier in the recent past, so these features probably owe their origin to glacial processes. The most impressive moraines are ridges of loose debris that parallel the two sides of the glacier and rise more than 50 meters above the icy surface. Clearly, the Athabasca Glacier is part of an actively changing landscape and not simply a static pile of ice.

You cannot resist the opportunity to take a tour to the top of the glacier. A specially equipped bus carries you and your friends onto the glacier where you set out on a short hike (Figure 18.1e). The first thing that you notice is the hard icy surface; it is slushy in a few places, but definitely not fluffy snow. The ice melts in the summer sun, and streams of water flow in shallow channels. You follow one stream to where it disappears into a deep opening in the glacier surface (Figure 18.1f). Perhaps this water plunges all of the way to the bottom of the ice and emerges in the streams that exit the front of the glacier.

Your vacation stop piques your interest in glaciers and glaciation. How do glaciers form? Does the ice, the meltwater streams, or both, transport the chaotic moraine sediment? How do glaciers modify the landscape? If the Athabasca Glacier retreated 1.5 kilometers upvalley in the last 150 years, where was it during the ice age? What causes an ice age? When will the next ice age occur?

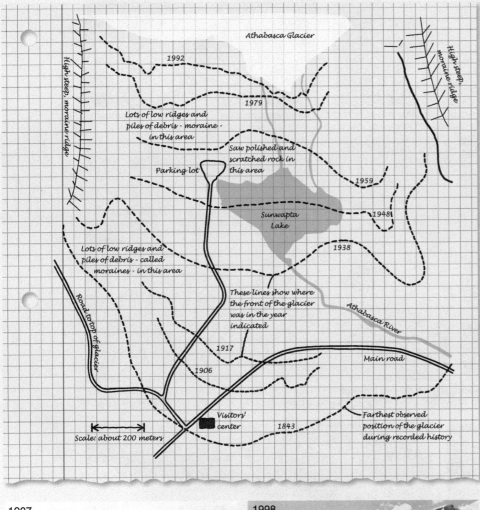

▲ Figure 18.2 **Overview of glacial features.** This map summarizes features observed in the field near the end of the Athabasca Glacier. The photos were taken from the same location at different times and show how the front of the glacier retreated upvalley during the twentieth century.

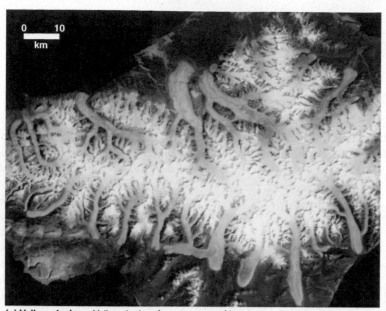

(a) Valley glaciers: Valley glaciers form tongues of ice descending into valleys eroded in snow-covered mountains on Bylot Island in northern Canada (left). Franz Josef Glacier in New Zealand (right) illustrates how valley glaciers are confined between bedrock valley walls.

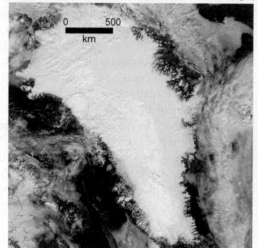

(b) Ice-sheet glaciers: This view of Greenland from space (left) shows that the large island is almost completely covered by a glacial ice sheet. The view from an airplane flying over Antarctica (right) shows that only the highest mountain peaks stick out above the kilometers-thick ice sheet that covers the continent.

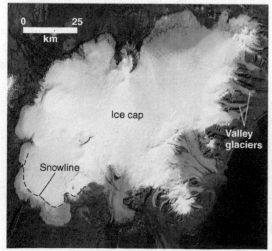

(c) Ice-cap glaciers: Part of the island of Iceland is covered by an ice cap that is visible in the satellite view (left). Glaciers also descend some valleys around the periphery of the ice cap. The ice is darker in this summer view where the previous winter snow fall has melted below the snowline. The field photo (right) shows the edge of the ice cap.

▲ **Figure 18.3 Visualize the types of glaciers.** Valley glaciers are confined between valley walls of rock. Ice sheets and ice caps flow outward in all directions across the landscape and are not confined by valleys. Ice sheets are larger than ice caps.

18.1 What Is a Glacier?

Geologists define a **glacier** as an accumulation of snow and ice that is thick enough to flow under its own weight. Observations across Earth's surface, such as those at the locations illustrated in **Figure 18.3**, show that some glaciers occupy rock-walled valleys, whereas others are continent-scale sheets of ice. These observations of the different sizes and shapes form an easy classification.

Types of Glaciers

Valley glaciers, such as the Athabasca Glacier, are long and narrow glaciers confined within bedrock valleys (see Figure 18.3a). Valley glaciers flow from the highest elevations to the lowest elevations, just like streams.

Other glaciers are not confined by valleys and can be subdivided based on size. **Ice sheets** (Figure 18.3b) are larger than 50,000 square kilometers in area, whereas **ice caps** are smaller, unconfined glaciers (Figure 18.3c). The two ice sheets on Earth today cover most of Antarctica and Greenland (see Figure 18.3b), but you will learn in Section 18.10 that ice sheets covered large parts of North America and Europe during the last ice age. Ice sheets and caps flow outward in all directions from their highest ice-surface elevations, regardless of the buried topography of the underlying bedrock.

Where Glaciers Form

The presence of glaciers at high latitudes, or in high mountains, where year-round temperatures are chilly, reveals the importance of a cold climate to form glaciers. However, the exact conditions required for glacier formation include more than coldness, because we also have to keep in mind the need to have lots of snow. Many decades of observations show that glaciers form where snow persists year-round, because more snow falls during the winter than melts during the following summer. If you live in western North America or have visited this mountainous part of the continent in the summer, then you have seen snow patches that remain throughout the year. A few of these snowy areas are large enough to qualify as glaciers.

In addition, heavy winter snowfall is essential to form glaciers. Very little snow will accumulate if the climate is dry during the winter. The snow that does accumulate melts away during the summer, even if temperatures rise above freezing for a short time. On the other hand, some snow may persist throughout the year where winter snowfall is very heavy, even in areas with relatively warm summers.

To understand where glaciers form, it is convenient to define the **snowline**, which is the elevation above which snow persists throughout the year. If a region does not have areas at elevations above the snowline, then it cannot have glaciers, regardless of how much snow falls in the winter. **Figure 18.4** shows how the snowline elevation changes along a route drawn from the North Pole to the South Pole and through the mountainous western parts of North and South America. The lowest snowline elevations are at the high, cold latitudes. You may find it surprising that the snowline dips slightly near the equator. This happens because the very moist equatorial climate delivers great accumulations of snow to high mountain peaks. This is why Cotopaxi, a tall volcano located in the Andes less than one degree from the balmy equator, has glaciers. The mountain is 5900 meters high, where most of the tropical moisture falls as snow rather than rain. The snowline elevation also rises slightly around 70 degrees north latitude because the polar regions are very dry.

Snowlines in the recent geologic past were much lower than today. Figure 18.4 shows the lower snowline elevations in the Americas about 21,000 years ago, during the last ice age. This older snowline is defined by glacial erosional and depositional features that you will learn to recognize in this chapter.

Sizes of Glaciers

The area covered by glaciers is easily measured by looking at maps. Some valley glaciers cover less than one square kilometer, whereas the Antarctic ice sheet blankets 13.5 million square kilometers. Volume, rather than area, is a more complete measure of glacier size and requires knowledge of the thickness of glaciers. Drilling holes through the ice and

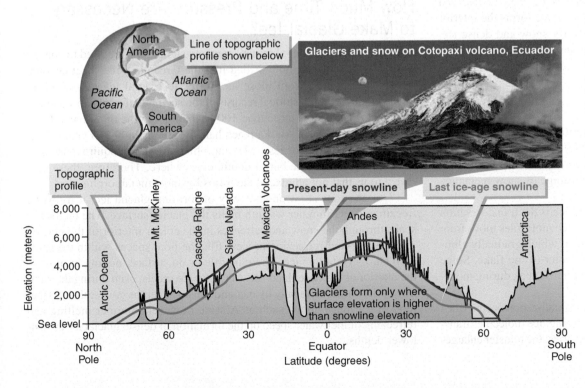

◀ **Figure 18.4 Graphing the snowline in the Americas.** The graph shows the elevations of the present-day snowline in comparison to land-surface elevations. Only Antarctica and high peaks in the Americas and are above the snowline, so these are the only places where glaciers form. The snowline elevation is high near the equator but tall mountains such as Cotopaxi, Ecuador, are higher still and have glaciers. The graph also shows that the snowline was much lower during the last ice age, so glaciers formed at lower elevations then than they do today.

echo-sounding methods, similar to those used to determine water depths in the ocean, provide geologists with many thickness measurements. These data show that valley glaciers are usually 50 to 300 meters thick, whereas large areas of the Antarctic ice sheet are more than 4 kilometers thick.

The total volume of glacial ice in Antarctica is about 29 million cubic kilometers. Imagine that volume of ice spread across all of North America—it would make a layer 1.25 kilometers thick, or more than three times the height of the Empire State Building. Seventy percent of Earth's fresh water is frozen in the Antarctic ice sheet.

Putting It Together—What Is a Glacier?

• A glacier is a mass of snow and ice that flows under its own weight.

• Valley glaciers move downslope between the rocky walls of valleys. Ice sheets and ice caps, different only in size, are thicker ice masses that flow radially outward from a central high point.

• Glaciers form where more snow falls during the winter than melts during the summer over a prolonged period of time. Both cold temperatures and abundant winter snowfall are required to form glaciers.

• The snowline is the elevation above where snow persists all year and where glaciers form. The snowline is lowest at high latitudes, where year-round temperatures are cooler.

18.2 How Does Glacial Ice Form?

We have said that glaciers occur only in areas of heavy snowfall, and you may be familiar with freshly fallen, fluffy snow. But observations in tunnels, in icebergs, and even on glacier surfaces such as the one you observed at Athabasca (Figure 18.1) show that hard, compact ice forms the interior of glaciers. What is the relationship between fluffy snow and dense ice, and how does that ice affect glacier movement and the ability of glaciers to sculpt scenic landscapes? Where does the ice come from?

Snow Metamorphism

Ice is a mineral that metamorphoses at temperatures and pressures that exist close to Earth's surface. Most of the dense, almost nonporous glacial ice results from metamorphism of snow, as depicted in **Figure 18.5** (also see Section 6.3 and Figure 6.8 for brief coverage of snow metamorphism).

The first stage in the transition from snow to ice is the transformation of irregular snowflakes into rounded snow grains. Ice molecules move from the outer points of the snowflake toward its center, which gradually eliminates the delicate points while thickening the center of the flake. Some rounding also happens as the delicate snowflake points break during movement or compaction.

Next, the rounded ice grains recrystallize. Recrystallization occurs when high pressure exists where adjacent grains are touching. Ice molecules transfer from one grain to another along the grain boundaries; the transfer enlarges

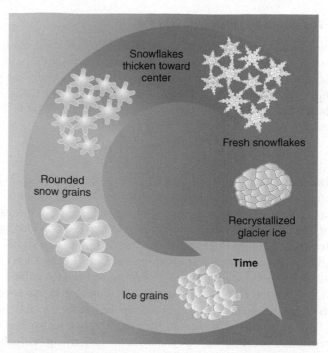

▲ **Figure 18.5 Snow metamorphoses to glacier ice.** The porosity of fresh-fallen snow is very high because of the irregular shape of the snowflakes. As the snowflakes are buried below more snow, the ice molecules move from the edges to the center of the flakes and eventually the snow transforms into rounded grains. Minor melting and refreezing during the summer converts the snow grains into small ice grains, which pack closer together than the original snowflakes. When the ice grains are buried to a depth of a few tens of meters, the pressure causes recrystallization into nonporous glacier ice.

some grains as others shrink, and eventually all of the air spaces fill with crystalline ice. The resulting interlocking ice crystals are several millimeters to a few centimeters across. The glacial ice is much denser (0.9 g/cm^3) than fresh snow (0.05 g/cm^3) but still less dense than water (1.0 g/cm^3).

How Much Time and Pressure Are Necessary to Make Glacial Ice?

Scientific drill holes bored into glaciers reveal the time required to convert snow to glacial ice. A hole drilled into the Greenland ice sheet encountered the transition from loose ice grains to glacial ice at a depth of 66 meters. Based on historical records of the accumulation of snow and ice, the ice at 66 meters is more than 100 years old. On the other hand, holes drilled into glaciers in Alaska reach hard ice that is only 3–5 years old at a depth of 15 meters or less. Metamorphism of snow requires the same pressure, and the same burial depth, everywhere. Therefore, these data reveal to us that not all glacial ice forms by snow metamorphism.

Observations show another way of forming glacial ice: it forms by freezing water. Summer warmth melts the glacier surface. The meltwater soaks through the snow and refreezes in the colder interior of the glacier. The melting-and-refreezing process fills the pore spaces within the snow with hard, crystalline ice. There is little or no surface melting in most of the interior of the Greenland and Antarctic ice sheets, even during the summer, so ice forms there only by metamorphism at depths greater than 60 meters. In warmer regions such as Alaska, however, summer melting and refreezing of meltwater speed up the formation of dense, glacial ice at shallower depths.

Putting It Together—How Does Glacial Ice Form?

• Glacial ice forms either by the metamorphic recrystallization of snow or by freezing of meltwater that soaks into the glacier.

18.3 How Does Ice Flow?

The critical feature that distinguishes a glacier from a persistent patch of snow and ice is the fact that glaciers flow. Furthermore, the flowing ice does geologically important work by eroding rock, as indicated by the scratched and polished surfaces at the Athabasca Glacier (Figure 18.1c) and further implied by the scooped-out shape of the valley (Figure 18.1a). The material eroded from one location is eventually deposited at another site to form, for example, the moraines you saw in front of the glacier (Figure 18.1d). Therefore, it is important for us to understand that we can prove that glaciers move and understand why the ice flows.

How Fast Do Glaciers Move?

Writers sometimes describe a very slow process as occurring at "glacial speed." This metaphor refers to the extremely slow pace of glacial motion. In fact, in the eighteenth century many people were skeptical that glaciers could erode rock until experiments in the Swiss Alps proved that glaciers actually moved. To prove glacial flow, stakes were driven into the surface of the ice and tracked year to year as they slowly traveled downslope.

Figure 18.6 shows data that document the flow of the Athabasca Glacier. Survey markers placed on the glacier surface show that the recent surface speed is not very impressive. If you patiently stood on the fastest part of the glacier, then you would move only half the length of a football field in one year.

The data also demonstrate variations in the velocity that need to be explained by how glaciers flow. For example, the surface velocity is fastest in the center of the glacier and slower at the edges, which is best explained as a result of greater friction against the rocky valley walls. The bending of a pipe inserted into the glacier (Figure 18.6) shows that the surface of the glacier is moving faster than the interior. This pattern of changing velocity with depth into the glacier also reflects the effects of friction, in this case where the glacier is in contact with the underlying rock. The velocity is not zero at the bottom of the glacier, however, which means that the glacier slides across the valley bottom. In fact, most of the motion in the Athabasca Glacier results from this basal slip rather than by flow within the ice.

Measurements at many glaciers reveal considerable variation in flow velocities. Surface velocities range from less than 2 meters per year to more than 8 kilometers per year. Some glaciers rapidly surge forward over weeks or months at rates as rapid as 80 meters per day. Glaciers in southeastern Alaska and the Swiss Alps yield the fastest velocities, and those in Greenland and Antarctica are the slowest. As we continue to explore how glaciers flow, we can learn why these variations exist.

Glacial Ice Is Weak Rock

Let's start by exploring how solid ice can flow. A key consideration is that glacial ice is much weaker than the common rocks in Earth's crust. Ice crystals have a well-developed cleavage that is similar to micas, and form thin sheets. When glacial ice is stressed, the ice crystals slide on these cleavage planes and the whole mass deforms plastically. The critical stress for plastic ice deformation is equal to the pressure at a depth of only a few tens of meters. This value contrasts sharply with a depth of about 10 kilometers to reach conditions for the plastic flow of granite in the continental crust.

Therefore, weak ice flows under the stress exerted by its own weight. This is why glaciers form only where winter snow accumulation exceeds

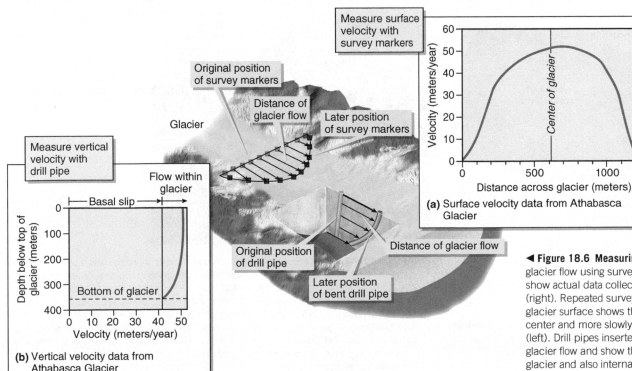

◄ **Figure 18.6 Measuring glacier flow.** Geologists measure glacier flow using survey markers and drill pipes. The graphs show actual data collected at the Athabasca Glacier in Canada (right). Repeated surveying of markers placed on the moving glacier surface shows that the glacier moves fastest in the center and more slowly where there is friction along valley walls (left). Drill pipes inserted into the glacier bend because of glacier flow and show that there is slip along the base of the glacier and also internal flow, which decreases downward toward the bottom.

summer melting; eventually the accumulating ice is thick enough to flow as a glacier.

Figure 18.7 illustrates observations that demonstrate plastic flow in the lower part of a glacier, whereas the upper brittle part breaks into large fractures. **Crevasses** are cracks in the brittle upper part of glacier caused by motion of the lower, plastically deforming part. Crevasses rarely persist below depths of about 50 meters.

Flow from Where Ice Builds Up to Where It Melts

A reason why glaciers flow relates to the accumulation and melting of ice. Adding ice at high elevation causes glaciers to flow downslope because of the pull of gravity. However, a growing glacier eventually descends to lower, warmer elevations where summer melting exceeds winter accumulation. A glacier, therefore, consists of two zones, which are illustrated in **Figure 18.8**:

1. The high-elevation **zone of accumulation**, where the winter snow accumulation is greater than the amount of snow that melts during the summer.

2. The low-elevation **zone of wastage** (also called the "zone of ablation"), where summer melting exceeds the winter snow accumulation.

Over the course of a year, mass is added to the glacier in the zone of accumulation, and it is

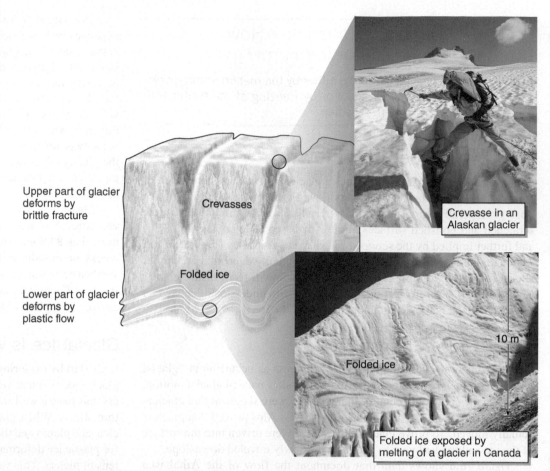

▲ **Figure 18.7 Brittle and plastic deformation of ice.** Brittle and plastic behavior of ice resembles rock deformation. Crevasses form by brittle fracture in the upper part of the glacier where pressure is low. Ice folds by plastic deformation in the lower part of the glacier where pressure is high.

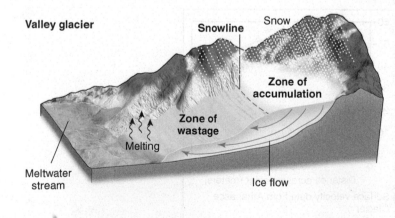

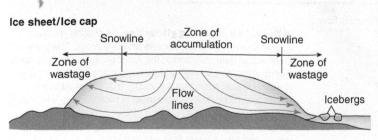

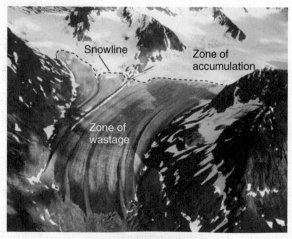

◄ **Figure 18.8 Zones of accumulation and wastage.** Zones of accumulation and wastage are separated by the snowline. Winter snowfall adds more mass to the zone of accumulation than is lost by summer melting. Mass is lost from the zone of wastage because summer melting and detaching icebergs is greater than the winter snowfall. Ice flows from the zone of accumulation to the zone of wastage and also flows downward in the zone of accumulation and upward in the zone of wastage. The photo shows these two zones for a glacier in Alaska. Persistence of snow above the snowline causes the zone of accumulation to have a bright white surface, whereas summer melting exposes darker, sediment-rich ice in the zone of wastage.

subtracted from the zone of wastage. Mass is also lost if icebergs form where a glacier terminates in a lake or the ocean. The snowline elevation, where there is no overall gain or loss in ice mass over the course of a year, forms the boundary between the zones of accumulation and wastage.

Ice flows from the zone of accumulation toward the zone of wastage. This is the same direction as flow from the head to the toe of a valley glacier, or from the center to the edge of an ice sheet (Figure 18.8). The ice also moves downward in the zone of accumulation, as new material is added on top each year. Ice deep in the zone of wastage moves upward as surficial ice melts. These movements mean that glacial ice flows toward the underlying ground surface above the snowline and away from the ground surface below the snowline. This pattern of motion will be important later on when we examine how glaciers erode rock and deposit sediment.

Figure 18.9 allows us to integrate some observations of glacier flow velocities, observed deformation of the ice surface, and the pattern of flow within the ice. The ice speeds up and stretches where it moves down steep slopes. The resulting tensional stress causes the upper, brittle part of the glacier to break into crevasses. Farther downslope, where the glacier slows down, the ice compresses and thickens, causing crevasses to close and, in some cases, forming thrust faults in the ice.

It is important at this point to consider something that may at first sound confusing—glacial ice always flows toward the edge of the glacier, even if that edge is retreating upslope over time. To illustrate this point, remember that observations prove that the Athabasca Glacier has retreated 1.5 kilometers *up* the valley over the last 150 years (Figures 18.1 and 18.2) even though measurements illustrated in Figure 18.6 show that the ice is flowing *down* the valley. So, ice always flows downslope regardless of whether the glacier toe is stationary, advancing, or retreating.

Figure 18.10 shows that comparing amounts of accumulation and wastage explains whether the front of the glacier remains stationary, advances, or retreats over time. The front of the glacier is stationary if the annual mass gained above the snowline exactly equals the mass lost below the snowline (see Figure 18.10a). If accumulation is greater than wastage, then the front of the glacier moves downslope because more ice flows

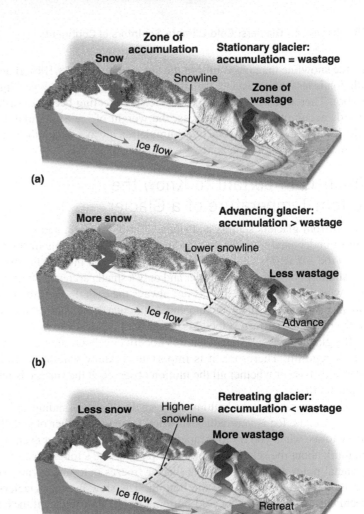

(a)

(b)

(c)

▲ Figure 18.10 Why glaciers advance, retreat, or remain stationary. Ice flows from the zone of accumulation to the zone of wastage. The front of the glacier remains stationary if the annual accumulation and wastage are equal. If accumulation exceeds wastage, then the front of the glacier advances. If wastage exceeds accumulation, then the front of the glacier retreats, even though ice flow is always toward the zone of wastage. Modern glaciers show all three behaviors, although retreating glaciers are more common. In the Alps of southern Europe, for example, the area covered by glacier ice is now 35 percent less than in 1850.

ACTIVE ART

Glacial Advance and Retreat. See the how a glacier moves when the front advances, retreats, or is stationary.

▼ Figure 18.9 How glacier surface features relate to flow velocity. The diagram shows how flow velocity usually increases where the ice flow is downward and decreases where the ice flow is upward. Flow is also faster where the underlying slope becomes steeper and the ice slows down where the slope angle decreases. Crevasses form where the glacier speeds up and causes the ice to stretch. The ice compresses where the glacier slows down, so crevasses squeeze closed and thrust faults form in the ice. The photo shows the opening and closing of crevasses in a New Zealand glacier where slope and flow velocity change.

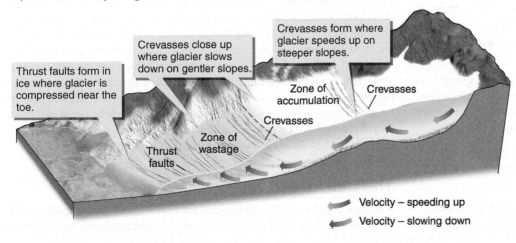

past the snowline than melts during the summer (Figure 18.10b). If accumulation is less than wastage, then the glacier toe retreats upslope (Figure 18.10c). Therefore, the Athabasca Glacier is retreating because wastage exceeds accumulation. Nonetheless, the weight of ice accumulating at high elevations within the zone of accumulation always causes the glacier to flow downslope into the zone of wastage.

Why It Is Important to Know the Bottom Temperature of a Glacier

Erosion from glaciers sculpts and shapes Earth's surface, and geologists want to understand how glaciers do this. The measurements of the bent pipe in the Athabasca Glacier (Figure 18.6) show that most of the motion is by sliding at the base of the glacier and only a small part results from plastic flow within the glacier. This is not always the case. In fact, some very slow-moving glaciers in Greenland and Antarctica do not slide at the base, and the measured surface velocity is due entirely to plastic flow within the ice. If a glacier does not slide, then it is not effectively eroding underlying rock or regolith. Therefore, it is important to know whether a glacier slides at its base or whether all the motion observed at the surface is internal plastic flow.

The temperature at the bottom of the ice distinguishes sliding and non-sliding glaciers. Ice is very adhesive when frozen to another object. This is easy to see when you try to scrape ice from a windshield (or even when you think about the schoolyard prank of placing one's tongue on a freezing metal pole). If the bottom of a glacier freezes onto underlying rock, then there is no ice motion at the bottom. The slow-moving glaciers in Greenland and Antarctica are cold-bottom glaciers. In contrast, if the basal temperature is at or above the melting temperature, then liquid water is present below the ice and sliding occurs. Fast-moving glaciers in Alaska and the Swiss Alps, along with the Athabasca Glacier, are warm-bottom glaciers. Furthermore, studies in tunnels underneath glaciers show that the velocity of warm-bottom glaciers is fastest when there is more water present at the base of the ice. The important thing to remember is that only glaciers with warmer bottom temperatures slide at the base, and these glaciers, or parts of glaciers, do the most geologic work.

Putting It Together—How Does Ice Flow?

• Flow velocities are fastest at the surface in the center of the glacier, and range from less than 2 meters/year to more than 80 meters/day. Velocity decreases downward through the glacier and toward the valley margins (in the case of a valley glacier).

• Variations in the velocity cause different parts of the glacier to experience tensional and compressional stresses. Tension causes brittle fracturing revealed by crevasses. Compression presses crevasses closed and may cause thickening and thrust faulting in the ice.

• Glacial ice flows under its own weight because it is a very weak rock that deforms plastically where it is more than a few tens of meters thick.

• Ice flows from the zone of accumulation to the zone of wastage, regardless of whether the toe of the glacier is stationary, advancing, or retreating. Flow is generally toward the bottom of the glacier in the zone of accumulation and toward the surface in the zone of wastage.

• If the basal temperature is at or above the melting temperature, then the glacier slides across underlying rock and regolith. Where the glacier freezes to underlying rock and regolith, it moves only by internal plastic flow.

18.4 How Do Glaciers Erode and Transport Sediment?

Geologists and naturalists have long been fascinated with the landforms of modern and ancient glacial landscapes. Study of these landforms, and the materials that compose them, reveals the processes of glacial erosion, transport of sediment, and deposition of sediment that are complemented by observations at active glaciers. **Table 18.1** lists and explains the terms applied to many of these landforms. Most of the words probably sound unfamiliar to you and originate in Europe, where the landforms were first described. Only some of these terms are used in this chapter, although all of them appear in the geologic literature.

Comparing Glaciers and Rivers

Glaciers are sometimes described as "rivers of ice." However, do glaciers erode and transport sediment the same way that streams do? It seems reasonable to expect that there are substantial differences in how solid ice and liquid water erode and transport sediment.

Observing active erosion and sediment transport is considerably more difficult for glaciers than for streams. Geologists study erosion and transport of stream sediment by direct observations in natural channels, or by indirect and controlled experiments in glass-walled laboratory channels (see Section 16.4). Glaciers, on the other hand, are opaque and move too slowly for direct observation of sediment movement. Nonetheless, geologists can link observations in natural tunnels below glaciers to field studies of glaciers that recently have retreated to reveal erosional features, such as the scratched and polished bedrock observed in front of the Athabasca Glacier (Figure 18.1c).

Glaciers Exert Shear Stress

Shear stress, the force exerted by a moving object, is a critical factor for explaining erosion by both streams and glaciers (see Section 16.3 for description of shear stress in a stream). Shear stress increases with the increasing weight of the flowing material. The weight of valley glaciers that are 50 to 300 meters thick or ice sheets more than 3 kilometers thick is huge compared to the weight of water in streams, which are usually less than 10 meters deep. Glaciers, therefore, exert much stronger shear stresses on their beds than do streams, and they are also much more effective agents of lateral erosion on valley walls. For comparison, the shear stress at the bottom of the modest-size Athabasca Glacier is four times greater than the shear stress exerted by flowing water near the mouth of the Mississippi River.

The Role of Meltwater

It is impossible to understand glacial erosion without also considering the meltwater present at the bottom of the glacier. On one hand, water reduces

TABLE 18.1 Names of Glacial Landforms

Name of Landform	Origin of Name	Formation Appearance	Illustration of Landform
Depositional landforms			
Moraine	A French word describing heaps of stony debris.	Any accumulation of nonbedded glacial sediment deposited at the base, front, or bottom of a glacier.	Figures 18.17, 18.25, 18.29
Kame	From the Scottish *comb*, describing a steep ridge.	Outwash alluvium deposited alongside or beneath a glacier that remained as a hill or ridge after the adjacent glacial ice melted away.	Figure 18.20
Esker	From the Irish *eiscir*, meaning ridge.	Long, steep-sided curving ridge of alluvium that filled an ice tunnel beneath a glacier and was left behind when the glacier melted.	Figure 18.20
Drumlin	From the Gaelic *druim*, meaning rounded hill.	A ridge of till, or less commonly outwash, with a streamlined shape molded by flow at the base of a glacier. The slope of the ridge is steepest in the direction from which the ice approached and is more gentle and tapered in the direction of glacier flow.	Figure 18.31
Erosional landforms			
Arête	A French word for fish bone.	A narrow, knife-edge ridge separating glacially eroded valleys or cirques.	Figure 18.24
Horn	An Old German word describing the sharp, bony projection from the head of an animal.	A very steep, pointy mountain peak sculpted between three or more cirques; epitomized by the Matterhorn in the Alps.	Figure 18.24
Roche moutonée	Derives from resemblance to eighteenth-century French wigs called moutonées, after the mutton fat used to hold them in place.	An elongate bedrock ridge eroded by a glacier and parallel to the direction of glacier movement. The ridge has a gently sloping, smooth abraded slope that faces in the direction of glacier approach and a steep, plucked slope that faces in the direction of glacier flow.	Figure 18.11b
Cirque	A French word meaning both circus and ring.	A steep-sided amphitheater depression or half bowl high on a mountain side at the head of a glacially eroded valley.	Figure 18.27
Tarn	Icelandic term for a small lake.	Commonly refers to a lake within a cirque where glacial erosion overdeepened the cirque floor to form an enclosed depression.	Figure 18.27

friction, which reduces the wearing down of rock beneath the glacier. On the other hand, water enhances basal sliding, which allows the glacier to move and erode. On the whole, the presence of water increases the erosion potential of a glacier. Water flowing at the base of the ice also carries away glacial-erosion products and erodes channels into the underlying bedrock or cuts tunnels in the overriding ice.

Erosion by Abrasion

Some glacial erosion occurs by abrasion (see Section 16.3 for descriptions of this process in streams). Ice has a Mohs hardness of 1.5, which is comparable to that of talc (see Figure 2.4), so ice is too soft to scratch many rocks. Glaciers use transported rock and mineral debris trapped within the ice as abrasive tools.

Sharp-edged rocks frozen into the bottom of a slow-moving glacier are the tools that scratch and gouge the underlying bedrock. **Figure 18.11** shows examples of the parallel **striations** that demonstrate abrasion at the base of a glacier by these frozen-in rocks. Abrasion generates a huge volume of very fine-grained rock dust, called **glacial flour**. If meltwater at the base of the glacier does not immediately wash away the glacial flour, then these very fine mineral particles grind along the bedrock surface and polish it smooth.

Glacial abrasion, therefore, is analogous to using sandpaper on wood. Coarse sand paper rapidly removes uneven bumps in wood, whereas fine sand paper produces a smooth polish. Rubbing coarse sand paper across a smooth, polished surface produces scratches.

Erosion by Plucking

Glaciers also pluck apart rock along preexisting joints and cracks in a very similar fashion to stream plucking. The plucking process is more erosive at the bottom of a glacier than at the bottom of a stream for four reasons:

1. The higher shear stress of flowing ice more readily disaggregates fractured rock.
2. Freezing and thawing of meltwater at the bottom of the glacier produces new rock fractures and further pries open existing ones.
3. The plastic glacier ice squeezes into fractures and pries the rock apart.
4. Water confined at the bottom of the glacier is pressurized, like water in a hose, which increases the ability of the flowing water to dislodge blocks of rock.

Plucking results in jagged rocky promontories that face into the direction of glacier flow, as shown in Figure 18.11.

(a) Glacial striations in limestone, Marblehead, Ohio

(b) Glacier-eroded rocky hill, Yosemite, California

▲ **Figure 18.11 Evidence of glacier erosion and plucking.**

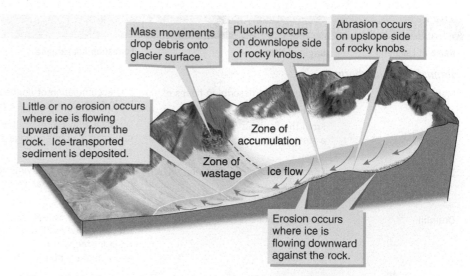

▲ **Figure 18.12 Where glacial erosion occurs.** Most glacial erosion occurs where ice flow pushes downward against the underlying rock and regolith. Ice abrades the upslope sides of rocky knobs beneath the glacier and plucks the downslope sides (also see Figure 18.11). Mass wasting along valley margins drops debris onto the ice surface. Deposition occurs where internal ice flow is upward away from the underlying rock in the zone of wastage.

Contributions from the Freeze-Thaw Process and Mass Movement

Not all of the erosive action is at the bottom of the glacier. What happens on the valley walls above a valley glacier, or along mountainsides that stick up through ice caps and ice sheets? Physical weathering by freeze and thaw (see Figure 5.3a for more about physical weathering) is very intense in glacial environments, especially during seasons where temperature oscillates daily between subfreezing and above freezing. The weathered rock drops as rock falls and rock slides onto the glacier (see Figure 15.30 for a dramatic example). Glaciers are efficient conveyor belts that collect and transport debris from mass movements along the valley walls.

Where Erosion Occurs

Figure 18.12 summarizes the types of locations where geologists observe glacial erosion. Erosion is most intense where the moving ice shoves against the rock. Erosion, therefore, is characteristic of the zone of accumulation where the ice flow is downward against the underlying rock and regolith (see Figure 18.8), and also where rock projects up into the ice, in either the zone of accumulation or wastage.

Erosion also is intense along the valley walls. The velocity variations across a valley glacier from side to side (Figure 18.6) reveal strong frictional resistance between ice and rock, which enhances plucking and abrasion at the edges of the glacier. Mass movements further modify the steep slopes above the glacier.

The Transport of Sediment on Top of, Within, and Below Glaciers

Glaciers shove and drag sediment at the base of the ice, carry it frozen within the ice, and piggyback it along the top of the ice. **Figure 18.13** illustrates these processes. Sediment eroded by abrasion and plucking freezes into the ice. Some debris remains at the base of the glacier but much of it eventually moves upward near rock obstacles and into the zone of wastage (Figures 18.9 and 18.13). When surface snow and ice melt in the zone of wastage, the sediment remains to give the glacier a dirty appearance (see Figure 18.8).

Mass-movement debris that drops onto the ice litters the tops of valley glaciers and also some ice sheets that move past higher mountain peaks. **Figure 18.14** shows how glaciers carry this debris as a moving ridge of boulder-rich sediment along the glacier margin, called a **lateral moraine**. Where two valley glaciers join, the lateral moraines combine to form a ribbon of sediment within the glacier, which is a **medial moraine**.

Meltwater moving beneath the glacier also transports large volumes of sediment. An example of sediment-laden meltwater occurs at the Athabasca Glacier (see Figure 18.1b). Recall that it was milky white. Why was that? The color of meltwater streams reflects the great abundance of suspended load, which is the glacial flour produced by abrasion.

The Rates of Glacial Erosion

Geologists calculate the total erosive power of glaciers by measuring the amount of glacial sediment deposited over a measured time interval. These calculations show that slow-moving, cold-bottom ice-sheet glaciers in Greenland and Antarctica lower land-surface elevation only by about 0.01 millimeter per year. At the other extreme, fast-moving, warm-bottom valley glaciers in southeastern Alaska erode downward and sideways about 10–100 millimeters per year. Although this erosion rate may seem rather small, the mass of produced sediment is gigantic. One glacier in southeastern Alaska covers an area only about twice the size of Boston, Massachusetts, but it erodes more than 200,000 metric tons of rock from every

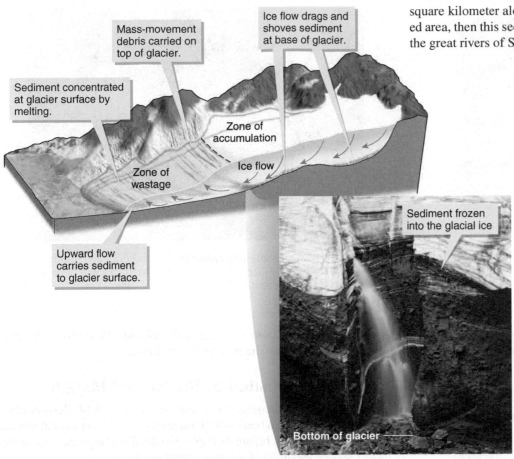

Sediment concentrated at glacier surface by melting.

Mass-movement debris carried on top of glacier.

Ice flow drags and shoves sediment at base of glacier.

Zone of accumulation

Zone of wastage

Ice flow

Upward flow carries sediment to glacier surface.

Sediment frozen into the glacial ice

Bottom of glacier

▲ **Figure 18.13 How glaciers transport sediment.** Glaciers push sediment along at the base, especially where flow is downward toward the bottom of the glacier. Sediment freezes into the ice, moves upward in the zone of wastage, and remains on the glacier surface when the surrounding ice melts. Mass-movement debris is carried along piggyback on top of the glacier.

▲ **Figure 18.14 Visualizing how lateral and medial moraines form.** Lateral moraines are ridges of accumulated rock-fall debris along the margin of a glacier. Medial moraines are ribbons of debris within the glacier. The view in this photo shows that medial moraines form where lateral moraines join at the junction of two glaciers.

square kilometer along its base every year. If you compare an equal eroded area, then this sediment load is more than 100 times greater than that of the great rivers of Southeast Asia (see Figure 16.7).

Putting It Together— How Do Glaciers Erode and Transport Sediment?

• Glacial erosion takes place by abrasion and plucking. Debris frozen into the moving ice abrades underlying rock. Plucking occurs when rocks disaggregate along pre-existing fractures, which are commonly enhanced by freeze and thaw of meltwater or by pressurized meltwater or ice that is injected into the cracks.

• Glacial erosion is greatest where internal ice flow is downward against the underlying rock at the base of the glacier and also along the valley walls for valley glaciers.

• Sediment is shoved along at the base of the glacier, frozen within the moving ice where it may be carried above the base of the glacier, and carried piggyback at the top of the glacier.

• Mass movements deliver sediment to the glacier surface, where it is then transported as lateral or medial moraines on valley glaciers.

18.5 How Do Glaciers Deposit Sediment?

Geologists recognize two types of sedimentary deposits near or beneath modern glaciers, and these deposits are also widespread in previously glaciated landscapes. Geologists understand these depositional processes from a combination of studies around active glaciers and studies undertaken where glaciers have melted away. **Figure 18.15** illustrates these contrasting sediment types, which are described as follows:

1. Sediment deposited directly by the glacier is **till**.
2. Sediment mostly eroded by glaciers and then carried away by meltwater streams is **outwash** because it is, literally, the sediment that washes out of the glacier.

Till is a very poorly sorted mixture of gravel, sand, and mud, and it typically lacks bedding (Figure 18.15). Moraines, for example, consist of till. The rock fragments within the till erode from all of the area covered by glacial ice. Ice sheets transport large boulders, some weighing hundreds of metric tons, more than 1000 kilometers from where they originated. **Figure 18.16** illustrates an example of these far-traveled, out-of-place rocks, called **erratics**. One piece of evidence for ice-age glaciers in the Midwestern United States is the presence of large erratics of ancient metamorphic rocks that match up with outcrops near Hudson Bay in Canada.

Glacial cobble:
faceted edges, striations

Stream cobble:
rounded, smooth

▲ **Figure 18.15 What glacial deposits look like.** Till is poorly sorted, nonbedded sediment deposited directly by a glacier. Outwash is moderately- to well-sorted, bedded alluvium deposited by meltwater streams. Stream cobbles in outwash are smooth and rounded. Glacial cobbles in till have faceted faces and edges, and commonly are striated.

Outwash is much better sorted and has more distinct bedding than till (Figure 18.15). The gravelly and sandy alluvium is commonly associated with thinly bedded mud deposited in lakes that form near the glacier.

The Sediment Left Behind

Till deposited beneath the glacier forms a bumpy sediment sheet of irregular thickness, called **ground moraine**. Ground moraine, illustrated in **Figure 18.17**a, can originate by any of three processes:

1. A glacier drags and pushes a large volume of sediment at the bottom of the ice, which increases the friction at the base of the glacier. The frictional resistance builds up to the point where the rocky debris lodges against the ground surface and is left behind as the ice continues to flow.

2. Rock fragments frozen into the ice move upward in the zone of wastage (see Figure 18.9), but fragments dragging along the base of the glacier are left behind.

▲ **Figure 18.16 What a glacial erratic looks like.** This conspicuously out-of-place boulder in Central Park, New York City, rests on very different rock with a glacier-abraded surface. The boulder is a glacial erratic eroded from a far-distant location and deposited here by an ice-age glacier.

3. Melting at the bottom of the glacier releases rock fragments that previously froze into the base of the flowing ice.

Sediment Deposited at the Glacier Margin

High ridges of till flank most glacier margins. Figure 18.14 illustrates lateral moraines. The high boulder-rich ridges that tower over the Athabasca Glacier (see Figure 18.1a) are examples of lateral moraines that rise to the former surface elevation of the now shrunken glacier.

End moraines, seen in Figure 18.17b, form at the leading snout of the glacier. Some end-moraine till is regolith that is bulldozed along in front of the moving glacier. Most end moraines, however, are deposited directly from the glacial ice, as explained in **Figure 18.18**. To fully understand this process, recall that ice flows even when the front of the glacier is stationary (Figure 18.10). When melting occurs each summer, sediment melts out along the snout or accumulates on top of the glacier and then slides or washes down to the front. When the glacier is stationary for decades to millennia, the internal flow delivers sediment to the same area like a big conveyor belt and builds up the moraine ridge over time.

Sediment Deposited by Meltwater Streams

Melting glaciers produce much larger volumes of water and sediment than would otherwise result from direct precipitation on the drainage basin. Streams that exit valley glaciers carry away hundreds of metric tons of sediment each day. Stream discharge fluctuates dramatically because melting mostly occurs when the temperatures are highest. Summer discharge, for example, may be 10 times greater than winter discharge, and flow on a warm summer afternoon may be twice as great as during the cooler evening just a few hours later. The large sediment load, wildly fluctuating discharge, and typical low abundance of streamside vegetation in cold climate zones all combine to favor the development of braided stream patterns (see Figure 16.20 to review factors that favor formation of braided streams). The large sediment loads typically overwhelm the transport ability of the streams, causing widespread deposition of outwash alluvium in river valleys, as shown in **Figure 18.19**.

Meltwater streams also deposit outwash beneath and alongside the glacier, as shown in **Figure 18.20**. Meltwater streams along the glacier margin erode sediment from the lateral moraine and from mass

◄ Figure 18.17 **What moraines look like.**

(a) Ground moraine is an uneven veneer of till deposited beneath a glacier and exposed when the glacier retreats or melts away. The continuous, low ridge in the ground moraine was shaped by glacier flow.

(b) End and lateral moraines are ridges of till that form along the glacier margin.

movements along the valley wall and redistribute it into a wedge of alluvium between the valley wall and the glacial ice (these features are also called "kames"; see Table 18.1). In other cases, meltwater on top of the ice commonly enters the glacier through crevasses, as observed at the Athabasca Glacier (Figure 18.1f). The water flows rapidly down steep chutes and conduits and eventually flows along the base of the ice. The flowing water may carve out a tunnel within the ice that is then partly filled with alluvium. When the glacier later melts away, the tunnel-filling sediment remains as a ridge (this landform is also called an esker; see Table 18.1).

Putting It Together—How Do Glaciers Deposit Sediment?

• Till is poorly sorted debris deposited directly from the glacier. Ground moraine is an uneven sheet of till deposited beneath the glacier and left behind when it melts. Ridges of till form lateral moraines alongside the glacier and end moraines at the downslope snout of the glacier.

• Outwash is better-sorted alluvium deposited by meltwater streams downslope, alongside, and beneath glaciers.

▶ Figure 18.18 **How end moraines form.** End moraines form at the front of a stationary glacier by a process analogous to a conveyor belt. Upward and forward ice flow in the zone of wastage carries sediment to the front of the glacier, where it is released from melting ice. The sedimentary fragments accumulate as a ridge at the front of the glacier. Glaciers that retreat at uneven rates leave behind many end moraines, each one of which represents a location where the glacier front was stationary for a period of time.

ACTIVE ART

Glacial Processes. See the growth and movement of a glacier along with glacial erosion and deposition.

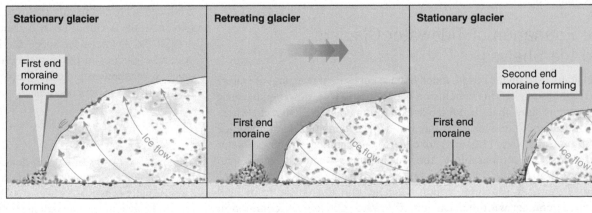

Stationary glacier | Retreating glacier | Stationary glacier

First end moraine forming | First end moraine / Second end moraine forming / First end moraine

Conveyor transport | Conveyor transport | Conveyor transport

Stationary conveyor | **Moving conveyor** | **Stationary conveyor**

▶ **Figure 18.19 What glacial-outwash rivers look like.** The photo on the left is a space shuttle view of outwash rivers draining from glaciers in southeastern Alaska into the Pacific Ocean. High-suspended-load transport of glacial flour turns the water gray and produces a sediment plume in the ocean. The photo on the right is an airplane view of the Sunwapta River, which drains glaciers in the Canadian Rockies. The braided channel pattern and milky water result from heavy sediment load and are common for outwash rivers. Notice the bus and automobiles on the road, which provide an impression of the large scale of the river.

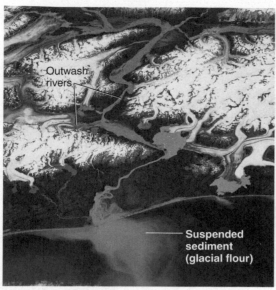

▶ **Figure 18.20 Meltwater streams alongside and beneath glaciers.** Outwash deposited alongside and below glaciers is left behind as benches and ridges after a glacier melts. The benches are also called kames and the ridges are sometimes called eskers (see Table 18.1).

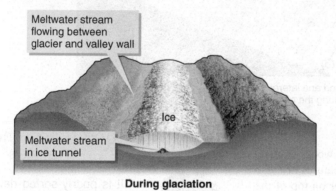
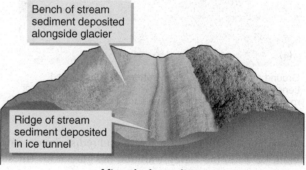

Meltwater stream flowing between glacier and valley wall

Meltwater stream in ice tunnel

Ice

During glaciation

Bench of stream sediment deposited alongside glacier

Ridge of stream sediment deposited in ice tunnel

After glacier melts away

18.6 What Happens When Glaciers Reach the Ocean?

Glaciers form on land but may flow into the ocean or into large meltwater lakes. Glacial ice and water interact in complex ways because ice is less dense than water.

The Formation of Tidewater Glaciers and Ice Shelves

Do glaciers float on the ocean, much like ice cubes float in a glass of water? You might think so. This conclusion is only partly true, however, because observations show that glacial ice does not immediately float right where the glacier snout moves into the water. The densities of glacial ice (0.9 g/cm^3) and water (1 g/cm^3) are close to one another in value, so water depth has to be comparable to the ice thickness—actually, more than nine-tenths of the ice thickness—before the glacier floats. **Figure 18.21** shows **tidewater glaciers**, which descend into the ocean from land and are in contact with the seafloor. Tidewater glaciers are common in southeastern Alaska, parts of northeastern Arctic Canada, and along the Greenland coast.

When a tidewater glacier moves into deeper water, it bobs up from the seafloor and floats on the water surface to form an **ice shelf**. **Figure 18.22**

illustrates the differences between ice shelves and **sea ice**, which is simply frozen seawater. Ice shelves are much thicker than sea ice and move primarily as a result of glacial flow rather than at the whim of ocean waves and currents. Ice shelves persist throughout the year, whereas a large proportion of high-latitude sea ice forms each winter and then melts in the summer.

Ice shelves are present along the margins of the Antarctic and Greenland ice sheets and locally in Arctic North America. Ice shelves form 44 percent of the Antarctic coastline and compose about 7 percent of the total area of the Antarctic ice sheet. On land, the predominantly cold-bottom Antarctic glaciers move very slowly, typically slower than 0.1 meter per year. At sea, however, the ice shelves float away from the coastline at velocities of 1–3 meters per year.

Making Icebergs

Icebergs are blocks of ice that detach from a glacier and float off into the ocean or a lake. Some icebergs break away from the front of tidewater glaciers, usually because of wave erosion that oversteepens the front of the glacier and causes mass movement (Figure 18.21). The largest icebergs break away from ice shelves. Ocean waves and currents stress the floating ice. Cracks form and eventually join up to completely separate blocks of the ice-shelf glacier, which then float free and move in directions determined by wind and ocean currents.

Glacier front collapsing to form icebergs

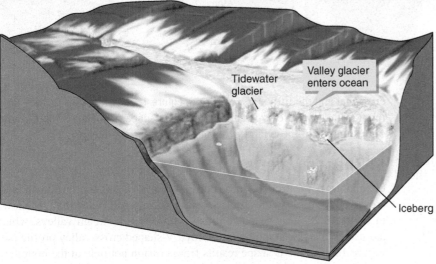

Tidewater glacier

Valley glacier enters ocean

Iceberg

◀ **Figure 18.21 What a tidewater glacier looks like.** This view from southeastern Alaska shows a tidewater glacier that originates on land and flows into the sea. The steep front of the glacier frequently collapses to form large icebergs. The diagram shows that the ice is in contact with the seafloor because the water is too shallow for the glacier to float.

◀ **Figure 18.22 What ice shelves and sea ice look like.** Ice shelves are thick, floating glaciers, whereas sea ice is thin sheets of frozen seawater. Both of these views are near the coast of Antarctica.

Ice shelf

Sea ice

Iceberg formation is an important wastage process for glaciers, especially in cold climates with minimal melting. Formation of icebergs along the Antarctic ice shelves accounts for 80 percent of the total glacier wastage for the entire continent. The largest observed iceberg-forming event occurred over 35 days in the summer of 2002, when 3250 square kilometers of an Antarctic ice shelf broke off into the South Atlantic Ocean, as shown in **Figure 18.23**.

Icebergs are very hazardous to ships. Large icebergs, some more than 50 kilometers long and 300 meters thick, float in the ocean for decades, so they cross shipping lanes far from the parent glacier. When ships collide with icebergs, the mass of the ice exerts sufficient stress against the hull to crush wood or rupture steel plates. Most of the iceberg is concealed below the water and may extend over a significantly greater area than the exposed iceberg above the water line. This is the literal meaning of the familiar expression for something more to come, "It's just the tip of the iceberg." In 1912 the *Titanic*, then the world's largest ship, sank after colliding with an iceberg. The accident occurred south of Newfoundland, Canada, more than 2000 kilometers away from the west coast of Greenland, where the iceberg originated.

January 31, 2002
50km

March 7, 2002
50km

► **Figure 18.23 An ice shelf breaks up to form icebergs.** These two satellite photos provide before and after views of the break-up of the Larsen B Ice Shelf on the Atlantic Ocean side of Antarctica. The total area of the disintegrated ice shelf is 3250 km², which you can compare to the 2717-km² area of Rhode Island. The average thickness of the ice shelf was 220 meters, and the breakup produced 720 billion metric tons of icebergs. The light blue areas between the large icebergs in the right photo are patches of seawater littered with countless smaller icebergs.

Putting It Together—What Happens when Glaciers Reach the Ocean?

• Glacial ice is less dense than water, but glaciers float only where water depth is more than nine-tenths of the ice thickness.

• Tidewater glaciers are in contact with the seafloor, whereas ice shelves float. Sea ice is simply thin sheets of frozen seawater and is not related to glaciers.

• Icebergs are large blocks of floating ice that break off of tidewater glaciers and ice shelves.

18.7 How Do Valley Glaciers Modify the Landscape?

Glaciers play an important role in forming landscapes. It is important to understand landscape modification by moving glaciers for two reasons:

1. Glaciers are not as common as streams, but they do more erosive work. In some places, glaciers are the dominant force creating the landscape. This is especially true in high, mountainous regions where cold temperatures and heavy snowfall produce glaciers that erode the rock and provide for isostatic feedbacks for further uplift (to review the relationship between erosion and uplift, see Section 13.3 and Figure 13.12).

2. Ice-age glaciers sculpted the primary landscape features over large areas of North America and northern Europe that are currently ice free. These phenomena are an indication of changing global climatic conditions during Earth history.

What would you expect to see following the retreat of a valley glacier? Five landforms are particularly distinctive of the current or former action of valley glaciers: (1) U-shaped valleys, (2) places where erosion overdeepened valleys to form lakes, (3) hanging valleys, (4) knife-edge ridges and pointed peaks, and (5) moraine ridges. The first four features are

erosional and the fifth is depositional. **Figure 18.24** illustrates the development of these landforms in a glaciated landscape, and **Figure 18.25** points out examples of these landforms at the Athabasca Glacier.

U-Shaped Valleys

Glacial valleys eroded into bedrock have a distinctive U-shaped, cross-valley profile, with very steep walls and a broad valley floor, as shown in **Figure 18.26**. This shape contrasts with bedrock stream valleys, which are usually narrow at the bottom with a V-shaped cross-valley profile (see Figure 18.24). The U shape results from erosion not only at the bottom of the valley, like in the case of a stream, but also along the sides of the valley where the glacial ice, in some cases hundreds of meters thick, grinds against rock.

Fjords are U-shaped, glacier-eroded valleys along coastlines that are now partly submerged beneath the sea to form long, deep, steep-walled bays. *Fjord* is a Norwegian word, and the Norway coast features dozens of these glacial landforms. Fjords are also present in North America along the Atlantic, Arctic, and Pacific Ocean coasts of Canada and in southern Alaska.

Overdeepened Valleys

The elevation of a stream-valley floor always decreases continuously in the down-valley direction. In contrast, the long profile of glacially eroded valleys commonly includes overdeepened places where the valley floor actually slopes inward to make a bowl. Water accumulates to form lakes within these eroded bowls after the glacier melts away, as shown in **Figure 18.27** (also see Figure 18.26).

Glaciers erode deeply into the bedrock and overdeepen the valley floor to form these low spots. Deep erosion is especially common at the upslope end of glaciated valleys, where the steeply eroded valley walls partially enclose a natural amphitheater called a **cirque**. Some cirque depressions fill with lakes (called tarns; see Table 18.1), and they are common landforms in the Rocky Mountains (Figure 18.27).

Geologists cannot directly observe the overdeepened erosional bowls beneath active glaciers, so they are uncertain of why glaciers erode more deeply at some locations. Exceptionally deep scour sometimes coincides

Before glaciation

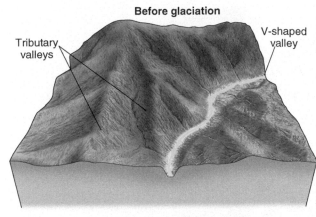

Tributary valleys

V-shaped valley

During glaciation

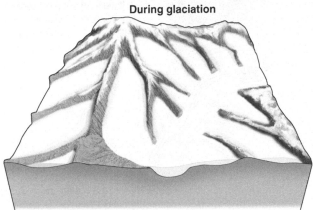

After glaciation

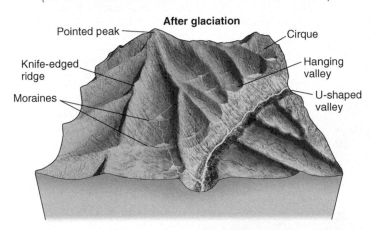

Pointed peak

Cirque

Knife-edged ridge

Hanging valley

Moraines

U-shaped valley

Mt. Everest

Glaciers carved the valleys around Mt. Everest, Earth's highest mountain. Notice the broad glacial valleys and narrow intervening ridges that rise to pointed peaks. This photo was taken by astronauts aboard the International Space Station.

The Teton Range, in Wyoming, features the knife-edged ridges, pointed peaks, and hanging valleys of a previously glaciated mountain range. The ridge in the bottom foreground is an end moraine.

◀ **Figure 18.24 Landscape modification by valley glaciers.** Valley glaciers transform V-shaped stream valleys into broad, steep-sided, U-shaped valleys. The valleys are separated by narrow, knife-edged ridges that rise to sharp, pointed peaks (the ridges and peaks are also sometimes called arêtes and horns; see Table 18.1). Some glacial valleys are eroded more deeply than others, leaving tributary valleys hanging in the sky with streams falling over high waterfalls.

◀ **Figure 18.25 The landscape modified by the Athabasca Glacier.** This photograph illustrates the common landscape features of valley glaciers that are seen at the Athabasca Glacier. Compare this view with the photos and map in Figures 18.1 and 18.2.

Hanging valleys

U-shaped valley

Lateral moraine

Lateral moraine

End moraines

End moraine

◄ **Figure 18.26 What glaciated valleys look like.** These photos show glaciated valleys in southeastern Alaska. The photo on the left illustrates the broad floor and steep sides that characterize glacially eroded, U-shaped valleys. The photo on the right illustrates how glaciated valleys along coastlines have submerged to form fjords as sea level rose after the ice age.

Profile of a river-eroded valley

Elevation decreases uniformly in the downstream direction.

Profile of a glacier-eroded valley

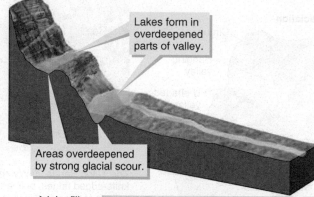

Lakes form in overdeepened parts of valley.

Areas overdeepened by strong glacial scour.

◄ **Figure 18.27 What overdeepened valleys looks like.** Lakes form in previously glaciated valleys where glaciers scour out deep bowls into the rock. A cirque is the overdeepened head of a glacial valley that is partly encircled by an amphitheater of steep mountain ridges.

Lakes fill overdeepened parts of glaciated valley; Misty Fjords National Monument, Alaska

A lake fills a cirque at the head of a glacial valley; Glacier National Park, Montana

with softer or highly fractured rock that is more easily plucked by the glacier. In other cases, however, the rock seems uniform throughout the valley. Intense local erosion by meltwater beneath the ice may also be a cause of overdeepening, especially where crevasses permit large amounts of water to flow to the base of the glacier. The increased water pressure at the base of the glacier near these meltwater-input locations enhances fracturing in the underlying rock and permits more plucking by the glacier. This explanation would also apply to deep erosion of cirques, because there is always an open gap at the head of a glacier, between the ice and the adjacent bedrock wall, where meltwater flows down to the base of the glacier. Overdeepening also happens where tributary glaciers join the main glacial flow, probably because the added ice from the glacier increases the stress exerted on underlying rock.

Hanging Valleys

Another feature that distinguishes stream and glacial valleys is the elevation where tributary valleys join the main valley (see Figure 18.24). Streams usually join together at the same elevation—that is to say, a tributary stream neither ponds into a lake because it joins a higher-elevation main stream, nor does it cascade as a waterfall into a lower-elevation main stream. In contrast, tributary valleys in glaciated landscapes commonly terminate high above the main valley. This is not obvious while ice fills the valleys because the ice surface is uniform, and only the basal elevations are substantially different (see Figure 18.24). Retreating glaciers in side valleys typically disconnect from the glaciers in the deeper main valleys, as seen at the Athabasca Glacier (Figure 18.25). When the glacier completely melts away, the tributary-valley floor hangs hundreds of meters above the main valley, resulting in spectacular waterfalls, as illustrated in **Figure 18.28**.

Hanging valleys erode because the main-valley glacier is thicker than the side-valley tributary glaciers. The main-valley glacier is thicker because it gains ice from each tributary that joins it. The shear stress is higher, and the depth of erosion is greater where the ice is thicker. This means that the thicker, main-valley glacier erodes more deeply than the thinner, side-valley glaciers.

Knife-Edged Ridges and Pointed Peaks

The ridges between widening U-shaped glacial valleys become narrower, until they rise steeply to very narrow, almost knifelike ridges (also known as arêtes; see Table 18.1 and Figure 18.24). Where several valleys slope radially away from a single mountain peak, the glaciers erode deeply into the central mountain. This leaves behind a very pointy pyramid (commonly called a horn). Earth's highest peak, Mount Everest, has the pointed pyramid shape that results from glacial erosion (Figure 18.24).

Lateral and End Moraines

Lateral and end moraines are recognizable alongside and in front of the Athabasca Glacier in Figure 18.25 and are also distinctive landforms in older glaciated landscapes, as seen in **Figure 18.29**. Ridges of till along the sides of valleys show the edges of former valley glaciers, and the heights of these lateral-moraine ridges above the valley floor provide good estimates of the thickness of the former glacier. End moraines are ridges of till that cross the valley and, along with lateral moraines or the rock walls of the valley, they may obstruct stream flow to produce a lake.

Putting It Together—How Do Valley Glaciers Modify the Landscape?

- Five erosional and depositional landforms distinguish glacially modified mountainous landscapes from stream-eroded landscapes.

- U-shaped valleys form by glacial erosion that not only deepens, but also widens valleys.

▲ **Figure 18.28 What a hanging valley looks like.** Bridal Veil Falls at Yosemite National Park, California, drops from a hanging valley. Ice-age glaciers eroded the Merced River valley, in the foreground, much deeper than its tributary valleys. The mouths of tributary streams are now much higher than the main river, so the water pours over waterfalls into the Merced valley.

◄ **Figure 18.29 Lateral and end moraines mark locations of former valley glaciers.** Glaciated valleys include lateral and end moraines of till deposited by the glacier. The extent of a former glacier in this valley in the Sierra Nevada, California, is indicated by the location of end moraines. The glacier was at least as thick as the top of the lateral moraines.

• Unusually deep erosion occurs at the cirque and at various locations along the valley. The overdeepened parts of the valley contain lakes after the glaciers melt away.

• Thick glaciers erode more deeply than their thinner tributary glaciers. After the glaciers melt, the different depths of erosion result in hanging valleys where tributary streams fall in tall waterfalls to join larger streams in the bottom of deeper valley floors.

• The widening of adjacent glaciated valleys results in the narrowing of intervening ridges and peaks. The resulting landscape consists of broad, scooped-out valleys separated by knife-edged ridges and pointed peaks.

• Lateral- and end-moraine ridges remain in glaciated valleys after glaciers melt away, and may impede streams, forming lakes on the valley floor.

18.8 How Do Ice Sheets Modify the Landscape?

Imagine a humongous retreating ice sheet—what kinds of features would you expect to see left behind? Some erosional and depositional features of huge ice-sheet glaciers simply are larger versions of those left by valley glaciers. However, because ice sheets are not confined in valleys, this means that ice sheets do not form features such as hanging valleys or lateral moraines.

Ice-sheet erosion occurs primarily beneath the zone of accumulation, where ice flow is directed downward to the glacier bed, whereas deposition takes place mostly beneath the zone of wastage. Some deposition also takes place within the original erosional zone as the ice melts in the waning stages of the ice age. Geologists' understanding of ice-sheet-modified landscapes comes largely from studies of North American and European landscapes resulting from the last ice age. **Table 18.2** compares landscape features from valley and ice-sheet glaciers.

Four landscape characteristics are particularly indicative of past ice-sheet glaciation: (1) large areas of scoured, plucked, and abraded rock surfaces; (2) large regions thickly covered with till; (3) streamlined ridges that parallel the direction of glacier movement; and (4) landscapes of countless lakes, ranging from small ponds to the largest lakes on Earth, formed by both erosional and depositional processes.

Figure 18.30 illustrates landscape features related to ice-sheet glaciation. The distribution and appearance of erosional and depositional ice-age glacial landforms in northeastern North America are shown in **Figure 18.31**.

Widespread Scour of Bedrock

Ice sheets dramatically reshape the ground surface because the shear stress exerted by ice that is more than a kilometer thick leads to considerable erosion.

TABLE 18.2 Comparison of Glacial Landscape Features

Valley Glaciers	Ice Sheets
Erosional Landscape Features	
U-shaped valleys and hanging valleys.	Large areas of scoured, plucked, and abraded rock surfaces.
Knife-edge ridges and pointed peaks.	Streamlined ridges that parallel the direction of glacier movement.
Places where erosion overdeepened valleys to form lakes and cirques.	Lakes, scoured from rock and ranging in size from small ponds to the largest lakes on Earth.
Depositional Landscape Features	
Till ridges forming lateral and end moraines and typically thin ground moraine.	Large areas thickly covered with ground moraine with multiple end moraines extending for hundreds of kilometers.
Outwash-stream deposits partly fill valleys and plains downslope of the glacier.	Outwash-stream deposits partly or completely fill valleys and form widespread plains below, above, and downslope of glacial till.
	Small kettle lakes form where ice blocks melt in till and outwash.

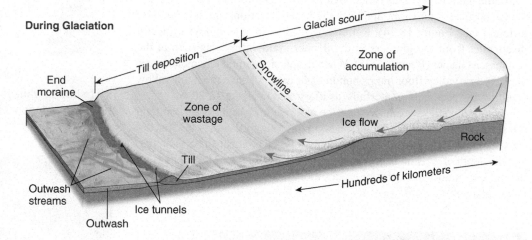

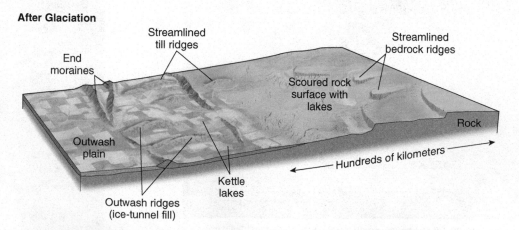

▲ **Figure 18.30 Landscape modification by ice sheets.** Moving ice sheets scour bedrock and deposit till over huge areas, with erosion coinciding with the zone of accumulation and deposition occurring within the zone of wastage. Outwash streams distribute sediment across vast plains beyond the ice-sheet margin. Lakes formed where water fills deep scours into rock, in kettle depressions left behind where ice melted in till and outwash, and within low spots in irregular ground-moraine and end-moraine topography. Ice movement erodes bedrock and till into elongate ridges (sometimes called "roche moutonée" and "drumlins"; see Table 18.1) that are streamlined parallel to the direction of ice flow.

Regolith is completely stripped away within the zone of accumulation, except where the ice froze to the ground surface, and the bedrock is highly abraded and plucked. Large ice sheets cover millions of square kilometers, so they erode a wide variety of rock types. Some areas are more susceptible to glacial erosion than others because of varying rock resistance to abrasion and plucking. The results, shown in Figure 18.31a, are ridges of hard rock interspersed with water-filled depressions scoured out of softer or more fractured rock. The total depth of erosion by at least a dozen ice sheets crossing eastern Canada over the last 3 million years averages out to about 200 meters.

Thick Till and Long Moraines

In contrast to the large areas of glacial erosion are the equally impressive areas covered by unsorted glacial till (example photos shown in Figures 18.31b–e). Till in eastern North America is typically 30–60 meters thick and is locally more than 200 meters thick near and south of the Great Lakes. Till is so thick and widespread that the bedrock geology over tens of thousands of square kilometers is known only from rocks encountered in wells and uncommon outcrops in the bottoms of the deepest stream valleys.

▶ Figure 18.31 Where ice-age erosion and deposition occurred in North America.

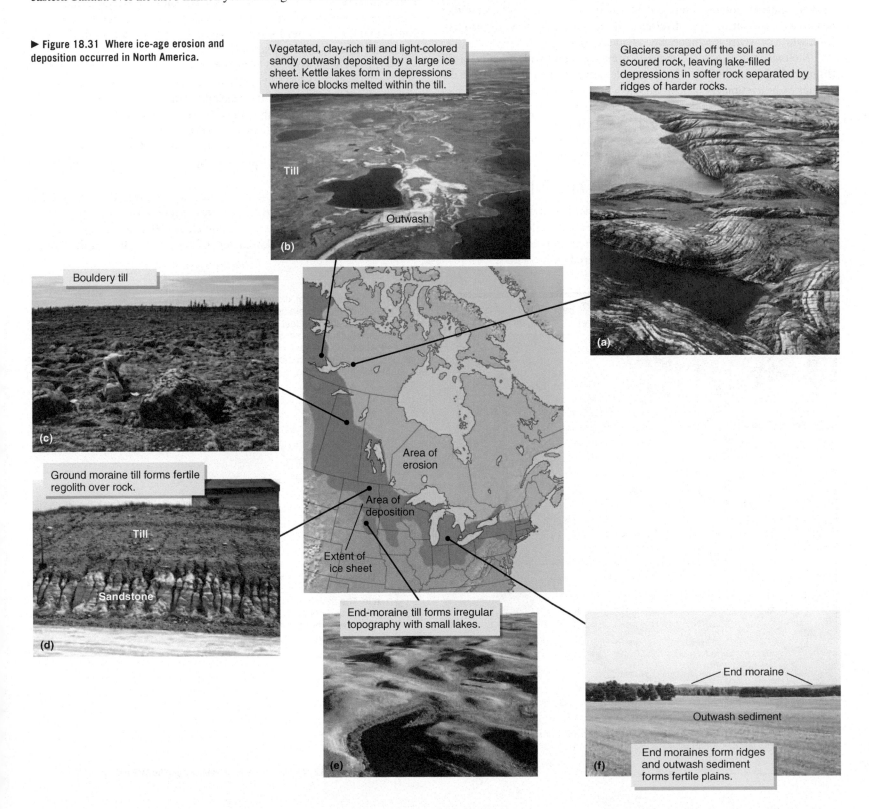

Vegetated, clay-rich till and light-colored sandy outwash deposited by a large ice sheet. Kettle lakes form in depressions where ice blocks melted within the till.

Till

Outwash

(b)

Glaciers scraped off the soil and scoured rock, leaving lake-filled depressions in softer rock separated by ridges of harder rocks.

(a)

Bouldery till

(c)

Ground moraine till forms fertile regolith over rock.

Till

Sandstone

(d)

Area of erosion

Area of deposition

Extent of ice sheet

End-moraine till forms irregular topography with small lakes.

(e)

End moraine

Outwash sediment

(f)

End moraines form ridges and outwash sediment forms fertile plains.

Landscapes are very uneven because of irregular till thickness and depressions, called **kettles**, where large blocks of ice melted within the till (or within related outwash alluvium).

End-moraine ridges can be followed across the landscape for hundreds of kilometers, as illustrated by **Figure 18.32**. The moraines are commonly 10 or more kilometers wide and rise no more than a few tens of meters above the surrounding land surface (Figure 18.31f). Ice sheets have a curving front, so the moraines share the same curving outline (see Figure 18.32a).

Each glacier generates multiple end moraines (see Figures 18.30 and 18.32). One end moraine forms at the farthest advance of the glacier. Other moraines form during glacial retreat, where the ice front was stationary for a while or where it temporarily advanced again.

Meltwater-stream deposits are commonly found with the glacial till (Figures 18.30 and 18.31f). These include sediment deposited beneath the glacier or in ice tunnels, glacial outwash that was overrun by an advancing ice sheet, or outwash deposited on top of till by a retreating ice sheet. Outwash deposits also extend into nonglaciated regions downslope of the original ice sheet. These deposits are so voluminous that they completely fill former stream valleys more than 100 meters deep.

Streamlined Features in Rock and Till

Advancing glaciers sculpt rock, till, and outwash alluvium into furrows and ridges that are elongate parallel to the direction of flowing ice. These erosional effects are also seen with valley glaciers, but they are most striking in size and extent in landscapes once covered by ice sheets.

The streamlined hills of till are tapered by the flowing ice and resemble a whale's back projecting above the water line. These ridges (sometimes called "drumlins"; see Table 18.1) not only provide evidence of past glaciation, but they also reveal the direction of glacier movement. Furrows and ridges eroded into till may persist for several kilometers, as though a giant comb was dragged across the landscape.

Glacially Formed Lakes

Lakes are very common features in glaciated landscapes. Some lakes mark locations where glaciers scooped out rock to leave a bowl that collected water (Figures 18.30 and 18.31a). Lakes occupy 20 percent of the land surface in the most extensively scoured areas of central Canada, a region with an average of more than 100 lakes within every 20-kilometer-by-20-kilometer square. Other lakes and ponds are kettles (Figures 18.30 and 18.31b). The irregular topography of till, especially within end moraines, commonly includes closed depressions that fill with water (Figures 18.30 and 18.31e). Minnesota's nickname, "Land of 10,000 Lakes," owes its origin to numerous lakes formed by both glacial erosion and deposition; most are kettles or correspond to low spots within the irregular till topography of moraines.

The Great Lakes are the most spectacular glacially eroded lakes on Earth. **Figure 18.33** shows these lakes, which cover an area of about 400,000 square kilometers and are as much as 400 meters deep. The lakes contain about 23,000 cubic kilometers of water—one-fifth of the world's

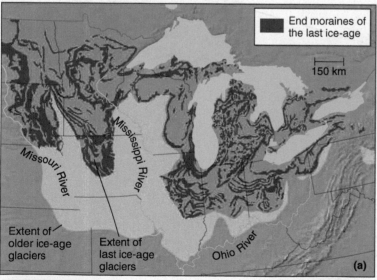

Ridges winding for hundreds of kilometers across the upper Midwest are end moraines left by a retreating ice sheet during the last ice age. Notice how the moraines outline the margins of ice-sheet ice tongues that carved out the Great Lakes. Older ice ages deposited till and moraine ridges even farther south than the last ice age.

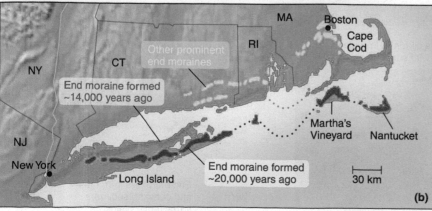

Ridges on Long Island are end moraines deposited during the last ice age. The moraines are partly submerged below the Atlantic Ocean farther east, but are traced to similar moraine ridges that form islands and peninsulas in southern Massachusetts.

◄ **Figure 18.32 Mapping the end moraines from the last ice age.**

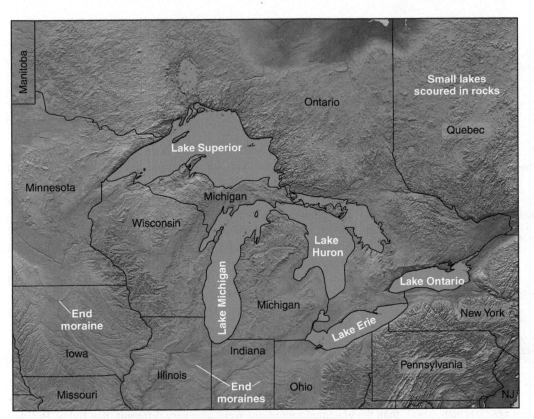

▲ Figure 18.33 Ice sheets shaped the Great Lakes region. This map consists of colored satellite radar images of North America that illustrate the irregular glacially scoured landscapes around and north of the Great Lakes in contrast to the smoother topography of glacial deposits with moraine ridges southwest of the Great Lakes. The Great Lakes are the most impressive examples of glacier scour, but also notice the countless small lakes in Canada that occupy scoured-out depressions in rock.

freshwater. Many advancing ice sheets progressively scoured out the Great Lakes in locations of easily eroded rock during the last 2 million years. The glaciers carved out of a dipping layer of soft evaporite sedimentary rock that encircles the lower peninsula of Michigan to form Lakes Michigan and Huron, along with western Lake Erie. Central and eastern Lake Erie and Lake Ontario formed where the ice sheets scooped out easily eroded shale. Lake Superior fills a deep depression where the ice sheets scoured out the sedimentary and volcanic rocks filling an ancient rift valley that is flanked by harder metamorphic rocks.

Putting It Together—How Do Ice Sheets Modify the Landscape?

• Ice-sheet glaciation scours bedrock and deposits till across hundreds of thousands of square kilometers.

• Rock, till, and outwash are sculpted by glacier flow into streamlined ridges and furrows that are elongate parallel to the direction of ice-sheet movement.

• Lakes are too numerous to count in glaciated landscapes. Lakes occupy deep scours into rock and irregular depressions within glacial till, some of which result from melting of ice originally deposited with the sediment. The Great Lakes were carved out of relatively soft rock by ice-age glaciers.

18.9 What Did North America Look Like During the Last Ice Age?

The last ice age began about 120,000 years ago, late in the Pleistocene epoch (see Figure 7.9 for a refresher on the geologic time scale). Spurts of glacial advance, alternating with minor retreats, continued over the next 100,000 years. The continent-scale ice sheets reached their largest extent about 21,000 years ago. Then the glaciers melted back very rapidly compared to their long period of fitful advance, and ice sheets retreated to their present, permanent holdout positions in Antarctica and Greenland by 6000 years ago.

North America looked very different during the last ice age than it does now. Not only did ice cover huge areas, but cooler climate also affected vegetation distribution beyond the glaciated areas. Continent outlines were different because of lower sea level, and large lakes existed in the now arid Southwest because of diminished evaporation during this cooler time.

The Extent of Glaciers

Figure 18.34 shows the extent of Northern Hemisphere glaciers at the peak of the ice age. The outline of glaciers is determined by looking at the distribution of glacial till and where end moraines are located (see Figure 18.32, for example). There were two great ice sheets in North America—Laurentide in the east and Cordilleran in the west. At their largest extent, the two ice sheets joined together across the plains of western Canada and also linked eastward to the Greenland ice sheet. The combined area of the Laurentide and Cordilleran ice sheets was 16 million square kilometers, which is roughly two-thirds the area of North America, and the total ice volume was similar to the modern Antarctic ice sheet. Ice caps formed locally in the higher elevations of the Rocky Mountains and Sierra Nevada and fed into valley glaciers that extended to the adjacent plains. Glaciers also formed along the high volcanic peaks of the Cascade Range where smaller remnant glaciers remain today. Landscapes illustrated in Figures 18.11, 18.16, 18.25–18.29, and 18.31 owe their origins to this ice age.

Where the Ice Sheets Came From

The Laurentide and Cordilleran ice sheets did not sweep southward from the North Pole, but they gradually grew outward from several initial ice caps in Canada (see the ice-flow arrows in Figure 18.34). The far Arctic north is very dry, so the thickest glacial ice accumulated farther south where there is more snow between latitudes 55–65° North. Large areas of central and northern Alaska were ice free (Figure 18.34) because snowfall accumulation was too low for glaciers to form.

Geologists estimate that the Laurentide ice sheet was more than 3 kilometers thick near Hudson Bay, as shown in **Figure 18.35**. Lobes of ice extending south of the Great Lakes into the United States were only 500–1000 meters thick.

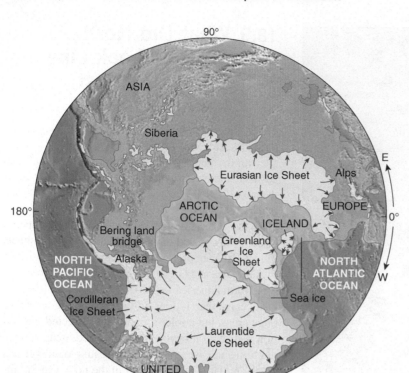

▲ **Figure 18.34 Visualizing the extent of glaciers during the last ice age.** This map shows the area covered by northern hemisphere glaciers 21,000 years ago, as seen looking down from above the North Pole. The mapped coastal outlines are different from today because sea level was lower when large volumes of water were stored in the glacial ice. The Bering land bridge, connecting Asia and North America, provided a path for human migration into the Americas from Asia.

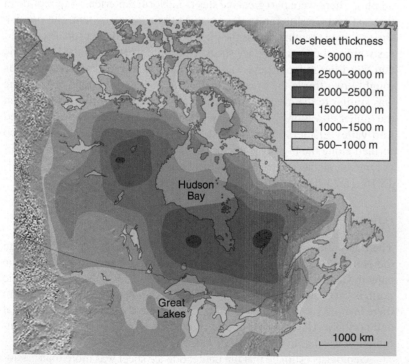

▲ **Figure 18.35 Visualizing the thickness of an ice-age ice sheet.** This map shows the estimated thickness of the Laurentide ice sheet. The thicknesses are calculated using equations that explain the thickness and physics of flow in the modern Antarctic and Greenland ice sheets. The ice was thickest in the areas south and west of Hudson Bay where the ice sheet started forming.

Lower Sea Levels and the Bering Land Connection

When glaciers expand on continents, sea level falls in the oceans. This happens because ice accumulation in glaciers temporarily removes water from the water cycle. The global ice-age glacier volumes were sufficient to lower sea level by about 100–120 meters during the peak of the ice age 21,000 years ago.

The exposure of the seafloor between Asia and Alaska when ice-age sea level was low is significant for the human geography of North and South America (see Figure 18.34). This "Bering land bridge" connected the continents where the shallow Bering Sea exists today. Asiatic people migrated to North and South America across the land bridge. There is no convincing evidence that Asiatic people arrived in Alaska during peak glacial conditions, about 21,000 years ago. The harshly cold climate conditions and the limited availability of game animals or edible plants were obstacles to migration into this area at the time of lowest sea level. However, the land bridge remained until about 12,000 years ago because sea level rose very slowly as the ice sheets melted. The first human migration into central Alaska occurred before 12,000 years ago, and migration along the coastline may have begun by 14,000 years ago.

Landscapes Beyond the Glaciers

North American landscapes south of the ice sheets and at elevations lower than the glaciated mountains were also very different than today. Braided rivers choked with outwash sediment flowed southward from the ice sheets and spilled from the glaciated mountains into the surrounding plains.

Most of the meltwater from the Laurentide ice sheet entered the Mississippi River valley between 21,000 and 14,000 years ago, and the discharge in the upper Mississippi was probably 6 to 8 times larger than it is today. River valleys filled with outwash sediment in the early stages of ice-sheet melting, and then the streams eroded valleys into those deposits when the outwash-sediment supply diminished. The older braided-stream deposits underlie large areas adjacent to the Mississippi valley, as shown in **Figure 18.36**.

The sediment-laden rivers flowed southward from the Laurentide ice sheet and eastward from the Rocky Mountains across a cold, windswept, sparsely vegetated landscape. The wind blew away fine sand and silt from the floodplains and bars of the outwash streams. This sediment built up sand dunes and a frosting of fine silt that now form the parent materials for fertile soils across most of the Mississippi River drainage basin. The wind-blown silt deposits are called **loess**, a German term for the same type of loose silt that is common along the Rhine River in Europe. **Figure 18.37** depicts the distribution of loess in the central United States, where successive glaciations left more than 20 meters of windblown silt near major rivers.

Ice-age pollen collected from lake and floodplain deposits reveal a very different distribution of vegetation than is seen today. Cold-climate mosses and lichens that exist today only along the Arctic Ocean coastline of Canada were the dominant vegetation much farther south in the Midwestern United States. Spruce trees now found near and north of the Great Lakes formed forests in the southeastern United States all the way to the Gulf of Mexico. The southward shift of cold-climate vegetation resulted from the globally cold conditions and the effect of chilling winds that swept southward from the high Laurentide ice sheet.

The deserts of the southwestern United States were awash in deep lakes during the last ice age, as shown in **Figure 18.38**. Many fault-block valleys in this region have no drainage outlets. Today, precipitation runoff flows into the valleys and then evaporates to leave salty deposits (see

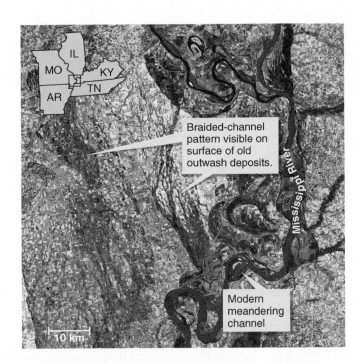

◀ **Figure 18.36 Braided outwash rivers in the Mississippi Valley.** This satellite image shows the meandering Mississippi River and a checkerboard pattern of farmland. The landscape west of the river is the upper surface of valley-filling outwash sediment deposited between approximately 8000 and 11,000 years ago. The braided pattern of the outwash is still clearly visible on this landscape.

Section 16.13). However, ancient beaches located high up on the adjacent mountainsides, and muddy lake deposits below the valley floors with remains of invertebrate animals that thrive in deep freshwater lakes reveal a different scene 21,000 years ago. The thick northern ice sheets produced weather patterns different from today that brought much more abundant winter snow and rain into this region. Evaporation was also less, compared to today, because of the cooler temperatures of the global ice-age climate. The largest ancient Southwestern lake was Lake Bonneville, which covered 51,800 square kilometers of northwestern Utah and adjacent Nevada and Idaho to depths as great as 305 meters (Figure 18.38). This huge lake has mostly evaporated in the warmer, drier post-ice-age climate, leaving behind the Great Salt Lake as a meager remnant along with extensive evaporites that make up the floor of the Bonneville Salt Flats. You may be familiar with the salt flats as the place where super-fast land vehicles are tested.

EXTENSION MODULE 18.1

Ice Age Lakes in the Great Basin. Learn the geologic evidence for deep lakes in the desert Great Basin during the last ice age.

Landscape Changes During Glacial Retreat

Figure 18.39 maps out the changing geography during the demise of the last ice age. The retreat of the great ice sheets is linked to important geologic features in North America.

The northward retreat of the glaciers from the northern United States exposed freshly eroded bedrock, recently deposited till, and a new landscape for stream drainage. **Figure 18.40** shows how glacial erosion and deposition completely rearranged the drainage basins over nearly a third of North America. Glaciers are long gone from this region today, but the locations of rivers, lakes, drainage divides, and even the directions that the rivers flow are a legacy of glacial modification of the landscape.

Huge lakes, mapped in Figure 18.39, formed during melting of the Laurentide ice sheet. The initial meltwater did not readily run off in rivers to the Atlantic Ocean because the land surface was depressed by the weight of the glacial ice. The area of depressed land extended outward more than 100 kilometers beyond the terminus of the ice sheet and caused the surface to slope

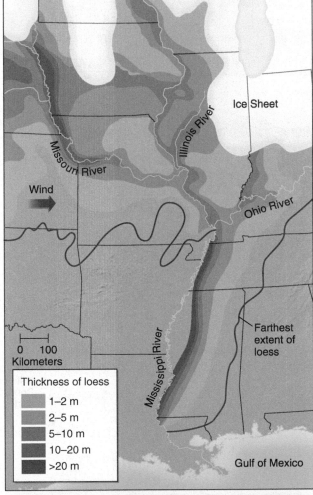

Thickness of loess

	1–2 m
	2–5 m
	5–10 m
	10–20 m
	>20 m

▲ **Figure 18.37 Ice-age loess deposited downwind of outwash rivers.** The map shows the variations in thickness of a blanket of wind-blown silt, called "loess," that accumulated alongside and downwind of the ice-age outwash rivers. The photograph shows a thick exposure of loess in northwestern Missouri.

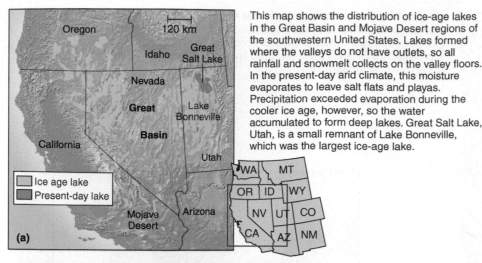

This map shows the distribution of ice-age lakes in the Great Basin and Mojave Desert regions of the southwestern United States. Lakes formed where the valleys do not have outlets, so all rainfall and snowmelt collects on the valley floors. In the present-day arid climate, this moisture evaporates to leave salt flats and playas. Precipitation exceeded evaporation during the cooler ice age, however, so the water accumulated to form deep lakes. Great Salt Lake, Utah, is a small remnant of Lake Bonneville, which was the largest ice-age lake.

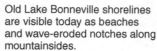

Old Lake Bonneville shorelines are visible today as beaches and wave-eroded notches along mountainsides.

Most of Lake Bonneville has dried up, leaving behind thick salt deposits that form the Bonneville Salt Flats. Notice the circled person for scale.

▲ Figure 18.38 Evidence for ice-age lakes in the Great Basin.

toward the glacier (see Section 13.3 for further description of isostatic subsidence and rebound caused by large glaciers). The big meltwater lakes have largely disappeared or are greatly reduced in size compared to 10,000–12,000 years ago (Figure 18.40) except for lakes filling deep glacier-eroded troughs in bedrock, such as the Great Lakes. However, peat bogs and wetlands remain over most of the region covered by the older, vanished lakes. Some of the lake water drained eastward through the St. Lawrence Valley when the Laurentide ice sheet retreated to the north, and some drained southward as the crust slowly rebounded upward in response to removal of the weight of the glacial ice. Nearly flat plains in the Upper Midwest of the United States and in southern Canada are former lake bottoms. Many large cities, including Chicago, Detroit, Toronto, and Cleveland, are built on these lake plains.

EXTENSION MODULE 18.2

Humongous Ice-Age Floods in the Pacific Northwest.
Learn about the incredible erosional and depositional features formed by ice-age floods more than 100 meters deep that rushed across the Northwestern United States.

Sea level rose while the glaciers melted. The initial rate of sea-level rise was faster than the rate of rebound uplift of the crust that had been depressed under the weight of more than 3 kilometers of glacial ice. As a result, seawater flooded southwestward along the St. Lawrence River valley past Montréal, Quebec, and the Hudson Bay shoreline was as much as 250 kilometers inland of where it is today (Figure 18.39). Eventually, the slow upward rebound of the crust raised much of this inundated area above sea level to establish the modern shoreline.

Muddy deposits of the former lakes and shallow seas present hazards and engineering challenges, especially in southeastern Canada. The young, poorly consolidated lake and marine clays settle unevenly under the weight of buildings and highways. The marine-clay deposits along the St. Lawrence Valley are particularly sensitive to landslide failure where wet clay slides off its foundation of glacial till or bedrock into stream valleys. The instability of these clay deposits makes the St. Lawrence Valley one of the most landslide-prone regions of the world. The most destructive of these events destroyed a Quebec village in 1971 and claimed 31 lives.

Implications for Soil Formation and Fertility

Glaciation partly explains soil characteristics in central North America. The Laurentide ice sheet scoured away the pre-existing soil and regolith over nearly half of the area that was covered by ice in Canada. Today, much of this area lacks soil or has only thin, immature soil because there has only been about 12,000 years of weathering to produce new regolith.

In contrast, very fertile soils are found in the western and southern areas of the former Laurentide ice sheet. Glacial till, locally thick outwash, and meltwater-lake sediment consist mostly of freshly ground-up rock debris eroded by the ice-age glaciers. These deposits either buried or replaced older soils resulting from tens of millions of years of weathering in a mostly humid climate. The mineral nutrients had been exhausted in the old soils by this long period of weathering, and the newly deposited unweathered and crushed rock allowed soil formation to start over in regolith that was rich in mineral nutrients. Particularly striking differences in soil fertility are seen today within the Ohio River drainage basin, where crop productivity is very high in glaciated Illinois, Indiana, and western Ohio but markedly less productive in nonglaciated Kentucky, to the south.

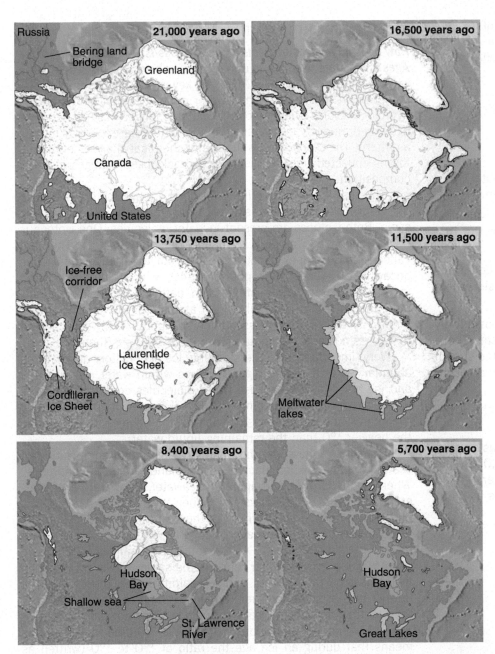

◄ **Figure 18.39 Mapping the glacial retreat.** These maps show the changing locations of glacial ice, meltwater lakes, and shoreline position since the peak of the last ice age, about 21,000 years ago. An ice-free corridor formed between the contracting Cordilleran and Laurentide ice sheets about 14,000 years ago and permitted southward migration of arctic inhabitants into the present United States. Large meltwater lakes around the edge of the Laurentide ice sheet eventually drained to the Atlantic Ocean as the ice sheet retreated. The land was depressed by the weight of the thickest ice so that the Hudson Bay lowlands and St. Lawrence River valley were briefly submerged by shallow seas until the land gradually rose back up above sea level.

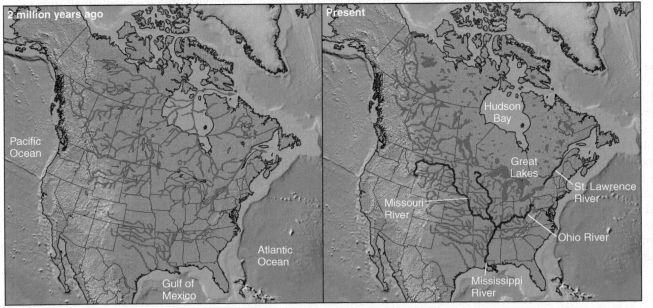

◄ **Figure 18.40 Glaciers rearranged river drainage.** When the ice ages were just beginning about 2 million years ago, most streams in central North America drained through what is now Hudson Bay. The Great Lakes did not exist. Repeated ice sheet glaciations rearranged this drainage pattern. Widespread glacier scour of bedrock created lakes across most of eastern and northern Canada and excavated the Great Lakes. The Missouri and Ohio Rivers occupy valleys close to the farthest southern extent of till and end moraines (see Figure 18.32a).

Fresh regolith for soil formation also characterizes most of the mid-continent plains of the United States, even south of the glaciated region. Fertile soils in this region form in thick loess deposits and valley filling outwash alluvium.

Putting It Together—What Did North America Look Like During the Last Ice Age?

• During the last ice age, which peaked about 21,000 years ago, glacial ice covered about two-thirds of North America.

• Sea level was 100–120 meters lower than it is today, producing a land connection between Asia and Alaska along which the first humans migrated to the Americas.

• Cooler and wetter climate in the southwestern United States allowed deep lakes to form within valleys of the Great Basin, where only shallow salty lakes and salt flats are present today.

• Outwash choked river valleys, causing widespread deposition by braided streams during the early stages of glacial retreat. These deposits now form stream terraces. Wind blowing across the stream valleys swept up fine-grained sediment and deposited it in adjacent areas as loess.

• Large lakes formed on the crust depressed by the weight of the thick Laurentide ice sheet.

• Soil formation throughout the glaciated region and in areas covered by outwash and loess has been taking place for less than 21,000 years within regolith that experienced little chemical weathering and thus had lost few mineral nutrients. The most fertile soils in North America are found on deposits related to the ice age.

18.10 How Do We Know . . . How to Determine When Ice Ages Happened?

Picture the Problem

How Many Ice Ages Were There, and When Did They Happen? The last ice age was very recent in geologic history and even within human history. Clearly, a future ice age and the accompanying cold global climate would have negative impacts. It is important, therefore, to know how many ice ages occurred in the past and how often they recur.

The extent and age of the last ice age are best known because it happened most recently and its deposits are present at the surface. In many places, however, successive layers of glacial till, separated by soils, indicate many glacial advances and retreats in the past. Each ice-age advance deposited till, and soil formed on the till after the glaciers retreated. Although the tills and soils record multiple ice ages, it is very difficult to piece together a complete history of the older glaciations because the last glacial advance substantially eroded the older deposits and highly modified the landscape. So how do geologists know about the number and timing of older ice ages? They turn to seafloor sediment, rather than glacial deposits on continents, to decipher a complete history of ice ages.

The Background of the Method

How Do Isotope Measurements of Marine Fossils Track the Timing of Ice Ages? There are two keys to understanding how geologists determine when the ice ages happened. The first key is a subtle change in water composition that occurs when water molecules move through the hydrologic cycle. Evidence of alternating changes in seawater composition during ice ages versus the intervening warm periods reveals a pattern. The second key is the use of chemical analyses of marine fossils collected from the seafloor to track these changes in water composition.

The first key to determining the long ice-age history relates to the behavior of two stable isotopes of oxygen within the hydrologic cycle. Both isotopes are present in water molecules. One isotope is ^{18}O (pronounced "oxygen 18"), and the other is ^{16}O. They are called *stable* isotopes because they do not experience radioactive decay, like carbon-14 or potassium-40, which were introduced in Chapter 7 for their usefulness in dating geological materials. More than 99 percent of all oxygen atoms are ^{16}O, and from your familiarity with atomic mass numbers (Figure 7.16), you can tell that in its nucleus ^{18}O contains two more neutrons than ^{16}O.

Very important to understanding ice-age history is that the abundances of the two oxygen isotopes within water, water vapor, and ice are different because water molecules containing ^{18}O are heavier than those containing ^{16}O. For example, when water partly evaporates, the water vapor has more of the lighter ^{16}O and less of the heavier ^{18}O than the remaining liquid water.

Figure 18.41 illustrates how the oxygen-isotope composition of seawater changes when glaciers advance and retreat on continents. When clouds form from evaporated seawater, the water vapor preferentially includes the lighter ^{16}O. When the vapor later condenses and precipitates over continents, the resulting liquid eventually finds its way back to the ocean through the hydrologic cycle, which returns the ^{16}O to the sea. As a result, the overall ratio of the two oxygen isotopes in seawater remains the same. However, a change in the isotope ratio occurs if the precipitation in the continental interior falls as snow and then remains on land as glacial ice rather than melting and flowing back to the sea. The snowfall in the continental interior contains more ^{16}O and less ^{18}O than seawater. This means that during an ice age the ratio of ^{18}O to ^{16}O (written as $^{18}O/^{16}O$) increases in seawater. When the ice sheets melt, this ratio in seawater decreases because the ^{16}O-rich water within the melting glacial ice returns to the ocean.

The second key to reconstructing the timing of past ice ages is the analyses of marine fossils to determine how the $^{18}O/^{16}O$ ratio in seawater changed back through time. Geologists rely on the nearly microscopic shells of a group of marine organisms called "foraminifera," an example of which is illustrated in Figure 18.41. Foraminifera secrete tiny shells composed of calcite. The $^{18}O/^{16}O$ ratio of the oxygen atoms in the calcite shells records the ratio of the seawater oxygen isotopes at the times when the foraminifera lived. Geologists separate fossil foraminifera shells from sediment layers beneath the seafloor and then measure their isotope ratios in the laboratory to obtain a history of the seawater $^{18}O/^{16}O$ ratio. The changes through time in the ratio of the oxygen isotopes in the ocean simultaneously reveal the history of growth and melting of ice sheets on continents.

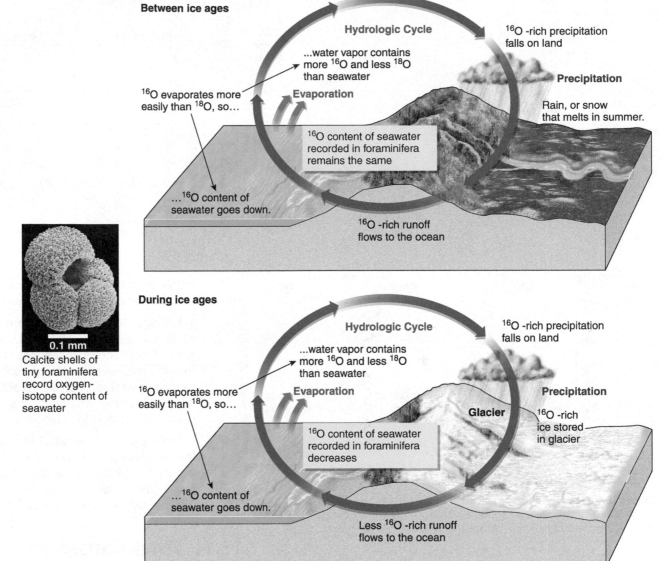

► Figure 18.41 How oxygen isotopes track ancient ice ages. Water molecules containing the ^{18}O and ^{16}O isotopes separate from each other within the hydrologic cycle. Water with ^{16}O more easily evaporates from the ocean, so rain and snow that fall on the continents contain more ^{16}O than seawater. During times between ice ages the ^{16}O-rich runoff from continents returns to the ocean. During ice ages the ^{16}O-rich water is partly stored in glacial ice, so the ^{16}O content of seawater decreases. The tiny calcite shells of fossil foraminifera record these variations in the oxygen-isotope content of seawater.

Calcite shells of tiny foraminifera record oxygen-isotope content of seawater

Collect the Data

When Did the Ice Ages Take Place? Figure 18.42 illustrates the records of oxygen-isotope ratios in foraminifera from sedimentary layers sampled from beneath the seafloor in the Atlantic and Pacific Oceans. Here are the key interpretations drawn from these data:

- The isotope ratios change in the same fashion at two widely separated locations. This is an important observation because it indicates that the measurements track changes in global ocean composition and not local changes.

- The $^{18}O/^{16}O$ ratio is very high in foraminifera deposited on the seafloor 21,000 years ago.

This is an important observation because it matches the record of the last ice age on the continents. Therefore, other time periods with similar high ratios are interpreted as older ice ages.

Geologists have obtained several dozen similar oxygen-isotope records from locations throughout the world oceans. These records extend back more than 100 million years in a few places. They suggest that small glaciers probably existed in the Mesozoic Era, most likely in Antarctica, which has been at the South Pole since Mesozoic time. The isotope data suggest that Antarctica glaciers expanded about 33 million years ago, during the middle of the Cenozoic Era. The data further indicate that significant glaciers first appeared in the Northern Hemisphere about 2.5 million years ago, and they have advanced and retreated dozens of times since then.

Insights

Do Ice Ages Have Rhythm? Geologists were very excited about the oxygen-isotope records when they were first constructed during the 1960s because the data revealed something more than just when the ice ages occurred. The data plots like the ones shown in Figure 18.42 look a little bit like electronic records of a beating heart. The peaks and troughs in the isotope values are not randomly distributed through time. Instead, there seems to be a definite beat, a rhythm to the changing climate. The peaks corresponding to the warmest periods occur about every 100,000 years. The

▶ **Figure 18.42 What oxygen-isotope data reveal.** These graphs summarize oxygen-isotope measurements in fossil foraminifera from two locations. Using the analysis presented in Figure 18.41, ice ages correspond to times when there is less ^{16}O in the fossil foraminifera, which also means less ^{16}O in the ocean. The occurrence of relatively low ^{16}O content 21,000 years ago corroborates this interpretation and permits recognition of earlier ice ages from the isotope data.

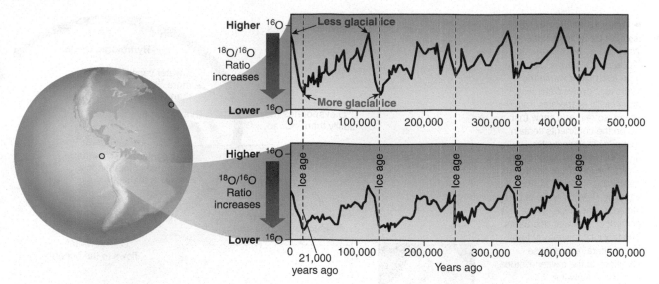

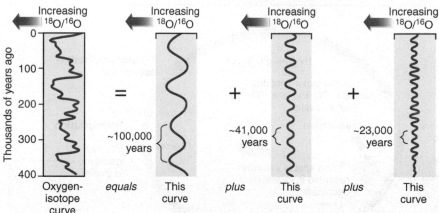

▲ **Figure 18.43 The oxygen-isotope record has rhythm.** The pattern of peaks and troughs in the oxygen-isotope curve is the sum of three more regular curves where consecutive peaks or troughs are separated by 100,000 years, 41,000 years, and 23,000 years.

intervening smaller peaks and troughs are also spaced fairly regularly at 23,000 years and 41,000 years. **Figure 18.43** shows how to explain the isotope curve: It is the sum of curves with rhythms of 23,000, 41,000, and 100,000 years.

The ice-age rhythm, which is explained in the next section, has turned out to be an important clue to what causes ice ages. Although not anticipated at the outset, the oxygen-isotope data collected to determine *when* ice ages occurred also opened up decades of exciting research into *why* ice ages occur.

EXTENSION MODULE 18.3

Ice Ages through Earth's History. *Learn the geologic evidence for ice ages that happened hundreds of millions and billions of years ago.*

Putting It Together—How Do We Know . . . How to Determine When Ice Ages Happened?

• The ratio of two stable isotopes of oxygen present in seawater molecules changes through time as glaciers advance and retreat on continents. The variations are caused by different behaviors

of the isotopes when water molecules move through the hydrologic cycle as liquid water, water vapor, and ice.

• Marine foraminifera secrete miniature calcite shells, in which the oxygen atoms record the oxygen-isotope composition of the ocean when the foraminifera lived. Fossil foraminifera sampled from sediment layers beneath the seafloor permit reconstruction of the oxygen-isotope composition of seawater, and hence the record of glaciation on continents, back through time.

• The oxygen-isotope ratios in foraminifera reveal dozens of ice-age glaciations in the northern hemisphere over the last 2.5 million years. These data also show rhythmic variation in glacial-ice volume that varies over times of approximately 23,000, 41,000, and 100,000 years.

18.11 What Causes Ice Ages?

There has to be a driving force or forces that fluctuate in a regular, rhythmic fashion to cause ice ages in order to explain the oxygen-isotope data. Oceanographers and geologists recognize that the persistent beats in the isotope data at 23,000, 41,000, and 100,000 years coincide with calculated periodic variations in Earth's rotation on its axis and in its orbit around the Sun. These variations were calculated during the early twentieth century by Milutin Milankovitch and are commonly called the **Milankovitch cycles**. Milankovitch predicted that these cycles should affect the amount of solar energy that reaches Earth, but prior to measuring the oxygen isotopes in foraminifera, there was no clear evidence to support his hypothesis.

Variations in Earth's Rotation and Orbit

Calculations of the gravitational pull of the Sun, Moon, and other planets on Earth's orbit and rotation reveal three important effects on climate. **Figure 18.44** illustrates these effects.

• Earth's rotation axis is tilted away from vertical. The tilt causes seasons, because the parts of Earth's surface and atmosphere that tilt toward the Sun receive more heat than areas that point away (see Figure 18.44a). The angle between a vertical line and the rotation axis is called the *obliquity*. The current tilt angle is 23.5°, but astronomers' calculations show that it varies from as little as 21.5° to as much as 24.5° and back about every 41,000 years.

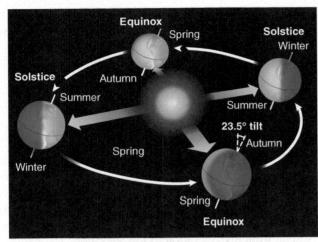

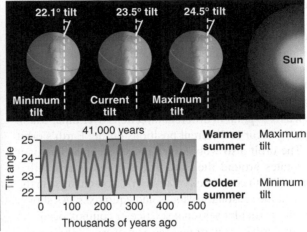

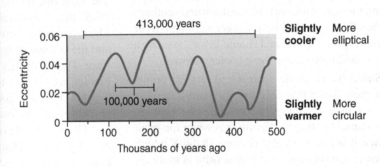

(a) Tilt of Earth's rotation axis

Earth's rotation axis is tilted at a 23.5 degree angle. The hemisphere that points toward the Sun receives more direct solar heating and is warmer than the hemisphere that points away from the Sun. The tilt, therefore, explains the seasons, and why northern hemisphere seasons are opposite from those in the southern hemisphere.

The tilt angle varies from low to high and back to low values every 41,000 years. Summer heating is greater and winter cooling is greater when the tilt angle is high. Summers are cooler, therefore, when the tilt angle is low. Persistence of year-round snow is favored by cooler summers.

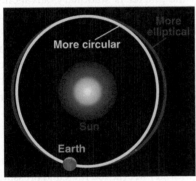

(b) Shape of Earth's orbit

Earth's orbit is slightly elliptical (or eccentric), rather than circular. When the orbit is more elliptical, the year-round solar heating from the Sun is slightly less than when the orbit is more circular. Extreme values of eccentricity recur every 100,000 years, and there is another cycle that recurs every 413,000 years, but the amount of temperature change is very small.

Wobble of the rotation axis

Earth's rotation axis wobbles just like a spinning top. This means that the North Pole points to a different direction in space at different times.

The orbit around the Sun also wobbles, similar to the looping of a hula hoop around your waist.

Wobble of the orbit

Earth is at the same position in its orbit for each solstice or equinox approximately every 23,000 years. When Earth is far from the Sun on the June 21 solstice, the northern hemisphere summer is coldest and the southern hemisphere summer (December 21) is warmest.

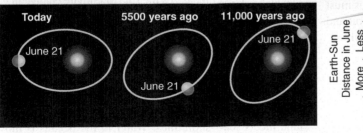

(c) Wobble of Earth's rotation axis and orbit

▲ **Figure 18.44 How Earth's rotation and orbit vary over time.** The Milankovitch cycles describe predictable changes in three aspects of Earth's orbit and rotation that also have predictable effects on climate.

- The shape of Earth's orbit is slightly elliptical. *Eccentricity* describes how elliptical an orbit is; a circular orbit has an eccentricity of zero. Earth's eccentricity varies over time between two very small values, 0.005 to 0.05 (Figure 18.44b). The most extreme oscillations between eccentricity values take place about every 413,000 years, but there are smaller variations that occur every 100,000 years.

- Earth wobbles as it rotates, much like a spinning top. This wobble causes the rotation axis to point in different directions over time, which causes the seasons to occur at different positions during Earth's orbit around the Sun. The orbit also wobbles, vaguely resembling a giant hula hoop as it rotates around the Sun. These wobbles are called *precession* and, acting with eccentricity variation, cause changes in the Earth-Sun distance at the seasonal solstices and equinoxes. As shown in Figure 18.44c, any particular seasonal solstice or equinox occurs at the same position in Earth's orbit around the Sun every 23,000 years.

Variations in Solar Heating of Earth's Surface

Figure 18.44 also shows how variations in tilt, elliptical orbit, and wobble affect Earth's climate.

- The poles receive the most summer sunshine and heating, and the least winter warming, when the obliquity (tilt angle) is high. As a result, the difference in summer and winter temperatures in both northern and southern hemispheres is greatest when the angle is highest. Cooler summers occur when the tilt angle is lowest (Figure 18.44a) and should favor the growth of glaciers because less snow will melt during these cooler summers.

- The difference in Earth-Sun distance at the nearest and farthest points in the orbit from the Sun increases when eccentricity increases. This causes very slight (<0.1 percent) variations in the total solar heat received by the planet. The annual heating is slightly less when the orbit is more elliptical than when it is more circular (Figure 18.44b), which means that ice ages are favored during times when the orbit is more elliptical.

- If Earth's orbit were perfectly circular, then it would not matter where the planet was in its orbit during each season of the year. However, the elliptical orbit and the precession (wobble) combine to produce significant climate effects. For example, if the northern hemisphere summer occurs when Earth is far from the Sun, then the summer temperatures are cool compared to when Earth is close to the Sun during the summer.

The match between the rhythm of the ice ages revealed by the oxygen-isotope ratios in ancient foraminifera with the Milankovitch cycles shows that climate change on the scale of tens of thousands and hundreds of thousands of years is mostly regulated by Earth's orbit and rotation. The climate changes predicted by the Milankovitch cycles do not explain all of the detailed oscillations in climate, so scientists know that other factors, like ocean circulation and changes in the composition of atmospheric gases must also play a role. However, it is clear that the variations in Earth's orbit and rotation explain when ice ages, and intervening warmer times, take place.

Putting It Together—What Causes Ice Ages?

- Variations in the obliquity of Earth's rotation axis, the eccentricity of Earth orbit, and the precession of the seasons around the orbit are predicted to cause rhythmic variations in the amount of solar heat received by the planet.

- These rhythms correspond to the rhythms in the oxygen-isotope record of glacial-ice volume, suggesting a cause and effect relationship.

Where Are You and Where Are You Going?

Glaciers are slowly flowing masses of ice that cover about 10 percent of Earth's surface. The ice forms from metamorphism of snow and freezing of snow meltwater. Glaciers form where temperatures are cold and precipitation is abundant so that winter snowfall exceeds summer melting. Eventually, enough snow and ice builds so that the ice flows under its own weight from high elevation to low elevation. The climate conditions for glacier formation are typically met in high mountains and at high latitudes. Valley glaciers are confined by valley walls in mountains. Ice caps and larger ice sheets are thicker glaciers that bury most or all surface topography and are not confined in valleys.

Ice always flows from the zone of accumulation to the zone of wastage regardless of whether the glacier is stationary, advancing, or retreating. The snowline defines the boundary between these two zones and it also marks the elevation above which winter snowfall exceeds summer melting. The relationship between the amount of accumulation above the snowline and the amount of wastage below it determines the position of the glacier front. The ice moves both by internal flow and by slip at the bottom of the glacier, as long as the glacier does not freeze to underlying rock and regolith. Ice undergoes brittle deforms near the surface, to form crevasses, and plastically below about 50 meters.

Glaciers erode rock and regolith because of the high shear stress exerted by thick ice. Liquid water also plays essential roles in glacier erosion. Freezing and thawing of water in cracked rock and pressurized injection of water into cracks disaggregates rock that is then plucked loose by glacier flow. Sediment eroded by glaciers is ultimately deposited as till left behind or melted out of the ice, or as outwash deposited by meltwater streams.

Glaciers sculpt and smooth landscapes and produce many unique landforms. Jagged peaks, knife-edge mountain ridges, broad U-shaped valleys flanked by hanging-valley waterfalls, expansive terrain pockmarked with lakes, moraine ridges, and fertile plains of glacially eroded debris are the hallmarks of glaciation. These glacial landscapes cover far more area than modern glaciers now cover, which reveals extensive modification of Earth's surface by flowing ice during past ice ages. The surface geology of most of North America reflects glacial processes that last peaked in intensity about 21,000 years ago.

The observation that present-day landscapes are inherited from past glacial ice ages emphasizes the role of climate change in geological processes. More than a dozen ice ages occurred over the last 2 million years. The chemical compositions of tiny marine organisms reveal that the ice ages and intervening warmer climate periods, such as the present day, alternate like clockwork. The timekeepers of climate change are variations in Earth's rotation and orbit around the Sun that affect the summer heating at high northern-hemisphere latitudes where the ice-age ice sheets are born.

You are now ready to study the important geologic boundary where land meets sea—shorelines. Water in the oceans is moved by wind and tides and not by the downward pull of gravity that governs flowing water in streams. Comprehension of how wind and tides move water leads to an understanding of how scenic coastal landscapes are produced, why hazardous coastal erosion occurs, and even why your favorite recreational beach exists. The highly variable features of shorelines are understood by combining your acquired knowledge of rock weathering, erosion, tectonic uplift, and subsidence with new knowledge of oceanic processes.

Consideration of glacial ice-age history offers another motivation for studying shorelines. Growth of ice-age glaciers implies a simultaneous drop in global sea level. Sea level rises when glaciers melt back during warm periods. Shorelines should show a record of shifting sea level that parallels the history of glacial advances and retreats. Sea level is currently rising, with critical implications for coastal cities.

Active Art

Glacial Advance and Retreat. See how a glacier moves when the front advances, retreats, or is stationary.

Glacial Processes. See the growth and movement of a glacier along with glacial erosion and deposition.

Extension Modules

Extension Module 18.1: Ice Age Lakes in the Great Basin. Learn the geologic evidence for deep lakes in the desert Great Basin during the last ice age.

Extension Module 18.2: Humongous Ice-Age Floods in the Pacific Northwest. Learn about the incredible erosional and depositional features formed by ice-age floods more than 100 meters deep that rushed across the Northwestern United States.

Extension Module 18.3: Ice Ages through Earth's History. Learn the geologic evidence for ice ages that happened hundreds of millions and billions of years ago.

Confirm Your Knowledge

1. What is a glacier? What features distinguish the three types of glaciers?
2. Use your own words to describe the zones of accumulation and wastage and the snowline? How does the snowline elevation change if climate changes?
3. What two conditions must be met in order for a glacier to form?
4. How does glacier ice form?
5. How does a glacier move and where does the movement take place? How quickly do glaciers move?
6. Glaciers are sometimes called "rivers of ice" but while it is easy to observe rivers, it is difficult to directly observe glacial flows, because the ice is opaque and moves very slowly. How do geologists overcome these obstacles to studying glacier movement, erosion, and deposition?
7. Explain how the front of a glacier can move upslope in a retreating valley glacier if the ice is always moving downslope.
8. Describe how glaciers erode rock and carry sediment.
9. Explain why glaciers are much more effective agents of erosion than streams, even though glaciers move much slower than streams.
10. Define "moraine." What are the types of moraines?
11. What are the two main types of sedimentary deposits formed near or beneath modern glaciers? How would you distinguish these two deposits types in an area that was glaciated in the past?
12. Which machine is more analogous to the deposition of sediment at the front of a stationary glacier, a bulldozer or a conveyor belt? Why?
13. How do icebergs form?
14. Explain the types of landscape features formed by valley-glacier erosion.
15. Describe the features left behind after an ice sheet retreats from a region.
16. During the most recent glacial advance, which peaked 21,000 years ago, how much of North America was covered with ice?
17. What were the consequences of ice-age climate for the parts of the continental US not covered in ice during the last ice age?
18. How do geologists determine the number of ice ages and when they happened?

Confirm Your Understanding

1. Write an answer for each question in the section headings.
2. Because glacial ice is a natural, consolidated aggregate of minerals it is considered to be a rock. Explain what type of rock it is: igneous, metamorphic, or sedimentary.
3. Explain why a glacial valley has a U-shaped cross section, whereas mountains stream valleys are usually V-shaped.
4. Why are waterfalls common in previously glaciated regions? Why are lakes very common in previously glaciated regions?
5. What were the conditions in your hometown during the peak of the last ice age, 21,000 years ago? Be sure to explain how you arrived at your interpretation.
6. How would you differentiate between glacial outwash streams and braided streams associated with an alluvial fan?
7. Explain how Milankovitch cycles explain when ice ages occur.
8. Take a few minutes to study the full-page photograph at the beginning of the chapter. Then, write a narrative that describes all of the visible features of glacial processes and landforms, explaining how each feature formed.

CHAPTER 19

Shorelines: Changing Landscapes Where Land Meets Sea

Why Study Shorelines?

Water covers 70 percent of Earth's surface, so humans have always had a practical and aesthetic relationship with the sea. Half of the global population lives within 60 kilometers of a shoreline. Fourteen of the 20 largest American cities are located along a coast, and coastal counties contain more than half of the population. Most readers of this book probably live near a coast or have at least visited a shoreline.

The advantages of coastal living are accompanied by the challenges of coping with active shoreline processes. Waves and tides erode the shore and transport huge quantities of sediment. These processes, along with changing sea level, modify the shape of the shoreline, damage or destroy structures, and affect the ability of ships to access ports. Increases in coastal population and urbanization require an understanding of how geologic processes shape coastal landscapes.

This chapter links the shaping and reshaping of shorelines to the processes that move water in the ocean.

After Completing This Chapter, You Will Be Able to

- Explain the origin of waves and tides.
- Apply knowledge of waves and tides to explain shoreline landscapes.
- Explain how human activities modify shorelines.
- Evaluate evidence for, and the consequences of, rising sea level.

Pathway to Learning

19.1 What Factors Determine the Shape of a Shoreline?

19.2 How Do Waves Form and Move in Water?

19.3 How Do Waves Form Shoreline Landscapes?

19.4 What Is the Role of Tides in Forming Coastal Landscapes?

Atlantic Ocean waves lap against a developed coastline in South Carolina

IN THE FIELD

Imagine taking a vacation drive along the varied landscapes of the Oregon shoreline, as seen in **Figure 19.1**. In some places, you see beaches (Figure 19.1a), as you would expect to, but in other locations, you see rocky cliffs that drop straight down into the ocean (Figure 19.1b). Steep-sided, rocky islands stand a short distance offshore, suggesting that erosion separates large rock outcrops from the mainland. You wonder why the shoreline alternates between rocky cliffs and sandy beaches.

The sounds of the coast also impress you. Waves crash incessantly against rocky cliffs, and water roars as it surges up onto a sandy beach, followed by gurgling sounds as the water pulls back into the sea. The ocean is a powerfully moving body of water, and you wonder what processes cause the water to move.

You can stroll on some beaches. Some beaches are well-sorted sand, whereas others, especially near rocky sea cliffs, consist of well-rounded and polished cobbles (Figure 19.1c). When the waves rush up onto the gravel beaches, you hear the clatter of the cobbles banging together, and you know right away why they are so well rounded. Broken seashells litter the beach, apparently transported in from offshore by waves. The waves clearly move sediment, and presumably erode rock and regolith, but how do these processes work?

The coastal highway crosses many streams flowing to the ocean. Towns located where the largest rivers meet the coast are home to commercial fisheries and major tourist attractions. You wonder why people construct "jetties" (long, parallel rock walls) that stick out into the ocean on either side of the river mouths (Figure 19.1d). The beaches often wall off the seaward end of the valley, as seen in Figure 19.1d, to provide a bay where fishing fleets are moored in quiet water away from the waves that pound the beach.

You linger in one of these coastal communities. During the several hours of your visit, you notice that the water level drops along the bayside waterfront, exposing barnacle-covered piers and bare, muddy banks (Figure 19.1e). You realize that the changing water level records a falling tide. What causes tides, and do they play a role in shaping shoreline landscapes?

Shorelines are dynamic places with varied landforms. How do waves and tides explain how coastal landscapes develop? How might these processes produce hazards to coastal communities and how do people use structures, such as the jetties, to diminish these hazards? These are critical topics to pursue in your study of shorelines.

(a) Sandy beach with incoming waves near Newport.

(b) Steep rocky cliff of basalt drops directly into the ocean without a beach; Cape Foulweather.

Crab, about 4 cm across

(c) Smooth, rounded pebbles and cobbles of basalt found on a small beach near where photo B was taken.

Jetties

(d) Yaquina Bay, where the Yaquina River enters the Pacific Ocean at Newport. The quiet water of the bay contrasts with the crashing white-capped waves on the beach on either side of the river mouth.

Water line at high tide

Barnacles attached to pier are submerged at high tide.

(e) Photos taken about 6 hours apart show the different water levels of high and low tides on a pier in Yaquina Bay.

▲ Figure 19.1 Views of shoreline landscapes along the Oregon Coast.

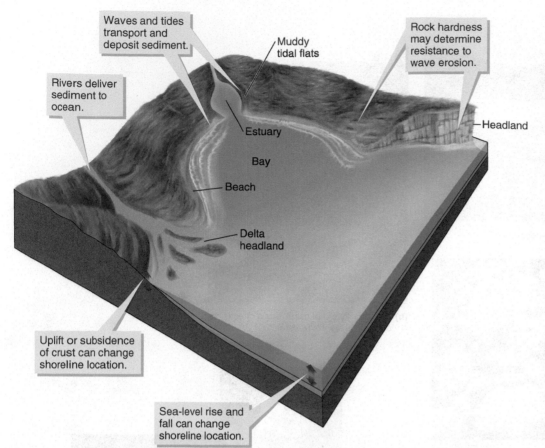

▲ **Figure 19.2 Factors that influence the appearance of coastal landscapes.**
Shorelines are rarely straight but alternate between headlands and bays. Variations in coastal landscapes relate to how easily waves and tides erode coastal rock and regolith, where rivers deliver sediment, and how shoreline position changes because of sea-level change or because of uplift and subsidence of crust.

19.1 What Factors Determine the Shape of a Shoreline?

Shorelines are the boundary between land and sea. The features illustrated in Figure 19.1 demonstrate that shorelines are not perfectly straight. **Figure 19.2** shows there are places where land juts out into the sea to form a **headland** (Figure 19.1b, for example). Headlands commonly have geographic names that include the words "cape" or "head." In contrast, other parts of the shoreline are deep recesses, called **bays** (see Figure 19.1d). Land may rise abruptly above the water in steep cliffs, whereas in most places there are beaches, which are low-sloping landforms composed of unconsolidated sediment moved by waves and tides (contrast Figures 19.1a and b).

Shoreline materials vary in their resistance to erosion, which is one contributing factor to irregular coastal outlines. The extents of lithification, abundance of fractures, and presence or absence of bedding or foliation planes are rock characteristics that determine strength and erodibility. Strong rocks compose many headlands (Figure 19.1b), whereas weak rock or regolith may form the shoreline along adjacent bays (Figure 19.2).

Stream erosion contributes to the formation of bays. Stream valleys partially submerge beneath the sea if the land subsides or sea level rises. **Estuaries** are submerged parts of stream valleys where freshwater and sea-water mix. The bay seen in Figure 19.1d is an estuary. Therefore, along

submerging coastlines the bays may be drowned river valleys, and the headlands are the ridges that separate adjacent valleys. In this case it is stream erosion rather than material properties that accounts for the irregular coastline.

The availability of loose sediment also influences the shoreline shape. Some beaches consist of local, wave-eroded sediment, as in the case where a gravelly beach alongside a headland contains only the rock types found on the headland (see Figure 19.1c, for example). In most locations, however, rivers deliver sediment to the ocean where waves and tides then spread the sediment along the coastline to form beaches. Deltas form headlands (see Section 16.5) where rivers build out the coastline by depositing more sediment than waves and tides can erode and redistribute (Figure 19.2).

Shoreline shape and location change over time (see Figure 19.2). Coastal erosion causes shorelines to retreat in the landward direction, whereas sediment deposition causes shorelines to advance seaward. Rise and fall in sea level, caused by changes in ocean-basin shape (see Section 13.4) or volume of water in the oceans (see Section 18.9), cause shorelines to shift landward or seaward. Shoreline shape also changes with uplift of the crust where former seafloor lifts above sea level or subsidence where land sinks beneath the sea.

Putting It Together—What Factors Determine the Shape of a Shoreline?

• Irregular shoreline shape results from variations in several factors: erodibility of rock, sediment delivery from rivers, sediment movement by waves and tides, changes in sea level, and uplift or subsidence of the crust.

• Streams may form estuaries or deltas where they meet the ocean. Estuaries are previously eroded stream valleys that submerge because of land subsidence or sea-level rise. Deltaic headlands form where sediment is deposited at the mouth of a stream.

19.2 How Do Waves Form and Move in Water?

It is hard to imagine a view of the ocean without seeing waves. The origin of ocean waves and the ability of waves to erode and transport sediment are essential to understanding shoreline landscapes.

What Are Waves?

Section 8.2 introduced the general idea that waves are disturbances that transport energy through a medium such as rock, air, or water. A more detailed look at the shape of waves and some of their properties will guide our understanding of how waves move through water.

Figure 19.3 shows how you can form waves by shaking the end of a rope up and down. Several features serve to describe the wave. Low troughs separate the high crests on the wave. The **wave height** describes the vertical distance between adjacent crest and trough, and the **wavelength** is the distance between successive crests, or successive troughs. The time that elapses for successive crests, or troughs, to pass a stationary point is the **wave period**. Typical ocean waves have wave heights between 0.5 and 20 meters, wavelengths of 10 to 200 meters, and periods of 5 to 20 seconds.

A wave moves through a medium, but particles in the medium return to the original location after the wave passes. For example, the wave moves from one end of the rope to the other, but no part of the rope travels from one end to the other (Figure 19.3). A typical ocean wave moves across 1000 kilometers of water in about a day, but it is the wave, and not the water, that travels that distance. Just as seismic waves represent a pulse of energy moving through Earth, ocean waves are disturbances in the water surface caused by the transfer of energy in the direction of wave movement.

ACTIVE ART

Properties of Waves. See how to recognize the parts of a wave and to measure the wave period.

How Water Waves Form

Coastal observers long ago realized that waves are highest during windy weather, which indicates that wind blowing across water produces waves. The friction that occurs where moving air touches water transfers some of the energy and momentum of the moving air to the water and places the water surface in motion. The physics of energy transfer from wind to water is a perplexing process, but several key relationships are clear from observations.

Increases in the velocity of the wind, the duration of the wind, and the distance that the wind is in contact with the water all cause increases in wave height, wavelength, and wave period. Wind speed and duration stir up the highest waves during prolonged windy storms, whereas fair weather allows the low waves of a calm sea. The importance of the distance that the wind blows over the water explains why even very high winds do not produce large waves on swimming pools or small ponds, even though hurricane winds blowing over the ocean generate waves that are more than 20 meters high.

Energy transfers from storm wind to the ocean surface, then moves across the ocean as waves, and finally arrives at the shoreline. When waves reach the shoreline, the energy performs work to move sediment and rock and part of the energy converts to sound waves that you hear as the "roar" of the surf.

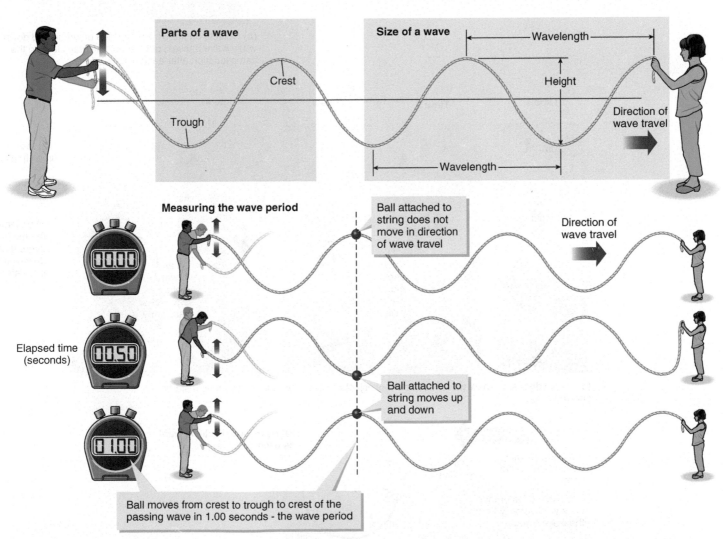

▲ **Figure 19.3 Describing the properties of a wave.** Shaking a rope produces waves that compare well with natural waves, and provides a useful analogy for understanding terms that describe the parts and size of a wave and the wave period.

How Waves Move Water

For seismic waves, or rope waves, the energy is introduced at a single point. Waves at sea are different because wind imparts energy to the ocean across the entire water surface. The amount of wave motion then decreases downward from the surface.

Figure 19.4 summarizes observations of how water moves when a wave passes. An object floating on the water surface moves up and forward as the crest of the wave goes by, and then it moves down and backward when the trough of the wave passes. The floating object, therefore, traces a circular path during one wave period. Particles suspended in the water move through circular orbits of decreasing diameter at deeper depths. There is no motion below a level called the **wave base**. The higher the wave is at the surface, the deeper the wave base. The lack of motion below wave base explains why submerged divers and

▶ **Figure 19.4 Visualizing water motion caused by a passing wave.**

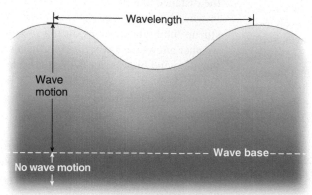

(a) Waves cause water motion only down to wave base.

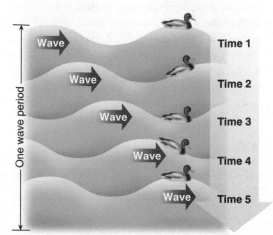

(b) Objects on the water's surface move up and down when wave passes, but the objects stay close to the same location after each wave period.

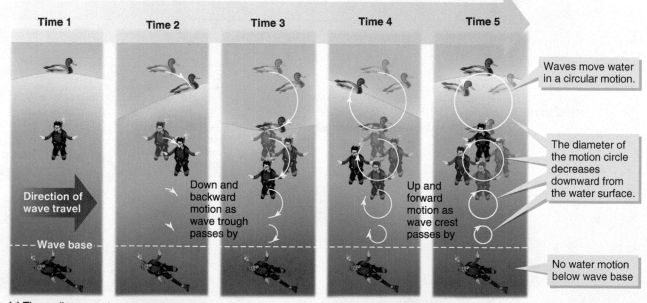

(c) These diagrams show wave movement of objects in water through the five time instants shown in (b).

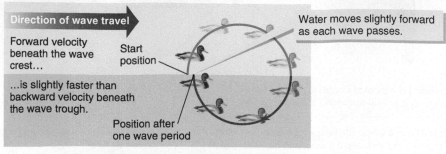

(d) Water moves slightly forward as the wave passes through.

submarines are unaffected by the waves that cause seasickness on surface vessels.

The circular wave motion of water imparts a small amount of forward movement of the water that is not seen in the rope wave. The motion beneath the passing wave decreases downward, so a particle of water moves farther when it moves up and forward than when it moves down and backward (Figure 19.4d). Therefore, water actually does move slowly in the direction of wave movement. This is an important departure from our earlier generalization about waves because when the wave reaches shore a mass of water is piled against the coast, which has important implications for erosion and sediment movement. High-velocity hurricane winds move such large masses of water to the shore that the resulting storm surges locally raise sea level by several meters and cause flooding many kilometers inland.

What Happens When Waves Approach the Shoreline?

Wave motion changes where waves approach shore through progressively shallower water. **Figure 19.5** illustrates that the changes start to take place where water depth equals the wave-base depth. When waves continue into shallower water, a decreasing volume of water must transport the same amount of energy. This causes many observed changes: Wave height increases, crests become narrower than troughs, wavelength decreases, and the forward velocity of waves decreases.

The changing shape of the waves in shallower water makes the particle motions become elliptical (Figure 19.5) rather than circular. The motion ellipses are flatter downward so that the motion at the seafloor is simply back-and-forth rather than circular (Figure 19.5). The continual to-and-fro motion on the seafloor shapes loose sand into ripples that symmetrically slope at the same angle toward shore and away from shore (see the photograph in Figure 19.5).

In still shallower water the waves become even steeper, and the forward wave velocity slows until the water at the crest moves landward faster than the wave itself. At this point, the wave forms a **breaker** where water in the wave crest collapses down and toward the beach, as seen in **Figure 19.6**. In most cases, the wave breaks into a noisy, foaming mass of water that surges up onto the beach (Figure 19.6a). Where the steepness of the seafloor

▲ **Figure 19.6 What breakers look like.** When a wave approaches the shore, the water in the steepening wave crest moves landward faster than water in the adjacent, landward trough so that the top of the wave collapses forward as a breaker. The breakers shown in (a) simply spill onto the beach. Excellent surfing waves, such as the one shown in (b), are much higher and break with an overhanging curl.

► **Figure 19.5 How wave motion changes close to shore.**

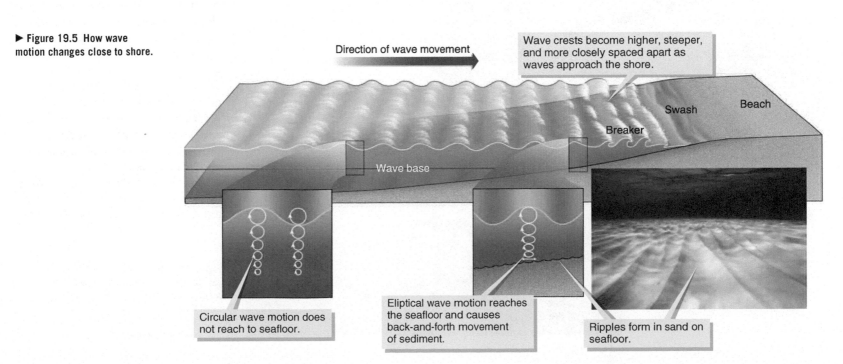

Direction of wave movement

Wave crests become higher, steeper, and more closely spaced apart as waves approach the shore.

Swash

Beach

Breaker

Wave base

Circular wave motion does not reach to seafloor.

Eliptical wave motion reaches the seafloor and causes back-and-forth movement of sediment.

Ripples form in sand on seafloor.

New South Wales, Australia

(a)

Cape Arago, Oregon

(b)

▲ **Figure 19.7 How waves end.** Waves end on the shore with water (a) swashing onto a beach after the wave breaks, or (b) crashing into headlands where waves meet the shore before breaking.

and the wave height are just right, however, the top of the wave curls over and plunges downward in the classic surfing wave (Figure 19.6b). When the wave breaks, the energy transfers from the water wave into sound waves and into mechanical energy that moves sediment. The remaining momentum carries the water up the beach until the pull of gravity slows it to a stop, and then the water flows back into the ocean, as illustrated in **Figure 19.7**a. Where steep sea cliffs drop directly into deep water, the wave does not slow sufficiently to break, so the full energy of the wave arrives with a deafening crash against the rock, as seen in Figure 19.7b.

If you watch waves along a beach, they always seem to come directly toward you, despite the fact that waves in the open ocean move in the direction that the wind blows. Given the variety of orientations of the shoreline and directions of blowing wind, how is it possible for the wave crests to be always parallel to the shore? Refraction, the change in direction of wave motion caused by a change in wave velocity, provides the answer to this question (check back to Section 8.2 for a refresher on refraction).

Figure 19.8 illustrates how and why waves refract in water. Where a wave moves near a shoreline, the part of the wave that is closest to land moves through shallower water than does the offshore part of the wave. The nearshore part of the wave touches bottom, steepens, and slows. The progressive shoreward decrease in wave velocity causes each wave to bend toward the shore so that the wave crests are parallel, or nearly parallel, to the shoreline when the waves break.

How Waves Generate Currents

Wave motion is forward and backward, but observations show that waves also generate water currents that move in just one direction, carrying boats, swimmers, and other objects in the water along the shoreline or even out to sea. **Longshore currents** move parallel to shore, whereas **rip currents** move water away from the coast.

Figure 19.9 explains the formation of longshore and rip currents. Where refracting waves approach the shoreline at an angle, the water is forced to move parallel to the coast and away from where the waves converge onto the shore. This flow of water along the shore defines the longshore current. Of course, each arriving wave contributes more water into the longshore current, and this water cannot simply pile up along the shore because gravity is pulling the excess water back out to sea. As a result, part of the current flows offshore as a rip current. Longshore currents may persist for tens of kilometers unless deflected seaward by a headland or pier. Rip currents are hazardous when the current speed heading

Beach   Breakers

Waves touch bottom, slow down, and bend parallel to shore.

Refracted waves

Wave crest   Waves

Waves

Waves

Wave base

Waves move at an angle to shore in deep water.

Waves refract to parallel the shore.

Wave crest

◄ **Figure 19.8 Water waves refract.** Waves refract in shallow water to approach with wave crests parallel to the beach. Waves bend parallel to the coast because the shallow part of each wave slows down where it makes contact with the seafloor close to shore.

ACTIVE ART

Water Wave Motion and Refraction. *See how waves move in water and refract along a shoreline.*

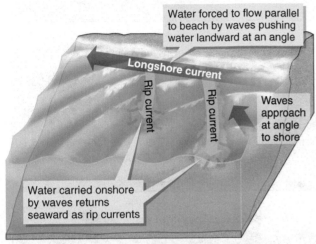

▲ **Figure 19.9 How longshore and rip currents form.** Waves approaching the shore at an angle force water to move along the shore in the direction of the wave motion. This along-shore water movement forms longshore currents. Some water that waves carry against the beach also moves offshore in rip currents.

▲ **Figure 19.10 Tsunami meets the shore.** A towering tsunami heads toward a beach in Thailand following a giant earthquake more than 1000 kilometers away near the Indonesian island of Sumatra. The rocky seafloor is normally submerged but was exposed as the tsunami wave trough reached the shoreline ahead of the wave crest in the background. Those people curious to examine the oddly exposed seafloor, but also lacking knowledge about tsunami, then had to run for their lives as the giant wave crest rushed in.

offshore is too great to swim against; swimmers encountering a rip current can be carried out to sea and drowned.

How Tsunami Waves Are Different from Other Waves

All that you have just learned relates to the normal wind-driven waves on the surface of oceans and lakes, but different processes must account for tsunami waves. Monstrous tsunami are far more deadly than storm waves, as was tragically demonstrated by the more than 230,000 lives lost in eleven nations surrounding the Indian Ocean on December 26, 2004. You learned a little bit about tsunami and their relationship to earthquakes in Section 11.8, and it is important to understand how and why they are different from regular waves.

Some simple observations illustrate key differences between tsunami and regular sea waves.

- Although tall storm waves commonly arrive on distant, sunny coastlines, they rarely arrive without warning. Instead, the wave height builds gradually over many hours or days. Tsunami, in contrast, commonly arrive without warning as a series of several waves of incredible magnitude that can be more than ten meters high, as illustrated in **Figure 19.10**.

- Instead of breaking on the beach, a tsunami wave comes onto land as an ever-rising surge of water that floods inland for more than a kilometer on gently sloping shorelines over a period of tens of minutes. The flood wave then quickly withdraws back to the sea only to be followed by the next wave several minutes to an hour later.

The suddenness, great height, longevity of the surging wave onto land, and the long time between waves are characteristic of tsunami.

Quantitative measurements also bear out the differences between tsunami and everyday waves as indicated by the following comparisons:

- Wind-generated waves typically have a period of 5 to 20 seconds, whereas tsunami periods are typically 10 minutes to 2 hours.

- Wind-generated waves have a wavelength of 10 to 200 meters, whereas tsunami wavelengths are in the range of 100 to 500 kilometers in the open ocean.

- Wind-generated waves rarely move faster than about 50 kilometers per hour, whereas tsunami reach peak speeds greater than 800 kilometers per hour (similar to a commercial jet airplane).

The differences between normal waves and tsunami relate to how they form. The fast-moving, long-wavelength tsunami require huge amounts of energy to place such a large volume of water in motion. Wind speeds would need to exceed 10,000 kilometers per hour to transfer the requisite energy that fuels a large tsunami. The long wavelengths also indicate that the entire thickness of the ocean water column, not simply a surface zone dragged along by wind, is moving. For an analogy, wind does not generate tsunami any more than wind is likely to blow water out of a swimming pool. To make a big wave that sloshes out of a swimming pool at a pool party, you need to get lots of people to jump into the pool at once. The resulting energetic wave results by displacing water out of the pool.

ACTIVE ART

Tsunami. See how an earthquake forms a tsunami and how a tsunami travels across the ocean.

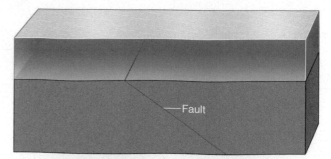

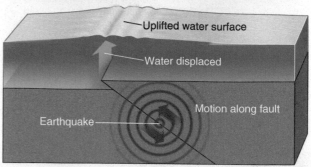

Similarly, tsunami represent the sudden displacement of water over the entire depth of the ocean. Fault movement during earthquakes is the most common generating mechanism, but landslides under water or into the sea from land, the formation of calderas on volcanic islands, and the ocean impact of large meteors from space (not witnessed by humans but recorded in ancient sedimentary rocks) also cause tsunami. The connection between tsunami and sudden geologic events has been evident for centuries as residents of coastal communities, still dazed by earthquake shaking or from witnessing a huge landslide or cataclysmic volcanic explosion were then swamped by giant waves that seemingly raced onto land from nowhere.

Figure 19.11 illustrates how earthquakes generate tsunami. Although rock movement is not directly observed during submarine earthquakes, we know that it must happen by drawing comparison to displacement of rocks along faults during earthquakes on land (Figure 11.27 illustrates examples). Motion along a fault that ruptures the seafloor also moves the overlying water. Upward displacement of the ocean conveys potential energy to the water column, and the water flows away from the uplift so that the ocean surface returns to its original condition. For example, fault motion during the 2004 Sumatra earthquake raised the seafloor and ocean surface by about 5 meters within an area approximately 400 kilometers long and 50 kilometers wide. This motion lifted more than 100 billion metric tons of water above sea level over a matter of a few minutes.

Highly energetic waves generated by an impulse, such as fault motion, have very large wavelengths and low wave heights compared to waves resulting from persistent input of modest surface energy by wind. Tsunami are notoriously difficult to recognize in the open ocean because in deep water the wave may only be 20 centimeters high and more than 100 kilometers across, as depicted in **Figure 19.12**. The wave is so broad compared

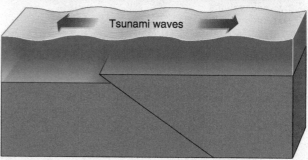

▲ **Figure 19.11 The relationship between earthquakes and tsunami.** Motion along a fault that ruptures the seafloor during an earthquake also displaces the overlying water and raises the ocean surface above sea level. The elevated volume of water spreads horizontally, which generates a tsunami.

Tsunami wavelength and velocity decrease when waves move from deep water to shallow water. Wave height increases dramatically. Wave period remains the same, which means that the water remains above normal sea-level for as long as an hour as each wave crest moves inland. The height, width, and velocity of tsunami causes severe destruction along low-lying coastlines, as illustrated by these satellite images obtained before and after the December 2004 tsunami struck Lhoknga, Indonesia.

◀ **Figure 19.12 Visualizing why tsunami are destructive.**

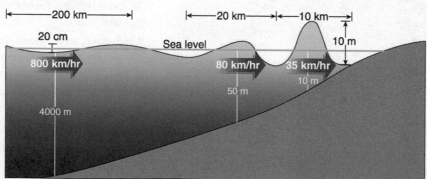

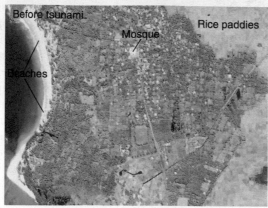

to its height that it is virtually invisible. Remember, too, as with any wave, that although the wave energy moves at jet-aircraft speed, the actual forward movement of the water is negligible in the open sea, which further decreases the likelihood of detection.

Like regular waves, however, tsunami slow down in shallow water close to shore. The immense energy is confined into a smaller volume of water, causing the ocean surface to rise as the wavelength shortens and the volume of water beneath each wave crest compresses into a smaller area above the seafloor. As Figure 19.12 shows, the fast-moving, low waves in the deep ocean build into roughly 10-meter-high, 5-kilometer-wide surges of water that move too fast to outrun, despite having decelerated from their peak speed. The actual dimensions of tsunami are hard to predict along coastlines because the height and wavelength depend on the shape of the shoreline and the slope of the submerged seafloor in addition to the total tsunami energy.

Putting It Together—How Do Waves Form and Move in Water?

• Wind blowing across water produces waves. High waves require strong winds, blowing for long time periods, across long distances of water.

• Wave motion decreases downward in the water to wave base, the depth below the water surface to which wave motion takes place. There is a small amount of overall water motion in the direction of wave movement.

• Wave height increases and wave velocity decreases where waves move into water that is shallower than wave base. Eventually, the wave crests move forward faster than the slowing waves, causing the waves to break.

• Waves refract where they simultaneously travel slowly in shallow water and more rapidly in deep water. Refracted waves turn parallel, or nearly parallel, to the shoreline.

• Waves move water shoreward, parallel to the coast as longshore currents, and then back to sea as rip currents.

• Displacement of the ocean water, usually by fault movement or landslides generates deadly tsunami. Tsunami have a longer wavelength, larger period, and faster velocity than much less energetic wind-generated waves.

19.3 How Do Waves Form Shoreline Landscapes?

Now that you have a fundamental understanding of what waves are and how they form, we can move on to consider how waves shape shorelines. Waves are the primary process that form and modify landscapes in most coastal regions.

How Waves Erode, Transport, and Deposit Sediment

The power of waves to do work in moving materials on a shoreline is most impressively seen where waves crash into rocky headlands (see Figure 19.7b). Storm waves exert incredible force against rocks, some equivalent to 10,000 kilograms weighing down on each square meter. These forces

pluck off large blocks of rock that were already separated by fractures and joints or that have fallen at the base of the cliff after mass movement. Erosion is most severe near the base of a cliff and may produce an overhang that is unstable and prone to failure by mass movement, as seen in **Figure 19.13**.

Figure 19.14 shows how wave refraction focuses erosion at headlands and favors deposition in adjacent bays along the shore. Incoming waves first encounter shallow water near headlands along highly irregular coasts. The waves refract and bend toward the front and sides of the protruding headland. Refraction directs more water toward the headland than toward the bays, so wave height is higher along the headland than in the bays. These processes focus wave erosion on the headlands. Then, longshore currents

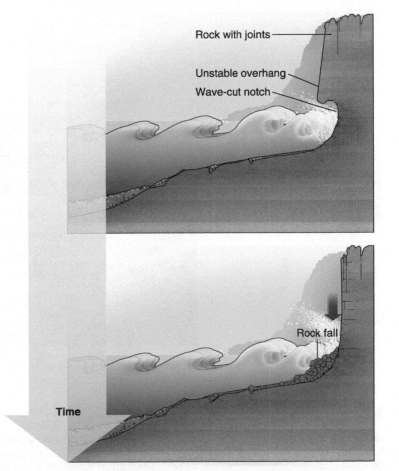

▲ **Figure 19.13 Why mass movements occur along coasts.** Wave erosion focuses at the base of cliffs to form a notch and an unstable overhang that eventually fails by rock fall.

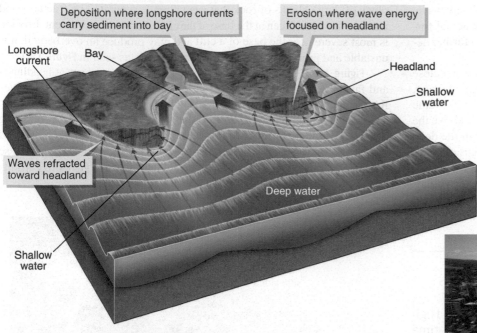

Deposition where longshore currents carry sediment into bay

Erosion where wave energy focused on headland

Longshore current

Bay

Headland

Shallow water

Waves refracted toward headland

Deep water

Shallow water

◀ **Figure 19.14 Refraction causes erosion on headlands and deposition in bays.**
Refraction bends waves toward headlands, which focuses wave energy to erode the headland. Longshore currents then carry sediment eroded from headlands into the adjacent bays, where wave heights are low.

Deposition to form a beach in the bay

Wave erosion on headland

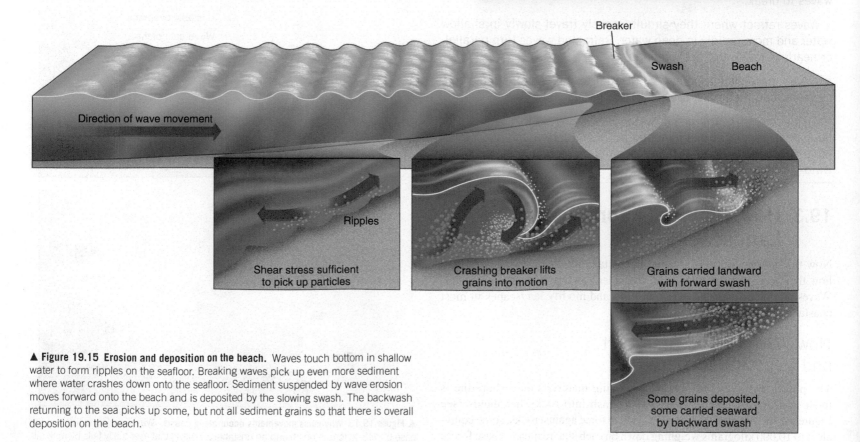

Breaker

Swash   Beach

Direction of wave movement

Ripples

Shear stress sufficient to pick up particles

Crashing breaker lifts grains into motion

Grains carried landward with forward swash

Some grains deposited, some carried seaward by backward swash

▲ **Figure 19.15 Erosion and deposition on the beach.** Waves touch bottom in shallow water to form ripples on the seafloor. Breaking waves pick up even more sediment where water crashes down onto the seafloor. Sediment suspended by wave erosion moves forward onto the beach and is deposited by the slowing swash. The backwash returning to the sea picks up some, but not all sediment grains so that there is overall deposition on the beach.

carry sediment eroded from the headlands toward the adjacent bays where wave height is much lower, and deposition occurs on the beach.

Observations of sediment movement by waves show that both erosion and deposition take place along a beach and on the adjacent shallow seafloor. **Figure 19.15** summarizes these observations. The to-and-fro motion of the water exerts shear stress on the seafloor that moves sediment, as is also demonstrated by the seafloor ripples in Figure 19.5. Sediment grains that are suspended in the water move toward land because of the small, landward component of water transport by the waves explained in Figure 19.4. The greatest seafloor erosion occurs where waves break and plunge downward onto unconsolidated sediment (Figure 19.15). The force of the water slamming down on the seafloor suspends the sediment and carries it forward as the wave swashes up onto the beach.

There are two reasons why the sediment-carrying capacity of the water diminishes as it swashes up the beach (Figure 19.15):

1. The velocity of the water decreases because it moves upslope against the force of gravity.
2. Some water percolates into the pore spaces between the sand or gravel particles on the beach, which decreases the amount of moving water available to carry sediment.

The water rises onto the beach, slows to a brief standstill, and then retreats back down to the sea. The backwash is less erosive than the initially energetic landward swash, but it still picks up the smallest grains and carries them back out to sea. In this fashion, waves tend to move the coarser sediment grains landward and deposit them on the beach while the finer particles move back offshore. This explains why clastic marine sediment closest to a shoreline is coarser grained than the sediment deposited below wave-base depths (for example, see Section 5.6 and Figure 5.20).

Some sediment also moves along the beach rather than up and down the beach slope, as illustrated in **Figure 19.16**. Sediment stirred up by the breaking wave moves up the beach perpendicular to the direction of wave movement, which is usually at an angle to the shoreline. The backwash, however, flows directly down the beach slope into the water because of the pull of gravity. The different paths of sediment grains during transport by swash and backwash causes **beach drift,** a zigzag movement of sediment along the beach. Some of the seafloor sediment stirred up by the waves is also transported parallel to shore by longshore currents (see Figure 19.8) rather than up onto the beach.

How Waves Shape Rocky Shorelines

Most rocky shorelines slowly erode in the landward direction. **Figure 19.17** illustrates progressive stages in this shoreline reshaping process. Wave erosion hollows out caves at the base of a headland. The caves may connect through a narrow headland to form an arch. As the arch widens, the unsupported rock roof of the arch collapses, which isolates the part of the original headland to produce a **sea stack**. The sea stack landform is simply a small rocky island that is located close to the shore (pictured in Figure 19.17; also see Figure 19.1b).

Another feature of rocky shorelines are nearly horizontal benches of rock that are commonly submerged and wave swept at high tide, is exposed to view when the tide is low. These benches are called **wave-cut platforms** (see Figure 19.17). The platform develops when waves break up rock and carry away rock fragments loosened by biologic and chemical processes. **Figure 19.18** illustrates weathering and erosion processes that affect coastal rock exposures. Sea urchins, sea anemones, clams, sponges, and other marine animals bore holes into rock in order to improve their

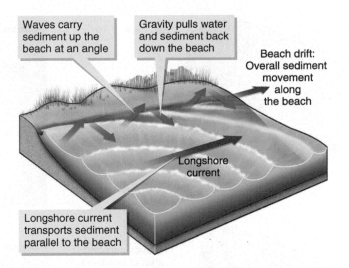

▲ **Figure 19.16 How sediment moves along the shoreline.** Longshore currents and beach drift transport sediment along the shoreline.

ACTIVE ART

Beach Drift and Longshore Current. See how beach drift and longshore current move sediment along the shoreline.

own resistance to being moved by waves. Other animals, such as snails and chitons, abrade rocks like miniature files as they graze on algae attached to the rock surface. Shallow pools of water left on the platform at low tide enhance rock weathering through growth of salt crystals in cracks by evaporation, or dissolution of soluble rocks such as limestone (see Section 5.1 for more about salt weathering). Sediment grains carried back and forth across the platform by waves also abrade the underlying rock.

How Waves Make Beaches

The definition of a beach is pretty clear to anyone who has been to a shore. A **beach** is the nonvegetated area of unconsolidated sediment that extends from the low-tide line to a landward line defined by a cliff, sand dunes, or permanent vegetation. **Figure 19.19** shows that a typical beach has a relatively steep part close to the water, called the **beach face**, and a flatter part, called the **berm**. The **berm crest** marks the boundary between the beach face and the berm. Sunbathers prefer the berm part of the beach because it is beyond the reach of most fair-weather waves, is nearly flat, and is likely composed of soft, fine sand.

Measurements by geologists show that the height of incoming waves and the size of the beach sediment determine the height of the berm crest and the slope of the beach face. **Figure 19.20**a shows sediment deposited by the swash and backwash all along the beach-face up to the berm crest. Sediment deposited on the beach face builds the beach out toward the ocean and up to the berm-crest elevation. The higher the waves, the greater the swash, and the higher the berm crest. Where the beach sediment is coarse grained and permeable, however, more of the swash soaks into the beach. As a result, gravel beaches tend to have steeper beach faces and lower berm heights than sand beaches experiencing the same incoming wave heights.

The shape of a beach changes through the year because wave height changes with the seasons (Figure 19.20b). High storm waves (usually during the winter) erode the lower part of the beach face and build a new berm at a higher elevation than the fair-weather berm. Fair-weather conditions (usually during the summer) lead to more beach deposition that builds out a new beach

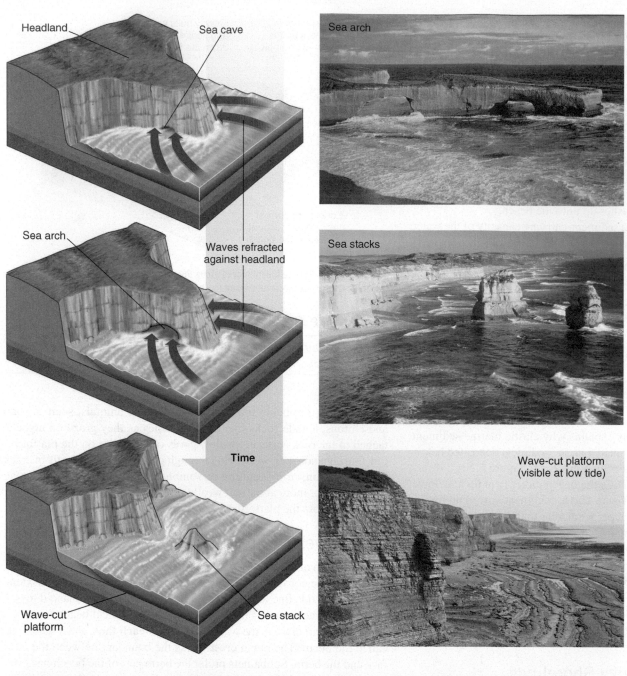

Headland

Sea cave

Sea arch

Waves refracted against headland

Time

Wave-cut platform

Sea stack

Sea arch

Sea stacks

Wave-cut platform (visible at low tide)

◀ **Figure 19.17 What rocky coastlines look like.** Wave erosion of rocky headlands produces caves that may connect beneath the headland to form a sea arch. Collapse of the sea arch then separates a sea stack from the mainland. Erosion at wave base cuts a platform in the rock.

Sea urchins bore holes into rock

(a)

Salt weathering produces pits in rock

(b)

◀ **Figure 19.18 How rock weathers at the shoreline.** (a) Some animals, such as sea urchins, bore holes into rock. Others, such as starfish and some snails, abrade rock surfaces as they scrape off algae and other organisms for food. (b) Evaporation of salty sea spray and tide pools exposed at low tide causes precipitation of salt crystals that disaggregate rocks. Waves erode the weathered rock fragments and leave behind pitted surfaces where water accumulates and accentuates the salt-weathering process.

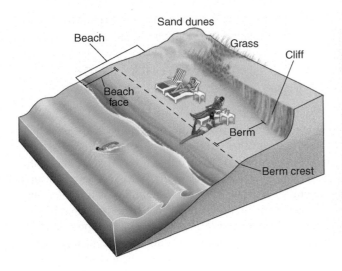

◀ **Figure 19.19 The shape of a beach.** The beach is a bare area of unconsolidated sediment that extends from the low-tide line to a landward boundary marked by a sea cliff, vegetation, or wind-blown sand dunes. The beach includes the seaward-inclined beach face and the near horizontal to slightly landward inclined berm. The berm crest separates the beach face and the berm.

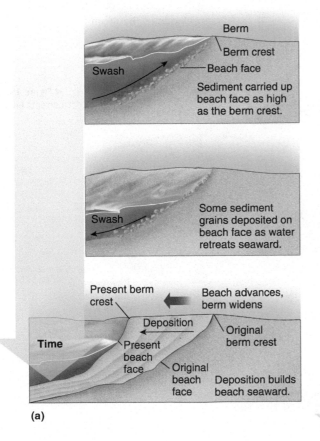

(a)

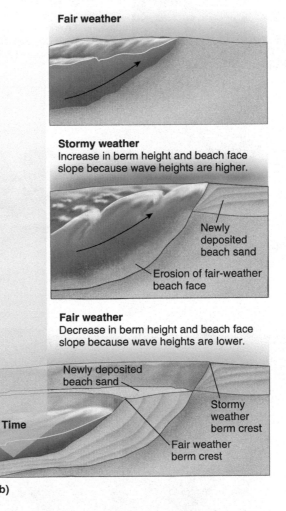

(b)

▲ **Figure 19.20 Visualizing how a berm forms and changes.** (a) At high tide wave swash carries sediment up the beach face as far as the berm crest. Sediment accumulates on the beach face so that the beach builds seaward over time to produce a wide berm. (b) The berm-crest height increases when high storm waves pound the beach. Under stormy conditions the beach face erodes, and deposition builds up a higher berm. When fair-weather conditions return, waves build a new beach with a lower berm.

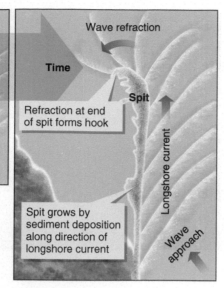

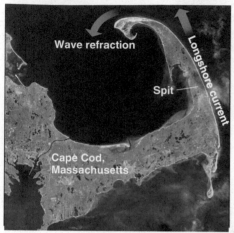

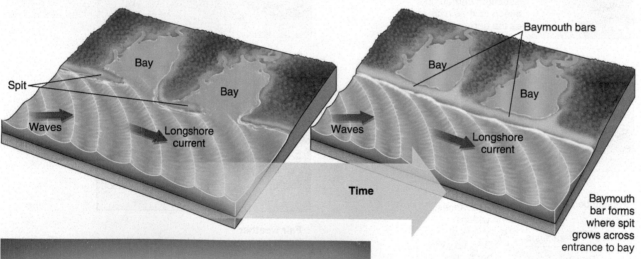

below the former storm berm. As a result, a single beach may exhibit both a winter and a summer berm.

How Longshore Currents Shape Shorelines

Erosion and deposition of sediment by longshore currents also produce distinctive landforms. **Figure 19.21** shows how longshore currents build **spits**, which are elongate sediment ridges that are attached to headlands and point in the direction of the longshore current. Wave refraction around the end of the spit causes sediment deposition that may form a prominent hook-shaped beach, such as that seen at Cape Cod, Massachusetts (Figure 19.21). **Figure 19.22** illustrates how longshore currents may build **baymouth bars**, which are spits that link one headland to the next and close off the intervening bay from the ocean.

Sediment erosion and transport by longshore currents is dramatically illustrated where structures are built perpendicular to the shoreline. **Figure 19.23** shows how **groins**, which are walls built perpendicular to the shoreline, trap sediment on the up-current side to widen a beach. The longshore current erodes the beach, however, on the down-current side of the groin.

◀ **Figure 19.21 How longshore currents build spits.** The longshore current carries sediment off the end of a headland to build a spit. Wave refraction around the end of the spit shapes the sediment into a hook. Cape Cod, Massachusetts, as seen from the International Space Station, is an example of a hooked spit.

◀ **Figure 19.22 How longshore currents build baymouth bars.**

Baymouth bars in The Hamptons, Long Island, New York

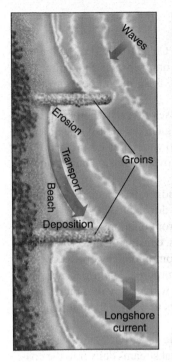

◀ **Figure 19.23 How groins modify beaches.** Groins project into the longshore current and interrupt sediment transport along the beach. Deposition widens the beach on the up-current side of the groin whereas the current erodes sand from the down-current side of the groin. The photo shows sediment erosion and deposition alongside groins on Lake Michigan in Chicago, Illinois. Wave processes in large lakes are the same as wave processes in the ocean.

Beach eroded on downcurrent side of groin

Wide beach deposited on upcurrent side of groin

Longshore current

Groins

ACTIVE ART

Effects of Groins and Jetties. See how groins and jetties influence shoreline deposition and erosion.

Sediment Sources for Beaches

Field observations reveal many sources for beach sediment.

- Waves erode some beach sediment from shoreline rock and regolith.
- Streams deliver large volumes of sediment to the ocean, where waves and longshore currents carry some of this sediment onto beaches.

- Waves erode sediment from the seafloor as indicated by seashells on beaches.

Figure 19.24 illustrates a variety of beach deposits. Many white-sand beaches contain mostly quartz (Figure 19.24a). Quartz is abundant in river sediment delivered to the sea because it is the weathering-resistant mineral that is most abundant in continental rocks (see Section 5.2). Quartz also lacks cleavage, so continuous movement and abrasion by waves do not easily break it. Black-sand beaches are common on basaltic volcanic islands (Figure 19.24b). Basalt does not contain quartz to produce white sand, but the fine-grained igneous rock is crushed into small sand-size grains by wave erosion along the shore. Some coastlines lack major rivers to introduce sediment to the coast, and wave energy may be too weak to significantly erode rock or regolith exposed at the shore. In these cases, the beach consists of broken seashells carried landward by the waves (Figure 19.24c). All beach-sediment grains are well rounded because of constant wave agitation that abrades the grains against one another.

Quartz-sand beach, North Carolina

(a) Sand weathered and eroded from continents contains abundant quartz, because quartz resists weathering and does not abrade or break easily during transport by streams.

(b) Black sand beaches on volcanic islands consist of wave-eroded fragments of basalt lava.

Basalt-sand beach, Hawaii

The Sediment Budget Determines Beach Growth or Loss

Historical observations show that the width of many beaches either increase or decrease. Whether a beach grows, shrinks, or is relatively unchanged over time is an important aspect of human interactions with dynamic

Shell-sand beach, Florida

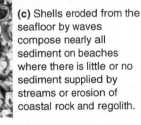

(c) Shells eroded from the seafloor by waves compose nearly all sediment on beaches where there is little or no sediment supplied by streams or erosion of coastal rock and regolith.

◀ **Figure 19.24 What beach sediment looks like.**

shorelines. It is reasonable to assume that beaches get narrower where erosion dominates and they grow wider where sediment deposition is taking place. Geologists put together a beach sediment budget in order to determine whether sediment is being gained or lost along a coastline in order to evaluate growth and loss of beaches.

Figure 19.25 illustrates the budget of sediment gains and losses along a coastline. Natural sediment gains to a section of shoreline include

- the amount of sediment delivered by longshore currents from the adjacent shoreline.
- the amount of sediment eroded from local sea cliffs and headlands.
- the amount of sediment delivered to the shore by rivers.
- the amount of sediment eroded by waves offshore of the beach and then transported onshore.

Rivers and coastal erosion are usually the largest sediment sources, including sediment spread out along the shore by longshore currents. Sediment gains, therefore, are largest near the mouths of major rivers and where easily eroded rock or regolith forms the shoreline. For example, erosion of steep exposures of thick, unconsolidated ice-age glacial sediments supplies abundant sand and gravel to the beaches of New England and Washington.

Natural sediment losses to a section of shoreline include

- the amount of sediment carried farther along the shore by longshore currents.
- the amount of beach sediment eroded by waves and carried offshore by rip currents or backflow of storm surges during storms.
- the amount of sand that wind blows off of the beach and onto adjacent coastal sand dunes.

Coastal sand dunes, explored further in Chapter 20, form by the persistent movement of wind across loose sand on the beach. Sand dunes are common near beaches and, in many places, form the highest elevations along the coast.

Comparing the sediment gains and losses reveals whether or not a beach grows or shrinks. If more sediment is gained than lost, then there is overall deposition, which causes the beach to widen toward the sea. If the losses are larger than the gains, then there is overall erosion and the beach becomes narrower.

Human activities affect the sediment budget. Groins cause sediment surpluses in up-current areas and deficits in down current areas of beach (Figure 19.23). Pumping or dredging sand from offshore and adding it to the beach artificially nourishes some beaches, causing them to widen even if naturally they would be eroding and becoming narrower. Beach sand and gravel are excavated for construction purposes, leading to narrower beaches, whereas in other cases beaches are widened by dumping of sediment that was dredged from adjacent harbors to deepen them for entry of large ships.

Some changes in sediment supply relate to activities farther inland that change the amount of sediment supply that streams deliver to the coast (see Section 16.11 for more information on human impacts on streams). Changes in land use from natural vegetation to agricultural or urban development add sediment to streams, whereas dams trap sediment in reservoirs that would otherwise be transported to the ocean where it would contribute to beach formation. Figure 19.26 illustrates twentieth-century changes in sediment delivery to the United States Atlantic coast. Urbanization increased sediment supply to beaches in the Northeast while construction of dams and development of agricultural practices that decrease soil erosion led to substantial decreases in sediment load south of Chesapeake

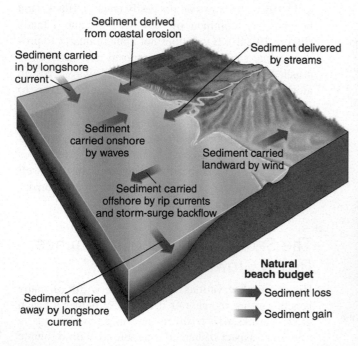

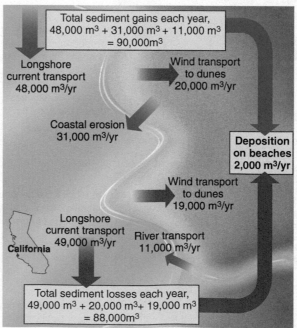

▲ Figure 19.25 Visualizing the beach-sediment budget. A beach grows or shrinks through time depending on the relative amounts of sediment carried to or eroded away from the part of the shoreline where the beach is located. An actual budget illustrated on the right shows that a volume of sediment that would fill almost 500 railroad boxcars is added to this 20-kilometer-long stretch of California coastline each year.

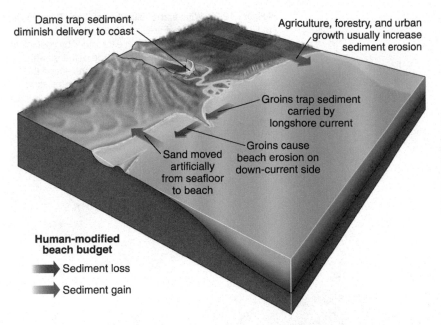

Dams trap sediment, diminish delivery to coast

Agriculture, forestry, and urban growth usually increase sediment erosion

Groins trap sediment carried by longshore current

Groins cause beach erosion on down-current side

Sand moved artificially from seafloor to beach

Human activities complicate beach budgets. Groins disrupt deposition and erosion by longshore currents. Sand may be artificially added or subtracted from the beach. Land-use changes and dam construction modify natural river sediment delivery to shorelines.

Human-modified beach budget

➡ Sediment loss

➡ Sediment gain

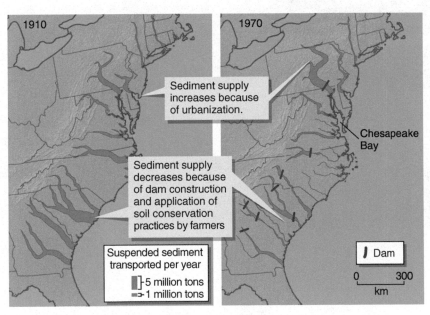

1910

1970

Sediment supply increases because of urbanization.

Chesapeake Bay

Sediment supply decreases because of dam construction and application of soil conservation practices by farmers

Suspended sediment transported per year

5 million tons
1 million tons

▮ Dam

0   300
km

Sediment supply to Atlantic beaches changed dramatically during the twentieth century.

The gray areas in the river valleys schematically represent the amount of sediment carried by each river and not the width of the stream.

▲ **Figure 19.26 Visualizing human modifications of the beach budget.**

Bay. Similar historic changes in sediment supply occurred on the Pacific Coast, where dam building in California reduced sediment delivery to 23 percent of the beaches in the state by an average of 25 percent.

How Barrier Islands and Tidal Inlets Form

Many coastlines include **barrier islands**, which are long, narrow ridges of land that form parallel to, but separate from, the mainland coast. Barrier islands form 13 percent of the world's coastlines, and there are about 300 such islands along the Atlantic and Gulf of Mexico coasts of the United States. Some of the most heavily developed and most valuable real estate in the country is located on barrier islands, in places such as Atlantic City, New Jersey; Miami Beach, Florida; and Galveston, Texas.

Figure 19.27 summarizes the features of barrier-island coastlines. Barrier islands consist of sediment rather than rock. Each barrier island is typically 10 to 100 kilometers long but usually less than 5 kilometers wide. Barrier coastlines consist of many islands that resemble beads on a necklace. Narrow **tidal inlets** separate the islands and focus tide-produced currents between the lagoon and open ocean. The highest elevations on barrier islands are usually sand dunes built by wind blowing across the beach on the seaward side of the island.

The landward side of the barrier island faces a lagoon or marsh that is largely unaffected by waves but may exhibit features caused by tides, which are discussed in the next section. **Lagoons** are shallow, calm-water bodies of water located landward of obstacles that absorb wave energy directed toward the shoreline. Example obstacles are barrier islands, spits, baymouth

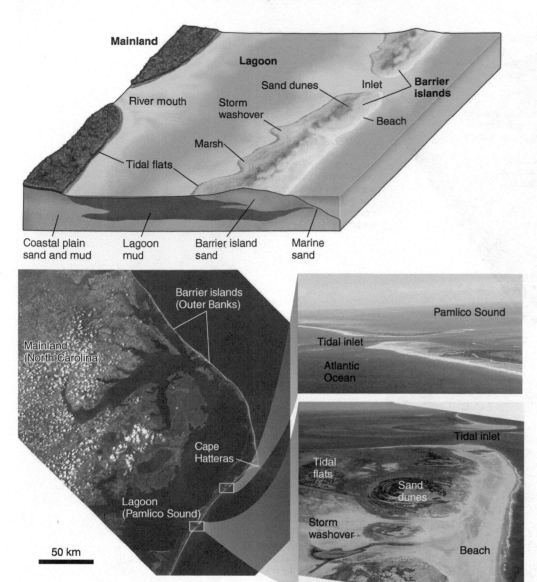

◄ **Figure 19.27 What barrier islands look like.** Barrier islands are long, low islands that are separated from the mainland by a quiet-water lagoon, and from each other by tidal inlets. Waves form a beach on the seaward side of the island, whereas tidal mud flats and marshes are present on the lagoon side. Wind blowing across the beach forms sand dunes along the high spine of the island. Storm waves wash across the islands and carry beach and dune sand into the lagoon. Barrier islands form long sections of the Atlantic Coast of the United States, including the pictured locations in North Carolina.

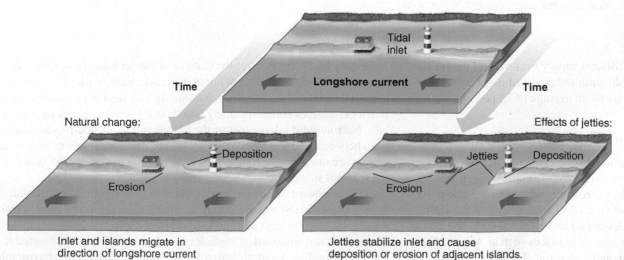

◄ **Figure 19.28 Why islands and tidal inlets migrate.** Longshore current erosion and deposition cause barrier islands and tidal inlets to migrate in the direction of longshore transport. Erosion occurs on the up-current ends of islands and deposition occurs on the down-current ends of islands. Erosion and deposition on opposite sides of an inlet cause the islands and inlet to migrate together without changing the inlet width. Engineers construct jetties to keep inlets open to ships. The jetty traps longshore-transported sediment, which causes growth of one island at the expense of the other.

bars, and offshore coral reefs. High storm waves readily cross the narrow, low barrier islands. Storm-wave erosion may breach the island to form a new tidal inlet and wash beach and dune sand into the quiet lagoon.

Barrier islands and intervening tidal inlets shift in the direction of the longshore current, as shown in **Figure 19.28**. Wave refraction and long-shore currents erode sediment from the up-current end of one island and deposit it on the down-current end of the next one. The process of erosion on one end of each island and deposition on the other end causes both the inlet and the islands to migrate.

Many inlets serve as shipping lanes between the open ocean and mainland ports in lagoons and estuaries. It is important to maintain these inlets in a stable position by building **jetties**, which are longer versions of groins built adjacent to an inlet. Jetties, visible in Figure 19.1d, are walls that keep the inlet from shifting with the longshore current. As illustrated in Figure 19.28, a jetty traps sediment on the up-current side, which keeps the barrier island from shifting into the tidal inlet. The trapping of sediment by the jetty may starve the next barrier island in the chain from sediment, causing it to retreat landward under the effects of wave erosion.

Putting It Together—How Do Waves Form Shoreline Landscapes?

- Wave motion touches the seafloor in shallow water, erodes loose sediment, and carries it landward where deposition occurs on beaches.

- Longshore currents carry sediment parallel to shore to produce spits and baymouth bars.

- Refracted-wave attack on rocky headlands gradually straightens the shoreline. Sea stacks and wave-cut platforms are common land-forms on eroded rocky coasts.

- Beach sediment comes from sediment carried to the ocean by rivers and from wave erosion of the seafloor and coast.

- Overall beach growth or shrinkage depends on long-term imbalance between the volume of sediment added or subtracted from the beach. Human activities affect the beach-sediment budget.

- Barrier islands are separated by tidal inlets and migrate along the coastline in the direction of longshore currents.

19.4 What Is the Role of Tides in Forming Coastal Landscapes?

Rising and falling tides are an essential process of dynamic, changing shoreline landscapes, as is apparent by your observations on the Oregon Coast in Figure 19.1e. The **tide** is the slow, up-and-down movement of sea level that occurs each day because of gravitational interactions of the Moon and Sun with Earth.

The change in sea level between low and high tide, called the **tidal range**, is less than 3 meters along most modern shorelines, but in some places is greater than 15 meters. Tides typically move water at slow velocities compared to waves. Exceptions to this generalization occur where the

mass of rising and falling water is constricted in inlets between islands or forced in and out of funnel-shaped estuaries. These tide-generated currents move at several meters per second.

Why Tides Exist

The attracting force of gravity between Earth, Moon, and Sun explains tides. Measurements of the gravitational force, dating back to Isaac Newton in the seventeenth century, indicate that the force exerted on one body of matter by another is larger when (a) the masses of the objects are larger and (b) when they are closer together. The Sun is the largest mass in our solar system, although it is far away from Earth. The Moon has a relatively small mass, but it is close to Earth and thus exerts a significant gravitational attraction. The Sun and Moon are the most important planetary bodies for calculating the gravity force that causes tides; all other planets are too small or too distant to have much effect.

Figure 19.29 shows how gravitational attraction produces ocean tides. All points on and within Earth are pulled toward the Moon by gravity (and the same force pulls the Moon toward Earth). For simplicity, think of Earth as a rigid sphere and that the magnitude of the gravitational attraction of the Moon on Earth is simply a force acting from the center of Earth toward the center of the Moon. Also for simplicity, imagine a uniform layer of ocean water covering Earth's surface (see Figure 19.29). Points on the ocean surface are attracted toward the Moon's center by varying amounts, with water on the side closest to the Moon tugged a little bit more than water on the far side, because the water closest to the Moon experiences a larger gravitational attraction than the water on the far side of Earth.

You can calculate the actual tidal displacement of the ocean by subtracting the distance of whole-Earth movement from the amount of ocean movement at the different points on the planet surface. As shown in Figure 19.29, the resulting displacement of the ocean relative to rigid Earth causes two tidal bulges, one on the side of Earth facing the Moon and the other on the side opposite the Moon. On the side facing the Moon, the ocean water is pulled farther toward the Moon than Earth is pulled, so the water surface rises. On the side opposite the Moon, Earth is pulled farther toward the Moon than the water surface is pulled, so the water surface rises relative to the rigid Earth surface. Water moves toward these rising tidal bulges in the ocean from elsewhere on the planet, where the tide is falling. Earth rotates under these two bulges in the water surface so that every location on the surface experiences two high tides, and two intervening low tides, each day.

Figure 19.29 also illustrates the gravitational effect of the Sun, which either adds to or subtracts from the tidal force exerted by the Moon. The high tidal range of **spring tides** occurs when Earth, Moon, and Sun are aligned, so that the gravitational attraction of the Sun on Earth adds to the attraction of the Moon. The lower tidal range of **neap tides** occurs when the Moon's gravitational attraction is oriented perpendicular to that of the Sun.

Many factors complicate this simple view of tidal generation. The rotation of Earth, variations in ocean depth, and the location of land distort water movement toward the tidal bulges. These factors cause some areas to experience only one high and one low tide each day, rather than two of each, and the magnitude of each successive high or low tide can be different.

Where Tides Affect the Shoreline

Tides rise and fall along every coast, but they have the most noticeable effect on landscapes where tidal range is high while the wave energy is low.

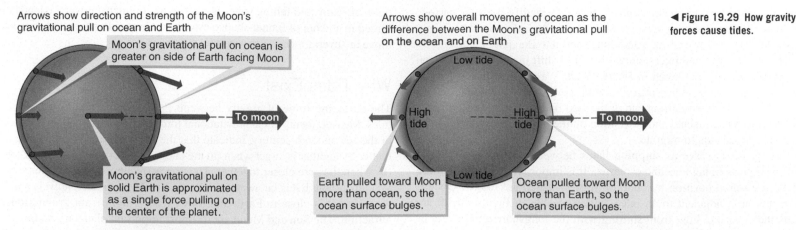

Arrows show direction and strength of the Moon's gravitational pull on ocean and Earth

Moon's gravitational pull on ocean is greater on side of Earth facing Moon

To moon

Moon's gravitational pull on solid Earth is approximated as a single force pulling on the center of the planet.

Arrows show overall movement of ocean as the difference between the Moon's gravitational pull on the ocean and on Earth

Low tide

High tide

High tide

To moon

Low tide

Earth pulled toward Moon more than ocean, so the ocean surface bulges.

Ocean pulled toward Moon more than Earth, so the ocean surface bulges.

◄ **Figure 19.29 How gravity forces cause tides.**

The Moon's gravity pulls the fluid ocean water independent of the solid Earth, which forms outward tidal bulges in the ocean—one facing the Moon and one on the opposite side of Earth. Gravitational attraction of water toward the high-tide bulges produces low tides elsewhere. Earth rotates below these areas of high and low tides to produce daily variations in water level.

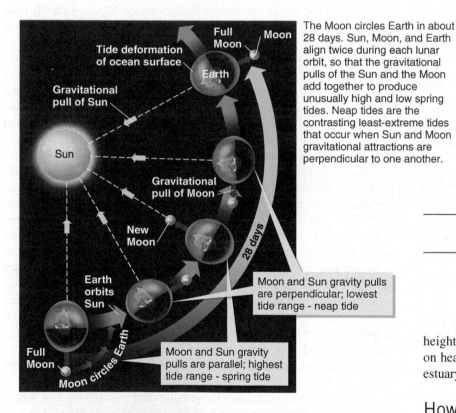

Full Moon   Moon

Tide deformation of ocean surface

Earth

Gravitational pull of Sun

Sun

Gravitational pull of Moon

New Moon

Earth orbits Sun

28 days

Full Moon

Moon circles Earth

Moon and Sun gravity pulls are perpendicular; lowest tide range - neap tide

Moon and Sun gravity pulls are parallel; highest tide range - spring tide

The Moon circles Earth in about 28 days. Sun, Moon, and Earth align twice during each lunar orbit, so that the gravitational pulls of the Sun and the Moon add together to produce unusually high and low spring tides. Neap tides are the contrasting least-extreme tides that occur when Sun and Moon gravitational attractions are perpendicular to one another.

ACTIVE ART

How Tides Work. *See how the positions of Earth, Moon, and Sun determine the rise and fall of the tides.*

Figure 19.30 illustrates that you can notice the difference between high and low tides on wave-swept beaches, but tidal changes are more obvious on shorelines where waves are small and wide beaches do not exist. In these cases, the prominent landform is gently sloping, muddy **tidal flats**, which are marshy or barren areas of land submerged at high tide and exposed at low tide (see Figure 19.30).

Some tidal flats are found along wave-dominated coastlines but exist in locations such as lagoons and estuaries that are protected from wave action. Lagoons are separated from the open ocean by barrier islands (Figure 19.27) or offshore coral reefs that absorb the incoming wave energy. Wind moving across the lagoon creates waves, but these are very small because the wind is in contact with the water over a very short distance, usually only a few kilometers. Estuaries are very deep embayments in a coastline where the ocean extends landward into a river valley (Figure 19.2). Wave

height decreases inland within an estuary because wave energy focuses on headlands flanking the estuary mouth and in shallow water along the estuary shoreline.

How Tidal Flats Form

The rising tide carries sediment landward and then leaves it behind on the tidal flat when the tide falls, as shown in **Figure 19.31**. The rising tide slows down as it moves up the gentle slope of the tidal flat and approaches the high-tide mark, where the velocity decreases to zero before water retreats back across the flat to the low-tide mark. Sediment transported by the rising water is deposited when the rising tide slows down. When the tide falls, the initial current velocity draining off of the tidal flat is too slow to pick up all of the sediment that settled out at high tide. These changes in the velocity and direction of flowing water cause more sediment to be carried landward during the rising tide than is carried seaward during the falling tide, so that each tidal cycle leaves a veneer of newly deposited sediment. The sediment is also coarser grained on the seaward side of the flat than on the landward side, because tidal currents transport the sediment from sea toward land, rather than from land toward sea.

High tide   **Tides on a Tidal Flat**   Low tide

◄ **Figure 19.30 Recognizing tides at the shore.** The width of exposed beach is the only visible difference on this Mexico beach between low and high tides. Tidal flats, such as this one in eastern Canada, include vegetated areas that are at or above the high tide level. Intervening unvegetated areas are submerged most of the time and when exposed at low tide are muddy surfaces crossed by small channels that drain pore water out of the exposed sediment.

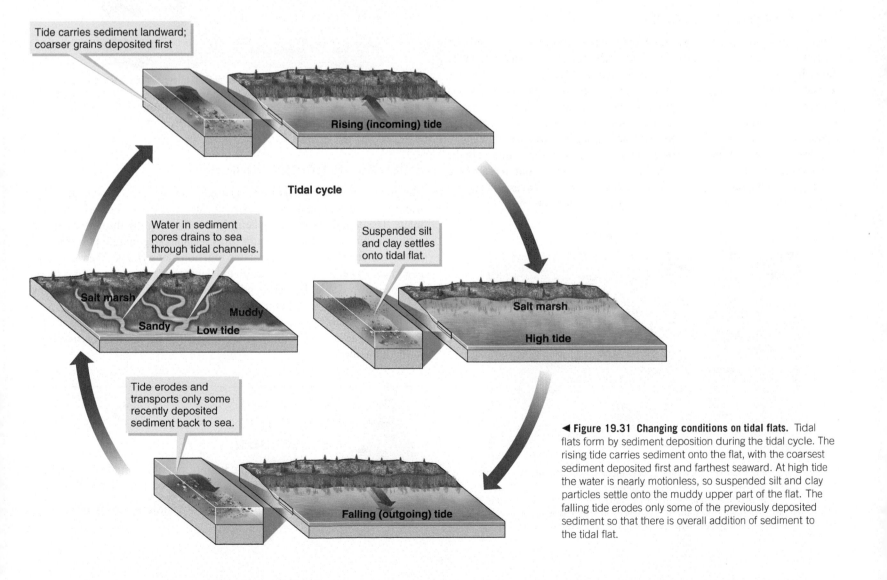

Tide carries sediment landward; coarser grains deposited first

Rising (incoming) tide

Tidal cycle

Water in sediment pores drains to sea through tidal channels.

Suspended silt and clay settles onto tidal flat.

Salt marsh

Muddy

Sandy   Low tide

Salt marsh

High tide

Tide erodes and transports only some recently deposited sediment back to sea.

Falling (outgoing) tide

◄ **Figure 19.31 Changing conditions on tidal flats.** Tidal flats form by sediment deposition during the tidal cycle. The rising tide carries sediment onto the flat, with the coarsest sediment deposited first and farthest seaward. At high tide the water is nearly motionless, so suspended silt and clay particles settle onto the muddy upper part of the flat. The falling tide erodes only some of the previously deposited sediment so that there is overall addition of sediment to the tidal flat.

Water draining seaward during the falling tide erodes channels in the tidal-flat surface. These **tidal channels** (sometimes called tidal creeks) resemble those caused by drainage of water off of hillslopes on land. Like stream channels, these tidal channels have tributaries and usually have very sinuous channel patterns consistent with the cohesive, muddy character of the sediment. Water may continue to flow in tidal channels at low tide because shallow ground water slowly seeps out of the saturated muddy sediment and into the channels.

The part of the tidal flat that is exposed at low tide lacks vegetation because few plants are tolerant of submergence in salt water during high tide. Grasses and low shrubs with limited tolerance of salt water may form salt marshes along the landward side of the tidal flat, which submerges only during spring tides. Mangrove trees form dense forests near the high-tide mark along some coasts.

Chemical sediment covers tidal flats in desert regions where there are no streams to bring clastic sediment to low-relief, wave-protected shorelines. An example is shown in **Figure 19.32**. Seawater evaporation on the tidal flat causes calcite and dolomite precipitation, partly by biologic processes in microbial mats that thrive in hot, damp conditions. Evaporation draws seawater to the surface above the high-tide mark, where it evaporates to form stark, barren deposits of evaporite minerals, such as gypsum and halite.

Tidal Processes at Inlets

Tides are also important in shaping barrier-island shorelines. Not only do tidal flats form on the lagoon side of a barrier (Figure 19.27), but **Figure 19.33** shows how tidal currents move sediment back and forth through the inlets between adjacent islands. Tidal currents are much stronger in the tidal inlets than on the adjacent tidal flats because the rising and falling tide funnels back and forth through narrow constrictions between the open ocean and lagoon. The rising tide carries sediment suspended by waves and longshore currents through the tidal inlet toward land and deposits the sediment in the quiet-water lagoon. The falling tide carries sediment back out to sea where it is deposited beyond the beach, or it may be eroded by waves and carried back onto the beach or carried away by longshore currents.

▲ **Figure 19.32 Chemical sedimentation occurs along desert coastlines.** The photo shows cracked mineral crusts of salt, gypsum, and dolomite formed on tidal flats by evaporation of seawater at low tide along the Persian Gulf in the country of Qatar. The hot, dry climate favors evaporite mineral crystallization. There is very little clastic mud or sand on the tidal flat because there are no rivers to deliver sediment to the ocean.

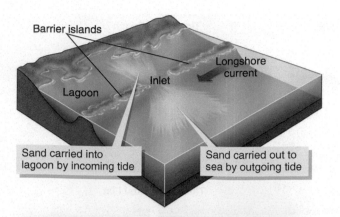

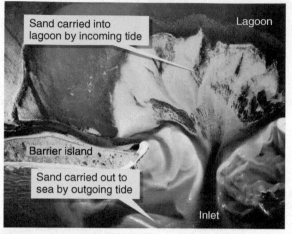

▲ **Figure 19.33 Recognizing tidal processes at tidal inlets.** Tidal currents enter and exit lagoons through the narrow tidal inlets between barrier islands. These focused currents transport sand to form submerged sand bars on either side of the inlet, as shown in the aerial photograph of an area in Long Island, New York.

Tides as an Energy Source

The persistent rise and fall of tides is used in some places to generate electricity. Tidal power plants are constructed as dams across estuaries or tidal inlets. The dam focuses the tidal currents to flow rapidly through a narrow artificial channel, where the water turns turbines to generate power in the same fashion that hydroelectric dams function along rivers. The rising tide generates power as it flows through the dam to enter the lagoon or estuary on the landward side. Rather than letting the water return through the dam at low tide, it is retained behind the dam until it can be released to generate electricity when power demand is high. The higher the tidal range, the more water can be held behind the dam to generate electricity by return flow through the turbines.

Putting It Together—What Is the Role of Tides in Forming Coastal Landscapes?

• Tides are periodic variations in sea-surface elevation caused by gravitational interactions of the Moon and Sun with Earth.

• Tidal influences on shorelines are most visible where wave energy is small.

• Tidal flats are gently sloping, muddy surfaces that are continually submerged at high tide and exposed at low tide. Some sediment carried landward by the rising tide is left behind when the tide goes out so that tidal flats gradually build upward and seaward over time. Evaporite minerals precipitate on tidal flats along arid coastlines.

• Tide currents funnel back and forth through tidal inlets between open ocean and lagoons on barrier-island coastlines. Tidal currents at inlets carry sediment both landward into the lagoon and seaward, where longshore currents may redistribute it onto barrier beaches. Tidal currents in inlets also power some electricity generation.

19.5 Why Does Shoreline Location Change Through Time?

Shorelines are dynamic places that change in appearance on relatively short time frames. Daily tides, seasonal variations in wave energy, powerful storms, and longshore transport of sediment produce these changes. Historical and geological observations also demonstrate slower, long-term changes that take place over centuries or even many millennia. Over geologic time scales these long-term changes are recorded by successive sedimentary layers that were deposited in progressively shallower or deeper water. (Check back to Section 5.6 and Figures 5.19 and 5.20 to review this sedimentary process.)

Processes That Change Shoreline Location

Figure 19.34 illustrates the processes the geologists observe to cause changes in shoreline location. If a point on land later submerges beneath the sea, or a point on the seafloor is later located on dry land, then it is possible to assume that sea level changed. It is important to keep in mind, however, that the measurement of changing sea level is made relative to a fixed location on land or at sea. This means that the submergence of land could be explained by global sea-level rise *or* subsidence of the land beneath the sea. Likewise, emergence of the seafloor happens either by global sea-level fall *or* seafloor uplift.

Relative sea-level change describes a shift in local shoreline position caused either by global sea-level fluctuation, local uplift and subsidence of crust, or a combination of these processes. Global **absolute sea-level change** mostly results from changes in the rates of creation of seafloor at mid-ocean ridges (discussed in Section 13.4), or from changes in the volume of glacial ice on continents (see Section 18.9). Tectonic and isostatic forces (described in Chapters 11–13) cause uplift and subsidence of the crust along shorelines that can account for relative sea-level change even if absolute sea-level is not changing.

Shoreline location can also change because of sediment erosion and deposition without a change in relative sea level (Figure 19.34). The largest seaward shoreline shifts occur where deltas build out into the ocean at the mouths of large, sediment-laden rivers (see Figure 16.16). If, on the other hand, the sediment losses outpace the gains in the sediment budget, and if coastal rock or regolith is easily eroded by waves, then wave erosion causes the shoreline to retreat landward.

Submergent and Emergent Shorelines

Figure 19.35 shows that the shape of shorelines reflects relative sea-level rise or fall. Relative sea-level rise submerges stream valleys below sea level to form a highly irregular shoreline of drowned-valley estuaries separated by headlands. Emergence, on the other hand, leaves behind wave-cut platforms and old beach deposits above sea level (also see Figure 13.13).

Global sea-level change, uplift or subsidence of crust, and deposition or erosion of sediment may operate simultaneously to affect the position of a local shoreline. The challenge is to separate out the role of

▼ **Figure 19.34 Why shoreline positions change through time.** Shorelines shift landward during relative sea-level rise, caused by global rise in sea level or local subsidence of the crust, or both. Shorelines shift seaward during relative sea-level fall, which results from global sea-level fall or uplift of the crust. Erosion and deposition also change shoreline position without any relative sea-level change, such as when deltas build land out into the sea.

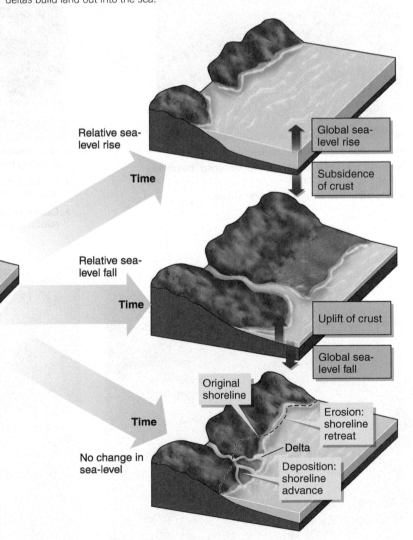

Change in shoreline position because...

Relative sea-level rise — Time → Global sea-level rise / Subsidence of crust

Relative sea-level fall — Time → Uplift of crust / Global sea-level fall

No change in sea-level — Time → Original shoreline / Erosion: shoreline retreat / Delta / Deposition: shoreline advance

Submergent shoreline

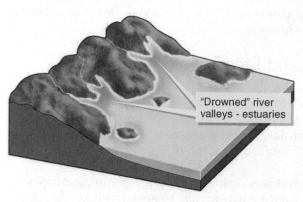

Emergent shoreline

Present shoreline features:

Former shoreline features:

Sea cliff

Beach

Sea cliff

Beach

Sea stack

Wave-cut platform

Sea stack

Wave-cut platform

Example: Maine coast

"Drowned" river valleys

Example: California coast

Uplifted wave-cut platform

- Shoreline results from relative sea-level rise.
- Lower parts of river valleys are "drowned" below sea level to form estuaries.
- Shoreline has a highly irregular shape.

- Shoreline results from relative sea-level fall; usually resulting from uplift of crust.
- Former shoreline and seafloor features are exposed above sea level.
- Old wave-cut platforms form flat benches along relatively straight shorelines.

◄ **Figure 19.35 What submergent and emergent shorelines look like.**

Shoreline before earthquake

Shoreline after earthquake

Dead coral and other marine organisms

◄ **Figure 19.36 Earthquakes change shorelines.** Shoreline modifications by tectonic uplift and subsidence are dramatically illustrated by changes resulting from a moment-magnitude 8.7 earthquake offshore of Sumatra, Indonesia in March 2005. The photo on the left shows an extensive area of seafloor that is now exposed as land surface after being uplifted approximately 2.5 m. About 150 km away, more than 1 m of subsidence during the earthquake partially submerged the village shown in the photo on the right.

each process. Present-day shoreline configurations are unquestionably affected by 20,000 years of absolute sea-level rise resulting from melting of the last ice-age glaciers (see Section 18.9). The evidence for absolute sea-level rise and its impacts are thoroughly explored in the next two sections. The remainder of this section demonstrates how geologists know that the other processes actively affect shorelines.

Evidence That Uplift and Subsidence Change Shorelines

Major earthquakes commonly result in uplift or subsidence along coastal areas, as shown in **Figure 19.36**. Over long time periods, tectonic forces have profound impacts on the location of shorelines.

Roman ruins near Naples, Italy, clearly show signs of changing relative sea level, as shown in **Figure 19.37**. Charles Lyell illustrated these ruins at the front of his 1830 text *Principles of Geology* because it is a dramatic example of dynamic Earth processes. The illustrated Roman building was built on dry land in the second century B.C.E. Holes bored into the marble columns by marine mollusks are present, however, at elevations as high as 6 meters above the present water line. This means that within a scant fraction of human history, the building was submerged below sea level and then raised back up to its present elevation. The Roman city was built within a restless volcanic caldera, so scientists can say in this case that moving magma in the crust was the cause of this down-and-up movement of the land surface.

Uplift and subsidence affect the Oregon shoreline that you visited in the field at the beginning of this chapter. Elevations of permanently marked locations along the shore were carefully surveyed at two times, 57 years apart. **Figure 19.38** shows a graph that depicts the survey results. Some locations rose in elevation during that time, while others sank. The data indicate that some areas of the coastline are experiencing tectonic uplift while other parts are subsiding. Subduction of the Juan de Fuca plate under the North American plate (see Figure 12.3 for locations of plate boundaries) accounts for this deformation.

Evidence That Sediment Deposition and Erosion Change Shorelines

Figure 19.39 illustrates an example of shoreline advance toward the sea where an estuary filled in with river sediment. This location, in northwestern Turkey, is significant because the geologic history is relevant to interpreting history presented in the classic epic *Iliad*, written by Greek poet Homer. *Iliad* chronicles the war fought for the conquest of Troy, about 3250 years ago. Homer described Troy as being close to the shoreline. Archaeologists thought they found the location of the ancient fortress city, but at a location nearly 7 kilometers from the coast. The long distance from the shore to the presumed location of Troy is inconsistent with Homer's text, leading some archaeologists to question whether the real location of Troy was known.

Geologists contributed to solving this problem by hypothesizing that the shoreline had shifted during the intervening 3250 years. Geologists extracted sediment samples from as deep as 50 meters below the river floodplains near where Troy was thought to be located and discovered marine sediment layers below the surface. The marine deposits are buried beneath sediment containing shells of estuarine animals and river deposits near the surface. Geologists combined the recorded depositional environments with radioactive-isotope dates on the shells (using the ^{14}C method mentioned in Section 7.7) to determine that Troy was once located along the shore of an estuary. The estuary gradually filled in with river sediment over a period of about 8000 years. Homer's description of Troy matches with the interpreted archaeological record of the city when considering the location of the shoreline 3250 years ago. The confusion about the location of Troy arose because sediment deposition drastically changed the geography of the coastline.

In contrast, **Figure 19.40** illustrates evidence of shoreline retreat because of sediment erosion. Coastal erosion is currently taking place along 80 percent of the Atlantic and Pacific shorelines in the conterminous United States. One significant cause of the erosion is a decrease is sediment supply to the coast resulting from construction of dams along rivers (Figure 19.26). The

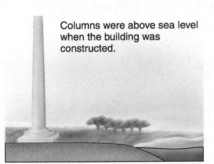
Columns were above sea level when the building was constructed.

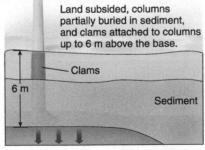
Land subsided, columns partially buried in sediment, and clams attached to columns up to 6 m above the base.

Clams

6 m

Sediment

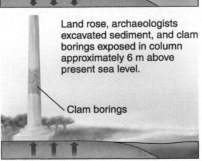
Land rose, archaeologists excavated sediment, and clam borings exposed in column approximately 6 m above present sea level.

Clam borings

Today

As illustrated by Lyell in 1830

Clam borings in marble columns

▲ **Figure 19.37 Evidence for historic, relative sea-level change in Italy.** Clam borings in the marble columns of a Roman ruin reveal relative sea level changes during the last 2200 years near Naples, Italy. This example of submergence and emergence of a shoreline resulting from deformation was first illustrated in Charles Lyell's geology textbook in 1830.

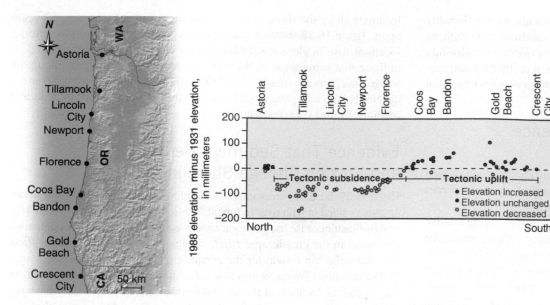

◀ **Figure 19.38 Measured uplift and subsidence along the Oregon Coast.** Surveys of permanently marked locations along the Oregon Coast in 1931 and 1988 show that some parts of the coast experienced uplift (relative sea level fall) and other parts subsided (relative sea level rise) between the surveys.

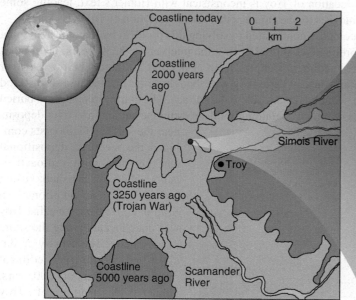

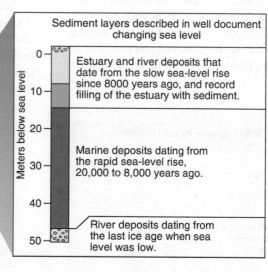

◀ **Figure 19.39 Historic shoreline change caused by sediment deposition.** Ancient Troy was on the shoreline of an estuary 5000 years ago but is now almost 7 kilometers from the ocean. Study of sediment layers show that an estuary filled with river sediment so that the coast gradually shifted away from Troy.

Sediment layers described in well document changing sea level

Estuary and river deposits that date from the slow sea-level rise since 8000 years ago, and record filling of the estuary with sediment.

Marine deposits dating from the rapid sea-level rise, 20,000 to 8,000 years ago.

River deposits dating from the last ice age when sea level was low.

◀ **Figure 19.40 Historic shoreline change caused by sediment erosion.** Erosion is reshaping the South Carolina shoreline, as dramatically illustrated by the Morris Island lighthouse. The lighthouse was on the beach in the 1940s but is now more than 400 meters from shore.

sand deposited behind the dams never reaches the ocean, which removes sediment from the beach budget and leads to coastal erosion.

Putting It Together—Why Does Shoreline Location Change Through Time?

• Shorelines shift with time because of global sea-level change, uplift or subsidence of the coast, and deposition or erosion of sediment by waves, currents, and tides.

• Emergent shorelines form by relative sea-level fall, which results from uplift, global sea-level fall, or both. Emergent shorelines typically have steep cliffs, and prominent flat benches that formed below sea level as wave-cut platforms.

• Submergent shorelines form by relative sea-level rise, which results from subsidence, global sea-level rise, or both. Submergent shorelines typically have low relief and have a highly irregular outline dominated by river valleys that drowned to form estuaries.

19.6 How Do We Know . . . That Global Sea Level Is Rising?

Statements about rising sea level are commonly seen in the news. How do geologists know that sea level is rising?

We can start with data from glacial geologic studies. Observations on land reveal the area and likely thickness of glacial ice on continents during the last ice age, which peaked about 21,000 years ago (see Figure 18.42). Using area and thickness, you can calculate the volume of the glacial ice and the volume of water contained in the ice rather than remaining in the ocean. This calculation shows that that sea level 21,000 years ago was approximately 100–120 meters lower than it is today. A reasonable hypothesis, therefore, is that sea level has been rising from this time of lower sea level. In testing this hypothesis it is important to keep in mind that while sea level rose after the last ice age, this does not mean that sea level is still rising now, nor does it tell us whether sea level rose at a uniform rate since the last ice age.

Picture the Problem

Why Is It Important to Document Changing Sea Level? Approximately 100 million people live around the world at locations less than 1 meter above sea level, so it is essential to know whether sea level is rising now and, if so, how fast it is rising. This information is necessary to determine the magnitude of the change and how to respond to it. Submergence of large coastal cities (such as New York City and New Orleans, Louisiana) would have potentially disastrous economic and social impacts.

Consider two approaches to test the hypothesis that sea level is rising and to determine the rate of sea-level change. One approach uses historic measurements of sea level. The second approach uses geologic data to see whether the historic measurements are consistent with sea-level changes measured over the longer time interval since the last ice age.

Examine the Evidence for Historic Sea-Level Change

What Changes in Sea Level Do Tide Gages Reveal? Tide gages have measured sea-level changes at major ports for more than a century. **Figure 19.41** is a graph of a tide-gage data collected at New York City. Averaging together the high- and low-tide elevations for each day and then averaging all of these daily values determines a single sea-level value for each year. Notice the large sea-level variation from one year to the next. Most of the variability relates to changes in weather conditions because strong winds and variations in atmospheric pressure influence the water-surface elevation. Despite these year-to-year fluctuations, there is a general trend of rising sea level since 1900, at a long-term average rate of 3 millimeters per year.

The problem with the tide-gage data is that tide gages measure *relative* sea-level rise. The data plotted in Figure 19.41 do not exclude the possibility that sea level is stationary or actually falling while the land around the New York City tide gage is sinking. Indeed, tide-gage data from around the world show tremendous variability in the rates of relative sea-level change, and many gages reveal relative sea-level fall rather than rise. To eliminate the effects of tectonic uplift and subsidence, geologists avoid using data from ports in tectonically active areas near plate boundaries.

There is still another problem, however, that relates to shifting mass on Earth's surface. During the ice age, the weight of ice pushed the continental crust down, while adjacent areas bulged slightly upward, much like the displacement observed when sitting on a water bed (Section 13.3 provides more detail on this process). Oceanic areas also changed elevation, because less water mass pushed down on the crust when sea level was lower during the ice age. Earth surface elevations are still adjusting to the removal of ice mass from some continents and the addition of water mass to the oceans. These active adjustments cause slow uplift and subsidence that affect measurements of relative sea level.

Tide gages cannot be used to separately measure change in land-surface elevation and change in water-surface elevation, so they

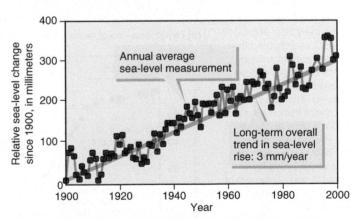

▲ **Figure 19.41 Tide gage data reveals sea-level rise in New York City.** Tide-gage data show that relative sea level rose about 300 millimeters (about the height of this page) during the twentieth century. Short-term variations from year to year mostly relate to weather variations. High atmospheric pressure depresses the water surface, and just a 3 percent decrease in air pressure causes the water surface to rise 300 millimeters. This means that the average, annual sea level is higher in years with weather conditions having generally lower air pressure and lower when there are more days with high air pressure. Long records are required, therefore, to average out these weather-related changes in sea level and isolate the actual relative sea-level change.

do not measure the ongoing change in global sea level. Geologists adjust for this shortcoming by calculating the uplift and subsidence effects resulting from moving the mass of glacial ice off of the continents and adding water mass to the global ocean. In combination with the tide-gage data, these calculations reveal that global sea level rose at an average rate of about 1.5 to 2 millimeters per year during the twentieth century.

The bottom line from this analysis is that global sea level is rising, but the rate of rise is very slow. There is also the uncertainty that results from not being able to directly measure the value of global sea-level change. The difference between estimates of 1.5 millimeter per year and 2 millimeters per year may seem insignificant, but the faster estimate is 33 percent larger than the slower one, so this uncertainty is not trivial.

Examine the Evidence for Prehistoric Sea-Level Change

What Changes in Sea Level Do Drowned Coral Reefs Reveal? Geologists use corals as "dipsticks" of former sea levels. *Acropora palmata* is a Caribbean Sea coral that commonly grows right up to the low-tide line and never in water that is deeper than 5 meters. However, dead specimens of this coral species form reefs that are submerged more than 120 meters *below* current sea level. A reasonable hypothesis,

illustrated in **Figure 19.42**, is that the deep, dead corals mark the locations of former reefs that grew near sea level and then drowned when sea level rose after the ice age. If this hypothesis is true, then the ages of dead corals should be progressively more recent at shallower depth, and the elevations of the corals would record elevations of former ocean surfaces as sea level rose. The rate of sea-level rise can be calculated as the slope of a line on a graph that plots the depth of corals and their ages. The necessary graph of data is provided in Figure 19.42.

The ages determined by using the ^{14}C dating method on samples of submerged coral extend back to the last ice age. The ages are older at greater depth, as predicted by the hypothesis. There is, however, still a problem of determining how much of the relative sea-level rise is global sea-level rise rather than local tectonic subsidence, because the samples come from islands close to the convergent boundary at the eastern edge of the Caribbean plate (see Figure 12.3 for the location of the plate boundaries). In this area, however, the presence of reefs more than 100,000 years old on dry land reveals the islands are uplifted by tectonic processes, which would cause relative sea-level fall rather than rise. Thus, the submerged dead reefs record absolute sea-level rise, and the amount of rise is adjusted in Figure 19.42 for the estimated amount of uplift determined from the still older reefs exposed on land.

▶ **Figure 19.42 Ancient corals reveal sea-level rise in the Caribbean.** (a) Researchers hypothesized that specimens of the coral *Acropora palmata* would reveal long-term sea-level change, because although living corals only live within 5 meters of the water surface, dead corals are seen at greater depths. The dead corals lived when sea level was lower and died when rising sea level submerged them deeper than 5 meters. This means that dead corals should be progressively older at greater depth. (b) Geologists used the ^{14}C radioactive-isotope dating to determine the ages of dead corals sampled from different water depths. The data confirm the hypothesis that corals are progressively older at deeper depths. The data also reveal (1) that 20,000 years ago sea level was about 120 m lower than present, and (2) the rate of sea-level rise decreased about 8000 years ago.

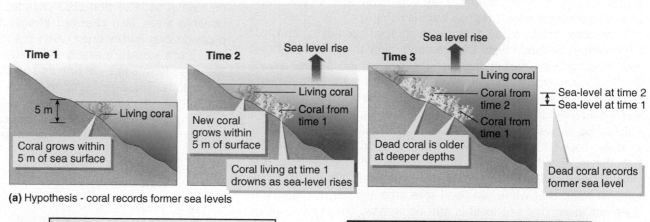

(a) Hypothesis - coral records former sea levels

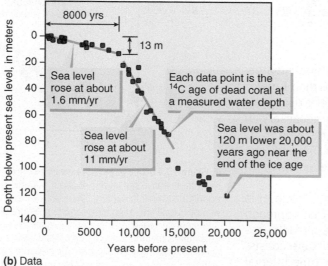

(b) Data

Acropora palmata

The Caribbean coral data also reveal changes in the rate of sea-level rise over time (Figure 19.42). Over the last 8000 years, sea level rose 13 meters, which equates to an average rise of 1.6 millimeters per year. Prior to 8000 years ago, however, the rate of sea-level rise was much faster, averaging 11 millimeters per year.

Integrate the Data Sets

Do the Results Support One Another? Consistency of results obtained in different ways is an important part of scientific data analysis. Scientists commonly approach the same problem with different data, collected by different methods, and used in different analyses with different assumptions and uncertainties. If the different data and approaches yield consistent results, then the methods and assumptions employed in the investigations are strongly supported.

In the case of reconstructing sea-level change, there are three pertinent data sets to compare. Figures 19.41 and 19.42 represent two of these data sets. The third relevant data set is the extent of ice-age glaciers, which is used to calculate the total sea-level change since 21,000 years ago (see Section 18.9). The Caribbean coral data indicate 120 meters of sea-level rise since the last ice age, which is consistent with the estimates of 100–120 meters determined from calculations of ice-age glacier volume. The coral data also suggest that sea level rose at about 1.6 millimeters per year during the last 8000 years. This rate of rise agrees with the modern estimated rate of between 1.5 and 2 millimeters per year as calculated from tide-gage measurements and modified to account for slow adjustments of the crust to shifting masses of glacial ice and ocean water. The three data sets are, therefore, consistent with one another.

Insights

What Are the Causes and Future of Sea-Level Rise? The data collection and analyses that establish sea-level rise as fact also lead to an evaluation of *why* sea level is rising and *how much* it might rise in the near future. It would seem simple enough to suggest that rising sea level results from melting glacial ice. There are other processes, however, that could cause global sea-level rise.

One cause of rising sea level is the expansion of ocean water as a result of global warming over the last century. Water expands when it warms up, so sea level rises as ocean temperature increases. A 1-degree-Celsius increase in global water temperature produces a roughly 2-centimeter increase in sea level. Averaged over the twentieth century, this process caused sea level to rise about 0.5 millimeter per year.

Is it possible that human interference with the hydrologic cycle also causes rising sea level? Ground water extracted for irrigation or municipal water supplies only partly returns to aquifers. Streams probably transport most of the extracted water to the ocean. Other increases in streamflow to the oceans result from land-use changes, such as urbanization and deforestation, which decrease water infiltration and increase runoff. Other activities decrease stream discharge, however, such as evaporation of water from reservoirs behind dams and irrigated agricultural land. Some water impounded in reservoirs also infiltrates to recharge ground water rather than flowing to the ocean, and the water that is stored in reservoirs is withheld from the ocean. The magnitudes of these human impacts to increase, or decrease, the amount of water reaching the oceans are very difficult to measure or estimate. Current best estimates suggest that it is more likely that human activities decrease, rather than increase, sea level.

Taking all of these observations together, it seems necessary to call upon melting glaciers in order to explain rising sea level. However, perhaps as much as one third of the twentieth century sea level rise was the result of water expansion in warming oceans.

How much and for how long will sea level continue to rise? After all, sea level fluctuated throughout Earth history with falling levels as well as rising ones. A clue comes from the presence of 125,000-year-old shoreline deposits exposed on land in many tectonically stable parts of the world. These ancient shorelines indicate a sea level for that time—when glacial records reveal ice sheet volume on continents that is not very different than at present—to be approximately 5–6 meters higher than today. One hypothesis, therefore, is that 5–6 meters of additional sea-level rise will occur before the next ice age begins. Unfortunately, this is a difficult hypothesis to test other than by waiting to see what actually happens.

Sea level rose at a rate of 1.5–2 millimeters per year for the last 8000 years, so a reasonable hypothesis is that it will likely continue to rise at 1.5–2 millimeters per year into the near future. Global sea level will likely rise 10 to 20 centimeters during the next century, and areas experiencing tectonic or isostatic subsidence can expect even more relative sea-level rise. If climate warming increases melting of ice sheets in Greenland and Antarctica, then the amount of sea-level rise will be greater, perhaps as much as 50 centimeters by 2100. The long dimension of this page is close to 30 centimeters, so this amount of sea-level rise may seem insignificant, but it will have profound effects in many coastal areas where surface slopes are very low.

Putting It Together—How Do We Know . . . That Global Sea Level Is Rising?

• Tide-gage records document historic relative sea-level rise and drowned corals document long-term relative sea-level rise since the last ice age.

• Separating global sea-level rise from relative sea-level rise affected by vertical movements of the lithosphere is difficult.

• Global sea level rose at an average rate of 1.5 to 2 millimeters per year through the twentieth century. Expansion of seawater because of global warming explains as much as one third of the rise. Melting glacial ice is most likely responsible for the additional sea-level rise.

19.7 What Are the Consequences of Rising Sea Level?

The shape of modern shorelines reflects 21,000 years of rising sea level, and the shorelines will continue to change as sea level rises further. Some of the expected results are as follows:

• Coastal areas will gradually submerge. Shoreline retreat by as much as 100 meters, roughly the length of a football field, can take place with as little as 10 centimeters of sea-level rise where the land surface is nearly flat.

- Islands will gradually shrink and perhaps disappear, with decreasing area to support populations and agriculture. There are more than 1000 inhabited islands on Earth with maximum elevations less than one meter above sea level. **Figure 19.43**, for example, shows how Key West will gradually submerge as sea level rises over the next few centuries. The Florida Keys are low-elevation islands composed mostly of coral reefs that were completely submerged 125,000 years ago.

- Coastlines will erode as wave energy focuses farther inland. This effect is most obvious during strong storms along shorelines composed of

easily eroded material. **Figure 19.44** shows evidence of coastal erosion during recent hurricanes, which was enhanced by rising sea level. Coastal erosion has important implications for the preservation of beaches, which are the biggest tourist attractions worldwide. Unless rivers supply sediment to shorelines in excess of the amount that is eroded, the shorelines will retreat landward.

- High tides will inundate increasingly larger land areas, flooding coastal wetlands along lagoons and estuaries and killing plants that are not salt-water tolerant. Tides also extend farther up river valleys, which slows river flow to the ocean and causes flooding along the stream banks.

- Coastal water tables rise as sea level rises and causes landward incursions of salty ground water that is undrinkable and unsuitable for irrigation.

The effects of sea-level rise on coastal landscape evolution and its effects on human structures and activities are particularly well documented for beaches and barrier islands, estuaries, deltas, and sea cliffs.

▼ **Figure 19.43 Visualizing the submergence of islands.** These maps show how islands in the Florida Keys will submerge if sea-level rise continues. At current local rates of relative sea-level rise, a 1-meter rise will occur over the next 250 years and 2 meters of submergence will occur by 500 years from now. Large areas of these and other small islands around the world are at elevations less than 1 meter above sea level.

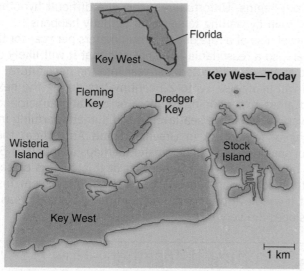

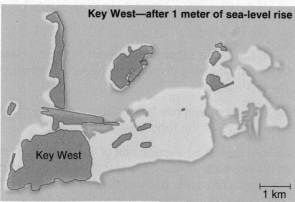

▼ **Figure 19.44 Coastal erosion during hurricanes.** These photos show progressive erosion of Dauphin Island, Alabama. The middle photo was taken right after the passage of Hurricane Ivan and the bottom photo was taken two days after Hurricane Katrina. The white arrow points to the same house in each photo, and highlights the erosion of the seaward (toward the bottom) side of the barrier island. The white areas are sand that washed over the island during each hurricane, notably filling a boat channel in the bottom photograph.

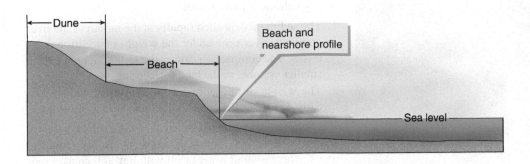

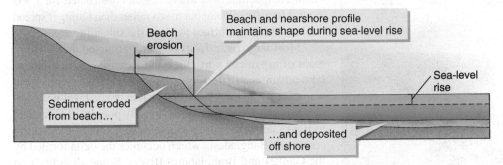

▲ Figure 19.45 Why sea-level rise causes beach erosion. Even small amounts of sea-level rise cause large amounts of beach erosion and shoreline retreat. The profile shape of the beach and shallow seafloor remains the same during sea-level rise, which causes large amounts of beach erosion.

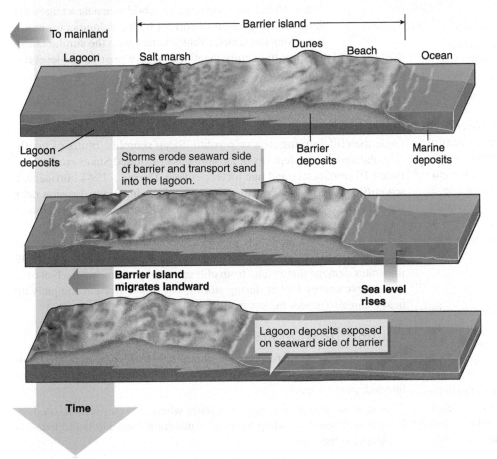

The Effects of Rising Sea Level on Beaches and Barrier Islands

Sea-level rise causes erosion of beaches and landward retreat of the shoreline, as shown in **Figure 19.45**. Wave energy focuses farther inland on the beach, and more of the beach submerges at high tide. Wave erosion sculpts a smooth, concave-up profile along the beach and into the shallow nearshore environment. This profile shifts upward and landward as sea level rises, so that each *centimeter* of sea-level rise commonly causes more than 1.5 *meters* of shoreline retreat.

Sea-level rise causes barrier islands to migrate landward. **Figure 19.46** shows how storms erode the seaward-facing beach and wash the sediment across the island and into the landward lagoon. This pattern of sediment erosion and deposition is also visible in Figure 19.44. The barrier islands on the United States east coast have migrated for thousands of years. Geologists reached this conclusion because the sandy beach and dune deposits forming the islands rest on top of older muddy lagoon deposits that were deposited landward of the islands when the islands used to be farther offshore. These older lagoon deposits are detected in wells drilled on the barrier island and are locally exposed on the seaward beaches when they erode during storms (Figure 19.46).

The barrier coastlines of New Jersey, Delaware, and Maryland are heavily populated, and homes are constructed on or just landward of the beach. Relative sea-level rise probably will cause about 50 meters of landward retreat of the shoreline in this region by 2050. This amount of erosion has the potential to cause considerable damage because the beaches are generally only approximately 25–30 meters wide.

The Effects of Rising Sea Level on Estuaries

Estuaries form when relative sea-level rise submerges low-gradient river valleys along coastlines. Modern estuaries, such as Chesapeake Bay, shown in **Figure 19.47**, formed during the most recent rise in sea level that began about 20,000 years ago. For estuaries to persist, the rate of sea-level rise to submerge the valley must exceed the rate of sediment deposition by the rivers, which tends to fill in the estuary. The estuary at Troy, for example (see Figure 19.39), formed when the rate of sea-level rise was rapid, prior to 8000 years ago (see Figure 19.42) and then filled in with sediment when the rate of rise decreased.

◄ Figure 19.46 Barrier islands migrate landward when sea level rises. Waves erode barrier-island beaches, and storms wash over the islands and carry sand into lagoons. These processes cause the islands to slowly back-peddle landward over muddy lagoon deposits. Oyster shells collected on Atlantic beaches are eroded from old lagoon deposits that are now exposed on the seafloor near the beach.

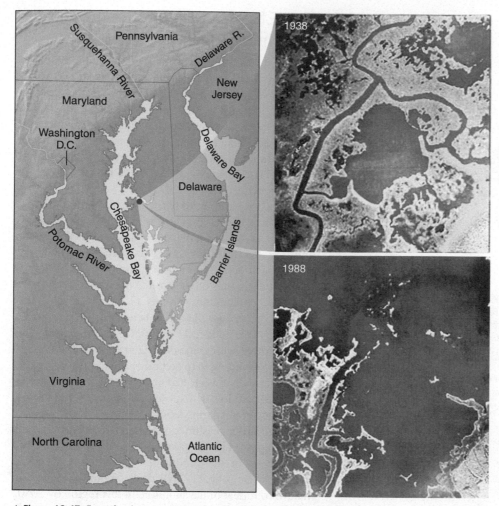

▲ **Figure 19.47 Estuaries form by submergence of river valleys.** Large bays on the Atlantic coast are estuaries formed by drowning of river valleys beneath the rising sea. Submergence of coastal wetlands is easily recorded over short historic time scales, as seen in these two aerial photographs taken 50 years apart.

Chesapeake Bay is an example of an estuary that continues to enlarge by submergence as sea level rises (see Figure 19.47). The progressive submergence of the tidal flats and coastal marshes has ecological consequences in addition to threatening bayside homes and roads. In a completely natural situation, the coastal marshes would simply shift landward as sea level rose. Along heavily populated coastlines, however, there usually is only a narrow band of wetland between the high-tide mark and roads and buildings. Human-built structures restrict landward shift of the marshes, so once the wetland submerges, it is gone.

The Effects of Rising Sea Level on Deltas

Deltas are headlands that build into the sea because sediment delivery by rivers outpaces shoreline retreat by sea-level rise. It might seem likely, therefore, that deltas are immune from the effects of sea-level rise, but this is not the case. Deltas cover tens of thousands of square kilometers, but sediment deposition occurs only on small parts of the delta at any one time. This means that while part of the delta builds seaward, the rest of the delta is a low-elevation plain that is susceptible to inundation by rising sea level.

Relative sea-level rise is commonly greater on delta coastlines than along the same shore that is more distant from the river mouth. Not only

is global sea level rising, but the land is also sinking. Deltas slowly subside for two reasons:

1. The sediment deposited rapidly at the mouth of the river compacts over time under the weight of additional deposits. The compaction causes the sediment to occupy a smaller volume, so the land surface subsides.
2. The weight of the sediment forming the delta isostatically depresses the crust, so the land slowly subsides.

Subsidence and submergence of delta coastlines are even greater where dams diminish sediment supply from the rivers, where ground water or oil withdrawal increases sediment compaction, and where levees constructed for flood control funnel sediment out to sea rather than letting it spread out over the delta surface to fill in the subsiding areas.

Most large river deltas on Earth attracted the development of urbanized port cities and extensive agriculture to take advantage of fertile soil and readily available fresh water for irrigation. A one-meter rise in sea level will submerge about 15 percent of the densely populated Nile delta in Egypt and will submerge about 10 percent of the entire country of Bangladesh, which occupies the delta formed by the Ganges and Brahmaputra Rivers. Some shorelines on the Mississippi delta in Louisiana retreat as fast as 20 meters per year, with annual submergence of an area equal in size to Washington, D.C. The city of New Orleans has subsided below sea level and is surrounded by walls to prevent flooding. **Figure 19.48** illustrates the flooding of historic Venice, Italy, at high tide, which results from the combination of global sea-level rise and subsidence near a large river delta. A comparison of current high-tide marks on buildings along the famous Venice canals with the similar marks visible in early eighteenth-century paintings shows a relative sea-level rise of about 70 centimeters between 1727 and 2002.

The Effects of Rising Sea Level on Sea Cliffs

The response of sea cliffs to sea-level rise depends on how easily waves erode the cliff-forming rock or regolith. Steep shoreline bluffs of poorly consolidated glacial deposits in the northeastern United States erode at between 10 centimeters and one meter per year. During a 1944 hurricane, a sea cliff on Long Island retreated 12 meters in a single day. Erosion rates for sea cliffs of hard granite, on the other hand, are imperceptibly slow at 1 millimeter per year, or less.

Most of the scenic, highly developed shoreline of California features homes built at the edge of sea cliffs eroded in sedimentary rocks. **Figure 19.49** illustrates damage that results from cliff erosion by storm waves. Not only is the wave energy higher during storms, but runoff from accompanying heavy rain also erodes the steep cliff faces.

Human Responses to Shoreline Erosion

There are three general responses to shoreline erosion caused by the combination of global sea-level rise, land subsidence, and changes in shoreline sediment budgets:

1. Armor the shoreline to hold it in place where it is currently located
2. Add sediment to eroding beaches to maintain their width and location
3. Abandon the coast

▲ **Figure 19.48 How relative rising sea level affects Venice.** Pedestrians wade in the ocean or stroll on elevated boardwalks when spring high tide floods historic Venice, Italy. Tidal flooding of Venice occurs because of the combination of global sea-level rise and subsidence of land beneath the city.

▲ **Figure 19.49 Sea-cliff erosion destroys homes.** Homes collapse onto the beach along this California shoreline because waves easily erode the soft sedimentary rock forming the sea cliff.

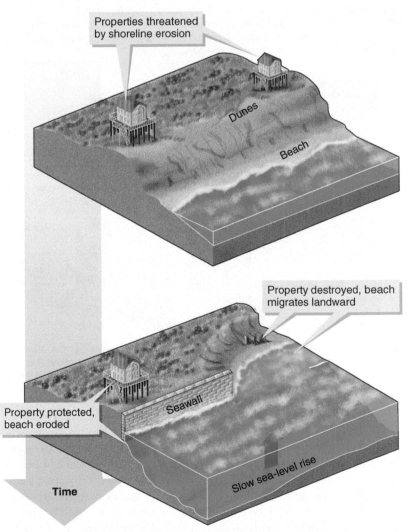

▲ **Figure 19.50 How seawalls protect property but not beaches.** Wave-resistant walls stop erosional retreat of shorelines and protect coastal buildings. However, as sea level rises, the beach erodes away on the seaward side of the wall, which diminishes recreational value. The beach remains if it is allowed to migrate landward with rising sea level, but then coastal erosion destroys unprotected buildings.

Armoring usually means building **seawalls** on the beach parallel to the shoreline as shown in **Figure 19.50**. Seawalls vary in height and are constructed from wood, plastic, concrete, rock, steel, junk cars, rubber tires, or sandbags. Resistance to storm-wave erosion is greatest for high walls constructed of strong materials. Seawalls do protect coastal property but do not necessarily save the beach, as seen in Figure 19.50. Wave erosion eventually carves away the beach until the waves crash against the seawall. Potential beach-sediment sources in coastal dunes or readily eroded sea cliffs are isolated behind the seawall, which diminishes the sediment budget for the beach.

Replenishing eroded beach sand is an alternative to building seawalls. Coastal communities heavily utilize this nourishment option to restore economically important recreational beaches, as shown in **Figure 19.51**. The procedure also maintains beaches in front of coastal properties so that wave energy is spent on the beach and does not threaten buildings. In some places, sand that was eroded and transported offshore is dredged onto barges that transport the sediment back to the beach, or sand and water are pumped from the seafloor and spread onto the beach. In other cases, beaches are replenished with sediment excavated from harbors that are filling with river deposits. Beach replenishment is very expensive and always temporary, because the conditions eroding the beach are not changed. Some resort beaches in New Jersey have been replenished more than 40 times since 1950.

▲ **Figure 19.51 Replenishing Miami Beach.** The photo on the left shows groins and seawalls protecting resort hotels on nearly beachless Miami Beach in the 1970s. The photo on the right shows a wide beach at the same location following beach replenishment efforts in the early 1980s. Millions of dollars were spent to pump wave-eroded sand onto the beach from the nearby seafloor.

New lighthouse location

Wave erosion threatens lighthouse despite protective groin

Lighthouse moving on rails

▲ **Figure 19.52 Retreating from coastal erosion.** The Cape Hatteras Lighthouse was moved inland in 1999 because it was at risk of being destroyed by coastal erosion. The lighthouse was more than 500 m from the shore when it was constructed in 1870.

Retreating in the face of rising sea level and coastal erosion is another option. Many beachfront homes that are threatened by erosion are simply jacked onto trucks and moved to more inland properties. In 1999, the historic Cape Hatteras, North Carolina, lighthouse was moved 884 meters inland, as shown in **Figure 19.52**, to avoid certain destruction by coastal erosion. The lighthouse was built more than 500 meters from the beach in 1870, but storm waves lapped at its base by the early 1980s. The cost of this relocation was 12 million dollars. Wholesale movement of coastal communities and large resort hotels is clearly not economically feasible, so seawalls and beach-replenishment projects continue as costly, but still less expensive, responses to the shifting shore. If sea levels continue to rise, however, then natural coastal processes will eventually destroy coastal developments with extraordinary economic costs.

Putting It Together—What Are the Consequences of Rising Sea Level?

• Sea-level rise threatens large coastal and island populations with submergence and the hazards of coastal erosion.

• Responses to coastal erosion on United States shorelines focus on armoring the shore against erosion or replenishing beach sand eroded by waves. These solutions are expensive and only temporary.

EXTENSION MODULE 19.1

Changing Shorelines in the Great Lakes. *Learn the causes for fluctuating water levels in the Great Lakes, and compare coastal erosion problems on Great Lakes shorelines with those that happen on ocean shores.*

Where Are You and Where Are You Going?

Wind-driven waves, tides, and associated currents are agents of change at shorelines. These processes create wide sandy beaches, steep rocky headlands, broad marshy tidal flats, and long chains of barrier islands. The diverse appearances of shoreline landscapes relate not only to waves, tides, and currents, but also to the erodibility of rock and regolith at the coast, uplift and subsidence of the crust, variations in sediment supplied by rivers, and absolute sea-level rise and fall.

Shoreline positions shift over short, human time scales and long, geologic time scales. Emergence of once-submerged seafloor occurs when absolute sea level falls or land rises because of tectonic and isostatic processes. Submergence of land beneath the sea occurs when absolute sea level rises or the land subsides. Shorelines also extend seaward, where rivers deliver large volumes of sediment that fill in bays and nearshore shallow areas.

Sea level rose at a rate of 1.5–2.0 millimeters per year during the twentieth century. Both melting of glacier ice and expansion of the warming ocean contribute to sea-level rise. However, some coastlines are rising faster than sea level because of tectonic uplift or isostatic uplift where glacial ice previously depressed the crust. Most heavily populated coastal areas and islands are, nonetheless, at risk of submergence and increasing threats of coastal erosion hazards as absolute sea level rises.

You have now completed your study of geologic processes and landscape modification related to water—flowing in streams, flowing underground, flowing as ice, driven by wind and tides. The next chapter examines wind as a geologic process. Wind produces waves, which play the major role in shoreline geology considered in this chapter. Wind is also important on land as an agent of sediment erosion that lowers landscape elevations and as an agent of deposition to form sand dunes that cover hundreds of thousands of square kilometers in deserts. The presence of sand dunes on barrier islands indicates, however, that wind is not just a geologic agent in deserts. In fact, regardless of where you live, you only have to look at the dust that settles in your home or classroom to contemplate the importance of wind.

Active Art

Properties of Waves. See how to recognize the parts of a wave and to measure the wave period.

Water Wave Motion and Refraction. See how waves move in water and refract along a shoreline.

Tsunami. See how an earthquake forms a tsunami and how a tsunami travels across the ocean.

Beach Drift and Longshore Current. See how beach drift and longshore current move sediment along the shoreline.

Effects of Groins and Jetties. See how groins and jetties influence shoreline deposition and erosion.

How Tides Work. See how the positions of Earth, Moon, and Sun determine the rise and fall of the tides.

Extension Module

Extension Module 19.1: Changing Shorelines in the Great Lakes. Learn the causes for fluctuating water levels in the Great Lakes, and compare coastal erosion problems on Great Lakes shorelines with those that happen on ocean shores.

Confirm Your Knowledge

1. Irregularities in coastal outlines are due, in part, to differences in the way shoreline materials erode. What causes these differences?
2. Where does the sand at a beach come from?
3. What factors can cause shoreline shape and position to change over time?
4. How do waves form? How do they travel? Why do they break?
5. Why do waves not form on swimming pools or small ponds?
6. Distinguish wave height from wavelength.
7. How do longshore currents cause sediment deposition or erosion along the shoreline?
8. Describe how construction of groins and jetties affect adjacent beaches.
9. What human activities happening tens or hundreds of kilometers from a coast can still end up affecting the shoreline sediment budget?
10. How and why do barrier islands and tidal inlets move over time?
11. Which has a greater influence on Earth's ocean tides, the Sun or the Moon? Why?
12. What is the difference between a spring tide and a neap tide? Do spring tides only happen in the spring?
13. What features would you look for as evidence that a shoreline is submergent or emergent?
14. Why is it so difficult to determine how fast sea level is rising?
15. What evidence would you present to convince a skeptic that sea level is rising?
16. What evidence exists to support the hypothesis that global sea level will rise 5 to 6 meters higher than it is today?
17. List the expected results if sea level rises by 1 meter.

Confirm Your Understanding

1. Write an answer for each question in the section headings.
2. Many shoreline shapes represent a dynamic balance between sediment delivered to the shore by streams and erosion from wave action. Explain how a beach or spit might change over time because of variations in these two processes.
3. What do ocean waves and seismic waves have in common. How do they differ?
4. Explain the transfer of energy in going from wind on the open ocean to the roar of the surf.
5. Explain why a floating object moves slowly in the direction of wave movement.
6. If you lived along the shoreline, on which side of a groin, relative to the longshore current, would you prefer to locate your home? Why?
7. Barrier islands commonly contain valuable real estate that is heavily developed. What is a long-term natural problem inherent to barrier islands?
8. What would the tides on Earth be like if Earth had two moons identical in size and orbit except that they were 180 degrees apart? What if they were 90 degrees apart?
9. Are all changes in relative sea level due to changes in the volume of water in the ocean? Why or why not?
10. Some geologists hypothesize that the tug and pull of lunar tidal forces are strong enough to trigger volcanic eruptions and earthquakes. Explain how you think this might happen. How would you test your hypotheses and what data would you need?
11. Look at the photos in Figure 19.44. What actions could have been taken to diminish the property loss that is evident in these views?

CHAPTER 20

Wind: A Global Geologic Process

Why Study Wind?

All processes that erode and transport Earth materials are geologically important and shape the landscape. Movement in the atmosphere—wind—is such a process. If you have stood on a sandy beach, or in a desert, or even near piles of dirt at a construction site on a windy day, then you have felt windblown sediment stinging your skin. Sand dunes are a familiar landform caused by wind transport of sediment across the ground. Just as your skin feels as though it is being sandblasted on a windy day in a sandy landscape, wind-blown particles abrade rock surfaces. Wind is a truly global process, visible not only in the desert sand dunes, but also in features in other environments that you may not have thought wind would affect. Beyond Earth, for example, vast dust storms periodically obscure the surface of Mars.

As a geologic process, wind also affects humans. Wind erosion removes topsoil, which diminishes agricultural productivity. Thousands of square kilometers of once vegetated land are converted into barren desert each year. Wind-blown sediment reduces visibility in populated areas and causes highway accidents. Dust consisting of tiny mineral particles, pollen, spores, and windborne microbes, remains aloft in the swirling atmosphere for months, causing hazy skies and respiratory ailments. More than 2 billion metric tons of fine mineral particles loft into the atmosphere from Earth's surface each year.

After Completing This Chapter, You Will Be Able to

- Explain why and how the atmosphere moves to cause wind.
- Describe how wind erodes, transports, and deposits sediment.
- Apply an understanding of wind processes to the formation of landscapes.
- Relate the geology of windblown dust to other elements of the Earth system.

Pathway to Learning

Windblown sand dunes invade farmers' fields in an Egyptian desert oasis.

IN THE FIELD

onsider two imaginary vacations to places where wind processes clearly are at work. Your first stop is a desert in the southwestern United States. Your second trip is to the Atlantic Coast.

From a mountain vantage point high above Death Valley National Park, you look over the stark landscape shown in **Figure 20.1**a. There is virtually no vegetation to obscure your view of steep, rocky cliffs; widespread fans of stream-transported sand and gravel at the mountain bases; and glaring white salt flats where rare accumulations of surface water dried up in the past.

Similar scenes appear throughout many areas of the western United States, and these areas commonly are described as "deserts." You conclude that the near absence of vegetation must relate to the lack of moisture to nourish plant growth. Indeed, Death Valley receives less than 5 centimeters of rain per year, which is pretty sparse compared to a greener place such as Chicago, Illinois, where 86 centimeters of precipitation falls each year.

Driving down from the mountains to the valley floor, you encounter an area of sand dunes (Figure 20.1b). The ridges of loose sand, resembling ocean waves frozen in place, cover more than 50 square kilometers of Death Valley, and the highest dunes are 213 meters tall (or taller than a 50-story building). Smaller undulating ripples stripe the soft and sandy dune surfaces. So much sand moves across the surface with each wind gust that for a moment the ground almost seems to flow like liquid. During more sustained gusts you notice that the ripples shift position in the direction of the wind.

Dunes form from sand blowing in the wind, but your curiosity and observations tell you there is more to the desert.

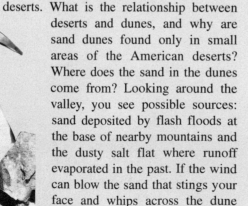

Sand dunes are part of many people's images of deserts, but in reality they constitute a very small part of this one in Death Valley. In fact, sand dunes cover the surface of only about 2 percent of North American deserts. What is the relationship between deserts and dunes, and why are sand dunes found only in small areas of the American deserts? Where does the sand in the dunes come from? Looking around the valley, you see possible sources: sand deposited by flash floods at the base of nearby mountains and the dusty salt flat where runoff evaporated in the past. If the wind can blow the sand that stings your face and whips across the dune

surfaces, then what is the fate of the even smaller particles, the dust grains not present in this great sand pile?

Figures 20.1c and d summarize your observations on the Atlantic Coast, at Jockey's Ridge State Park, located on a barrier island on the North Carolina coast. You came here to enjoy the beach but are surprised to see that Jockey's Ridge is a sand dune. In fact, at a height of about 30 meters, Jockey's Ridge is the highest sand dune in the eastern United States.

Why do tall sand dunes form along the shoreline in the humid, rainy southeast? One reason might be that there is a lot of sand that is not covered by vegetation. The beach, especially wide at low tide, consists of loose sand that swirls around your bare legs in the wind just as it did at Death Valley. A strong wind blows, just as it does in Death Valley, making Jockey's Ridge a favorite spot for kite flying and hang gliding. The persistent winds and the soft landing spots provided by sandy beaches and dunes along the barrier islands of North Carolina are what attracted the Wright brothers to Kitty Hawk, approximately 15 kilometers from Jockey's Ridge, to successfully test their first powered airplane.

Compared to the dunes in Death Valley, the dunes at Jockey's Ridge cover a very small area. You notice that the bare, active coastal dunes form a narrow band alongside the beach but merge landward into grass-covered ridges that have the shapes of sand dunes. These ridges must be dunes that formed at an earlier time, but why did they stop moving and become stable landforms for plant growth?

These two field experiences offer a few new perspectives on sand dunes. Sand dunes are not just features of deserts, and they do not even cover very much of the desert Southwest. The common variables for sand dunes at the two places you observed seem to be

- blowing wind
- lots of loose sand
- very little vegetation, at least where the active dunes are present.

Your simple list inspires new questions, however. Why does the air move to begin with? What factors of geology and climate determine the effectiveness of erosion by wind, or how loose sand for the wind to blow is even present, or how lack of vegetation commonly characterizes dune environments? What other landforms, besides sand dunes, owe their origin to blowing wind? If sand dunes are landforms that occur where wind deposits sand, then what landforms are characteristic of places where wind erodes sand? How far does wind transport sediment?

▶ Figure 20.1 **Wind shapes landscapes in deserts and along coasts.**

(a) The view of Death Valley from Dante's View shows a desolate desert, dominated by white salt flats where lakes evaporated in the past.

(b) Big sand dunes, with rippled surfaces, cover part of Death Valley.

(d) Persistent wind attracts hang-gliding and kite-flying enthusiasts to Jockey's Ridge State Park.

(c) The big costal sand dunes at Jockey's Ridge are only a kilometer away from the densely vegetated landscape more typical of the southeastern United States.

20.1 Why Does Wind Blow?

Before examining evidence of the geologic work done by wind, it is essential to understand why air moves to create wind. It is also important to understand the factors that determine how strongly the wind blows and the direction it blows from.

What Is Wind?

Wind is motion in the atmosphere. Movement of gas molecules in the atmosphere occurs for the same basic reason that motion occurs within the solid Earth (described in Chapter 10)—wind results from atmospheric convection. Dense air sinks to Earth's surface and displaces less dense air upward for the same reason that less dense mantle moves up as denser mantle sinks. This observation makes it essential for us to consider the factors that determine variations in atmosphere density.

Temperature is one variable that determines the density of the atmosphere, as illustrated in **Figure 20.2**. Heating causes the atmospheric gases to expand so that warm air is less dense than cool air. Density differences cause air to rise over regions of atmospheric heating and to sink in cool regions. Air also moves along Earth's surface from the cool regions to the warm regions as dense, cool air displaces warm, less dense air. It is this lateral motion that we experience as wind.

Water vapor in the atmosphere contributes to contrasting densities between adjacent volumes of air (see Figure 20.2). Humid air is less dense than dry air at the same temperature because the water molecule has a lower mass than the nitrogen and oxygen molecules that compose most of the atmosphere (see Figure 9.14 for data on atmosphere composition). Moist, low-density air lifts from Earth's surface whereas dry air of similar temperature sinks. The temperature and moisture effects on air density commonly work together, because warm air can hold more moisture than cold air.

Convection in the atmosphere explains the variations in air pressure that figure prominently in weather forecasts. Air pressure is low where the air rises away from the surface and high where denser air sinks against the surface. This means that surface winds tend to move from areas of high surface pressure toward areas of low surface pressure (see Figure 20.2). Pressure differences also cause the air circulation to complete a loop in the upper atmosphere. Sinking air reduces air pressure in the upper atmosphere

because mass is moving downward. On the other hand, rising air compresses air mass into the upper atmosphere, which increases air pressure compared to regions of sinking air.

Convective motion in the atmosphere is much faster than convection within Earth's mantle. Primarily, this difference results from the fact that the viscosity of air is negligible compared to the viscosity of mantle rocks. Average wind speeds are 10–20 kilometers per hour, and hurricane winds can exceed 200 kilometers per hour. These rates of convective movement hugely exceed the motion in the mantle driven by the same fundamental physical process at a few centimeters per year.

Why are winds stronger during storms? Wind speed relates to the pressure difference between adjacent regions of high and low air pressure—the greater the difference in air pressure, the faster the wind speed. Windy storms and hurricanes coincide with areas of extremely low air pressure where warm, moisture-rich air rises rapidly. The large difference in air pressure between the storm region, which has extremely low pressure, and the adjacent areas of higher pressure are one cause of high-velocity, damaging wind.

Wind direction and speed vary from day to day in response to the shifting locations of high- and low-pressure regions in the atmosphere, but observations of how particular wind directions dominate different areas on a yearly basis reveal a global pattern of atmosphere motion. For example, in the conterminous United States the wind blows most often from the west toward the east. We describe the wind direction in terms of where the wind blows from, rather than toward, so the prevailing wind in the conterminous United States is westerly. In contrast, easterly winds are most common in Hawaii. Dominant wind direction also varies with season in some locations. At Jockey's Ridge, for example, westerly winds are most common, but winter storms usually bring wind from the east or northeast. The seasonal flip-flop in wind direction moves the coastal dunes back and forth so that, instead of blowing primarily landward or seaward, they remain relatively permanent features. Clearly, our understanding of how wind works is incomplete until we determine why wind commonly blows from different directions at different locations, and at different times of year.

Constructing Global Wind Patterns

Figure 20.3a shows a simple view of atmosphere convection to begin your investigation of the directions of blowing wind. You will soon see that this scheme is unrealistic, but it is easiest to start with a simple idea and then add reality that is more complex. The first observation is that the equatorial region receives more direct solar heating than the poles. Therefore, it is reasonable to conclude that dense, cool air descends at the poles and moves toward the equator where less dense, warm air ascends. Said in another way, air moves from surface high pressure near the poles toward a belt of surface low pressure along the equator. In this view, northerly winds dominate in the northern hemisphere; southerly winds dominate in the southern hemisphere. However, you do not see the expected pattern of prevailing westerly and easterly winds observed in different parts of the United States and across the globe.

We can add more reality, as shown in Figure 20.3b, by including the effect of moisture transport on atmosphere convection. Warm air rising from the low-pressure belt at the equator is also rich in water vapor, because (a) warm temperatures favor evaporation of surface moisture, and (b) warm air holds more water vapor than cold air. This initially warm, moist air mass cools, however, as it moves higher in the atmosphere.

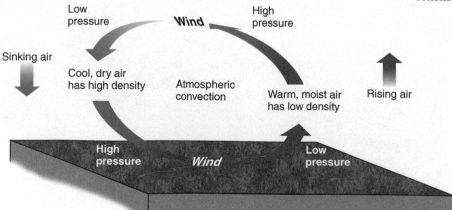

▲ **Figure 20.2 Why the atmosphere convects.** Air density depends on temperature and moisture content. Warm air is less dense than cool air, and moist air is less dense than dry air. This means that cool, dry air sinks, while warm, moist air rises. The vertical movement of air produces regions of low and high air pressure. Air moves horizontally from regions of high pressure toward regions of low pressure. This horizontal motion is wind.

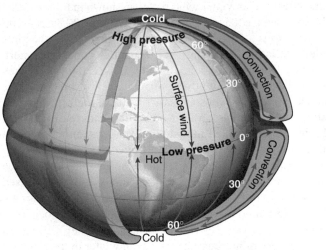

(a) Simple temperature-driven convection

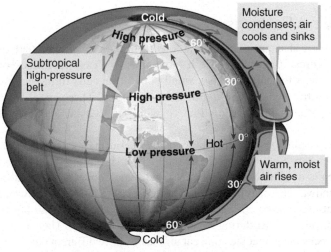

Moisture condenses; air cools and sinks

Subtropical high-pressure belt

Warm, moist air rises

(b) Adding the effect of moisture on atmospheric circulation

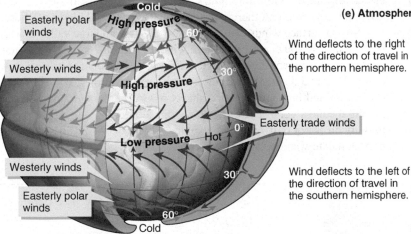

Easterly polar winds

Westerly winds

Easterly trade winds

Westerly winds

Easterly polar winds

Wind deflects to the right of the direction of travel in the northern hemisphere.

Wind deflects to the left of the direction of travel in the southern hemisphere.

(c) Adding the Coriolis effect on atmospheric circulation

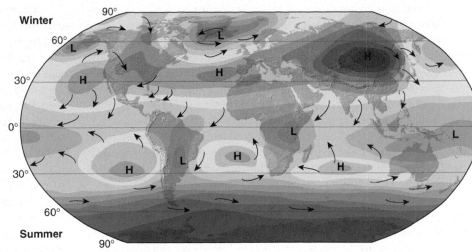

(d) Atmospheric circulation in January

Air pressure

Low (L)　　　High (H)

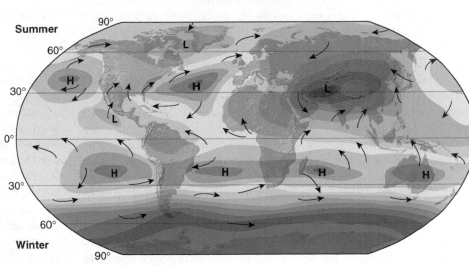

(e) Atmospheric circulation in July

◀ **Figure 20.3 Understanding global wind patterns.** If air circulation resulted only from simple convection caused by greater warming of the atmosphere at the equator, then Earth's wind pattern would resemble that seen in (a). In reality, the situation is more complicated because evaporation and condensation of water vapor also affect air density, which causes air to sink toward the surface to produce the subtropical high-pressure belts shown in (b). You must also consider Earth's rotation, which causes the Coriolis effect that deflects the air currents, as shown in (c). There are also seasonal differences in the distribution of high- and low-pressure areas on Earth's surface because solar heating of land and ocean surfaces is not equal. These seasonal variations in surface heating and air pressure cause different wind directions, which are depicted in parts (d) and (e).

ACTIVE ART

Global Wind Patterns. See the processes that determine global wind patterns.

This cooling of rising air is caused by the upward decrease in air pressure that results from the decrease in the weight of overlying atmosphere as the air moves toward space. It may not be obvious why a change in pressure should cause a change in temperature. Laboratory experiments show that whenever a solid, liquid, or gas is compressed, its temperature rises, while its volume decreases. The total amount of heat energy does not change, but it is distributed over a smaller volume, causing the temperature to rise. Likewise, temperature decreases whenever a solid, liquid, or gas expands. You have experienced this phenomenon if you ever noticed an aerosol can become cold during use. The gas in the can cools when it expands as some of the gas is released. This cooling effect also happens as rising air experiences lower pressure.

Of great importance to our model for wind, the cooling air cannot hold the water vapor that it absorbed at the warmer surface. Cooling, therefore, causes the water vapor to condense as clouds and rain. This means that the originally warm, moist, low-density air is now both cooler and drier, which are properties that increase its density.

The simple convection between equator and poles (Figure 20.3a) is cut short, therefore, by sinking of cool, dry air only 30 degrees north and south of the equator (Figure 20.3b). The descending air forms high-pressure belts at these subtropical latitudes and then flows both northward and southward. Descending cold air near the poles forces another convection loop at high latitudes. This pattern is more realistic than Figure 20.3a, but it still predicts only northerly and southerly winds.

The final step toward envisioning *real* global wind patterns is to add in the effect of Earth's rotation. Rotation causes all objects that move over Earth's surface or through the atmosphere to experience a horizontal drift called the **Coriolis effect** (in honor of Gaspard-Gustave de Coriolis, the nineteenth-century French engineer who mathematically quantified it). The Coriolis effect shifts moving objects to the right of their initial path in the northern hemisphere and to the left in the southern hemisphere. Figure 20.3c adds Coriolis deflections to the schematic wind pattern produced by convection. For example, air moving southward from the northern hemisphere subtropical high-pressure zone deflects to the right (toward the west) to produce easterly winds, whereas air moving northward from this high-pressure belt deflects to the right to produce westerly winds. The picture is almost complete: The Coriolis effect works with convection to produce westerly and easterly winds. What remains unexplained, however, is why the wind directions do not remain the same year-round.

EXTENSION MODULE 20.1

How the Coriolis Effect Works. *Learn more about how Earth's rotation causes moving objects to follow curving, rather than straight-line, paths.*

Figures 20.3d and 20.3e show the actual global wind directions and the locations of measured high- and low-pressure regions in the atmosphere during the seasonal solstices. The actual circulation differences between seasons and the general pattern predicted in Figure 20.3c relate to two factors:

1. Solar heating of the surface and atmosphere differs between summer and winter. This means that the locations of high- and low-pressure regions and the directions of winds between them shift with the seasons.

2. Land and ocean surfaces heat unevenly. Water absorbs and holds solar heat much more efficiently than does rock or soil. Incoming heat is distributed through large depths of ocean water but only penetrates a short distance beneath the land surface. This means that, compared to oceans, the land surfaces reflect more heat into the atmosphere during summer months, which produces low-pressure areas of rising warm air above the land.

The overall effect is that winter wind patterns, shown in Figures 20.3d and 20.3e, closely resemble the ideal case shown in Figure 20.3c. Subtropical high pressure forms a fairly continuous belt that encircles the globe close to 30 degrees latitude, and this belt separates zones of easterly and westerly winds. During the summer months, however, lower air pressure over the continents breaks up the continuous belt of high pressure. Very low summer air pressure over southeastern Africa, southern Asia, and northern Australia draws in moist air from adjacent oceans to cause torrential seasonal rains on land, called **monsoons**. Weaker summer-monsoon wind and rain also occur in the southwestern United States and Mexico.

We can now explain the global wind patterns as the summed-up effect of convection and Earth's rotation, plus complications caused by the seasons and uneven heating of land and sea. Easterly winds in Hawaii and westerlies in the conterminous United States are explained by location at different latitudes. The seasonally variable winds at Jockey's Ridge are consistent with a pattern of northeasterly winds in winter in the southeast United States (Figure 20.3d) and westerly winds in summer (Figure 20.3e).

Local Wind Patterns

Surface features also cause local variations from the global wind pattern shown in Figures 20.3d and e. High mountains, for example, block or deflect wind near the surface. Also, small convection cells form because of differential heating and cooling of Earth's surface. These small-scale circulation patterns produce distinctive winds in coastal and glacial regions.

Along coastlines, light winds blow landward during the day and seaward at night, especially during the summer. This air circulation between land and sea is a miniature version of global-scale convection. The seaward and landward breezes form as illustrated in **Figure 20.4** because water absorbs heat much more efficiently than does rock or soil. The air above land is, therefore, warmer during the day than air above water. The temperature differences mean that the adjacent air masses have different densities, which result in different air pressures. Cool air moves landward from the ocean and displaces the warm air upward. Water also stores heat more effectively during nighttime hours, whereas rock and soil surfaces cool down rapidly. At night, therefore, the wind reverses and blows offshore.

Strong surface winds commonly flow away from large valley glaciers and ice sheets. You can explain these winds as relating to the cold air temperatures above areas covered by ice and snow. This cold, dense air flows down slope and displaces warmer air at lower elevations. Winds blowing downward and outward from ice sheets in Greenland and Antarctica are among the most persistently strong winds recorded on Earth. The rapidly moving air is not only cold, but also dry, because cold air does not transport abundant water vapor.

Putting It Together—Why Does Wind Blow?

• Wind is motion of Earth's gaseous atmosphere.

• Convection causes atmospheric motion. Cool, dry air is denser than warm, moist air. This means that cool, dry air moves downward and displaces warm, moist air upward.

• Global zones of easterly and westerly winds alternating with latitude result from the combination of atmosphere convection and the Coriolis effect caused by Earth's rotation. Irregular heating and

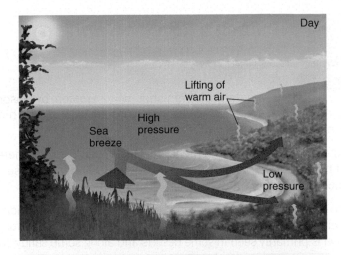

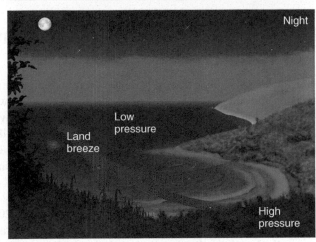

▲ **Figure 20.4 Daily changes in wind direction between land and sea.** Warming of the air over land during the day causes air to rise, which forms low pressure over land. Sea breezes blow onshore toward the region of low pressure. At night, the land cools rapidly and the overlying air mass becomes cool and dense; meanwhile, the water and its overlying air mass remain warm. This causes the region of low pressure to shift over the ocean, which results in offshore land breezes.

cooling of continents and oceans along with seasonal temperature variations add additional complexity to global atmospheric motion.

• Local wind patterns result from local variations in topography and from irregular heating and cooling of Earth's surface.

20.2 Where Is Wind an Influential Process in the Landscape?

We know that wind blows everywhere on Earth, but the impact of wind on landscapes is more obvious in places such as Death Valley and Jockey's Ridge. Based on similar field experiences around the globe it is apparent that wind is of geologic importance where three conditions are met:

1. Strong winds blow frequently.
2. Vegetation is sparse or absent.
3. Small, loose, dry particles are abundant on the surface.

Examples of wind erosion and sediment transport meeting these three criteria are shown in **Figure 20.5**. Wind power is the most straightforward

▲ **Figure 20.5 Wind erosion and sediment transport.** Dust storms occur at a variety of scales. NASA Space Shuttle astronauts took the top photo of a dust storm stretching more than 100 kilometers across Afghanistan. The bottom photo shows a red haze of sand and dust swept from the northern Arizona landscape on a windy spring day. The rocks in this area are 150-million-year-old sand-dune sandstones. Fine sand grains separate from the rocks during weathering and provide a bountiful source of sand for modern wind to transport. All three conditions for effective wind transport are illustrated: very little vegetation, strong winds, and abundant fine sediment.

factor; wind must blow strongly and often in order to be influential. The other two conditions are a little more complex and require further consideration.

Wind Is Effective Where Vegetation Is Sparse or Absent

You can observe that vegetation diminishes the geologic effectiveness of wind for two reasons:

1. Plant roots bind surface particles so that they cannot be picked up by the wind.
2. Plants rise above the land surface and absorb and deflect the force of the wind so that surface particles are sheltered from potential wind erosion.

The presence of sand dunes in some deserts and along coastal beaches is partly explained by the lack of vegetation to hold down the sediment or to slow the near-surface winds.

The sparseness or absence of vegetation in deserts (see Figure 20.1) results from a lack of moisture. By definition, **deserts** are regions where annual precipitation is less than 25 centimeters. The lack of moisture means that it is difficult for plants to grow. Plants that are adapted to desert climates are spaced far apart in comparison to the dense forests of wetter climate zones. Widely spaced plants do not impede the ability of wind to erode and transport surface particles. It is also important to remember that although deserts are commonly thought of as very hot, temperature is not part of the definition; some deserts are cold, such as in Antarctica.

The sparseness of vegetation in most coastal regions, such as Jockey's Ridge (Figure 20.1b), is explained by shoreline processes, rather than lack of rain. Unlike deserts, there is no rainfall limit for formation of coastal dunes, which exist in places that may receive a meter or more of rainfall each year. Sandy beaches lack vegetation because plants do not grow within the area swept by saltwater each day by the waves and tides or seasonally by storms. In addition, lush plant growth rarely exists landward of the beach along coasts that experience strong sea breezes. This is because the moving salty air dries out plant tissue and typically restricts plant growth to grasses that are tolerant of dry, salty conditions.

Wind Is Effective Where Surface Particles Are Small

Wind is not as effective as water in picking up surface sediment. Air density is only 1/800th the density of water, so shear stresses exerted on the surface by blowing wind are considerably less than for flowing water. Water also exerts a buoyancy effect on particles that make them easier to pick up and move. The buoyancy effect exists because water density is greater than one-third the density of most minerals, so mineral grains weigh less than two-thirds as much in water as they do in air. Grains are easier to move in water than air because the grains effectively weigh less in water than in air.

These limitations of wind as an agent of sediment erosion and transport mean that it is uncommon for mineral grains larger than about 0.5 millimeter, which is sand size, to move with the wind. If the surface has abundant particles smaller than 0.5 millimeter, then wind can shape the landscape by picking up and moving the sediment. On the other hand, if there is no loose unconsolidated sediment or the particles are coarser than 0.5 millimeter, then the landscape is unaffected by wind regardless of a lack of vegetation or the presence of strong winds. Gravel that is too coarse for wind to move covers large expanses of desert surfaces. This is why evidence for the erosion and modification of the landscape by the wind is not seen everywhere in desert climates, because many desert surface materials are too large for wind to move.

What landscapes have abundant fine-grained sediment that wind can transport? Strong winds readily move any recently deposited fine-grained, dry material that is not covered by vegetation. Stream floodplains and bars expose sand and finer-grained sediment. Sandy beaches are another candidate for wind erosion, as seen at Jockey's Ridge. Dried-up lakebeds, which are common in the western United States, are a source of fine sediment that includes evaporite-mineral particles.

Putting It Together—Where Is Wind an Influential Process in the Landscape?

• Vegetation diminishes wind erosion by absorbing wind energy that would otherwise reach the surface and by binding together loose sediment with roots.

• Wind rarely picks up mineral grains larger than 0.5 millimeter, so wind modifies landscapes only where there is abundant, loose, fine-grained sediment on the surface.

• The combination of strong wind, sparse or absent vegetation, and abundance of fine sediment that is a necessary condition for wind erosion is primarily seen in some deserts and along some sandy coasts.

20.3 What Determines the Locations of Deserts?

Deserts are the largest regions that meet the three requirements for wind to function as an important geologic process. Although windblown sand deposits like the dunes in Death Valley and at Jockey's Ridge cover only approximately 6 percent of Earth's land surface, 97 percent of dune deposits are found in deserts. Therefore, to further our understanding of wind as a geologic agent we need to understand where deserts are located and how they form.

Figure 20.6 shows the global location of deserts. We can apply knowledge gained from Section 20.2 (especially Figures 20.3d and 20.3e) to explain why deserts are found in these locations. **Figure 20.7** illustrates that desert regions lack moisture because they are located where (1) dry air descends to the surface, (2) the air is too cold to hold moisture, (3) mountains block moist air from oceans, (4) moist ocean air moves away from land, or (5) some combination of these conditions exists.

Subtropical Deserts

Most of the deserts depicted on Figure 20.6 are located close to 30 degrees north and south latitude. These include the famous Sahara Desert of northern Africa, the Arabian Desert, the Kalahari Desert of southern Africa, large expanses of Australia, and the deserts straddling the border between the United States and Mexico. These deserts coincide with the subtropical high-pressure belts within Earth's atmospheric circulation (Figure 20.3). Air descending to the surface at these latitudes is dry, having previously lost its moisture to heavy tropical rainfall closer to the equator.

As the descending air compresses against Earth's surface, the air also warms up (Figure 20.7a), in the reverse of the pressure-related temperature change that caused the air to cool when it first rose in the equatorial low-pressure belt. The results of this warming are staggeringly high temperatures, including the United States record of 57°C (134°F) at Death Valley and the global record of 58°C (136°F) in North Africa. The high air temperatures add to the drying effect because warm air readily absorbs moisture rather than releasing it as rain and snow.

Rain-Shadow Deserts

Deserts commonly form on the downwind side of mountain ranges, regardless of latitude (Figure 20.7b). Winds moving toward mountains are

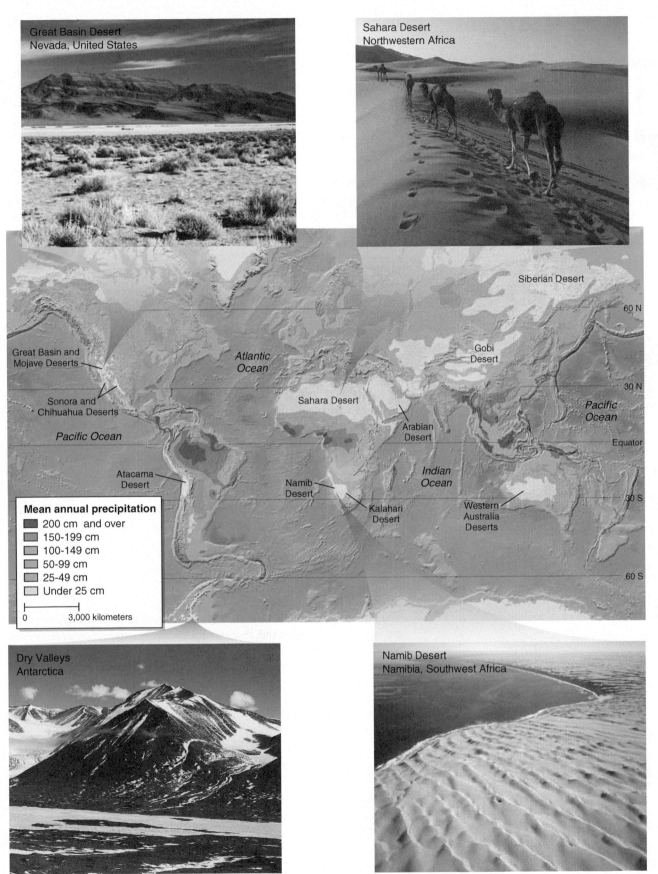

Great Basin Desert
Nevada, United States

Sahara Desert
Northwestern Africa

◀ Figure 20.6 Where deserts form. The map of mean annual precipitation shows dramatic variability in moisture delivered to different areas of continents. Deserts are those areas that receive less than 25 centimeters of precipitation, on average, each year. The largest deserts cluster near the high-pressure belts at 30 degrees north and south latitude, in the cold and dry polar regions, within continental interiors downwind of mountains, and in narrow strips along some coastlines.

Siberian Desert

60 N

Great Basin and
Mojave Deserts

Atlantic
Ocean

Gobi
Desert

30 N

Sonora and
Chihuahua Deserts

Sahara Desert

Pacific
Ocean

Pacific Ocean

Arabian
Desert

Equator

Atacama
Desert

Indian
Ocean

Namib
Desert

30 S

Mean annual precipitation
- 200 cm and over
- 150-199 cm
- 100-149 cm
- 50-99 cm
- 25-49 cm
- Under 25 cm

Kalahari
Desert

Western
Australia
Deserts

0   3,000 kilometers

60 S

Dry Valleys
Antarctica

Namib Desert
Namibia, Southwest Africa

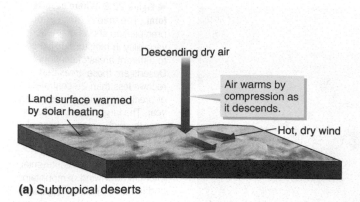

(a) Subtropical deserts

Descending dry air

Air warms by compression as it descends.

Land surface warmed by solar heating

Hot, dry wind

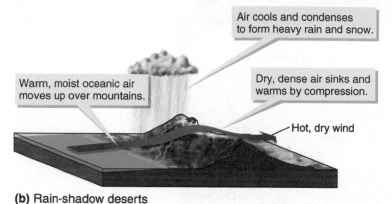

Air cools and condenses to form heavy rain and snow.

Warm, moist oceanic air moves up over mountains.

Dry, dense air sinks and warms by compression.

Hot, dry wind

(b) Rain-shadow deserts

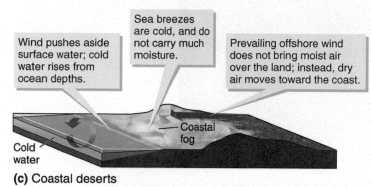

Sea breezes are cold, and do not carry much moisture.

Wind pushes aside surface water; cold water rises from ocean depths.

Prevailing offshore wind does not bring moist air over the land; instead, dry air moves toward the coast.

Coastal fog

Cold water

(c) Coastal deserts

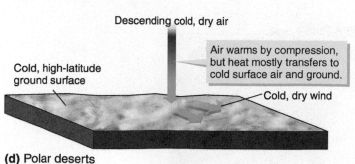

Descending cold, dry air

Cold, high-latitude ground surface

Air warms by compression, but heat mostly transfers to cold surface air and ground.

Cold, dry wind

(d) Polar deserts

▲ **Figure 20.7 How deserts form.** Dry air at Earth's surface causes the formation of deserts. These diagrams show four processes that produce dry air masses.

forced upward over the high topography. The lifted air expands and cools, which causes condensation of moisture as rain or snow in the mountains. The now denser cool and dry air descends the downwind side of the mountains, where it warms by compression against the surface and has been depleted of moisture. The regions downwind of mountains are called rain-shadow

deserts because the rain and snow falls on the upwind side of the mountain range and leaves a shadow zone deprived of precipitation on the downwind side.

The rain-shadow effect contributes to the origin of desert conditions in the western United States. High mountain ranges in the coastal states of Washington, Oregon, and California intercept moisture carried landward from the Pacific Ocean by prevailing westerly winds. As a result, rain-shadow deserts parallel the east side of the mountain ranges through Nevada, eastern Oregon, and eastern Washington (Figure 20.6).

The great deserts of central Asia are also separated from ocean moisture by high mountains. The problem here is exaggerated compared to North America because the interior of the huge Asian continent is much farther from oceans. The farther that moist air masses move inland from oceans, the greater the likelihood that moisture condenses and precipitates as rain or snow, making sure continental interiors remain very dry.

Coastal Deserts

Some deserts exist along shorelines, even though the atmosphere obtains most of its moisture by evaporation over oceans. The Atacama Desert of western South America (Figure 20.6) is one example of a huge coastal desert. The world record for the longest period without rainfall is 14 years and 4 months at a coastal village in northern Chile. Other coastal deserts include Baja California, the northwestern Sahara, the Namib Desert in southern Africa, and the northwest coast of Australia.

How can deserts form in such close proximity to ocean moisture? The key observation for each coastal desert listed above is that all are located on the west edges of continents at latitudes where the prevailing wind is from the east. This means that atmospheric circulation carries dry continental air offshore rather than bringing moist oceanic air onshore (Figure 20.7c). The offshore-directed wind also blows relatively warm surface water away from shore, so that colder water from deep in the ocean moves to the surface. The cold ocean water has a refrigerating effect on the surface air, so that even when sea breezes move onshore, the air is cold and dry.

Polar Deserts

The polar regions of North America, Greenland, Siberia, and Antarctica are as dry as the Sahara and, in some places, rarely experience air temperatures above the freezing point of water. These cold deserts, all located above 60 degrees latitude, provide contrasts to the stereotypical view of scorching hot deserts. However, it is the coldness of these regions that explains their dryness: Cold air cannot hold significant moisture, so climate conditions at these high latitudes are dominated by the downward flow of cold, dry air (Figure 20.7d).

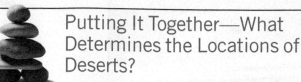

Putting It Together—What Determines the Locations of Deserts?

• Deserts form any place where atmospheric circulation deprives a region of moisture.

• Deserts form where dry air descends in high-pressure belts, where mountains intercept moisture carried landward from oceans, where wind predominantly blows offshore, and where air is extremely cold.

20.4 How Does Wind Pick Up and Transport Sediment?

You can use what you have learned about stream and glacier erosion as a starting point for understanding how wind moves sediment (look back to Sections 16.3 and 18.4 for information about stream and glacier erosion). There are both similarities and differences in how wind picks up sediment compared to erosion by flowing water and glaciers. Erosion happens because a moving mass of water, ice, or air exerts sufficient force on the surface to cause particles to move along with the current. Shear stress is the force that moves the particles. The shear stress is greatest where the slope is steeper and the thickness (and thus the weight) of ice, water, or air is greater.

Wind also exerts shear stress, but it is not very useful to think of slope and weight as part of the process. After all, air does not weigh very much, and it is difficult to calculate what thickness of atmosphere is moving with the wind. Gravity is the force that causes water, ice, and air to move. For water and ice, gravity pulls the moving mass down slope, so that the force is always exerted downward across the landscape. However, the situation is different for air. Gravity pulls denser air against Earth's surface, where it then moves toward areas of lower pressure. The direction the wind blows can be uphill or downhill, or even along flat surfaces. Geologists, therefore, can best estimate the shear stress of wind by measuring wind velocity at different heights above the ground surface, rather than measuring slopes and weights. Wind velocity at the sediment surface is highest where the surface is smooth and there is no vegetation.

Getting Particles in Motion

Fast winds are needed not only to move large sediment grains, but also to move very small grains. This observation, partly intuitive and partly confusing at first glance, is based on the same principle that applies to how flowing water picks up sediment (look back to Section 16.3 for more information). Large grains weigh more than small grains, so it makes sense that large grains only move in the strongest winds. Even then, wind is rarely capable of moving grains larger than 0.5 millimeter across, which is within the defined range of sand (0.063–2.0 millimeters). Surprisingly, even smaller grains do not move more easily than sand. Measurements show that sediment grains progressively smaller than about 0.1 millimeter also need increasingly faster winds to move. Small sediment grains, especially silt and clay particles composed of clay minerals, are very cohesive because of electrostatic attractions between the small particles. In order for wind to pick up these small particles, the wind has to exert a shear stress that exceeds both the weight of the particles and the cohesion that holds them together.

Observations in the field and in wind-tunnel experiments show that many of the particles transported by wind are not directly picked up by wind but are instead knocked loose from the surface by the impact of other, already moving particles. Grains about 0.1 millimeter across are the easiest for wind to pick up, and their movement requires a surface wind speed of only about 5 kilometers per hour, which is equivalent to average walking speed. When these small grains bounce along the surface with the blowing wind, they knock loose larger or more cohesive grains that were motionless. Wind-tunnel experiments show that each time a bouncing sand grain hits the surface it kicks up about 10 more sediment grains. Once propelled into the air, many of these grains are sufficiently small for the wind to carry. Larger grains fall back to the surface, where they kick loose some

▲ Figure 20.8 Lifting particles off the ground. Fine dust is easily transported by light breezes but is not as easily picked up from the ground because these fine particles tend to be cohesive. In this photo, the truck tires, not than wind, dislodged the fine particles that are then transported by the breeze. Dust is not blowing elsewhere in this view, which means that the wind is capable of transporting the dust but cannot erode the fine particles.

more grains while the tiniest grains drift off as dust. The end result is that most of the grains carried by wind are dislodged from the surface by impacts of a smaller proportion of bouncing windblown grains, rather than initially picked up by wind blowing over the ground surface.

Figure 20.8 illustrates a familiar example of how wind-induced shear stress, alone, is not enough to pick up very small sediment particles. The billowing cloud of dust that frequently follows a vehicle on a dirt road or a tractor in a field demonstrates that wind velocity is frequently sufficient to carry the small particles but lacks the ability to set the particles in motion. The only place that these particles move is where the additional shear stress exerted on the surface by rotating tires or moving farm implements is adequate to overcome the cohesive properties of the particles that resist motion.

Observations also show that dry particles are easier to erode than moist particles. Water in pore spaces exerts a cohesive force on adjacent sediment particles that resists wind erosion. This is one reason why wind erosion is most prevalent at times when ground surfaces are dry.

Transporting Particles

Figure 20.9 summarizes observations of sediment transported by wind. Whether picked up directly by wind or knocked loose by bouncing grains, the transport of sediment particles is similar to particle movement in flowing water. Grains move by either rolling or bouncing along the ground, or they are completely suspended in the moving air. Field observations and wind-tunnel experiments show that small sand grains between 0.1 and 0.3 millimeter across typically bounce along the surface. Larger grains tend to roll along the ground.

What happens to particles smaller than 0.1 millimeter? The wind suspends these tiny particles in the atmosphere. **Dust** is the term commonly applied to these suspended particles. Some dust particles settle to the surface when the wind is calm, but grains smaller than 0.02 millimeter across

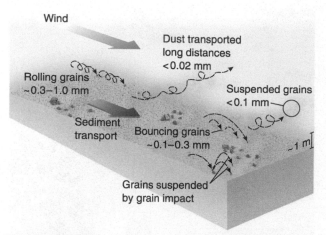

▲ **Figure 20.9 How wind transports sediment.** Wind-transported sediment grains roll, bounce along the ground, or are carried in suspension in whirling air currents. Bouncing grains commonly set more particles in motion when they impact the ground surface. Wind only moves sand and finer-grained sediment. The finest dust particles may travel thousands of kilometers.

depressions sheltered from the wind. Additional deposition produces a low sand pile that becomes the nucleus of a growing dune.

Figure 20.11 shows how sand dunes move. The downwind side of the dune is usually steeper than the upwind side. Wind erodes sand on the upwind side, which faces into the current. The wind carries an eroded sand grain to the top of the dune, where it then rolls, bounces, or settles onto the downwind side that is sheltered from the wind by the dune crest. Over long periods of erosion on the upwind side of a dune and deposition on the downwind side, a dune noticeably migrates in the direction of the prevailing wind. When you examine cross-beds in excavations, you can see the former positions of the steep, downwind side of the migrating dune (Figure 20.11; also look back at Figure 5.17 for more details on the movement of dunes to produce cross-bedding).

ACTIVE ART

How Wind Moves Sediment. See how blowing wind moves sediment.

remain suspended in even the most imperceptible breeze and may travel thousands of kilometers. Explosive volcanic eruptions demonstrate the efficiency of wind transport of tiny particles, as illustrated in **Figure 20.10**. Some volcanic eruptions eject volcanic ash and gases more than 20 kilometers into the atmosphere. The gases condense into tiny liquid droplets that, along with the smallest ash particles, may travel around the world many times and remain aloft for more than a decade.

Bouncing sand grains form moving ripples and dunes (Figure 20.1). Sand grains moved by the wind place additional grains in motion each time they bounce on the surface. All of these grains move along the surface in the direction that the wind blows. The ripple ridges, therefore, are similar to small mounds of dirt swept across the floor in front of a broom. The taller sand dunes build up because of variations in surface wind velocity that cause alternating areas of sediment erosion and deposition. Bouncing sand grains may come to rest against an obstacle, such as a plant or cobble, and in surface

Astronauts in the International Space Station took this picture of ash and gases erupting from Mt. Etna, on the Italian island of Sicily, in 2002. Ash from this eruption fell 560 km away in northern Africa.

Satellite instruments tracked the global spread of wind-blown condensed-gas droplets following the June 1991 eruption of Mt. Pinatubo, in the Philippines. Most of the gas droplets were sulfur dioxide, which absorbs solar heat and caused approximately 2-degree Celsius drop in average global temperature. The sulfur dioxide also mixed with water vapor to form sulfuric acid, which etched airplane windows.

Increasing condensed-gas droplets

▶ **Figure 20.10 Carrying fine volcanic particles long distances.**

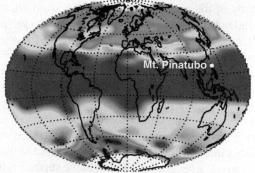

Before eruption

June 15 - July 25, 1991

August 23 - September 30, 1991

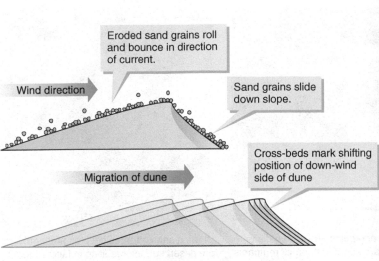

▲ Figure 20.11 How sand dunes migrate.

ACTIVE ART

How Sand Dunes Move. See how sand dunes migrate over time to form cross-bedding.

Putting It Together—How Does Wind Pick Up and Transport Sediment?

• Wind erodes particles where the wind velocity exerts sufficient stress to exceed the weight and cohesion of grains, which resist movement. Cohesion results either from electrostatic attraction between clay and silt particles or from moisture.

• Grains moved by the wind dislodge stationary particles and place them in motion, too.

• Wind transports grains by rolling and bouncing along the ground and as suspended dust.

• Bouncing sand grains form ripples and dunes. Dunes migrate by simultaneous erosion (on the upwind side of the dune) and deposition (on the steeper downwind side). Cross-beds show the changing position of the downwind side of the migrating dune.

20.5 How Does Wind Shape the Landscape?

Sand dunes are the most familiar and probably the most obvious features of wind-modified landscapes. Dunes are depositional features, and the sand has to first erode from somewhere. There are many landscape features that reveal the importance of wind erosion.

Landscapes Eroded by Wind

Observations, such as those illustrated in **Figure 20.12**, show that wind erodes landscapes by two different processes:

1. **Deflation** (from Latin, meaning "to blow away") is the process through which landscape elevations are lowered as wind removes fine particles.

2. Abrasion by windblown particles erodes exposed rock and regolith surfaces. This sandblasting effect reshapes rock outcrops and slowly wears away sediment grains that are too large for wind to pick up.

Evidence of wind erosion, and then deposition of eroded particles, is not only widespread on Earth; some of these same distinctive features

Wind erosion

Deflation

Abrasion

Pans

Yardangs

Ventifacts

Deflation pans central New Mexico

Wind

Dunes

Evaporite deposits

Pans

Wind erodes fine sediment and deposits it to form a dune. Deflated area, called a pan, may temporarily fill with surface runoff or ground-water seepage. Evaporite minerals crystallize when water evaporates.

Yardangs, Egypt

Blowing sand abrades soft rock or regolith and wind deflates the eroded particles. Yardangs are abraded bedrock remnants within the deflated landscape, and are elongated in the direction of the prevailing wind with pedestal-and-cap shapes resulting from greatest erosion closest to the ground.

Ventifacts, Western United States

Blowing sand abrades cobbles and boulders on the desert surface resulting in faceted rocks called ventifacts.

▲ Figure 20.12 Features resulting from wind deflation and abrasion.

appear in images sent to Earth by NASA's Mars rovers, such as seen in **Figure 20.13**. Clearly, wind also shapes landscapes on other planets.

Deflation commonly scours out elliptical or circular areas of easily eroded sediment to leave distinctive depressions called **pans** (Figure 20.12). A low dune commonly forms along the downwind side of the pan, showing that most of the deflated sediment only travels a short distance before

it is deposited. In regions of fluctuating water table, pans form during dry periods and then partly fill with infiltrated ground water when the water table rises to the bottom of the pan. Some pan depressions also accumulate surface runoff. Evaporation of water deposits a crust of evaporite minerals, such as gypsum and halite, to produce a playa on the floor of the pan (go back to Section 16.13 for more information on playas).

Ventifacts are loose rocks that show evidence of abrasion by wind-blown sand (see Figures 20.12 and 20.13). The term originates from Latin roots that mean "made by the wind." Persistent sandblasting abrades smooth planar surfaces in rocks that are shaped like polished facets on a gemstone. The facets commonly meet along sharp edges so that the rock resembles

This football-size rock has the faceted faces and sharp edges characteristic of ventifacts.

Low sand dunes, less than 1 m tall, sweep across the floor of Endurance Crater.

▲ Figure 20.13 Recognizing wind action on Mars. Photos returned by NASA's Mars Exploration Rovers *Spirit* and *Opportunity* in 2004 clearly show evidence of wind processes on the Martian surface. Ventifacts and sand dunes on Mars are identical to features on Earth.

the outline of a brazil nut. The multiple abraded facets result from varying wind directions and shifting of the rock over time.

Deflation and abrasion work together to produce a distinctive landform called a **yardang**, which is a wind-parallel ridge of soft rock or slightly consolidated sediment that remains after surrounding material is eroded (see Figure 20.12). The term originated in Turkey, but this landform is recognized on all continents. Most yardangs are less than 5 meters high and typically no more than 10 meters long and are very steep sided. Larger yardangs are more than 100 meters high, several kilometers long, and are scattered across several hundred thousand square kilometers of the great deserts of North Africa and central Asia.

Can wind erosion substantially lower surface elevations? Wind erosion on a large scale is as clearly evident as are the erosive effects of flowing water and glaciers. From a theoretical standpoint, it seems reasonable that wind should be able to erode deeply as long as the surface materials are loose and sufficiently fine-grained for deflation. However, deflation cannot continue below the water table because wind will not blow away cohesive, moist sand. In desert regions, however, soft, loose, unsaturated sediment may be hundreds of meters thick. Is it possible for wind to lower landscapes by hundreds of meters? The presence of yardangs as high as 200 meters suggests that wind may erode deeply, but it is also unclear whether all this erosion relates to ongoing desert sand blasting or whether it may have partly occurred by an earlier episode of stream erosion. Nonetheless, some geologists hypothesize that deflation explains nearly 20,000 square kilometers in the Sahara Desert with depressions below sea level. It is difficult to prove, however, that streams or tectonic processes did not also play roles in making these large depressions. Although features such as pans, ventifacts, and yardangs result from wind erosion, the scale at which wind lowers surface elevations remains unclear.

The Variety of Sand Dunes

Actively moving, windblown sand forms the most distinctive landforms of deserts and coastlines. In some places, as shown in **Figure 20.14**a, the sand simply accumulates in low sheets or forms piles at the base of shrubs that block the wind to cause deposition. Larger sand dunes are the most prominent and well-known deposits resulting from wind transport of sediment. Dunes in coastal regions are conspicuous for their towering height above adjacent beaches (Figure 20.1c), but they rarely cover large areas. Many deserts, on the other hand, feature vast areas of sand

dunes, such as the expanse pictured in Figure 20.14b, where the seemingly endless, wavelike undulations of the dune crests and troughs suggest waves on the ocean. Sand dunes are a part of everyone's first impressions of what a desert should look like, but dunes actually cover less than 20 percent of the arid-zone surface of Earth and only 2 percent of desert landscapes in the United States.

A survey of dunes would show that they appear in a variety of shapes and sizes. **Figure 20.15** illustrates and summarizes the characteristics of five common types of dunes. The observations and measurements summarized in **Figure 20.16** show that the abundance of sand for wind to move and the uniformity of wind direction are the primary factors that determine dune shape.

Blankets of Loess

What happens to fine particles that remain suspended in the wind while sand accumulates and moves slowly in dunes? It turns out that windblown silt covers nearly as much land surface as is covered by sand dunes. Although these silt deposits are tens to hundreds of meters thick, the silt does not form impressive landforms comparable to sand dunes. Instead, the silt particles accumulate gradually, mostly in vegetated landscapes where wind velocity is slow enough for small particles to settle to the ground. Surface elevations gradually rise because of sediment deposition, but the shape of the landscape is preserved as if covered by fallen snow. Geologists use a German term, *loess*, to describe on-land deposition of wind-borne sediment that is mostly silt size.

Figure 20.17 outlines areas on Earth where loess forms continuous blankets that are many meters thick. The loess deposits in central Asia and South America are thicker and contain more mixed-in sand in close proximity to adjacent deserts. This geographic relationship of loess and deserts confirms that silt blows out of deserts and accumulates on adjacent landscapes with greater vegetation cover and lower wind velocities. Thick loess deposits in the central and northwestern United States and in Europe tend to be thicker and sandier close to river valleys (look back at Figure 18.37, for example). Most of this loess accumulated during and shortly after the last ice age (peaking about 21,000 years ago) when cold dry winds blew off the glaciers and across river floodplains that were covered with large volumes of sediment carried by meltwater streams. During ice-age climatic conditions the river valleys were less vegetated and experienced stronger

(a)

(b)

◄ **Figure 20.14 Sand deposits at different scales.** (a) Some sand dunes are very small, such as these low dunes that formed around desert shrubs. (b) Other dunes are hundreds of meters high and cover tens of thousands of square kilometers, as seen in this view of the Namib Desert in Namibia.

Parabolic dunes form where wind erosion attacks a barren area in a largely vegetated landscape, commonly near a beach. A depression forms where erosion is most intense and the sand is trapped by vegetation to accumulate as a dune. The U-shape resembles the graph of a parabola with the depression in the center and the two ends of the parabola pointing upwind.

Barchan dunes are crescent shaped dunes that form on desert surfaces where gravel or rock are more abundant than wind-transported sand. The word derives from an ancient Turkish word describing sandy hills. Barchans resemble parabolic dunes except that the ends of the crescent barchans point downwind.

Transverse dunes form where sand supply is abundant and dunes continuously carpet large areas. The term "transverse" emphasizes that the curvy dune crests are mostly transverse (perpendicular) to the prevailing wind direction. These particular dunes consist of gypsum sand grains at White Sands National Monument, New Mexico.

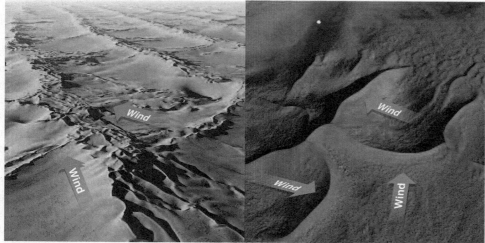

Linear dunes are long sand ridges oriented parallel to the prevailing wind direction. Linear dunes typically form where sand supply is more limited than where transverse dunes form and where wind direction is more variable than regions characterized by barchans.

Star dunes are radiating sand ridges resulting from highly variable wind directions crossing large areas of readily eroded sand.

◀ **Figure 20.15 Visualizing types of sand dunes.**

winds than are recorded today, so wind processes were more influential then than now.

Loess forms rich soil. Most surface loess accumulated within the last 20,000 years and is not extensively weathered, so mineral-nutrient content is much higher than for regions where soil formation has been ongoing for much longer. The fine grain size of the silt is ideal for retaining infiltrating ground water and allows crops to grow with little or no irrigation, even where climatic conditions border on arid. Farmers must exercise care when cultivating loess soils, however, because runoff easily erodes the silt grains where the loess is not covered with vegetation, which causes high soil-erosion rates.

Desert Pavements: Formed by Sediment Deflation or Accumulation?

One reason that sand dunes are rare in some deserts is the presence of closely spaced pieces of gravel to form a smooth surface called **desert pavement**. **Figure 20.18** illustrates typical characteristics of desert pavements, including that the pavement is usually only one or two stones thick and overlies silt. Desert pavements are an important control on wind erosion because pavement particles are too large for wind to erode, and they form a surface armor that protects underlying fine sediment from erosion. On the other hand, where stream erosion or human activities remove parts

◀ **Figure 20.16 The factors that determine dune shape.** The shapes of sand dunes depend on the availability of sand for the wind to move and the uniformity of wind direction.

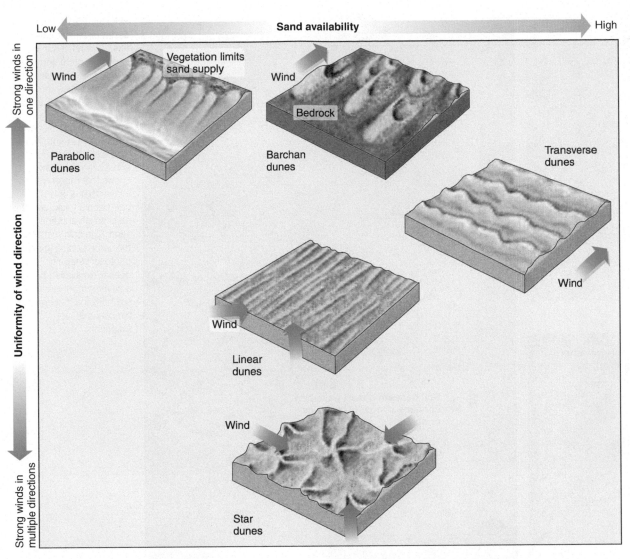

Sand availability

Low ← → High

Uniformity of wind direction

Strong winds in one direction

Strong winds in multiple directions

Wind

Vegetation limits sand supply

Parabolic dunes

Wind

Bedrock

Barchan dunes

Transverse dunes

Wind

Wind

Linear dunes

Wind

Star dunes

◀ **Figure 20.17 Visualizing where loess is found.** The map shows the occurrences of thick deposits of windblown silt, called loess. Thin, discontinuous loess deposits occur in other locations not shown here. The photographs show typical loess outcrops. The cohesive silt particles form deposits that erode and excavate to leave steep slopes.

Eastern Washington

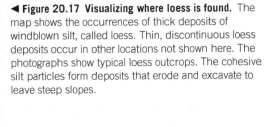

Western Iowa

China

Desert pavement

Silt and fine sand underlies the desert pavement

Pebble and cobble desert pavement on surface

Soil beneath desert pavement contains almost no pebbles or cobbles

◄ **Figure 20.18 What desert pavement looks like.** Low-relief surfaces in deserts are commonly covered with closely spaced pebbles and cobbles that resemble a cobblestone pavement. The top-left photo shows a pavement of limestone pebbles in Nevada and the photo below it shows that this gravel is a layer that is only one-pebble thick, and rests on silt and fine sand. The coin rests on the same pebble in each photo for reference. A deeper excavation below a pavement reveals desert soil horizons that contain almost no gravel fragments.

of the thin gravel surface layer, wind can easily erode the exposed silt. How is desert pavement formed?

The role of wind to produce desert pavement is controversial. **Figure 20.19**a illustrates a deflation hypothesis for the origin of desert pavement. This hypothesis states that wind deflates sand and dust particles within poorly sorted surface deposits that also include gravel. The abundance of gravel fragments on the surface increases through time as finer particles blow away. Eventually the surface is completely covered with coarse particles that the wind is unable to move. The hypothesis is consistent with worldwide occurrence of pavements in deserts where wind is clearly an important geologic process. This hypothesis does not, however, readily explain two features:

1. Silty layers, in some cases tens of centimeters thick, immediately underlie the pavement armor but contain few if any gravel fragments (Figure 20.18). Therefore, deflation of the fine-grained sediment will not leave behind a layer of gravel.

2. Some geologic data reveal that all of the pavement clasts have been at the surface for the same period of time, whereas deflation requires exposure of clasts at different times as enclosing fine particles gradually blow away to expose each clast.

Figure 20.19b depicts an alternative hypothesis for how sediment might accumulate to form desert pavement. This hypothesis simultaneously addresses the two weak points in the deflation hypothesis. Clasts projecting above the surface of an initially coarse deposit trap windblown dust particles. The dust sifts into cracks between surface clasts, probably aided by infiltrating rainwater. The dust includes silt- and clay-size particles that swell and shrink during wetting and drying events. Wetting also dissolves soluble minerals in the dust, which then precipitate when the sediment dries. The shrinking and swelling, along with mineral dissolution and precipitation, shift the surface clasts around and open cracks in the accumulating fine-grained layer, and this action allows newly arriving dust particles to accumulate. Slow, gradual accumulation of dust progressively lifts the

(a) Forming desert pavement by wind erosion

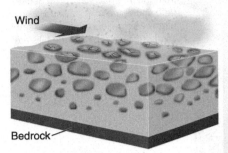

Wind

Bedrock

Mixture of gravel, sand, and silt.

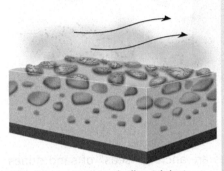

Wind blows away sand, silt, and dust, which leaves gravel behind.

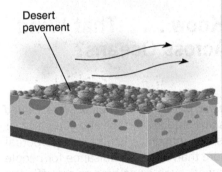

Desert pavement

Gravel completely covers the surface producing a pavement that diminishes further wind erosion.

(b) Forming desert pavement by wind deposition

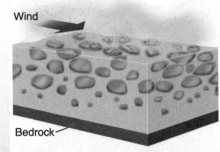

Wind

Bedrock

Dust accumulates on surface, especially where trapped against pebbles and cobbles.

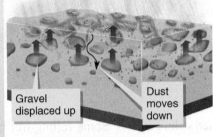

Gravel displaced up

Dust moves down

Rain washes the fine dust downward. Soluble minerals dissolve and then precipitate below the surface when the infiltrating rainwater evaporates. Accumulation of dust, shrinking and swelling of clay minerals, and precipitation of minerals lift the gravel fragments to the surface.

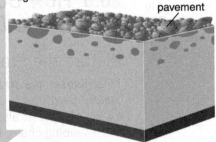

Desert pavement

Continuous gravel pavement eventually accumulates at the surface above a thickening layer of finer sediment.

Time

◀ **Figure 20.19 How to form desert pavement.** Geologists debate the origin of desert pavement. (a) This popular hypothesis attributes pavement formation to erosion. (b) This more recent hypothesis attributes pavement formation to deposition, and is more consistent with observed characteristics of most desert pavements, which overlie thick deposits of fine-grained sediment without much gravel (Figure 20.18).

surface gravel fragments higher and higher above where they started. This hypothesis is more complicated than the simple deflation idea, but it is more consistent with observed data. The more likely sediment-accumulation hypothesis also reveals that development of the landscape was due to wind deposition rather than to wind erosion.

Wind-Formed Landscapes of the Recent Past

Geologists have long recognized rolling vegetated hills of sand that resemble sand dunes except for the fact that they are covered in vegetation and do not move. The conclusion, then, is that there are many places on Earth where wind shaped the landscape in the recent past when climatic conditions were more favorable for producing sand dunes than at present. Rolling hills across parts of the southeastern and central United States are inactive

sand dunes where relatively recent vegetation growth has stopped sand movement and stabilized the dunes.

Inactive sand-dune deposits are especially common in the High Plains region. Rivers carry abundant sand eastward from the Rocky Mountains onto the semiarid, almost treeless plains where wind blows the sand out of the river valleys and across the countryside. Present climatic conditions are just sufficiently moist to support adequate grass cover to reduce significant wind erosion and transport of sand. Geologic data show that a drier climate in the recent past reduced the grass cover so that sand dunes formed.

The Sand Hills of northwestern Nebraska are the largest sea of dunes in this region, and are illustrated in **Figure 20.20**. The sandy hills are actually grass-covered sand dunes that cover 50,000 square kilometers and are as high as 80 meters (roughly equivalent to a 20-story building). Geologists

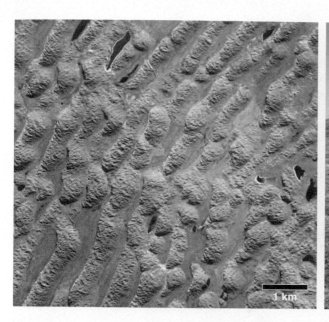 

◀ **Figure 20.20 Sand dunes of the past.** The photo taken by astronauts from the International Space Station (left) shows the rolling Sand Hills of northwestern Nebraska that resemble sand dunes except that the hills are vegetated, as seen in the closer view (right). The Sand Hills are ancient sand dunes that are vegetated and stationary landforms current climatic conditions.

hypothesize that these dunes formed during a past ice age when sediment supply in nearby rivers was greater than today and cold, dry winds swept southward from nearby glacial ice sheets and eastward from the cold glacial valleys in the Rocky Mountains. Careful studies show, however, that regardless of when the dunes first formed, they have been active much more recently than the demise of the ice-age glaciers. Measured wind velocities in the Sand Hills are sufficiently strong to move sand about half of the year. The prairie grasses covering the old dunes are the only factor that keeps the dunes from moving. Long periods of drought, including one episode that occurred only 200–300 years ago, reduce the distribution of grasses so that the dunes reactivate as moving features on the landscape.

Putting It Together—How Does Wind Shape the Landscape?

• Wind erosion occurs by deflation and abrasion. Pans are low areas formed by deep deflation, ventifacts are abraded rocks on the desert surface, and yardangs are peculiarly shaped ridges that are sandblasted and streamlined by blowing sand.

• Dunes are large sand piles moved by wind, with steep slopes that face toward the direction of movement. The variety of sand dune shapes relates to the abundance of sand, the abundance of vegetation, and the consistency of wind direction.

• Deposits of windblown silt, called "loess," blanket large areas and provide fertile soil.

• Desert pavements look like they are a covering of gravel fragments left behind by deflation of smaller particles. It is more likely, however, that these pavements form where dust accumulates beneath gravel particles that have always been at the surface.

• Some rolling-hill landscapes are ancient "seas" of sand dunes. Although the dunes are now overgrown by stabilizing vegetation, the dunes actively moved in the past when vegetation was sparse or absent because of drier climate.

20.6 How Do We Know . . . That Wind Blows Dust Across Oceans?

Picture the Problem

What Causes the Haziest Summer Days in the Southeastern United States? Drag your finger across any outdoor surface and you discover a layer of dust. It also finds its way indoors—dusting is a perpetual housekeeping chore. Dust is more than a simple nuisance for people who suffer from respiratory ailments such as asthma or are afflicted by dusty allergens. Dust is, therefore, monitored as a measure of air quality in the United States and other countries. Dust is also routinely monitored as part of an effort to maintain clear skies with long-distance visibility in environmentally sensitive areas, such as many national parks and wildlife refuges. Some of the tiny particles filtered from air-quality samples are airborne pollution from smokestacks and vehicles, but many are tiny mineral fragments. You have seen dust everywhere, but have you ever wondered where it comes from?

Air-quality data collected across the United States repeatedly show that the Southeast is the dustiest region during summer months. Periods of highest dust concentrations, called dust episodes, coincide with or shortly follow even more severe dusty haze conditions in the Caribbean islands south of Florida. This pattern contrasts with data collected during the spring, which shows higher dust concentrations in the western part of the country. Dust originating from the dry semiarid and arid landscapes in the West is logical, but where does dust come from to obscure summer skies in the humid Southeast and Caribbean?

Develop a Hypothesis

Does the Dust Come from Africa? Wind transports dust. Mineral dust erodes from landscapes where wind erosion is effective, such as deserts. Consideration of wind directions and desert locations may reveal a possible source for summer dust in the Southeast.

Figures 20.3d and 20.3e show that the southeastern United States is located within a zone of slightly shifting subtropical high pressure. Most winter wind blows from the northeast, but the summer flow is from the south, as easterly winds move away from high pressure over the Atlantic Ocean and then northward toward lower pressure above the warming North America continent. This influx of warm moist air from the Caribbean keeps the southeastern United States humid, despite the fact that the region is located at the same latitude as deserts in the western United States and Africa. The easterly summer winds from Africa might carry dust from the Sahara westward into the Caribbean and then northward into the eastern United States. If this hypothesis is true, then some air-quality concerns in the Southeast relate to processes of wind erosion on another continent.

Test the Hypothesis

How Is Dust Tracked to Its Source? As one test of the hypothesis, consider the satellite image in **Figure 20.21**, which shows dust blowing across the Atlantic Ocean from northwestern Africa. Hazy-sky conditions in the Caribbean coincide with these dusty outbursts from the Sahara. Satellite images, therefore, provide one way of directly tracking Saharan dust by direct observation. This technique has limitations however, because dust concentration decreases during transport as grains slowly settle to the ocean or onto islands. When the dust is more dispersed or dilute, it is not always clear on the satellite images that the African dust moves as far as the southeastern United States.

Figure 20.22 shows measured dust concentrations in the eastern United States during a 2-week period. The maps show the airborne concentration of extremely small dust particles, less than 25/10,000ths of a millimeter across, which remain suspended in moving air for a long time and travel long distances. The mapped changes in dust concentration clearly show that the dust came from the south and, it turns out, at the same time that a Saharan dust episode affected the Caribbean.

Scientific conclusions are strongest when corroborated by different methods and data. Tracking dust by satellite images and measurement of dust concentration support the hypothesis that the haziest summer days in the Southeast result from windblown dust originating in Africa. If the dust could be "fingerprinted" to be from the Sahara, then there would be even stronger evidence to support the hypothesis. Air-quality filters in the Southeast show summer dust is reddish, because of abundant iron-oxide minerals, whereas spring dust is gray. The color difference suggests a composition difference that may help track different dust sources during the two seasons.

Figure 20.23 highlights seasonal differences in dust concentration and chemical composition that we seek to explain. During the spring, dust concentration is highest in the Southwest. So, it seems that spring dust originates in the dry western United States and moves with the prevailing westerly wind to affect much of the country. Also important to notice is that the composition of the fine dust

Clear day in the Virgin Islands

Hazy day in the Virgin Islands

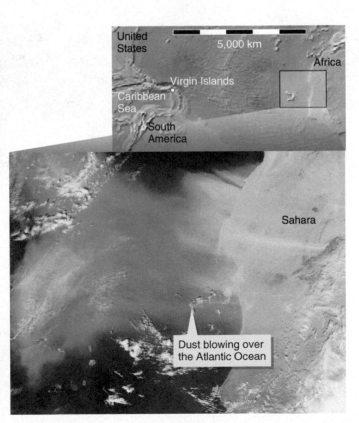

United States

5,000 km

Africa

Virgin Islands

Caribbean Sea

South America

Sahara

Dust blowing over the Atlantic Ocean

◄ **Figure 20.21 Dust out of Africa.** The satellite image on the right shows Sahara Desert dust blowing eastward over the Atlantic Ocean. During these dust events, skies are hazy in the Caribbean Sea, which is more than 5000 km downwind from Africa.

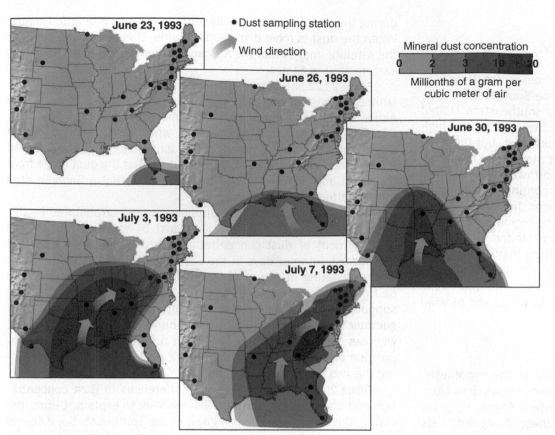

• Dust sampling station

⇨ Wind direction

Mineral dust concentration

| 0 | 2 | 3 | 10 | 20 |

Millionths of a gram per
cubic meter of air

◄ **Figure 20.22 Measuring dust concentrations in the
eastern United States.** Measurements at specially
designed sampling stations track the northward
movement of wind-borne mineral dust during a
summer dust episode in the eastern United States.

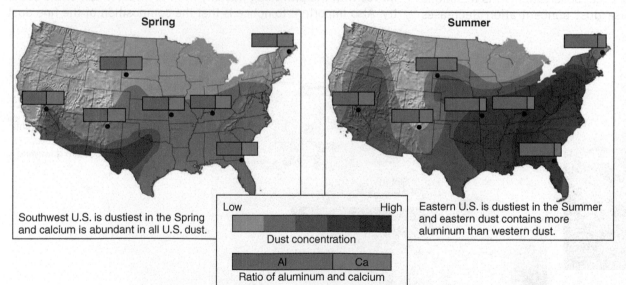

Southwest U.S. is dustiest in the Spring
and calcium is abundant in all U.S. dust.

Eastern U.S. is dustiest in the Summer
and eastern dust contains more
aluminum than western dust.

Low   High

Dust concentration

| Al | Ca |

Ratio of aluminum and calcium

◄ **Figure 20.23 Seasonal differences in
dust concentration and composition.**
Spring dust concentrations are high in
the southwestern United States and
summer dust concentrations are highest
in the east. Comparison of aluminum
and calcium contents shows that dust
from the southwest, in both spring and
summer, is higher in calcium than is the
summer dust in the Southeast. Calcium-
rich dust blows off the surface of dry
playa lakes in the western United States,
but summer dust in the Southeast is rich
in aluminum-rich clay minerals from
Africa.

particles is similar everywhere during the spring, suggesting a single source of dust. However, when a summer dust episode affects the Southeast, the dust composition changes in those areas where the dust concentration is highest. Summer dust in the Southeast has more aluminum and less calcium than dust blowing out of the western deserts. The high aluminum reflects high clay-mineral content, whereas high calcium levels indicate abundant calcite or gypsum in the dust. Dust from the western United States deserts contains calcite and gypsum deflated from dried-up playa lakes, as shown in Figure 20.23. Summer dust in the Southeast lacks this calcium-rich "fingerprint" of western United States dust. Instead, the abundant clay, revealed by the very high aluminum content, indicates a desert source that lacks numerous playa lakes. Lakes are extremely rare in the Sahara, so Saharan dust is high in aluminum and low in calcium. Saharan dust is also iron rich, which accounts for the red color of summer dust.

Insights

What Is the Fallout of Dust from Distant Places? By using the compositional fingerprint as a guide, scientists estimate that Saharan dust episodes occur in the Southeast about three times each summer and, on average, last for about 10 days. About 30 percent of the continental United States is hazy with African dust during the typical dust episode. Perhaps as much as one billion metric tons of African dust crosses the Atlantic each year.

The dust episodes not only further emphasize the importance of wind as a geologic process, but they also explain some biologic phenomena. "Red tides" in coastal marine waters commonly coincide with dust episodes that provide mineral nutrients for tiny plankton. Unusually large concentrations of tiny red plankton cause the red discoloration of the seawater. This plankton poisons fish and shellfish and causes illness in humans who eat the poisoned seafood. Red tides, therefore, cause closure of fisheries and recreational beaches. Fortunately, the red plankton usually are not abundant because of a lack of nutritional iron in the seawater. However, Saharan dust delivers iron during dust episodes so that the plankton temporarily flourish. In addition to tiny mineral grains, the African dust includes microbes that are linked to widespread disease and death of coral throughout the Caribbean. About 40 million tons of African dust also arrive in South America every year. The delivery of mineral nutrients in the dust explains the fertility of soil in the Amazon rainforest, which would otherwise have long ago lost these nutrients to mineral dissolution in the tropical climate.

The western United States also received distant dust from Asian deserts. Satellite images show large dust clouds moving with prevailing westerly winds from central Asia, across the Pacific Ocean, and completely across North America. The largest of these Asian episodes so far recorded was in April 2001; it doubled the dust in the North American sky during the six days it took for the dust cloud to cross the continent.

The geologic importance of wind seems obvious when standing among the dunes of Death Valley or Jockey's Ridge. Now you know that wind is a globally important mover of sediment, with impact on human health and natural ecosystems.

Putting It Together—How Do We Know . . . That Wind Blows Dust Across Oceans?

• Deserts are the primary source of dust, which wind transports from continent to continent across wide oceans.

• Saharan dust blows into the Caribbean and eastern United States during the summer. Atmospheric scientists track the dust clouds by satellites. A chemical fingerprint distinguishes Saharan dust from dust originating in the western United States deserts.

• Dust not only diminishes visibility and inflames respiratory ailments, but also causes changes in marine ecology, such as red tides and the demise of coral reefs.

Where Are You and Where Are You Going?

Wind joins with glacial ice, water flowing in streams, and water moved by waves and tides, as an important geologic process that modifies Earth's surface. Like water and ice, wind erodes the surface and deposits the resulting sediment to form distinctive landforms and landscapes.

Wind blows everywhere; its direction and velocity are wind-determined by variations in air pressure from place to place. The varying air pressures, in turn, result from large-scale atmospheric circulation driven by a combination of convection and Earth's rotation. These air-pressure and wind patterns also determine the locations of dry regions called deserts.

Wind is most effective at modifying Earth's surface where small, loose grains are abundant on the surface, vegetation is sparse, and wind blows strongly. This combination of conditions is most commonly met in deserts. It can occur in other settings, however, especially coastal areas adjacent to wide, sandy beaches. Rolling hills in some locations are ancient sand dunes that are now stabilized by vegetation but were active during earlier drier times when plant cover was more sparse.

Sand dunes are the most recognizable landform produced by blowing wind, but most deserts contain only local regions of blowing sand. Where sand dunes are present, they reveal a wide variety of shapes determined by the abundance of sand, the sparseness of vegetation that inhibits sand movement, and the consistency of wind directions. Deflated pans, sand-blasted yardangs, and abraded and pitted ventifacts reveal evidence of wind erosion, on Mars as well as on Earth. Gravel-size fragments cover large areas of desert landscapes and are too large for wind to move. Pavement-like surfaces of gravel armor over layers of fine silt that gradually accumulate beneath the larger stones.

Wind also transports and deposits fine dust. These small particles move along with the wind and may accumulate far from their original sources. Thick windblown silt blankets of loess weather to produce soil that is notable for high nutrient content and excellent water retention. Some loess accumulates downwind of deserts. Other loess links to river valleys, where streams deposit silt that is then picked up and moved farther across the landscape by wind. Loess is especially thick and widespread near rivers that carried large volumes of sediment and were swept by strong, cold, dry winds during and shortly after the last ice age. Even smaller dust particles, only thousandths of a millimeter across, are carried across oceans from the great deserts of North Africa and central Asia. Although springtime dust in the United States comes mostly from windy deserts and dried-up lakebeds, it is sometimes supplemented by dust that crosses the Pacific Ocean from Asia. The dustiest summer days are in the southeastern United States, where dust is delivered by wind blowing westward from the Sahara.

Wind is not only an agent of geologic change, but also a feature of daily weather and the longer-term characteristics of weather that we call climate. Climate changes over geologic time, as indicated by stable sand dunes where once there was desert and evidence for glaciers that once covered much of North America. There is much attention given in the news media and political speeches to the phenomenon of global warming as climate change on a human time frame. In the final chapter of the text you will learn about the evidence for global warming, evaluate the role of human activities on climate change, and examine the ongoing climate change from the deep time perspective of geology.

Active Art

Global Wind Patterns. See the processes that determine global wind patterns.

How Wind Moves Sediment. See how blowing wind moves sediment.

How Sand Dunes Move. See how sand dunes migrate over time to form cross-bedding.

Extension Module

Extension Module 20.1: How the Coriolis Effect Works. Learn more about how Earth's rotation causes moving objects to follow curving, rather than straight-line, paths.

Confirm Your Knowledge

1. What is wind?
2. Wind exists because of density differences in the atmosphere. What factors determine differences in atmospheric density?
3. How does air pressure relate to wind direction?
4. Define the "Coriolis effect." How does it affect moving objects in the northern and southern hemispheres?
5. What are monsoons? Why do they occur?
6. Local winds are more variable than predicted global patterns in Figure 20.3c. Explain the causes of these local variations.
7. What conditions are necessary for the formation of sand dunes?
8. What is a desert?
9. What are the factors that determine the locations of deserts?
10. Compared with water, what are the limitations of wind as an agent of erosion?
11. Why is it just as difficult for the wind to pick up dust as it is for it to pick up sand?
12. How do particles move with the wind?
13. How would you recognize and distinguish between landscapes affected by wind erosion and deposition?
14. What is a desert pavement? How do desert pavements probably form? What evidence supports this interpretation?
15. What evidence supports the hypothesis that most of the dust that contributes to hazy summer skies in the southeastern United States originates in Africa?

Confirm Your Understanding

1. Write an answer for each question in the section headings.
2. What would you expect to be the prevailing winter wind directions for the following locations?
   a. Auckland, New Zealand, Lat: 36 degrees 55 minutes S
   b. Brasília, Brazil, Lat: 15 degrees 47 minutes S
   c. Cape Town, South Africa, Lat: 33 degrees 56 minutes S
   d. Des Moines, Iowa, Lat: 40 degrees 55 minutes N
   e. Kabul, Afghanistan, Lat: 34 degrees 31 minutes N
   f. Lagos, Nigeria, Lat: 06 degrees 27 minutes N
   g. Paris, France, Lat: 48 degrees 52 minutes N
3. Although Mark Twain never actually said the "coldest winter I ever spent was a summer in San Francisco," he did say that "you can never go without a coat in the summer in the city of San Francisco." Using what you know about the general behavior of local wind patterns, explain why the summer in San Francisco is so cold compared to nearby places in California at the same latitude.
4. How can some windy regions show no evidence of wind erosion?
5. If you lived adjacent to an unpaved road that generated a lot of dust in your home, what could you do to diminish the dust (besides paving the road)?
6. Dust in the southeastern United States is thought to come from Africa. How can this be, when the prevailing wind pattern at this latitude is westerly?
7. Locate your hometown or college town in Figures 20.3d and 20.3e. Explain the factors that determine the prevailing wind direction and the seasonal variations in wind direction.
8. Write a paragraph that explains how wind affects or relates to all aspects of the Earth system: the geosphere, hydrosphere, atmosphere, and biosphere.
9. Compare convection inside Earth with convection in the atmosphere. How are they similar? How are they different?
10. Look back to Figure 10.6. Given your additional learning about convection in Chapter 20, explain the slopes of the lines between points A and A′, and between B and B′ in Figure 10.6.

CHAPTER 21

Global Warming: Real-Time Change in the Earth System

Why Study Climate Change?

We hear about "global warming" almost daily. The news sounds like impending doom: Climate change is causing killer heat waves; making sea level rise to flood coastal communities; worsening droughts; and intensifying hurricanes. Most scientists attribute this apparent catastrophe to human activities, whereas some politicians, commentators, and bloggers, supported by a minority of scientists, say that this is a natural change. Earth has been ever-changing for billions of years without the presence of humans; so, how can we separate natural and human causes of changing climate?

Geologists participate in climate-change research because many geologic processes relate to climate: water supplies, floods, the expansion and retreat of glaciers, and the rise and fall of sea level. Also, the geologic record provides a perspective of climate change extending back millions of years.

Skepticism is central to science, and you may be skeptical of the causes of global warming. What is the evidence for global warming? What is the evidence that humans have a hand in the alleged warming? How certain are the conclusions based on this evidence? Here you will examine the evidence, the logic of the conclusions accepted by most Earth scientists, and the degree of certainty attached to those conclusions. In the process, we will present new data and will incorporate many concepts learned in the previous chapters of this text.

After Completing This Chapter, You Will Be Able to

- Evaluate the evidence for global warming.
- Explain the natural and anthropogenic (human-related) processes that cause climate change.
- Explain how climatic conditions are interpreted for times longer than human measurements.
- Evaluate the evidence that human activities contribute to global warming.

Pathway to Learning

21.1 What Is the Evidence for Global Warming?

21.2 What Is the Geologic Record of Climate Change?

21.3 Why Does Climate Change?

These photos of Muir Glacier in Glacier Bay National Park, Alaska, were taken from the same location in 1941 (top) and 2004 (bottom), and demonstrate dramatic glacial retreat in response to warmer temperatures.

21.5 How Do We Know . . . That Humans Cause Global Warming?

21.4 What Natural Processes and Human Activities Affect Global Temperature?

21.6 How Will Climate Change in the Future?

IN THE LAB

Imagine the following scenario: In your morning history class you learn that the Northwest Passage was a long-desired sea route through the Arctic Ocean along the northern coast of North America. This route was explored in order to establish trade between northwestern Europe and Asia without the long trip around Africa or South America (remembering that the Panama and Suez canals did not exist before the twentieth century). Many explorers sought to find this northern route between the late sixteenth and late nineteenth centuries. All of them were turned back by thick sea ice, or their ships became frozen into the ice, leading to disastrous results. Modern ice-breaker ships have plowed through the sea ice in the summer months, but even then this is possible only by hugging the shore, where the water is too shallow for large merchant ships. Almost as an aside, your history professor mentions that recent climate change has reduced the sea ice, opening the Northwest Passage for a few months; perhaps, she suggests, it will be permanently open in the near future, leading to new maritime shipping routes.

Then, you go to your anthropology class where you are studying the Inuit, the indigenous people of the Canadian Arctic. The Inuit regularly travel and hunt on sea ice, so they observe variations in the ice from year to year and pass down their observations from generation to generation. In recent years they have become increasingly concerned that the ice is thinner than observed in their personal experience or oral history. As a result, the ice is less safe to travel across and the hunting season is shorter. Permanent snow patches used as sources of drinking water during dry summer months and as places to store butchered game are disappearing. Glaciers have noticeably retreated in single human lifetimes, and meltwater streams run faster, diminishing fishing opportunities.

It is hard to avoid the conclusion that the observations encountered in your history and anthropology classes are related. You think that these recent changes in the Arctic are important examples of human activities affected by changes in the Earth system. You decide to visit your geology professor to learn more about the disappearing ice in the far north. Your instructor suggests that you do an independent-study project to document what changes are taking place and whether they are unusual.

Your first discovered resource is a satellite image, reproduced in **Figure 21.1**a, which shows the Northwest Passage free of sea ice in September 2007. Is this evidence of global warming? It clearly is an unusual situation because the passage has been blocked with sea ice, even in the summer, since the beginning of explorations more than 400 years ago. Can you use additional satellite images to determine how much the ice cover fluctuates from year to year? The problem, which is evident in Figure 21.1a, is the difficulty of distinguishing sea ice from clouds.

You continue your research and learn that the remote sensing of Earth's surface by satellites is not limited to normal photographs. Some satellites measure heat that is released from Earth's surface into space as natural microwaves. The microwave energy output of land, water, and ice are very different, and the microwaves pass through clouds without being absorbed. This means that detection of microwaves by satellites permits the mapping out of ice, water, or land, regardless of cloud cover. An example of such a map appears as Figure 21.1b, and an artistic rendition of a view of the Arctic from space is shown in Figure 21.1c. It is striking to see how little sea ice existed in 2007 compared to the average conditions over nearly 30 years (Figure 21.1b). From many such maps you are able to build the graph in Figure 21.1d that allows you to interpret changes in the area covered by sea ice over time. Notably, the 2008 sea-ice extent is greater than in 2007, and the small area of ice cover in 2007 was very unusual. However, despite year-to-year variations in sea-ice cover, the overall trend is toward a decrease in Arctic ice cover over time.

However, a more important question comes to mind: How meaningful are 30 years of data? Perhaps there are natural oscillations that happen over longer periods. When one looks at climate across all of geological time there are clearly big changes, such as those from ice age cold to intervening warm spells, each of which is thousands of years long. Just how do geologists and climatologists reconstruct the history of climate change? Moreover, why is it that so many scientists accept the hypothesis that not only is Earth warming up, but also that most of the warming results from human activity?

▶ Figure 21.1 **Visualizing changes in Arctic sea ice.**

21.1 What Is the Evidence for Global Warming?

If you had the task of testing the hypothesis that Earth's surface temperatures are rising, how would you do it? The first thing that probably comes to mind is to collect thermometer measurements everywhere in the world over a long period of time and see whether they show an increase in global air temperature. A good scientific test, however, goes even further and proposes related hypotheses to test. In other words, if the records do show

that temperature is rising, then other phenomena should be observed that corroborate the temperature data. For example, crop-growing seasons should get longer, animal migrations should happen at different times, glaciers should melt, sea level should rise, and polar sea ice should contract to cover a smaller area (as you have already seen in Figure 21.1). In this section, we will explore what the temperature records show and how they are constructed. Then we will check the interpretation of the temperature records against two geological observations—changes in glacier size and sea level.

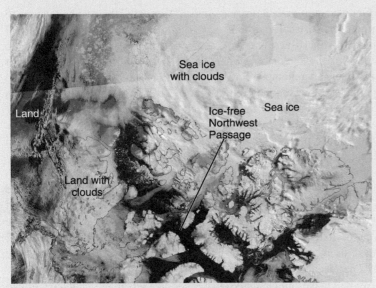

(a) Satellite view of the Canadian Arctic in September 2007 shows open water in the Northwest Passage for the first time in recorded history.

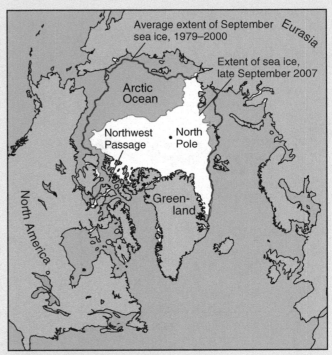

(b) A map of sea-ice extent in September 2007 based on satellite measurements of microwave radiation emitted from Earth's surface.

(c) An artist's rendition of the Arctic as seen from space in September 2007.

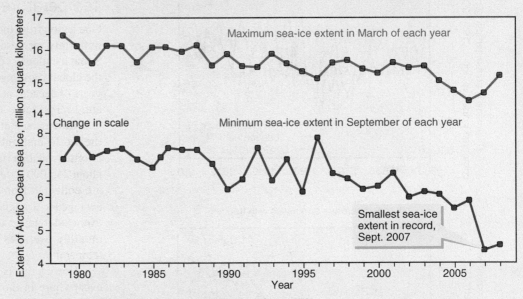

(d) Graph of area covered by Arctic sea ice by year since 1979.

The Evidence from Local Temperature Records

Figure 21.2a displays instrumental temperature records for West Palm Beach, Florida; Albuquerque, New Mexico; and Dickinson, North Dakota. Comparing the three graphs shows slight temperature increases at all locations in recent years but direct comparison is challenging because the average temperatures are so different at the three cities. Location explains these differences. Dickinson is much farther north than the other two cities,

so it experiences cooler temperatures. West Palm Beach is at sea level, whereas Albuquerque is more than 1500 meters above sea level; air temperatures decrease at higher elevation.

We can get around this comparison difficulty by calculating how much the annual temperature each year differs from the average for all measurements at each city. This departure from the average is graphed in Figure 21.2b. All three temperature records are now on the same graph, but it is still difficult to see overall trends because the scattering of a few unusually cool or warm years produces a very spiky plot.

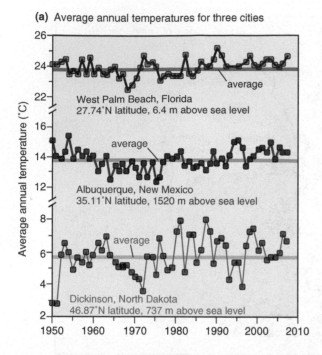

(a) Average annual temperatures for three cities

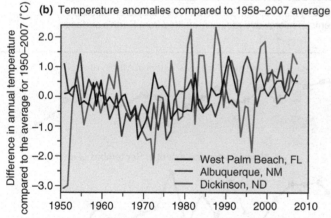

(b) Temperature anomalies compared to 1958–2007 average

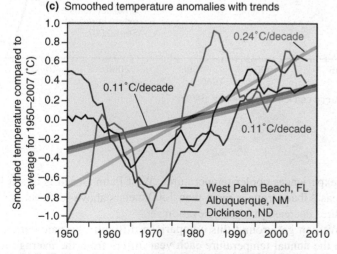

(c) Smoothed temperature anomalies with trends

▲ **Figure 21.2 Ways of studying temperature change over time.** Comparison of temperature measurements from different locations is easier when variations in temperature are compared to average values, fluctuations are mathematically smoothed out, and long-term trends are approximated by lines.

Our next step in data processing is to "smooth" the graphed values to eliminate the distracting spikes. This is done by averaging the annual temperatures in 10-year intervals. For example, the value plotted for the year 1980 is the average of values from 1976–1985; the value plotted at 1981 is the average temperature for 1977–1986, and so on. The smoothed temperature values are graphed in Figure 21.2c. This graph also includes best-fit lines among all of the smoothed data points for each city in order to see any long-term trends in temperature change. This longer-term view describes the climate of each city rather than daily to yearly variations in the local weather.

Now we can better see similarities and differences. All three cities experienced overall warming since 1970. West Palm Beach and Albuquerque also show a cooling trend from 1950 to 1960, while Dickinson experienced warming during this period. The best-fit lines show that for the entire 58-year period, West Palm Beach and Albuquerque experienced a 0.11°C increase in average temperature per decade and Dickinson warmed by 0.24°C per decade.

We will make similar analyses of other data sets in this chapter. Climate scientists start by looking at annual variations, convert these to departures from the average values, smooth the data to better show climate variability during longer time spans, and examine overall trends for longer periods.

The Evidence from Global Temperature Records

One lesson from the data graphed in Figure 21.2 is that temperatures do not uniformly increase or decrease during the same times at every location. Some locations experience warming, while others are cooling. To address the global warming hypothesis, we must look not only at trends over many years or decades at individual locations, but also trends in average temperature across large areas.

Climate scientists collect data from all over the world. They divide the world map into rectangular boxes that are 5° of latitude by 5° of longitude in area (in the continental United States, each box has an area of about 245,000 square kilometers). Temperature data for a particular month are collected from every measurement location within a box and then averaged to a single value for that rectangle. Then, all of the box values are averaged together to define the global temperature for the month, and the monthly averages are combined to calculate the annual global mean temperature.

However, it is not easy to come up with temperature measurements everywhere in the world. The total number of measurement locations was very small in the mid-nineteenth century but it does increase to about 3000 locations after 1950. The spacing of data locations on land is very dense in populated parts of the world, such as the United States, southern Canada, Europe and Japan, but is very sparse over the interior of South America, Africa, and all of Antarctica. Data for the oceans consist of sea-surface-temperature measurements recorded by merchant and naval ships, mostly along the main shipping routes. Water, rather than air, temperatures are recorded because shipboard measurements of air temperature are more variable than the water temperatures and data analysis shows that sea-surface temperature corresponds closely to the average air temperature.

Figure 21.3 maps out the resulting temperature trends in each boxed area of Earth's surface for 1979–2005, a time when the data were very abundant. Almost all areas of the globe show evidence of warming during this period, especially in the northern hemisphere.

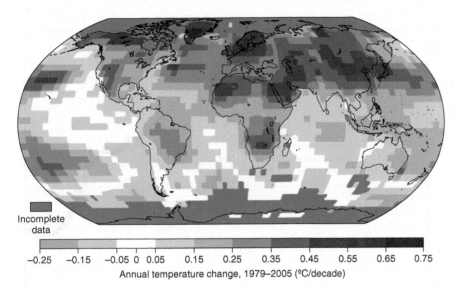

◀ **Figure 21.3 How temperature is changing around the world.** The map shows how much temperature has changed between 1979 and 2005 within rectangular areas that are 5° of latitude wide and 5° of longitude tall.

compared to later years (shown in Figure 21.4b). Nevertheless, when we include all of these uncertainties into the plot (Figure 21.4a), the overall warming since 1850 and especially since 1950 is clearly demonstrated and cannot be explained by uncertainty.

The Supporting Evidence from Glacier Observations

If the global temperature is rising, then we should expect to see evidence that glaciers are melting. The pair of

We can take the average temperature for each grid box in each year and make a graph of global temperature change as shown in **Figure 21.4**a. Like the record of Arctic sea-ice variation, there are ups and downs in the curve, but there is an overall warming trend. Drawing a trend line through all of the data shows an overall average temperature increase of 0.05°C per decade between 1850 and 2000. However, the overall slope of the temperature-change line is even steeper in later times, indicating that the rate of warming is increasing. Eleven of the twelve years from 1995 through 2006 are ranked among the twelve warmest years since 1850. A total temperature increase of about 0.7°C (1.3°F) in the last century may not seem very significant to you. However, when we examine temperature change over geologic time in Section 21.2 you will see that this actually is a very rapid temperature rise.

It is important to ask, however, how certain we can be that these data represent real changes and not errors in measurement or biases introduced by how the measurements were averaged. Climatologists have carefully examined the sources of these uncertainties, and the list of these sources is quite extensive. There can be errors in reading the thermometers, changes in the type of thermometer used, variations in the times of day when the temperature is measured, and even changes in the design of the instrument enclosures that contain the thermometers. Over time, methods of shipboard sea-surface temperature measurements changed from drawing up water in wood buckets or canvas bags to measuring the temperature of the water pumped in to cool the engines. Even if none of these sources of individual measurement uncertainties existed, there would still be uncertainties introduced by the process of averaging the values to obtain a single temperature for a grid box. This uncertainty relates to the number of stations inside the box, how evenly spaced the stations are within the box, and how much the climate conditions vary within the box (for example, from coastal to mountain conditions).

All of these uncertainties are greater for earlier measurements than for more recent ones. Particularly important are the more widely spaced measurements in the nineteenth century,

(a) Average annual global temperature with uncertainty and overall trends

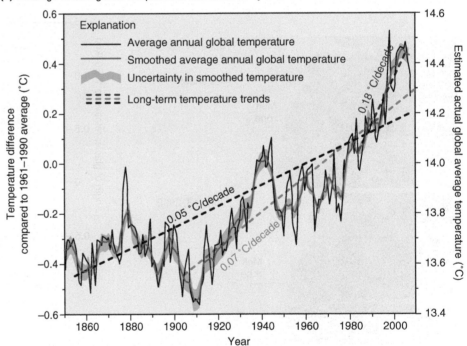

(b) Percentage of Earth surface where measurements are used to calculate average global temperature

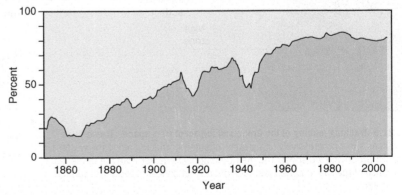

▲ **Figure 21.4 Measured changes in global temperature.**

photos at the beginning of this chapter is dramatic evidence of the rapid retreat of one Alaskan glacier in only 60 years. **Figure 21.5** shows recent evidence of enhanced melting of the Greenland ice cap. Nevertheless, what about data that cover a longer time period that could be more closely matched to the temperature record graphed in Figure 21.4?

There are many historical records that locate the downslope edges of glaciers, especially in the Alps and northwestern North America. An example record for a Norwegian glacier is shown in **Figure 21.6**a. When data such as these are compiled for 169 glaciers from around the world and then averaged together, a global picture emerges of glacier advance from sometime prior to 1600 into the mid-1800s and glacier retreat since that time (Figure 21.6b). The expansion of ice during the eighteenth and early nineteenth centuries coincides with a time of cooler temperatures in the northern hemisphere that is commonly referred to as the Little Ice Age, although glacier advances during this time were trivial in comparison to that of the last "real" ice age 21,000 years ago (Figure 18.34).

(a)

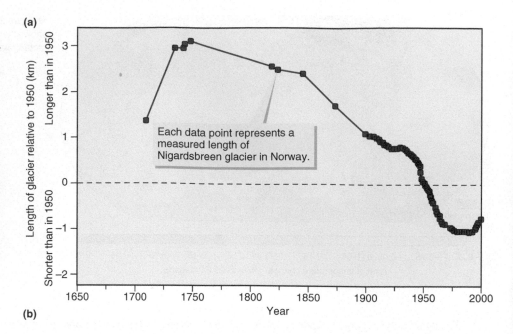

Each data point represents a measured length of Nigardsbreen glacier in Norway.

(b)

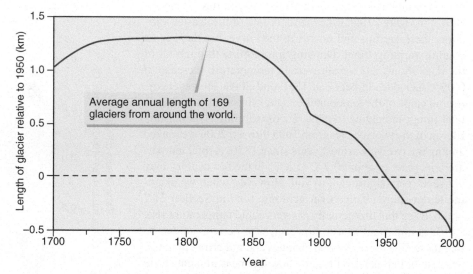

Average annual length of 169 glaciers from around the world.

▲ Figure 21.6 **Measurements show that glaciers are retreating.**

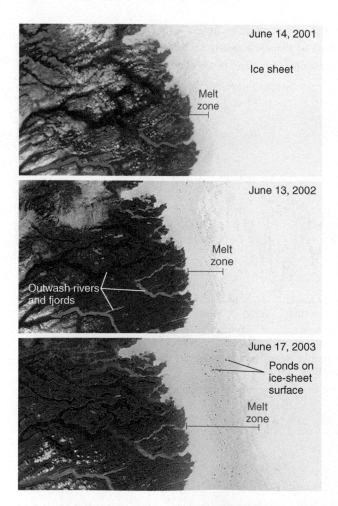

June 14, 2001

Ice sheet

Melt zone

June 13, 2002

Outwash rivers and fjords

Melt zone

June 17, 2003

Ponds on ice-sheet surface

Melt zone

▲ Figure 21.5 **Tracking melting of the Greenland ice sheet from space.** These three satellite images show progressively larger areas of summer melting along the western margin of the Greenland ice sheet. The melt zone appears darker than the unmelted glacier because of meltwater ponds and rock debris that concentrates at the surface as the ice melts.

The compilation of changing glacier lengths (Figure 21.6b) is consistent with global rise in temperature since the mid-nineteenth century (Figure 21.4). The consistency of the two data sets adds credibility to the methods used to construct both climate records.

The Supporting Evidence from Sea-Level Measurements

Two aspects of global warming should cause rising sea level. The most obvious is that warming melts glacier ice, which adds water to the ocean. The second process is the warming of the ocean itself, which causes the water to expand.

The history of sea-level change is mostly known from tide-gauge measurements in harbors located around the world. Section 19.6 explores how these data are used to reconstruct a history of changing sea level. Careful analysis is necessary because the water levels recorded by tide gauges are also affected by local, short-term variations in weather conditions and, in some locations, long-term uplift or subsidence of the coastline. Satellite measurements of sea-level elevation using sophisticated radar devices began in 1993. The satellite data measure changes in water elevation all over the world oceans, whereas tide gauges record changes only along coastlines.

Figure 21.7 shows how sea level has changed since 1870. As with the temperature record, the uncertainties in sea-level elevation are greater in the earlier part of the record than in more recent years. As predicted by the temperature reconstruction (Figure 21.4), sea level has risen throughout this time period.

Not only has sea level risen since 1870, but the rate of sea-level rise also increased in the mid twentieth century, which correlates with increased warming. Since 1960, the average rate of sea-level rise has been about 1.8 millimeters per year with an uncertainty of 0.5 millimeter per year (1.8 ± 0.5 mm/yr). Shipboard measurements of water temperature at many locations and at many depths in the world ocean have recorded warming, too. The warming is detected as deep as 3000 meters and in the upper 250 meters of the ocean has been changing 0.05–0.1°C per decade almost everywhere in the world since 1955, which is as far back as a sufficient number of measurements exist for making an analysis. Using these temperature data and laboratory measurements of how much seawater expands as it is heated, oceanographers have determined that this ocean warming would cause 0.5–1.0 mm/yr of sea-level rise since 1960. Measured melting of glaciers, including the large ice caps in Greenland and Antarctica, also contributes 0.5–1.0 mm/yr of sea-level rise. Although there are large uncertainties in the magnitudes of sea-level rise attributed to water expansion and glacier melting, they are a reasonable match to the actual measured sea-level rise.

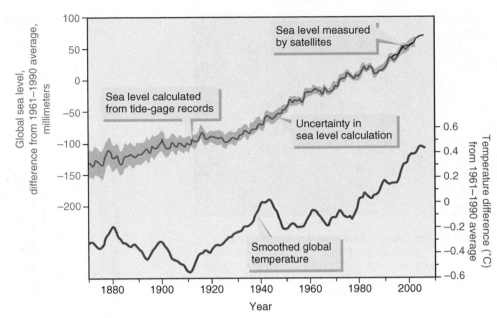

▲ Figure 21.7 Comparing changing global sea level to changing global temperature.

Putting It Together—What Is the Evidence for Global Warming?

• Global temperature change is determined from annual variations in temperatures averaged over large rectangular areas of Earth's land and sea surface. Since 1850 some places have cooled but most have warmed. Changes in global temperature increase or decrease over short times, but the overall trend since 1850 has been toward increasing rates of warming.

• Glaciers around the world advanced in the eighteenth century but have, overall, retreated since 1850, which is consistent with global warming.

• Sea level has been rising at faster rates since the mid-twentieth century, which is consistent with increasing rates of global warming.

21.2 What Is the Geologic Record of Climate Change?

The record of global warming over the last 150 years is well established by temperature measurements and is consistent with observations of natural phenomena, such as decreased ice cover in the Arctic, glacier retreat, and rising sea level. However, the data examined so far do not provide what a geologist would call a long-term perspective on climate change. How do we know whether the recent warming is unusual or part of natural fluctuations? How much has temperature varied over millennia or even longer geologic time spans? If we are to understand the importance of natural versus human causes for global warming, then it is important to examine a longer record of global temperature.

How Paleoclimate Is Reconstructed

The recording and interpreting of past climate conditions constitutes the field of paleoclimatology. Reconstructing a record of past temperatures is very challenging. After all, there are no thermometer records of temperature in America 500 years ago, let alone 5000, 5 million, or 500 million years ago. In the absence of instrumental measurements, paleoclimatologists seek other data that are stand-ins for temperature. These stand-ins are measurable properties in older biological or geological materials that are known from laboratory studies to vary with temperature, so that ancient temperatures can be estimated without being directly measured. A data record that substitutes for direct instrumental measurements is called a **proxy**.

To understand how proxies work, let's examine tree rings, which are commonly used to reconstruct paleoclimate. **Figure 21.8** shows the annual growth rings in a tree. Because a tree grows one ring each year, it is very simple to count the rings inward from the outer bark and know which ring grew during a particular year. It is common to locate trees that are more than 500 years old and the oldest living trees are about 5000 years old. Tree

▲ **Figure 21.8 Tree rings are proxy records of climate change.** This cut pine tree shows annual growth rings in the wood of the tree trunk. Light-colored wood forms early in the growing season and the dark wood grows late in the growing season. A tree ring for one year consists of the combination of light and dark wood.

rings contain a record of environmental conditions at the time the tree grew. Paleoclimatologists do not need to cut down a tree to measure and collect wood from each ring; instead they drill out a narrow cylinder of wood from the tree without doing serious harm.

The width of annual rings varies within a tree, which means that trees grow more during some years than during others. Careful measurements of tree growth show that many factors affect growth, such as moisture, temperature, fire, disease, and soil nutrients. Furthermore, the relative importance of these factors varies from species to species and within a species from one location to another.

To build a proxy-temperature record it is important to know that temperature is the major factor determining the annual increment of tree growth in conifers growing at high elevations and high latitudes. In these two environments moisture is almost always adequate for healthy trees and plays very little role in different rates of growth from year to year. The correlation to temperature is even stronger when the measured density of wood that forms late in the summer growing season is combined with measurements of annual ring width. **Figure 21.9** shows the results of a proxy temperature record obtained from studying 404 trees at six localities in the Canadian Rocky Mountains.

Figure 21.9a is an especially critical graph because it compares the proxy temperature estimates to nearby instrumental temperature measurements. If tree rings provide a reliable proxy record, then the tree-ring temperature estimates should correspond to actual temperature measurements for those years where both records exist. There is no perfect match between these two temperature records, but some differences are expected because of (a) the subordinate role of factors other than temperature in tree growth, and (b) the fact that the instrumental measurements were not made at the exact same locations where the trees are growing. Nonetheless, the same general increasing and decreasing variations exist in both the proxy and instrumental temperature values.

The generally good correlation of tree-ring properties and measured temperature lends confidence to looking at the longer period of proxy temperature measurements plotted in Figure 21.9b. Smoothing out the annual variability provides the best view of general trends in temperature for 1000 years, and also diminishes the influence of other factors that may

(a) Comparison of measured temperatures to proxy temperatures estimated from tree-ring analyses

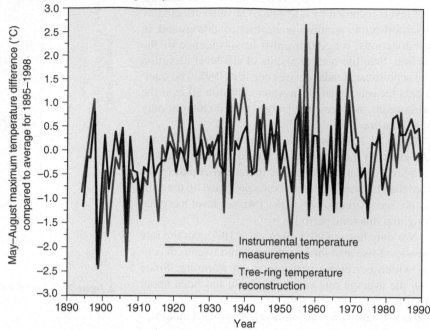

(b) Proxy temperature record for the last 1000 years in the Canadian Rocky Mountains, based on tree-ring analyses

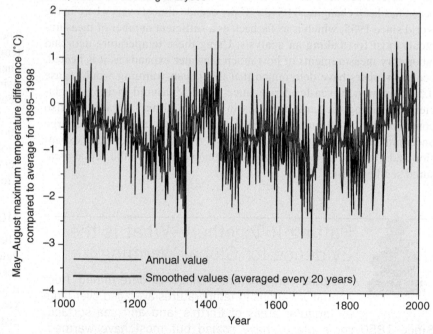

▲ **Figure 21.9 Tree-ring proxy records of temperature in the Canadian Rocky Mountains.**

influence the proxy temperature values. Based on this graph, it is reasonable to conclude that growing-season temperatures in the late twentieth century in the Canadian Rockies are among the highest in the last 1000 years, and comparable warmth likely occurred only around 1050 and 1400.

Paleoclimatologists employ many proxy records to record ancient climates. Some records are qualitative observations in rocks, such as ancient deposits attributed to glaciers or desert sand dunes, and climatic interpretations of fossils, such as plant leaf shapes that are adapted to different temperature and precipitation. Other proxies are quantitative and based

▶ **Figure 21.10 Temperature change across geologic time.** The graph schematically shows overall climatic cooling since the Mesozoic based on many different proxy data sets. Notice the abrupt fluctuations between colder glacial ice ages and warmer interglacial times during the last million years.

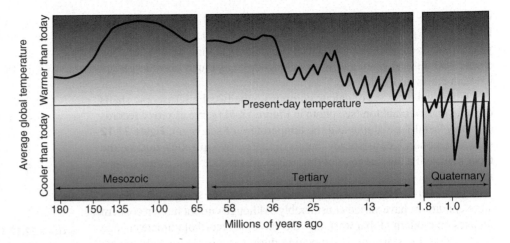

on geochemical measurements; for example using oxygen-isotope ratios in tiny calcite-secreting plankton to determine past glacial-ice volume (explained in Figure 18.41). Combining many different proxy data sets provides the most complete picture of past climates. However, just like the instrumental records of temperature and sea level, the proxy temperature estimates are more uncertain in earlier times than more recently.

Climate Change Before the Quaternary Period

The power of paleoclimate-proxy data is particularly apparent when we see temperature reconstructions over long, geologic time intervals. **Figure 21.10** graphs the estimated temperature change since the Jurassic Period, the time of the dinosaurs. Notice that the vertical axis on the graph lacks any numbers. This is because the uncertainties are very large when evaluating proxy data from rocks this old, so although the relative variations in temperatures from one increment of time to another are usually clear, the actual mean-annual global temperature value is elusive. The best estimates of global temperature at the beginning of the Cenozoic, 65.5 million years ago, suggest conditions 5–10°C warmer than today.

Two things stand out from the temperatures graphed in Figure 21.10. First, global temperature was warmer during most of the last 180 million years than it is today. Overall, global climate experienced a cooling trend during the Cenozoic. Second, the graph is very smooth for the older part of the record and dominated by fluctuating peaks and troughs in more recent times. Part of this difference relates to the resolution of the climate record. Some climate records provide proxy temperature values for each year (tree rings, for example) extending back to about 800,000 years. Beyond that time, the uncertainty in the age of the material that provides the estimated temperature value may be tens of thousands of years, or even larger. As a result, these values are averaged together to show general smooth trends for the earlier part of the graph. However, part of the spikiness of the later part of the record seems to realistically portray unusually wide swings in the climate system, which we will explore a bit more below.

We can combine these generalized temperature trends with the longer geologic record of large ice-sheet glaciers. Glaciers produce distinctive landforms and deposits that survive in the geologic record. Most easily recognized are the chaotic deposits of glacial till (Figure 18.15) and striations scratched into outcrops by the passing ice that encloses protruding rocks (Figure 18.11). The photo in **Figure 21.11** shows that ancient till deposits overlying striated bedrock provide records of glacial ice ages deep in the geologic past. Geologists use paleogeographic reconstructions based on plate tectonics to relocate each glaciated region to its place on the globe when the glaciers existed. Then, the extent of glacial advance away from polar regions and toward lower latitudes can be plotted, as done in Figure 21.11, to identify the coldest times in global climate history. This plot shows that the cooling experienced during Cenozoic time coincides with an increasing build up of glacial ice, which has been permanent in Antarctica for at least the last 12 million years (and perhaps much longer). During the Quaternary Period glaciers periodically advanced over large

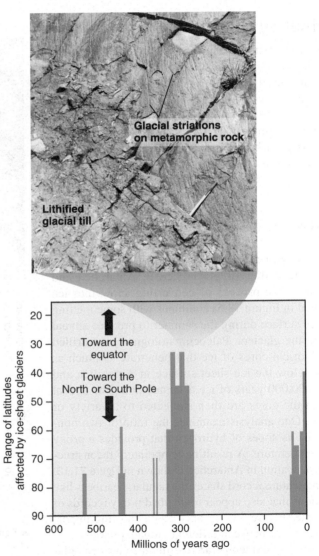

▲ **Figure 21.11 Reconstructing past ice ages.** The photograph shows lithified glacial till resting on striated rock as evidence of glacial deposition and erosion, respectively, in South Africa during the late Paleozoic. Similar evidence of ancient ice sheets is found in many parts of the world in rocks of various ages. Using plate tectonic principles to restore continents to their ancient locations (see Figure 12.43, for example) allows geologists to describe the scale of ancient ice ages according to the lowest latitude where the glacial evidence is found; the results are shown on the graph.

regions of the Northern Hemisphere (see Figure 18.34, for example). The late Cenozoic glacial extent is unusual when compared to the 540-million-year Phanerozoic rock record. Only a prolonged cold interval centered on a time 300 million years ago shows evidence of as much or more glacial ice than exists on Earth today.

Chemical analyses of calcite composing the fossil microscopic shells of some marine plankton (see photo in Figure 18.41) serve as proxy records of both glacial-ice volume and the temperature of sea water. **Figure 21.12** shows a general consistency between these two proxies during the Cenozoic, with the oceans cooling while glacier ice has, overall, been increasing.

The record of climate change over geologic time shows that temperature conditions have varied considerably. Although current human concern focuses on modern global warming, the geologic record of climate change implies that this warm up is happening during one of the overall coolest times in at least 250 million years.

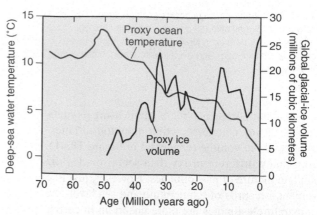

▲ **Figure 21.12 Comparing proxy records of ocean temperature and glacial ice.** Geochemical analyses of calcite secreted by plankton that then accumulated as fossils in deep-sea sediment provide proxy records of both an unsteady cooling of the world ocean and unsteady growth of ice sheets on land during the Cenozoic.

Fluctuating Quaternary Climate

Quaternary climate is characterized by extreme fluctuations, which are schematically illustrated in Figure 21.10 and reconstructed in detail from many parts of the world. The cold times were the ice ages, the latest one reaching its peak about 21,000 years ago. The timing of ice ages and intervening warm periods is explained in Section 18.10. Ice ages have reached their peak about every 100,000 years, like a very consistent clock, for the last 800,000 years. The warmest of the interglacial times between ice ages persist for 5,000 to 30,000 years before temperatures start to cool down (Figure 18.42). The transitions from ice ages to interglacial warmth usually requires no more than 10,000 years.

The longest records of Quaternary climate change come from studies of glacial ice in Antarctica and Greenland. The process of converting snow into ice (described in Figure 18.5) combines with some melting of the ice surface during the summer to produce annual layers in the glaciers. Paleoclimatologists have drilled out cylindrical cores of ice that penetrate as much as 3.2 km below the ice-sheet surface in Antarctica and sample 800,000 years of ice accumulation. The annual layers in the cores are then subjected to a variety of analyses. One analysis examines the ratio of two non-radioactive isotopes of hydrogen that provides a proxy for air temperature. A resulting temperature reconstruction for a location in Antarctica is shown in **Figure 21.13**. This temperature record shares the regular, heartbeat-like fluctuations that also appear in graphed proxy records of ice volume (Figure 18.42), once again showing important consistency between different climate proxies.

The maximum temperature difference between glacial and interglacial times in Antarctica is about 10°C. Other temperature proxies collected around the world suggest that the temperature variation is less extreme at lower latitudes, and also less variable in seawater, so that the global surface temperature during a Quaternary ice age is probably only about 5°C cooler than during interglacial (between ice age) times. Most of that 5° warm-up since the end of the last ice age was spread out over about 10,000 years. Notice, therefore,

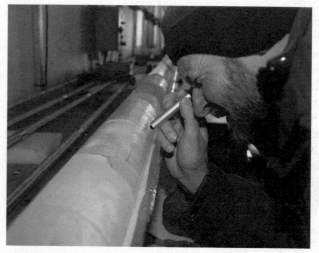

(a) A paleoclimatologist uses a magnifying lens to examine annual layers in an ice core from Antarctica.

(b) The graph shows a proxy air temperature record based on nearly 5800 measurements of hydrogen isotopes from samples cut from the ice core shown in (a).

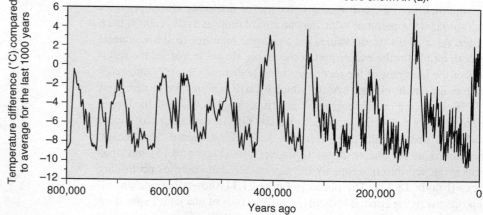

▲ **Figure 21.13 A long history of Antarctic temperature.** Analyses of hydrogen isotopes from an ice core drilled more than 3.2 kilometers downward through the Antarctic ice sheet provide an 800,000-year long proxy record of temperature.

that the approximately 0.7°C rise in global temperature just since 1900 (Figure 21.4), which seems small, is actually a very substantial change when placed in this longer, geologic-time perspective.

The Quaternary climate records also provide insightful evidence of very abrupt shifts in temperature that happened separately from the regular swings between glacial and interglacial conditions. The most dramatic examples occurred around 12,900 years ago. As the world was warming out of the previous ice age, many locations experienced an abrupt step backwards to ice-age coldness. This cold snap was first recognized in Scandinavia. Here, sedimentary layers consist of glacial till overlain by mud containing wood and pollen remains of a temperate forest that is then abruptly followed by layers lacking evidence for trees but, instead, contain pollen of mountain avens. Mountain avens is an Arctic wildflower that lives today only in the coldest tundra environments. Therefore, the sedimentary layers record a post–ice-age warm up that abruptly reversed to colder conditions. The cold period is named the Younger Dryas by paleoclimatologists in recognition of the distinctive wildflower, whose Latin name is *Dryas octopetala*.

The proxy data sets shown in **Figure 21.14** illustrate two important characteristics of the Younger Dryas. First, the cold period lasted only about 1300 years. Second, the Younger Dryas began and ended very abruptly. The temperature drop to ice-age conditions took less than a century. The roughly 10°C warm-up in Greenland at the end of the Younger Dryas happened in less than a decade. However, even though the records of the Younger Dryas are abundant in the northern hemisphere, it was not clearly a global event. Proxy temperature data from Antarctica indicate warming at this same time, and measurements of expansion and retreat of glaciers in South America and New Zealand are not consistent with a global cool down during the Younger Dryas. Although probably not a time of global cooling, the Younger Dryas tells us that the natural climate system is capable of very rapid change.

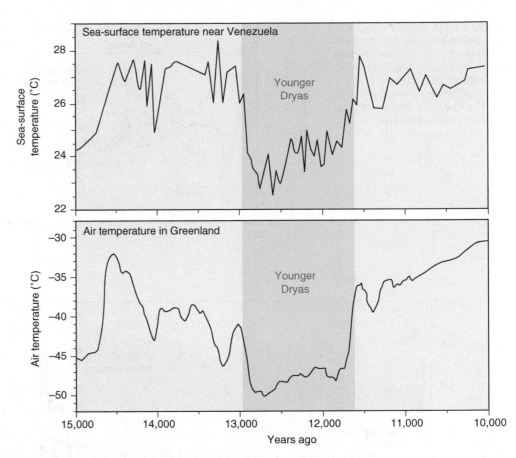

▲ **Figure 21.14 Documenting the Younger Dryas cold spell.** Proxy temperature records from fossil organisms collected offshore of northern Venezuela and from an ice core in Greenland are among many from northern hemisphere localities that show evidence of the Younger Dryas cold interval. Notice that these temperature records show that the cold spell began and ended very abruptly.

The Climate of the Last 1000 Years

While these long-term geologic perspectives on climate change provide important insights to natural temperature fluctuations, the last millennium arguably provides the best record for comparison to modern-day global warming.

Historical records from northern Europe suggest significant temperature oscillations during the last thousand or so years. For example, much of this region experienced warm weather from about CE 800 to 1200. Conditions were sufficiently mild in Greenland that the Vikings colonized the coastal areas starting in CE 986, finding an adequate growing season for raising some crops and livestock along with ice-free harbors that allowed fishing. Historical accounts also refer to cultivation of pomegranates, figs, and olives as far north as the Rhine Valley of Germany, where modern-day winters are too cold for these plants to survive. This Medieval Warm Period, named for its overlap with the European Medieval historical period, was followed by cooler temperatures, including the glacier advances of the Little Ice Age between about 1600 and 1850 (Figure 21.6b). As Greenland cooled down, the Viking settlements were stressed by crop-killing frosts and harbors that sometimes remained icebound throughout the entire summer; Greenland villages were gradually abandoned between 1350 and 1500.

Tree-ring records of northern hemisphere temperatures also register the Medieval Warm Period and Little Ice Age. This can be seen to some extent in the Canadian Rockies (Figure 21.9), where temperatures between 1650 and 1850 were generally cooler than between 1000 and 1200. A better picture is obtained, however, when combining many tree-ring records over large areas in order to average out local climate variability. **Figure 21.15** shows such a temperature reconstruction based on the study of tree rings from 14 locations in North America and Eurasia. Because tree rings provide reliable insights into temperature variability only from generally cool regions, nearly all of the data used for making Figure 21.15 come from locations at latitudes higher than 45 degrees North.

Notably, the historical accounts, tree-ring data, and other proxies supporting the existence of the Medieval Warm Period and Little Ice Age come from very few places and almost entirely in the northern hemisphere. If we are to compare this longer temperature record to recent *global* warming we need a *global* proxy record. **Figure 21.16** plots the most recent such global temperature reconstruction, back to CE 500, with a comparison to instrumental measurements since 1850. Even though the uncertainty is quite large for the proxy temperatures, the warming trend since 1850 is much more abrupt than any record of the last 1000 years. In addition, the global temperatures of the last 15 years are quite likely the warmest of this entire period. The Medieval Warm Period, when examined with global, rather than just northern hemisphere, data does not appear to have been as warm as the present.

Examining paleoclimatic records at different time and geographic scales puts the modern-day global warming in perspective. Surface temperature has fluctuated throughout Earth history and when considering all of

▶ **Figure 21.15 Regional tree-ring temperature reconstructions for the last 1200 years.** Tree-ring measurements from 14 locations at high latitudes in the northern hemisphere provide a basis for comparing recent warming to the Medieval Warm Period and Little Ice Age conditions recorded in historical documents.

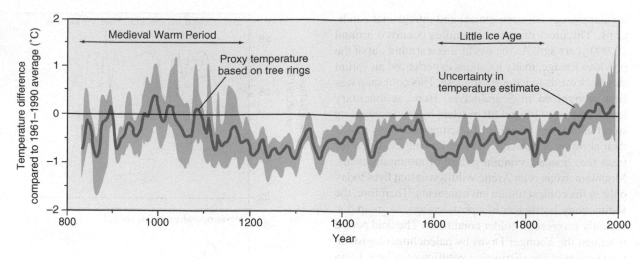

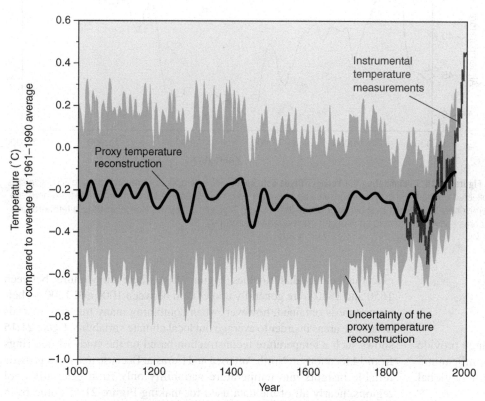

▲ **Figure 21.16 Global temperature changes for the last millennium.** A combination of many different temperature proxies from localities around the world, on land and in the sea, provides long-term reference for comparison to the shorter record of instrumental temperature measurements. The uncertainties of the temperature reconstruction are very large but the proxy temperatures closely match the actual measured temperatures for the times where both records exist. Temperatures recorded in the last decade of the twentieth century, and probably since 1950, have no precedent in the previous 1000 years.

Phanerozoic time, we live in a relatively cool period (Figures 21.10–21.12). Quaternary time, however, shows dramatic fluctuations on shorter time frames, and Earth is currently in a relatively warm interglacial period. Although current northern hemisphere temperatures may not differ very much from the Medieval Warm Period, this earlier warming was apparently not a global phenomenon. Therefore, when looking at the global record, the current warming appears unprecedented within the last 1000 years.

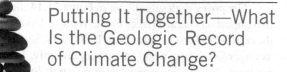

Putting It Together—What Is the Geologic Record of Climate Change?

• Measurements of physical and chemical properties of biologic and geologic materials that are known to vary with temperature provide proxy records of paleoclimate back through geologic time.

• The natural climate system, operating prior to the appearance of humans on Earth, exhibits considerable variations, including very abrupt cooling and warming, such as that defining the Younger Dryas cold snap between 12,900 and 11,600 years ago.

• Current global warming is unprecedented during the last 1000 years and is occurring during an interglacial period within a trend of long-term cooling of global temperatures during the Cenozoic.

21.3 Why Does Climate Change?

The evidence for climate change on long, geologic time scales and short, historic time scales is clearly documented. Our next task is to understand how these changes in temperature, at both time scales, can happen. Air temperature is a measure of the amount of heat energy present in the atmosphere. Therefore, in order to explain temperature changes we need to understand the way heat moves into, through, and out of the atmosphere.

The Energy Budget

The Sun is the source of the heat energy that determines air temperature and drives all aspects of the climate system, including winds and precipitation patterns. Satellites orbiting Earth detect about 342 W/m² of heat energy inbound from the Sun. This quantity of heat is about 4500 times greater than the heat flow that arrives at the surface from inside Earth (Figure 10.5).

► **Figure 21.17 Comparing incoming and outgoing energy waves.** The incoming shortwave energy from the Sun has a different wavelength than the longwave energy radiated from Earth's surface. This energy, felt as heat and some of it seen as visible light, occupies only a small part of the spectrum of energy wavelengths exhibited by different forms of electromagnetic energy.

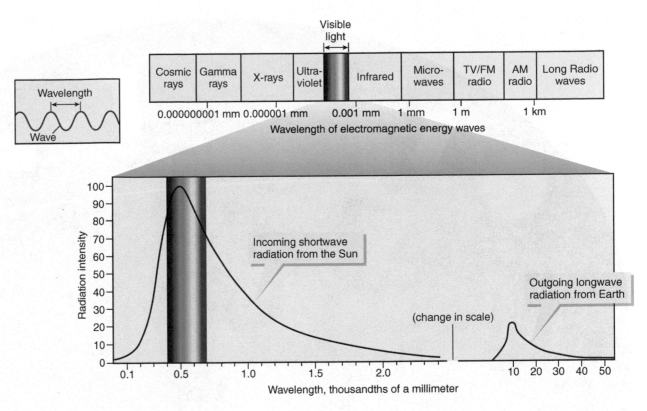

Therefore, even though we placed considerable emphasis on Earth's internal heat in previous chapters, we can ignore it when studying climate.

However, energy is not only absorbed at the surface and in the atmosphere, but it is also emitted back into space from both the surface and the atmosphere. Over long periods and averaged over Earth's entire surface, the incoming energy is balanced by outgoing energy. We can use the phrase **energy budget** to describe this balancing of incoming and outgoing energy in much the same way that a financial budget tracks incoming revenues and outgoing expenses.

How does the energy budget relate to the surface temperature? The essential thing to understand in answering this question is that the measured characteristics of the outgoing energy are different from the incoming energy. Energy radiated by the Sun to Earth and radiated back to space from Earth is in the form of waves. The wavelengths of the radiant energy (distance from one wave crest to another) are extremely tiny compared to ocean waves (described in Chapter 19) or earthquake waves (described in Chapters 8 and 11). **Figure 21.17** shows that the wavelengths of incoming solar radiation are different from the outgoing heat emissions. The incoming short-wavelength (shortwave) radiation is mostly in the form of visible light along with ultraviolet radiation that causes sunburn. The outgoing long-wavelength (longwave) radiation is similar to the heat that you feel when sitting close to a fire or place your hand near the stove. A very small amount of the emitted longwave energy is also in the form of microwaves that have a wavelength of about 1 millimeter; detection of these naturally emitted microwaves is critical to the mapping of sea ice, as you saw at the beginning of the chapter.

With this understanding of the differences between incoming and outgoing heat energy, we can explore the total energy budget for the atmosphere, which will be our key to understanding the factors that change climate at various time scales. **Figure 21.18** is a pictorial representation of the energy budget. The illustration contains a lot of information but if we concentrate on the variables that can change so as to also affect Earth's surface temperature, we can focus on just three factors.

Identifying Factors That Change in the Energy Budget

One factor that clearly would cause global surface temperature change is variations in the amount of incoming solar energy. If the total incoming energy increased or decreased, then surface temperatures would warm up or cool down accordingly.

A second factor that would cause changes in surface temperature is variations in the amount of solar energy that is reflected back into space. The tendency of material to reflect rather than to absorb solar energy is called the **albedo**. High-albedo materials, such as white cloud tops, snow and ice, city streets and buildings, and bare ground, reflect some incoming shortwave energy back into space. Low-albedo materials, such as water and most vegetation, absorb rather than reflect this energy. If surface materials absorb the solar energy, they then contain more heat that can be radiated into the near-surface atmosphere, which causes it to warm up, too. Measurements of the incoming solar radiation by satellites compared with measurements at Earth's surface show that only about 49 percent of the incoming shortwave energy is absorbed at the surface by rocks, soil, water, plants, snow, and ice. About 20 percent is absorbed by gases and particles in the atmosphere. The remaining 31 percent is reflected back into space; therefore Earth's albedo is said to be 31 percent. Figure 21.18 shows that most of the albedo relates to clouds and aerosols (tiny fluid droplets in the atmosphere) that reflect solar energy, and about one-third of Earth's albedo is determined by properties of the planet's surface. So, if the amount of clouds or aerosols in the atmosphere change or if the distribution of bare rock, vegetation, water, snow, and ice change on Earth's surface change, then the albedo and surface temperatures would also change.

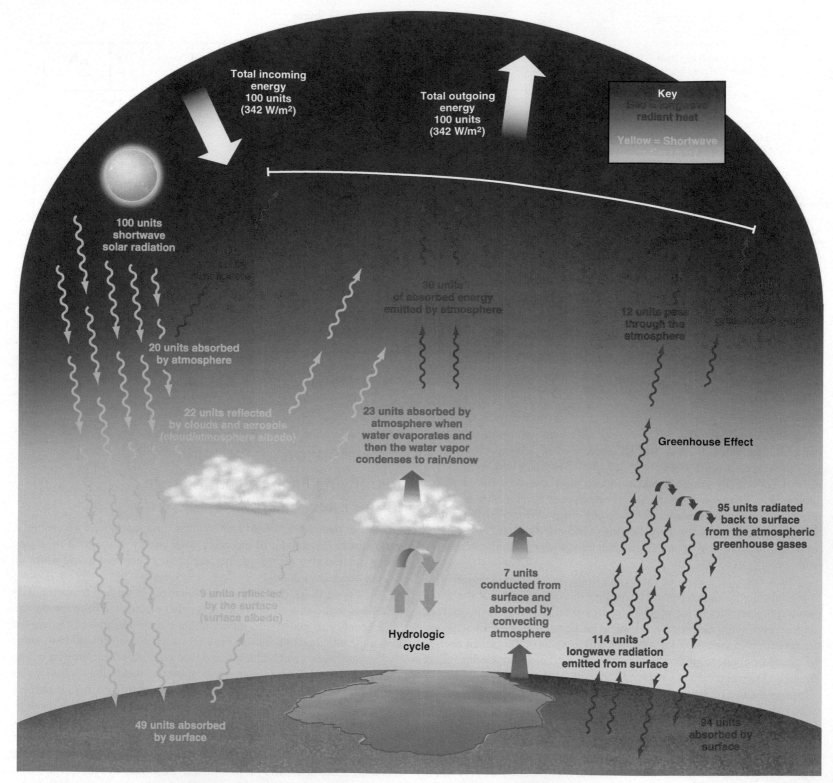

▲ **Figure 21.18 Visualizing Earth's atmospheric energy budget.**

The third factor that would change the temperature relates to the trapping of heat near Earth's surface, which is illustrated on the right side of Figure 21.18. Gases in the atmosphere tend to be much more transparent to shortwave radiation than to longwave radiation. Nitrogen (N_2) and oxygen (O_2), which make up 99 percent of the atmosphere, neither absorb the incoming shortwave nor outgoing longwave radiation to any significant

extent. However, other atmospheric gases do interact with the incoming and outgoing radiation, as shown in **Figure 21.19**.

Ozone is the only atmospheric gas that has a significant effect on incoming shortwave, particularly ultraviolet, radiation. You may have heard concerns about decreased ozone concentrations in the atmosphere because of reactions between ozone and human-introduced gases used as

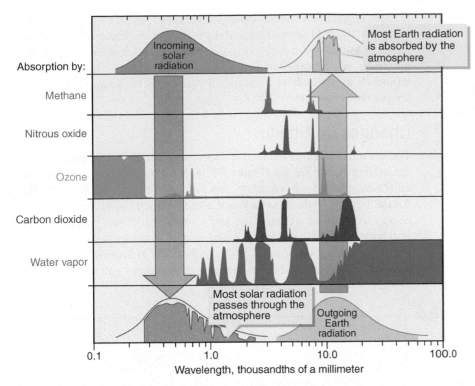

▲ **Figure 21.19 How atmospheric gases absorb radiant energy.** Some atmospheric gases absorb energy at the same wavelengths as incoming solar radiation or outgoing Earth radiation. Most solar radiation passes through the atmosphere to warm Earth's surface. However, most outgoing Earth radiation is intercepted by the greenhouses gases, of which water vapor, carbon dioxide, nitrous oxide, and methane are the most significant. When comparing this diagram to Figure 21.17, notice that the horizontal axis of this graph is logarithmic rather than linear.

refrigerants, propellants in aerosol-spray cans, and a variety of manufacturing uses. The depletion of ozone threatens to cause increased risk of sunburn and skin cancer because more ultraviolet radiation is reaching Earth's surface now than in the past. Ozone depletion has been reversed by international agreement to reduce emission of the harmful human-introduced gases.

More important to our interest in temperature change is the observation that outgoing longwave radiation is absorbed by many atmospheric gases, most notably water vapor, carbon dioxide, methane, and nitrous oxide. These gases, even though they are present in trace amounts, absorb approximately 89 percent of the energy that is emitted from Earth's surface and then release most of it back toward the surface and only a small part into space.

You have likely heard of this backward reflection of heat emitted from Earth's surface as the **greenhouse effect**, and the gases responsible for this effect are known as the "greenhouse gases." The name comes from the similar effect in a glass-roofed greenhouse. In greenhouses, most of the solar heat passes through the roof, but the heat radiated from the soil and plants inside does not penetrate glass and so is reflected back into the greenhouse, keeping the inside temperature much warmer than the outside temperature. Similarly, satellites measure the temperature at the top of our atmosphere to be about −16°C, whereas the average surface temperature is +15°C; the difference is due to the greenhouse effect. Earth's surface would be uninhabitable without the greenhouse warming provided by our blanket-like atmosphere.

Considerable attention has focused on relating recent global warming to increasing concentrations of greenhouse gases in the atmosphere.

Although water vapor is an abundant greenhouse gas, it readily reaches saturation in the atmosphere and is returned to the surface as rain and snow. Therefore, the hydrologic cycle (Figure 16.2) keeps the water-vapor concentration from changing very much. This is why carbon dioxide, methane, and a few other gases are tracked more closely as indications of changing the greenhouse effect.

Ocean Circulation and Climate

We also need to think about the effects of ocean circulation on surface temperature. The energy budget determines the total amount of energy received by Earth but ocean circulation moves that heat around the planet. The Sun's rays reach the low latitudes more directly through most of the year than at high latitudes. This difference accounts for variations in temperature with latitude. However, temperatures are not the same at all locations at the same latitude, as we might expect. For example, London, England, which has an average annual temperature of about 10°C, is at the same latitude as Hudson Bay, Canada, which has an average annual temperature of −5°C.

Most of this otherwise unexpected temperature difference is explained by ocean circulation. Water has a very low albedo, so it absorbs solar energy. However, water does not sit still; the oceans have a complex circulation system that is mostly driven by the wind blowing over the surface and partly caused by density variations in the seawater resulting from combinations of water temperature and saltiness (salinity). Circulation of water volumes that have different temperatures can, therefore, transport heat around on Earth's surface. In the case of London, ocean currents bring warm, equatorial water into the North Atlantic, which keeps coastal areas in this region much warmer than would otherwise be expected at this high latitude. If the circulation system changes, then so will surface temperatures, regardless of the energy budget.

Putting It Together—Why Does Climate Change?

• Climate is fundamentally driven by energy from the Sun.

• Changes in the energy budget at Earth's surface are primarily affected by changes in the solar energy received by Earth, albedo, and atmosphere composition. Variations in ocean circulation affect the distribution of heat on Earth's surface.

21.4   What Natural Processes and Human Activities Affect Global Temperature?

Some aspects of the energy budget, ocean circulation, or both have to be changing to account for global warming. What changes can we detect and which ones are part of natural fluctuations in the Earth system and which one are attributed to human activities? We also have to keep in mind that some changes will increase temperature, while others decrease temperature. In the end, all the contributing factors that affect temperature must be taken into account to determine whether the net, overall result is warming or cooling.

Changes in Solar Energy

Research shows that the amount of solar energy received by Earth is determined by two different processes, which operate at different time scales. One process is variation in the heat output of the Sun. The other is variations in Earth's orbit and rotation on its axis that cause fluctuations in how much of that heat is received by Earth.

Measuring variations in the Sun's output has to be made in space, because heat measurements at Earth's surface are also influenced by albedo and greenhouse gases. The necessary direct measurements of incoming solar heat by satellites have been collected only since 1978. These data are shown in **Figure 21.20** and indicate that change in solar-heat output corresponds directly to the cyclic fluctuation in the number of visible sunspots, which alternates from high numbers (maxima) to low (minima) and back to high values every 11 years. However, the change in solar energy is very small, differing by only about one-tenth of 1 percent between sunspot minima and maxima.

Of greater interest to the warming experienced over the last several centuries is whether there are variations in solar energy over longer times than the 11-year sunspot cycle. Unfortunately, scientists have not yet discovered a very reliable proxy for solar energy, so describing any long-term energy variation becomes a theoretical exercise based on the physics of the Sun's "furnace" and observations of brightness variations among other stars. Several different estimates of past solar energy radiated by the Sun have been developed. The most recent model estimate is used to make the reconstruction shown in Figure 21.20. A long-term increase in solar energy output is apparent in this graph although it is largely hidden by the larger 11-year variability. This total overall trend is, however, very small and equates to about 0.1 W/m^2 increase in solar heating since 1700.

Solar energy received from the Sun is also affected by variations in Earth's orbit and rotation. These cyclic variations, called Milankovitch cycles, are fully explained in Figure 18.44. Variations in the eccentricity (departure from a circle to an ellipse) of the orbit, at a 100,000-year cycle; tilt of the rotation axis (obliquity) at a 41,000-year cycle; and wobble of the rotation axis, at approximately 23,000 years, affects both the total amount of solar energy reaching Earth and the differences in heating between summer and winter and between northern and summer hemispheres. Evidence that the Milankovitch cycles are an important natural control of climate change is revealed by

changes in glacial ice volume and air temperature (compare Figure 18.42 and Figure 21.13) that follow a repeated rhythmic pattern, which corresponds to the three Milankovitch cycles (Figure 18.43). These changes, however, are important only for explaining climate variability over time spans of tens of thousands to hundreds of thousand of years; they do not predict a global temperature change since 1700.

Changes in Albedo

The influence of albedo to produce short-term effects is illustrated by the fluctuating Arctic sea ice (Figure 21.1). Weather during the summer of 2007 was notable for more sunny days and fewer clouds than normal in the Arctic. Fewer clouds meant a lower albedo, which permitted more solar energy to reach the surface. Unusually strong winds and ocean currents also broke up ice that had become very thin by warming in previous years, which exposed more seawater. As a combination of melting and breaking up of ice by strong winds and currents exposed more water, the very low

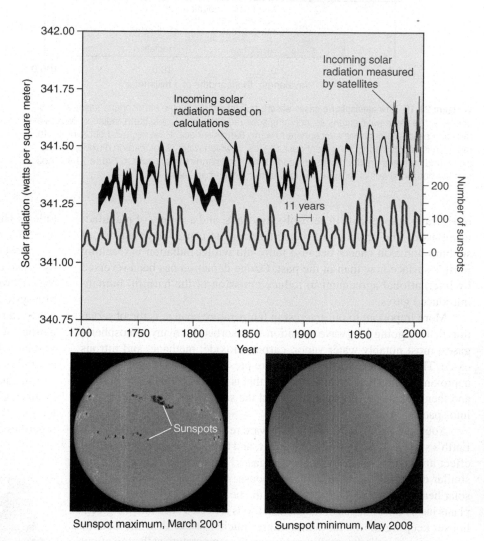

▲ **Figure 21.20 Solar-energy output varies with the sunspot cycle.** Telescope observations of the sun show that the number of sunspots varies in a cycle that is about 11 years long (ranging between 9 and 14 years). Solar-energy output changes with the sunspot cycle. Direct satellite measurements of solar radiation exist only since 1978. The black curve shows calculated solar radiation based on the physics of how stars such as our Sun are known to work. The thickness of the line reflects uncertainty in the energy output, especially at earlier times. Notice that there has been an overall, very small increase in overall solar-energy output independent of the sunspot cycle; energy received during recent sunspot minima is about 0.1 W/m^2 higher than in the 1700s, even though the number of observed sunspots during minima has stayed the same.

albedo of the water increased the amount of heat absorbed at the surface, which increased the sea-ice melting.

Notice in this example that melting was enhanced by the melting ice, which is then described as a positive feedback that increased the warming that was already underway. It is likely that a contrasting negative feedback kicked in during 2008. With record melting of freshwater sea ice during 2007, the saltiness, also called **salinity**, of the Arctic Ocean was diminished. The saltier the water, the colder the temperature required to form sea ice. Summer 2007 melting lowered the salinity so it was easier to reform sea ice during the following winter, which probably contributed to the upward spike in March 2008. These feedback relationships among processes make it very difficult to consider all of the factors that cause warming or cooling.

Long-term, rather than year-to-year, variations in cloud cover across the globe might be expected as a result of global warming, because warmer temperatures cause more evaporation and more cloud formation. The calculated global effect of increased cloud albedo is a decrease in solar energy received at Earth's surface, or a cooling effect, of about 0.2 W/m^2 since 1850.

The largest albedo changes that have occurred during the time of measured global warming are related to human land use. Large forested regions were cleared to produce crops and develop pastures for livestock. For example, only 6–7 percent of the global land surface was cultivated or used for pasture in 1750 compared to 39 percent in 1990. Cropland and pasture vegetation has a lower albedo than forested landscapes, especially when the ground is bare between growing seasons. So, this shift in land use has increased the solar energy that is reflected into space. Current calculations, however, suggest that the total impact of land-use changes is a decrease in solar energy absorbed at the surface of only 0.2 W/m^2.

Changes in Atmosphere Composition by Human Activities

The changes in solar energy and albedo during the last 300 years are very small compared to the 342 W/m^2 of solar energy that reaches the top of the atmosphere. For this reason, greenhouse-gas concentration is the factor that receives the most attention in the current debate on climate change. Therefore, it is critical that we understand why scientists focus on this factor.

You might wonder how these gases, measured in quantities such as parts per million (1/10,000th of 1 percent) or parts per billion (1/10 millionth of a percent), could possibly change Earths surface temperature. It turns out that the heat-trapping ability of these gas molecules really does have a dramatic effect on temperature. For example, an increase in carbon dioxide of only 100 parts per million would have about 20 times more heating effect than the estimated increase in solar-energy output since 1700.

Of the various greenhouse gases whose concentrations have increased in recent history, it is most worthwhile to concentrate on carbon dioxide and methane. These two gases account for more than 90 percent of changes in the greenhouse effect. Both gases are artificially released into the atmosphere by the extraction and burning of fossil fuels. The fossil fuels (coal, natural gas, and oil) consist of carbon or complex hydrogen and carbon compounds. Energy is released during burning by breaking the bonds between carbon atoms and between carbon and hydrogen atoms. The "free" carbon atoms then immediately bond with oxygen in the atmosphere to form carbon dioxide. Methane is the dominant component of natural gas, and it is commonly found with coal and oil. Methane, therefore, is released directly into the atmosphere from coal mines and some oil drilling operations. Methane is also released by agricultural activity, especially from rice paddies and by the digestive processes of ruminant livestock, such as cows and sheep, and also from decaying garbage in landfills.

Direct measurements of greenhouse concentrations in the atmosphere go back only to the 1960s but our data go back much farther by measuring gas levels inside air bubbles trapped in glacial ice. We already considered the usefulness of chemical analyses of ice cores drilled out of glaciers to provide proxy temperature data (Figure 21.13) and to track changes in global glacial-ice volume (Figure 18.42). Another important feature of glacial ice is the air trapped as bubbles when porous snow compacts into ice. These bubbles are natural samples of the ancient atmosphere that can be analyzed in the laboratory to measure greenhouse-gas concentrations long before humans were consuming fossil fuels. **Figure 21.21**a shows that

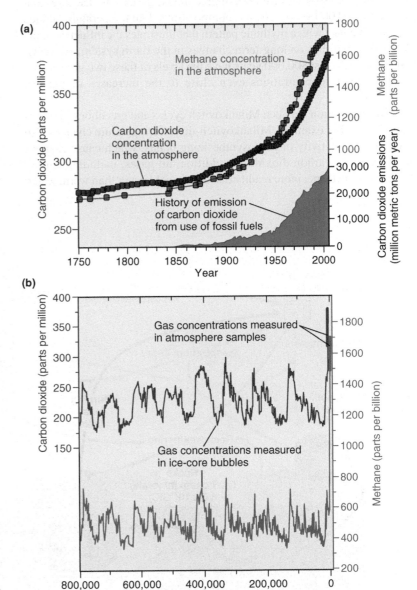

▲ **Figure 21.21 Changing concentrations of greenhouse gases.** (a) Historic levels of carbon dioxide and methane, measured directly since the 1960s and from ice-core air bubbles at earlier times, increase parallel to the rising consumption of fossil fuels (coal, oil, and natural gas). (b) A longer record of greenhouse-gas levels is represented by analyses of air bubbles in a long ice core drilled in Antarctica. The rhythmic fluctuations in gas concentrations match with the proxy temperature record from the same core (Figure 21.13). Both temperature and gas concentrations vary in accordance with the Milankovitch cycles, which also control global glacial-ice volume at this time (Section 18.11).

carbon dioxide and methane levels have clearly risen since 1750 and very closely follow the pattern of increasing consumption of fossil fuels. This similarity in pattern supports the conclusion that fossil-fuel consumption accounts for the recent rise in the concentration of these greenhouse gases in the atmosphere. Land-use changes also contribute to increasing greenhouse-gas concentrations shown in Figure 21.21a, because (a) forests remove more carbon dioxide by photosynthesis than does the vegetation introduced by humans, and (b) the cleared trees are either burned, which creates carbon dioxide, or rot, which creates methane. Even cement production, which was introduced in Section 6.1 as an example of a degassing reaction, releases 1.2 billion metric tons of carbon dioxide into the atmosphere every year.

Figure 21.21b shows a record of greenhouse gases from ice cores that goes back 800,000 years. The gas compositions, like temperature and ice volume, fluctuate with a rhythmic pattern that illustrates the influence of the Milankovitch cycles on long-term changes in the Earth system. However, at no time in the last 800,000 years have the levels of these two greenhouse gases reached concentrations even close to the increases of the last 250 years.

The connection between Milankovitch cycles and greenhouse gases is complicated. For example, Milankovitch-driven temperature changes also affect the productivity of photosynthesizing plants, which cause changes in atmospheric carbon dioxide. In addition, all gases, including carbon dioxide and methane, more readily dissolve in cold water than warm water,

so their greatest abundances shift back and forth between ocean and atmosphere as climate shifts.

In fact, not all of the carbon dioxide emitted during the last few centuries of industrialization with fossil-fuel energy sources has stayed in the atmosphere. About one-third of the carbon dioxide has been consumed by increasingly productive photosynthesizing organisms, which thrive in a carbon-dioxide-rich atmosphere. Another third of this added greenhouse gas has been absorbed into seawater, analogous to how carbon dioxide dissolves into artificially produced carbonated beverages. Recalling that the mixture of carbon dioxide and water makes a weak acid (Section 5.1), it should come as no surprise that measurements show that the world oceans are gradually becoming acidic. Because calcite dissolves in this weak acid (Figure 5.4b) there have also been measurable impacts on calcite-secreting organisms, such as coral in recent decades.

How Greenhouse-Gas Concentrations Relate to the Carbon Cycle

The movement of carbon through the geosphere (as fossil fuels in rocks, for example), biosphere, and atmosphere explains changes in important greenhouse gases over both historic and geologic time frames. **Figure 21.22** shows where carbon is stored in the Earth system, and schematically illustrates how carbon cycles between these storage reservoirs. The vast

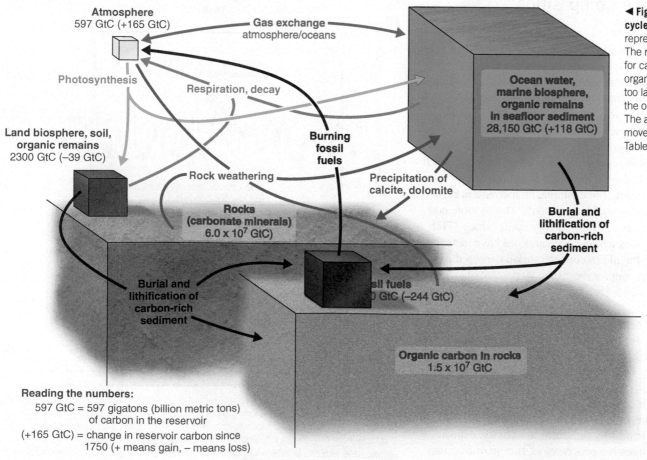

◄ **Figure 21.22 Visualizing the carbon cycle.** Each box in this diagram represents a storage reservoir of carbon. The reservoirs are drawn to scale except for carbon stored as minerals and organic matter in rocks, which are much too large to show. Fossil fuels are part of the organic carbon reservoir in rocks. The arrows show the processes that move carbon between the reservoirs and Table 21.1 describes these processes.

Atmosphere
597 GtC (+165 GtC)

Gas exchange
atmosphere/oceans

Photosynthesis

Respiration, decay

Land biosphere, soil, organic remains
2300 GtC (−39 GtC)

Burning fossil fuels

Rock weathering

Ocean water, marine biosphere, organic remains in seafloor sediment
28,150 GtC (+118 GtC)

Precipitation of calcite, dolomite

Rocks (carbonate minerals)
6.0×10^7 GtC

Burial and lithification of carbon-rich sediment

Burial and lithification of carbon-rich sediment

sil fuels
GtC (−244 GtC)

Organic carbon in rocks
1.5×10^7 GtC

Reading the numbers:
597 GtC = 597 gigatons (billion metric tons) of carbon in the reservoir
(+165 GtC) = change in reservoir carbon since 1750 (+ means gain, − means loss)

TABLE 21.1 Some Processes That Add or Subtract Atmospheric Carbon Dioxide

Processes Adding Carbon Dioxide to the Atmosphere	Processes Subtracting Carbon Dioxide from the Atmosphere
More important on time scales of tens to thousands of years (human time scale):	
Respiration of carbon dioxide by most organisms while breaking down organic compounds to obtain energy	Photosynthesis by plants and some one-celled organisms removes carbon dioxide to produce organic tissue and oxygen
Burning fossil fuels (coal, oil, and natural gas) releases carbon atoms that bond with oxygen in the atmosphere to produce carbon dioxide	Decades-long cooling of the oceans absorbs carbon dioxide from the atmosphere because gases are more soluble in water as temperature decreases
Methane released from wetlands, rice paddies, and landfills partly reacts with oxygen to produce carbon dioxide	
Land clearing for development and agriculture usually reduces the photosynthetic removal of carbon dioxide, which causes levels to increase	
Decades-long warming of the oceans releases carbon dioxide to the atmosphere because gases are less soluble in water as temperature increases	
More important on time scales of millions of years (geologic time scale):	
Decreased burial, and therefore increased decay, of organic carbon originating from dead plants and marine plankton, or weathering of previously buried organic carbon	Increased burial, and therefore decreased decay, of organic carbon originating from dead plants and marine plankton, some of which ultimately forms fossil fuels
Metamorphic reactions that include degassing of carbon dioxide	Weathering reactions that break down silicate minerals consume carbon dioxide
Long-term climatic cooling decreases rock weathering rates especially if expanding glaciers bury rock so that it is not weathered; diminished weathering decreases consumption of carbon dioxide by weathering reactions, which causes levels to increase	Long-term climatic warming increases rock weathering rates, increasing consumption of carbon dioxide by weathering reactions, which causes levels to decrease
Increased volcanic activity releases larger volumes of carbon dioxide from the mantle	Decreased volcanic activity decreases the volume of carbon dioxide that degasses from the mantle, which lowers levels
Long-term warming of the oceans releases carbon dioxide to the atmosphere because gases are less soluble in water as temperature increases	Long-term cooling of the oceans absorbs carbon dioxide from the atmosphere because gases are more soluble in water as temperature decreases

majority of Earth's carbon is in rocks, mostly within the atomic structure of calcite and other carbonate minerals and partly as buried organic matter. The organic carbon also exists in the large reservoir of carbon represented by living plants and animals. As these organisms die and decay, the resulting organic carbon oxidizes into carbon dioxide or is buried in sediment. The fossil fuels form from the buried organic carbon found in rocks, and this reservoir of carbon would have remained mostly isolated from the atmosphere if humans did not extract and burn coal, oil, and natural gas to create an unnatural source of carbon dioxide.

Table 21.1 lists some of the processes that either add or decrease carbon dioxide in the atmosphere as a result of the carbon cycle. Some of these processes, such as burning of fossil fuels, solubility of carbon dioxide in the ocean that changes with water temperature, and plant photosynthesis and respiration, affect atmospheric carbon dioxide over human lifetimes. Others, such as rock weathering and metamorphism tend to be longer, geologic-timeframe controls on atmospheric composition.

To illustrate this long-term relationship between carbon dioxide and climate, let's examine why many geologists attribute most of the climate variations illustrated in Figures 21.10 and 21.11 to changes in atmospheric carbon dioxide levels. Ice cores provide carbon dioxide concentration only back to 800,000 years ago. Proxy records of earlier carbon-dioxide levels are revealed in geochemical data collected from rocks. The proxy records are combined with computer-model calculations that are based on a number of geological data sets, such as burial of carbon in sedimentary rocks of different ages, variations in volcanic activity, and variations in sediment accumulation, which also reflects changes in rock weathering that consumes carbon dioxide (Table 21.1). Both the proxy data and the

computer-model estimates of carbon dioxide levels have large uncertainties. Nonetheless, as shown in **Figure 21.23**, these estimates of carbon dioxide levels do show large scale increasing and decreasing trends in carbon dioxide that are consistent with the proxies of ancient temperature.

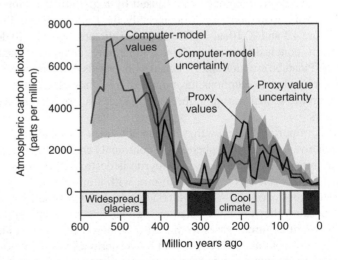

▲ Figure 21.23 Carbon dioxide levels fluctuated during geologic history. Carbon dioxide levels in the atmosphere for the last 575 million years are estimated from proxy chemical patterns preserved in rocks and by computer models that consider changes in carbon storage among rocks, oceans, and the atmosphere that are implied by features in rocks. Both estimates have large uncertainties but are consistent with one another. The times of estimated low carbon-dioxide levels correlate with geologic evidence of large-scale glaciers on land or unusually cool water temperatures in the oceans, as shown by the bars along the bottom of the graph.

The Effects of Volcanic Eruptions on Atmosphere Composition

Section 9.4 introduces the importance of volcanic eruptions for contributing gases to Earth's atmosphere, and carbon dioxide is a significant component of volcanic gases (Figure 9.14, Table 21.1). How important are the gases released by volcanic eruptions to cause natural climate change? Averaging measured carbon dioxide emissions from many volcanoes and then adjusting this value by comparison to the total volume of lava and pyroclastic material erupted on Earth each year, leads to the conclusion that volcanoes contribute about 65 *million* metric tons of carbon dioxide to the atmosphere on an annual basis. This is a tiny fraction when compared to the estimated 32 *billion* metric tons of carbon dioxide that enters the atmosphere each year because of fossil fuel consumption (Figure 21.21a). Nonetheless, at times in the geologic past when rates of seafloor spreading were faster than today, there was much more carbon dioxide released from volcanoes along mid-ocean ridges. Intense volcanic activity was an important contributor to the higher carbon dioxide levels 100–150 million years ago (Figure 21.23), compared to today.

Although the amount of carbon dioxide erupted by volcanoes is not causing global warming, volcanoes do change atmosphere composition in ways that cause global cooling. Volcanic ash shields Earth's surface from incoming solar radiation like shade from an umbrella, but this effect is very short lived because the ash particles settle back to the surface within a few weeks or months after an eruption. More important are the gaseous sulfur compounds ejected very high into the atmosphere during very explosive eruptions. The sulfurous gas condenses into very tiny aerosol droplets that remain airborne for many years and encircle the globe (Figure 20.10) to have a widespread impact on atmosphere composition. The sulfurous aerosols absorb solar energy in the upper atmosphere, which cools Earth's surface. About two or three times every century there is an eruption that is large enough to affect global climate. The most famous example was in 1816, known as the "year without a summer" across northern Europe and eastern North America because of killing frosts in every month of the year. This unusually cool summer was caused by a gigantic eruption of the Tambora volcano in Indonesia. More recently, the 1991 eruption of Pinatubo (see Figure 4.1 and 20.10) also affected global climate. **Figure 21.24** demonstrates this conclusion because global temperatures declined immediately after the Pinatubo eruption and remained at low levels for more than 3 years. Therefore, volcanic eruptions have more of a cooling effect than being a contributor to global warming; the cooling effect is large, but lasts only a few years.

It is important to know that human activities also add cooling sulfurous aerosols to the atmosphere. Coal commonly contains pyrite, an iron-sulfide mineral, as an impurity. The pyrite is destroyed when the coal is burned, which releases the sulfur to mix with atmospheric oxygen and form sulfurous aerosols. Although residing in the atmosphere for shorter times because they are not ejected as high as volcanic eruption plumes, the release of sulfurous gases from coal-fired power plants and industrial facilities is a continuous process with longer-term effects on atmosphere temperature. Sulfurous emissions have been curtailed in North America and most of Europe in an effort to reduce air pollution and the acid rain that results when the sulfurous gases mix with water vapor to create sulfuric acid. At the same time, however, sulfur emissions are rising in increasingly industrialized India and China. Nonetheless, global sulfurous gas contributions to the atmosphere decreased by almost 20 percent between 1970 and 2000. Ironically, the improvement of air quality by decreasing

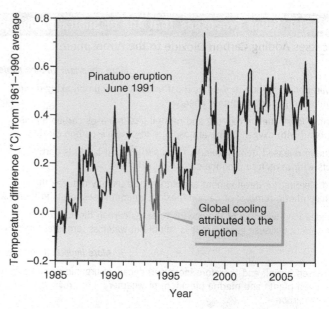

▲ **Figure 21.24 Volcanic eruptions cause global cooling.** Global temperature, plotted here as monthly averages, decreased immediately following the 1991 eruption of Pinatubo in the Philippines, and remained unexpectedly low for about three years.

sulfurous-gas emissions has likely contributed indirectly to global warming by removing an effect that had partly countered the influence of increased greenhouse-gas emissions.

Balancing the Energy Budget

We have seen that a variety of natural and human-related processes have either increased or decreased the total heat that is felt at Earth's surface in recent centuries. **Figure 21.25** illustrates these changes in the energy budget

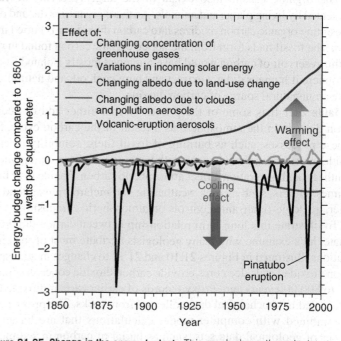

▲ **Figure 21.25 Change in the energy budget.** This graph shows the calculated changes in heat energy reaching Earth's surface since 1850 as a result of natural processes (incoming solar energy and volcanic eruptions) and human activities (greenhouse-gas concentrations, land-use changes, pollution haze, and aerosols).

since 1850. Although not shown, there are uncertainties associated with each of the lines that are graphed in Figure 21.25, especially for times prior to 1900. The greatest certainty is associated with the effects of greenhouse gas concentrations, which is also the biggest change in the budget entries. Volcanic eruptions have substantial, well known cooling effects, but these effects last for only a few years. The other factors in the energy budget have, more or less, either increased or decreased overall from 1850 to 2000. These factors play an important role in Section 21.5, where we will see the results of computer models that use these factors to calculate global temperatures for comparison to the actual instrumental measurements.

Changes in Ocean Circulation

Can ocean circulation have a significant impact on global warming? Among the factors that cause global temperature changes, it is important to consider warming or cooling that may relate to shifts in ocean circulation. We will examine two examples. One relates to the previously mentioned current that brings warmth to London and other locations in the North Atlantic Ocean. The second example is the fluctuation between El Niño and La Niña weather conditions.

Figure 21.26 illustrates, in a generalized way, an important pattern of circulation observed in the Atlantic Ocean. This circulation pattern is not caused by wind blowing across the water but is, instead, convection caused by density variations in seawater.

Temperature and salinity are the two factors that affect seawater density. Cold water is denser than warm water, and salty water is denser than freshwater. Atlantic Ocean surface water absorbs heat from the Sun at the tropical latitudes near the equator. This warming decreases the water density, but the warmth also enhances evaporation, which makes the water saltier. Measurements show that the warming effect is more important on the overall density, allowing the unusually salty water to remain at or close to the surface as it moves northward. Arriving in the far North Atlantic, the surface water encounters southward moving cold air from the Arctic (Figure 20.3). The warm surface water conducts and radiates heat into this cold air, accounting for the warmth experienced on land in this region, including in London. As the water transfers its heat and cools down, it becomes very dense both because it is cold and because it is salty. The added density from the salinity is now important, because the deeper water is also cold and it is the combination of coldness and saltiness that causes the surface water to abruptly sink deep into the North Atlantic Ocean. The cold, salty water, named North Atlantic Deep Water, then flows southward to complete the convection loop.

Just like subducting plates drive plate motion (Figure 12.42), the sinking formation of North Atlantic Deep Water drives the whole current system in a conveyor-belt-like motion. If the cold, salty, sinking part of the current loop does not exist, then the conveyor belt stops and the warm surface current will not move northward.

Paleoclimatologists hypothesize that changes in this conveyor-belt current system caused the Younger Dryas cold snap, described in Section 21.2. Studies of the remains of microscopic marine organisms in the North Atlantic that date to Younger Dryas show that the cold snap began with abrupt freshening of surface water. The evidence is variations in the chemistry of the microscopic shells as well as variations in species that are particularly sensitive to changes in salinity. Scientists have not reached a consensus on the source of the freshwater, and it may have come from several sources. Wherever the freshwater came from, everyone agrees that it originated as large volumes of meltwater runoff from North America, and

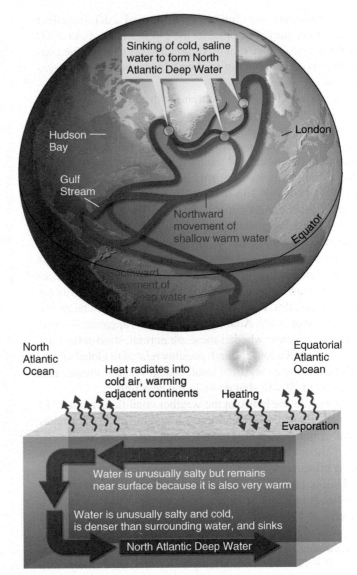

▲ **Figure 21.26 Conveyor-belt circulation affects North Atlantic climate.** The map shows a conveyor-like pattern of shallow, warm-water currents and deep, cold-water currents that operates in the Atlantic Ocean. The schematic cross-section through the ocean shows that the northward-moving shallow current is both unusually warm and salty because of heating and evaporation near the equator. Farther north, the current radiates heat into cold air adjacent to northwestern Europe, producing unusually warm air temperatures at this high latitude. The combination of high salinity and decreasing water temperature increases the water density, causing it to sink and form North Atlantic Deep Water. The sinking action is what causes the conveyor-belt circulation to operate; otherwise warm currents would not flow so far north in the Atlantic Ocean.

ACTIVE ART

Ocean Circulation. See how wind-driven surface currents affect water temperatures, and also the conveyor-belt circulation that includes North Atlantic Deep Water.

possibly Eurasia, coincident with the retreat of the ice-age glaciers. When this low-density fresh water flowed out onto the surface of the North Atlantic, the salinity was greatly reduced. This meant that the surface water was no longer dense enough to sink and the conveyor belt stopped. No warm water moved into the North Atlantic and the northern hemisphere continents abruptly cooled and the climate change caused a domino effect that ultimately impacted most of the globe. After about 1300 years the surface-water salinity returned to normal and the conveyor belt restarted, bringing an abrupt end to the Younger Dryas cold snap.

The relationship of ocean currents to the Younger Dryas is an important example of how quickly global climate can change because of a perturbation in one small area. It is also notable, and somewhat ironic, that this cold period was a direct result of overall warming; the melting of the northern hemisphere ice sheets produced the freshwater that shut down the conveyor belt and caused abrupt cooling. This scenario, in greatly exaggerated form, was the basis for the 2004 movie *The Day After Tomorrow*.

Could ongoing global warming cause sufficient melting of freshwater sea ice in the Arctic and glaciers in Greenland so that a Younger-Dryas-like cold snap happens again? This ironic twist of fate is certainly possible. Reliable measurements of surface-water salinity and current velocities have been available only for a couple of decades. A decrease in the current strength related to the formation of North Atlantic Deep Water is apparent in these data, but it is unclear whether these are normal, short-term fluctuations or part of a longer trend, possibly related to global warming. Either way, the measured small changes do not suggest that the conveyor belt will shut down in the near future.

The oscillation between the weather conditions named "El Niño" and "La Niña" is a real-time, observable result of changes in ocean circulation. Similar to the North Atlantic conveyor circulation, these two weather patterns originate in one location, in this case offshore of western South America, but have global impact. Unlike the conveyor belt, the El Niño-La Niña current change is also intimately linked to atmospheric circulation.

Figure 21.27 illustrates how changes in combined ocean and atmospheric circulation cause climate change. Driven by the easterly trade winds (explained in Figure 20.3), the currents in the equatorial Pacific Ocean normally move westward away from South America and toward Australia. A relatively weak counter current produces an eastward return flow of water. The easterly trade winds blow offshore from South America, and this allows upwelling of cold, deep water to the surface. The overall result is a strong variation in sea-surface temperature from relatively cool in the eastern equatorial Pacific to much warmer in the west.

El Niño describes a weather pattern where the trade winds are unusually weak, allowing more eastward current flow and unusually warm conditions along the coast of South America. This weather pattern was first described by residents of coastal Peru who named it El Niño (Spanish for "little boy") because the unusual warmth was particularly noticeable near Christmas. The warmer water temperatures also cause warmer air temperatures and greater evaporation. As a result, El Niño conditions in coastal South America, Central America, and even the Southwestern United States are unusually wet, because the increased evaporation from the warmer ocean feeds into a more vigorous hydrologic cycle that also increases rainfall.

La Niña (little girl) describes a contrasting pattern where the equatorial counter current is unusually weak, which causes anomalously cool water, and drier precipitation patterns, along and near the South American coast. The fluctuations between El Niño and La Niña, with intervening periods of average, or

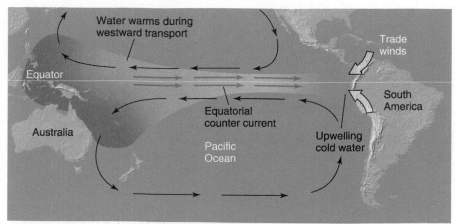

Normal conditions: Trade winds move water westward in the tropics, allowing cold water from the deep ocean to upwell at the surface near South America. The water then warms up as it moves toward Australia. The equatorial counter current moves some of this water back toward the east.

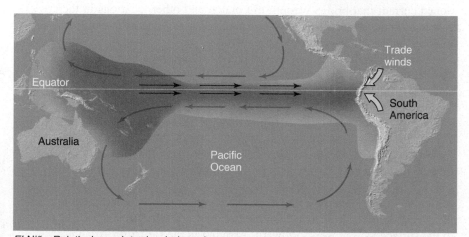

El Niño: Relatively weak trade winds and a stronger, eastward-flowing equatorial counter current decreases westward transport of warm surface water and supresses upwelling of cold water along the west coast of South America to produce unusually warm water in the equatorial Pacific Ocean.

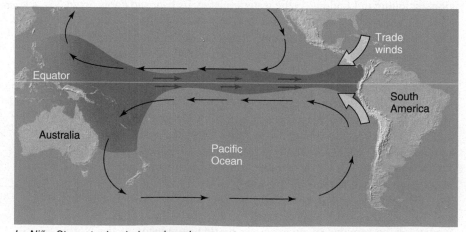

La Niña: Strong trade winds and weak equatorial counter current increases upwelling of cold water along South American coast, producing unusually cool water in the equatorial Pacific Ocean.

▲ **Figure 21.27 Contrasting El Niño and La Niña circulation and sea-surface temperature patterns.**

ACTIVE ART

El Niño and La Niña. See how changing atmospheric and oceanic circulation systems in the Pacific Ocean affect weather conditions.

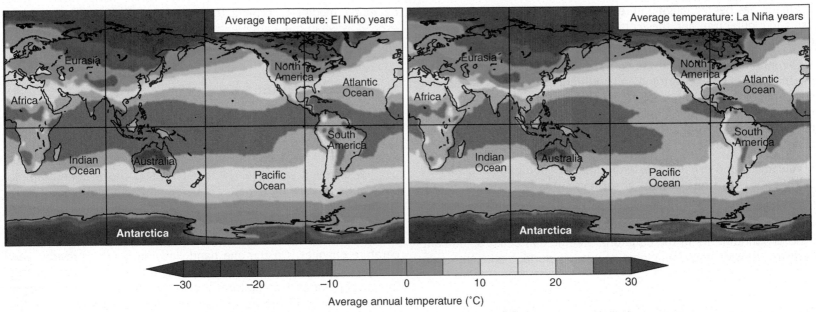

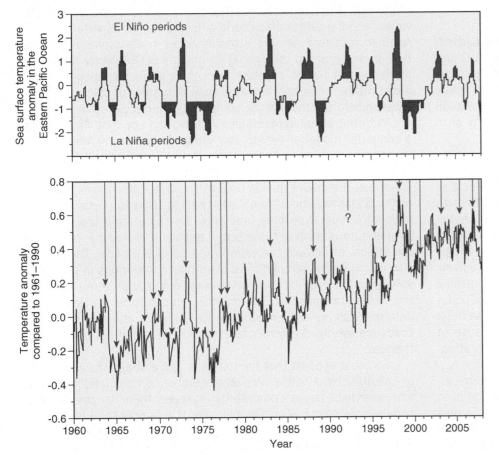

▲ **Figure 21.28 Global temperatures change during El Niño and La Niña years.** These maps compare the annual global temperatures averaged over several El Niño and La Niña years. The most obvious differences are in the Pacific Ocean but subtle differences are observed in most parts of the world.

normal, conditions relate to processes operating in just one area of the world. However, the changes in Pacific Ocean surface temperatures upset general atmospheric circulation to cause variations in temperature, precipitation, or both that are measured around the world.

Given our particular interest in air temperature, **Figure 21.28** shows the average variations in global surface temperatures during El Niños and La Niñas between 1955 and 2003. The major differences are seen in the Pacific Ocean, but when averaged across the globe, the world is generally warmer during El Niños and cooler during La Niñas. This result is best seen by comparing the record of sea-surface temperatures off the coast of Peru with the instrumental record of global temperature, depicted in **Figure 21.29**. We can now see that most of the undulations between relatively cooler and warmer times along the overall global-warming trend of the last few decades (Figures 21.4 and 21.28) are explained by this cyclic variation in circulation that begins in the equatorial Pacific Ocean.

Putting It Together—What Natural Processes and Human Activities Affect Global Temperature?

• Variations in Earth's orbit and rotation are the major cause of changes in solar energy reaching Earth. These variations cause climate changes on time spans of tens of thousands to a hundred thousand years, which explains oscillations between glacial and interglacial climates but not historic global warming.

• Variations in the energy output of the Sun are not well known but are probably very small.

▲ **Figure 21.29 Global temperatures oscillate with El Niño and La Niña.** El Niño and La Niña conditions are defined by extremes in sea-surface temperatures measured off the western coast of South America (top graph). The role of these equatorial circulation changes on global climate is demonstrated by the correlation, shown by arrows, between El Niños and La Niñas with warm and cold spikes, respectively, on the global temperature curve (bottom graph). The only failed correlation is the global cool interval that corresponds to the 1992–1993 El Niño (red question mark). The warming expected with this El Niño was erased by cooling caused by the 1991 eruption of Pinatubo (see Figure 21.24). Notice that while El Niño warming and La Niña cooling provide an explanation for global temperature variations over periods of several years, this cycle does not explain the overall warming after 1970.

- Albedo has been changing in response to historic changes in land use and cloud cover and these changes may cause small amounts of cooling.

- Greenhouse-gas concentrations have increased dramatically during the time of global warming and correlate to increases in the consumption of carbon-rich fossil fuels. This factor is the largest factor that increases heat in the energy budget. Variations in atmospheric carbon dioxide may be the primary driver of climate change over geologic time scales, too.

- Volcanoes, and to a lesser extent human consumption of coal, add significant quantities of heat absorbing sulfurous aerosols to the atmosphere that could cause short periods of dramatic cooling.

- Changes in convective circulation of heat in the North Atlantic and wind-driven currents in the equatorial Pacific can cause significant hemisphere-wide and even global changes in surface temperature.

21.5 How Do We Know . . . That Humans Cause Global Warming?

Define the Problem

What Parts of Global Warming Are Natural Versus Being Human Caused?
Very few people who examine the data plotted in Figures 21.3 and 21.4 and who understand how those data are obtained dispute the evidence for global warming during recent decades. Instead, many of the disputes that you may be aware of relate to the extent to which we should attribute this climate change to natural variability or to human activity.

Our consideration of the energy budget reveals that some variability must be natural. Examples include variations in solar energy received at the top of the atmosphere and brief cooling effects of very large volcanic eruptions. Another natural variability, although unrelated to the energy budget, is the changing ocean currents, such as the oscillations between El Niño and La Niña conditions.

When examining the effects of human activities on the energy budget, we need to consider that some of these activities have likely worked to lower temperatures rather than raise them. Increasing surface albedo caused by land-use change and sulfur-aerosol pollution in the atmosphere are examples of cooling effects. These must be included when considering the warming expected because of substantial measured increases of greenhouse gases in the atmosphere (Figure 21.21).

The correlation of fossil-fuel consumption with atmospheric carbon dioxide levels (Figure 21.21a) and the correspondence of both of these trends to measured temperature increase (compare Figures 21.4a and 21.21a) can be used to make a compelling argument that human activity is causing global warming. Scientists do not assume, however, that just because two measurements vary in similar ways at similar times that one factor causes change in the other. Correlation suggests a connection but this could be coincidence, so rigorous science goes further. The more convincing step is to hypothesize a clear physical explanation for how changing one factor causes changes in others *and* that the amount of change is consistent with the observed measurements. To test the hypothesis, scientists commonly use computer models to explore these cause-and-effect relationships, especially when there are many variables that need to be simultaneously considered.

Build a Computer Model

How Do We Convert the Energy Budget into Global Temperature?
Computer models are the outputs of elaborate computer programs that describe natural processes by mathematical equations and use measurements to establish single values, or ranges of possible values, for different variables. Section 10.4 also provides an example of computer modeling to better understand geological processes.

In order to make a model of temperature change that can be compared to measured values, such as those plotted in Figure 21.4, the changes in the energy budget need to be converted into temperature changes. On the one hand, laboratory work provides very straightforward equations for converting the input of heat energy into a value for air temperature. The variability of the energy-budget factors shown in Figure 21.25 can, therefore, be used to make the model. On the other hand, we know that the actual air temperature at the surface depends on time and location. For example, we have to take into account the temperature differences among the seasons of the year, the temperature differences related to latitude and elevation, and the substantial albedo differences of land, water, and ice. Another important variable is that changes in atmosphere temperature are not uniform throughout its roughly 30 kilometer thickness, so the model has to consider how changes in the energy budget affect different levels of the atmosphere in order to calculate a surface temperature that can be compared to surface measurements.

With these requirements in mind, we can see that constructing a computer model of year-to-year global temperature change must share some features with the methods used to construct the global temperature record from measurements shown in Figure 21.4. Therefore, the models divide up Earth's surface into boxes, typically at the 5°-latitude-by-5°-longitude spacing used to average out the regional temperature measurements. By calculating the expected temperature in each of these boxed regions it is possible to include the effects of latitude, elevation, and surface albedo. The models also calculate the temperatures at different elevations in the atmosphere within each box and an average monthly air temperature at the surface for each box. The monthly temperatures are averaged to determine an annual temperature for each box and then all of the boxes are averaged together to estimate the global temperature for the year.

So far, it appears that the model is just a straightforward series of calculations, albeit a very large number of calculations (that is why scientists use a computer!). However, there are many other complexities to consider. For example, if a temperature increase is calculated in the Arctic, then sea ice may melt, exposing more water, and changing the albedo value within one or more boxes, which changes the air temperature. As another example, the model must include how the ocean temperature changes through time because heat moves back and forth between the air above land and sea on both daily and seasonal time frames. As yet another example, warming would increase evaporation that may increase cloud cover in some areas, which changes albedo in a way that might then cause cooling. And, we still have not considered the effects of ocean circulation changes that reflect changes in wind strength and direction that are also connected to variations in atmospheric heating. Keeping these examples in mind tells us that the best models are those that account for all of these feedbacks and domino effects between the variables in the climate system.

There is another critical factor to consider—uncertainty. For all that scientists have learned in recent decades about the energy-budget factors summarized in Figure 21.25, not all these factors are equally well known. The best understood values, especially on an annual basis over the last century, are greenhouse-gas concentrations and sulfur-aerosols erupted by volcanoes. The sum total of the albedo effects of land-use change are fairly well known but the year-to-year variations are not known. Solar-energy received at the top of the atmosphere is a significant part of the energy budget and has been directly measured for only 30 years. Considerable controversy exists among geologists, astronomers, and climatologists about how the Sun's output has changed over several centuries. The relatively small overall increase in solar heating shown in Figure 21.25 is a recent estimate but it is not the only one. The implications for computer models is that it is worthwhile to run the model more than once using different values where values are uncertain.

Evaluate the Results

What Factors Explain Global Warming? Figure 21.30 summarizes the critical results of the computer modeling exercise designed to answer our question about natural and human causes of temperature change. Two graphs are presented and each one compares model-calculated global temperature from 1900 to 2005 with the measured temperature. One graph (Figure 21.30a) shows the results of computer models that consider all of the energy-budget changes during this time period (Figure 21.20). The other graph (Figure 21.30b) shows the results if human-caused changes in the energy budget are left out of the calculations. Model results in this second graph keep the greenhouse-gas concentrations at levels that predate the burning of fossil fuels. For example, carbon dioxide concentration is set at a constant value of 280 parts per million, which is the actual value around the year 1800 (Figure 21.21a). The changes in the energy budget affected by land-use changes and sulfur-aerosol air pollution are also excluded when calculating the temperature changes associated with natural processes.

An important observation in both graphs is that there are many simulated temperature-change curves. The tangled-string appearance of lines in both graphs results from plotting up the calculations from several different computer models and using different energy-budget values where those values are uncertain. The different models use different approaches to including all of the interactions and feedbacks among climate variables. Some models include the calculated effects of El Niño and La Niña, while others do not. By using different computer models and different values for uncertain variables, the graphs in Figure 21.30 provide the best estimates of the uncertainty in calculating the natural and human causes of climate change.

The results graphed in Figure 21.30 distinguish natural and human causes of global warming. If the human causes are left out of the computer calculations, the measured temperature change from 1900 to about 1970 can be explained just by the natural factors, the

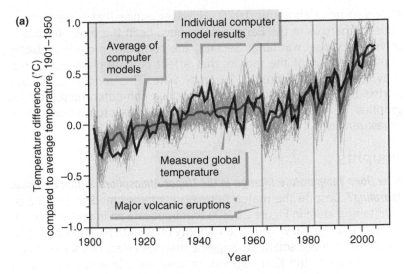

(a)

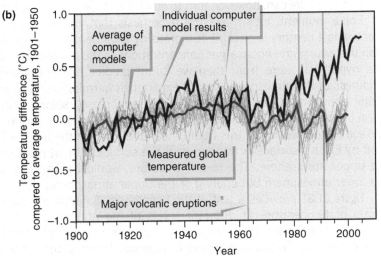

(b)

▲ **Figure 21.30 Computer models separate natural and human causes of global warming.** (a) Comparison of measured temperature to global temperature calculations from 58 simulations using 14 computer models and applying all natural and human-caused variations in heat energy at Earth's surface. (b) Comparison of measured temperature to global temperature calculations from 19 simulations using 5 computer models and applying only natural variations in heat energy at Earth's surface and assuming constant, 1850 values for greenhouse-gas concentrations. Natural processes fail to account for global warming after about 1970. Notice the significant effect of major volcanic eruptions on measured and modeled temperature.

increasing output of heat from the sun interrupted occasionally by the cooling effect of volcanic eruptions. There are some years where measured temperature during this period was cooler or warmer than the average of the model calculations but, for most years, the measured temperature falls somewhere within the range of uncertainty represented by the different model results (Figure 21.30b). In contrast, the increase in global temperature after 1970 exceeds even the most extreme computer-model results that contain only the natural processes. In fact, these computer models calculate a small amount of cooling in the latter part of the twentieth century. However, when the human impacts are included the computer models do match the measured temperatures (Figure 21.30a).

Two conclusions can be drawn. First, natural processes, primarily solar energy arriving at the top of the atmosphere and volcanic eruptions, were the most important drivers of temperature changes observed up until about 1970. However, since that time human-caused changes in the Earth system have been the most important drivers of global warming. Second, the computer models probably capture enough of the factors that affect climate changes so that they can be used to predict future climate. This conclusion is supported by the observation that the various computer-model results graphed in Figure 21.30a are all close to the actual temperature measurements.

Insights

How Does Temperature Change in the Upper Atmosphere Relate to Global Warming? Despite the match between computer models and measured temperature in Figure 21.30a that implicate an important human influence on global warming, a minority of scientists remains skeptical. Most of these scientists suggest that the effect of varying solar-energy input to the Earth system is inadequately accounted for in these models. We can, however, turn to other data to test this criticism.

For a moment, let's make the hypothesis that global warming over the last century is a result of increasing solar heat output rather than increasing greenhouse-gas concentrations in the atmosphere. If this hypothesis were true, then we could predict that the whole thickness of the atmosphere should experience warming because all of the atmosphere would be affected by the increased solar energy. If, on the other hand, greenhouse gases are controlling the measured surface air temperatures, this would mean that more of the heat radiated by Earth's surface is trapped near the surface without reaching the upper atmosphere. The result, therefore, would be warming of the lower atmosphere but cooling of the upper atmosphere.

Figure 21.31 provides a test of the hypothesis. Satellite measurements of temperature at different levels of the atmosphere show warming below about 15 kilometers and cooling in the upper atmosphere since 1978. This cooling trend is inconsistent with the hypothesis that variations in solar energy cause the recent global warming. Instead, the cooling is consistent with the importance of increasing concentrations of greenhouse gases because of human activities that is incorporated in the computer-model results in Figure 21.30a.

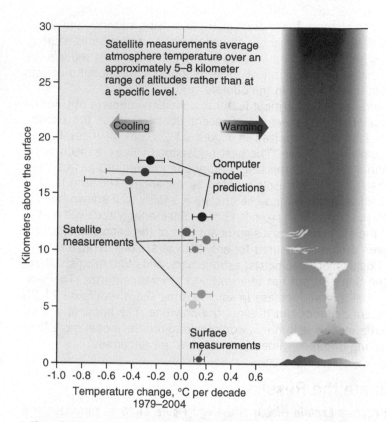

▲ **Figure 21.31 The upper atmosphere is cooling.** Satellite measurements of atmosphere temperature, averaged over various atmosphere altitudes, show that the upper atmosphere has cooled in recent decades while warming occurred at lower altitudes and at the surface. Cooling in the upper atmosphere is predicted by computer models that include changing values of gases in Earth's atmosphere during this same time. Increasing greenhouse gases trap heat in the lower atmosphere and ozone depletion has decreased the amount of incoming solar energy that is absorbed everywhere in the atmosphere, leading to the observed cooling above about 15 kilometers.

Putting It Together—How Do We Know . . . That Humans Cause Global Warming?

• Computer model reconstructions of year-to-year global temperature change explain warming from 1900 to about 1970 only by natural causes, mostly a small increase in overall energy output from the Sun interrupted by brief introductions of cooling aerosols to the atmosphere by violent volcanic eruptions.

• Computer models can account for the observed warming since 1970 only by including human impacts on the energy budget, especially the increasing production of greenhouse gases.

• Cooling of the upper atmosphere during the last 30 years while near-surface temperatures have risen also is consistent with greenhouse warming of Earth's surface and is inconsistent with alternative hypotheses ascribing global warming to inadequately known variations in solar energy output from the Sun.

21.6 How Will Climate Change in the Future?

In bringing our consideration of global warming to a close, it is important to consider what the implications of recent climate change hold for the future. It is important to keep in mind that although we focused on global temperature in this chapter, temperature changes also link to changes in precipitation, lengths of seasons, severity of storms, and patterns of wind and ocean currents. For that reason, we can be more complete by thinking in terms of global climate change, rather than just global warming.

Computer Models of Future Climate

Hundreds of scientists have contributed to our current understanding of climate change. Periodically, the body of research pertinent to climate change is reviewed by the Intergovernmental Panel on Climate Change (IPCC), which was established in 1988 by organizations within the United Nations to evaluate the risk of climate change caused by human activities. Working groups within the IPCC, each consisting of dozens of scholars from around the world, evaluate the evidence for ongoing climate change, the prospects for future climate change, and potential social and economic impacts of these changes. The four multivolume assessment reports published by the IPCC (in 1990, 1995, 2001, and 2007) are important resources for understanding not only the evidence supporting global warming, but also predictions for the future.

The ability of computer models to reproduce measured temperatures (Figure 21.30) supports the validity of using these models to predict future climate change. This approach, however, is not without criticism. For one thing, geoscientists are primarily scholars of past changes on Earth, so the tools for predicting future changes are much more uncertain. Or, stated another way, hindsight is always more certain than looking ahead. In addition, it is important to keep in mind that modeling climate change into the future assumes that we fully understand the interactions and feedbacks among climate-change variables, including those that might prohibit an endless rise in global temperature. Equally important, given the demonstrated importance of greenhouse gases on climate, are uncertain assumptions about future greenhouse-gas emissions.

Figure 21.32 summarizes possible future-temperature scenarios adopted by the scientists who authored the 2007 IPCC assessment report. The four scenarios are presented as calculated temperature changes that start in 2000. Three of the scenarios make different assumptions about future increases in greenhouse-gas emissions and the fourth examines a prediction if greenhouse-gas emissions did not rise above the 2000 level. Notice that even with the unrealistic assumption of no increase in greenhouse-gas emissions, the computer models predict about 0.2°C of additional warming until about 2020. This warming happens mostly because the heat absorbed in the oceans during the late twentieth century would radiate back into the atmosphere during future decades. The predictions for various future emission scenarios produce the same result for the next 30 years, with a rate of temperature rise that is at least twice as large as the warming observed during the twentieth century. By 2100, the temperature predictions are different, however, ranging from about 1.5°C to almost 4°C warmer than in 2000.

One way of evaluating these somewhat shocking results, is to see whether the model-based predictions made in earlier IPCC reports were consistent with temperature changes over the few years elapsed since those reports were issued. **Figure 21.33** explores this comparison. Although each of the first three IPCC reports was issued in a different year, the computer-model projections were all made relative to starting conditions in 1990. The actual measured temperatures fall within the uncertainties of each prediction. This suggests that the short-term predictive ability of these computer models is good, although it does not test the validity of predictions made for many decades or a century.

▶ **Figure 21.33 Comparing projected and measured temperatures.** The first three IPCC assessment reports included global-temperature projections using temperature in 1990 as a starting point. Each successive projection was based on computer models that included increasing understanding of climate behavior, which produced progressively less model uncertainty in projected temperatures for the period 1990–2005. Each successive model also projected less global warming over this short interval. The actual measured temperatures are close to the projections with a slight tendency for the later models to under predict the actual measured warming.

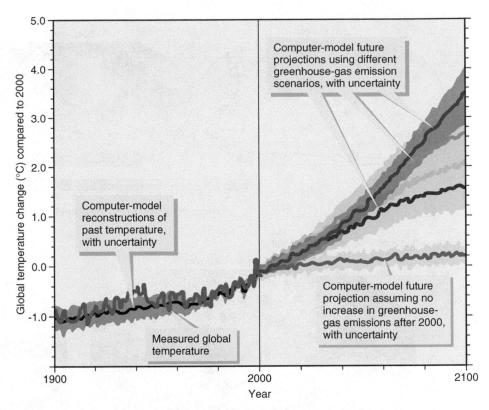

▲ **Figure 21.32 Projections of future global warming.** Existing computer models that closely reconstruct actual global temperature changes for the twentieth century (left side of graph) are used to project twenty-first–century temperature change (right side of graph) based on different scenarios of global economic growth, population growth, and energy innovation that will affect future greenhouse-gas emissions.

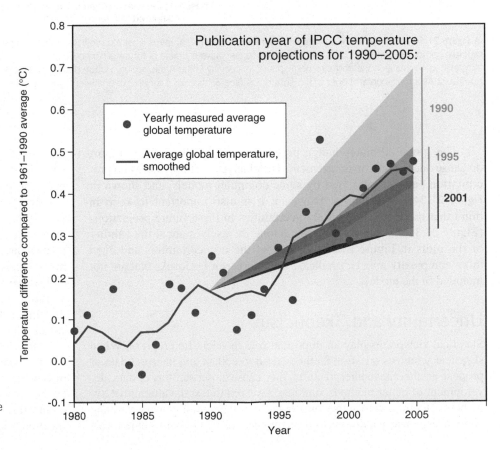

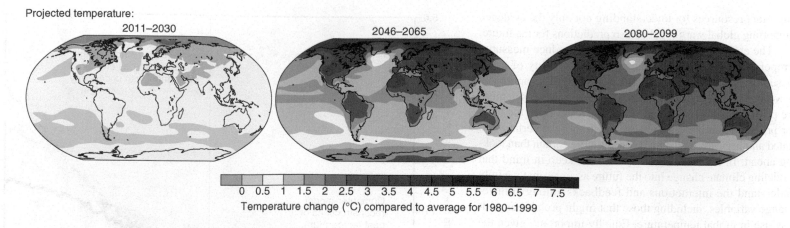

Projected temperature:

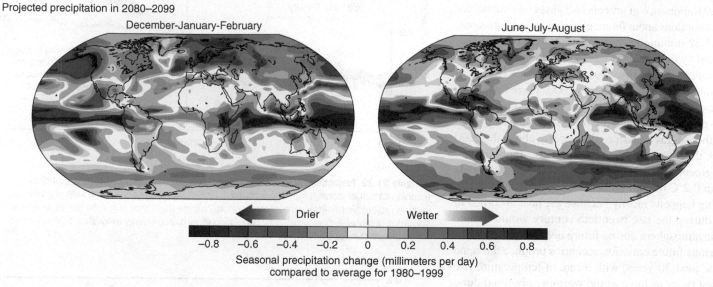

Projected precipitation in 2080–2099

▲ **Figure 21.34 What the future climate might be like.** These global maps illustrate projected future temperature and precipitation patterns based on averaging the results of multiple computer climate models that all used a conservative estimate of future greenhouse emissions. This global warming forecast predicts greatest twenty-first–century temperature increase in the northern hemisphere, especially in the Arctic region. Future precipitation is projected to be wetter near the equator and areas affected by the Southeast Asia monsoon and drier in the subtropics (see Section 20.1 for an explanation of global wind patterns that affect precipitation patterns).

The global-climate-change predictions in the 2007 IPCC report do cause one to pause in amazement. The changes in temperature and precipitation patterns predicted by some computer models, and shown in **Figure 21.34**, are stunning. However, it is also important to keep in mind that there are substantial uncertainties in these future projections (Figure 21.32). Scientists disagree among themselves about the validity of the plots in Figure 21.34 because of these uncertainties and even more complexity may be produced by unforeseen feedbacks that are not included in the models.

Uncertainty and Skepticism

Skeptical viewpoints play an important role in scientific progress. When skeptical scientists question methods used to collect and interpret data or propose an alternative interpretation, they cause closer scrutiny of data, development of better methods, and new research to test the alternative hypotheses. The new work either further supports the prevailing view, winning wider acceptance and diminishing skepticism, or an alternative hypothesis

turns out to provide the better explanation, causing research to move in a new direction. For example, plate tectonics was highly criticized when it was first proposed in the 1960s and its development into a widely accepted theory took many years of work, spurred on by testing hypotheses. Unfortunately, when scientific debates concern topics that are of wide interest outside the scientific community, such as climate change, there are many nonscientists who misunderstand the significance of these disagreements. Those who do not understand how science is done and progresses and whose values are in some way placed in conflict with the prevailing scientific viewpoint regrettably misinterpret the debates as an indication that this challenging viewpoint is based on poor science. Decision makers can use this misinterpretation of scientific progress as a basis for inaction.

However, it is the remaining challenges to understand how Earth works that motivate geologists to do more research. Whether it is the mysteries of how sand is deposited in the deep sea (Section 5.7) or why hot spots exist (Section 12.6), the uncertainty in the measurement of absolute ages of minerals (Section 7.8), competing ideas about the origin of Earth's water (Section 9.5) or desert pavements (Section 20.5), or the abandonment of continental drift in favor of plate tectonics (Section 12.1), scientists accept disagreements and uncertainty as what science is all about rather than as examples of the failure of science to find the "right answer." Single correct answers can be elusive and what seems like a satisfactory explanation now may be contradicted by evidence that remains to be discovered. There is no question that scientists, including geologists, have a lot to learn about climate change that may impact predictions for the future. The questions you must answer for yourself are these:

- Is climate change taking place?
- If so, are humans partly responsible for the changing climate?
- And, if so, what actions are justified for the future?

This chapter has provided evidence that you can evaluate to answer the first two questions. The last question requires additional contemplation of social and economic questions and your own personal values. However, now that you know the scientific evidence for global warming, we hope that you will continue to explore this issue on your own in the future.

Putting It Together—How Will Climate Change in the Future?

- Computer models predict a doubling of the rates of warming over the next 30 years but differ on longer-term forecasts, suggesting increases ranging from 1.5° to 4°C by 2100.

- The predictions of future warming are very uncertain and may not adequately include complicated interactions among components of the Earth system that affect temperature, and can only speculate about future greenhouse-gas emissions.

- Uncertainties in the predictions and disagreement among scientists of what future climate changes will occur is interpreted by some people to indicate the inability of science to assist decision making about human activities that may affect global warming. Alternatively, these disagreements and uncertainties can be viewed as typical of scientific progress.

Where Are You and Where Are You Going?

"Global warming" is the concept that the global annual temperatures on Earth have been steadily warming for more than a century because of both natural and primarily human-derived causes. There is spirited public debate as to whether global warming is taking place, and if so, the extents to which humans are responsible. It is because of this debate that students of geoscience should understand and evaluate the pertinent data.

The evidence for ongoing global warming is derived from compiling annual variations in temperatures averaged over Earth's land and sea surface. Not only do these analyses show warming of about 0.7°C since 1850, but rates of warming also increased during the twentieth century. The increasing temperatures are corroborated by measurements of decreasing sea ice in the Arctic, the retreat of melting glaciers, and the rise of global sea level.

The geologic record provides a perspective on climate change extending back millions of years, which is important for efforts to distinguish natural and human effects on global warming. Because temperature measurements do not exist for the ancient past, paleoclimatologists use proxies, including tree rings and data collected from sediments and ice cores, to infer ancient climates. This record shows that current warming is taking place within a warm interval between ice ages, which from the long-term view is happening during one of the overall coolest times in the last half-billion years of Earth history. Proxy data sets sometimes show warming in one location while cooling occurs in another and that the magnitudes of warming and cooling can be different at different locations. These regional differences must be averaged together in order to document global warming or cooling. Current global warming is unprecedented in the paleoclimate record of the last 1200 years, and this recent warmth stands out beyond the uncertainties related to measurement errors or interpretation of proxy records.

Climate is fundamentally driven by energy received from the Sun. Energy radiated from Sun to Earth differs in wavelength from the energy radiated into space from Earth's surface and atmosphere. Differences in radiant energy wavelength affect how energy is absorbed and reflected from different surfaces. Changes in the amount of solar energy reaching Earth, albedo of the atmosphere and surface, and concentrations of atmospheric gases are factors that change the overall energy budget. Another contributing factor is variation in ocean circulation, which distributes heat on Earth's surface. Variations in Earth's orbit and rotation are the major cause of changes in solar energy reaching Earth over long times, causing fluctuations between glacial and interglacial conditions, but do not explain historic warming. Very small increases in solar-energy output during the current warming trend are documented, but are not known with a high degree of certainty. Albedo has changed in response to historic changes in land use and cloud cover, and these changes likely cause small amounts of cooling. Concentrations of greenhouse gases, such as carbon dioxide and methane, have increased dramatically during the time of global warming and correlate to increased consumption of carbon-rich fossil fuels. Sulfurous aerosols emitted during volcanic eruptions and from human consumption of coal cause short periods of cooling. The circulation of heat with convective ocean currents in the North Atlantic and wind-driven currents in the equatorial Pacific, which cause El Niño and La Niña climatic extremes, also contribute to changes in global surface temperatures.

Computer models of the climate system can explain warming from 1900 to approximately 1970 by natural causes, but temperature increases observed since 1970 must include human impacts on the energy budget, especially the increasing production of greenhouse gases. Skeptics of the human impact on global warming suggest that the effect of varying solar-energy input to the Earth system is inadequately accounted for in these models and is the primary, and natural, driver of global warming. If this was true, then we could predict that the whole thickness of the atmosphere should experience warming because all of the atmosphere would be affected by the increased solar energy. This is not the case, as satellite measurements since 1978 of temperature at different levels of the atmosphere show warming below about 15 kilometers and cooling in the upper atmosphere. These observations are consistent with greenhouse warming of Earth's surface rather than variations in solar-energy output.

Forward-looking computer models predict a doubling of the rates of warming within the next 30 years and a total global temperature rise as great as 4°C by 2100. These predictions are uncertain as they may not adequately include complicated interactions among components of the Earth system that affect temperature, and require speculation about future greenhouse-gas emissions. Uncertainty of predictions is one reason global warming is a controversial subject.

Scientists from many disciplines (including atmospheric scientists, oceanographers, astronomers, and biologists, as well as geologists) provide critical scientific data for evaluating the existence and causes of global warming. Many sciences must contribute knowledge and expertise in order to decipher such a complex problem. The ongoing global warming debate is, therefore, an excellent example of the role of geology within studies of the entire Earth system. The debate itself is an example of how science progresses through the proposing and testing of alternative hypotheses within an overall framework of skepticism until all that remains are the conclusions that best explain observations and measurements. Regrettably, many people do not understand that disagreement and pursuit of alternative explanations are parts of doing good science and, instead, interpret the scientific debate about global warming as a basis for indecision about the future implications. Global warming is but one example of how scientists evaluate problems that are significant to humanity and potentially provide answers for our future as we continue to pursue our understanding of how Earth works.

Active Art

Global Warming. See how differences in the wavelength of incoming and outgoing radiant heat cause warming of the atmosphere close to Earth's surface.

Ocean Circulation. See how wind-driven surface currents affect water temperatures, and also the conveyor-belt circulation that includes North Atlantic Deep Water.

El Niño and La Niña. See how changing atmospheric and oceanic circulation systems in the Pacific Ocean affect weather conditions.

Confirm Your Knowledge

1. Describe the global warming concept in your own words.
2. What is the difference between weather and climate?
3. What does it mean to "smooth" graphed values? Why do climate scientists "smooth" temperature data?
4. How do climate scientists collect temperature data from all over the world to calculate the annual global mean temperature?
5. Why does temperature data collected in more recent times have less uncertainty than for the past?
6. What evidence from glaciers and sea-level measurements corroborates global warming?
7. What is a "proxy"? Describe some of the proxies of ancient climate conditions.
8. Describe the Younger Dryas period, the evidence for its existence, and the hypothesis for its origin.
9. List and briefly describe the components of the energy budget.
10. Why are the differing wavelengths of incoming and outgoing radiant heat important for understanding the energy budget?
11. What is the greenhouse effect? List the greenhouse gases.
12. Which changes in the energy budget over the last 150 years are natural and which are caused by human activities? Be sure to note whether each change would cause cooling or warming of the atmosphere near Earth's surface.

13. Which processes contribute to increasing carbon dioxide and methane in Earth's atmosphere?
14. How do climate scientists determine carbon dioxide concentrations for the ancient atmosphere?
15. Describe the carbon cycle on Earth with an emphasis on how processes within the cycle affect the carbon-dioxide concentration in the atmosphere.
16. How does burning coal contribute to both increasing and decreasing global temperature?
17. How do volcanic eruptions affect Earth's temperature?
18. How do temperature and salinity affect seawater density? Why are these relationships among temperature, salinity, and density important for understanding climate in the North Atlantic region?
19. Describe and explain the causes of El Niño and La Niña.
20. How do computer models contribute to separating natural from human-caused global warming?
21. How do climate scientists estimate future temperature changes?

Confirm Your Understanding

1. Write an answer for each question in the section headings.

2. What predictions can you make from the hypothesis that global temperatures are rising? Think of predictions beyond those discussed in this chapter.

3. Why is it important to recognize and include uncertainties when dealing with data? Use uncertainties described and graphed in this chapter, and others, as examples in your answer.

4. A friend examines Figure 21.4 and argues that a temperature increase of about 0.7°C in the last 150 years is so small that global warming is certainly insignificant and perhaps even nonexistent. What is your response?

5. How can climate scientists be sure that temperature proxies are valid for reconstructing ancient climate?

6. Change of any kind should be described in terms of the time over which the change takes place. Describe global climate change on these time scales: (a) during the last 180 million years; (b) during the Cenozoic, (c) during the Quaternary, (d) during the last 21,000 years; (d) during the last 1000 years, (e) during the last century; (f) during the last decade.

7. Why is London, England, relatively warm even though it is located at the same high, cold latitude as Hudson Bay, Canada?

8. Some climate scientists feel that humans have affected climate and greenhouse gases in the atmosphere for at least 5000 years, dating back far before consumption of fossil fuels. Based on what you have learned in this chapter, how could the rise of agricultural societies have produced this longer impact on climate?

9. Explain how the amount of solar energy received at the top of Earth's atmosphere varies on both the short term (decade and century) and the long term (thousands to hundreds of thousands of years).

10. There are many atmospheric gases that contribute to greenhouse warming, so why do climate scientists focus on carbon dioxide and methane and not on more abundant water vapor?

11. Explain how volcanic eruptions contribute to variations in Earth's temperature through geologic time and on historic time frames.

12. You are talking with family and friends and tell them that global climate change is (or is not) caused by humans. Explain your point of view to them using scientific data.

13. Use Figure 21.32 to draw a conclusion about global warming during the twenty-first century.

14. If global temperature increases as much as 4° C by 2100, what global effects might you anticipate as a result?

APPENDIX A

Periodic Table of the Elements

Abundance of elements in Earth's crust

- $> 10\%$
- $1{-}10\%$
- $0.1{-}1\%$
- $0.001{-}0.1\%$
- $1 \times 10^{-6}{-}0.001\%$
- $< 1 \times 10^{-6}\%$
- Not present in nature

Legend cell: 22 — Atomic number; Ti; Titanium; 47.88 — Atomic weight

Element	No.	Name	Weight
H	1	Hydrogen	1.008
He	2	Helium	4.003
Li	3	Lithium	6.941
Be	4	Beryllium	9.012
B	5	Boron	10.81
C	6	Carbon	12.01
N	7	Nitrogen	14.01
O	8	Oxygen	16.00
F	9	Fluorine	19.00
Ne	10	Neon	20.18
Na	11	Sodium	22.99
Mg	12	Magnesium	24.31
Al	13	Aluminum	26.98
Si	14	Silicon	28.09
P	15	Phosphorus	30.97
S	16	Sulfur	32.07
Cl	17	Chlorine	35.45
Ar	18	Argon	39.95
K	19	Potassium	39.10
Ca	20	Calcium	40.08
Sc	21	Scandium	44.96
Ti	22	Titanium	47.88
V	23	Vanadium	50.94
Cr	24	Chromium	52.00
Mn	25	Manganese	54.94
Fe	26	Iron	55.85
Co	27	Cobalt	58.93
Ni	28	Nickel	58.69
Cu	29	Copper	63.55
Zn	30	Zinc	65.39
Ga	31	Gallium	69.72
Ge	32	Germanium	72.64
As	33	Arsenic	74.92
Se	34	Selenium	78.96
Br	35	Bromine	79.90
Kr	36	Krypton	83.80
Rb	37	Rubidium	85.47
Sr	38	Strontium	87.62
Y	39	Yttrium	88.91
Zr	40	Zirconium	91.22
Nb	41	Niobium	92.91
Mo	42	Molybdenum	95.94
Tc	43	Technetium	(98)
Ru	44	Ruthenium	101.1
Rh	45	Rhodium	102.9
Pd	46	Palladium	106.4
Ag	47	Silver	107.9
Cd	48	Cadmium	112.4
In	49	Indium	114.8
Sn	50	Tin	118.7
Sb	51	Antimony	121.8
Te	52	Tellurium	127.6
I	53	Iodine	126.9
Xe	54	Xenon	131.3
Cs	55	Cesium	132.9
Ba	56	Barium	137.3
La	57	Lanthanum	138.9
Hf	72	Hafnium	178.5
Ta	73	Tantalum	180.9
W	74	Tungsten	183.9
Re	75	Rhenium	186.2
Os	76	Osmium	190.2
Ir	77	Iridium	192.2
Pt	78	Platinum	195.1
Au	79	Gold	197.0
Hg	80	Mercury	200.6
Tl	81	Thallium	204.4
Pb	82	Lead	207.2
Bi	83	Bismuth	209.0
Po	84	Polonium	(209)
At	85	Astatine	(210)
Rn	86	Radon	(222)
Fr	87	Francium	(223)
Ra	88	Radium	(226)
Ac	89	Actinium	(227)
Rf	104	Rutherfordium	(261)
Db	105	Dubnium	(262)
Sg	106	Seaborgium	(266)
Bh	107	Bohrium	(264)
Hs	108	Hassium	(269.1)
Mt	109	Meitnerium	(268)
Ds	110	Darmstadtium	(271)
Uuu	111	Unununium	(272)
Uub	112	Ununbium	(277)
UUq	114	Ununquadium	(289)

Lanthanides

Element	No.	Name	Weight
Ce	58	Cerium	140.1
Pr	59	Praseodymium	140.9
Nd	60	Neodymium	144.2
Pm	61	Promethium	(145)
Sm	62	Samarium	150.4
Eu	63	Europium	152.0
Gd	64	Gadolinium	157.3
Tb	65	Terbium	158.9
Dy	66	Dysprosium	162.5
Ho	67	Holmium	164.9
Er	68	Erbium	167.3
Tm	69	Thulium	168.9
Yb	70	Ytterbium	173.0
Lu	71	Lutetium	175.0

Actinides

Element	No.	Name	Weight
Th	90	Thorium	232.0
Pa	91	Protactinium	231
U	92	Uranium	238.0
Np	93	Neptunium	(237)
Pu	94	Plutonium	(244)
Am	95	Americium	(243)
Cm	96	Curium	(247)
Bk	97	Berkelium	(247)
Cf	98	Californium	(251)
Es	99	Einsteinium	(252)
Fm	100	Fermium	(257)
Md	101	Mendelevium	(258)
No	102	Nobelium	(259)
Lr	103	Lawrencium	(262)

APPENDIX B

Ions and Natural Isotopes that Are Significant to Geologic Studies

Element (symbol)	Negative Ions	Native Element	Positive Ions	Isotopes (most abundant, *radioactive*)
Aluminum (Al)		Al	Al^{3+}	^{26}Al, ^{27}Al
Argon (Ar)		Ar		^{36}Ar, ^{40}Ar
Arsenic (As)	As^{3-}		As^{3+}	^{75}As
Barium (Ba)			Ba^{2+}	^{136}Ba, ^{137}Ba
Boron (B)			B^{3+}	^{11}B
Calcium (Ca)			Ca^{2+}	^{40}Ca
Carbon (C)	C^{4-}	C	C^{4+}	^{12}C, ^{13}C, ^{14}C
Chlorine (Cl)	Cl^{1-}			^{35}Cl, ^{37}Cl
Chromium (Cr)		Cr	Cr^{2+}, Cr^{3+}	^{52}Cr
Copper (Cu)		Cu	Cu^{1+}, Cu^{2+}	^{63}Cu, ^{65}Cu
Fluorine (F)	F^{1-}			^{19}F
Gold (Au)		Au	Au^{1+}	^{197}Au
Helium (He)		He		^{3}He, ^{4}He
Hydrogen (H)	H^{1-}		H^{1+}	^{1}H, ^{2}H, ^{3}H
Iridium (Ir)		Ir	Ir^{4+}	^{191}Ir, ^{193}Ir
Iron (Fe)		Fe	Fe^{2+}, Fe^{3+}	^{54}Fe, ^{56}Fe, ^{57}Fe
Lead (Pb)			Pb^{2+}, Pb^{4+}	^{204}Pb, ^{206}Pb, ^{207}Pb, ^{208}Pb, ^{210}Pb
Lithium (Li)			Li^{1+}	^{7}Li
Magnesium (Mg)			Mg^{2+}	^{24}Mg, ^{25}Mg, ^{26}Mg
Manganese (Mn)			Mn^{2+}, Mn^{3+}, Mn^{4+}	^{55}Mn
Mercury (Hg)		Hg	Hg^{1+}	^{198}Hg, ^{199}Hg, ^{200}Hg, ^{201}Hg, ^{202}Hg
Molybdenum (Mo)			Mo^{4+}, Mo^{6+}	^{92}Mo, ^{95}Mo, ^{96}Mo, ^{98}Mo, ^{100}Mo
Neodymium (Nd)			Nd^{3+}	^{142}Nd, ^{143}Nd, ^{144}Nd, ^{146}Nd
Nickel (Ni)		Ni	Ni^{2+}, Ni^{3+}	^{58}Ni, ^{60}Ni
Nitrogen (N)	N^{1-}	N_2	N^{5+}	^{14}N, ^{15}N
Oxygen (O)	O^{2-}	O_2		^{16}O, ^{17}O, ^{18}O
Phosphorous (P)			P^{5+}	^{31}P
Platinum (Pt)		Pt	Pt^{2+}	^{194}Pt, ^{195}Pt, ^{196}Pt
Potassium (K)			K^{1+}	^{39}K, ^{40}K, ^{41}K
Radon (Rn)		Rn		^{218}Rn, ^{219}Rn, ^{220}Rn, ^{222}Rn
Rubidium (Rb)			Rb^{1+}	^{85}Rb, ^{87}Rb
Samarium (Sm)			Sm^{3+}	^{147}Sm, ^{148}Sm, ^{149}Sm, ^{152}Sm, ^{154}Sm
Silicon (Si)			Si^{4+}	^{28}Si, ^{29}Si, ^{30}Si
Silver (Ag)		Ag	Ag^{1+}	^{107}Ag, ^{109}Ag
Strontium			Sr^{2+}	^{86}Sr, ^{87}Sr, ^{88}Sr
Sulfur (S)	S^{2-}	S	S^{2+}, S^{4+}, S^{6+}	^{32}S, ^{33}S, ^{34}S, ^{36}S
Thorium (Th)			Th^{4+}	^{230}Th, ^{231}Th, ^{232}Th, ^{234}Th
Tin (Sn)		Sn	Sn^{2+}	^{116}Sn, ^{118}Sn, ^{120}Sn
Titanium (Ti)			Ti^{4+}	^{48}Ti
Uranium (U)			U^{4+}, U^{6+}	^{234}U, ^{235}U, ^{238}U
Zinc (Zn)			Zn^{2+}	^{64}Zn, ^{66}Zn, ^{68}Zn

APPENDIX C

Common Conversions

Length

	Inch (in)	Feet (ft)	Mile (mi)	Centimeter (cm)	Meter (m)	Kilometer (km)
1 in =	1	0.0833	0.0000158	2.54	0.0254	0.0000254
1 ft =	12	1	0.0001894	30.48	0.3048	0.0003048
1 mi =	63,360	5,280	1	160,934.4	1,609.344	1.609344
1 cm =	0.39370	0.0328084	0.00000621	1	0.01	0.00001
1 m =	39.370	3.28084	0.000621	100	1	0.001
1 km =	39,370	3,280.84	0.621	100,000	1,000	1

Area

	Square Inch (in^2)	Square Feet (ft^2)	Square Mile (mi^2)	Square Centimeter (cm^2)	Square Meter (m^2)	Square Kilometer (km^2)
1 in^2 =	1	0.0069444	2.491×10^{-10}	6.452	0.0006452	6.452×10^{-10}
1 ft^2 =	144	1	3.587×10^{-8}	929.03	0.092903	9.2903×10^{-8}
1 mi^2 =	4.014×10^9	2.788×10^7	1	2.59×10^{10}	2.59×10^6	2.59
1 cm^2 =	0.1550	0.0010764	3.861×10^{-11}	1	0.0001	1×10^{-10}
1 m^2 =	1,550	10.764	3.861×10^{-7}	10,000	1	0.000001
1 km^2 =	1.55×10^9	1.0764×10^7	0.3861	1×10^{10}	1×10^6	1

Volume

	Cubic Inch (in^3)	Cubic Feet (ft^3)	Cubic Meter (m^3)	Quart (qt)	Liter (l)	Gallon (Gal.; U.S.)
1 in^3 =	1	0.0005787	0.0000164	0.017316	0.0163871	0.004329
1 ft^3 =	1,728	1	0.0283168	29.9220779	28.3168467	7.4805195
1 m^3 =	61,023.74	35.3147	1	1,056.6882	1,000	264.1721
1 qt =	57.75	0.0334201	0.0009464	1	0.946353	0.25
1 l =	61.02374	0.0353147	0.001	1.0566882	1	0.2641721
1 gal =	231	0.1336806	0.0037854	4	3.7854	1

Mass and Weight

	Ounce (oz)	Pound (lb)	Short Ton (T)	Gram (g)	Kilogram (kg)	Metric Ton (t)
1 oz =	1	0.0625	0.0000313	28.3	0.0283	0.0000283
1 lb =	16	1	0.0005	453.6	0.4536	0.0004536
1 T =	32,000	2,000	1	907,184.7	907.1847	0.9071847
1 g =	0.035274	0.0022046	0.00000110	1	0.001	0.000001
1 kg =	35.274	2.2046	0.00110	1,000	1	0.001
1 t =	35,274	2,204.6	1.10	1×10^6	1,000	1

Temperature

To convert from Fahrenheit (F) to Celsius (C): $°C = (°F - 32°)/1.8$
To convert from Celsius (C) to Fahrenheit (F): $°F = (°C × 1.8) + 32°$

Fahrenheit (°F)	Celsius (°C)	
32	0	Freezing point of water at 1 atmosphere pressure
50	10	
68	20	
86	30	
104	40	
122	50	
140	60	
158	70	
176	80	
194	90	
212	100	Boiling point of water at 1 atmosphere pressure
572	300	
932	500	
1292	700	
1652	900	
1832	1000	
3632	2000	
5432	3000	
7232	4000	
9032	5000	
10832	6000	

GLOSSARY

A horizon: A surficial soil horizon that contains organic matter mixed with mineral and rock fragments.

Abrasion: Erosion of rock by the impact of sediment particles carried by water, wind, or glacial ice.

Absolute age: Specific time when an event occurred or a particular number of years that have elapsed since an event occurred.

Absolute sea level change: Global change in sea level caused either by changes in the volume of water in the oceans or changes in average ocean depth.

Accreted terrane: See *exotic terrane*.

Achondrite: Stony meteorite that lacks chondrules and formed by igneous processes.

Aftershock: One of many earthquakes that follow a larger-magnitude mainshock in the same region.

Albedo: The amount of light and heat reflected by a planetary object or by particular surfaces on the object.

Alluvial fan: Fan-shaped mass of alluvium that forms where stream flow abruptly exits a confined channel onto an unconfined valley floor; commonly forms at the base of a mountain range.

Alluvial stream: Stream that flows on loose sediment (alluvium).

Alluvium: Sediment deposited by streams.

Amphibolite: Metamorphic rock composed primarily of amphibole minerals, typically also contains plagioclase feldspar. Usually formed by metamorphism of mafic or intermediate volcanic rocks.

Andesite: Aphanitic igneous rock solidified from intermediate-composition magma.

Angle of repose: The maximum angle of a stable slope determined by friction, cohesion, and the shapes of the particles in the mass of loose materials that make up the slope.

Angular unconformity: An unconformity defined by a sharp boundary between intervals of layered rocks (either sedimentary beds or lava flows) that are inclined at different angles.

Angular velocity: The velocity of circular motion around a fixed center point, which is described as the angle of rotation divided by the time over which the rotation occurred (e.g., degrees per year).

Anthracite: The hard variety of coal produced by metamorphism of sedimentary coal.

Anticline: Arch-shaped fold where limbs dip away from the hinge line; oldest rocks are exposed in the center of the fold when the rocks are eroded.

Aphanitic: Igneous-rock texture resulting from rapid crystallization of magma to form very small crystals that are generally invisible to the unaided eye.

Apparent polar wander path: The apparent path of motion by a magnetic pole that is implied by paleomagnetic data collected from rocks of different ages. The word "apparent" emphasizes that it only appears like the pole is wandering, because the data instead require motion of the continents where the rocks were collected.

Aquifer: A body of rock or regolith with sufficient porosity and permeability to provide water in useful quantities to wells or springs.

Arkose: Sandstone containing at least 25 percent feldspar in addition to abundant quartz.

Artesian: An adjective describing ground water that is confined under pressure such that the water naturally rises toward the surface in a well or spring.

Assimilation: Melting of rock adjacent to magma, which modifies the magma composition.

Asteroid: Irregularly shaped, rocky planetesimal primarily found in a belt orbiting the Sun between Mars and Jupiter.

Asthenosphere: A weak layer of Earth's mantle below the lithosphere that may contain small amounts of magma.

Atomic mass number: The sum of the number of protons and the number of neutrons in the nucleus of an atom.

Atomic mass unit: Equal to the mass of a proton, or 1.673×10^{-24} gram; abbreviated AMU.

Atomic number: The number of protons in the nucleus of an atom. Each element has a unique atomic number.

Atom: The smallest unit of matter that engages in chemical reactions and cannot be chemically broken down into simpler components.

Axial plane: An imaginary surface separating rocks on each side of a fold.

Axis: An imaginary line that represents where the axial plane intersects an original plane in the rocks, such as a bedding plane.

B horizon: The soil horizon notable for colors and textures that indicate accumulation of minerals that are not present in the parent material. Typically found below the A horizon and above the C horizon.

Bar: A body of sediment exposed above a low water level in a stream but that is submerged and moves at higher flow.

Barchan: Crescent-shaped sand dune that forms on desert surfaces where wind-transported sand is not abundant. Barchans resemble parabolic dunes except that the ends of the crescent on a barchan point in the downwind direction.

Barrier island: Long, narrow sandy ridge of land that forms parallel to, but separate from, the mainland coast.

Basalt: Aphanitic mafic igneous rock.

Base level: The elevation down to which a river has the ability to erode its bed everywhere along its course.

Basin: Depression caused by subsidence of the crust and typically accumulating thick sediment and sedimentary rocks. May also refer to an eroded depression; also see *drainage basin*.

Batholith: Large body of intrusive igneous rocks formed by the crystallization of a magma chamber beneath the surface.

Bauxite: Multicolored ore of aluminum hydroxide minerals that is the primary source of commercial aluminum; commonly found in modern or ancient rainforest soils.

Baymouth bar: A spit that links one headland to the next and closes off the intervening bay from the ocean.

Bay: Deep recess along a shoreline.

Beach drift: The along-shore transport of sediment on a beach resulting from alternating oblique up-beach transport by swash and directly offshore transport by backwash.

Beach face: Relatively steep part of a beach close to the water.

Beach: Low-sloping, nonvegetated area of unconsolidated sediment moved by waves and tides that extends from the low-tide line to a landward line defined by a cliff, sand dunes, or permanent vegetation.

Bedding: Distinctive layering caused by layer-by-layer deposition of sedimentary materials at Earth's surface.

Bedload: Large grains that roll, slide, and bounce along the bottom of a stream.

Bedrock: Large, continuous exposures of solid rock, whereas the term "rock" by itself could describe a small fragment that you pick up from the ground.

Bedrock stream: Stream that flows on solid rock.

Berm: A typically flat part of a beach formed landward of the sloping beach face.

Berm crest: Linear boundary between the relatively steep beach face and the nearly horizontal berm.

Bituminous coal: The most common lithified variety of sedimentary coal.

Body wave: Seismic (earthquake) wave that passes through the interior of Earth. See *P wave* and *S wave* as examples.

Bomb: Pyroclastic fragment more than 64 millimeters across.

Bond: The force that holds atoms together to form molecules.

Braided: Describes an alluvial stream notable for abundant bars that divide the stream flow into threads that separate and rejoin around the bars; so named for the pattern of water flow around the bars that resembles braided rope or hair.

Breaker: A water-surface wave that becomes so steep that the crest outraces the rest of the wave and collapses forward.

Breccia: A clastic sedimentary rock consisting of angular fragments larger than 2 millimeters across.

Brittle: Describes deformation characterized by rock fracture. Contrast with *plastic*.

C horizon: The lowest soil horizon consisting of the least-weathered material, which commonly is the parent material for soil formation.

C-type asteroid: Asteroid with low albedo and chemical composition similar to the Sun but with slightly lower abundances of gaseous elements.

Cabochon: A round or oval gem with a simple curved top, commonly used to display the beauty of a gem.

Caldera: Circular or elliptical depression at Earth's surface formed when the roof of a magma chamber collapses because a significant volume of magma was removed from the chamber without immediate replenishment from deeper levels.

Cementation: The filling of pore spaces with precipitated minerals to produce an interlocking framework with clastic sediment grains; a step in lithification of sediment to form sedimentary rock.

Centrifugal force: An apparent, not real, force that accounts for the change in direction of motion of an object on the surface of a rotating object.

Chemical equilibrium: The condition in a chemical reaction when the concentrations of reactants and products no longer changes, even though the reaction continues with reactants forming products and products breaking down into reactants.

Chemical reaction: The coming together of atoms or molecular compounds that results in a change.

Chemical sediment: Sediment formed by precipitation of chemical compounds from water.

Chemical weathering: Dissolution of some minerals and formation of new minerals and dissolved ions as a result of water and oxygen reacting with minerals.

Chert: Chemical sedimentary rock composed of microscopic quartz crystals formed by chemical precipitation. Also a biologic sedimentary rock formed by accumulation of silica-rich microorganisms.

Chondrite: Stony meteorite containing many tiny spheres of silicate minerals, called chondrules, with textures that indicate crystallization from molten droplets.

Cinder: Lapilli (2–64 millimeters across) of basaltic or andesitic composition. Also called scoria.

Cinder cone: A conical accumulation of loose volcanic cinder lapilli and bombs around a central crater. Also called a scoria cone.

Cirque: Steeply eroded walls at the upslope end of glaciated valleys that partially enclose a natural amphitheater, sometimes occupied by a lake.

Clastic sediment: Residue of particles that remains after rocks weather.

Clastic: Sedimentary-rock texture describing particles formed by weathering of preexisting rocks.

Cleavage: Pattern of breakage along smooth planes in a mineral and the shape of the resulting fragments. Also defines the separation of metamorphic rock into thin sheets.

Coal: Sedimentary rock composed almost entirely of the compacted remains of fossil plants.

Cohesion: Attraction of particles to each other at the atomic level, caused by opposite electrostatic charges on adjacent particles.

Comet: A planetesimal formed mostly of ice, with subordinate rock, and that partly vaporizes to form a long gaseous tail in proximity to the sun.

Compaction: Decrease in volume caused by partial or complete elimination of pore spaces between sediment particles, usually caused by the weight of overlying sediment; a step in lithification of sediment to form sedimentary rock.

Composite volcano: Volcano composed of interlayered lava flows and pyroclastic deposits and typically with slopes generally steeper than 25 degrees.

Compression: A stress caused by oppositely directed forces that shorten or decrease the volume of materials.

Conduction: Process of transferring heat by contact between two surfaces of different temperatures; heat transfers without motion of matter.

Cone of depression: A cone-shaped depression in the water-table surface that forms where wells discharge ground water. The water-table surface slopes in all directions toward the well so that the amount of ground water flowing to the well is equal to the amount pumped from the well.

Confined aquifer: Ground water in a permeable layer that is between impermeable layers (confining beds) that prevent shallow and deeper ground waters from mixing.

Confining bed: Low-permeability material that restricts the movement of ground water into or out of adjacent aquifers.

Conglomerate: A clastic sedimentary rock composed primarily of rounded gravel-size particles (greater than 2 millimeters across).

Contact metamorphic rock: Rock changed (metamorphosed) by heat and fluid in close proximity to magma intrusions or lava flows.

Contact metamorphism: Metamorphism near the contacts of igneous intrusions or beneath erupted lava flows, produced primarily by heat and fluid flow in the region adjacent to the magma or lava.

Continental drift: Hypothesis of German meteorologist Alfred Wegener that all continents were once joined as a single continent, which he named Pangea, from where they drifted to their current positions.

Convection: Process of simultaneously transferring heat and matter by movement of fluid or plastically deforming rock because of density contrast; denser and typically colder material sinks while less dense and typically warmer material rises.

Convergent plate boundary: Curving zone on Earth's surface where plates collide nearly head-on into one another, compressing the lithosphere and causing subduction of one plate beneath the other.

Core: The central region of Earth composed primarily of iron metal and consisting of a molten liquid outer part and a solid inner part.

Coriolis effect: The tendency of objects in motion on or above Earth's surface to appear to be deflected to the right in the northern hemisphere and to the left in the southern hemisphere because of Earth's rotation.

Correlation: Demonstrated equivalence of rocks exposed in different locations.

Covalent bond: Union formed when two or more atoms mutually share electrons.

Craton: Low-elevation, tectonically stable interior region of a continent, exposing either very ancient Precambrian rocks or relatively thin coverings of sedimentary rocks resting unconformably on older Precambrian rocks.

Creep: Very slow flow of rock or regolith detected only by dislocation or bending of features at the surface.

Crevasse: Crack in the brittle upper part of a glacier caused by motion of the lower, plastically deforming part.

Cross-bed: Sedimentary structure of inclined layers within a bed and formed by shifting dunes and ripples. Cross-beds dip in the direction of current transport.

Crust: Outermost concentric layer of Earth composed mostly of silicate minerals and containing more silicon and aluminum than the underlying mantle.

Crystal face: Smooth, flat surface with regular geometric shape that forms part of the outer surface of a mineral specimen.

Cutbank: A steep bank eroded on the outside of a bend in a stream channel where flow is fast and deep.

Dacite: Aphanitic igneous rock solidified from felsic magma with a composition between andesite and rhyolite.

Darcy's law: The discharge of fluid through pore spaces is proportional to the area of the flow and to the hydraulic gradient. The law is an equation used to compute the quantity of water flowing through an aquifer.

Daughter isotope: An elemental isotope produced by decay of a radioactive parent isotope.

Debris avalanche: A very rapid flow of rock, regolith, vegetation, and sometimes ice.

Debris flow: A flow of regolith and water that behaves like a high-viscosity fluid.

Declination: Angle made by the two lines that connect a point on Earth's surface to the magnetic north pole and to the geographic north pole.

Decompression melting: A process of magma formation where rock partly melts because pressure decreases at a nearly constant high temperature. Usually happens by movement of mantle rocks toward Earth's surface.

Deflation: The erosion and transport of loose particles by wind, which lowers surface elevation.

Dehydration metamorphism: High-temperature and -pressure metamorphism where water-bearing minerals react to form minerals that lack water. Water is released by these reactions.

Delta: Landform protruding outward from a coastline and produced by sediment deposition where a stream enters a lake, reservoir, or sea.

Density: A measure of how compact a substance is, and mathematically defined by the mass divided by the volume of the substance.

Desert pavement: Closely spaced gravel fragments that cover barren, rocky deserts to form a smooth surface.

Desert: Region where annual precipitation is less than 25 centimeters.

Dike: A tabular, steeply inclined igneous intrusion that cuts across sedimentary layers, if present.

Diorite: Phaneritic igneous rock solidified from intermediate-composition magma.

Dip: Angle between an imaginary horizontal plane and the planar margin of a geologic feature; used with the measurement of strike to describe the orientation of any planar geologic feature such as a rock layer, fault, or margin of an igneous intrusion.

Dip-slip fault: Fault along which rocks move parallel to the dip direction of the fault plane.

Discharge: The volume of fluid that passes a location within an interval of time (e.g., cubic meters per second).

Disconformity: An unconformity defined by a sharp erosional boundary between intervals of sedimentary or volcanic rocks where layers above and below the boundary are parallel to one another.

Dissolution: Chemical reactions where bonds break between atoms or molecules that then disperse in water.

Dissolved load: The chemical ions dissolved in stream water.

Distributary channel: One of several branching streams formed by separation of flow from a single channel around bars and vegetated islands of stream-deposited sediment. Most commonly form on deltas and alluvial fans.

Divergent plate boundary: Linear or curving zones where plates move apart from one another and new lithosphere forms.

Divide: A relatively high ridge that separates the drainage basin of one stream from adjacent drainage basins.

Dolostone: Chemical sedimentary rock composed of the calcium and magnesium carbonate mineral, dolomite.

Drainage basin: Area from which a stream gathers water; can be used to describe the size of a stream.

Dune: A curving ridge of loose sediment, taller than 1 centimeter, which moves along with water or wind currents.

Dust: Wind-transported particles smaller than 0.1 millimeter.

Dynamo: Device that generates electric current by rapidly rotating a large magnet inside coils of electrically conductive wire.

E horizon: The highly leached soil horizon that forms in some cases below the A horizon and defined by the absence or near-absence of organic matter, easily weathered minerals, or weathering products like clay, oxide, and hydroxide minerals.

Earthquake: A release of stored energy resulting from the breaking and sudden movement of stressed rock.

Eclogite: Very high-grade metamorphic rock that lacks water-bearing minerals, is dominated by garnet and sodium-rich pyroxene (which gives the rock a blue coloration), and sometimes includes minor quartz.

El Niño: A weather pattern where the trade winds in the eastern Pacific Ocean are unusually weak, allowing more eastward current flow and unusually warm conditions along the coast of South America; typically correlates to warmer global temperatures. The opposite condition is called La Niña.

Elastic: Describes deformation in which a rock returns to its original dimensions after stress is removed, much the way a rubber band returns to its original shape after stretching and letting it go.

Elastic limit: Value of applied stress at which rock deformation is permanent and cannot be reversed; the value of increasing stress where rocks break or flow.

Elastic rebound theory: The nonpermanent bending of rock caused by strain on either side of a locked fault and which is recovered after the fault breaks during an earthquake.

Element: Chemical substance that cannot be split into simpler substances.

Elongation: The stretching strain resulting from tensional stress.

End moraine: A ridge of till deposited at the leading snout of a glacier.

Energy: A measure of the ability to do work.

Energy budget: The numerical balancing of the incoming energy to Earth's atmosphere and the outgoing energy from Earth to space.

Epicenter: Point on Earth's surface directly above the focus of an earthquake.

Erratic: Rock transported by a glacier and deposited where similar rocks are not present.

Estuary: Submerged part of a coastal stream valley where freshwater and seawater mix.

Euler pole: The pole of rotation that describes the motion of an object on the surface of a sphere.

Evaporite: Chemical sedimentary rock formed by minerals, such as halite and gypsum, that crystallize when water evaporates, causing ions to bond together.

Evapotranspiration: The transfer of water vapor from Earth's surface into the atmosphere by evaporation of moisture from rock or soil and by the transpiration of moisture from plant leaves.

Exfoliation: Deformation in which rocks expand as overlying rock is eroded, forming joints parallel to the ground; in some cases, the rock displaced by expansion peels away in thin layers.

Exotic terrane: Blocks of crust added to a continent, such as most of far western North America, and that consist of rocks that do not resemble rocks of the same age in adjacent blocks or the rest of the continent.

Extrusive (or volcanic) rock: Igneous rock formed by eruption of lava flows and pyroclastic materials onto Earth's surface.

Facet: Planar surface that is artificially cut on a gemstone to accentuate luster and transparency.

Fall: A mass movement where material detaches from a steep slope and then free falls through the air, or bounces and rolls downslope.

Fault: A fracture plane along which rock or regolith is displaced.

Fault scarp: Cliff or low step in the ground surface caused by displacement along a fault.

Felsic: Describes igneous rocks composed mostly of quartz, sodium-rich plagioclase, and potassium feldspars, and the magmas that these rocks crystallize from; derived from the words *fel*dspar and *sili*ca.

Fjord: Glacier-eroded valley along a coastline that is partly submerged beneath the sea to form a long, deep, steep-walled bay.

Flood: Overflow of water beyond the banks of a stream that occurs when discharge is too large to be contained within the channel.

Flood wall: An artificial wall typically constructed of concrete or steel along the banks of a river to confine high-discharge flows to the channel and protect the floodplain from flooding.

Floodplain: The land surface adjacent to a stream channel that is constructed by stream erosion and deposition and that is inundated during floods.

Floodway: An area of little or no development alongside a stream that is wide enough and low enough to carry the predicted discharge of the 100-year flood.

Flow: A mass movement of rock, regolith, or both, which behaves like a high-viscosity liquid.

Focus: Location of an earthquake inside Earth (also called the hypocenter).

Foliation: Planes of minerals formed in response to stress; a feature of many metamorphic rocks.

Footwall: Rock or regolith that exist below a fault plane.

Foreshock: One of many relatively low-magnitude earthquakes that precede a larger mainshock in the same region.

Fossil: Remains of an organism preserved in rock. Fossils may consist of the original mineral matter secreted by an organism, petrified organic material, or an impression left behind after most or all of the organic material has been destroyed.

Fossil fuel: Combustible energy source such as coal, oil, or natural gas formed from organic matter buried with sediment.

Fractional crystallization: Process by which the composition of a melt changes through time because minerals that form early during crystallization differ in composition from the magma and are physically separated from the magma.

Fracture: Nonuniform breakage of a mineral to leave an uneven surface, in contrast to smooth cleavage planes. Rock breakage that characterizes brittle deformation.

Fracture zone: Linear to slightly curved boundary between oceanic lithosphere of different ages and elevations within the same lithospheric plate. Fracture zones connect to transform plate boundaries that separate mid-ocean-ridge segments.

Friction: Force that opposes motion between two objects in contact with one another.

Gabbro: Phaneritic, mafic igneous rock.

Gemstone: Mineral, rock, or organic substance that has value based on beauty, color, luster, transparency, durability, and rarity.

Geographic information system (GIS): Computer-software tool that collects, stores, retrieves, analyzes, and displays data referenced to specific locations as distinct diagrammatic layers.

Geologic cross section: Diagram showing the interpretation of subsurface geology based on surface measurements and sometimes based on rock samples recovered from wells.

Geologic map: Map that portrays the distribution and orientation of rock types, locations of faults and folds, ages of rocks, and locations and nature of contacts between rock types.

Geologic time scale: Established chronological order of time intervals in geologic history.

Geology: The science of the origin, composition, structure, and history of Earth.

Geothermal gradient: The increase in temperature with increasing depth beneath Earth's surface.

Glacial flour: Very fine-grained particles resulting from abrasion of bedrock surfaces by rocks frozen into the bottom of glaciers; produces a distinctive milky discoloration in glacial-melt water streams.

Glacial rebound: Isostatic adjustment of surface elevation resulting from melting of glacial ice. When ice weight is added to the crust, the underlying crust subsides and the adjacent area bulges; when the ice melts, the originally depressed areas rise up and the bulges sink.

Glacier: An accumulation of snow and ice that is thick enough to flow under its own weight.

Gneiss: High-grade metamorphic rock defined by foliation of parallel compositional layers of light-colored (e.g., quartz, feldspar) and dark-colored (e.g., biotite, amphibole, pyroxene, garnet) minerals.

Graded bed: A sedimentary bed defined by a gradual variation in grain size from coarse at the bottom to fine at the top.

Graben: Block of crust displaced downward along normal faults.

Granite: Phaneritic, felsic igneous rock.

Gravity: A mutually attractive force between objects that depends on the distances between objects and their masses.

Greenhouse effect: The warming of Earth's surface that results from gases in the atmosphere (notably including carbon dioxide, water vapor, and methane) that permit solar energy to reach Earth's surface but stop radiated heat from going back into space; so named because these gases behave similar to glass in a greenhouse that raises temperature to permit year-round plant growth.

Greenstone: Low-grade metamorphic rock that contains abundant green minerals, usually with chlorite (iron-magnesium mica) as the primary constituent, along with green amphibole, feldspar, and quartz; typically forms by metamorphism of volcanic rocks.

Groin: A wall built perpendicular to the shoreline to trap sediment transported by longshore currents.

Ground moraine: Till deposited beneath a glacier to form a bumpy sediment sheet of irregular thickness.

Ground water: Water below Earth's surface that moves slowly through pore spaces and fractures within regolith and rock.

Half-life: Time interval required for half of the radioactive parent-isotope atoms to decay to form an equal number of daughter-isotope atoms.

Hanging wall: Rock or regolith that exist above a fault plane.

Hardness: A measure of the resistance to scratching a mineral surface.

Hard water: Water with high concentrations of dissolved ions, usually calcium and magnesium ions

from dissolution of calcite and dolomite in limestone and dolostone. These ions commonly reach concentrations in the water where they precipitate and clog pipes with rings of calcite, dolomite, or other minerals.

Head: An elevation that is proportional to the total energy of a fluid. In ground-water hydrology, head is the elevation that water rises in a well and describes the potential and pressure energy at the point where the well opens to the aquifer.

Headland: Place where land juts into the sea or a lake; sometimes called a cape.

Headwaters: Source of a stream at high elevation close to a divide.

Heat flow: Total amount of heat escaping through the surface of Earth and originating within the planet, mostly from radioactive decay.

High-grade metamorphism: Metamorphism occurring at temperatures higher than 600°C.

Hinge line: The imaginary line drawn along a deformed layer where the dip direction changes (also called the axis).

Hornfels: Any very hard, nonfoliated, metamorphic rock composed mostly or entirely of microscopically small crystals.

Horst: Block of crust displaced upward along normal faults.

Hot spot: An area of intense volcanic activity not explained by plate-boundary processes. Hot spots form where asthenosphere rises beneath lithospheric plates.

Hydraulic conductivity: A quantity that is proportional to the rate at which the fluid can move through a permeable material. Hydraulic conductivity is determined by both the properties of the fluid and the material that the fluid flows through.

Hydrologic cycle: A concept describing the movement of liquid water and water vapor through all parts of the Earth system.

Hydrolysis: Chemical reaction between a solid compound and water to produce a solid compound, which contains water molecules, and dissolved ions.

Hydrothermal metamorphic rock: Rock resulting from reaction of hot fluid with a preexisting rock.

Hydrothermal metamorphism: Metamorphic process resulting from hot water circulating through pore spaces and cracks in preexisting rock.

Hypothesis: Testable prediction about a natural process that can be checked by collecting data.

Ice cap: Broad glacier not confined by topography and with an area less than 50,000 square kilometers.

Ice sheet: Broad glacier not confined by topography and with an area greater than 50,000 square kilometers. Ice sheets currently cover most of Antarctica and Greenland.

Ice shelf: The floating part of a glacier that moved from land into deep water.

Iceberg: Block of glacial ice that detaches from a glacier and floats in the ocean or a lake.

Igneous rock: Rock that crystallized from molten material originating inside Earth.

Inclination: Angle between the magnetic field force line and Earth's surface.

Incompressibility: A measure of how material resists changing volume when subjected to high pressure (also known as the bulk modulus).

Index minerals: Minerals whose presence in metamorphic rocks allow the estimation of the pressure and temperature of rock formation.

Intensity: Measure of earthquake violence. The Mercalli Intensity Scale, denoted by Roman numerals I to XII, describes the extent to which people feel a quake, damage to structures, and secondary effects such as landslides.

Intermediate: Describes igneous rocks (e.g., andesite, diorite), and the magmas they crystallize from, that have a composition in between that of mafic and felsic rocks and magmas.

Intrusive rock: Igneous rock formed from magma injected into pre-existing rocks below the surface.

Iron meteorites: Meteorites mostly composed of iron and nickel metal.

Ionic bond: Atomic bond formed by the attraction of oppositely charged ions to one another to balance their charges.

Ion: Charged atom resulting from the gain or loss of one or more electrons so that the number of protons and electrons are unequal. A negatively charged ion is called an anion and a positively charged ion is a cation.

Isograd: The boundary on a geologic map between zones defined by different metamorphic index minerals. The boundary, therefore, indicates where a metamorphic reaction took place to consume or add an index mineral in the rocks.

Isotope: One of two or more atoms of the same element that have the same number of protons but different numbers of neutrons.

Jetty: A wall built where a channel enters the sea or a lake in order to keep the channel from filling with sediment transported by longshore currents. Jetties are commonly built in pairs on either side of a harbor entrance or tidal inlet.

Joint: Fracture in rock where little or no displacement has occurred.

Karst topography: Landscape pockmarked by sinkholes or a highly irregular landscape of high-standing rock towers and intervening depressions and valleys; there may be only limited surface stream flow as a result of deranged drainage patterns caused by rock dissolution.

Kettle: Depression where large blocks of ice melted within till or outwash; typically filled with water to form a pond or lake.

La Niña: A weather pattern where the eastward equatorial counter current in the Pacific Ocean is unusually weak, which causes anomalously cool water, and drier precipitation patterns, along and near the South American coast. Global temperatures are usually cooler at the same time. The opposite condition is called El Niño.

Lagoon: Shallow, relatively quiet water body on the landward side of an obstacle such as a barrier island, spit, baymouth bar, or offshore coral reef that absorbs wave energy directed toward the shoreline.

Lahar: Indonesian term describing the rapid flow of water and loose debris down steep slopes of volcanoes.

Lapilli: Pyroclastic fragments ranging in size from 2 millimeters to 64 millimeters across (singular: lapillus).

Lateral moraine: Boulder-rich sediment that forms a ridge along the margin of a valley glacier.

Lava: Molten material that erupts from a volcano and solidifies to form extrusive igneous rock.

Lava dome: A steep-sided lava flow, commonly almost as high as it is wide, and caused by the extrusion of extremely viscous lava.

Lava flow: Molten material extruded at, and flowing away from, a volcano.

Law: Scientific description of how nature is observed to behave.

Left-lateral strike-slip fault: Strike-slip fault across which features are displaced to the left (also called a sinistral strike-slip fault).

Limb: One side of a fold where all of the rocks dip in the same direction.

Limestone: Chemical sedimentary rock composed of calcite.

Linear dune: Long, narrow sand ridge oriented parallel to the prevailing wind direction; usually formed where sand supply is limited and wind direction is variable.

Liquefaction: The transformation of loosely packed, water-saturated sediment into a fluid mass. Commonly occurs when grains settle during earthquake ground shaking, which displaces the water in the intervening pore spaces upward such that the water pressure moves the grains apart and the whole sediment-water mixture loses strength.

Lithic sandstone: Sandstone composed mostly of sand-size rock fragments, rather than individual mineral fragments.

Lithification: The process of transforming loose sediment into sedimentary rock by compaction and cementation.

Lithosphere: Outer strong shell of Earth consisting of the crust and uppermost mantle.

Loess: Deposit of windblown silt.

Longshore current: A current that moves parallel to shore and forms where wave crests do not approach exactly parallel to the shoreline.

Low-grade metamorphism: Metamorphism occurring at temperatures approximately between 200°C and 400°C and at pressures less than 4 kilobars.

Low-velocity zone: The part of the upper mantle where seismic-wave velocity does not increase systematically with greater depth; the top of this zone typically defines the boundary between the lithosphere and the asthenosphere.

Luster: The nature of light reflection from mineral surfaces.

M-type asteroid: A bright asteroid consisting of metallic iron.

Mafic: Describes igneous rocks composed mostly of pyroxene, calcium-rich plagioclase feldspar, and olivine (basalt, gabbro) that have relatively high *ma*gnesium and iron (*ferric*) contents and a relatively low silicon content.

Magma: Molten material formed and residing within Earth.

Magnetic field: A region where lines of force act to move charged particles. The field is produced around a magnet or around a conductor carrying an electric current.

Magnetic reversal: An abrupt flip of Earth's magnetic poles that occurs at irregular time intervals.

Magnitude: Measure of earthquake size related to the amount of energy released by the earthquake and the amplitude of the waves recorded on a seismogram.

Mainshock: Largest of a sequence of earthquakes that is preceded by lower-magnitude foreshocks and followed by lower-magnitude aftershocks.

Mantle: The mostly solid but generally weak silicate zone of Earth below the crust and above the core.

Marble: Nonfoliated metamorphic rock composed of calcite and formed by the metamorphism of limestone.

Mass movement: Gravity-driven downslope motion of rock and regolith (also called mass wasting).

Massive: Describes rocks that lack layering or bedding.

Matter: Anything that has mass and occupies space.

Meandering: Tendency of a stream channel to gradually shift position across a valley as a result of simultaneous erosion and deposition on opposite stream banks over time.

Medial moraine: A ribbon of sediment within a glacier caused where lateral moraines combine at and downslope of the junction of two valley glaciers.

Medium-grade metamorphism: Metamorphism occurring approximately between temperatures of 400°C to 600°C.

Metallic bond: Bond formed where electrons freely roam around a number of different atoms, typically of the same element, giving the electrons a mobility that accounts for the ability of metallic substances to conduct electricity.

Metamorphic facies: An association of metamorphic minerals that are stable together over a defined range of temperature and pressure conditions.

Metamorphic rock: A rock distinguished from a preexisting rock by a change in the minerals that comprise it, or a rearrangement of the existing minerals, or both, as a result of reactions that occur at high temperature, high pressure, or in the presence of hot fluid, or a combination of all three.

Metamorphism: Processes within Earth that produce solid state mineralogical, chemical, and textural changes that alter the appearance of preexisting rocks.

Meteorite: Object from space that lands on Earth.

Meteor: Object from space that approaches Earth (called a meteorite after it lands on Earth).

Mid-ocean ridge: Long, continuous submerged volcanic mountain chain winding through Earth's oceans; coincide with divergent plate boundary.

Migmatite: Rock that resembles gneiss except that light-colored bands have the igneous-crystallization texture of granite resulting from high-grade temperature, pressure, and fluid conditions that cause partial melting, whereas dark layers reveal metamorphic crystal growth and recrystallization.

Milankovitch cycles: Periodic variations in Earth's rotation on its axis and in its orbit around the Sun that affect the amount of solar energy that reaches Earth.

Mineral: A naturally occurring solid with a definite, or only slightly variable chemical composition, and an ordered atomic structure formed mostly, but not entirely, by inorganic processes.

Mohorovičić discontinuity (Moho): The location of an abrupt increase in seismic velocity where mafic to felsic igneous and metamorphosed igneous rocks of the crust are underlain by mantle peridotite. The Moho is typically encountered 5–20 km deep beneath ocean basins, is thinnest near mid-ocean ridges, and is thickest (25–75 kilometers) below continents.

Mohs Hardness Scale: Relative scale for describing mineral hardness, ranging in value from 1 to 10, with the hardest mineral, diamond, having a hardness of 10.

Molecule: Substance composed of two or more atoms.

Moment magnitude: The most commonly used earthquake-magnitude scale that relies on the calculation of the seismic moment rather than the height of waves traced on a seismogram.

Monsoon: Torrential seasonal rains caused by seasonal changes in air pressure that bring warm, moist air from oceans over land.

Moraine: Heap of stony debris deposited along the margins of, or beneath, a glacier.

Mouth: Lowest elevation along a stream where it enters another stream, a lake, or the ocean.

Mudcrack: Open crack in muddy sediment or filled crack in mudstone or shale caused by contraction of clay minerals that absorb water and swell when wet, but then lose water, shrink, and crack while drying. Many cracks form and intersect to define polygon-shaped blocks of sediment.

Mudstone: Clastic sedimentary rock composed primarily of mud (silt and clay); called shale if the rock easily splits into thin sheets.

Natural levee: A ridge of sediment alongside a stream channel produced by deposition of sediment adjacent to the channel during floods.

Neap tide: The lowest high tide or highest low tide that occurs twice during a month. The relatively low tidal range results from the subtractive effects of gravitational pull exerted by the Moon and Sun.

Nebula: Cloud of gas and dust in space.

Nonconformity: An unconformity defined by a sharp boundary between plutonic-igneous or metamorphic rocks and overlying younger sedimentary or volcanic rocks.

Normal fault: A fault formed where the hanging-wall block moves downward compared to the foot-wall block.

Normal-polarity interval: Time interval when Earth's magnetic field is oriented as it is today, such that a compass needle points toward the north magnetic pole.

Normal stress: Force applied perpendicular to a surface.

Nuclear fission: A process where an atomic nucleus breaks into two roughly equal-sized atoms; results in the release of large amounts of energy.

Nuclear fusion: The combining (fusing) of two atomic nuclei to form a new element; results in the release of large amounts of energy.

Nucleus (atomic): The center of the atom. The nucleus contains one or more protons, which are particles with a positive electrical charge, and usually one or more neutral neutrons.

O horizon: The surface soil horizon that consists only of organic matter without minerals.

Oblique-slip faults: Faults along which rock movement includes a component along the dip direction of the fault plane and a component along the strike direction of the fault plane.

Obsidian: Volcanic glass of felsic composition that is typically opaque, commonly dark gray to black or brown.

Ore: Rock containing important metallic elements that must be extracted from the minerals by metallurgical processing that breaks the mineral bonds.

Outwash: Sediment eroded by glaciers that is then carried away by meltwater streams.

Oxbow lake: A highly curved lake (resembling an oxbow on a yoke) formed on a river floodplain when erosion cuts off a meander loop of a sinuous stream channel.

Oxidation: Chemical reaction between a substance and oxygen that produces new substances.

P wave (primary wave): Seismic body wave that displaces material in the same direction that the wave is moving, which causes alternating squeezing and stretching of the material as the wave passes.

pH: A measure of the acidity or alkalinity of a solution. Explicitly, pH is the logarithm of the hydronium-ion concentration. A solution with a pH of 7 is neutral, pH values less than 7 are acidic, and pH values greater than 7 are alkaline.

Paleomagnetism: The ancient orientation of Earth's magnetic field that is recorded in rocks.

Pan: Wind-eroded depression in regolith, commonly circular or elliptical, and commonly elongate parallel to the prevailing wind direction.

Parabolic dune: Crescent-shaped dune with an overall shape that resembles the graph of a parabola with a depression ("blowout") in the center and the two ends of the parabola pointing in the upwind direction.

Parent isotope: Radioactive isotope that decays through time to a daughter isotope.

Parent material: Rock or regolith from which a soil forms.

Peat: Black, organic-rich soil that contains more plant residue than mineral grains.

Pegmatite: Unusually coarse-grained igneous rock with crystals ranging in size from several centimeters to several meters.

Perched ground water: A locally saturated region above the water table formed where an impermeable layer impedes downward infiltration.

Peridotite: Phaneritic ultramafic igneous rock that is uncommon at Earth's surface but composes almost the entire upper mantle.

Period: The fundamental division of time on the geologic time scale, with boundaries defined by the presence of key fossils.

Permeability: The ability of a porous material to permit fluid flow through the material.

Phaneritic: Igneous rock texture defined by coarse, easily visible mineral crystals formed by slow crystallization.

Phyllite: Moderately high-grade metamorphic rock that forms from slate; mica grains in phyllite are coarser than those in slate and generate a silky sheen from the reflection of light from the parallel mica cleavage surfaces.

Physical weathering: Disaggregation of rocks by mechanical processes.

Planet: A celestial body that (1) orbits the Sun, (2) has sufficient mass and therefore gravitational pull to form into a nearly spherical shape, and (3) cleared out all other objects in the neighborhood of its orbit, hence completing the accretion process.

Planetary accretion: The process of planetary formation and growth by the gradual accumulation of small objects to make larger ones.

Planetary differentiation: Process of physical separation of relatively dense compounds toward the center of a spinning, growing planetesimal or planet, while less dense compounds remain near the surface.

Planetary embryo: A planetary object about one-hundredth to one tenth the size of Earth and formed by accretion of planetesimals.

Planetesimal: A small, solid object in space representing the first stage in accretion within the solar nebula. Planetesimals grow by cohesion between particles or by gravitational attraction between particles ranging in size from fine dust to as much as a few kilometers in diameter.

Plastic: Permanent deformation where rock flows rather than breaking.

Plate: One of several discrete, rigid to semi-rigid, roughly 100-km-thick slabs that make up Earth's lithosphere.

Plate tectonics: Theory that Earth's lithosphere is not seamlessly continuous but is broken into discrete pieces that move slowly relative to one another and change in size over geologic time.

Playa: A dry lake bed characterized by a dusty salt flat of evaporite minerals.

Plucking: A process of erosion by rivers or glaciers where rock outcrops break into fragments along pre-existing fractures.

Plume: One of many hypothesized narrow columns of unusually hot mantle that rise by convection from the core–mantle boundary; many hot spots may be plumes.

Plunging fold: Folded rock where the fold axis is not horizontal but is inclined (plunges) downward.

Pluton: General term for bodies of igneous rock that form in varying shapes and sizes form where magma solidifies beneath the surface.

Plutonic (or intrusive) rock: Igneous rock formed where magma solidifies below Earth's surface.

Pluvial lake: A lake that exists only during a period of exceptionally high rainfall; typically refers to ice-age lakes that formed in what are today mostly dry enclosed drainage basins in the western United States.

Point bar: A place where sediment accumulates on the inside of a bend in a stream, where the water flow is slow and shallow.

Polarity: The orientation of Earth's magnetic field, which is described as either normal or reverse polarity.

Polymorph: One of two or more minerals with identical chemical composition but with different arrangements of atoms.

Pores: Open spaces in a rock between mineral grains or open spaces between particles in regolith.

Porosity: Percentage of the total volume of regolith or rock composed of pores.

Porphyritic: Igneous rock texture defined by some large crystals surrounded by smaller crystals.

Potential energy: Energy that an object possesses because of its elevation. The potential energy that causes objects to move is greater for objects at high elevation than for objects at low elevation. Objects move from areas where they possess high potential energy to areas where they possess low potential energy.

Principle: A guiding concept that consistently works to describe the natural world.

Principle of cross-cutting relationships: Geologic features that cut across rocks (e.g., faults, igneous intrusions) must have formed after the rocks that they cut through.

Principle of faunal succession: Fossils serve to determine the relative age of enclosing rocks because there is a consistent chronologic sequence of fossil animals through geologic time.

Principle of inclusions: Objects enclosed in rock must be older than the time of rock formation.

Principle of isostasy: Surface elevations are adjusted by uplift and subsidence to maintain a condition where low-density rock floats on denser, underlying rock so that the pressure is everywhere the same at the base of any thick vertical column through the crust and upper mantle.

Principle of lateral continuity: Sedimentary beds are continuously deposited over large areas until encountering a barrier that limits their deposition.

Principle of original horizontality: Sedimentary beds are horizontal or near horizontal when deposited.

Principle of superposition: Where rocks are found in layers, one above the other, the lowest rock formed first, with each successively higher layer being younger than the one below.

Principle of uniformitarianism: Geologic processes and natural laws now operating on and within Earth have acted throughout geologic time; the logic and method by which geologists reconstruct past events.

Protostar: An early stage of star formation where dust and gases have concentrated into a dense central cloud but there is insufficient heat to initiate hydrogen fusion as a heat source, as in a true star.

Province: A discrete block of fault-bounded Precambrian crust composed of igneous and metamorphic rocks that formed during a particular time interval.

Proxy: A data record that substitutes for direct instrumental measurements; usually refers to estimates of climatic conditions prior to collection of temperature and precipitation measurements.

Pumice: Lightweight, highly vesicular, felsic volcanic fragments; typically pyroclastic fragments of lapilli or bomb size.

Pyroclastic: A class of fragmental volcanic rocks formed when explosions break apart magma and eject the resulting liquid drops and blobs that quickly quench into glass and fall to the ground.

Pyroclastic-fall deposits: Deposits consisting of pyroclastic fragments that are ejected high above a volcano, drift downwind, and then settle to the ground. The deposits exhibit a uniform decrease in particle size with increasing distance, up to one or two thousand kilometers, from the source volcano.

Pyroclastic-flow deposits: Poorly sorted mixtures of ash, lapilli, and bombs deposited by avalanches of pumice and ash flowing down the slope of a volcano; commonly thinner and finer grained at greater distances from the source.

Quartz sandstone: Sandstone composed primarily (more than 90 percent) of quartz, suggesting an environment where chemical weathering destroyed all of the other minerals in the source rock.

Quartzite: Nonfoliated metamorphic rock consisting of recrystallized quartz; formed by metamorphism of quartz sandstone.

Radiation: The process of energy transport, including heat, in the form of waves or particles, such as light.

Radioactive decay: The process by which unstable (radioactive) isotopes transform to new elements by a change in the number of protons and neutrons in the nucleus.

Radioactivity: Energy released when atoms of one element are transformed into atoms of another element resulting from processes that change the number of protons and neutrons in the nucleus.

Rayleigh number: The ratio of the gravity force that drives convection to the viscous force that resists convection. Calculating the Rayleigh number for a substance experiencing a temperature gradient assesses whether heat transfers by convection or conduction.

Recharge: Water added to the saturated zone.

Recrystallization: The transfer of atoms from one part of a crystal to another part of the same crystal or to a different crystal; generally causes an increase in the size of some crystals at the expense of others and commonly changes crystal shape.

Recurrence interval: The time interval between the occurrences of a type of event. Usually used in relation to the probability that a river floods each year; a recurrence interval of 100 years means that there is a one-in-one-hundred (0.01) probability of a flood with a specified discharge each year.

Reflection: The phenomenon where waves bounce off a boundary.

Refraction: The bending of a wave caused by a change in wave velocity.

Regional metamorphic rock: Rock changed by metamorphic processes affecting an entire region, such as occurs near a convergent plate boundary.

Regional metamorphism: Metamorphism over large areas not related to specific igneous intrusions or sources of hydrothermal fluid. Typically related to the formation of mountain belts near subduction zones, and involving progressively increasing temperature- and pressure-driven mineralogical and textural changes to rock.

Regolith: Unconsolidated remains of weathered rock overlying solid rock, some of it transported as sediment and some of it remaining as fragments found above or near the source bedrock.

Relative age: Ordering of objects or features from oldest to youngest; establishing the age of one thing as older or younger than another.

Relative sea level change: Shifting shoreline position caused either by absolute sea level fluctuation, uplift and subsidence of crust, or a combination of these processes.

Relief: Difference in elevation between two locations.

Reserve: That part of a naturally occurring resource that are economic to use at present.

Resource: A concentration of useful natural materials that are economic to extract and produce now or in the foreseeable future.

Reverse fault: A fault inclined at an angle steeper than 45 degrees and formed where the hanging wall rock moves upward compared to the footwall.

Reversed polarity interval: Time interval when Earth's magnetic field is oriented the opposite of the present orientation so that compasses would point toward the South Pole rather than toward the North Pole.

Rhyolite: Aphanitic igneous rock solidified from felsic magma.

Rift valley: Long valley occupying graben blocks of crust displaced downward along normal faults.

Right-lateral strike-slip fault: Strike-slip fault across which features are displaced to the right (also called a dextral strike-slip fault).

Rigidity: A measure of how much force is needed to change the shape of a solid object without changing its volume.

Rip current: Ocean current that moves water directly offshore and away from the coast.

Ripple: Curving ridge of loose sediment, shorter than 1 centimeter high, which moves along with water or wind currents, or moves back and forth beneath oscillating water waves.

Rock cycle: A sequence of processes and products that relate each rock type to the others, and that describes rocks as continuously forming from pre-existing rocks.

Rotation: Motion where an object turns around a pivot point or rotation axis. Points on an object move along curving paths and each point moves a different distance.

Rounding: A characteristic of the shape of clastic-sediment grains that describes the extent to which the edges and corners of grains abrade during transport by wind or water.

S-type asteroid: A high-albedo asteroid rich in metallic iron along with iron- and magnesium-rich silicates.

S wave (secondary wave): Seismic body wave that displaces material at right angles to the direction of wave motion; only travels through solids.

Salinity: Measure of the salt content of water.

Sandstone: Clastic sedimentary rock composed of sand grains that are 1/16th to 2 millimeters across.

Saturated zone: The area below the water table where all pores and fractures are filled with water.

Schist: A shiny, mica-rich, foliated metamorphic rock that forms from phyllite exposed to higher temperatures and pressures; consists of parallel mica crystals that are coarse enough to be seen with the naked eye.

Scientific method: Process used to systematically and objectively examine and explain a problem or observed phenomenon.

Scoria: See *cinder*.

Sea ice: Frozen seawater.

Sea stack: Small rocky island close to a shoreline and left behind by the collapse of an arch during preferential erosion of a headland by waves.

Seafloor spreading: The process whereby oceanic lithosphere is pulled apart along the crests of mid-ocean ridges as new lithosphere forms and fills the gap where the plates separate.

Seamount: Submarine volcano that does not reach the ocean surface to form an island.

Seawall: Structure of wood, plastic, concrete, rock, steel, junk cars, rubber tires, or sandbags built on a beach parallel to the shoreline to hold the shoreline in place against wave erosion.

Sediment load: Particles carried by a stream as bedload and suspended load.

Sedimentary rock: Rock consisting of the particulate and precipitated dissolved products of the weathering of older rocks.

Sedimentary structure: A physical feature, such as cross-bedding, produced in sediment at the time it is deposited, or shortly after deposition and before lithification into rock.

Seismic moment: Measure of the amount of energy released during an earthquake based on the strength of the rock that broke during the earthquake, the area of the fault plane that ruptured, and the distance that rocks moved on either side of the fault.

Seismic tomography: Method of determining the internal structure of Earth by mapping locations where seismic waves travel slightly faster or slightly slower than in adjacent rock at the same depth.

Seismic waves: Elastic energy waves that pass through and along the surface of Earth following an earthquake or explosion.

Seismogram: Record from a seismometer.

Seismometer: An instrument that detects seismic-wave motion and amplifies the wave pattern to make a record.

Serpentinite: Metamorphic rock composed almost entirely of serpentine, a hydrous magnesium-rich silicate; typically forms by metamorphism of peridotite.

Shadow zones: Areas at Earth's surface where P, S, or both P and S waves are not recorded.

Shale: A type of mudstone in which alignment of clay minerals cause the rock to break in thin sheets.

Shear: A deformation, or strain, where adjacent parts of a rock slide parallel to their plane of contact without overall shortening or elongation.

Shear stress: Force applied parallel to a surface.

Shield volcano: Volcano consisting of very thin and widespread basaltic lava flows and with gentle slopes typically less than 15 degrees.

Shortening: The strain resulting from an applied compressive stress that decreases the distance between points, decreases the volume of a body, or both.

Silica tetrahedron: The basic building block of the silicate mineral crystal structure, consisting of four oxygen atoms surrounding and bonded to a silicon atom.

Sill: A tabular, commonly horizontal or near-horizontal igneous intrusion that usually forms by injection of magma between sedimentary layers.

Sinkhole: A depression on Earth's surface caused by collapse of surface rock and regolith into a large underground cavity that formed by ground water dissolution of rock.

Slate: A metamorphic rock consisting mostly of fine-grained mica, which causes the rock to separate along parallel rock-cleavage planes; commonly produced by metamorphism of shale or other mudstone.

Slide: A mass movement where rock and regolith move downslope in contact with a surface of rupture, which separates moving material from stationary material.

Slump: A type of slide where rock or regolith move by rotation along a curved surface of rupture.

Snowline: The elevation above where snow persists throughout the year. The snowline separates zones of accumulation and wastage in a glacier.

Soft water: Water with low concentrations of dissolved ions, usually artificially produced from hard water by using water softeners—special filtration devices that chemically remove calcium and magnesium ions.

Soil: The mostly unconsolidated to loosely consolidated surface residue that results from weathering of rock as minerals interact with water and organisms. The resulting mixture of mineral and organic constituents has different physical or compositional properties, or both, than the material from which it was derived.

Soil horizons: Layers of soil, each with physical or compositional properties that distinguish them from adjacent layers.

Solar nebula: A disk-shaped spinning cloud formed by the inward collapse of a region of gas and dust in space.

Solar wind: The streams of charged particles produced by fast moving protons and electrons blasting outward from the Sun.

Sorting: Describes the range in grain size of a clastic sedimentary deposit or rock, with well sorted specimens containing mostly one grain size and poorly sorted sediment containing a wide range of grain sizes.

Specific gravity: The ratio of the mass of a substance compared to the mass of the same volume of water; comparable to density.

Speleothem: A mineral deposit in a cave caused by mineral precipitation from water.

Spit: An elongate sediment ridge formed by long-shore currents and that projects from headlands in the direction of the longshore current. Wave refraction around the end of the spit causes sediment deposition along the tip and landward side of the spit and may form a prominent hook-shaped beach.

Spring: Place where ground water emerges onto the surface.

Spring tide: The highest high tide or lowest low tide that occurs twice a month. The relatively high tidal range results from the additive effect of the gravitational pull exerted by the Moon and Sun.

Stage: The measured elevation of the water surface in a stream. The stage is defined relative to an arbitrarily selected elevation that is typically close to the elevation of the deepest part of the stream channel.

Stalactite: A conical or cylindrical speleothem developed downward from the roof of a cave by precipitation of minerals from dripping water; usually composed of calcite.

Stalagmite: A conical or cylindrical speleothem developed upward from the floor of a cave by precipitation of minerals from dripping water; usually composed of calcite.

Star dune: Sand dune consisting of curving ridges of sand that radiate from the center and resulting from highly variable wind directions crossing large areas of readily eroded sand.

Stones: A class of meteorites, including chondrites and achondrites, with compositions similar to rocks and minerals found on Earth.

Stony irons: A class of meteorites that contain approximately equal volumes of metal and silicate minerals.

Strain: The measurable deformation resulting from stress.

Stratification: See *bedding*.

Streak: Color of the residue left behind from scratching a mineral on a non-glazed porcelain plate.

Stream: Flowing water that moves through a channel and simultaneously transports dissolved and particulate products of rock weathering.

Stream gage: Device that measures water level at a location in a stream where the channel cross section has been carefully surveyed; stream gage data are used to calculate stream discharge.

Stream power: A measure of the ability of a stream to do work. A simple way to calculate stream power averaged over a unit area of stream bed is to multiply the shear stress and average flow velocity.

Strength: Measure of the amount of stress a material can endure before it fails, either by breaking or flowing.

Stress: The magnitude of force divided by the area over which the stress is applied.

Striation: A scratch on a bedrock surface commonly created by sharp-edged rocks frozen in the bottom of

a glacier as the glacier slowly moves in a straight-line path.

Strike: Compass orientation of a line produced by the intersection of an imaginary horizontal plane with an inclined plane such as a tilted bed, fault plane, or edge of an intrusion; used with dip to describe the orientation of any planar geologic feature.

Strike-slip fault: A fault where rocks are displaced by horizontal movement along the strike direction of the fault plane.

Structural geology: The study of deformation in rocks at a microscopic to regional (hundreds of square kilometers) scale.

Subduction: The process by which a lithospheric plate descends beneath a neighboring plate into the deeper mantle.

Surface of rupture: Planar or curved face along which moving material of a slide separates from stationary material.

Surface waves: Seismic (earthquake) waves on Earth's surface that decrease in intensity with depth.

Suspended load: The fine-grained sediment intimately mixed with the water and flowing above the bed of a stream.

Syncline: Trough-shaped fold where limbs dip toward the hinge line; youngest rocks are present in the center of the eroded fold.

Talus: Piles of loose, fallen rock, debris, and earth found at the base of a steep slope.

Tectonics: The field of study that encompasses rock deformation at a regional to global scale.

Tension: A stress caused by oppositely directed forces that elongate or increase the volume of materials.

Terraces: Step-and-bench landforms, including stream terraces alongside and above a river channel.

Theory: Established, thoroughly tested, generally accepted explanation for an observed natural phenomenon that is supported by a substantial body of data.

Thrust fault: A fault dipping at an angle less than 45 degrees and formed where the hanging wall rocks move upward compared to the footwall.

Tidal channel: Channel eroded by water draining seaward during the falling tide; sometimes called a tidal creek.

Tidal creek: See *tidal channel*.

Tidal current: Fast-moving current that forms where the rising and falling tide is constricted to inlets between islands, or forced in and out of funnel-shaped estuaries.

Tidal flat: Gently sloping, muddy, marshy, or barren area of land submerged at high tide and exposed at low tide.

Tidal inlet: Narrow body of water between barrier islands that connects a lagoon with the open ocean.

Tidal range: Change in sea-level elevation between low and high tide.

Tide: The slow rhythmic, alternating rise and fall in the surface of the ocean caused by the gravitational pull of the Moon and Sun on the oceans and rigid Earth.

Tidewater glaciers: Glaciers that descend into the ocean from land and are in contact with the seafloor.

Till: Very poorly sorted mixture of boulders, cobbles, gravel, sand, and mud deposited directly by a glacier.

Tillite: A poorly sorted sedimentary rock interpreted to have originated as glacial till.

Tonalite: Phaneritic igneous rock solidified from felsic magma with a composition between diorite and granite.

Transform plate boundary: Zones where lithospheric plates slide past one another with neither creation nor destruction of lithosphere.

Transition zone: The region of the mantle between 410 and 660 kilometers below the surface where minerals undergo changes between peridotite upper mantle and the lower mantle composed of high-pressure minerals.

Translation: Motion where the orientation of the object does not change while it moves, and all points on an object move the same distance.

Transpiration: Process whereby water in plant leaves and stems is released as water vapor.

Transverse dune: A linear or curving sand dune with a crest oriented perpendicular to the prevailing wind direction; forms where sand supply is abundant and wind direction varies only slightly.

Travertine: A porous variety of limestone that typically forms where calcite precipitates from water around springs.

Tributary: A relatively small stream that flows into a larger stream.

Tsunami: Fast-moving, long-wavelength sea waves caused by sudden displacement of the seafloor, typically caused by submarine fault motion or landslide.

Turbidite: Sedimentary deposit created by a turbidity current, typically consisting of a graded bed with a sharp, eroded base and ripple-formed cross-bedding.

Turbidity current: Mixture of sediment and water that flows along the seafloor or a lake bottom.

Ultramafic: Describes igneous rocks with a relatively low silica content, dominated by olivine and pyroxene and lacking quartz or significant feldspar.

Unconformity: Breaks in the continuity of the geologic record between rock units. The absence of rocks representing some interval of time that usually results from erosion.

Unsaturated zone: The region above the water table where pores are partly filled with water and partly filled with air. Water moves downward through the unsaturated zone to join ground water in the saturated zone.

Valley glaciers: Elongate glaciers confined between bedrock valleys.

Van der Waals forces: Uneven distribution of electrical charges around a neutral molecule that exert slight attractive and repulsive forces.

Ventifact: A loose rock with faceted, planar surfaces abraded by wind-blown sand.

Vesicles: Cavities in volcanic rocks occupied by gas when the rock solidified.

Viscosity: The property of a fluid that describes its resistance to flow.

Volcanic ash: Pyroclastic fragments less than 2 millimeters across.

Volcanic (or extrusive) rock: Rock originating from eruption of molten material at Earth's surface, including the products of lava flows, pyroclastic falls, and pyroclastic flows.

Volcanic neck: A nearly cylindrical pipe-shaped intrusion of plutonic rock or tuff that marks the feeding conduit of a volcano and later exposed by erosion.

Volcano: Hill, ridge, or mountain formed by the accumulation of lava flows and pyroclastic deposits around the conduit, or vent, from which they were erupted.

Wadati-Benioff zone: The inclined zone of earthquakes foci characteristic of subduction zones at convergent plate boundaries.

Water table: The surface that marks the top of the ground water and forms the boundary between the saturated and unsaturated zones.

Wave: Disruption that moves through a medium as a pulse of energy with little if any overall transport of the medium in the direction of wave movement.

Wave base: The depth to which oscillatory wave motion persists downward below the water surface.

Wave height: Distance between the adjacent crest and trough of a wave.

Wave period: The time elapsed between the passage of two successive peaks or troughs of a wave past a point.

Wave-cut platform: Nearly horizontal bench eroded by waves along a rocky coast, commonly submerged at high tide and exposed to view at low tide.

Wavelength: The distance between two successive peaks or troughs of a wave.

Weathering: Deterioration of rocks at Earth's surface, more systematically defined as the response of the geosphere at its interface with the atmosphere, hydrosphere, and biosphere to reduce rocks into loose particles while dissolving some minerals and producing new ones.

Welded tuff: A pyroclastic deposit that forms a rock because the hot pyroclastic fragments compacted and welded to one another because of the weight of overlying and rapidly accumulating pyroclastic material.

Wet melting: A process of magma generation resulting from the introduction of water into hot rock, which lowers the melting temperature of the rock.

Wind: Movement of gas molecules in the atmosphere caused by convection.

Yardang: A wind-parallel ridge of soft rock or slightly consolidated sediment that remains after surrounding material is eroded by wind abrasion.

Yield strength: The point at which rock deformation is permanent and cannot be reversed by decreasing the stress.

Zone: Used in the study of metamorphic rocks to indicate the area on a map where a particular metamorphic index mineral is present in the rocks.

Zone of accumulation: The high-elevation zone of a glacier where the winter snow accumulation exceeds summer melting.

Zone of wastage: The low-elevation zone of a glacier (also called the zone of ablation) where melting exceeds the winter snow accumulation.

CREDITS

Chapter 1
Chapter opening, Richard Cummins/SuperStock; 1.1a, Peter Carsten/National Geographic Image Collection; 1.1b, Gary/Fotolia; 1.1c, Melvin Grubb/Grubb Photo Service, Inc; 1.2a, Michal Novotny/Fotolia; 1.2b, John Wang/Getty Images; 1.2c, Jim Wark/Peter Arnold; 1.4, Based on maps from the Association of Bay Area Governments; 1.5a, EyeWire Collection/Getty Images; 1.5b, Colin Keates © Dorling Kindersley, Courtesy of the Natural History Museum, London; 1.5c, Dirk Wiersma/Photo Researchers; 1.7a, Martin Bond/Photo Researchers; 1.7b, Marli Miller, University of Oregon; 1.8, Dr. Ross W. Boulanger; 1.12, James A. Sugar/Corbis/Bettmann

Chapter 2
Chapter opening, Javier Trueba/MSF/Photo Researchers; 2.1a, Kevin Schafer/Corbis/ Bettmann; 2.1b, Gary A. Smith; 2.2a, Mark A. Schneider/Photo Researchers; 2.2b, Klaus Guldbrandsen/Photo Researchers; 2.3, Gary A. Smith; 2.5a, Aurora Pun; 2.5b, Aurora Pun; 2.7a, Aurora Pun; 2.7b, Aurora Pun; 2.8, Gary A. Smith; 2.10a, Mark A. Schneider/Photo Researchers; 2.10b, Arnold Fisher/Photo Researchers; 2.10c, Mark A. Schneider/Photo Researchers; 2.17, Clive Streeter © Dorling Kindersley; 2.21a, Dr. David Barber, University of Essex; 2.23a, Zbynek Burival/Shutterstock; 2.23b, Mark A. Schneider/Photo Researchers; 2.25b, Charles D. Winters/Photo Researchers; 2.26a, Dennis Tasa, Tasa Graphic Arts, Inc; 2.26b, Harry Taylor © Dorling Kindersley; 2.27, Greg C. Grace/Alamy; 2.29, Ted Foxx/Alamy; 2.30a, Martin Land/Photo Researchers; 2.30b, Dirk Wiersma/Photo Researchers; 2.30c, Photolibrary.com

Chapter 3
Chapter opening, Tyler Stableford/Getty Images; 3.1a, Bert Sagara/Getty Images; 3.1b, Anton Foltin/Shutterstock; 3.1c, Gary A. Smith; 3.1d, Gary A. Smith; 3.2a, Bruce Forster © Dorling Kindersley; 3.2b, Robert & Jean Pollock/Science Source; 3.2c, Gary A. Smith; 3.2d, Michael Nolan/Robert Harding World Imagery/Alamy; 3.3a, Aurora Pun; 3.3b, Gary A. Smith; 3.4a, Douglas Peebles, Flirt Collection/ Photolibrary.com; 3.4b, D.A. Swanson/Hawaiian Volcano Observatory, U.S. Geological Survey; 3.4c1, Andrew Alden; 3.4c2, Aurora Pun; 3.4d, Aurora Pun; 3.7a, Gary A. Smith; 3.7b, Alan Kearney/Getty Images; 3.9a, Steve Austin/Papilo/Corbis; 3.9b, Craig Aurness/Corbis/Bettmann; 3.11, U.S. Geological Survey

Chapter 4
Chapter opening, Agence France Presse/Getty Images; 4.1a, Rainer Albiez/Fotolia; 4.1b, U.S. Air Force; 4.1c, Craig Aurness/CORBIS; 4.1d, Gary A. Smith; 4.2a, Albert J. Copley/Getty Images; 4.2b, Andreas Einsiedel © Dorling Kindersley; 4.2c, Aurora Pun; 4.2d, Gary A. Smith; 4.2e, Grace Davies/Omni-Photo Communications; 4.3b, Breck P. Kent/Animals Animals/Earth Scenes; 4.3c, Breck P. Kent; 4.3d, Tyler Boyes/Fotolia; 4.3e, Photo courtesy of Donna Tucker; 4.3f, E.R. Degginger/Photo Researchers; 4.3g, Edward Kinsman/Photo Researchers; 4.3h, Harry Taylor © Dorling Kindersley; 4.3i, Aleksandr Volkov/Fotolia; 4.3j, Property of Charles E. Jones; 4.4a, J.P. Lockwood/U.S. Geological Survey/U.S. Department of the Interior; 4.4b, Carver Mostardi/Alamy; 4.4c, Gary A. Smith; 4.4d, Aurora Pun; 4.5, After E. J. Tarbuck and F. K. Lutgens, 2005, *Earth: An Introduction to Physical Geology,* 8th ed., Prentice Hall; 4.6a, Photo courtesy of Karl E. Karlstrom; 4.6b, Marli Bryant Miller; 4.7, Michael Collier; 4.8a, Dan Suzio/Photo Researchers, Inc; 4.8b, J.D. Griggs/Hawaiian Volcano Observatory, U.S. Geological Survey; 4.9a, Michael T. Sedam/CORBIS; 4.9b, Katie Dickinson/Shutterstock; 4.10a, Chris Luneski/Alamy; 4.10b, Steve Schilling, Cascades Volcano Observatory, U.S. Geological Survey; 4.11a, Gary A. Smith; 4.11b, Gary A. Smith; 4.12a, Michael Collier; 4.12b, J.D. Griggs/U.S. Geological Survey/U.S. Department of the Interior; 4.12c, Vacclav/Fotolia; 4.12d, Michael Clynne/U.S. Geological Survey/U.S. Department of the Interior; 4.13b, J. D. Griggs/Hawaiian Volcano Observatory, U.S. Geological Survey; 4.14b, Francois Gohier; 4.24, Paul Chesley/Getty Images; 4.25, AP/Wide World Photos; 4.26, Archives de l'Academie des Sciences; 4.27, Associated Press; 4.29, Lee Prince/Shutterstock; 4.30a, Michael N. Spilde; 4.30b, Jacana/Photo Researchers; 4.30c, Getty Images

Chapter 5
Chapter opening, © Yva Momatuik & John Eastcott/Minden Pictures; 5.1a, Gary Smith; 5.1b, Lucido Studio Inc./Corbis RF; 5.1c, Gary A. Smith; 5.1d, Gary A. Smith; 5.1e, Gary A. Smith; 5.2a, Gary A. Smith; 5.2b, Marli Miller; 5.3a, granitepeaker/Fotolia; 5.3c, David Nunuk/Photo Researchers; 5.4a, Aurora Pun; 5.4b, Dan Guravich/Photo Researchers; 5.4c, Photo courtesy of John Bloch; 5.5a, Gary A. Smith; 5.5b, © David Muench/David Muench Photography Inc; 5.5c, Jess Alford/Getty Images; 5.7a, Marli Miller; 5.7b, Marli Miller; 5.7c, Gary A. Smith; 5.9a, Charles R. Belinky/Photo Researchers; 5.9b, Colin Keates © Dorling Kindersley, Courtesy of the Natural History Museum, London; 5.9c, Dr. B. Booth/GeoScience Features Picture Library; 5.9d, Dr. B. Booth/GeoScience Features Picture Library; 5.9e, Gary Ombler/Dorling Kindersley; 5.10a, Joyce Photographics/Photo Researchers; 5.10b, Edward Kinsman/Photo Researchers; 5.10c, Harry Taylor © Dorling Kindersley; 5.10d, MarcelClemens/Shutterstock; 5.10e, anetlanda/Fotolia; 5.11a, Colin Keates © Dorling Kindersley, Courtesy of the Natural History Museum, London; 5.11b, Colin Keates © Dorling Kindersley; 5.11c, Charlie Ott/Photo Researchers; 5.11d, Joy Spurr/Photoshot Holdings Ltd; 5.12a1, Tim Mead/Photolibrary; 5.12a2, Spike Walker/Getty Images; 5.12b, Dave Houseknecht/U.S. Geological Survey/U.S. Department of the Interior; 5.13, Data from U.S. Dept. of Energy, Energy Information Agency; 5.14a, David R. Frazier/Photo Researchers; 5.14b, Gary A. Smith; 5.14c, MODIS/NASA Headquarters; 5.14d, Gary A. Smith; 5.15a, Adam Sylvester/Photo Researchers; 5.15b, Gary A. Smith; 5.16a, Gary A. Smith; 5.16b, Gary A. Smith; 5.17a1, Hugh Sitton/ Getty Images; 5.17a2, Richard Hamilton Smith/Corbis/Bettmann; 5.17c1, Marli Miller; 5.17c2, Gary A. Smith; 5.18a, Aurora Pun; 5.18b, Aurora Pun; 5.18c, Gary A. Smith; 5.19, Ralph Lee Hopkins/Photolibrary.com; 5.20, After Levin, *The Earth through Time,* 4th ed., Saunders; 5.21a, Marli Miller; 5.21b, Gary A. Smith; 5.21c, Gary A. Smith; 5.22b, "Turbidity currents as a cause of graded bedding," by P. H. Kuenen and C. I. Migliorini, *The Journal of Geology,* volume 58 (1950); 5.22c, "Turbidity currents as a cause of graded bedding," by P. H. Kuenen and C. I. Migliorini, *The Journal of Geology,* volume 58 (1950); 5.23, Map after D. J. W. Piper et al., 1988, Geol. Soc. America Spec. Pap. 229, pp. 77–92; 5.24a, Map after Levin, *The Earth through Time,* 4th ed. (Saunders); 5.24b, Ronald C. Blakey; EOC5.1, NASA/JPL/Cornell; EOC5.2, Gary A. Smith

Chapter 6
Chapter opening, Kenneth Murray/Photo Researchers; 6.1a, Dr. J. Alcock/Penn State Abington College; 6.1b, Dr. J. Alcock/Penn State Abington College; 6.1c, Property of Charles E. Jones; 6.7, After Davidson, Reed, and Davis, *Exploring Earth,* 1st ed, 1997, Prentice Hall; 6.8, After Duff, *Holmes' Principles of Physical Geology,* 4th ed. 1993, Chapman and Hall; 6.9a, Colin Keates © Dorling Kindersley, Courtesy of the Natural History Museum, London; 6.9b, Jim Wehtje/Getty Images; 6.10b, Doug Martin/Photo Researchers; 6.11b, Marli Miller; 6.13a, Aurora Pun; 6.13b, Aurora Pun; 6.15, W. D.Yardley, W. S. MacKenzie, and C. Guilford, Atlas of Metamorphic Rocks and their Textures, 1990, Longman Scientific and Technical, Essex; 6.16, After A. D. Edgar, *Experimental Petrology: Basic Principles and Techniques,* 1973, Clarendon Press; 6.17, Data from M. J. Holdaway and B. Mukhopadhyay, 1993, *American Mineralogist,* vol. 78, p. 298; 6.18, Data from M. J. Holdaway and B. Mukhopadhyay, 1993, *American Mineralogist,* vol. 78, p. 298; 6.20a, Property of Charles E. Jones; 6.20b, Harry Taylor © Dorling Kindersley; 6.20c, Property of Charles E. Jones; 6.20d, Richard M. Busch; 6.20e, Property of Charles E. Jones; 6.20f, Property of Charles E. Jones; 6.20f, 6.20g, Harry Taylor © Dorling Kindersley; 6.20h, Property of Charles E. Jones; 6.20i, Martin Shields/Photo Researchers; 6.20j, Aurora Pun; 6.20k, Aurora Pun; 6.20l, Harry Taylor © Dorling Kindersley; 6.21, After Marshak, *Earth: Portrait of a Planet,* Norton, 2001; 6.22, Aurora Pun; 6.24a, After Nagy and Parmentier, 1982, *Earth and Planetary Science Letters,* vol. 59, pp. 1–10; 6.24b, After Davidson, Reed, and Davis, *Exploring Earth,* 1st ed, 1997, Prentice Hall; 6.26, Ken MacDonald/Science Photo Library/Photo Researchers; 6.27, After Humphirs et al., 1995, Nature, vol. 377, p. 713; 6.28, After Duff, *Holmes' Principles of Physical Geology,* 4th ed. 1993, Chapman and Hall

Chapter 7
Chapter opening, Marli Miller; 7.2, Gary A. Smith; 7.3, Edward Kinsman/Photo Researchers; 7.4a, Gary A. Smith; 7.4b, Fletcher & Baylis/Photo Researchers; 7.6a, Stuart Wilson/Photo Researchers; 7.8, After C. L. and M. A. Fenton, *Giants of Geology,* Doubleday, 1952; 7.9, After *International Stratigraphic Chart,* 2004, by the International Commission on Stratigraphy; 7.11a, Marli Miller; 7.12a, Gary A. Smith; 7.13a, Gary A. Smith; 7.18a, Data from Libby, W. F., 1961, Radiocarbon Dating, *Science,* vol. 133, pp. 621–629; 7.18b, Data from Williams, I. S., Compston, W., and Chappell, B. W., 1983, Zircon and monazite U-Pb systems and the histories of I-type magmas, Berridale batholith, Australia, *Journal of Petrology,* vol. 21, pp. 76–97; 7.23, Data from Beckinsale, R. D., and Gale, N. H., 1969, A reappraisal of the decay constants and branching ratio of 40K, *Earth and Planetary Science Letters,* vol. 6, pp. 289–294

Chapter 8

Chapter opening, The Natural History Museum, London; 8.1a, Patrick Lynch/PH ESM; 8.1b, Patrick Lynch/PH ESM; 8.3, Gary A. Smith; 8.4a, Bradley Hacker; 8.4b, Photo courtesy of Aaron Yoshinobu; 8.6a, David Mack/Photo Researchers; 8.8, Russell D. Curtis/Photo Researchers

Chapter 9

Chapter opening, JPL/Texas A&M/Cornell/NASA; 9.1a, NASA; 9.1b, NASA Headquarters; 9.2a, NASA Headquarters; 9.2b, NASA/JPL/Space Science Institute; 9.4, C. R. O'Dell/NASA; 9.5, After W. K. Hamblin and E. H. Christiansen, *Exploring the Planets,* 1990, Macmillan; 9.6, Photo Researchers; 9.9, After M. J. Drake and K. Righter, 2002, Determining the composition of the Earth, *Nature,* vol. 416, pp. 39–44; 9.10, After G. Schubert, D. L. Turcotte, and P. Olson, 2001, *Mantle Convection in the Earth and Planets,* Cambridge University Press; 9.11, D. A. Seal/NASA; 9.12, After M. J. Drake and K. Righter (2002), Determining the composition of the Earth, *Nature,* vol. 416, pp. 39–44; 9.13, Ron Miller/Black Cat Studios; 9.14, Data for volcanoes from H. Sigurdsson, editor, 2000, *Encyclopedia of Volcanoes,* Academic Press. Atmospheric data from G. Faure, 1991, *Principles and Applications of Inorganic Geochemistry,* Macmillan Publishing Company

Chapter 10

Chapter opening, Pierre Vetsch/Geological Survey of South Africa; 10.1, Cadre Design; 10.5, Map courtesy of Henry Pollack, University of Michigan; 10.8, Modified from G. Schubert, D. L. Turcotte, and P. Olson, 2001, *Mantle Convection in the Earth and Planets,* Cambridge University Press; 10.9, After B. Travis, P. Olson, and G. Schubert, 1990, The transition from two-dimensional to three-dimensional planforms in infinite-Prandtl-number thermal convection, *Journal of Fluid Mechanics,* vol. 216, pp. 71–91; 10.10a, Dr. Paul Tackley, University of California, Los Angeles; 10.10b, Dr. Paul Tackley, University of California, Los Angeles; 10.12b, Cordelia Molloy/Photo Researchers; 10.13, Map after N. Olsen, 2007, Will the magnetic North Pole move to Siberia? *EOS,* v. 88, p. 293. Graph data from the National Geophysical Data Center; 10.14, Richard Megna/ Fundamental Photographs, NYC; 10.16, After J. Bloxham and D. Gubbins, 1989, The evolution of the Earth's magnetic field, *Scientific American;* 10.17, Images courtesy of Gary Glatzmaier, after Glatzmaier, G. A., 2002, Geodynamo simulations—how realistic are they? *Annual Reviews of Earth and Planetary Science,* vol. 30, pp. 237–257; 10.18, Images courtesy of Gary Glatzmaier, after Glatzmaier, G. A., 2002, Geodynamo simulations—how realistic are they? *Annual Reviews of Earth and Planetary Science,* vol. 30, pp. 237–257

Chapter 11

Chapter opening, Copyright © Louis Maher; 11.1a, Gary A. Smith; 11.1b, Jon Price/ Nevada Bureau of Mines and Geology; 11.1c, Photo courtesy of John W. Geissman; 11.1d, Gary A. Smith; 11.2b, Jarno Gonzalez Zarraonandia/istockphoto; 11.4a, Breck P. Kent; 11.4b, Bernhard Edmaier/SPL/Photo Researchers; 11.5, LesPalenik/Shutterstock; 11.6a, Gary A. Smith; 11.6b, Marli Miller; 11.8a, Fletcher & Baylis/Photo Researchers; 11.8b, Educational Images Ltd/Custom Medical Stock Photo Newscom; 11.8c, © Marli Miller, University of Oregon; 11.8d, Marli Miller; 11.9, After L. A. Woodward and R. L. Ruetschilling, 1976, *Geology of San Ysidro Quadrangle, New Mexico,* New Mexico Bureau of Mines and Mineral Resources Geologic Map 37; 11.11, After E. J. Tarbuck and F. K. Lutgens, 2005, *Earth: An Introduction to Physical Geology,* 8th ed., Prentice Hall; 11.12, Brian J. Skinner; 11.14, After G. H. Davis, 1984, *Structural Geology of Rocks and Regions,* Wiley and Son; 11.15, B. Willis/U.S. Geological Survey/Geologic Inquiries Group; 11.16, After S. Judson and S. M. Richardson, 1995, *Earth: An Introduction to Geologic Change,* Prentice Hall; 11.17, M. S. Patterson/Mt. Stromlo & Siding Spring Observatories; 11.19, After M. P. Billings, *Structural Geology,* 3rd ed., Prentice Hall; 11.20, After M. P. Billings, *Structural Geology,* 3rd ed., Prentice Hall; 11.22a, NASA/JPL; 11.22b, Marli Miller; 11.22c, Lloyd Cluff/Corbis; 11.22d, Paul Breeding/Maryland Geological Survey; 11.24, Gary A. Smith; 11.25b, Michael Collier; 11.26, A. Jones, L. Siebert, P. Kimberly, and J. F. Luhr, 2000, *Earthquakes and Eruptions,* Smithsonian Institution, Global Volcanism Program Digital Information Series GVP-2; 11.27a, Lloyd Cluff/CORBIS; 11.27b, U.S. Geological Survey, Denver; 11.28, Data from the Southern California Earthquake Data Center; 11.29, Thomas McGuire; 11.30b, David Schmidt; 11.31, After U.S. Geological Survey Fact Sheet 014-03; 11.32, After L. Brewer, 1992, Preliminary damage and intensity survey, *Earthquakes and Volcanoes,* v. 23, pp. 219–226; 11.33, After B. A. Bolt, 1993, *Earthquakes and Geological Discovery,* Scientific American Library; 11.35, After IRIS Consortium, 2002, How often do earthquakes occur? Education and Outreach Series No. 3; 11.36a, Asahi Shimbun/SIPA Press; 11.36b, U.S. Geological Survey/U.S. Department of the Interior; 11.36c, Ryan Pile/Corbis; 11.36d, Photo courtesy of Dr. Peter W. Weigand and Dr. Sandra L. Jewett; 11.37a, After G. Plafker and J. P. Galloway, 1989, Lessons Learned from the Loma Prieta, California, Earthquake of October 17, 1989, U.S. Geological Survey Circular 1045; S. E. Hough, 2002, *Earthshaking Science: What We Know (and Don't Know) about Earthquakes,* Princeton University Press; 11.37b, Loma Prieta Collection, Earthquake Engineering Research Center, University of California, Berkeley; 11.38a, After F. I. Gonzales, May 1999, Tsunami!, *Scientific American,* pp. 58–65; 11.38b, DigitalGlobe; 11.38c, U.S. Geological Survey, Denver; 11.39, U.S. Geological Survey

Chapter 12

Chapter opening, Getty Images; 12.1a1, Patrick Lynch/PH ESM; 12.1a2, Patrick Lynch/PH ESM; 12.1b1, Patrick Lynch/PH ESM; 12.1b2, Patrick Lynch/PH ESM; 12.2, After A. Wegener, 1915, Die Entstehung der Kontinente und Ozeane, Vieweg, Braunschweig; 12.3, Data from A. Jones, L. Siebert, P. Kimberly, and J. F. Luhr, 2000, *Earthquakes and Eruptions,* Smithsonian Institution, Global Volcanism Program Digital Information Series, GVP-2; 12.5a, iStockphoto/Thinkstock; 12.5b, jack stephens/Alamy; 12.6, Data from A. Jones, L. Siebert, P. Kimberly, and J. F. Luhr, 2000, *Earthquakes and Eruptions,* Smithsonian Institution, Global Volcanism Program Digital Information Series, GVP-2; 12.7, Data from A. Jones, L. Siebert, P. Kimberly, and J. F. Luhr, 2000, *Earthquakes and Eruptions,* Smithsonian Institution, Global Volcanism Program Digital Information Series, GVP-2; 12.9, After E. J. Tarbuck and F. K. Lutgens, 2002, *Earth: An Introduction to Physical Geology,* 7th ed., Prentice Hall; 12.10a, Georg Gerster/Photo Researchers; 12.10b, After J. A. Karson and P. A. Rona, 1990, Block-tilting, transfer faults, and structural control of magmatic and hydrothermal processes in the TAG area, Mid-Atlantic Ridge 26°N., *Geological Society of America Bulletin,* vol. 102, no. 12, pp. 1635–1645; 12.13, After R. D. Müller, W. R. Roest, J. Y. Royer, L. M. Gahagan, and J. G. Sclater, 1997, Digital isochrons of the world's ocean floor, *Journal of Geophysical Research,* vol. 102, pp. 3211–3214; 12.14, After (top) R. P. von Herzen and S. Uyeda, 1963, Heat flow through the eastern Pacific Ocean floor, *Journal of Geophysical Research,* vol. 68, pp. 4219–4250; (bottom) W.C. Pitman, 1978, The relationship between eustasy and stratigraphic sequences of passive margins, *Geological Society of America Bulletin,* vol. 89, pp. 1389–1403; 12.15, After (top) W. C. Pitman, 1978, The relationship between eustasy and stratigraphic sequences of passive margins, *Geological Society of America Bulletin,* vol. 89, pp. 1389–1403; (bottom) S. Stein and C. A. Stein, 1996, Thermo-mechanical evolution of oceanic lithosphere: Implications for the subduction process and deep earthquakes, *American Geophysical Union Monograph* 96, pp. 1–18; 12.16, Courtesy of Roderick Brown, University of Glasgow; 12.17b, Nicholas Parfitt/Getty Images Inc; 12.18, After S. Marshak, 2001, *Earth: Portrait of a Planet,* Norton; 12.19a, National Geophysical Data Center, NOAA; 12.19b, Alex Blackburn Clayton/Photolibrary.com; 12.19c, Photo courtesy of Tobias Fischer; 12.21, Data from A. Jones, L. Siebert, P. Kimberly, and J. F. Luhr, 2000, *Earthquakes and Eruptions,* Smithsonian Institution, Global Volcanism Program Digital Information Series, GVP-2; 12.22, After R. van der Hilst, 1995, Complex morphology of subducted lithosphere in the mantle beneath the Tonga trench, *Nature,* vol. 374, pp. 154–157; 12.23, After D. D. Blackwell, R. G. Bowen, D. A. Hull, J. Riccio, and J. L. Steele, 1982, Heat flow, arc volcanism, and subduction in northern Oregon, *Journal of Geophysical Research,* vol. 87, pp. 8735–8754; 12.25, After E. J. Tarbuck and F. K. Lutgens, 2002, *Earth: An Introduction to Physical Geology,* 7th ed., Prentice Hall; 12.26b, Joseph A. Dellinger; 12.27a, G.K. Gilbert/U.S. Geological Survey, Denver; 12.27b, Lloyd Cluff/Corbis/Bettmann; 12.27c, Map after R. G. Stanley, 1987, New estimates of displacement along the San Andreas fault in central California based on paleobathymetry and paleogeography, *Geology,* vol. 15, pp. 171–174; 12.29, After V. Courtillot, A. Davaille, J. Besse, and J. Stock, 2003, Three distinct types of hotspots in the Earth's mantle, *Earth and Planetary Science Letters,* vol. 205, pp. 295–308; 12.30a, NASA Headquarters; 12.30b, Aurora Pun; 12.34, After V. Courtillot, A. Davaille, J. Besse, and J. Stock, 2003, Three distinct types of hotspots in the Earth's mantle, *Earth and Planetary Science Letters,* vol. 205, pp. 295–308; 12.36, After D. F. Argus and R. G. Gordon, 1991, No-net-rotation model of current plate velocities incorporating plate motion model NUVEL-1, *Geophysical Research Letters,* vol. 18, pp. 2039–2042; 12.38, NASA Jet Propulsion Laboratory, GPS Time Series; 12.39, After D. F. Argus and R. G. Gordon, 1991, No-net-rotation model of current plate velocities incorporating plate motion model NUVEL-1, *Geophysical Research Letters,* vol. 18, pp. 2039–2042; NASA Jet Propulsion Laboratory, GPS Time Series; 12.40, After R. A. Bennett, J. L. Davis, and B. P. Wernicke, 1999, Present day pattern of Cordillean deformation in the Western United States, *Geology,* vol. 27, pp. 371–374; 12.42, After D. W. Forsyth and S. Uyeda, 1975, On the relative importance of the driving forces of plate motion, *Geophysical Journal of the Royal Astronomical Society,* vol. 43, pp. 163–200; 12.43a, Dr. Ronald Blakey, Northern Arizona University-Flagstaff; 12.43b, Dr. Ronald Blakey, Northern Arizona University-Flagstaff; 12.43c, Dr. Ronald Blakey, Northern Arizona University–Flagstaff; 12.43d, Dr. Ronald Blakey, Northern Arizona University-Flagstaff; 12.43e, Dr. Ronald Blakey, Northern Arizona University–Flagstaff; 12.43g, Dr. Ronald Blakey, Northern Arizona University–Flagstaff; 12.43f, Dr. Ronald Blakey, Northern Arizona University–Flagstaff

Chapter 13

Chapter opening, Jim Wark/Photolibrary.com; 13.1a, Bobbé Christopherson; 13.1b, Thomas Hallstein/Alamy Images; 13.1c, Diane Miller/Monsoon Images/ Photolibrary.com; 13.2, After P. R. Pinet, 1992, *Oceanography: An Introduction to the Planet Oceanus,* West Pub; 13.8, After J. Jackson, 2002, Strength of the continental lithosphere: Time to abandon the jelly sandwich? *GSA Today,* vol. 12, no. 9, pp. 4–10; 13.13a, Photo by Dr. John Riley from his book Flora of the Hudson Bay Lowland and its Postglacial Origins, Courtesy of NRC Research Press; 13.13b, Map after A. B. Watts, 2001, *Isostasy and Flexure of the Lithosphere,* Cambridge University Press, with GPS data from NASA/Jet Propulsion Laboratory; 13.15, After R. D. Muller and others, 2008, Long-term sea-level fluctuations driven by ocean-basin dynamics, *Science,* vol. 319, pp. 1357–1362; 13.18a, Map courtesy of the National Geophysical Data Center/NOAA; 13.18b, SuperStock, Inc; 13.20a, EyeWire Collection/Getty Images; 13.20b, AIRPHOTO—Jim Wark; 13.23, After S. Marshak, 2001, Earth, Portrait of a Planet, Norton; 13.24, After A. Nur and Z. Ben-Avraham, 1982, Oceanic plateaus, the fragmentation of continents and mountain building, Journal of Geophysical Research, vol. 87, pp. 3644–3661; 13.26a, Marli Miller; 13.26b, Natural Selection/Bill Byrne/Photolibrary.com; 13.27a, © Robert Hildebrand; 13.27b, Marli Miller; 13.28, After K. E. Karlstrom, S. S. Harlan, M. L. Williams, J. McLelland, J. W. Geissman, and K. I. Ahall, 2001, Long-lived (1.8-0.8 Ga) Cordilleran-type orogen in southern Laurentia, its extensions to Australia and Baltic, and implications for refining Rodinia, *Precambrian Research,* vol. 111, pp. 1–30; 13.30, After D. R. Lowe, 1992, Major events in the geological development of the Pre-cambrian Earth, in J. W. Schopf and C. Klein, eds., *The Proterozoic Biosphere: A Multi-Disciplinary Study,* Cambridge University Press

Chapter 14

Chapter opening, Georg Gerster/Photo Researchers, Inc.; 14.1a1, Gary A. Smith; 14.1a2, USDA/Natural Resources Conservation Service; 14.1b1, Photo courtesy of Leslie McFadden; 14.1b2, Photo courtesy of Leslie McFadden; 14.2, USDA/Natural Resources Conservation Service; 14.4a, USDA/Natural Resources Conservation Service; 14.4b, Marbut Collection, Soil Science Society of America, Inc; 14.7, Photo courtesy of N. C. Brady and R. Weil, 2002, *The Nature and Properties of Soils,* 13th ed., Prentice Hall, Upper Saddle River, NJ, with permission of R. Weil; 14.8, After M. N. Machette, 1985, Calcic Soils of the Southwestern United States, Geological Society of America Special Paper 203, pp. 1–21; 14.9, U.S. Department of Agriculture; 14.11a, After N. C. Brady and R. R. Weil, 1996, *The Nature and Properties of Soils,* 11th ed., Prentice Hall; 14.11b, USDA/Natural Resources Conservation Service; 14.11c, Jeff Vanuga/USDA/Natural Resources Conservation Service; 14.11d, keller/Fotolia; 14.14, After Natural Resource Conservation Service; 14.15a, After Natural Resource Conservation Service; 14.15b, USDA/Natural Resources Conservation Service; 14.15c, USDA/Natural Resources Conservation Service; 14.15d, USDA/Natural Resources Conservation Service; 14.16a, After Natural Resource Conservation Service; 14.16b, Grahame McConnell/Photolibrary.com; 14.16c, © 2004 Theodore Gray www.element-collection.com; 14.17, USDA/Natural Resources Conservation Service; 14.18, Randall J. Schaetzl, Michigan State University; 14.19, Randall J. Schaetzl, Michigan State University; 14.20b, Gary A. Smith; 14.21, Data from L. H. Gile and R. B. Grossman, 1979, *The Desert Project Soil Monograph,* U.S. Dept. of Agriculture, Soil Conservation Service; 14.22, Data from L. H. Gile and R. B. Grossman, 1979, *The Desert Project Soil Monograph,* U.S. Dept. of Agriculture, Soil Conservation Service; 14.23, After B. H. Wilkinson and B. J. McElroy, 2007, The impact of humans on continental erosion and sedimentation, *GSA Bulletin*; vol. 119, pp. 140–156; 14.24a, Lynn Betts/USDA/Natural Resources Conservation Service; 14.24b, Lynn Betts/USDA/Natural Resources Conservation Services; 14.24c, USDA/Natural Resources Conservation Service; 14.25a, USDA/Natural Resources Conservation Service; 14.25b, USDA/Natural Resources Conservation Service; 14.26, USDA/Natural Resources Conservation Service; 14.27b, Lynn Betts/USDA/Natural Resources Conservation Service; 14.28, Tim McCabe/ USDA/Natural Resources Conservation Service; 14.29, Erwin Cole/USDA/Natural Resources Conservation Service

Chapter 15

Chapter opening, Michael D. Kennedy/U.S. Navy; 15.1a1, Dr. David F. Walter; 15.1a2, Edwin L. Harp/U.S. Geological Survey/U.S. Department of the Interior; 15.1b1, J. R. Stacy/U.S. Geological Survey/U.S. Department of the Interior; 15.1b2, Aurora Pun; 15.1c, U.S. Geological Survey/U.S. Department of the Interior; 15.1d, AP Wide World Photos; 15.2, After D. J. Varnes, 1978, Landslides, Analysis and Control, Transportation Research Board Special Report 176, National Research Council, pp. 1–33; 15.3a, Jim Cole/AP Wide World Photos; 15.3b, AP Wide World Photos; 15.4, U.S. Geological Survey/U.S. Department of the Interior; 15.5a, Marli Miller; 15.5b1, Photo courtesy of Richard Young; 15.5b2, Photo courtesy of Richard Young; 15.6, After S. Marshak, 2001, *Earth: Portrait of a Planet,* Norton; 15.7a, Photo courtesy of Michael Dolan and Michigan Technological University; 15.7b, Susan Cannon/U.S. Geological Survey/U.S. Department of the Interior; 15.8a, D. Bradley/National Oceanic and Atmospheric Administration; 15.8b, Colorado Geological Survey/NGDC; 15.9b, U.S. Geological Survey/U.S. Department of the Interior; 15.19, Andrea Holland-Sears/U.S. Forestry Service; 15.20a2, Terry Donnelly/Getty Images Inc; 15.20b2, Gary A. Smith; 15.21a, Lloyd S. Cluff; 15.21b, Lloyd S. Cluff; 15.22b, V.V. Bertero/National Information Service for Earthquake Engineering; 15.23, Thomas Pierson/U.S. Geological Survey; 15.24b1, Gary Rosenquist; 15.24b2, Gary Rosenquist; 15.24b3, Gary Rosenquist; 15.24b4, Gary Rosenquist; 15.24b5, Gary A. Smith; 15.30, Institute of Geological & Nuclear Sciences Ltd.

Chapter 16

Chapter opening, Adriel Heisey/Adriel Heisey Photography; 16.1a1, EyeCatchLight/Fotolia; 16.1a2, Liz Hymans/CORBIS; 16.1b, Michael Maples/U.S. Army Corps of Engineers, Headquarters; 16.5b, James Bartolino/U.S. Geological Survey/U.S. Department of the Interior; 16.7a, After M. A. Summerfield, 1991, *Global Geomorphology,* Longman Scientific and Technical; 16.7b, Photo courtesy of NASA and the US/Japan ASTER Team; 16.8a, Gary A. Smith; 16.8b, Stephen St. John/National Geographic/Photolibrary.com; 16.11b, Photo courtesy of Robert S. Anderson; 16.12b1, Photo courtesy of Keith J. Tinkler; 16.12b2, Photo courtesy of Keith J. Tinkler; 16.13b, Gary A. Smith; 16.15b, Marli Miller; 16.16b, Photo courtesy of NASA and the US/Japan ASTER Team; 16.17, After M. Church, 1992, Channel morphology and typology, in *The Rivers Handbook, Hydrological and Ecological Principles,* Blackwell Scientific Publications, vol. 1, pp. 126–143; 16.19, After D. F. Ritter, R. C. Kochel, and J. R. Miller, 2002, *Process geomorphology,* 4th ed., McGraw Hill, and L. B. Leopold, 1994, *A View of the River,* Harvard University Press; 16.23b, Altitude (Yann Arthus-Bertrand)/Peter Arnold; 16.24b, AP Wide World Photos; 16.25a, AP Wide World Photos; 16.25b, Kimberly White/REUTERS/Corbis/Bettmann; 16.26a, Earth Satellite Corp./Photo Researchers; 16.26b, Earth Satellite Corp./Photo Researchers; 16.27, David Greedy/Getty Images; 16.32, Marli Miller; 16.36a, Richard T. Nowitz/CORBIS; 16.36b, Grant Meyer; 16.39b, Geordie Torr/Alamy; 16.39c, NASA/JPL; 16.40b, Carroll Claver/Photolibrary.com; 16.40c, Marli Miller; 16.41a, Michael A. Clynne, Volcano Hazards Team/U.S. Geological Survey; 16.41b, Robert L. Schuster/U.S. Geological Survey; 16.42b, NASA

Chapter 17

Chapter opening, William H. Mullins/Photo Researchers; 17.1a1, Photo courtesy of Laura Wilson. www.LauraWilson.com; 17.1a2, National Park Service; 17.1b1, Associated Press; 17.1b2, Jim Tuten/Black Star/Stock Photo/Black Star; 17.2b1, Patrick Lynch/PH ESM; 17.2b2, Patrick Lynch/PH ESM; 17.2b3, Patrick Lynch/PH ESM; 17.3b, steheap/Fotolia; 17.4b, Russell Burden/Image Stock Imagery/Photolibrary.com; 17.4c, Lothar Schroter/Mauritius/ Photolibrary.com; 17.8b, Robert Shedlock/U.S. Geological Survey/U.S. Department of the Interior; 17.9, Map by USGS; 17.12, Data from USGS; 17.13, After Winter, T. C., Harvey, J. W., Franke, O. L., and Alley, W. M., 1998, *Ground Water and Surface Water: A Single Resource,* USGS Circular 1139; 17.14a, After Winter, T. C., Harvey, J. W., Franke, O. L., and Alley, W. M., 1998, *Ground Water and Surface Water: A Single Resource,* USGS Circular 1139; 17.14b1, Copyright Larry Fellows, Arizona Geological Survey. Image courtesy of American Geological Institute, ImageBank, http://www.earthscienceworld.org/ imagebank; 17.14b2, Copyright Larry Fellows, Arizona Geological Survey. Image courtesy of American Geological Institute, ImageBank, http://www.earthscienceworld.org/ imagebank; 17.16, After M. K. Hubbert, 1940, The theory of ground-water motion, *Journal of Geology,* vol. 48, pp. 785–944, and J. A. Tóth, 1963, A theoretical analysis of ground-water flow in small drainage basins, *Journal of Geophysical Research,* vol. 68, pp. 4375–4387; 17.18a, After G. A. Waring, 1965, Thermal springs of the United States and other countries of the world: A summary, USGS Professional Paper 492; 17.18b1, Aurora Pun; 17.18b2, Aurora Pun; 17.18b3, Susan Leavines Harris/Photo Researchers, Inc; 17.25, After D. R. LeBlanc, S. P. Garabedian, K. M. Hess, L. W. Gelhar, R. D. Quadri, K. G. Stollenwerk, and W. W. Wood, 1991, Large-scale natural gradient tracer test in sand and gravel, Cape Cod, Massachusetts. 1. Experimental design and observed tracer movement, *Water Resources Research,* vol. 27, pp. 895–910; 17.25b, Denis R. LeBlanc/U.S. Geological Survey/U.S. Department of the Interior; 17.26, After D. R. LeBlanc, S. P. Garabedian, K. M. Hess, L. W. Gelhar, R. D. Quadri, K. G. Stollenwerk, and W. W. Wood, 1991, Large-scale natural gradient tracer test in sand and gravel, Cape Cod, Massachusetts. 1. Experimental design and observed tracer movement, *Water Resources Research,* vol. 27, pp. 895–910; 17.29, USGS/Department of Interior; 17.30, Photo courtesy of © Avonsoft Water Treatment Ltd; 17.31, After J. R. Craig, D. J. Vaughan, and B. J. Skinner, 2001, *Resources of the Earth,* 3rd ed., Prentice Hall; 17.31b, The Natural History Museum, London; 17.35a, Tony Waltham/Photolibrary.com; 17.35b, After McKnight and D. Hess, 2004, Physical

20.21a2, Courtesy of Virginia Garrison, USGS; 20.21b, Photo courtesy of NASA/MODIS; 20.22, After K. D. Perry, T. A. Cahill, R. A. Eldred, and D. D. Dutcher, 1997, Long-range transport of North African dust to the eastern United States, *Journal of Geophysical Research*, vol. 102, pp. 11225–11238; 20.23a, Data provided by the IMPROVE Program (Interagency Monitoring of Protecting Visual Environments)

Chapter 21

Chapter opening, (a) William O. Field/National Snow and Ice Data Center/World Data Center for Glaciology, (b) Bruce F. Molnia/National Snow and Ice Data Center/World Data Center for Glaciology; 21.1a, Terry Haran/National Snow and Ice Data Center/World Data Center for Glaciology; 21.1b, Data from National Snow and Ice Data Center; 21.1c, NASA/Goddard Space Flight Center Scientific Visualization Studio; 21.1d, Data from National Snow and Ice Data Center; 21.2, Data from NOAA/National Climatic Data Center; 21.3, After IPCC, 2007, *Climate change 2007: The physical basis*, Cambridge; 21.4, Data from HadCRUT3v, Climatic Research Unit, Hadley Centre, UK Met Office; uncertainty envelope after Brohan, P., J. J. Kennedy, I. Harris, S. F. B. Tett, and P. D. Jones, 2006: Uncertainty estimates in regional and global observed temperature changes: a new dataset from 1850. *J. Geophysical Research* 111, D12106, doi:10.1029/2005JD006548; 21.5a, ODIS Rapid Response Team, NASA Goddard Space Flight Center; 21.5b, ODIS Rapid Response Team, NASA Goddard Space Flight Center; 21.5c, ODIS Rapid Response Team, NASA Goddard Space Flight Center; 21.6, After Oerlemans, J., 2005, Extracting a climate signal from 169 glacier records, Science, v. 308, pp. 675–677; 21.7, After Church, J. A., and White, N. J., 2006, A 20th century acceleration in global sea-level rise, Geophys. Res. Lett., v. 33, L01602, doi:10.1029/2005GL024826; temperature data from HadCRUT3v, Hadley Centre; 21.8, Peter Ryan/Photo Researchers; 21.9, Data from Luckman, B. H., and R. J. S. Wilson, 2006, Canadian Rockies Summer Temperature Reconstruction. IGBP PAGES/World Data Center for Paleoclimatology Data Contribution Series #2006-011. NOAA/NCDC Paleoclimatology Program; 21.10, After Duff, *Holmes' Principles of Physical Geology*, 4th ed., 1993, Chapman and Hall; 21.11a, Gary A. Smith; 21.11b, After Royer et al., 2004, CO_2 as a primary driver of Phanerozoic climate, GSA Today, v. 14, pp. 4–10; 21.12, After C. H. Lear, H. Elderfield, P. A. Wilson, 2000, Cenozoic deep-sea temperatures and global ice volumes from Mg/Ca in benthic foraminiferal calcite, Science, v. 287, pp. 269–272; 21.13a, Copyright © Eurelios/Photo Researchers; 21.13b, Data from J. Jouzel and V. Masson-Delmotte, 2008, EPICA Dome C Ice Core 800kyr

deuterium data and temperature estimates, doi:10.1594/PANGAEA.683655; 21.14, Data from Lea, D. W., et al., 2003, Cariaco Basin Foraminiferal Mg/Ca and SST Reconstruction, IGBP PAGES/World Data Center for Paleoclimatology Data Contribution Series #2003-067 and Alley, R. B., 2004, GISP2 Ice Core Temperature and Accumulation Data. IGBP PAGES/World Data Center for Paleoclimatology Data Contribution Series #2004-013. NOAA/NGDC Paleoclimatology Program; 21.15, Data from Esper, J., et al., 2003, Northern Hemisphere Extratropical Temperature Reconstruction, IGBP PAGES/World Data Center for Paleoclimatology Data Contribution Series #2003-036. NOAA/NGDC Paleoclimatology Program, Boulder, CO, USA; 21.16, Data from Mann, M. E. and P. D. Jones, 2003, 2,000 Year Hemispheric Multi-Proxy Temperature Reconstructions, IGBP PAGES/World Data Center for Paleoclimatology Data Contribution Series #2003-051. NOAA/NGDC Paleoclimatology Program and HadCRUT3v, Climatic Research Unit, Hadley Centre, UK Met Office; 21.17, After McKnight and D. Hess, 2004, *Physical Geography: A Landscape Appreciation*, 7th ed., Prentice Hall; 21.18, After J. T. Kiehl and K. E. Trenberth, 1997, Earth's Annual Global Mean Energy Budget, *Bull. Amer. Met. Soc.*, vol. 78, pp. 197–208; 21.20a, Data from Y. M. Wang and others, 2005, Modeling the sun's magnetic field and irradiance since 1713, *Astrophys. Jour.*, vol. 625, pp. 522–538, and National Geophysical Data Center, Solar Data Services; 21.20b, NASA/ESA SOHO; 21.20c, NASA/ESA SOHO; 21.21, After D. Lüthi and others, 2008, High-resolution carbon dioxide concentration record 650,000–800,000 years before present, *Nature*, vol. 453, pp. 379–382; L. Loulergue and others, 2008, Orbital and millennial-scale features of atmospheric CH_4 over the past 800,000 years, *Nature*, vol. 453, pp. 383–386; Compendium of Data on Global Change/Oak Ridge National Laboratory; 21.23, After Royer et al., 2004, CO_2 as a primary driver of Phanerozoic climate, *GSA Today*, vol. 14, pp. 4–10; 21.24, Data from HadCRUT3v, Climatic Research Unit, Hadley Centre, UK Met Office; 21.25, After IPCC, 2007, *Climate change 2007: The physical basis*, Cambridge; 21.28, NOAA Earth System Research Laboratory; 21.29, Data from HadCRUT3v, Climatic Research Unit, Hadley Centre, UK Met Office; 21.30, After IPCC, 2007, *Climate change 2007: The physical basis*, Cambridge; 21.31, Data from IPCC, 2007, *Climate change 2007: The physical basis*, Cambridge; and D. Shindell, 2008, Cool zone, *Nature Geoscience*, vol. 1, pp. 85–86; 21.32, After IPCC, 2007, *Climate change 2007: The physical basis*, Cambridge; 21.33, After IPCC, 2007, *Climate change 2007: The physical basis*, Cambridge; 21.34, After IPCC, 2007, *Climate change 2007: The physical basis*, Cambridge

INDEX

A

A horizon, in soil, 328–29, 330, 331
A'a, 65
Abrasion
  Glacial erosion by, 449–50
  Rivers channel erosion by, 382
Accreted terrane, 317–319
Acid rain, 558
Age of rocks. *See* Rocks, age of
Airy, George, 305
Albedo, 551, 554
Alluvial fans, streams, 384
Alluvial streams, 380
Alluvium, streams, 380
Aluminum
  aluminum-for-silicon ion substitution, 35, 37
  aluminum oxide in forest/rainforest soils, 336–37, 339
  aluminum oxide in soil B and E horizons, 329, 331
  aluminum-silicate mineral stability, 125–26
  bauxite (aluminum ore), 38, 339
  in Earth's composition, 35
  in eastern *vs.* western US dust, 534–35
  in feldspar minerals, 37
  in garnets, 38
  in granite, 72
  ion size (relative), 33, 35
  in magma, 60
Amethyst, 26, 32
Amphibole/amphibolite
  amphibole group, description/composition of, 36, 37
  appearance of (typical), 37, 129
  characteristics of, 131
  as coarse-grained, foliated rock, 128
  in composite banding/foliation, 121
  in continental crust, 170
  in gneiss, 128, 130
  in hydrothermal metamorphism, 133
  as index mineral, 127
  in mafic igneous rocks, 132
  oxidation reactions in, 91
  water molecules in, 70, 73, 201
  weathering products of, 91
Andalusite
  in igneous intrusions (Nevada), 134
  as index mineral, 127, 134
  in metamorphic rocks, 133
  as polymorph mineral, 125
  stability of, 125–26
Andesite
  appearance of (typical), 60, 61
  in composite volcanoes, 67, 74
  formation of, 63
Angle of repose, 360
Antarctica
  Antarctic temperature, age/history of, 547, 548
  glacial movement, speed of, 445, 448
  glaciers, plastic flow of, 448
  ice sheet, appearance of (typical), 442
  ice sheet, area/volume of, 443–44
  ice sheet, melting of, 445
  ice shelves along, 454, 455, 456
  iceberg formation at, 445
  See also Glaciers/glaciation; Ice; Ice age
Anticline, 232–234
Aphanitic rocks
  aphanitic, defined, 62
  appearance of (typical), 60, 61
  formation of, 62

and magma cooling rate, 63
  vs. phaneritic rock, 62
Aquifer
  defined, 416
  contamination, spread of, 428
  coastal, 430–31
  confined, 421–22
  types and properties of, 416–17
Aquitards, 416
Arête, glacier, 449
Arkose (sandstone), 96, 97
Artesian, 423
Ash and pumice, volcanic, 62, 63, 78–79
Asthenosphere
  decompression melting of, 269, 271
  description of, 13, 184
  formation of mid-ocean ridges, 14, 73–74, 269, 271, 274
  heat flow (at convergent plate boundaries), 279–820
  heat flow (at divergent plate boundaries), 274, 275
  in mantle convection, 17, 292
  in mantle-plume hypothesis, 285–86
  melting of, and geothermal gradient, 71
  and ocean water depth, 313
  peridotite in, 271
  seismic tomography in, 278–79, 282
  seismic wave velocity in, 183–84, 185, 186, 188
  in structure of mantle, 13
  in subduction zones, 14, 74, 201, 278, 281
  and surface topography, 316, 322
  viscosity of (and mantle convection), 214, 218
  See also Convection, mantle; Mantle
Atmosphere

  atmospheric composition and human activities, 555–56
  atmospheric gas radiant energy absorption, 552–53
  convection of, 17, 516
  early atmosphere, formation of, 203–4, 207
  energy budget, balancing, 552, 558–59
  greenhouse effect, 553
  ozone depletion, 552–53
  sulfur compound pollution of, 39, 131
  upper atmosphere cooling, effects of, 564
  volcanic gases and atmospheric composition, 204, 558
  water contained in (cubic kilometers), 205
  water-vapor concentration and hydrologic cycle, 553
  weathering, role in, 88, 91, 92
  windblown dust and atmospheric haze, 532–35
  See also Global warming; Wind
Atomic structure
  Basics of an Atom (Extension Module 2.1)
  defined, 26
  See also Bonds, chemical; Crystals/crystallization; Hardness; Ions
Avalanches. *See* Landslides; Mass Movements

B

B horizon, in soil, 328–30, 331
Barrier islands, shorelines, 493–95
Basalt
  basalt bombs, 62
  basalt dike (Grand Canyon), 64
  basalt flows (typical, columnar), 65, 78
  in black-sand beaches, 491
  and calcite-rich soil, 341
  in composite volcanoes, 67
  crystals in, 60
  and fractional crystallization, 75, 80, 201–2
  and hydrothermal metamorphism, 136
  melting of, 69–70
  metamorphosis to eclogite, 131
  mica in, 135, 136
  olivine in, 69–70
  partially-melted peridotite in, 201
  in subduction zones, 73–74
  typical appearance of, 61
  See also Igneous rock; Lava

Base level, streams, 387, 400
Batholith, formation of, 65
Bauxite, 38, 339
Baymouth bars, 490
Bays, shorelines, 478
Beaches. *See* Shorelines
Bedding, 49, 102
Bedload, streams, 383
Bedrock, 328
Bedrock streams, 380
Benioff, Hugo, 278
Bering land bridge, 464, 468
Biotite
  appearance of, 25
  in banding/foliation of metamorphic rock, 121, 128
  conversion to vermiculite, 91
  description, composition of, 36
  in granite, 60, 73, 90, 92
  as index mineral, 127
  in intermediate-range igneous rocks, 62
  as iron- and magnesium-rich mineral, 37, 39
  in mafic igneous rock, 132
  in metamorphism of shale, 130
  in non-foliated metamorphic rock, 129
  potassium-argon dating of, 158
  in schist and gneiss, 130
  in sedimentary rocks, 86–87
  in volcanic ejecta, 47
  weathering products of, 91
Black smokers, 135
Body waves (P and S waves) 173
Bombs, pyroclastic, 62, 63
Bonds, chemical
  breaking of (dissolving halite), 28–29
  in calcite, 29
  covalent, 27
  ions in, 27, 29, 33
  metallic, 27–28
  in quartz, 30
  van der Waals force, 28
  in water molecule, 28
Bowen, N.L., 72–73
Bowen's Reaction Series (Extension Module 4.1)
Braided streams, 375, 388, 389, 405, 454, 465
Breccia, 96

C

C horizon, in soil, 329, 330, 331
Calcite
  atomic bonding in, 29, 34
  atomic structure of, 28–30
  calcite-cemented soil particles, 45, 332, 334
  calcite-encased bottle (Oklahoma), 411
  calcite-rich soil, origins of, 341–43
  caliche, 332, 334
  chemical formula for, 26
  cleavage planes in, 24
  color variation in, 25
  crystal form of, 22, 23
  in deep-sea sediment (paleoclimate proxy record), 548
  dissolution of, 90, 556
  in foraminifera shells (and ice-age timing), 468–69
  Geochemistry of Calcite (Extension Module 5.3)
  iron in, 32
  reaction with hydrochloric acid, 30
  See also Soil
Calderas, 68–69
Caliche (calcite-cemented soil particles), 332, 334
Carbon cycle, 556–57
Carbon dioxide
  in global warming, 557
  in magma, 60, 76
  in metamorphosis of olivine, 122–23, 125
  See also Chemical weathering
Cementation, mineral, 45, 93–94
Chemical bonds. *See* Bonds, chemical